ELECTRO-OPTICAL DISPLAYS

OPTICAL ENGINEERING

Series Editor

Brian J. Thompson
Provost
University of Rochester
Rochester, New York

1. Electron and Ion Microscopy and Microanalysis: Principles and Applications, *Lawrence E. Murr*
2. Acousto-Optic Signal Processing: Theory and Implementation, *edited by Norman J. Berg and John N. Lee*
3. Electro-Optic and Acousto-Optic Scanning and Deflection, *Milton Gottlieb, Clive L. M. Ireland, and John Martin Ley*
4. Single-Mode Fiber Optics: Principles and Applications, *Luc B. Jeunhomme*
5. Pulse Code Formats for Fiber Optical Data Communication: Basic Principles and Applications, *David J. Morris*
6. Optical Materials: An Introduction to Selection and Application, *Solomon Musikant*
7. Infrared Methods for Gaseous Measurements: Theory and Practice, *edited by Joda Wormhoudt*
8. Laser Beam Scanning: Opto-Mechanical Devices, Systems, and Data Storage Optics, *edited by Gerald F. Marshall*
9. Opto-Mechanical Systems Design, *Paul R. Yoder, Jr.*
10. Optical Fiber Splices and Connectors: Theory and Methods, *Calvin M. Miller with Stephen C. Mettler and Ian A. White*
11. Laser Spectroscopy and Its Applications, *edited by Leon J. Radziemski, Richard W. Solarz and Jeffrey A. Paisner*
12. Infrared Optoelectronics: Devices and Applications, *William Nunley and J. Scott Bechtel*
13. Integrated Optical Circuits and Components: Design and Applications, *edited by Lynn D. Hutcheson*
14. Handbook of Molecular Lasers, *edited by Peter K. Cheo*
15. Handbook of Optical Fibers and Cables, *Hiroshi Murata*
16. Acousto-Optics, *Adrian Korpel*
17. Procedures in Applied Optics, *John Strong*
18. Handbook of Solid-State Lasers, *edited by Peter K. Cheo*
19. Optical Computing: Digital and Symbolic, *edited by Raymond Arrathoon*
20. Laser Applications in Physical Chemistry, *edited by D. K. Evans*
21. Laser-Induced Plasmas: Physical, Chemical, and Biological Applications, *edited by Leon J. Radziemski and David A. Cremers*

22. Infrared Technology Fundamentals, *Irving J. Spiro and Monroe Schlessinger*
23. Single-Mode Fiber Optics: Principles and Applications, Second Edition, Revised and Expanded, *Luc B. Jeunhomme*
24. Image Analysis Applications, *edited by Rangachar Kasturi and Mohan M. Trivedi*
25. Photoconductivity: Art, Science, and Technology, *N. V. Joshi*
26. Principles of Optical Circuit Engineering, *Mark A. Mentzer*
27. Lens Design, *Milton Laikin*
28. Optical Components, Systems, and Measurement Techniques, *Rajpal S. Sirohi and M. P. Kothiyal*
29. Electron and Ion Microscopy and Microanalysis: Principles and Applications, Second Edition, Revised and Expanded, *Lawrence E. Murr*
30. Handbook of Infrared Optical Materials, *edited by Paul Klocek*
31. Optical Scanning, *edited by Gerald F. Marshall*
32. Polymers for Lightwave and Integrated Optics: Technology and Applications, *edited by Lawrence A. Hornak*
33. Electro-Optical Displays, *edited by Mohammad A. Karim*

Additional Volumes in Preparation

Mathematical Morphology in Image Processing, *edited by Edward Dougherty*

Polarized Light: Fundamentals and Applications, *Edward Collett*

Opto-Mechanical Systems Design: Second Edition, Revised and Expanded, *Paul R. Yoder, Jr.*

Rare Earth Doped Fiber Lasers and Amplifiers, *edited by Michel Digonnet*

ELECTRO-OPTICAL DISPLAYS

EDITED BY
MOHAMMAD A. KARIM
The Center for Electro-Optics
The University of Dayton
Dayton, Ohio

Marcel Dekker, Inc. New York • Basel • Hong Kong

Library of Congress Cataloging-in-Publication Data

Electro-optical displays / edited by Mohammad A. Karim.
    p.   cm. -- (Optical engineering ; v. 33)
  Includes bibliographical references and index.
  ISBN 0-8247-8695-5
  1. Information display systems.   2. Electro-optical devices.
I. Karim, Mohammad A.   II. Series: Optical engineering (Marcel Dekker, Inc.) ; v. 33
TK7882.I6E36   1992
621.38--dc20   92-19407
                                                                                     CIP

TK
7882
.I6
E36
1992

This book is printed on acid-free paper.

Copyright © 1992 by Marcel Dekker, Inc. All Rights Reserved.

Neither this book nor any part may be reproduced or transmitted in any form or by any means, electronic or mechanical, including photocopying, microfilming, and recording, or by any information storage and retrieval system, without permission in writing from the publisher.

Marcel Dekker, Inc.
270 Madison Avenue, New York, New York 10016

Current printing (last digit):
10 9 8 7 6 5 4 3 2 1

PRINTED IN THE UNITED STATES OF AMERICA

About the Series

The series came of age with the publication of our twenty-first volume in 1989. The twenty-first volume was entitled *Laser-Induced Plasmas and Applications* and was a multi-authored work involving some twenty contributors and two editors: as such it represents one end of the spectrum of books that range from single-authored texts to multi-authored volumes. However, the philosophy of the series has remained the same: to discuss topics in optical engineering at the level that will be useful to those working in the field or attempting to design subsystems that are based on optical techniques or that have significant optical subsystems. The concept is not to provide detailed monographs on narrow subject areas but to deal with the material at a level that makes it immediately useful to the practicing scientist and engineer. These are not research monographs, although we expect that workers in optical research will find them extremely valuable.

There is no doubt that optical engineering is now established as an important discipline in its own right. The range of topics that can and should be included continues to grow. In the "About the Series" that I wrote for earlier volumes, I noted that the series covers "the topics that have been part of the rapid expansion of optical engineering." I then followed this with a list of such topics which we have already outgrown. I will not repeat that mistake this time! Since the series now exists, the topics that are appropriate are best exemplified by the titles of the volumes listed in the front of this book. More topics and volumes are forthcoming.

Brian J. Thompson
University of Rochester
Rochester, New York

Preface

Electro-optical displays serve perhaps as the most significant user–machine interface in this age of information. Demand for displays continues to grow and diversify as more and more application areas become identified. The roots of display technology go back to such disciplines as electronics, chemistry, physics, electro-optics, communications, psychology, vision, and many more. It is this aspect of diversity that often poses a stumbling block for integration of the total field of display technology. This book attempts to explain the principles, applications, and issues pertaining to some of these electro-optical displays. Each chapter has been developed independently by one or more authors who have working experience in research and development of the display area in question.

A total of twenty-eight workers from five countries participated in this massive project. Their backgrounds are also very diverse, with affiliations in government, industry, university, and consulting firms. The twenty chapters of this book have been organized into four parts: display fundamentals, display systems, evaluation of displays, and display issues.

The first part comprises five chapters. Chapter 1 introduces the fundamentals pertaining to the intensifier tube and cathode ray tube technologies. This is followed by a chapter in which David Armitage reviews liquid crystal (LC) properties relevant to display devices, and discusses matrix-addressing and photo-addressing issues, polymer-dispersed LCs, dynamic scattering and smectic-A scattering devices, and ferroelectric LC devices. In Chapter 3, William den Boer, F. C. Luo, and Zvi Yaniv introduce the aspects pertaining to microelectronics in active matrix LC displays and image sensors. The use of micro-electronic devices, such as transistors and diodes on glass elements, the limitation of the silicon wafer size, and applications are discussed. Alan Sobel next introduces flat-panel displays and discusses the basic problems common to all such displays—selection nonlinearity, the large number of devices and drive circuits required, and the design of the output devices. Various technologies are discussed in detail, including LCDs, gas discharge displays, electroluminescent displays, LED displays, and vacuum-fluorescent displays. In Chapter 5, Guo-Guang Mu, Zhi-Liang Fang, Xu-Ming Wang, and Yu-Guang Kuang introduce the encoding and decoding schemes for producing color image display using black-and-white film.

Part II consists of five chapters. Karen E. Jachimowicz introduces projection display technologies by describing CRT, LV, and laser projection techniques in Chapter 6. The principle of operation, the issues pertaining to image blending, and major components of each system type are analyzed and the application areas are identified. In Chapter 7, Larry F. Hodges studies time-multiplexed stereoscopic displays based on LC shutter technology. Factors affecting viewing of stereoscopic display systems are discussed. Interaction and effectiveness of binocular parallax cues in isolation and when combined with other cues such as perspective and occlusion are also reviewed. Peripheral vision display (PVD) is introduced next by Harry M. Assenheim. The discussion includes human factors in vision perception, techniques, hardware, and flight testing of commercial PVDs. Holographic heads-up displays (HUD) are studied by Robert B. Wood in Chapter 9. The chapter includes a rather general description of HUDs, HUD optics, holographic fabrication, optimization of optics, and applications in wide-field-of-view fighter aircraft, automobiles, and commercial aircraft. Finally, Phillip J. Rogers and Marvin H. Freeman discuss biocular display optics in Chapter 10. Characteristics of the eyes as they influence optical displays, methods of optimization, and considerations for use with image intensifiers, CRTs, stereoscopic and panoramic systems are considered in this chapter.

The third part also comprises five chapters. In Chapter 11, Abdul A. S. Awwal introduces the modulation transfer function (MTF) as well as the dynamic modulation transfer function for evaluating nondiscrete displays such as CRTs and intensifier tubes. In Chapter 12, Abdul A. S. Awwal discusses the seriousness of various image defects that are to be observed in nondiscrete displays and identifies means for correcting such image degradation. John C. Feltz next uses MTF analysis, in Chapter 13, to characterize image degradation in discrete displays and imagers. In particular, this chapter deals with charged-coupled device (CCD) based displays. Terrence S. Lomheim and Linda S. Kalman carry this a step further and study digital hardware simulation of CCD sensor systems. This chapter includes a discussion of two-dimensional polychromatic sensor MTF, signal and noise models, effects of digital video transmission including one- and two-dimensional video data compression. Finally, in Chapter 15, Jerry Silverman and Virgil E. Vickers introduce several algorithms for global (monotonic)

PREFACE

mapping of wide dynamic IR imagery to digitized bits for display. Their chapter discusses issues germane to real-time implementation and locally adaptive contrast enhancement.

The last part of the book, consisting of five chapters, examines display issues. Clarence E. Rash and Robert W. Verona begin by considering the human factor issues of image intensification and FLIR systems. Topics such as visual allusions, color perception, depth perception, and visual acuity pertaining to intensifier tubes and FLIR systems are discussed. In Chapter 17, Celeste M. Howard examines color control issues in digital displays. Problems of digital color control in CRT, light-valve projector, and LCDs are reviewed, and then the chapter discusses device-independent color rendering, color appearance at low light levels, maximization of luminance contrast and colorfulness, and gamma correction as well as examples of dome and helmet-mounted displays. Maxwell J. Wells and Michael Haas, in Chapter 18, discuss the human factors of helmet-mounted displays and sights. The chapter treats two issues in depth: the effects of whole-body vibration and the effects of field-of-view size. Julie M. Lindholm treats perceptual effects of sampling and discusses the corresponding reconstruction process in Chapter 19. Finally, in Chapter 20, Donald L. Moon brings about issues pertaining to display integration. In particular, integration technologies such as bussing structures, addressing, interfaces, computer hardware and software, and graphics are discussed from a systems point of view.

The chapters of this book are both original and tutorial in their scopes. This, I believe, will allow the text to be used for both advanced graduate courses as well as research and tutorial applications.

Mohammad A. Karim

Contents

About the Series   *iii*
Preface   *v*
Contributors   *xv*

Part I   Display Fundamentals

1. **Intensifier and Cathode-Ray Tube Technologies**   1
   Mohammad A. Karim and A. F. M. Yusuf Haider

2. **Liquid-Crystal Display Device Fundamentals**   19
   David Armitage

3. **Microelectronics in Active-Matrix LCDs and Image Sensors**   69
   William den Boer, F. C. Luo, and Zvi Yaniv

4. **Flat-Panel Displays**   121
   Alan Sobel

5. **Color Image Display with Black-and-White Film**   187
   Guo-Guang Mu, Zhi-Liang Fang, Xu-Ming Wang, and Yu-Guang Kuang

Part II   Display Systems

6. **Projection Display Technologies**   211
   Karen E. Jachimowicz

7. Stereoscopic Display 291
 Larry F. Hodges

8. Peripheral Vision Displays 311
 Harry M. Assenheim

9. Holographic Head-Up Displays 337
 Robert B. Wood

10. Biocular Display Optics 417
 Philip J. Rogers and Michael H. Freeman

Part III Evaluation of Displays

11. Standardization of Nondiscrete Displays 447
 Abdul Ahad S. Awwal

12. Restoration of Dynamically Degraded Images in Displays 475
 Abdul Ahad S. Awwal

13. Discrete Display Devices and Analysis Techniques 495
 John C. Feltz

14. Analytical Modeling and Digital Simulation of Scanning Charge-Coupled Device Imaging Systems 513
 Terrence S. Lomheim and Linda S. Kalman

15. Display and Enhancement of Infrared Images 585
 Jerry Silverman and Virgil E. Vickers

Part IV Display Issues

16. The Human Factor Considerations of Image Intensification and Thermal Imaging Systems 653
 Clarence E. Rash and Robert W. Verona

17. Color Control in Digital Displays 711
 Celeste McCollough Howard

18. The Human Factors of Helmet-Mounted Displays and Sights 743
 Maxwell J. Wells and Michael Haas

19. Perceptual Effects of Spatiotemporal Sampling 787
 Julie Mapes Lindholm

20.	**Electro-Optic Displays—The System Perspective** *Donald L. Moon*	**809**

Index *839*

Contributors

David Armitage Lockheed Missiles & Space Company, Inc., Palo Alto, California

Harry M. Assenheim Georgetown, Ontario, Canada

Abdul Ahad S. Awwal Wright State University, Computer Science and Engineering Department, Wright State University, Dayton, Ohio

William den Boer Ovonic Imaging Systems, Inc., Troy, Michigan

Zhi-Liang Fang Nankai University, Tianjin, China

John C. Feltz Systems Research and Applications Corporation, Arlington, Virginia

Michael H. Freeman Optics & Vision Ltd., Clwyd, United Kingdom

Michael Haas Harry G. Armstrong Aerospace Medical Research Laboratory, Wright-Patterson Air Force Base, Ohio

A. F. M. Yusuf Haider Department of Physics, The University of Dhaka, Dhaka, Bangladesh

Larry F. Hodges Georgia Institute of Technology, College of Computing, Graphics, Visualization & Usability Center, Atlanta, Georgia

Celeste McCollough Howard   *University of Dayton Research Institute, Aircrew Training Research Division, Armstrong Laboratory, Williams Air Force Base, Arizona*

Karen E. Jachimowicz*   *Honeywell, Inc., Phoenix, Arizona*

Mohammad A. Karim   *The Center for Electro-Optics, The University of Dayton, Dayton, Ohio*

Yu-Guang Kuang   *Academia Sinica, Changehun, China*

Linda S. Kalman   *The Aerospace Corporation, Los Angeles, California*

Julie Mapes Lindholm   *University of Dayton Research Institute, Aircrew Training Research Division, Armstrong Laboratory, Williams Air Force Base, Arizona*

Terrence S. Lomheim   *The Aerospace Corporation, Los Angeles, California*

F. C. Luo   *Ovonic Imaging Systems, Inc., Troy, Michigan*

Donald L. Moon   *The University of Dayton, Dayton, Ohio*

Guo-Guang Mu   *Nankai University, Tianjin, China*

Clarence E. Rash   *U.S. Army Aeromedical Research Laboratory, Fort Rucker, Alabama*

Philip J. Rogers   *Pilkington Optronics, Clwyd, United Kingdom*

Jerry Silverman   *Rome Laboratory, Hanscom Air Force Base, Massachusetts*

Alan Sobel   *Consultant, Evanston, Illinois*

Robert W. Verona   *Universal Energy Systems, Fort Rucker, Alabama*

Virgil E. Vickers   *Rome Laboratory, Hanscom Air Force Base, Massachusetts*

Xu-Ming Wang   *Nankai University, Tianjin, China*

Maxwell J. Wells   *Logicon Technical Services, Inc., Dayton, Ohio*

Robert B. Wood   *Hughes Flight Dynamics, Inc., Portland, Oregon*

Zvi Yaniv   *Ovonic Imaging Systems, Inc., Troy, Michigan*

---

*Current affiliation: Motorola, Inc., Tempe, Arizona

1
Intensifier and Cathode-Ray Tube Technologies

Mohammad A. Karim

*The Center for Electro-Optics,
The University of Dayton,
Dayton, Ohio*

A. F. M. Yusuf Haider

*The University of Dhaka,
Dhaka, Bangladesh*

1.1 INTRODUCTION

The human eye is a very astute and remarkably adaptable detector, but it is not free of physical shortcomings. Its spectral sensitivity is more appropriate for image perception under daylight conditions (Karim, 1990), and at best only about one photon in 12 falling on the retina is registered by the brain. The limitations to its applicability are controlled only by the density of detector elements on the retina and by its inherent aberrations. A portion of this limitation can be subdued with the assistance of instruments such as binoculars or a telescope by increasing the angular magnification but at the expense of the total viewing angle. At low light levels such as at night, however, the total number of detected quanta is so minimal that the photon noise resulting from its random fluctuations limits the minimum detectable image information. In addition, the quantum efficiency of the eye at night is on the order of about 1%.

Direct-view electronic image intensifiers can be used to rectify the problems that otherwise plague devices such as a binocular by accumulating a larger fraction of the available photons and using these more effectively. It must, however, be noted that the eye adapts to dark in several ways. First, the pupil dilates at low luminance levels, with pupil diameter increasing from 2 mm to about 8 mm to accumulate more light from the scene. The color-sensitive cones, which are more active in the daylight, become inoperative in starlight. The color-insensitive rods, which are overloaded in the daylight, slowly rebound in starlight and take over the detection role. However, because signals from a number of rods can be integrated in time, the overall signal is enhanced at the expense of the ability to resolve detail. These adaption schemes are responsible for improving the acuity of night vision.

Image intensifier tubes are electro-optical devices capable of detecting, intensifying, and shuttering optical images in the near-ultraviolet, visible, and near-infrared, as well as in the X-ray and gamma regions of the electromagnetic spectrum for displaying optical images. In addition to brightness gain, image intensifier tubes can be used to furnish viewfield zoom by simple electronic means. They are also coupled to television pickup tubes to enhance the sensitivity of low-light-level television systems. The functions performed by the direct-view intensifier tubes include (1) conversion of the two-dimensional image formed on the sensor surface into an electron image in parallel, (2) point-by-point intensification of the electron image, and (3) conversion of the intensified electron image formed on the display surface into a visual image in parallel. With minor exceptions, all image intensifiers make use of a photoemitter to accomplish the aforementioned three objectives. Whereas intensifier tubes take low-level light as the input that they intensify to make the input visible to human eyes, the electronic displays take only electrical signals and transform them into visual imagery in real time suitable for human interpretation.

Of the many different types of electronic displays available and those that are still being developed, cathode-ray tube (CRT) based display is perhaps the most versatile (Tannas, 1985). In general, CRTs serve as interfaces between users and machines and are sufficiently dynamic that they present information and can retain that information using refresh or memory schemes. Whereas intensifier tubes play a compelling role in applications such as night vision, night blindness aid, astronomy, electron microscopy, high-speed light shutters, and medical research, the CRT-based displays are famous for video and high-resolution applications including many in the consumer and entertainment markets. This chapter deals with the physics and engineering aspects of both of these devices, which are at the heart of the many other displays and display issues to be addressed in the subsequent chapters.

1.2 IMAGE INTENSIFIER TUBE BASICS

In order to achieve photon-noise-limited perception performance, it is necessary to meet two related objectives. First, each of the detected photons having given rise to an excited electron must go through an intensification process so that the resulting multiple photons at the viewing screen will, upon absorption by the retina, be sufficient to register an event. Second, background effects not related to the original two-dimensional image must not contribute to extraneous registrations. These two requirements can be met if the detection process of the device is of the photoemissive type. The detected photons may then give rise to individual photoelectrons that in vacuum can be accelerated by an external field to acquire sufficient energy to release a large number of photons. Later these photons may be made to hit a fluorescent surface so as to become visible to the human eye.

Classical telescopes serve a similar but limited purpose by means of a large-diameter objective lens and an ocular lens sharing a common focal plane. The scene at infinity is collected by the objective lens, which forms the intermediate image at the focal plane of the objective and ocular lenses. The small-diameter ocular lens forms the final image at infinity and thus becomes registered by the eye. An objective lens whose diameter is n times that of the pupil admits n^2 times as much light as does the unaided eye (Csorba, 1985). However, since the dimensions of the retinal image are increased by a factor equal to the magnification, the visual acuity is also enhanced by a factor equal to the magnification without any change in the scene brightness but at the expense of angular field of view, which is much diminished by a factor equal to the telescope power.

1.3 IMAGE INTENSIFIER TUBE TYPES

A single-stage image intensifier tube consists of an image sensor (usually a photocathode), an electron lens (Hawkes and Kasper, 1989), and a phosphor screen. The simplest version of an image tube (Fig. 1.1) is referred to as either biplanar or proximity focused. In it the screen and photocathode are positioned in very close proximity. This design ensures the presence of a homogeneous axial electric field inside the tube, making the tube relatively insensitive to voltage fluctua-

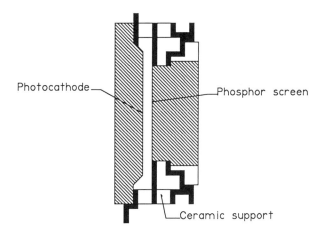

Figure 1.1 A biplanar image tube.

tions, and improves visual acuity without any loss of field of view. Such image tubes are generally distortion-free. The resolution and modulation transfer function of this image tube is often given in terms of the image plane displacement r from perfect focusing as

$$r = 2d(v_{lat}/V)^{1/2} \tag{1.1}$$

where d is the separation between the image and object planes, v_{lat} is the initial lateral velocity of the electron in electron volts, and V is the applied voltage (Freeman, 1973).

Since, in general, v_{lat} is less than 1 eV and the maximum allowable field strength is on the order of 10^5 V/cm, an r value of 10^{-3} cm can be achieved. It should be noted that the resolution is relatively independent of the initial longitudinal velocity v_{long} of the electron. The disadvantage of this type of image tube is that it does not have the resolution other types have. The high electric fields required tend to cause cold-electron emission, which contributes to shot noise and reduces contrast in the image (Biberman and Nudelman, 1971). In addition, the close spacing required makes the preparation of the photocathode difficult.

Figure 1.2 shows an electromagnetically focused single-stage image tube that superimposes a uniform magnetic field axially with the electrostatic field gradient. It provides the best resolution capability of any electro-optic image tube provided the relationship between the magnetic and electrostatic fields is maintained accurately. The set of homogeneous axial electric and magnetic fields produced by the accelerator rings and the magnet results in a focused erect image of the photocathode on the phosphor screen. The magnetic field in effect forces the electron to maintain a tighter trajectory. Emitted electrons traverse helical paths under the influence of the fields such that all electrons emitted from a point come together to form image points periodically after each complete cycle. For optimum focusing, the spacing d in centimeters is given by

$$d = 10NV^{1/2}/B \tag{1.2}$$

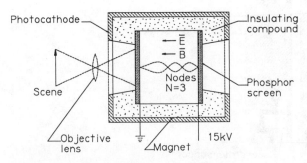

Figure 1.2 Electromagnetically focused image tube.

where B represents the magnetic flux density in gauss and N is the number of nodes between the object and image planes (Freeman, 1973). In the magnetically focused image tube the value of r is decreased from that given by Eq. (1.1) by a factor of $(4v_{long}/V)^{-1/2}$. The resolution can be improved further by slightly increasing d.

A large number of image intensifier tubes, such as the one shown in Figure 1.3, use electrostatic lenses as a focusing mechanism. Electrostatic focusing of electrons is achieved by superimposing a radially symmetric field on the longitudinal electric field. Such an electrostatic lens forms a first-order image and is analogous to a glass lens with variable index of refraction. It consists of a photocathode on one fiber-optic faceplate and an aluminized phosphor screen on the other. The input faceplate allows image transmission from an exterior plane to a suitable photocathode within the vacuum envelope. The aluminized screen, on the other hand, increases the light output per photoelectron and prohibits optical feedback to the cathode. Depositing the phosphor onto the other fiber-optic plate ensures the transmission of the image onto a second photocathode for a second-stage amplification or outside for immediate viewing.

An objective lens collects the scene information and focuses onto the photocathode of the tube. The intensified output surface is positioned at the focal plane of the ocular lens. Photoelectrons are emitted from the photocathode surface in a number proportional to the incident illumination. Subsequently the electro-optical lens system positioned inside the vacuum tube accelerates and focuses these electrons onto corresponding locations of the fluorescent phosphor screen. Each of the electrons releases a large number of photons, thus creating the intensified output image. Unfortunately, electrostatic lenses contribute to severe aberrations, astigmatism, and radial distortion. The inner surface of the fiber-optic plate is curved to minimize distortion. To prevent feedback of phosphor light to the photocathode, the cone-shaped electrode is blackened and the inner surface of the phosphor screen is aluminized.

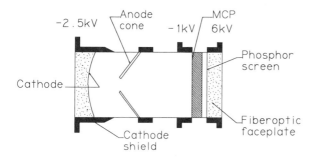

Figure 1.3 Schematic of an electrostatically focused image intensifier tube.

In this setup, the fluorescent screen potential is set at a very high value (in the kilovolt range) relative to the photocathode surface. For adequate focusing action, the image surface tends to have a curvature relative to the objective surface. The semitransparent photocathode converts the input light energy to an equivalent electric energy because of the photoelectric effect. The homogeneous axial electric field or a field with focusing properties accelerates the generated photoelectrons and raises their kinetic energy values. These photoelectrons finally strike the fluorescent screen, and their kinetic energy gives rise to an inverted light output via cathodoluminescent effect.

The photon gain of the image intensifier tube can be increased further by cascading two or more of the aforementioned intensifier units as shown in Figure 1.4a. In a cascaded system, a thin mica interstage coupler is used to support the fluorescent screen and the photocathode in close proximity to each other. Electrostatic focusing with approximately unity magnification is used in each of the units. A current gain of about 100 may be obtained across this mica target during the conversion of photoelectron beam energy into light and then that of light into photoelectron beam energy.

A three-stage cascaded system may provide gain in excess of 10^6. The electron gain from the first unit to the second unit is found to be proportional to the product of the applied voltage of the first unit, the light output of the first unit, the cathode sensitivity of the second unit, and a correction factor for spectral matching. An alternative method is to put all three stages in one vacuum envelope where the stages are partitioned by a thin membrane of glass or mica with phosphor on one side and the photocathode on the other. This technique is, however, more adaptable to magnetically focused tubes because of the curvature constraints. Another alternative approach involves the use of a microchannel plate (MCP) and a voltage (600–1000 V) applied across the plate as shown in Figure 1.4b.

Microchannel plates are current-amplifying optical devices having very high gain, low noise, fast time response, high spatial resolution, and relatively low power consumption. Accordingly, upon striking the MCP, the number of photoelectrons increases by a factor of 1000 or more. These electrons undergo further gain as they approach the phosphor screen. An MCP is a two-dimensional array of glass fibers (termed microchannels) of 10–50 μm inside diameter fused together to form a thin disk such that they have length-to-diameter ratios of between 40 and 100. A typical MCP has 10^4–10^7 channels that act as miniature electron multipliers. The inside surface of these microchannels is covered with a resistive electron emission film electrically connected to both input and output electrodes of the channel plate. The preferred electrical resistance is achieved by tapering off the metallic oxides of the glass by heating in hydrogen. Electrical contacts are made to each end of the channel by evaporating nichrome electrodes (Woodhead and Ward, 1977).

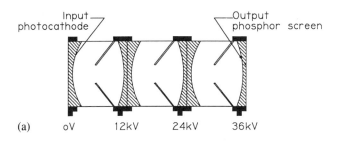

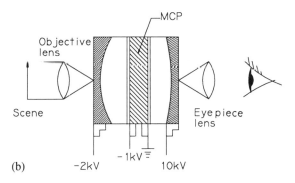

Figure 1.4 Image intensification using (a) modular cascade scheme and (b) microchannel plate.

When a photoelectron with some energy associated with potential between the photocathode and the input of the MCP enters one of the channels, it strikes the wall of the channel and generates secondary electrons. Under the influence of the electric field, the secondaries in turn accelerate through the channel, repeating the same process until the final stream of electrons exits the channel. An electron avalanche will result from the process provided the channel material has a secondary emission coefficient greater than unity. Applied voltage and the ratio of the length to the diameter of the channel determine the MCP gain. Usually, these microchannels are not positioned exactly perpendicular to the input and output surfaces but rather are biased at an angle of about 5°. This small bias angle ensures that the electrons first impact near the channel entrance and reduces light feedback from the phosphor screen. MCPs multiply the number rather than the energy of the photoelectrons. The MCP-based intensifier tube is particularly significant because of its extremely small size. Image inversion for this type of tube is normally achieved by using fiber twists through an angle of 180°.

In the arrangement illustrated in Figure 1.3, focusing is achieved by restoring the components of the field gradient normal to the electron trajectories. This effect thus contributes to a crossover of the electron trajectory, ultimately providing an inverted electron image. Consequently, this scheme may lead to problems, particularly when interfacing one electro-optical lens section either to another as is usually the case in many image intensifier tubes or to an objective lens. In the first-generation intensifier tubes, the curvatures of the two surfaces were accommodated by the use of fiber-optic faceplates. Subsequently, in the second-generation high-gain single-envelope intensifier tubes employing a high-gain MCP such as the one shown in Figure 1.5, the image surface is flattened and the corresponding effect is compensated by introducing additional focusing electrodes. This design followed from the understanding that a single high-gain tube would diminish the loss of modulation transfer occurring in the cascaded tubes.

The second-generation tube consists of a spherical cathode and a conical anode with the anode aperture placed a little off from the center of curvature of the photocathode. This arrangement is able to generate a flat inverted image of the photocathode on the input of the MCP. To provide optimum magnification, the MCP is placed at a certain distance beyond the anode aperture. The MCP intensifies the photoelectron image, which is then focused (through a proximity scheme) on the phosphor screen and an output fiber-optic screen. The second-generation intensifier tube has a radiant power comparable to the gain of a three-stage cascaded image tube and a length only slightly more than that of a single-stage first-generation tube. However, the second-generation image tube is also not without problems. In particular, the resolution drops off from the center to the edge, and the associated crossover contributes to a pincushion type of image distortion. It may be noted that the resolution at the center of the second-generation intensifier tube can be as good as that in the magnetically focused image tubes. In practice, however, the image plane is moved slightly inwards so as to provide for optimum resolution in regions beyond the center of the tube.

An alternative second-generation intensifier tube, referred to as the wafer tube, consists of a screen fiber-optic plate, an MCP, and a photocathode fiber-

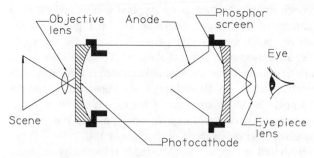

Figure 1.5 The second-generation image intensifier inverter tube.

optic faceplate wedge. The typical luminous gain of the wafer tube is on the order of 10^4. The second-generation image intensifiers have a fairly high and therefore poor noise figure, due to the statistics of the electron multiplication process within the channels. Again, the proximity focus configuration results in the production of positive ions in the high-charge regions of the channels, which in turn causes the destruction of photocathodes and therefore reduces cathode quantum efficiency.

Most of the second-generation wafer tubes use multialkali photocathodes. An improved version of these wafer tubes, however, makes use of GaAs/AlGaAs photocathodes as in the third-generation image intensifier tubes. The GaAs photocathodes have spectral response in the 0.8–0.9 μm range. The night sky spectral irradiance in this range is such that the photon flux is about six times as great as in the visible region. This increases the luminous sensitivity of the wafer tube photocathodes from 300 μA/lm to about 1100 μA/lm. In spite of this, ionic interaction with the MCP readily contaminates these third-generation photocathodes. This has forced the use of an ultrathin ion barrier membrane over the MCP input to allow the movement of primary electrons but inhibit the flow of ions between the MCP and the cathode. Usually an aluminum oxide membrane of 30–50 Å thickness capable of allowing electrons with energies in excess of 200 eV is used for this purpose.

1.4 IMAGE INTENSIFIER TUBE CHARACTERISTICS

There are three fundamentally important intensifier tube characteristics: brightness gain, equivalent background input (EBI) photocathode sensitivity, and modulation transfer function (MTF). The brightness gain is normally defined by the ratio of the output brightness in foot-lamberts (fL) divided by the input illumination on the photocathode in foot-candles (fc). For a consistent brightness gain measurement, a tungsten source operating at 2854 K is normally used as the input light source. However, this gain measurement is valid only as long as the output is Lambertian in angular distribution. Over a relatively small angle around the optical axis such as the acceptance angle of most eyepieces, such a Lambertian approximation is usually valid. When the output is a photographic plate, for example, it is more meaningful to identify the brightness gain in terms of its apparent ASA rating improvement.

Equivalent background input is a measure of the residual illumination on the screen in the absence of any input. Normally a nonzero quantity, EBI (in lumens per square centimeter) is caused mainly by the ion noise feedback, light feedback, thermal emission from the photocathode, and field emission. For most intensifier tubes, the typical value of EBI is 2×10^{-11} l/cm² or less.

The MTF is of particular importance as it allows one to determine the limiting resolution of an intensifier tube (Csorba, 1981; Jenkins, 1981). This measurement is not unique to intensifier tubes but applies equally to all imaging devices.

Very simply, the MTF describes how the modulation of 50–50 duty cycle opaque and transparent lines of a pattern of a given size is degraded by the imaging system. At higher spatial frequencies, the image contrast decreases until the pattern can no longer be distinguished. The spatial frequency at which this occurs is called the limiting resolution. This is a very useful concept because the MTF of an imaging system is also the product of the MTFs of its components.

A function of spatial frequency, the MTF describes how the imaging components affect signal modulation. For example, an MCP-based image tube is composed of a large number of discrete elements, and therefore the corresponding resolution is determined by the array geometry and its dimensions. Further, the electrons emerging from the channels spread out before reaching the screen, which in turn reduces the resolution of the device (Woodhead and Ward, 1977). In addition, there is the effect of phosphor persistence, which may be a limitation for certain intensifier tube applications. Phosphors tend to have long decay times and background buildup, which can be explained in terms of two simultaneously occurring time constants: prompt or fluorescent emission and delayed or phosphorescent emission (Torr, 1985). This results in an image-retention effect that contributes to degraded images of objects in motion (Awwal et al., 1991; Gao et al., 1990; Sandel et al, 1986). Because of persistence, the light emitted in a single frame does not uniquely specify the object during that frame. In subsequent chapters of this book, the concept of MTF will be used repeatedly for evaluating various display systems.

1.5 CATHODE-RAY TUBES

The electronic display is expected to be dynamic in that it must depict information in almost real time but also be able to hold that information using refresh schemes until new information is acquired. The primary display, aside from small alphanumerics, is the CRT, which is deeply entrenched in our day-to-day video world. In spite of competing technologies such as flat-panel displays, it is unlikely that the CRT will be totally replaced. It is by far the most common display device found in both general and special-purpose use. The cost factor and the trend toward using color displays of higher and higher resolution are the key factors guaranteeing the CRT's longevity. The CRT has satisfactory response speed, resolution, design, and life. Besides, there are very few electrical connections, and CRTs can present more information per unit time at a lower cost than any other display technology. A CRT display is generally categorized as having electrostatic or magnetic deflection, monochromatic or color video, and single or multiple beams.

Figure 1.6a shows the schematic of a CRT in which the cathodoluminescent phosphors are used at the output screen. Cathodoluminescence refers to the emission of radiation from a solid when it is bombarded by a beam of electrons. The electrons are generated by thermionic emission from a cathode and are directed onto a spot on the screen by means of a series of deflection plates held at various potentials. The beam is accelerated toward the screen by using an anode voltage of 20 kV or more. The viewing screen is coated with a phosphor that emits light when struck by a beam of electrons.

The electrons leaving the thermionic cathode, from what is known as a crossover gun CRT (Fig. 1.6b), first enter the cathode lens, which is present within the region bounded by the cathode and the two grid electrodes (Fig. 1.6a). The beam-forming electric field is introduced by subjecting the accelerating electrode (grid 2) to a positive potential with respect to the cathode. Either a zero or an appropriate negative potential is applied to G_1. Therefore, the generated electrons are converged to a crossover just in the vicinity of aperture G_2 in front of the cathode. This beam-forming strategy generates the highest current density at the crossover, thus producing a virtual image of the cathode that is generally more intense than the cathode itself. The crossover is next imaged by the prefocusing lens, as shown in Figure 1.6b and is then directed toward the focusing lens. The strength of the focusing lens is adjusted to obtain an image of the crossover at the phosphor screen. Alternatively, in the case of a laminar flow gun CRT, the electrons emitted from the thermionic cathode tend to move in streamline-like trajectories, although many still cross over, until finally they converge to a focus at the phosphor screen. The electron emission density falls off rapidly with radial distance in the case of the crossover gun CRT, but it is relatively uniform in the case of the laminar flow gun CRT. Ideally, therefore, the latter type of CRT provides between 1.3 and 2.1 times the current from the same cathode area.

The electron beam is sequentially scanned across the screen in a series of lines by means of deflecting fields acting orthogonally to the direction of the electron trajectory. The imagery is created on the raster as it is traced out. Typically two fields are interlaced to form a complete picture, with the raster lines of one field traced between those of the other. Thus, if a complete refreshing cycle takes t_r time, only odd-numbered lines are scanned during the first $t_r/2$ period, and the even-numbered lines are scanned during the remaining half. Consequently, our eyes treat the refreshing rate as if it were $2/t_r$ Hz instead of only $1/t_r$ Hz. The display is refreshed generally 60 times a second to avoid having a flickering image. In applications that involve very high resolution images containing thousands of scanning lines, lower refresh rates are used.

Although CRT displays consist of 525 scan lines in the United States, in other countries the number is 625. The bulb consisting of the electron gun, the deflectors, and the screen is maintained air-free for the purpose of having an electron beam and a display area. The video signals to be displayed are applied to both the electron gun and the deflectors in synchronization with the scanning signals. The deflected electrons strike the CRT phosphor screen, causing the phosphors

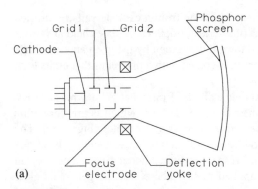

Figure 1.6 (a) Schematic CRT system (b) schematic of the beam-focusing lens; (c) CRT screen.

at that location to emit light. It is interesting to note that both absorption and emission distributions of phosphors are bell-shaped but the distribution peaks are relatively displaced in wavelength. Compared with the absorption distribution, the emission distribution peaks at a higher wavelength. This shift toward the red end of the spectrum is used to convert ultraviolet radiation to useful visible radiation. In particular, the CRT illumination caused by the cathodoluminescent phosphors is a strong function of both current and accelerating voltage V and is given by

$$L_e = Kf(i)V^n \tag{1.3}$$

where K is a constant, $f(i)$ is a function of current, and n ranges between 1.5 and 2.0. With a larger accelerating voltage, electrons penetrate further into the phosphor layer, causing more phosphor cells to irradiate.

The factors taken into consideration in selecting a particular phosphor are decay time, color, and luminous efficiency. Table 1.1 lists some of the more common phosphors and their characteristics. Note that even though a phosphor may have a higher radiant efficiency in the red while another may have a lower radiation efficiency in the green, the latter may have a more desirable luminous efficiency curve. Usually, the phosphor screen consists of a thin layer (~5 μm) of phosphor powder placed between the external glass face plate and a very thin layer (~0.1 μm) of aluminum backing as shown in Figure 1.6c. The aluminum backing prevents charge buildup and helps redirect light back toward the glass plate. The aluminum backing is thin enough that most of the electron beam energy can get through it. A substantial amount of the light that reaches the glass at normal or nearly normal angles is transmitted. However, a portion of the incident light (beyond the critical angle of incidence) may be totally internally reflected at the glass–air interface, and some of it may again be totally internally reflected at the phosphor–glass interface. Such physical situations yield a series of concen-

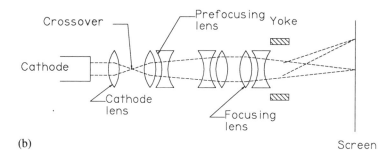

(b)

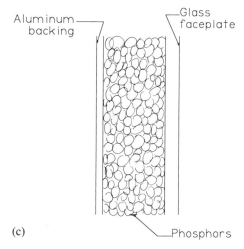

(c)

Figure 1.6 (Con't.)

tric circles of reduced brightness instead of producing one bright display spot. The combination of diffused display spots results in a display spot that has a Gaussian distribution profile.

Of the many available methods, the most common one for introducing color in a CRT display involves the use of a metal mask and three electron guns, each corresponding to a primary phosphor granule (red, blue, and green), as shown in Figure 1.7. The three electron guns are positioned around the longitudinal axis 120° apart, so that while each of the electron beams is passing through a particular mask hole it strikes a particular primary phosphor dot. There are two more variations of this design, in both of which the electron guns are positioned in line instead of in a delta configuration and the screen consists of a periodic array of vertical phosphor stripes (red, green, and blue). In one the mask is formed of slits, and in the other it is formed of slots as shown in Figures 1.7b and 1.7c. All three beams are deflected simultaneously. In addition, the focus elements for the

Table 1.1 Selected Phosphors and Their Characteristics

Phosphor	Peak wavelength (nm)	Color	Efficiency	90% Decay time (msec)
$CaWO_4$	430	Blue	3.4	0.03
ZnS:Ag	460	Blue	15.2	0.03
Zn_2SiO_4:Mn	525	Yellow-green	4.7	24
(Zn, Cd)S:Ag	530	Yellow-green	18.5	0.06
ZnS:CU	542	Yellow-green	12.4	0.10
La_2O_2S:Tb	543	Green	11.8	1
Gd_2O_2S:Tb	544	Green	10.2	1
YVO_4:Eu	617	Orange	12.0	9
Y_2O_2S:Eu	627	Red	13.1	0.90

three guns are connected in parallel so that a single focus control is sufficient to manipulate all beams. The three primary dots are closely packed in the screen so that proper color can be generated for each signal. Misalignment of the three beams causes a loss of purity for the colors. In any event, compared to the monochrome display, the CRT color reproduction process involves a loss of resolution to a certain degree because the primary phosphor cells are physically disjointed.

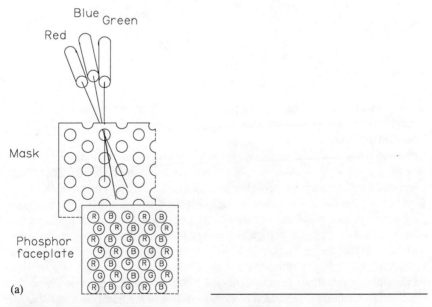

Figure 1.7 Color CRT systems in (a) delta configuration, (b) in-line configuration but with aperture mask, and (c) in-line configuration but with slotted mask.

INTENSIFIER AND CRT TECHNOLOGIES

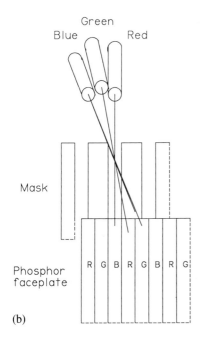

(b)

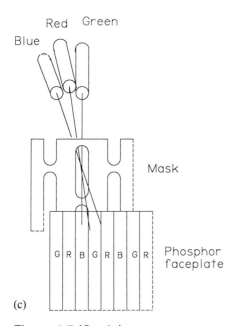

(c)

Figure 1.7 (Con't.)

1.6 CRT RESOLUTION AND CONTRAST

The resolution of a CRT is determined by the electron gun characteristics and by the optical properties of the phosphors used. Again, the overall electron gun performance is controlled by the current density distribution and is limited by several nonnegligible factors such as spot size magnification, cathode loading, lens aberrations, and thermal energy distribution. The lateral magnification m for a CRT is given by

$$m = (v/u)\,(V_c/V_s)^{1/2} \qquad (1.4)$$

where v and u, respectively, represent the image and object distances and V_c and V_s, respectively, represent cathode and screen potentials (Tannas, 1985). A decrease in modulation corresponds to an increase in resolution. The resolution of the CRT can thus be improved by increasing either the screen potential or the object distance of the gun.

To achieve high CRT resolution, it is necessary to have the highest current density at the focused spot. Again, the current density at the focused spot is directly proportional to the current density J_0 at the cathode. Assuming a Maxwellian velocity distribution for the emitted electrons and neglecting both space charge and aberrations, the current density is given by

$$J = \begin{cases} J_o/m^2, & m = 1 \\ J_o[(V_s e + 1)/kT]\sin^2 a, & m \ll 1 \end{cases} \qquad (1.5)$$

where k is the Maxwell–Boltzmann constant, T is the absolute temperature in kelvins, e is the charge of an electron, and a is half the apex angle of convergence at the focused spot (Tannas, 1985). For most CRTs, the magnification m is between 0.45 and 1. It must be emphasized that further increasing J will not necessarily increase the image current density. In fact, beyond a certain threshold value for J, space charge will begin to limit the image current density.

The way thermal effects begin to influence the CRT performance can be inferred by realizing that cathode loading J_0 is a parameter that superlinearly increases with temperature. Accordingly, it can be estimated that for those CRTs where m is close to unity, resolution becomes enhanced at higher temperatures. However, for those CRTs where m is very small (as in the case of higher resolution CRTs), if cathode loading is held constant, an increase in temperature will decrease the resolution.

In addition to electron gun characteristics, the phosphor particle size and screen thickness also limit the CRT resolution. Because of optical scattering, conventional phosphor powders limit screen resolution to less than that which can be realized at lower currents. Resolution is normally enhanced by reducing particle size (by cataphoretic deposition schemes) and screen thickness but at the expense of phosphor luminous efficiency. Phosphor persistence is another factor

INTENSIFIER AND CRT TECHNOLOGIES

that limits the performance of a CRT when it is employed to display images of fast-moving objects. The overall CRT display specifications thus directly correspond to phosphor type, persistence (Torr, 1985), and luminous efficiency. The ability of the eye to resolve detail depends on image contrast and brightness. The higher the luminance and contrast beyond a specified limit, the better the resolution of the eye. However, as luminance increases in the CRT, the contrast generally decreases.

Contrast is expressed in terms of the ratio of the luminance of bright and dim regions of the display. A particular phosphor-related effect, referred to as halation, plays a significant role in limiting the contrast ratio. This can be described as follows. Light rays that enter the glass at normal or nearly normal angles will be transmitted mostly undeterred. However, when the angle of incidence exceeds the critical angle, light will be totally internally reflected and returned to the inner surface of the CRT. The reflected light along with a part of it that may undergo further reflections will give rise to a series of concentric circles of diminishing brightness (Jenkins, 1981). This process of halation is able to reduce the contrast ratio from 50:1 to a value between 10:1 and 15:1. The value of the contrast ratio, however, can be improved by almost a factor of 2 by bonding a neutral density filter to the CRT faceplate. Other means of improving the contrast ratio or MTF involve the use of either powderless screens fabricated by vapor-phase deposition or fiber-optic windows that include absorbing material between the fibers to stop the process of halation.

1.7 DISCUSSION

The major use of image intensifier tubes is in night vision. However, in recent years, more and more image tubes have turned up in a host of nonmilitary applications, to the point where they represent an important case of technology transfer. Maintaining and achieving a contrast is an important consideration in the design of image tubes. In general, image tube performance is affected by light scatter, stray reflections, and inadequate component matching. The ongoing research and developmental work involves studying high-efficiency photocathodes with a spectral response extending further into the infrared as well as increasing the sensitivity of the next generation of image tubes.

The CRT is available in numerous versions and is capable of so many design trade-offs that it is almost impossible to rate one version against another. These design trade-offs include characteristics such as resolution, luminance, contrast ratio, video and deflection bandwidths, tube and phosphor luminous efficiencies, depth/diagonal ratio, size, duty cycle, life, cost per resolution element, and weight. Side by side, flat-panel display technologies are now very mature and support displays such as flat CRTs, electroluminescent displays, plasma panels, liquid-crystal displays, and many more. Competition between flat-panel and

CRT displays is very strong, but CRTs seem to endure. The CRT is so entrenched in the industrial and commercial world that it seems to be irreplaceable. However, the other display technologies also offer compelling features and newer application areas. The remaining chapters in this book discuss many of those challenging displays and display aspects.

REFERENCES

Awwal, A. A. S., A. K. Cherri, M. A. Karim, and D. L. Moon (1991). Dynamic modulation transfer function of a display system, *Appl. Opt. 30:* 201–205.
Biberman, L. M., and S. Nudelman, Eds. (1971). *Photoelectronic Imaging Devices,* Vol. 2, Plenum, New York.
Csorba, I. P. (1981). Modulation transfer function (MTF) of image intensifier tubes, *SPIE 274:* 42–49.
Csorba, I. P. (1985). *Image Tubes,* Howard W. Sams, Indianapolis, IN.
Freeman, C. F. (1973). Image intensifier tubes, *SPIE 42:* 3–13.
Gao, M. L., S. H. Zheng, and M. A. Karim (1990). Restoration of dynamically degraded gray level images in phosphor based display devices, *Opt. Eng. 29:* 878–882.
Hawkes, P. W., and E. Kasper (1989). *Principles of Electron Optics,* Academic, San Diego, CA.
Jenkins, A. J. (1981). Modulation transfer function (MTF) measurements on phosphor screens, *SPIE 274:* 154–158.
Karim, M. A. (1990). *Electro-Optical Devices and Systems,* PWS-Kent, Boston, MA.
Sandel, B. R., D. F. Collins, and A. L. Broadfoot (1986). Effect of phosphor persistence on photometers with image intensifiers and integrating readout devices, *Appl. Opt. 25:* 3697–3704.
Tannas, L. E., Jr., Ed. (1985). *Flat-Panel Displays and CRTs,* Van Nostrand Reinhold, New York.
Torr, M. R. (1985). Persistence of phosphor glow in microchannel plate image intensifiers, *Appl. Opt. 24:* 793–795.
Woodhead, A. W., and R. Ward (1977). The channel electron multiplier and its use in image intensifier, *Radio Electron. Eng. 47:* 545–553.

2
Liquid-Crystal Display Device Fundamentals

David Armitage

*Lockheed Missles & Space Company, Inc.,
Palo Alto, California*

2.1 INTRODUCTION

The liquid-crystal display (LCD) industry has grown rapidly over the past 20 years, and many device structures have evolved (Schadt, 1989). This chapter provides a brief description of the basic liquid-crystal properties relevant to LCDs and a guide to the important device structures. Several books and reviews are available that provide more detailed accounts (de Gennes, 1974; Priestly et al., 1974; Chandrasekhar, 1977; Blinov, 1983; Bahadur, 1984, 1990; L. A. Beresnev et al., 1988).

Liquid crystals are an intermediate phase existing between crystalline solids and isotropic liquids. The intermediate character gave rise to the term *mesophase*, but attempts to popularize this terminology have failed, possibly because of the dramatic conflict in the original description *liquid crystal*. Some revision of terminology has taken place over the years, but with no significant change in the LCD literature.

Liquid crystals provide a unique combination of fluidity and order that facilitates structural reorganization. Biological processes mediated by liquid crystalline phases are an interesting structural example. Electro-optical devices exploit the optical and dielectric anisotropy arising from order, together with the easy structural deformations associated with fluidity.

2.2 LIQUID-CRYSTAL PHASES

Ordinary liquids generally possess some short-range order on a molecular scale but lack any macroscopic or long-range order. This is described as the *isotropic phase* in liquid-crystal terminology. At sufficiently high temperature, any liquid-crystal material will become isotropic. The transition temperature to the isotropic state is called the *clearing temperature* because of the drastic reduction in bulk light scattering at this point.

Below the clearing temperature, various degrees of ordering in the fluid produce the liquid-crystal phases also known as *ordered fluids*. The solid-crystal phase is associated with the periodic lattice structures of positional order. This regular crystal periodicity is lost in the fluid phase. However, orientational order and diffuse structures are consistent with fluidity.

The liquid-crystal phases were first identified by their interesting birefringent textures observed with the polarizing microscope. The term *nematic* (Greek, thread) is derived from the threadlike structures that appear in this phase. A variant of the nematic phase is described as *cholesteric,* simply because the first materials demonstrating this property were cholesterol derivatives. The term *smectic* (Greek, soap) is associated with the soaplike character of the early materials.

Liquid-crystal materials are also broadly divided into lyotropic and thermotropic types. Lyotropics are solutions that exhibit a phase transition with concentration change and temperature. Thermotropic materials are not solutions, and their phase transitions are determined by temperature. Lyotropics are of importance in biology and related fields but of no value in current display technology.

2.2.1 Nematic Phase

The nematic liquid-crystal (NLC) phase is the simplest of the liquid-crystal structures. Orientational order exists in the presence of translational disorder. The structure of a uniaxial nematic is sketched in Figure 2.1, where typical liquid-crystal rodlike molecules are indicated, with dimensions on the order of 2 nm. The orientation axis is labeled by a unit vector **n**, known as the nematic director. The NLC is not polar, and $+\mathbf{n}$ is physically the same as $-\mathbf{n}$. Biaxial ordering is rare, and *biaxial nematics* appear to be of academic interest only.

The molecular orientational order gives rise to anisotropic properties such as birefringence. The nematic molecules undergo translational and rotational motions such that the orientational order is a statistical property subject to local fluctuations. The corresponding refractive fluctuations scatter light and limit the transmission of bulk nematics. However, the short optical path in LCDs makes light scattering insignificant.

The degree of NLC order is quantified by an orientational order-parameter S,

LCD DEVICE FUNDAMENTALS

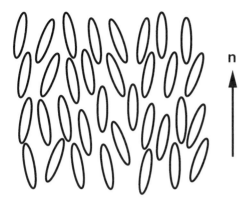

Figure 2.1 Nematic liquid-crystal phase.

$$S = \frac{1}{2}(3\langle\cos^2\theta\rangle - 1) \tag{2.1}$$

where θ is the fluctuation in polar angle relative to **n**, and $\langle\cos^2\theta\rangle$ is a molecular average. For complete order, $\theta = 0$, implying $S = 1$; in the totally random case, $\langle\cos^2\theta\rangle = 1/3$, giving $S = 0$ for the isotropic liquid. Typically S may approach 0.8, well below the clearing point, and decreases to approximately 0.4 near the discontinuous phase transition to the isotropic state.

2.2.2 Cholesteric Phase

The cholesteric liquid-crystal (CLC) phase is a nematic-like phase consisting of chiral molecules. The chiral component of molecular interaction produces a helical twisted structure of pitch p, characteristic of the cholesteric phase, as shown in Figure 2.2. The similarity of the cholesteric and nematic phases has led to the term *chiral nematic* (N*) as an alternative to the historic term cholesteric. The term twisted nematic is reserved for device structures.

The addition of soluble chiral molecules to an NLC phase produces a helical twisted structure identical to the CLC in macroscopic properties. The helical pitch is inversely proportional to the chiral component at low concentration. Molecular chirality may be right-handed (dextro) or left-handed (levo), and their effects on the helical pitch cancel. A *racemic* mixture has equal components of dextro and levo components and provides an untwisted NLC.

The periodicity associated with the CLC helix can diffract visible light to produce bright colors. The periodicity is temperature-dependent, causing the color to change dramatically with temperature. Temperature sensing with color change

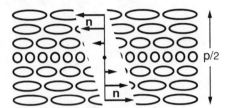

Figure 2.2 Cholesteric liquid-crystal phase.

is an important application of CLCs. Short-pitch cholesterics demonstrate *blue phases* close to the isotropic transition temperature. These phases are of considerable academic interest but of little value in device work at this stage of development. The cholesteric liquid crystal shows a richer variety of birefringent texture than the nematic. This texture is the basis of some device applications, where switching between scattering and clear states is employed.

2.2.3 Smectic Phase

The smectic liquid-crystal phases possess some diffuse positional order in addition to orientational order. The additional smectic order forms the layered structure shown for the smectic-A (S_A) phase in Figure 2.3. For some materials the NLC to S_A transition is a continuous or second-order phase transition, which gives some indication of the diffusivity of the layers indicated in Figure 2.3. For simplicity, the smectic layers and intralayer structure are sharply defined in the figures, but physically the structure is diffuse and is subject to thermal fluctuations. The layered order in the S_A phase can be described as a one-dimensional density wave of period d, generating an order-parameter $\tau = \langle \cos(2\pi z/d) \rangle$. The orientational fluctuations are greatly reduced in the S_A phase compared to the NLC. This is revealed by the reduction of light scattering in bulk single S_A crystals, which are transparent on visual inspection.

In the smectic-C (S_C) phase, the molecules have a polar angle tilt (θ) relative to the layer normal, as shown in Figure 2.4. The azimuth orientation of the molecular axis in the smectic plane is degenerate. The weak orientational elasticity allows nematic-like fluctuations of the degenerate azimuth angle while preserving the layer thickness. A single-crystal S_C scatters light, and like the nematic has limited transparency.

Molecular chirality in the S_C produces a twist in the azimuth tilt direction of successive layers, as shown in Figure 2.5. The chiral phase is indicated S_C^* and is the simplest phase to support ferroelectricity. An S_C phase can be converted to an S_C^* phase by addition of a soluble chiral component. The chirality may be right- or left-handed, and twist can be canceled, as discussed earlier in the cholesteric case.

LCD DEVICE FUNDAMENTALS

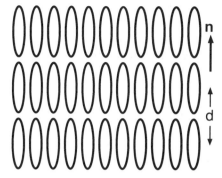

Figure 2.3 Smectic-A liquid-crystal phase.

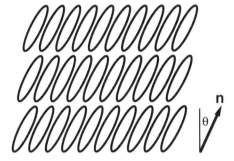

Figure 2.4 Smectic-C liquid-crystal phase.

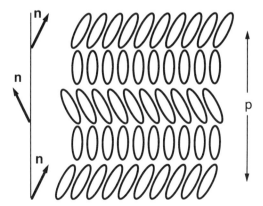

Figure 2.5 Chiral smectic-C liquid-crystal phase.

There are further smectic phases, which involve ordering effects in the plane but are of little interest to device work. The study of liquid-crystal phases continues to reveal unexpected behavior, such as the chiral S_A and the discotic phases. Although device applications are not obvious in these cases, the possibility of new phases and new device concepts should not be ignored.

2.2.4 Ferroelectric Phase

Nematic symmetry was originally derived from an observed nonpolar NLC state; therefore, by definition, an NLC is not ferroelectric. The symmetry conditions are altered in a distorted region of nematic, allowing polar effects in that region. The local polarization associated with the distortion is known as the *flexoelectric* effect. Flexoelectric effects are generally insignificant in LCD behavior at this stage of development; however, some device potential has been demonstrated (Barberi et al., 1989).

The flexoelectric properties of nematics led to the prediction of ferroelectricity in the S_C^* phase (Meyer et al., 1975; Meyer, 1977). Symmetry considerations show that S_C^* is the simplest phase allowing ferroelectricity. Figure 2.6 illustrates some important aspects of the ferroelectric S_C^* phase. The spontaneous polarization (P_S) vector lies in the smectic layer plane and is orthogonal to the local molecular director **n**. Relative to the smectic layer, **n** is described by a polar tilt angle θ and a degenerate azimuth angle ϕ. For successive smectic layers, ϕ increases and **n** spirals with helical pitch p around a cone angle 2θ. The polarization follows this rotation, where the helical P_S pattern results in zero macroscopic polarization.

The ferroelectric S_C^* phase can be achieved by doping the S_C phase with chiral molecules having suitable dipole moments (Kuczynski and Stegemeyer,

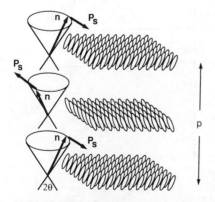

Figure 2.6 Ferroelectric chiral smectic-C.

1980). This is the basis of ferroelectric liquid-crystal (FLC) device materials development, where temperature range, viscosity, polarization, and tilt angle must be engineered. The additional freedom provided by chiral doping is an important advantage. Ferroelectricity is identified with molecular chirality, and the helical structure is a by-product of this chirality in liquid crystals. A nontwisted ferroelectric state can be achieved in a suitable mixture at a fixed temperature, where the chiral element is retained while canceling the helical manifestation. Complete suppression of the helical pitch is not required for FLC device materials, as the device thickness is substantially less than the pitch (Clark and Lagerwall, 1980). P_S values on the order of 100 nC/cm^2 have been achieved.

The molecular tilt angle θ is an order parameter of the S_C^* phase and decays on approach to the S_A phase transition. The P_S value is determined by ordering of molecular dipole moments and is therefore proportional to the molecular tilt angle. Close to the transition temperature, soft-mode behavior is important, and an electric field induced S_C molecular tilt known as the *electroclinic effect* is observed in the S_A temperature range (Garoff and Meyer, 1977).

2.3 LIQUID-CRYSTAL PROPERTIES

2.3.1 Elasticity

The minimum free energy of the NLC state corresponds to uniform orientation throughout the entire volume, analogous to the perfect single-crystal state in a solid. Elastic response to shearing forces is not possible in a fluid. Distortions in nematic orientation are opposed by forces much weaker than solid-state elasticity. The light-scattering behavior and low energy requirements of LCDs are related to the weak elasticity.

The NLC thermodynamic free energy can be expanded in spatial derivatives of **n**, where the coefficients in this expansion are identified with elasticity. Symmetry restrictions eliminate first-order terms and reduce the number of bulk curvature, (second-order) elastic constants to three independent terms:

$$\text{Bulk free energy} = (1/2)[K_1(\nabla \cdot \mathbf{n})^2 + K_2(\mathbf{n} \cdot \nabla \times \mathbf{n})^2 + K_3(\mathbf{n} \cdot \nabla \mathbf{n})^2] \quad (2.2)$$

The coefficients K_1, K_2, and K_3 are known as the Frank elastic constants and correspond to splay, twist, and bend deformations, respectively, as shown in Figure 2.7. Higher order terms in the free-energy expansion are associated with higher order elastic coefficients, which are of academic interest only. In semiquantitative work, considerable simplification follows from the approximation $K_1 = K_2 = K_3 = K$. In LCD NLCs at room temperature, K is on the order of 10 pN. The elasticity weakens with declining order related to increasing temperature such that $K \propto S^2$.

In smectics the elastic force controlling the layer periodicity is much stronger

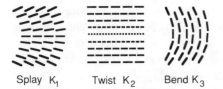

Figure 2.7 Nematic elastic constants.

than the orientational elasticity. This disparity in elasticity gives rise to defect structures where conservation of layer number restricts layer curvatures. The resulting birefringent texture, characteristic of the smectic phases, is described as *focal conic*. The conservation of layer number forbids bend and twist deformations in the S_A director field. However, the degenerate tilt angle in S_C crystals does allow some nematic-like elastic deformations in director field **n**.

2.3.2 Viscosity

The viscous behavior of liquid crystals is anisotropic; moreover, flow and orientation are coupled. A flow process induces a favored director orientation, and conversely a field-induced reorientation of the director induces a flow effect. A detailed analysis of the NLC involves five independent viscous coefficients. The behavior is usually approximated by a single rotational viscous coefficient (γ) in semiquantitative device work.

$$\text{Viscous torque} = \frac{\gamma \partial \theta}{\partial t} \quad (2.3)$$

The flow and rotation viscous interaction can be taken into account by a modified rotation viscosity (γ^*) (Pieranski et al., 1973; van Sprang and Koopman, 1988). At room temperature the LCD NLC flows with a viscosity about an order of magnitude greater than that of water.

The smectic viscosity is much lower for shear processes along the layers, relative to shear that disturbs the layer structure. The S_C degenerate tilt rotation has a viscous response similar to that of NLCs. At room temperature the bulk smectic generally resists flow with a greaselike level of viscosity.

The viscosity rises exponentially with decreasing temperature, imposing a corresponding increase in LCD response time. High viscosity at low temperature limits the range of video-rate LCDs to near-room-temperature operation. At a given temperature the viscosity is determined by molecular structure, and great effort has been made to synthesize low-viscosity NLCs for video-rate LCDs. At this stage of development only marginal improvements in NLC viscosity are anticipated.

2.3.3 Disclinations

Deviations from a solid crystal lattice are described as *dislocations*. The liquid-crystal analog is a sudden change in orientation, giving rise to the term *disclination* (e.g., de Gennes, 1974; Chandrasekhar and Ranganath, 1986). The stability of disclinations is determined by topological constraints and boundary conditions. In LCDs disclinations are usually defect structures that spoil the appearance and should be avoided. However, in smectic scattering devices the disclination structure is essential to the device function.

The disclination structure must minimize the elastic energy and satisfy the imposed topological constraints. In the nematic two-dimensional wedge (also known as a screw), disclinations provide a tractable example, where the nematic director orientation ϕ is linearly related to the polar angle θ,

$$\phi = s\theta + \theta_0 \tag{2.4}$$

Within the constraint of Eq. (2.4), the elastic free energy given in Eq. (2.2) is minimized. The director rotates $2s\pi$ around any circuit enclosing the singular disclination line, as shown in Figure 2.8 for equal elastic constants. The strength of the disclination is given by s, which must be an integer or half-integer. Low s values are energetically favored. The continuum approximation breaks down in the disclination core region, where the distortion approaches the molecular scale. Disclination points are observed in nematic droplets and are of practical interest in polymer-dispersed LCDs.

The smectic layer structure gives rise to a texture or defect structure described as focal-conic domains. The interlayer elasticity is much stiffer than the intralayer elasticity. Therefore, the distorted structure favors layers of equal thickness with common centers of curvature. Constant layer thickness and matching curvature can be accommodated in a space-filling structure containing related line

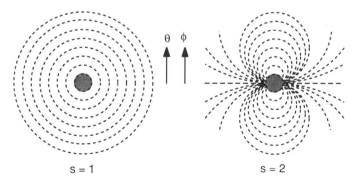

Figure 2.8 Nematic disclinations in two dimensions.

singularities of elliptical and hyperbolic form, hence the term focal-conic. An example of focal-conic structural cross section is shown in Figure 2.9. The focal-conic texture scatters light effectively and is the basis of certain LCDs. Appropriate boundary conditions allow the scattering focal-conic texture to be stored indefinitely. The helical periodicity of the cholesteric phase implies layerlike behavior, giving rise to focal-conic domain structures. However, in cholesteric textures the focal-conic aspect of the disclination lines is less developed.

The development of FLC devices has revived interest in smectic textures, particularly the S_C^* phase. The *zigzag* defect is a literal description of a defect first observed in FLC research devices. After much study the nature of the zigzag defect has been related to the mismatched *chevron* structures shown in Figure 2.10. The defect surface formed near the cell center at the chevron "V" is unique in being a disclination surface rather than a disclination line (Clark and Rieker, 1988).

2.3.4 Surface Alignment

Special conditions exist at the surface or interface region of a liquid crystal that determine the local orientation. The surface-induced orientation propagates to influence the bulk liquid-crystal orientation. The use of surface treatments to control the alignment in LCDs is a critical aspect of the technology (e.g., Cognard, 1982; Uchida, 1990).

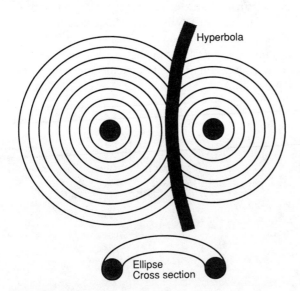

Figure 2.9 Focal-conic structure.

LCD DEVICE FUNDAMENTALS

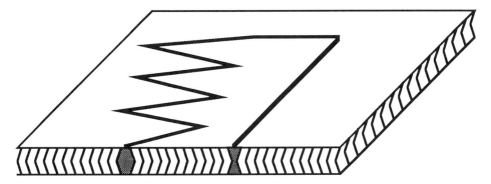

Figure 2.10 Chevron structure in S_C^* showing zigzag disclinations.

Surface alignment studies are complicated by the problems of reproducibility inherent in surfaces. A spurious monomolecular surface layer can drastically influence the orientation of liquid crystals in contact with the surface. Therefore, the observations of surface effects sometimes vary from laboratory to laboratory. However, some surface alignment techniques are reliable and easy to implement.

There are two basic alignments that the nematic director can adopt relative to the surface: perpendicular (*homeotropic*) and parallel (*homogeneous*). When applied to orientation, the terms homeotropic and homogeneous are confusing because their literal meaning is identical. The term *perpendicular alignment* is unambiguous. However, the term *parallel alignment* should be qualified by an azimuth direction in the plane. Moreover, nonuniform parallel alignment is possible, which explains the origin of the term *homogeneous* to distinguish uniform parallel alignment.

Strong anchoring is a term used to describe a surface alignment that is undisturbed by bulk distortions or applied electric or magnetic fields. The director is fixed at the surface and provides a boundary condition on any equations describing the director distribution. In *weak anchoring* the director has a preferred alignment, but the orientation can change in response to bulk distortion and applied fields. *Anchoring strength* is usually approximated by the expression

$$W = W_0 \sin^2 \theta \tag{2.5}$$

where θ is the angular deviation from the favored surface orientation and W_0 is a surface energy parameter describing the anchoring strength.

The short-range interactions of interface chemistry will determine the preferred molecular surface alignment. Longer range elastic interactions will also influence the alignment. If a grating-like surface topology favoring parallel alignment is provided, it can be shown that the anisotropic elasticity of the nematic promotes alignment parallel to the grating (Berreman, 1972). Figure 2.11 illus-

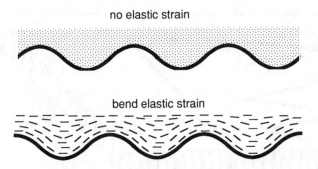

Figure 2.11 Surface grating topology favoring alignment along the grating.

trates the bend elastic distortion for alignment orthogonal to the grating; alignment along the grating is achieved without elastic distortion. The elastic energy varies inversely with the cube of the grating period, and calculations show that submicrometer periods are required for effective alignment. This simple physical theory explains how unidirectional rubbing of a surface can provide uniform parallel alignment by virtue of the induced grating-like microgrooves.

Other factors are involved in the rubbing process. A favorite alignment technique of the LCD industry employs polyimide coating followed by unidirectional rubbing with a polishing cloth or roller. The microgrooving associated with the rubbing is also accompanied by some polymer chain alignment. The polymer chain alignment induces a corresponding NLC alignment and appears to be the dominant mechanism (e.g., Geary et al., 1987; Ishihara et al., 1989).

Uniform parallel or perpendicular alignment is inadequate for a nematic LCD, where a *tilt bias* is required to control the director electric field response. The need for tilt bias will become clear when field effects and device structures are discussed. A rubbed polymer surface provides a uniform alignment, with the nematic director tilted 1–20° out of the surface plane. The tilt magnitude is determined by the polymer composition and rubbing details, while the alignment is along the rub direction.

Anisotropic surfaces providing liquid-crystal alignment can be prepared by oblique evaporation methods, sometimes described as the *Janning method* (Janning, 1972). Electron microscopic studies of the obliquely evaporated surfaces reveal the origin of the nematic alignment effect (Goodman et al., 1977; Armitage, 1980). Small-angle deposition (SAD) of silicon oxide, at a glancing angle of approximately 5° to the substrate surface, promotes tilted columnar growth toward the evaporant source. A nematic that aligns parallel to silicon oxide will align parallel to the columnar direction to minimize the surface and elastic energy. The resulting director tilt angle is uniform in a given sample but varies from 20° to 50° according to the mixture of oxides and deposition conditions.

Medium-angle deposition (MAD) of silicon oxide at an angle of approximately 30° to the substrate surface promotes a corrugated surface topology. The surface structure is similar to wind or water patterns on sand, where the corrugations run orthogonally to the evaporation direction. The nematic director aligns along the corrugations in the same way as the grating surface. The alignment is parallel to the surface, without tilt bias. For given substrate and source positions, the SAD- and MAD-induced alignments are orthogonal.

Tilt bias can be controlled by sequential SAD and MAD depositions, where the substrates are rotated 90° prior to the second deposition so as to reinforce the alignment direction. The tilt bias achieved in this method can be varied continuously according to the deposition thicknesses (e.g., Kahn, 1977). Oblique evaporation methods are used in low-production high-quality LCDs, but large-scale production favors low-cost rubbed polymeric film methods.

Surfactants provide a simple and effective method of controlling the surface chemistry. The surfactant is applied to the surface in liquid or vapor form and is generally designed to bond to oxide surface sites. The surfactant presents a chemical moiety to the liquid crystal that influences the alignment. Many surfactants are known that induce either perpendicular or random parallel nematic alignment. Surfactants can be combined with oblique evaporation methods to provide a tilt bias relative to perpendicular alignment (e.g., Cognard, 1982).

The Langmuir–Blogett method of creating oriented surface films has been applied to liquid-crystal alignment (Ikeno et al., 1988). However, this is essentially a laboratory technique and is difficult to apply to production methods. The alignment requirements of the supertwisted NLC and FLC devices have revived interest in alignment techniques. Control of the strength of the surface alignment forces and surface tilt are important aspects of some devices.

2.3.5 Dielectric Response

The NLC dielectric response is anisotropic, with uniaxial symmetry. The dielectric constants perpendicular and parallel to **n** are labeled $\varepsilon_\perp$ and $\varepsilon_\parallel$, respectively. The dielectric anisotropy, which may be positive or negative, generally increases with order S and is written

$$\Delta\varepsilon = \varepsilon_\parallel - \varepsilon_\perp \qquad (2.6)$$

The dielectric constants decrease with increasing frequency, as some components of the induced dipole moment are related to molecular motion. At room temperature, in the audio-frequency range, typical LCD NLC relative permittivities and anisotropies are on the order of 5. At optical frequency the dielectric response determines the refractive indices $n_\perp$ (ordinary) and $n_\parallel$ (extraordinary) perpendicular and parallel, respectively, to the director or optic axis. The birefringence (Δn) is always positive and increases with order S,

$$\Delta n = n_\parallel - n_\perp \tag{2.7}$$

For LCD NLC materials the refractive index is approximately 1.5, and the birefringence is on the order of 0.1.

The magnetic susceptibility anisotropy follows a similar pattern and is written $\Delta\chi = \chi_\parallel - \chi_\perp$. The magnetic response is of little interest in devices but is often used in experiments for freedom in orientation control, and without the current flow associated with electric fields. The smectic phases are usually described by the above expressions. However, the weak biaxiality of the S_C^* phase can be significant in device work (Elston, et al., 1990).

2.3.6 Applied Field Effects

An applied electric field contributes an additional term (F_E) to the liquid crystal free-energy density,

$$F_E = -(1/2)\Delta\varepsilon(\mathbf{E}\cdot\mathbf{n})^2 \tag{2.8}$$

which favors alignment along the electric field (**E**) direction for positive dielectric anisotropy $+\Delta\varepsilon$, and orthogonal to **E** for $-\Delta\varepsilon$. Physically, the induced dipole moment reacts with the applied field to produce a torque, which therefore varies as E^2. The torque and resulting director alignment are independent of **E** polarity. The orientational effect of an electric field is the basis for most LCDs. The magnetic field behavior is described by a similar expression involving magnetic field and susceptibility.

The FLC spontaneous dipole moment **P**s also contributes to the free energy density a term

$$F_E = -\mathbf{P}s\cdot\mathbf{E} \tag{2.9}$$

which favors orientation of **P**s along **E**, hence **n** orthogonal to **E**. Physically, the field acts on **P**s to produce a torque that varies as the first power and hence depends on **E** polarity. The ability to provide positive or negative torque according to the applied field polarity is a critical advantage in FLC devices.

In nematic LCDs the interplay between elastic and electric torques determines the operation. Threshold devices are based on the Freedericksz transition (e.g., de Gennes, 1974). It is instructive to examine the equations governing the response of the NLC cell shown in Figure 2.12, which has strong perpendicular alignment (boundary condition $\theta = 0$ at planar surfaces) and negative dielectric anisotropy. The free energy expressions (2.2) and (2.8) can be differentiated to give a torque balance equation between electric and elastic torques, where **n** is a function of z only (e.g., Deuling, 1978):

$$\frac{(K_1 \sin^2\theta + K_3 \cos^2\theta)\, d^2\theta}{dz^2} + \frac{(K_1 - K_3)\sin\theta\cos\theta\, d\theta}{dz}$$
$$= -E^2(z)\,\Delta\varepsilon\sin\theta\cos\theta \tag{2.10}$$

LCD DEVICE FUNDAMENTALS

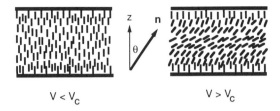

Figure 2.12 Freedericksz transition.

In the limit of small distortions, $E(z)$ is constant, and the initial condition $\theta(z) = 0$ is maintained below a critical field or voltage,

$$V_c = \pi[K_3/(-\Delta\varepsilon)]^{1/2} \tag{2.11}$$

Above the critical voltage V_c, the nematic deforms and $\theta(z)$ changes. The response can be approximated in the limit of low voltages, but otherwise computational work is required. The cell thickness is reflected in both the elastic and electric terms, making the critical voltage independent of thickness. For LCD NLCs, the critical voltage is generally on the order of 1 V.

For the parallel alignment ($\theta = 90°$) and positive anisotropy, the critical voltage becomes

$$V_c = \pi(K_1/\Delta\varepsilon)^{1/2} \tag{2.12}$$

The voltage-induced tilt direction is degenerate for precise planar or perpendicular alignment, generally resulting in tilt domains. A tilt bias of about 1° is sufficient to remove the ambiguity and avoid the domains without significant change in V_c.

The dynamics of the Freedericksz transition are complex, with higher spatial gradients of the director field responding at faster rates. A crude approximation that has some validity in device work takes the form (e.g., Blinov, 1983)

$$\tau_r = \gamma d^2/(\Delta\varepsilon V^2 - K\pi^2) \tag{2.13}$$

$$\tau_d = \gamma d^2/K\pi^2 \tag{2.14}$$

where τ_r and τ_d are exponential time constants describing the rise and decay rates for a cell of thickness d. The expression for τ_r greatly underestimates the response time when the nematic director and electric field approach orthogonal or parallel alignment. This follows from the vanishing of the electric torque under these conditions. The rise time can be reduced by increased drive voltage, but fast decay favors a thin cell structure. Decay times on the order of 10 msec are achieved in 4 μm thick cells.

The helical twist of the CLC can be unwound by a critical electric field (e.g., de Gennes, 1974),

$$E_c = (\pi^2/p)(K_2/\Delta\varepsilon)^{1/2} \tag{2.15}$$

This has LCD applications, where the effect is sometimes described as the field-induced CLC-to-NLC phase transition, although it is not a phase transition in the thermodynamic sense. The unwinding field can be used to switch from a scattering to a nonscattering state.

A similar unwinding of the S_C^* is related to a critical field,

$$E_c = (\pi^4/4)(B_3/pP_s) \tag{2.16}$$

where B_3 is the effective elastic coefficient (Meyer, 1977; Beresnev et al., 1988). The value of the unwinding field is of interest in the distorted-helix FLC device.

2.3.7 Electrochemistry

The electrical resistivity of field-effect LCD materials is very high, on the order of 10 GΩ·cm or more. Ionic conductivity is responsible for bulk charge transfer. The ions and liquid-crystal solvent form an electrolytic solution. This electrolyte and the contacting electrodes form an electrochemical cell. In the isolated state, a constant electrochemical potential is achieved in the system by ion migration. The equilibrium distribution of ions will depend on the detailed properties of the electrodes and solution. In the electrode surface region the structure depends on the species present, adsorption, and *double-layer* behavior. The double layer refers to the first solvated ionic layer at closest approach to the electrode, together with an outer diffuse ionic layer.

A direct current flow requires an applied voltage and a charge-exchange mechanism between the electrode and electrolyte. The mechanism is oxidation at the anode and reduction at the cathode, caused by a shift in electrode potential. If the *redox* processes are not completely reversible, the chemistry of the cell changes with current flow. At issue is the electrochemical stability of a liquid crystal and the electrode material in the presence of ionic impurities.

The minimum shift in electrode potential to oxidize or reduce a liquid-crystal molecule can be determined by electrochemical methods and is usually in the range of 0.5–2 V. A dc voltage applied to the cell drives the bulk ions to the electrode regions, screening the bulk liquid crystal from the applied field. All of the applied voltage is taken up by the double-layer electrode region. Therefore, dc voltages on the order of 1 V can induce electrochemical reactions in liquid crystals. The low-voltage electrochemical activity in NLCs was a severe problem in dynamic scattering devices, which depend on current flow. This led to the development of reversible redox dopants with lower electrochemical potentials designed to protect the NLC (e.g., Sussman, 1974; Barret et al., 1976a, 1976b).

Electrochemical effects are minimized by ac operation of the cell. Ion depletion of the bulk liquid crystal is eliminated to prevent electric field concentration in the electrode regions. Electrochemical processes that take place close to the

LCD DEVICE FUNDAMENTALS 35

electrodes are generally reversed as the electrode polarity is rapidly reversed at each cycle. Contaminants that undergo electrochemical change, such as water, can in turn react with the liquid crystal and electrodes to degrade material stability. Even with ac excitation, stabilizing additives with low and reversible redox potentials should protect against any residual electrochemical activity (Kohlmuller and Siemsen, 1982). With increasing voltage, the electric field approaches the electrochemical activity level, and ultimately dielectric breakdown.

The electrochemical behavior of an LCD is important in lifetime studies, where irreversible changes known as electrochemical degradation can take place. The electrochemical products generally raise the conductivity level and are deposited on the electrode surfaces, changing the alignment conditions. Lifetime tests are conducted on commercial LCDs to determine limitations on cell voltage, frequency, dc component, and purity. However, the results are inevitably proprietary. In the higher voltage devices it is generally found that an upper voltage level of 50 V at 100 Hz is acceptable, whereas low-voltage devices tolerate lower frequencies. The dc component of the ac drive voltage is usually made less than 50 mV to provide an adequate safety margin against electrochemical reactions.

2.3.8 Material Stability

Materials designed for LCDs must meet certain stability requirements (e.g., Schadt, 1988). Chemical and electrochemical stability are important, particularly in the presence of inevitable contaminants, such as water. Photochemical stability must accommodate stray UV light, and intense visible light in projection applications.

A wide temperature range must be accommodated. Crystallization at low temperatures produces voids, phase segregation, and deterioration in alignment on reheating into the liquid-crystal phase. The freezing point is most easily depressed by mixing several components to form a eutectic composition.

The elastic, dielectric, and optical properties are all functions of temperature. An LCD performance is vulnerable to temperature changes, but mixtures can be designed to compensate to some extent for the anticipated temperature effects. The liquid-crystal material in an LCD is always a mixture of compounds designed to suppress freezing and to optimize the performance of the particular device.

Nematic materials are well developed, but significant improvements are anticipated for specific applications. The active matrix LCDs are particularly demanding in resistivity levels >100 GΩ·cm to maintain the pixel charge at low duty cycles. Smectic materials are still evolving, as are the devices, and considerable material advances are expected.

2.4 DISPLAY MECHANISMS

2.4.1 Display Requirements

There are many LC effects that might be used in display devices, but the options are restricted by the display requirements. Adequate visual contrast may be as low as 2:1 or approach 100:1, depending on the application. Brightness and optical efficiency are needed in addition to contrast, particularly in outdoor or other bright environments. The viewing angular range, or off-axis viewing angle, is important in direct-view LCDs. Basic black-and-white operation is valued where color is provided by filters. A strongly nonlinear response, or sharp electro-optic characteristic, is critical in passive-matrix addressing. Tolerance to manufacturing fluctuations is a cost consideration. Power dissipation limits the display size and applications. Adequate response time is essential in video applications. Color and gray-level ranges are critical in TV-type applications. Stability and long lifetime are indispensable.

2.4.2 Nontwisted Nematics

A uniform and perpendicularly aligned NLC cell of thickness d, having $-\Delta\varepsilon$, is positioned between crossed polarizers as shown in Figure 2.13 for the relaxed and voltage-driven states. In the relaxed state, linearly polarized light propagates along the optic axis of the NLC and emerges unchanged, to be blocked by the analyzer. A voltage above the threshold given by Eq. (2.11) deforms the NLC so that optical propagation is at an oblique angle to the optic axis, producing elliptical polarization, having some transmission through the output analyzer. The effect is described as *tunable birefingence* or *electrically controlled birefringence* (ECB). The effective birefringence (Δn_{eff}) is related to the local nematic tilt angle θ and can be approximated in the limit of small Δn by

$$\Delta n_{eff} = \Delta n \sin^2 \theta \tag{2.17}$$

Most LCDs depend on effective birefringent changes associated with optic axis reorientation by electric fields, but the term ECB is identified with nontwisted structures.

The ECB NLCD was among the earliest field-effect devices to be investigated and has an extensive literature (e.g., Kahn, 1972). The contrast and brightness are maximized by linearly polarized input light oriented at 45° to the NLC tilt direction. The tilt direction is controlled by a small tilt bias built into the surface alignment. The ideal transmission intensity (T) can be expressed in terms of the birefringent phase retardation ($\delta = 2\pi d\, \Delta n_{eff}/\lambda$) between the linearly polarized orthogonal modes of equal amplitude:

$$T = \sin^2 \delta/2 \tag{2.18}$$

LCD DEVICE FUNDAMENTALS

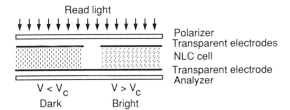

Figure 2.13 Electrically controlled birefringence modulation.

In the relaxed or off state, $\delta = 0$ and $T = 0$, which applies to all wavelengths. In the driven on state, maximum $T = 1$ when $\delta = (2n + 1)\pi$, and $T = 0$ when $\delta = 2n\pi$, as indicated in Figure 2.14. The transmission is wavelength (λ)-dependent, as over the visible range $\delta \sim 1/\lambda$. However, low order (low n) makes the wavelength effect tolerable. Thickness uniformity is an essential requirement.

For off-axis viewing, δ changes in relation to apparent cell thickness and NLC tilt angle. In the undistorted state, for off-axis viewing at polar angle θ and azimuth ϕ, the transmission can be approximated using Eqs. (2.17) and (2.18):

$$T = \sin^2 2\phi \, \sin^2[(\pi/\lambda)(\Delta nd \sin\theta \tan\theta)] \tag{2.19}$$

For minimum $\Delta nd \approx \lambda$, the off-axis dark transmission at $\phi = 45°$ and $\theta = 20°$ is $T = 15\%$, limiting the contrast ratio to 7:1. Increasing θ to 30° gives $T = 62\%$, with a corresponding fall in contrast to 1.6. This simple calculation, which ig-

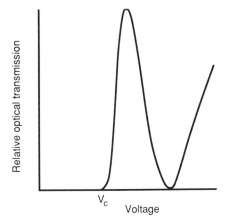

Figure 2.14 ECB voltage–transmission characteristic.

nores refraction at the cell surfaces, indicates how rapidly the contrast can disappear with viewing angle.

In general, the phase retardation as a function of applied voltage must be evaluated numerically, but in the low-voltage limit it can be approximated (e.g., Deuling, 1978) and simplifies for small birefringence,

$$\delta = \frac{(V/V_c - 1)(2\Delta nd/\lambda)}{K_1/K_3 - \Delta\varepsilon/\varepsilon_\perp} \tag{2.20}$$

The development of suitable NLCs with negative dielectric anisotropy has led to a revival of interest in ECB devices (e.g., Sterling et al., 1990). For passive matrix addressing, a sharp turn-on characteristic is required which is achieved in ECB devices by the following conditions implicit in Eq. (2.20): small pretilt from normal $<1°$; maximum retardation $d\Delta n$; small elastic ratio K_1/K_3; small $-\Delta\varepsilon/\varepsilon_\perp$; small V_c. A thin cell favors fast response according to Eqs. (2.13) and (2.14), and a wider viewing angle from Eq. (2.19), indicating that some compromise in d is required. The off-axis viewability can be improved by the addition of an external compensating layer having negative birefringence.

In active matrix addressing, ECB has advantages in higher contrast and viewing angle over twisted nematic structures (Clerc and Deutsch, 1987). The response speed of ECB is adequate for video applications. Details of the dynamical behavior are given by Labrunie and Robert (1973); and Pieranski et al. (1973). The response sharpness that can be achieved with ECB facilitates large-scale passive matrix addressing. However, some sacrifice in speed is required for sharpest response. A recent development employs electric field fringing effects to control the tilt direction. Generation of fringe fields allows bidirectional tilts in each pixel, which compensate to improve off-axis viewability (Clerc, 1991).

The alternative $+\Delta\varepsilon$ NLC configuration employing uniform parallel alignment (with small pretilt) can be analyzed in a similar manner, resulting in a similar expression,

$$\delta = \frac{(V/V_c - 1)(2\Delta nd/\lambda)}{K_3/K_1 + \Delta\varepsilon/\varepsilon_\perp} \tag{2.21}$$

Here we see that the ratio K_3/K_1 is inverted relative to Eq. (2.20). Parallel alignment has a major disadvantage in maximum retardation off state, which is ill defined because of cell thickness tolerances. A low retardation state can be approached at high voltages, but surface NLC effects limit the minimum retardation (Wu and Efron, 1986). The advantage of parallel alignment lies in a wide choice of NLCs, providing large retardation and fast response (e.g., Wu, 1990).

The response speed of a parallel aligned nematic LCD can be reduced by a bias voltage substantially above the critical level, which prevents complete relaxation. Under these conditions the cell activity is dominated by thin surface regions, giving a much faster response than the actual cell thickness would imply and suggesting the term *surface mode* (Fergason, 1980; Fergason and Berman, 1989). Dynamic range can be traded for speed to give frame rates approaching 1 kHz (Armitage et al., 1987). A rotation of one cell substrate through π radians to give parallel, rather than antiparallel, aligned surfaces generates the π *cell*. The opposed tilts at the π-cell surfaces tend to cancel retardation changes with viewing angle in the appropriate plane, providing this configuration with an advantage in off-axis viewing (Bos et al., 1985).

The frame speed can also be increased by choosing a nematic with $\Delta\varepsilon$ that changes sign as the frequency is increased from a low to a high value (e.g., Schadt, 1982). A driven on state and a driven off state are achieved by switching the frequency of the drive voltage. However, it is difficult to find suitable nematics with temperature-insensitive behavior. Moreover, high-frequency operation implies excessive power dissipation.

The fastest response is achieved by varying the direction of the applied electric field to achieve drive on and drive off. This implies a complication in the electrode structure, but speeds on the order of 10 μsec have been demonstrated (Channin and Carlson, 1976). A similar effect has been used to provide edge enhancement in a photoaddressed imaging device (Armitage and Thackara, 1989).

2.4.3 Twisted Nematics of 90° and 45°

The propagation of light in *twisted nematic* (TN) and helical structures provided by CLCs has a rich literature (e.g., de Gennes, 1974). In general, two orthogonal, counterrotating elliptically polarized optical modes are propagated, with orientation along the ordinary and extraordinary NLC optic axis, respectively. For twist Φ over distance d, in the Mauguin limit,

$$\Delta nd \gg \Phi\lambda/\pi \qquad (2.22)$$

the optical mode ellipticity collapses, becoming linearly polarized along the ordinary and extraordinary axes. Therefore, linearly polarized input light will follow the nematic twist in a form of waveguiding.

The thickness uniformity required and wavelength sensitivity of the ECB LCD led to the development of the 90° twisted nematic LCD as shown in Figure 2.15 (Schadt and Helfrich, 1971). The substrate alignments are orthogonal, and cholesteric doping removes the twist degeneracy. Waveguiding in the TN cell rotates light polarized along (or orthogonal to) the input nematic director through 90°, independent of cell thickness and wavelength. An applied electric field orients the nematic director in the direction of optical propagation, concentrating the twist in the central region of the cell and eliminating the waveguide rotation of the input light. In the voltage-activated state, the cell is similar to two orthogonal birefringent plates in series, which preserves the input polarization on transmission, independent of cell thickness and wavelength. Therefore the 90° twisted

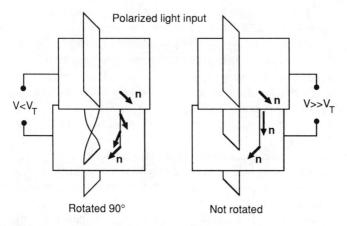

Figure 2.15 90° twisted nematic cell.

nematic cell can act as an optical switch in conjunction with input and output polarizers.

Detailed analysis for twist angle Φ, taking into account output interference effects between normal modes of finite ellipticity, gives the conditions for minimum transmission (Gooch and Tarry, 1975),

$$\Phi[1 + (\pi 2d\,\Delta n/\Phi\lambda)^2]^{1/2} = m\pi/2, \quad m \text{ even} \tag{2.23}$$

Fastest response is achieved at minimum d, which for the 90° twist cell ($\Phi = \pi/2$) gives $m = 2$ and $2d\,\Delta n/\lambda = \sqrt{3}$, as favored by active-matrix LCDs. A sharper electro-optic characteristic suitable for passive matrix addressing favors $m = 4$. Subsidiary maximum transmissions <10% occur for odd m values but are not significant in LCDs.

Unlike ECB, the TN optical response begins at a significantly higher voltage than the deformation threshold V_T, where a substantial deformation is required in order to inhibit waveguiding. The general form of the transmission–voltage characteristic is shown in Figure 2.16. The insensitivity of the characteristic to wavelength provides black/white contrast and accommodates color filters.

The NLC twisted through angle Φ undergoes a Freedericksz transition similar to the untwisted case (V_c), but at a different critical voltage (V_T) (e.g., Raynes, 1986).

$$V_T = V_c\left[1 + \left(\frac{\Phi}{\pi}\right)^2 \frac{K_3 - 2K_2}{K_1} + 4\frac{K_2}{K_1}\left(\frac{d}{p}\right)\left(\frac{\Phi}{\pi}\right)\right]^{1/2} \tag{2.24}$$

where cholesteric doping is represented by the d/p term, which is usually negli-

LCD DEVICE FUNDAMENTALS

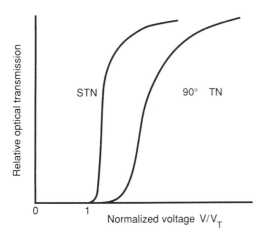

Figure 2.16 Response characteristic of twisted nematic and supertwisted nematic liquid crystals.

gible in the 90° twist case. For passive matrix addressing, the NLC and cell design are optimized for sharp response, off-axis viewing, and speed. The zero-twist response, Eq. (2.21), indicates for sharp response: large Δnd, small K_3/K_1 and $\Delta\varepsilon/\varepsilon_\perp$, which is confirmed by computational work in the 90° twist case. However, speed and off-axis viewing favor small d; therefore, some compromise is required.

The twist cell has a time response that differs from the nontwisted case and can be accounted for by substituting $K = K_1 + (K_3 - 2K_2)/4$ in Eq. (2.14) (Jakeman and Raynes, 1972). Therefore the response time can be minimized by manipulation of the elastic ratios as well as magnitudes.

The photoaddressed reflective light valve favors a 45° twist cell to avoid the inclusion of a polarizer adjacent to the reflector (Grindberg et al., 1975). The dark-background state is achieved by the waveguiding mode, where the reflected light polarization is unchanged from the input polarization. With increasing voltage the twist is broken and an ECB effect results, giving rise to the term *hybrid field effect*. The performance for this particular application can be optimized as indicated for the 90° twist cell and the ECB effect.

The detailed response of twist cells can be determined by numerical analysis. The dynamic response of the 90° twist cell has been computed, with detailed treatment of viscous effects (e.g., Berreman, 1975). An unusual bump on the dynamic electro-optic characteristic known as "optical bounce" was explained in terms of viscous flow orientation interaction, described as *backflow* phenomena.

2.4.4 Supertwisted and Lower-Twisted Nematics

In recent years the behavior of nematics twisted through a variety of angles has been explored to considerable advantage. Bistability is a 360° twist cell containing a cholesteric-doped nematic, with high-tilt alignment, was reported and related to potential improvements in passive matrix addressing, although the response was slow (Berreman and Heffner, 1980, 1981). The sharpness of the NLC cell electro-optic response is indicated by the deformation response of the director tilt angle (θ_m) at the cell center ($z = d/2$) to applied voltage. The low-voltage approximation to the response is expressed in terms of the cell twist (Φ) and voltage threshold (V_T) given by Eq. (2.24) (Raynes, 1986).

$$\theta_m^2 = 4(V/V_T - 1)/[F(K) + \Delta\varepsilon/\varepsilon_\perp] \qquad (2.25)$$

$$F(K) = \frac{(K_3/K_1)\{1 - (\Phi/\pi)^2[K_3/K_2 + (1 - 4\beta + \beta^2)K_2/K_3 + 2\beta - 1]\}}{1 + (\Phi/\pi)^2[K_3 - 2K_2(1 - \beta)]/K_1} \qquad (2.26)$$

The importance of cholesteric doping is recognized in the cell thickness to helical pitch ratio parameter $\beta = 2\pi d/p\Phi$. The sensitivity of the voltage response in Eq. (2.25) increases with twist, becoming infinite near 270° and bistable beyond, when the denominator in Eq. (2.25) approaches zero. Therefore, optimizing the twist is of critical importance to passive matrix addressing, where a sharp characteristic is favored. Figure 2.16 shows the sharp response of a *supertwisted nematic* (STN) in comparison with the normal 90° TN. The transmission of the STN increases rapidly close to the deformation voltage, unlike the excess voltage required in the TN case. Gray-scale production in passive matrix addressing requires some compromise in sharpness of the response.

Operation of the STN LCD at large enough cell thickness for the Mauguin limit to apply does not produce a sharp electro-optic response. The director tilt in the center region of the cell does change sharply at the critical voltage, but in the surface regions of the cell sufficient twist remains to rotate the polarization, preventing optical switching until substantially higher voltages (Scheffer and Nehring, 1985). Therefore the STN is operated with a thin cell beyond the Mauguin limit, making the response sensitive to d and λ. The polarizer and analyzer positions are rotated away from the NLC surface alignment directions, to optimize brightness and contrast. The off-axis polarizer positions imply ECB effects in addition to twisted nematic behavior.

The elastic term $F(K)$ is generally negative in the STN case; consequently, the voltage sensitivity in Eq. (2.25) increases with increase in K_3/K_1. This is opposite to the 90° twist cell, where $F(K)$ is generally positive. The voltage sensitivity increases with higher twist, lower pretilt, lower d/p, lower K_2/K_1, and lower $\Delta\varepsilon/\varepsilon_\perp$, as can be deduced from Eqs. (2.25) and (2.26). The optical transmission of the STN in the zero-voltage state can be treated analytically, and approximations derived for the optimum cell thickness (Raynes, 1987a). The analytic formulations express the influence of all the parameters and provide some shortcuts to optimum conditions. However, the detailed behavior and optimization of STN LCDs favors a computational approach (e.g., Scheffer and Nehring, 1985, 1990; van Sprang and Koopman, 1988).

Cholesteric doping is essential in stabilizing the high-twist direction in STN cells over the alternative lower energy low-twist direction—for example, $+270°$ rather than $-90°$. However, the development of a cholesteric *striped texture* can disrupt the STN function (Chigrinov et al., 1979; G. A. Beresnev et al., 1989; Scheffer and Nehring, 1990). The striped texture can be prevented by high-tilt alignment $>5°$ (e.g., Scheffer and Nehring, 1984, 1990). Commercial fabrication of STN LCDs requires low-cost high-tilt alignment (e.g., Becker et al., 1986; Nehring et al., 1987) or NLC materials accommodating low-tilt alignment (G. A. Beresnev et al., 1989).

When the STN LCD is optimized for contrast and brightness, the sensitivity to wavelength generates an unwanted coloration with white-light illumination. *Yellow mode* operation gives a yellowish-green birefringence color against black; rotation of one polarizer gives the *blue mode*, with a bright colorless appearance against a dark purplish-blue color (Scheffer and Nehring, 1984, 1985, 1990).

Black-on-white contrast is an advantage, particularly for color filter applications, and can be achieved by the *optical mode interference* (OMI) effect (Schadt and Leenhouts, 1987). The OMI has a 180° twist, with the input polarization along the NLC input director and orthogonal output polarizer. The operation is substantially below the Maugin limit, and Δnd is set to give optimum ellipticity of the normal mode. The behavior of the OMI is related to normal-mode behavior rather than a direct birefringent effect. Therefore the OMI is much less sensitive to wavelength and cell thickness than an STN, which is partly dependent on direct birefringence. Unfortunately, the OMI LCD trades these advantages for brightness, where the transmission is approximately half that of the standard STN LCD (Raynes, 1987b).

A solution to the STN color problem is provided by the double-layer (D-STN) method (Katoh et al., 1987), where a second (nonaddressed) STN cell of countertwist is positioned to optically compensate the originally addressed STN device. However, compensation using polymeric retarding films is gaining in popularity because of lower cost and bulk (Fujimura et al. 1991; Miyashita et al., 1991).

The dynamics of the STN cell have been calculated and compare well with experimental data (van Sprang and Koopman, 1988). For voltage operating points well away from the transition voltage, the response time is comparable with that of the nontwisted or 90° twist cells. In the transition region the response time increases in relation to the narrow range of voltage (Scheffer et al., 1985). The STN trades sharp electro-optic response or multiplexibility for response time. The material parameters should be optimized to avoid hysteresis effects for faster response time (Waters et al., 1985). The conditions for optimizing sharpness and speed have been reported, which give response <150 msec (Asano et al., 1986; Kondo et al., 1991).

The *lower-twisted nematic* (LTN) has less than 90° twist, and the associated crossed polarizers are oriented symmetrically about the twist angle. The electrooptic characteristic is not as voltage-sensitive as the 90° case. This deterioration in switch-on sharpness is a disadvantage in passive matrix addressing but favors improved gray-scale stability in active matrix addressing (Leenhouts et al., 1987).

The rapid evolution of the STN LCD has generated several equivalent or closely related terms: *supertwisted birefringence effect* (SBE), *highly twisted birefringence effect* (HBE), and *optical mode interference* (OMI). The terminology appears to be consolidating in the term STN, and a recent review has clarified the issues (Scheffer and Nehring, 1990). An analysis of all the twisted NLCDs in the field-off case has provided some general conclusions regarding optimum conditions (Ong, 1988; Lien, 1989, 1990).

2.4.5 Dynamic Scattering

The *dynamic scattering mode* (DSM) in NLC requires relatively low resistivity material (<10 GΩ·cm) and favors negative dielectric anisotropy. The NLC orientation is influenced by ionic current flow and dielectric torque. At a threshold on the order of 10 V, a striped pattern known as the *Williams domains* appears. Further increasing the voltage generates the DSM, a turbulent state that scatters light strongly. The DSM is a form of electrohydrodynamic instability and is of some academic interest.

The first LCDs were based on DSM (Heilmeier et al., 1968). Ionic current flow is essential to DSM; therefore, it is difficult to avoid electrochemical degradation. Excessive power dissipation and high voltage levels were other disadvantages that led to the eclipse of DSM by field-effect devices. The DSM has been recently reviewed by Bahadur (1990).

Interest in the DSM was revived by similar effects observed in S_A phases (Coates et al., 1978). The smectic phase has the advantage that the electrically induced scattering texture is stored when the voltage is removed. Moreover, the scattering texture can be electrically erased with higher frequency voltages on the order of 1 kHz. These attributes are important in large-scale passive matrix addressing, which was effectively demonstrated (Crossland and Cantor, 1985).

Any electrochemical degradation associated with DSM in smectics is mitigated by stored scattering, which allows infrequent electrical addressing. The voltage levels required are excessive but might be reduced by further materials development. However, development of DSM smectics has been eclipsed by the rise of STN, FLC, and PDLC devices. Current interest in DSM smectics is related to laser thermal addressing, where a scattering background can be created by DSM (e.g., Daley et al., 1989).

2.4.6 Cholesteric Devices

The CLC focal-conic texture provides effective light scattering and can be switched to a clear state by an applied electric field, which unwinds the pitch. The scattering and clear states are bistable if the cell thickness, CLC pitch, and surface alignment are optimized (Greubel, 1974). The bistable response allows large-scale matrix addressing. A number of LCD demonstrations have been made, but commercial products are slow to emerge. The scattering effect favors bright projection screen applications. A 640×400 pixel multicolor projection display with 2 sec frame time has been demonstrated (Yamagishi et al., 1988). More recently a 5×10^6 pixel electronic overhead projector system with 5 sec addressing has been demonstrated (Yabe et al., 1991). The slow response and limited contrast ratio are disadvantages in these devices.

The *circular dichroism* associated with the cholesteric spiral is of interest in filter applications (e.g., Blinov, 1983). A CLC with right-handed helix can reflect right-handed circularly polarized light (RCP) while transmitting left-handed circularly polarized light (LCP), and vice versa. The conditions of normal incidence on a parallel-aligned CLC cell (*Grandjean texture*) produce maximum reflection when the optical wavelength in the CLC is the same as the pitch, that is, vacuum wavelength $\lambda_o = p(n_\perp + n_\parallel)/2$. The behavior is related to Bragg reflection, but the higher order terms are absent. The reflection band has a wavelength range $\Delta\lambda = p\Delta n$. The reflected wave has the unusual property of being the same handedness as the incident wave.

A notch filter can be constructed by placing in series two CLC cells of opposite handedness turned to the same reflection wavelength. Unpolarized light can be regarded as a sum of RCP and LCP light, and both components are reflected by the combined notch filter device. The use of a liquid-crystal polymer in this application reduces the temperate sensibility of the filter (Tsai et al., 1989).

Unpolarized light can be efficiently polarized by a CLC filter in conjunction with an ordinary mirror. If the CLC filter transmits RCP, then the reflected LCP can be converted to RCP by further reflection from the ordinary mirror. Therefore, a system can be designed to convert unpolarized light completely to circularly polarized light. The circular polarization can be changed to linear polarization by means of a simple $\lambda/4$ retarder. Proper design of the CLC filters allows color separation in addition to polarized light conversion, which is important in projection LCD applications (Belayev et al., 1990; Schadt and Funfschilling, 1990). The operation of TN cells in circular polarized light is also discussed by Belayev et al. (1990) and Schadt and Funfschilling (1990).

2.4.7 Polymer-Dispersed Liquid Crystals

A polymer-dispersed liquid crystal (PDLC) is a distribution of liquid-crystal microdroplets in a polymer matrix. *Nematic curvilinear aligned phase* (NCAP) is an alternative term reflecting the unusual interface alignment. The nematic droplet size can be adjusted to provide a strong scattering effect at visible wavelengths. An applied electric field orients the nematic droplets such that with suitable conditions the scattering virtually disappears (Fergason, 1985; Doane et al., 1986). The display potential has led to a rapid development of PDLCs (Doane, 1990, Vaz, 1989).

The scattering/clear switching is illustrated in Figure 2.17, where the nematic ordinary refractive index matches that of the polymer matrix. Light propagating along the oriented nematic axis is not scattered because of refractive index matching. When the voltage is removed, the nematic relaxes and the droplet orientation becomes random. The resulting refractive index mismatch reasserts the scattering state.

The lack of polarizers in a scattering type of display implies an advantage in optical efficiency. A small droplet implies fast viscoelastic relaxation, which is achieved in PDLCs without the mechanical problems of a thin nematic cell. Large-area LCDs (on the order of square meters) can be fabricated with PDLC technology. However, part of the applied voltage is dropped across the unproductive polymer matrix, resulting in additional voltage requirements for PDLCs. The electro-optic response is far from sharp, and the PDLC is quite unsuited to passive matrix addressing.

There are two approaches for the preparation of a PDLC. The first is based on heterogeneous solutions such as emulsions or in microencapsulation procedures. The NLC is emulsified in a polymer solution, or colloid, so that the NLC droplets are formed before the polymer matrix gels. The emulsion is coated onto a sub-

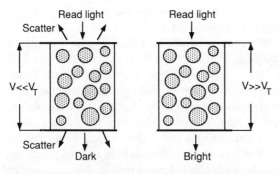

Figure 2.17 Polymer-dispersed liquid crystal.

strate, and excess solvent is removed by drying. The resulting film contains a dispersion of NLC droplets, where droplet density and size are controlled by the process details (Drzaic, 1986).

The alternative approach relies on controlled phase separation from a homogeneous solution. In polymerization-induced phase separation, the NLC is dissolved in a low molecular weight prepolymer solution, and phase separation is induced during the polymerization process as the NLC becomes less soluble (Smith and Vaz, 1988). Polymerization is achieved by thermal, chemical, or photochemical processes. Thermally induced phase separation is possible by cooling an NLC in a thermoplastic solution at elevated temperature (West, 1988). Solvent-induced phase separation is also effective, where the NLC and polymer have a common solvent (Doane et al., 1988). The droplet size and distribution are determined by process details such as phase separation rate.

The distribution in voltage between the inactive polymer matrix and active liquid-crystal dispersion is often a dominant factor in the switching voltage level. The dispersion/matrix inpedance ratio determines the electric field distribution, which is strongly dependent on frequency in general. The droplet director topology is influenced by director alignment on the polymer matrix, interface energy, and elastic energy. The droplet shape is also important (Drzaic and Muller, 1989).

Various droplet shapes and director distributions are shown in Figure 2.18. The threshold electric field and response time for director deformation in PDLC have been calculated under various conditions to generate the following expressions (Erdmann et al., 1989, 1990; Doane et al., 1988; Doane, 1990). For a spherical droplet of radius R, with perpendicular boundary conditions and radial director distribution as in Figure 2.18a, the deformation electric field can be approximated as

$$E = \frac{V}{d} \approx \frac{1.3}{R} \left(\frac{\rho_b}{\rho_{lc}} + 2 \right) \left(\frac{K}{\Delta \varepsilon} \right)^{1/2} \qquad (2.27)$$

where V is voltage applied across material thickness d, ρ_b/ρ_{lc} is the ratio of polymer to NLC resistivity, and $K/\Delta\varepsilon$ is the ratio of NLC elastic constant to dielectric anisotropy. The radial distribution is energetically unstable for droplet radius $R > 18d_e$, where the degenerate axial distribution indicated in Figure 2.18b is favored, and $d_e = K/W_o$ is a surface extrapolation length indicating the range of the surface interaction, as determined by the ratio of elastic to surface energy terms (Erdmann et al., 1990).

The spherical droplet director distribution is degenerate for tangential boundary alignment, as shown in Figures 2.18d and 2.18e. In an ellipsoidal droplet the degeneracy is removed, and director alignment energetics favor a bipolar point-defect configuration with predominant major axis orientation, as shown in Figure

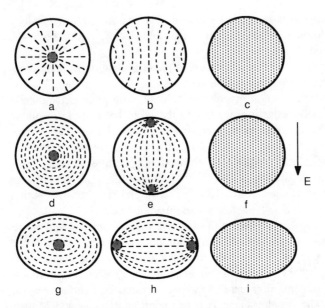

Figure 2.18 Droplet shape and director distribution. Perpendicular alignment: (a) radial, (b) axial. Tangential alignment: (d) and (g) confocal, (e) and (h) bipolar. Electric field aligned: (c), (f), and (i).

2.18h. On removal of the electric field, the director arrangement shown in Figure 2.18h. is re-formed. An expression for the threshold field can be written

$$E = \frac{V}{d} \approx \frac{1}{3a}\left(\frac{\rho_b}{\rho_{lc}}+2\right)\left[\frac{K(l^2-1)}{\Delta\varepsilon}\right]^{1/2} \tag{2.28}$$

where a is the length of the semimajor axis, and l is the ellipse aspect ratio.

Lower electric fields are always associated with lower resistivity ratios, higher dielectric anisotropy, and larger droplets. In Eq. (2.28) the aspect ratio of the ellipse is an important parameter in lowering the field requirement.

The response time is expressed in terms of a rotational viscosity γ, and for the radial spherical droplet it is written

$$\tau^{-1} \approx \frac{9\Delta\varepsilon E^2}{\gamma(\rho_b/\rho_{lc}+2)^2} + \frac{(2\pi)^2 K}{\gamma R^2} \tag{2.29}$$

while for the elliptical droplet the response time is

$$\tau^{-1} \approx \frac{9\Delta\varepsilon E^2}{\gamma(\rho_b/\rho_{lc}+2)^2} + \frac{K(l^2-1)}{\gamma a^2} \tag{2.30}$$

Fast decay in zero electric field favors low viscosity, small droplets, strong

LCD DEVICE FUNDAMENTALS

elasticity, and a large elliptic aspect ratio. A fast rise time requires, in addition, a strong dielectric torque and efficient field transfer to the NLC. In general, some compromise between high speed and low voltage is achieved for LCD applications.

Strong scattering in the visible spectrum requires droplet diameters on the order of 1 μm (Nomura et al., 1990). The mismatch in refractive index between polymer and liquid crystal should be large, implying a high NLC birefringence. The clear state is dependent on closely matching the NLC ordinary index with the polymer matrix over the required temperature range. Off-axis clear-state viewing is compromised by the NLC birefringence. However, if the matrix is formed from a properly oriented NLC polymer of matching birefringence, the off-axis clarity can be maintained over wide angles (Doane, 1990; Doane et al., 1990).

The development efforts expended over the past five years are beginning to provide display products (Drzaic et al., 1989). Television displays are of dominant interest, where PDLCs compare favorably with existing twisted nematic LCDs. The PDLC has advantages in speed, viewing angle, and assembly processes (Macknick et al., 1989). The demonstration of a full-color projection TV PDLC display indicates the current rate of progress (Kunigita et al., 1990).

2.4.8 Ferroelectric Surface-Stabilized Device

In the surface-stabilized ferroelectric liquid crystal (SSFLC) LCD, the FLC cell alignment surfaces play a critical role in stabilizing the FLC configuration while allowing a rapid switching effect (Clark and Lagerwall, 1980). The structure of the SSFLC cell is shown in Figure 2.19. Parallel surface alignment promotes *bookshelf geometry*, in which the smectic layers are perpendicular to the electrode surfaces. The helical smectic layers are perpendicular to the electrode surfaces. The helical twist of the FLC is suppressed when the cell spacing is significantly less than the pitch of the helix. The molecular director n and polarization

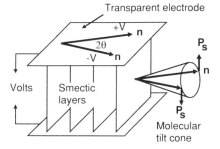

Figure 2.19 Surface-stabilized ferroelectric liquid-crystal cell.

P_S are fundamentally coupled, such that in the undistorted state P_S is perpendicular to the cell surface. If the polarity switches direction, the molecular director switches through twice the S_C^* tilt angle (2θ) in the surface plane, while the smectic layer structure remains intact. The switching process is envisioned as a rotation of the molecules in the smectic plane, around the tilt cone angle. The bistable states are separated by an elastic and surface energy barrier, which is overcome by an applied field, where polarity switching of the SSFLC rotates the optic axis of the cell by 2θ. The cell transmission intensity through crossed polarizers can be written

$$T = sin^2\ 4\theta\ sin^2(\pi \Delta nd/\lambda) \tag{2.31}$$

The optimum FLC tilt angle $\theta = 22.5°$, and birefringence colors are minimized in white-light conditions for first-order operation, when $\Delta nd \approx 0.26$ μm. The latter condition implies a thin cell with $d \approx 2$ μm, which also favors high switching speed, wide viewing angle, and stability.

The switching speed of the SSFLC is much faster than that of comparable NLC devices because of the first-order interaction ($P \cdot E$) of the spontaneous polarization with the applied electric field. Moreover, reversal of the applied field provides bidirectional drive. Practical switching speeds of about 10 μsec have been achieved. Bistability provides storage independent of control voltages and enables large-scale passive matrix addressing.

Director distortions in the NLC decay exponentially with distance, but the smectic layer structure propagates defects more extensively (de Gennes, 1974). Consequently, the smectic layered structure is, in general, more difficult to align than the NLC. Direct alignment of the smectic phase on cooling from the isotropic phase is possible (Patel et al., 1984). However, the most uniform alignment is achieved through alignment of a precursor long-pitch N*LC, which undergoes an S_A phase transition and finally an S_C^* transition with decreasing temperature. The SSFLC cell is prone to zigzag defects that reduce the contrast. The nature of the zigzag defect has become clearer in recent years and is identified with the chevron SSFLC cell structure shown in Figure 2.8 (Rieker et al., 1987; Clark and Rieker, 1988). Changes in chevron direction, as shown in Figure 2.8, give rise to the visible zigzag defects. The zigzag defect can be avoided by surface conditions that favor one chevron direction or by elimination of the chevron structure.

High-tilt alignment and parallel arrangement of the surfaces such that the alignment directions mimic a chevron and thereby favor that chevron direction in the FLC are successful in eliminating zigzag defects (Bos and Koehler/Beran, 1988; Yamamoto et al., 1989). In low-tilt rubbed polymer aligned SSFLC cells, the zigzag defects can be removed by a high electric field treatment, which also appears to eliminate the chevron structure in favor of bookshelf geometry (Patel et al., 1989; Hartmann and Luyckx-Smolders, 1990). FLC materials are being developed that favor the bookshelf geometry over the chevron structure and thereby avoid zigzag defects (Takanishi et al., 1990).

The surface alignment conditions of the SSFLC are unusual in having to accommodate a rotation of 45° at or in the surface region of the FLC. A strong surface interaction can disturb the bistability by favoring one of the states. Polar surface interactions are significant in the SSFLC and break the symmetry of the opposed surface states. More experimental work is required to clarify details of the surface behavior.

The threshold switching voltage of an SSFLC cell is quite low at approximately 0.1 V. The switching speed follows a square-law dependence on voltage at low voltage, turning to linear dependence at higher voltage. The square-law regime is related to nucleation and growth mechanisms, while linear dependence is expected for bulk switching (Clark et al., 1983). The bulk switching speed (τ) for an applied electric field (E) is limited by the viscosity (γ) associated with the rotation and can be approximated by

$$\tau = \gamma/P_s E \qquad (2.32)$$

High speed implies high P_s. However, bistability can be compromised by excessive P_s. Ionic impurities migrate under the influence of the internal polarization field and can create a counter field limited in discharge rate by ionic mobility. This effect can be prevented by alignment layers of sufficient electrical conductivity (Chieu and Yang, 1990).

The NLC has a substantial Freedericksz threshold of about 1 V, which is an advantage in LCD addressing. In the SSFLC device the low static threshold is compensated for by a *dynamic threshold,* for addressing purposes. The switching and latching of a bistable state is controlled by the dynamic threshold $V\tau_m$, the pulse amplitude and duration product, where τ_m is the minimum pulse duration for latching or memorization at voltage V. In the linear regime, $V\tau_m$ is constant; however, $\tau_m > \tau$ given by Eq. (2.32).

Gray-scale response of a bistable device is generally achieved by a halftone or dither mechanism. In the time domain the SSFLC can be modulated at frequencies beyond the visual response to generate gray scale. In the space domain the pixel can be subdivided to generate gray scale. A combination of time/space dither has been used to generate adequate gray scale in a matrix-addressed panel without excessive complications (e.g., Lagerwall et al., 1989; Dijon, 1990). A spatial dither can be achieved by charge-control operation without the complication of additional electrodes. If a pixel of given area A is supplied with charge Q, only a fraction $Q/A2P_s$ of the area can switch before the pixel voltage falls to zero. Therefore, by controlling the pixel charge, rather than the pixel voltage, a large gray-scale range is possible (e.g., Hartmann, 1989).

The SSFLC has been described in a simple ideal form that can be approached with careful cell preparation. In practice, many alternative structures have been identified in the SSFLC cell geometry. The bulk molecular director may deviate from surface parallelism, giving rise to an apparent reduction in cone switching angle. Twisted structures can form that reduce contrast ratio (e.g., Fukuda et al., 1989; Dijon, 1990). A detailed physics of the SSFLC operation is beginning to emerge but is far from complete.

2.4.9 Soft-Mode FLC

The electroclinic effect in S_A is related to the soft-mode behavior of the tilt angle, which is an order parameter in the phase transition. *Soft-mode ferroelectric liquid crystal* (SMFLC) is an alternative term used in FLC device physics (e.g., Anderson et al., 1989). The tilt angle θ in the S_A is linearly related to the applied field E,

$$\theta = \mu E/\alpha(T - T_c) \qquad (2.33)$$

and the response time τ is independent of field but limited by a rotational viscosity λ,

$$\tau = \lambda/\alpha(T - T_c) \qquad (2.34)$$

where μ and α are material constraints and $T - T_c$ is the temperature above the S_C^*–S_A transition. The response time is submicrosecond well above T_c, but θ is small. Close to T_c the θ response is more complicated and approaches the level needed for display devices, but the speed is comparable with that of SSFLC devices. The transmission through crossed polarizers follows Eq. (2.31).

The SMFLC device lacks bistability and memory but provides gray scale. The superior alignment properties associated with S_A operation can provide high contrast despite low brightness. Further materials development is in progress.

2.4.10 Distorted Helix FLC

If the S_C^* helical pitch is smaller than the optical wavelength, the optical propagation is described by a refractive index that is an average over the local director configuration. An applied electric field distorts the helix and modifies the configuration average to change the effective birefingence, as illustrated in Figure 2.20 for parallel surface alignment (Ostrovski and Chigrinov, 1980; L.A. Beresnev et al., 1988). The effective optic axis turns clockwise or counterclockwise according to the direction of the applied field. The development of suitable materials has opened interesting possibilities for the distorted helix ferroelectric (DHF) device (Beresnev and Blinov, 1989; Funfschilling and Schadt, 1989). The transmission through crossed polarizers follows Eq. (2.31).

The DHF provides gray scale at voltage on the order of 1 V with response speed on the order of 100 μsec, provided the pitch is not unwound. If the voltage is increased sufficiently to unwind the DHF pitch, the relaxation rate is much

LCD DEVICE FUNDAMENTALS

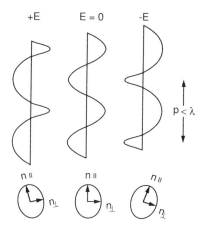

Figure 2.20 Distorted helix FLC.

slower. Uniform alignment is difficult and limits the contrast at this stage of development. Existing materials lack an N* precursor phase, where uniform alignment can be established. Moreover, a long-pitch N* phase may be inconsistent with the short pitch essential in the S_C^* phase. Alignment can be achieved by a shear motion between the cell substrates similar to the pioneer work in SSFLC devices (Funfschilling and Schadt, 1989). The DHF can show bistability or memory (Funfschilling and Schadt, 1990).

2.4.11 Dichroic Dyes

Dichroic dyes have an absorption coefficient that depends on molecular orientation. The orientation of a *guest* dichroic dye can be controlled by dissolving the dye in a *host* liquid crystal. The original guest–host LCD used an aligned NLC, with electric field control similar to ECB (Heilmeier and Zanoni, 1968). The dye absorption is generally higher for light polarized in the long direction of the molecule. The cell and polarizer are arranged to electrically rotate the NLC director and dye orientation along or away from the optical polarization.

A further development employed a CLC host that was arranged to accept all polarization directions, eliminating the polarizer (White and Taylor, 1974). An applied voltage unwound the CLC helix and aligned the director and dye in the direction of minimum absorption.

All of the liquid-crystal devices that are based on birefringent effects can function as guest-host dye LCDs. The advantages are a wider viewing angle, better uniformity, higher brightness in the absence of polarizers, color contrast, and improved appearance. These advantages should be weighed against the increased viscosity and slower response speed, dye stability problems, and broadening of the electro-optic response curve (or lower multiplexibility). The latter problem was instrumental in the development of STN LCDs (Waters et al., 1985).

The transmission of light polarized parallel $(T_\parallel)$ and perpendicular $(T_\perp)$ to the optic axis is expressed as

$$T_\parallel = \exp[-(2S+1)\alpha_o d] \tag{2.35}$$

$$T_\perp = \exp[-1(1-S)\alpha_o d] \tag{2.36}$$

where α_o is the dye attenuation constant in an isotropic liquid, d is the cell thickness, and S is the order parameter of the dye, which approximates the host value. High-order parameter hosts are required, which generally favors high clearing point NLCs.

2.5 ADDRESSING

2.5.1 NLC Passive Matrix Addressing

Direct drive is the simplest addressing circuit, where each element of the display is directly wired to a voltage source. Simple alphanumeric LCDs such as digital displays may use direct-drive addressing. The limitations of direct drive become obvious as the pixel count increases. A 100×100 pixel array requires 10,000 directly wired connections; in addition to the complexity there is not enough surface space for the wiring. Clearly, some method of multiplexing is necessary.

A multiplexing scheme based on writing that is equivalent to an X–Y coordinate arrangement is referred to as *matrix addressing,* where $N \times N$ pixels are addressed using $2N$ peripheral connections, as indicated in Figure 2.21 for a 5×5 array. This vast simplification in writing is achieved by pixels sharing elec-

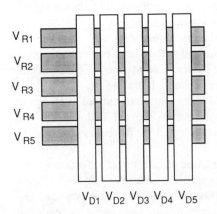

Figure 2.21 Matrix addressing.

LCD DEVICE FUNDAMENTALS

trodes, but at the expense of possible addressing confusion or crosstalk. The nonlinear response of the liquid crystal attenuates crosstalk in *passive matrix addressing*, also known as *direct-drive matrix addressing*. A nonlinear circuit element is incorporated at each pixel to suppress crosstalk in *active matrix addressing*.

A complete frame is written in time T, by sequentially activating the X (strobe) lines, while feeding the Y (data) lines in parallel. The dwell time per line cannot be more than T/N. In passive matrix addressing the required dwell time determines the minimum frame time. Active matrix addressing permits charge storage at each pixel, allowing the frame speed to approach the LC response speed.

A bistable FLC pixel can switch in <100 μsec dwell time, providing a fast frame rate with unlimited frame storage time. However, an NLC pixel is not bistable and relaxes toward the initial state between addressing pulses. A continuous NLC LCD requires a frame rate in excess of the relaxation rate, implying a frame-averaging effect in the pixel response. The NLC response described in Eq. (2.8) follows an E^2 dependence on applied electric field, implying that an NLC pixel will respond according to the mean-square addressing voltage. The NLC mean-square response provides the basis for optimization of the passive addressing scheme.

The passive matrix addressing of mean-square responding LCDs has been analyzed by Alt and Pleshko (1974), who introduce an LC nonlinearity parameter P in terms of the rms voltages for the selected (V_s) and nonselected (V_{ns}) states, relative to the optical threshold voltage (V_{th}):

$$P = (V_s - V_{ns})/V_{th} \tag{2.37}$$

The P value determines the limit on the number of addressable rows (N_{max}) and the corresponding voltages. For $P \ll 1$ the expressions simplify:

$$N_{max} = P^{-2} \tag{2.38}$$
$$V_D/V_{th} = (P/2)^{1/2} \tag{2.39}$$
$$V_R/V_{th} = (2/P)^{1/2} \tag{2.40}$$
$$V_s/V_{ns} = 1 + P \tag{2.41}$$

where $\pm V_D$ is the data pulse on the Y line and V_R is the strobe pulse on the X line, giving $V_s = V_R + V_D$ and $V_{ns} = V_R - V_D$.

Large-scale passive matrix addressing of mean-square-responding LCs requires device structures with small P, that is, well-defined sharp thresholds. Temperature stability, off-axis viewing, and structural uniformity impose additional restrictions on N_{max}. The standard 90° TN LCD is generally limited to 100 rows or less. It is basically the number of rows that is restricted, and larger arrays can be formed by a partitioning of the row structure (e.g., Kaneko, 1986). However, the numbers of drivers and interconnections increase. The STN and ECB LCDs have achieved 400-row operation at this stage of development.

2.5.2 SSFLC Passive Matrix Addressing

The bistable nature and speed of the SSFLC implies large-scale passive matrix addressing at video rate. The dynamic threshold $V\tau_m$ is the basis for matrix multiplexing. The matrix-addressing circuitry must optimize the switching of selected pixels over nonselected pixels. Kirchhoff's loop sum-zero voltage law limits the selection ratio $V_s/V_{ns} < 3$. The nonselect voltage V_{ns}, while below the dynamic threshold, still induces some electo-optic response that reduces the displayed contrast. The subthreshold response can be reduced by employing an FLC with negative dielectric anisotropy ($-\Delta\varepsilon$), where the dielectric mean-square drive opposes the dc polar drive (Le Pesant et al., 1985, Elston et al., 1990).

Polarity reversal is essential to FLC addressing and implies separate frame cycles for bright- and dark-state pixels, doubling the overall frame rate. A single switching cycle, bright to dark or vice versa, is sometimes described as a *slot*. An overall frame-addressing process is then indicated as four-slot, two-slot, etc. A pixel dc voltage component should be avoided for stability and electrochemical reasons. Elimination of direct current is an extra complication in the required addressing waveforms that favors the multislot schemes. The polarity, dc restrictions, mean-square considerations, and dynamic threshold are factors influencing the complexity of SSFLC addressing. Developments have been reviewed recently (Lagerwall et al., 1989; Dijon, 1990).

A four-slot scheme is shown in Figure 2.22, where the switching dark to bright or vice versa is determined by the phase of the dc-compensated pulse cycle (Harada et al., 1985). The strobe-row and data-column pulse cycles each have phase-reversed alternatives, implying four possible combinations of pixel voltage. There is an additional pixel voltage level associated with the strobe-row zero-voltage state. The five levels of pixel voltage are shown in Figure 2.22, where the dark field and bright field are written successively. The overall frame time is four times the intrinsic FLC switching speed.

The frame speed can be increased by writing the dark state on several lines simultaneously. This method allows the frame speed to approach twice the intrinsic speed. Alternatively, the dc compensation of the row drive can be sacrificed without serious consequences because of the low duty cycle (e.g., Matsumoto et al., 1989).

An example of current achievement in FLC passive matrix addressing is a 14 in., 1280×1120 pixel screen, refreshed at 6 Hz, for computer display applications (Kanbe et al., 1991).

2.5.3 Electron Beam Addressing and Photoaddressing

A more sophisticated version of direct drive is achieved in electron beam addressing. If each element of the LCD can store charge, then the e-beam can scan a complete frame similar to a CRT display. Charge storage is natural in a highly

LCD DEVICE FUNDAMENTALS 57

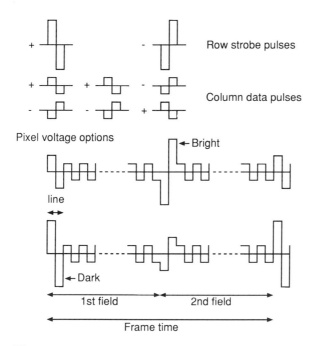

Figure 2.22 SSFLC matrix-addressing waveforms.

resistive LC, but the LC vapor pressure is not compatible with electron tube vacuum. Beam-addressed LCDs have been demonstrated in which the LC is isolated by a thin insulating membrane (Haven, 1983), or more recently a *charge transfer plate* (Hirsch et al., 1988). A photocathode or microchannel plate can provide an electronic "image" that addresses all the LCD elements in parallel and could provide an image intensifier for a projection display. Flat-panel displays are favored over vacuum tube devices; therefore the e-beam addressing of LCs is of interest only in special applications.

Photoaddressing is an alternative to e-beam addressing that avoids the vacuum tube handicap. The general device structure is shown in Figure 2.23. The photoreceptor may be a photoconductor or photodiode structure that converts the write light into a charge pattern to address the LC. A dielectric mirror isolates the write and read sides, allowing optical amplification with an intense readout light. Photoaddressed LCDs are characterized by resolution, contrast, photosensitivity, uniformity, and speed.

The resolution is described by a modulation transfer function (MTF), relating write-light sinusoidal spatial frequency modulation (f) to optical output intensity modulation (I).

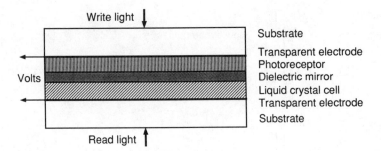

Figure 2.23 Photoaddressed LCD.

$$\text{MTF}(f) = (I_{max} - I_{min})/(I_{max} + I_{min}) \tag{2.42}$$

The MTF approaches unity at sufficiently low frequency, and approaches zero at high frequency. The spatial frequency (f) is generally quoted in cycles/mm or line pairs (lp)/mm. The resolution is fundamentally limited by the electric field fringing effects of the addressing charge accumulated at the photoreceptor–mirror interface. For an LC thickness d and photoreceptor thickness b, the upper limit to resolution at the 50% MTF value can be approximated (e.g., Armitage et al., 1989) as

$$f = \frac{1/d + 1/b}{4\pi} \tag{2.43}$$

The importance of thin liquid-crystal and photoreceptor layers is clear from Eq. (2.43). The practical upper limit of Eq. (2.43) is 100 lp/mm, but at this high value, charge migration in the photoreceptor will also limit resolution (Armitage et al., 1989). The limiting resolution is sometimes quoted and usually means the resolution at 10% MTF. Photoaddressing is capable of extremely high information content. Resolutions approaching 100 lp/mm are achievable over dimensions of 5 cm or more, that is, 10^8 pixel equivalency.

The device design and operating conditions require some compromise and may trade sensitivity for resolution and speed. Many photoactive materials have been used in LCD addressing (Vasil'ev et al., 1983). The slow response speed of photoconductors such as CdS led to the development of silicon photodiode addressing (Efron et al., 1985; Sayyah et al., 1989). Recent developments have favored amorphous silicon hydrogen alloy (a-Si:H) photoaddressing (Ashley and Davis, 1987; Williams et al., 1988; Sterling et al., 1990) including *p-i-n* diode structures (Moddel, 1991; Li et al., 1989).

Photoaddressing can be achieved by an optical scanning system or by a subsidiary display device such as a CRT. In the latter case the LCD acts as an optical amplifier and is used in projection displays. An analog gray scale and the scan-

LCD DEVICE FUNDAMENTALS

ning address mode for an a-Si:H photoaddressed FLC video projector LCD have recently been presented (Bone et al., 1991). Photoaddressed LCDs are of value in two-dimensional optical processing applications, where the term *spatial light modulator* (SLM) is employed (e.g., Armitage, 1989).

2.5.4 Thermal Addressing

A combination of heat pulses and electric field pulses can switch between scattering and clear states in an LC, which is the basis of the thermally addressed LCD. A complete background scattering state can be set by an overall heat pulse. Total erasure of the scattering state is achieved by an applied electric field. Therefore, negative or positive images are readily displayed. The S_A is favored over the CLC because of stronger scattering and the long-term stability of the scattering or clear state.

Thermal addressing has been incorporated into X–Y matrix addressing as a means of isolating the addressed row and avoiding crosstalk. The selected row is identified by an electrical heat pulse to the addressing electrode. During cooling an electric field is applied between the selected row and activated data columns to induce a clear texture. The inactive data columns (having no electric field) allow a scattering texture to form (e.g., Hockbaum, 1984). Development interest in this approach has waned because of the power requirements and limited frame speed.

Thermal addressing activity now centers on heat pulses delivered by laser beam, as indicated in Figure 2.24. This is also a form of optical addressing and has the advantage of large information content (Kahn 1973; Dewey, 1984). Commercial systems are available with 10^8 pixels (Kahn, 1989). Gray scale is achieved by modulating the thermal or voltage pulse during the write process. Frame speed is well below video rate, and slow-speed, high-resolution applications such as printing are favored.

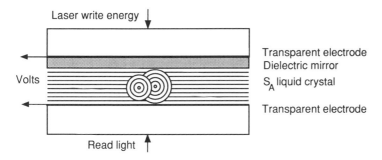

Figure 2.24 Optical-beam thermal addressing.

Considerable effort has been made to optimize materials and cell structure (Armitage, 1981). Smectic mixtures are designed to have a small nematic range before transforming to the isotropic phase. The nematic range allows easier control of selective erase. The smectic cell is operated close to the nematic transition temperature, so that a small heat pulse raises the local temperature through the nematic and into the isotropic phase. The rapid cooling inherent to a small volume freezes in the smectic scattering texture. A large dielectric anisotropy engineered into the smectic mixture allows complete erasure at sufficient electric field. Selective erasure of part of the image is achieved at a lower electric field by simultaneous laser-beam heating of the region.

ACKNOWLEDGMENTS

A critical reading of the manuscript by F. J. Kahn and M. Bone of Grayhawk Systems and V. Sethma of Kaiser Electronics is gratefully acknowledged.

References

Alt. P. M., and P. Pleshko (1974). Scanning limitations of liquid-crystal displays, *IEEE Trans. Elect. Dev. ED-21*:146.
Anderson, G., I. Dahl, L. Komitov, S. T. Lagerwall, K. Skarp, and B. Stebler (1989). Device physics of the soft-mode electro-optic effect, *J. Appl. Phys. 66*:4983.
Armitage, D. (1980). Alignment of liquid crystal on obliquely evaporated silicon oxide films, *J. Appl. Phys. 51*:2552.
Armitage, D. (1981). Thermal properties and heat flow in the laser addressed liquid-crystal display, *J. Appl. Phys. 52*:1294.
Armitage, D., J. I. Thackara, and W. D. Eades (1989). Photoaddressed liquid crystal spatial light modulators, *Appl. Opt. 28*:4763.
Armitage, D., and J. I. Thackara (1989). Photoaddressed liquid crystal edge-enhancing spatial light modulator, *Appl. Opt. 28*:219.
Armitage, D., J. I. Thackara, W. D. Eades, M. A. Stiller, and W. W. Anderson (1987). Fast nematic liquid crystal spatial light modulator, *SPIE 824*:34.
Asano, K., K. Arai, and S. Nishi (1986). SBE liquid crystal displays with improved response time, *Jpn. Display* Proc. 6th Int. Disp. Res. Conf. *1986*:392.
Ashley, P. R., and J. H. Davis (1987). Amorphous silicon photoconductor in a liquid crystal spatial light modulator, *Appl. Opt. 26*:241.
Bahadur, B. (1984). Liquid crystal displays, *Mol. Cryst. Liq. Cryst., Spec. Topics X 109*:3.
Bahadur, B. (Ed.) (1990). *Liquid Crystals: Applications and Uses, Vols. 1–3*, World Scientific, Teaneck, NJ.
Barberi, R., M. Boix, and G. Durand, (1989). Electrically controlled surface bistability in nematic liquid crystals. *Appl. Phys. Lett. 55*:2506.
Barret, S., F. Gaspard, R. Herino, and F. Mondon (1976a). Dynamic scattering in ne-

matic liquid crystals under dc conditions. 1. Basic electrochemical analysis, *J. Appl. Phys. 47:*2375.
Barret, S., F. Gaspard, R. Herino, and F. Mondon (1976b). Dynamic scattering in nematic liquid crystals under dc conditions. II. Monitoring of electrode processes and lifetime investigation, *J. Appl. Phys. 47:*2378.
Becker, M. E., R. A. Kilian, B. B. Kosmowski, and D. A. Mlynski (1986). Alignment properties of rubbed polymer surfaces, *Mol. Cryst. Liq. Cryst. 132:*167.
Belayev, S. V., M. Schadt, M. I. Barnik, J. Funfschilling, N. V. Malimoneko, and K. Schmitt (1990). Large aperture polarized light source and novel liquid crystal display operating modes, *Jpn. J. Appl. Phys. 29:*L634.
Beresnev, G. A., S. V. Belyaev, V. G. Chigrinov, and N. A. Khokhov (1989). Steepness of transmission–voltage curve and domains in supertwisted materials, *Proc. 9th International Display Research Conf. (Japan Display '89)*. Paper P2–5.
Beresnev, L. A., and L. M. Blinov (1989). Electro-optical effects in ferroelectric liquid crystals, *Ferroelectrics 92:*335.
Beresnev, L. A., L. M. Blinov, M. A. Osipov, and S. A. Pikin (1988). Ferroelectric liquid crystals, *Mol. Cryst. Liq. Cryst. 158A:* 3.
Berreman, D. W. (1972). Solid surface shape and the alignment of an adjacent nematic LC, *Phys. Rev. Lett. 26:*1683.
Berreman, D. W. (1975). Liquid crystal twist cell dynamics with backflow, *J. Appl. Phys. 46:*3746.
Berreman, D. W., and W. R. Heffner, (1980). New bistable cholesteric liquid-crystal display, *Appl. Phys. Lett., 37:*109.
Berreman, D. W., and W. R. Heffner, (1981). New bistable liquid-crystal twist cell, *J. Appl. Phys., 52:*3032.
Blinov, L. M. (1983). *Electro-Optical and Magneto-Optical Properties of Liquid Crystals,* Wiley, New York.
Bone, M., D. Haven, and D. Slobodin (1991). Video-rate photoaddressed ferroelectric LC light valve with gray scale, *SID 22:*254.
Bos, P. J., and K. R. Koehler/Beran (1988). A method for director alignment of SMC* devices, *Ferroelectrics 85:*15.
Bos, P. J., P. A. Johnson, K. R. Koehler/Beran (1985). The Π-cell: a new, fast liquid-crystal optical switching device, *Mol. Cryst. Liq. Cryst. 113:*329.
Chandrasekhar, S. (1977). *Liquid Crystals.* Cambridge Univ. Press, London.
Chandrasekhar, S., and G. S. Ranganath (1986). The structures and energetics of defects in liquid crystals, *Adv. Phys. 35:*507.
Channin, D. J., and D. E. Carlson (1976). Rapid turn-off in triode gate liquid crystal devices, *Appl. Phys. Lett. 28:*300.
Chieu, T. C., and K. H. Yang, (1990). Effect of alignment layer conductivity on the bistability of surface-stabilized ferroelectric liquid crystal devices, *Appl. Phys. Lett. 56:*1326.
Chigrinov, V. G., V. V. Belyaev, S. V. Belyaev, and M. F. Grebenkin (1979). Instabilities of cholesteric liquid crystals in an electric field, *Sov. Phys. JETP 50:*994.
Clark, N. A., and S. T. Lagerwall (1980). Submicrosecond bistable electro-optic switching in liquid crystals, *Appl. Phys. Lett. 36:*899.

Clark, N. A., and T. P. Rieker (1988). Smectic-C "chevron," a planar liquid-crystal defect: implications for the surface-stabilized ferroelectric liquid-crystal geometry, *Phys. Rev. A. 37:*1053.

Clark, N. A., M. A. Handschy, and S. T. Lagerwall (1983). Ferroelectric liquid crystal electro-optics using the surface stabilized structure, *Mol. Cryst. Liq. Cryst. 94:* 213.

Clerc, J. F. (1991). Vertically aligned liquid-crystal displays, *SID 91 Dig.,* 22:758.

Clerc, J. F., and J. C. Deutsch (1987). Birefringent mode in a thin liquid crystal layer for T.F.T. displays, *Proc. Eurodisplay, Lond.* 111.

Coates, D., W. A. Crossland, J. H. Morrissey, and B. Needham. (1978). Electrically induced scattering textures in smectic A phases and their electrical reversal, *J. Phys. D 11:*2025.

Cognard, J. (1982). Alignment of nematic liquid crystals and their mixtures, *Suppl. 1, Mol. Cryst. Liq. Cryst.*

Crossland, W. A., and S. Cantor (1985). An electrically addressed smectic storage device, *SID 85 Dig.,* 124.

Daley, R., A. J. Hughes, and D. G. McDonnell (1989). Light on dark laser addressed smectic liquid crystal projection displays, *Liq. Cryst. 4:*685.

de Gennes, P. G. (1974). *The Physics of Liquid Crystals,* Clarendon, Oxford.

Deuling, H. J. (1978). Elasticity of nematic liquid crystals, in *Solid State Physics, Suppl.* 14 (L. Liebert, Ed.), Academic, New York, p. 77.

Dewey, A. G. (1984). Laser-addressed liquid crystal displays, *Opt. Eng. 23:*230.

Dijon, J. (1990). Ferroelectric LCDs, in *Liquid Crystals: Application and Uses,* vol. 1 (B. Bahadur, Ed.), World Scientific, Teaneck, NJ, p. 305.

Doane, J. W. (1990). Polymer dispersed liquid crystal displays, in *Liquid Crystals: Applications and Uses,* Vol. 1 (B. Bahadur, Ed.), World Scientific, Teaneck, NJ, p. 361.

Doane, J. W., N. A. Vaz, B. G. Wu, and S. Zumer (1986). Field controlled light scattering from nematic microdroplets, *Appl. Phys. Lett.* 48:269.

Doane, J. W., J. L. Golemme, J. B. West, J. B. Whitehead, Jr., and B.-G. Wu (1988). Polymer dispersed liquid crystals for display applications, *Mol. Cryst. Liq. Cryst. 165:*511.

Doane, J. W., J. L. West, J. B. Whitehead, and D. S. Fredley (1990) Wide-angle-view PDLC displays, *SID 90 Dig.,* 224.

Drzaic, P. S. (1986). Polymer dispersed nematic liquid crystal for large area displays and light valves, *J. Appl. Phys. 60:*2142.

Drzaic, P. S., and A. Muller (1989). Droplet shape and reorientation fields in nematic droplet/polymer films, *Liq. Cryst. 5:*1467.

Drzaic, P. S., R. Wiley, and J. McCoy, (1989). High brightness and color contrast displays constructed from nematic droplet/polymer films incorporating pleochroic dyes, *SPIE,* 1084:41.

Efron, U., J. Grinberg, P. O. Braatz, M. J. Little, P. G. Reif, and R. N. Schwartz (1985). The silicon liquid-crystal light valve, *J. Appl. Phys. 57:*1356.

Elston, S. J., J. R. Sambles, and M. G. Clark, (1990). The mechanism of ac stabilization in ferroelectric liquid-crystal-filled cells, *J. Appl. Phys.,* 68:1242.

Erdmann, J., J. W. Doane, S. Zumer, and G. Chidichimo (1989). Electrooptic response of PDLC light shutters, *SPIE 1080*:32.

Erdmann, J., S. Zumer, and J. W. Doane (1990). Configuration transition in a nematic liquid crystal confined to a small spherical cavity, *Phys. Rev. Lett. 64*:1907.

Erdmann, J. H., S. Zumer, B. G. Wagner, and J. W. Doane (1990). Director configuration and configuration transitions in PDLC material, *SPIE 1257*:68.

Fergason, J. L. (1980). Performance of a matrix display using surface mode, *Biennial Display Res. Conf., IEEE*, 177.

Fergason, J. L. (1985). Polymer encapsulated liquid crystals for display and light control applications, *SID 85 Dig.*, 60.

Fergason, J. L., and A. L. Berman (1989). A push/pull surface-mode liquid-crystal shutter: technology and applications, *Liq. Cryst. 5*:1397.

Fujimura, Y., T. Nagatsuka, H. Yoshima, and T. Shimomura (1991). Optical properties of retardation films for STN-LCDs, *SID Dig.*, 739.

Fukuda, A., Y. Ouchi, H. Arai, H. Takano, K. Ishikawa, and H. Takezoe (1989), Complexities in the structure of ferroelectric liquid crystal cells. The chevron structure and twisted states, *Liq. Cryst. 5*:1055.

Funfschilling, J., and M. Schadt (1989). Fast responding and highly multiplexible distorted helix ferroelectric liquid-crystal displays, *J. Appl. Phys. 66*:3877.

Funfschilling, J., and M. Schadt (1990). Short-pitch bistable ferroelectric LCDs, *SID 90 Dig., Las Vegas*, 106.

Garoff, S., and R. Meyer (1977). Electroclinic effect at the A-C phase change in a chiral smectic liquid crystal, *Phys. Rev. Lett. 38*:848.

Geary, J. M., J. W. Goodby, A. R. Kmetz, and J. S. Patel (1987). The mechanism of polymer alignment of liquid-crystal materials, *J. Appl. Phys. 62*:4100.

Gooch, C. H., and H. A. Tarry (1975). The optical properties of twisted nematic liquid crystal structures with twisted angles $\leq 90°$, *J. Phys. D 8*:1575.

Goodman, L. A., J. T. McGinn, C. H. Anderson, and F. Digeronimo (1977). Topography of obliquely evaporated silicon oxide films and its effect of liquid-crystal orientation, *IEEE Trans. Elect. Dev. ED24*:795.

Greubel, W. (1974). Bistability behavior of texture in cholesteric liquid crystals in an electric field, *Appl. Phys. Lett. 25*:5.

Grindberg, J., A. Jacobson, W. P. Bleha, L. Miller, L. Fraas, D. Bosewell, and G. Meyer (1975). A new real-time non-coherent to coherent light image converter: the hybrid field effect liquid crystal light valve, *Opt. Eng. 4*:217.

Harada, T., M. Taguchi, K. Iwasa, and M. Kai (1985). An application of chiral smectic-C liquid crystal to a multiplexed large-area display, *SID 85 Dig.*, 131.

Hartmann, W. J. A. M. (1989). Charge-controlled phenomena in the surface-stabilized ferroelectric liquid-crystal structure, *J. Appl. Phys. 66*:1132.

Hartmann, W. J. A. M., and A. M. M. Luyckx-Smolders (1990). The bistability of the surface-stabilized ferroelectric liquid-crystal effect in electrically reoriented chevron structures, *J. Appl. Phys. 67*:1253.

Haven, D. A. (1983). Electron-beam addressed liquid-crystal light valve, *IEEE Trans. Elect. Dev. ED-30*:489.

Heilmeier, G. H., and L. A. Zanoni (1968). Guest–host interactions in nematic liquid crystals—a new electro-optic effect, *Appl. Phys. Lett. 13:*91.
Heilmeier, G. H., L. A. Zanoni, and L. A. Barton (1968). Dynamic scattering: a new electrooptic effect in certain classes of nematic liquid crystals, *Proc. IEEE 56:*1162.
Hirsch, P. W., I. Farber, and C. Warde (1988). Development of an *e*-beam charge-transfer liquid crystal light modulator, *Spatial Light Modulators Appl. Dig. 8:*11.
Hockbaum, A. (1984). Thermally addressed smectic liquid crystatl displays, *Opt. Eng. 23:*253.
Ikeno, H., A. Oh-saki, M. Nitta, N. Ozaki, Y. Yokoyama, K. Nakaya, and S. Kobayashi (1988). Electrooptic bistability of a ferroelectric liquid crystal device prepared using polyimide Langmuir–Blogett orientation films, *Jpn. J. Appl. Phys. 27:*L475.
Ishihara, S., H. Wakemoto, K. Nakazima, and Y. Matsuo (1989). The effect of rubbed polymer films on the liquid crystal alignment, *Liq. Cryst. 4:*669.
Jakeman, E., and E. P. Raynes (1972). Electro-optic response times in liquid crystals, *Phys. Lett. 39A:*69.
Janning, J. L. (1972). Thin film surface orientation for liquid crystals, *Appl. Phys. Lett. 21:*173.
Kahn, F. J. (1972). Electric-field-induced orientational deformation of nematic liquid crystals: tunable birefringence, *Appl. Phys. Lett. 20:*199.
Kahn, F. J. (1973). IR-laser-addressed thermo-optic smectic liquid-crystal storage displays, *Appl. Phys. Lett. 22:*111.
Kahn, F. J. (1977). Capacitive analysis of twisted nematic liquid crystal displays, *Mol. Cryst. Liq. Cryst. 38:*109.
Kahn, F. J. (1989). Photographic-quality, storage-type, liquid-crystal spatial-light-modulator system for optical information processing applications, *SPIE 151:*558.
Kanbe, J., H. Inoue, A. Mizutome, Y. Hanyuu, K. Katagiri, and S. Yoshihara (1991). High performance, large area FLC display with high graphic performance, *Ferroelectrics 114:*3.
Kaneko, E. (1986). Directly addressed matrix liquid crystal display panel with high information content, *Mol. Cryst. Liq. Cryst. 139:*81.
Katoh, K., Y. Endo, M. Akatsuka, M. Ohgawara, and K. Sawada (1987). Application of retardation compensation; a new highly multiplexable black-white liquid crystal display with two supertwisted nematic layers, *Jpn. J. Appl. Phys. 26:*L17784.
Kohlmuller, H., and G. Siemsen (1982). DC voltage-resistant liquid crystal mixtures for field effect displays, *Siemens Forsch-u. Entwickl-Ber. 11:*229.
Kondo, S., T. Yamamoto, A. Murayama, H. Hatoh, and S. Matsumoto (1991). A fast-response black and white ST-LCD with a retardation film, *SID 22:*747.
Kuczynski, W., and H. Stegemeyer (1980). Ferroelectric properties of smectic C liquid crystals with induced helical structure, *Chem. Phys. Lett. 70:*123.
Kunigita, M., Y. Hirai, Y. Ooi, S. Niiyama, T. Asakawa, K. Masumo, H. Kumai, M. Yuki, and T. Gunjima (1990). A full color projection TV using LC/polymer composite light valves, *SID 90 Dig.,* 227.
Labrunie, G., and J. Robert (1973). Transient behavior of the electrically controlled birefringence in a nematic liquid crystal, *J. Appl. Phys. 44:*4869.
Lagerwall, S. T., N. A. Clark, J. Dijon, and J. F. Clerc (1989). Ferroelectric liquid crystals: the development of devices, *Ferroelectrics 94.*1205.

Leenhouts, F., M. Schadt, and H.-J. Fromm (1987). Electro-optical characteristics of a new liquid-crystal display with an improved gray-scale capability, *Appl. Phys. Lett.* *50:*1468.

Le Pesant, J. P., J. M. Perbet, B. Mourey, M. Haring, G. Decobert, and J. C. Dubois (1985). Optical switching of chiral smectic C at room temperature, *Mol. Cryst. Liq. Cryst. 129:*61.

Li, W., R. A. Rice, G. Moddel, L. A. Pagano-Stauffer, and M. A. Handschy (1989). Hydrogenated amorphous-silicon photosensor for optically addressed high-speed spatial light modulator, *IEEE Trans. Electron. Dev. 36:*2959.

Lien, A. (1989). Optimization of the off-states for single-layer and double-layer general twisted nematic liquid-crystal displays, *IEEE Trans. Electron. Dev. 36:*1910.

Lien, A., (1990). The general and simplified Jones matrix representations for the high pretilt twisted nematic cell, *J. Appl. Phys.*, 67:2853.

Macknick, B., P. Jones, and L. White, (1989). High resolution displays using NCAP liquid crystals, *SPIE*, 1080:169.

Matsumoto, S., H. Hatoe, and A. Murayama (1989). Matrix liquid-crystal display device technologies, *Liq. Cryst. 5:*1345.

Meyer, R. B. (1977). Ferroelectric liquid crystals: a review, *Mol. Cryst. Liq. Cryst. 40:*33.

Meyer, R. B., L. Lieber, L. Strzelecki, and P. Keller (1975). Ferroelectric liquid crystals, *J. Phys. (Paris) Lett. 36:*L-69.

Miyashita, T., Y. Miyazawa, Z. Kikuchi, H. Aoki, and A. Mawatari (1991). A multicolor-wide-viewing-angle STN-LCD with multiple retardation films, *SID 91 Dig.*, 743.

Moddel, G. (1991). Optically addressed spatial light modulators which incorporate an amorphous silicon photosensor, in *Physics and Applications of Amorphous and Microcrystalline Semiconductor Devices — Optoelectronic Devices* (J. Kanicki, Ed.), (Chapter 11, p. 369–412). Artech House, Norwood, MA.

Nehring, J., H. Amstutz, P. A. Holmes, and A. Nevin (1987). High-pretilt polyphenylene layers for liquid-crystal displays, *Appl. Phys. Lett. 51:*1283.

Nomura, H., S. Suzuki, and Y. Atarashi (1990). Electrooptic properties of polymer films containing nematic liquid crystal microdroplets, *Jpn. J. Appl. Phys. 29:* 522.

Ong, H. L. (1988). Origin and characteristics of the optical properties of general twisted nematic liquid-crystal displays, *J. Appl. Phys. 64:*614.

Ostrovski, B. I., and V. G. Chigrinov (1980). Linear electro-optic effect in chiral smectic C liquid crystals, *Kristallografiya 25:*560.

Patel, J. S., T. M. Leslie, and J. W. Goodby (1984). A reliable method of alignment for smectic liquid crystals, *Ferroelectrics 59:*137.

Patel, J. S., S.-D. Lee, and J. W. Goodby (1989). Electric-field-induced layer reorientation in ferroelectric liquid crystals, *Phys. Rev. A 40:*2854.

Pieranski, P., F. Brochard, and E. Guyon (1973). Static and dynamic behavior of a nematic liquid crystal in a magnetic field, Part II: dynamics, *J. Phys. 34:*35.

Priestly, E. B., P. J. Wojtowicz, and P. Sheng Ed. (1974). *Introduction to Liquid Crystals,* Plenum, New York.

Raynes, E. P. (1986). The theory of supertwist transmissions, *Mol. Cryst. Liq. Cryst. Lett.* 4:1.
Raynes, E. P. (1987a). The optical properties of supertwisted liquid crystal layers, *Mol. Cryst. Liq. Cryst. Lett* 4:69.
Raynes, E. P. (1987b). The optical properties of the optical mode interference supertwist display, *Mol. Cryst. Liq. Cryst. Lett.* 4:159.
Rieker, T. P., N. A. Clark, G. S. Smith, D. S. Parmar, E. B. Sirota, and C. R. Safinya (1987). "Chevron" local layer structure in surface-stabilized ferroelectric smectic-C cells, *Phys. Rev. Lett.* 59:2658.
Sayyah, K., M. S. Welkowsky, P. G. Reif, and N. W. Goodwin (1989). High performance single crystal silicon liquid crystal light valve with good image uniformity, *Appl. Opt.* 28:4748.
Schadt, M. (1982). Low-frequency dielectric relaxation in nematics and dual-frequency addressing of field effects, *Mol. Cryst. Liq. Cryst.* 89:77.
Schadt, M. (1988). The twisted nematic effect: liquid crystal dispalys and liquid crystal materials, *Mol. Cryst. Liq. Cryst.* 165:405.
Schadt, M. (1989). The history of the liquid crystal display and liquid crystal material technology, *Liq. Cryst.* 5:57.
Schadt, M., and J. Funfschilling (1990). New liquid crystal polarized color projection principle, *Jpn. J. Appl. Phys.* 29:1974.
Schadt, M., and W. Helfrich (1971). Voltage-dependent optical activity of a twisted nematic liquid crystal, *Appl. Phys. Lett.* 18:127.
Schadt, M., and F. Leenhouts (1987). Electro-optical performance of a new, black-white and highly multiplexable liquid crystal display, *Appl. Phys. Lett.* 50:236.
Scheffer, T. J., and J. Nehring (1985). Investigation of the electro-optical properties of 270° chiral nematic layers in the birefringence mode, *J. Appl. Phys.* 58:3022.
Scheffer, T. J., and J. Nehring, (1984). A new, highly multiplexable liquid crystal display, *Appl. Phys. Lett.*, 45:1021.
Scheffer, T. J., and J. Nehring (1990). Twisted nematic and supertwisted nematic mode LCDs, in *Liquid Crystals: Applications and Uses, Vol. 1* (B. Bahadur, Ed.), World Scientific, Teaneck, NJ, p. 231.
Scheffer, T. J., J. Nehring, M. Kaufmann, H. Amstutz, D. Heimgartner, and P. Eglin (1985). 24 × 80 character LCD panel using the supertwisted birefringence effect, *SID 85 Dig.*, 120.
Smith, G. W., and N. A. Vaz (1988). The relation between formation kinetics and microdroplet size of epoxy-based polymer-dispersed liquid crystals, *Liq. Cryst.* 3:543.
Sterling, R. D., R. D. Te Kolste, J. M. Haggerty, T. C. Borah, and W. P. Bleha (1990). Video-rate liquid-crystal light-valve using an amorphous silicon photoconductor, *SID 90 Dig.*, 327.
Sussman, A. (1974). Electrochemistry in nematic liquid-crystal solvents, in *Inroduction to Liquid Crystals* (E. B. Priestly, P. J. Wojtowicz, and P. Sheng, Eds.). Plenum, New York, p. 319.
Takanishi, Y., Y. Ouchi, H. Takezoe, A. Fukuda, A. Mochizuki, and M. Nakatsuka (1990). Spontaneous formation of quasi-bookshelf layer structure in new ferroelectric liquid crystals derived form a naphthalene ring. *Jpn. J. Appl. Phys.* 29:L984.
Tsai, M. L., S. H. Chen, and S. D. Jacobs (1989). Optical notch filter using thermotropic liquid crystalline polymers, *Appl. Phys. Lett* 54:2395.

Uchida, T. (1990). Surface alignment of liquid crystals, in *Liquid Crystals: Applications and Uses, Vol. 3* (B. Bahadur, Ed.), World Scientific, NJ.
van Sprang, H. A., and H. G. Koopman (1988). Experimental and calculated results for the dynamics of oriented nematics with twist angles from 210° to 270°, *J. Appl. Phys. 64*:4873.
Vasil'ev, A. A., I. N. Kompanets, and A. V. Parfenov (1983). Progress in the development and applications of optically controlled liquid crystal spatial light modulators (review), *Kvant. Elek. 10:*1079 (*Sov. J. Quant. Elect. 13:*689).
Vaz, N. A. (1989). Polymer dispersed liquid crystal films: materials and applications, *SPIE 1080:*2.
Waters, C. M., E. P. Raynes, and V. Brimmell (1985). Design of highly multiplexed liquid crystal dye displays, *Mol. Cryst. Liq. Cryst. 123:*303.
West, J. L. (1988). Phase separation of liquid crystals in polymers, *Mol. Cryst. Liq. Cryst. 157:*427.
White, D. L., and G. N. Taylor (1974). New absorptive mode reflective liquid-crystal display device, *J. Appl. Phys. 45:* 4718.
Williams, D., S. G. Lathan, C. M. J. Powels, M. A. Powell, R. C. Chittick, A. P. Sparks, and N. Collings (1988). An amorphous silicon chiral smectic spatial light modulator, *J. Phys. D. 21:*S156.
Wu, S.-T. (1990). Nematic liquid crystals for active optics, in *Optical Materials, A Series of Advances,* Vol. 1 (S. Musikant, Ed.), Marcel Dekker, New York, p. 1.
Wu, S. T., and U. Efron (1986). Optical properties of thin nematic liquid crystal cells, *Appl. Phys. Lett. 48:*624.
Yabe, Y., H. Yamada, T. Hoshi, T. Yoshihara, A. Mochizuki, and Y. Yoneda (1991). A 5-M pixel overhead projection display utilizing a nematic-cholesteric phase-transition liquid crystal, *SID 22:*261.
Yamagishi, Y., M. Iwasaki, T. Yoshihara, A. Mochizuki, Y. Koike, and M. Haraguchi (1988). A multicolor projection display using nematic-cholesteric liquid crystal, *IEEE Int. Display Res. Conf. Dig.,* 204.
Yamamoto, N., Y. Yamada, and K. Mori (1989). Ferroelectric liquid crystal layer with high contrast ratio, *Jpn. J. Appl. Phys. 28:*524.

3
Microelectronics in Active-Matrix LCDs and Image Sensors

William den Boer, F. C. Luo, and Zvi Yaniv

*Ovonic Imaging Systems, Inc.,
Troy, Michigan*

3.1 ACTIVE-MATRIX LIQUID-CRYSTAL DISPLAYS

Large-area microelectronics on glass substrates has become an established and growing technology in the past few years. Thin-film transistors (TFTs) and diodes of amorphous silicon and polycrystalline silicon are now commonly used in contact image sensors and flat-panel displays. As a result, liquid-crystal displays (LCDs) in particular have increased in size, resolution, and performance and have become a viable contender to replace cathode-ray tubes.

Liquid-crystal displays based on the twisted nematic (TN) effect were developed in the early 1970s and were first applied as simple displays with several digits in watches and calculators. Dot matrix displays with a one-line-at-a-time driving method improved the resolution in the 1980s, and with the advent of supertwist displays screens with up to 400 lines appeared on the market.

For a description of the fundamentals of liquid crystals, including the operation principles of TN cells, the reader is referred to Chapter 2. Twisted nematic LC cells are operated with an ac voltage. The transmission vs. voltage curve is shown in Figure 3.1. When the number of multiplexed lines N in the display is increased, the ratio between ON and OFF voltage decreases according to the formula (Alt and Pleshko, 1974)

$$\frac{V_{on}}{V_{off}} = \left(\frac{\sqrt{N}+1}{\sqrt{N}-1}\right)^{1/2} \tag{3.1}$$

As N increases, the voltage ratio and the contrast ratio decrease. Supertwist displays have a steeper transmission–voltage curve and therefore a higher degree of multiplexibility. Contrast ratios of 15–20 have been achieved in 400-line supertwist displays. For both conventional TN and supertwist displays, however, the transmission curves depend strongly on viewing angle, as illustrated in Figure 3.1, and contrast ratios drop off rapidly for off-axis viewing.

To circumvent the reduction of contrast with increasing number of lines, Lechner (1970) proposed to incorporate a switch at each picture element in a matrix display so that the voltage across each pixel could be controlled independently and the same high contrast ratio of 100 or more obtained in simple, direct-driven displays could in principle also be achieved for high-information-content displays.

The switch can be either a diode or a transistor. Displays based on this principle are called *active matrix* LCDs (AMLCDs). Their fabrication requires the deposition and patterning of various metals, insulators, and semiconductors on glass substrates, comparable to the processing of integrated circuits. Brody et al. (1973) constructed the first AMLCD with CdSe thin-film transistors (TFTs) as the switching elements. In 1981 the first AMLCD with thin-film metal–insulator–metal (MIM) diodes as the pixel switches was reported (Baraff et al., 1981). MIM diodes seemed particularly attractive for this application because

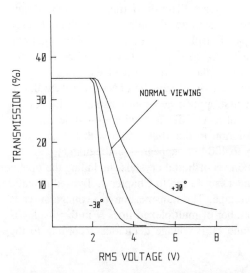

Figure 3.1 Transmission vs. rms voltage across pixel in a twisted nematic liquid crystal display for normal viewing and viewing under angles of $+30°$ (from above) and $-30°$ (from below).

they are relatively simple to fabricate and have current–voltage characteristics that are symmetric with respect to oppose polarities. This type of switch is therefore compatible with the ac drive of LCDs. Semiconductors such as CdSe are not compatible with standard processing in the microelectronics industry, which uses mainly silicon as the semiconductor material. Advanced photolithographic and etching processes have been developed over the years for silicon devices, and this technology is not readily applicable to CdSe TFTs.

Poly-crystalline silicon (poly-Si) and amorphous silicon devices do not suffer from this drawback and were developed for use in AMLCDs in the early 1980s. The first LC pocket television marketed in 1984 used a poly-Si TFT active matrix (Morozumi, 1984). Most poly-Si processes require high-temperature processing and therefore use expensive quartz substrates, but they offer the potential of integrating the driving electronics on the glass substrate.

In the 1970s plasma-deposited amorphous silicon (a-Si) was developed as a relatively high quality thin-film semiconductor material. Amorphous silicon can be easily deposited on large-area inexpensive glass substrates at a temperature below 300°C and can be doped p-type and n-type. PIN diodes of amorphous silicon for solar cells were developed in 1975, and their rectification ratio was improved to allow application as switches in AMLCDs (Seki et al., 1983).

LeComber et al. (1979) developed the first TFT with a-Si as the semiconductor material and suggested as one of its applications the active matrix LCD. In the 1980s several companies, in particular in Japan, developed a-Si TFT LCDs, mainly for pocket TVs with 3–5 in. screens. Prototype displays with a diagonal size of 15 in. (Wada et al., 1990) were demonstrated.

In the first half of this chapter (Sections 3.2–3.4), the operation principles and fabrication of active matrix LCDs are described. The advantages and drawbacks of various types of active matrix arrays are discussed.

3.2 OPERATION AND FABRICATION OF ACTIVE MATRIX LCDs

3.2.1 Requirements for ON and OFF Currents

In Figure 3.2 a cross-sectional view of an active matrix LCD with TFT switches is shown. The TFT array on the bottom glass substrate controls the rms voltage on each pixel, while the top substrate has a continuous transparent indium tin oxide (ITO)layer and, for color displays, patterned color filters.

The pixel switch in an active matrix LCD is required to charge the LC pixel in the select time T_s and hold the charge during the frame time T_f while the other rows in the matrix are scanned. The LC pixel can be represented, in a first approximation, by a simple capacitor C_{lc}. Ideally, the switch has zero impedance in the ON state and infinite impedance in the OFF state. In practice, ON currents in the microampere range and OFF currents in the picoampere range suffice.

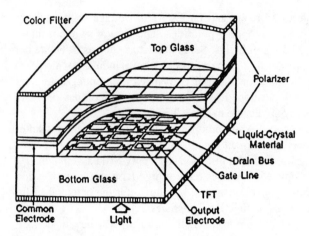

Figure 3.2 Cross-sectional view of TFT LCD.

The requirements for ON and OFF current can be readily calculated:

$$I_{on} > 2C_1 C_{pixel} V_{on} N/T_f \tag{3.2}$$

where C_{pixel} is the total pixel capacitance (including an auxiliary storage capacitor, if present). V_{on} is the voltage for the electrically activated (ON) state of the LC, and N is the number of rows in the display. The factor 2 appears in Eq. (3.2) to allow charging from $-V_{on}$ to $+V_{on}$ and vice versa, so that an ac voltage without dc component is applied across the liquid crystal. C_1 is a margin factor that ensures that the charging of the pixel is saturated.

The select time T_s is related to the frame time T_f by

$$T_s = T_f/N \tag{3.3}$$

The requirement for the OFF current is

$$I_{off} < C_2 C_{pixel} \Delta V_{lc}/T_f \tag{3.4}$$

where ΔV_{lc} is the allowable voltage drop across the liquid crystal due to the leakage current of the device and C_2 is another margin factor.

As an example we take a display with 1000 rows ($N = 1000$), $C_{pixel} = 1$ pF, $V_{on} = 5$ V, $T_f = 16$ msec (for a 60 Hz refresh rate), and $C_1 = 5$. We get

$$I_{on} > 3.1 \; \mu A$$

For the OFF current we get, assuming $\Delta V_{lc} = 0.1$ V and $C_2 = 0.2$,

$$I_{off} < 1.2 \; pA$$

MICROELECTRONICS IN LCDS AND IMAGE SENSORS

The calculations for ON and OFF current are valid for any type of switch used in AMLCDs, including TFTs and diodes.

3.2.2 TFT Displays

As TN LCDs require two transparent substrates sandwiched between polarizers, opaque crystalline silicon substrates cannot be used for this type of display.

Thin-film transistors (TFTs) with amorphous silicon, polycrystalline silicon, and CdSe film as the semiconductor material have been developed to function as the pixel switch at the periphery of each otherwise transparent pixel. TFTs in active matrix arrays operate in the enhancement mode. The layout and cross section of a TFT are schematically shown in Figure 3.3. TFT current–voltage characteristics can, in a first approximation, be described by the conventional MOSFET equations (Sze, 1969):

$$I_{sd} = \mu \, C_g \frac{W}{L} V_{sd} (V_g - V_{th} - \frac{V_{sd}}{2}), \qquad V_{sd} < V_g - V_{th} \qquad (3.5)$$

$$I_{sd} = \mu \, C_g \frac{W}{2L} (V_g - V_{th})^2, \qquad V_{sd} > V_g - V_{th} \qquad (3.6)$$

where W is the width and L is the length of the conducting channel of the TFT, μ is the field-effect mobility, C_g is the gate capacitance per unit area, V_g is the gate

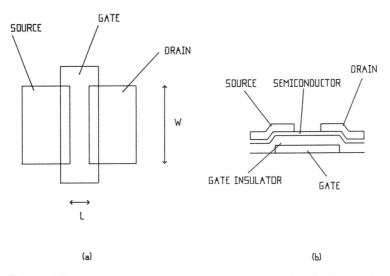

Figure 3.3 (a) Schematic layout and (b) cross section of a thin-film transistor.

voltage, V_{th} is the threshold voltage of the TFT, V_{sd} is the source-drain voltage, and I_{sd} is the source-drain current.

Figure 3.4 shows the characteristics of an optimized amorphous silicon TFT. The ON current and OFF current are in the microampere and picoampere range, respectively, as required. Amorphous silicon has been most successful for high-resolution large-area displays owing to its low dark conductivity and relatively easy fabrication on large-area glass substrates. Amorphous silicon TFTs with three different structures are used (Figure 3.5). In inverted staggered TFTs of the n^+ back-etch type, the gate metal is deposited and patterned first, followed by the deposition of the gate insulator (often Si_3N_4) and a-Si undoped and n^+ layers by plasma-enhanced chemical vapor deposition (PECVD). The a-Si layer is patterned into islands at the TFT locations and at the crossover area of select (gate) and video (source) lines.

After deposition and patterning of the source-drain metal, the n^+ layer in the channel is removed by etching, and a passivation layer is deposited to protect the channel. The n^+ layer provides a low-resistance ohmic contact for source and drain and suppresses hole injection at negative gate voltages.

In inverted staggered TFTs of the trilayer type, the process steps are the same as those of the back-etch type until the growth of the amorphous silicon. Deposition of the undoped a-Si layer is followed by a second insulator deposition. The

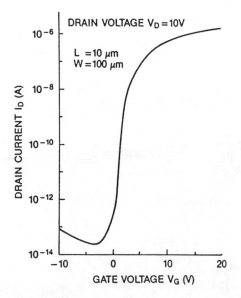

Figure 3.4 Characteristics of optimized a-Si TFT.

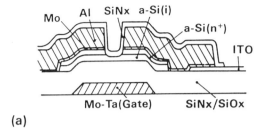

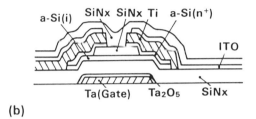

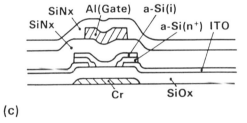

Figure 3.5 Cross sections of three different types of a-Si TFT's. (a) Inverted staggered (back-channel-etched); (b) inverted staggered (trilayered); (c) normal staggered. [From Oana (1989), reprinted with permission from Elsevier Science Publishers.]

second insulator, usually Si_3N_4, is patterned to define the channel area, and the n^+ layer and source-drain metal are subsequently deposited and patterned.

The trilayer-type TFT is more reproducible, as the active semiconductor layer is sandwiched between two insulators at all times. The a-Si layer can also be made thinner in this configuration, which makes the TFT less light-sensitive.

In the normal staggered type of TFT (Matsushita et al., 1987) a light shield layer is deposited and delineated first. Source and drain metal are than deposited and patterned, followed by the deposition and patterning of a-Si and gate insulator. The gate metal is deposited and patterned on top. The light shield is required to protect the TFT channel from light and eliminate photoleakage currents in the TFT.

In all three processes an additional deposition and patterning step is usually necessary for the ITO pixel electrode. Metals such as Ta, Cr, MoTa, and Al are used for the gates and select lines. Tantalum and MoTa can be anodically oxidized to provide an extra gate insulator, which virtually eliminates the occurrence of crossover shorts between select and video lines (Katayama et al., 1988; Dohjo et al., 1988). For source and drain and video lines, Ti, Mo, Al, or Al/Cr are frequently employed.

The TFTs are called staggered because the source and drain overlap the gate metal by a few micrometers. This is required to avoid current crowding at the edge of source and drain contacts and excess series resistance, which would reduce the ON current of the TFT.

Amorphous silicon TFTs have a field-effect mobility of 0.3–1 cm^2/(V·sec) and a threshold voltage V_{th} of about 2 V.

The mobility of poly-Si TFTs is much higher, around 50 cm^2/(V·sec). Poly-Si TFTs usually have a top gate configuration and can be fabricated in a low-temperature (600°C) process on hard glass substrates or in a high-temperature process (1000°C) on quartz substrates.

In Figure 3.6 an example of a low-temperature poly-Si TFT fabrication process is shown (Morozumi et al., 1986). First a 1500 Å phosphorus doped poly-Si layer is deposited on a hard glass substrate and patterned as source-drain electrodes. Then a very thin layer (typically 250 Å) of undoped poly-Si is deposited by low-pressure chemical vapor deposition at 600°C and patterned. An ITO film is then deposited and patterned into data lines and pixel electrodes. The gate insulator of 1500 Å thick SiO$_2$ is deposited by thermal CVD. Finally, the Cr gate electrodes are sputtered and delineated. For the high-temperature poly-Si TFT the gate insulator is thermally grown at about 1000°C, and source-drain doping

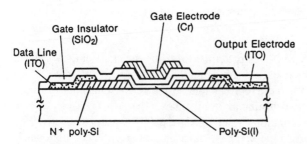

Figure 3.6 Cross section of low-temperature poly-Si TFT. [From Morozumi et al. (1986), reprinted with permission from Society for Information Display.]

is accomplished by ion implantation. Figure 3.7 shows the characteristics of a typical low-temperature poly-Si TFT (Morozumi et al., 1986).

One of the major advantages of poly-Si TFTs is their potential for integrated row and column driver circuits, which can significantly reduce the number of interconnections from the display substrate to the external electronics. For interlaced TV operation the row drivers have to operate at 16 kHz, which requires a TFT mobility of about 0.5 cm^2/(V·sec). This can be achieved by both a-Si and poly-Si TFTs. For the column drivers, however, the operating frequency has to exceed 8 MHz for monochrome and 23 MHz for color panels. A minimum mobility of 10 cm2/(V·sec) is required, which excludes a-Si.

In Figure 3.8 the operation of a TFT LCD is illustrated. Four pixels are shown, each with a storage capacitor C_{st} in parallel to the LC capacitance C_{lc}. The auxiliary capacitor, usually equal to 2–5 C_{lc}, is added to suppress the effects of LC leakage currents, which can cause nonuniform gray-scale performance, especially at elevated temperatures. In Figure 3.8 the storage capacitor for a pixel on select row i is connected to the adjacent gate line $i-1$, which was selected earlier.

It is also possible to connect C_{st} to a separate grid of common lines that is

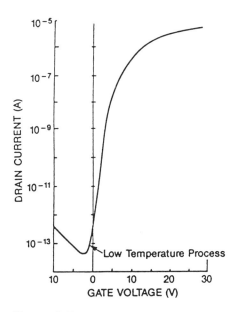

Figure 3.7 Characteristics of low-temperature poly-Si TFT. [From Morozumi et al. (1986), reprinted with permission from Society for Information Displays.]

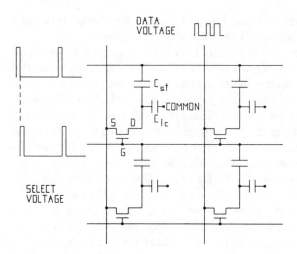

Figure 3.8 Operation principle of TFT LCD.

patterned interdigitated with the select lines (Tomita et al., 1989) or the video lines (Wada et al., 1990). In all cases the insulator that is used for the gate dielectric can also be used as the storage capacitor dielectric, so no extra processing is needed to fabricate the capacitor. The ITO counterelectrodes on the top plate are at a common potential.

The TFT display is operated in a one-row-at-a-time method. When the gates in one row are selected, the video information for that row is supplied at the source lines. All rows are sequentially selected, and during the nonselect time (the hold time) the charge on the pixel is retained by the combined capacitance of LC pixel and C_{st}.

In Figure 3.9a the equivalent circuit of a pixel including parasitic capacitances C_{ds} (between source and drain) and C_{gd} (between gate and drain) are shown. C_{ds} and C_{gd} affect the LC voltage. C_{gd}, in particular, is important and causes a negative voltage step ΔV_{pixel} when the gate is switched off (Figure 3.9b):

$$\Delta V_{\text{pixel}} = \frac{C_{gd}}{C_{gd} + C_{st} + C_{lc}} \Delta V_g \qquad (3.7)$$

where ΔV_g is the voltage swing on the gates. The common voltage at the top plate has to be adjusted so that a pure ac voltage without dc component appears across the liquid crystal. C_{gd}, which is approximately proportional to the gate-drain overlap and depends on the patterning accuracy of the metals, has to be constant across the display area to be able to offset the dc voltage.

If the dc component cannot be eliminated in some sections of the display,

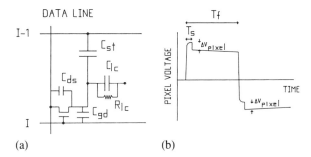

Figure 3.9 (a) Equivalent circuit of pixel with TFT including parasitic capacitances; (b) pixel voltage vs. time.

image retention can occur due to the charging of passivation and alignment layers. In addition, for a 60 Hz refresh rate, the dc voltage causes visible flicker at 30 Hz, half the refresh rate.

Flicker is often eliminated by a line-inversion driving method, in which the data voltage polarity is inverted every other row or column in addition to every frame. Since the flicker in adjacent pixels is then 180° out of phase, the eye perceives an area of the display as flickerless, although the individual pixels can still show flicker. Line inversion solves the flicker problem but does not eliminate image retention.

For color displays, red, green, and blue color filters are patterned on the top plate. In the interpixel area on the top plate, an opaque material is deposited and patterned. This "black matrix" improves the contrast ratio as it prevents light leakage in the areas between the pixels.

3.2.3 MIM Diode Displays

Baraff et al. (1981) proposed to use MIM diodes as the pixel switches in active matrix LCDs. The current–voltage characteristics of MIM diodes used in LCDs usually obey the Poole–Frenkel equation (Frenkel, 1938)

$$I = kV \exp(b\sqrt{V}) \qquad (3.8)$$

where I is the current, V is the applied voltage, and k and b are constants.

MIM diodes are bidirectional switches and have approximately symmetrical current–voltage characteristics. Diodes of Ta_2O_5 are relatively easy to fabricate. Films of Ta are first sputtered. After patterning, the tantalum is anodized in a dilute electrolyte such as citric acid, phosphoric acid, or ammonium tartrate. During the anodic oxidation, part of the Ta film is converted into Ta_2O_5. For an anodizing voltage between 20 and 50 V a thickness of 300–700 Å is obtained, suitable for application in LCDs.

The anodic oxidation process is self-limiting in the sense that the anodization current decays to practically zero and growth of the oxide film saturates at a fixed anodizing voltage (Maissel and Glang, 1970). Excellent film uniformity can therefore be obtained. The process is also self-healing. The anodization current density is higher at pinholes and weak points in the oxide film, and as a result virtually pinhole-free films are obtained.

The metal for the counterelectrode on top of the Ta_2O_5 is selected to obtain symmetric current–voltage characteristics (Morita et al., 1990). Titanium, chromium, and aluminum give good results whereas molybdenum, silver, and indium tin oxide (ITO) produce asymmetric curves with partially rectifying behavior.

The nonlinearity factor b for Ta_2O_5 is about 4, not high enough for LCDs with a high information content. Alternative insulators have therefore been investigated. One of them is off-stoichiometric silicon nitride (SiN_x) produced by PECVD (Suzuki et al., 1986). The ratio x of nitrogen to silicon in the film can be continuously varied from $x = 0$ for amorphous silicon to $x = 1.33$ for stoichiometric silicon nitride, which is a good insulator. This is done by varying the gas ratio NH_3/SiH_4 or N_2/SiH_4 in the plasma. By adjusting x and the thickness of the film, the resulting MIM diode can be tailored for LCD applications. The I–V characteristics of SiN_x diodes can also be fitted to the Poole–Frenkel equation. The factor b increases with increasing nitrogen content in the film and ranges from 4.5 to 6, significantly higher than for Ta_2O_5 MIM diodes. In Figure 3.10 the I–V characteristic of a typical SiN_x diode with an area of 50×50 μm is shown.

In a diode active matrix display, rows and columns are usually on opposite glass substrates, eliminating the possibility of crossover shorts. This is illustrated in Figure 3.11 for an MIM diode display. The select waveforms are applied to the

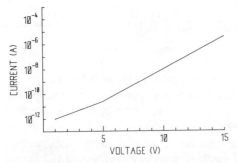

Figure 3.10 Current–voltage characteristics of 50 μm × 50 μm SiN_x MIM diode.

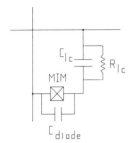

Figure 3.11 Circuit diagram of a pixel in a MIM diode display.

rows, and the data voltages are applied to the columns. The diodes can be connected either to the row lines or to the column lines. The diode has a parasitic capacitance, and since diode and LC capacitance are in series, any voltage change on the columns or rows will be partially absorbed by the LC capacitance (Figure 3.11). This capacitive charge transfer depends on the ratio of device capacitance to LC capacitance. To reduce crosstalk from the column lines it is important to keep this ratio small.

MIM diode displays can be operated with waveforms very similar to those used in directly multiplexed, "passive matrix" displays. In Figure 3.12 the ad-

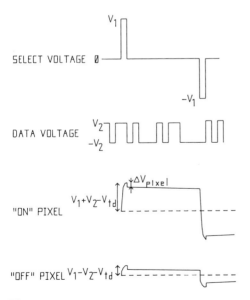

Figure 3.12 Addressing waveforms and pixel voltage for a MIM diode LCD.

dressing waveforms and pixel voltage are shown. For an activated pixel the total voltage across LC and diode is $V_1 + V_2$ during the select time. The LC pixel will then charge to $V_1 + V_2 - V_{td}$, where V_{td} is the threshold voltage of the MIM diode (the voltage at which the current is approximately 10^{-8} A). A nonactivated pixel will charge to $V_1 - V_2 - V_{td}$. V_1 and V_2 are selected to obtain the highest contrast ratio in the display and depend on the nonlinearity factor b and on N and T_f. During the next frame the polarities of select and data voltages are reversed to obtain an ac voltage across the pixel.

In practice, MIM diode characteristics are not perfectly symmetric with respect to opposite polarities. When the driving waveforms are symmetric, this can result in a dc voltage across the liquid crystal. By applying a dc offset to the data waveform or the select waveform, the dc component can be eliminated, provided the diode $I-V$ curves and asymmetry are the same across the entire display area.

Since the LC pixel and the MIM diode are in series, the threshold voltage V_{td} has to be uniform and stable across the entire display area to obtain uniform performance, especially for gray-scale operation. Different gray levels can be obtained by modulating the amplitude or the pulse width of the data voltage V_2.

The fabrication process of Ta_2O_5 MIM diode arrays is illustrated in Figure 3.13. Tantalum is deposited and patterned first. After anodizing the Ta to obtain 300–700 Å of Ta_2O_5, the top metal, usually Cr or Ti, is deposited and patterned. The overlap area between the Ta and the top metal defines the device area. Finally, ITO is deposited and patterned to form the pixel electrodes in the third photolithographic step. On the countersubstrate, ITO lines are patterned; these are usually the rows and are addressed with the select pulses. The video information is then supplied to the Ta column lines on the active substrate.

It is also possible to use the Ta lines as select rows and the ITO on the top substrate as column lines for data signals.

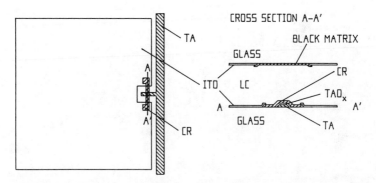

Figure 3.13 Pixel layout and cross section in a Ta_2O_5 MIM diode LCD.

Since Ta_2O_5 has a high dielectric constant of 22–25, the MIM device area has to be kept small to satisfy the condition $C_{diode}/C_{lc} < 0.1$. Photolithographic limitations on large-area substrates, however, make it difficult to reduce their size to less than 4×4 μm. To circumvent this problem attempts have been made to fabricate lateral MIM devices, in which the sidewall of the Ta pattern is used for the device (Morozumi et al., 1983) and the device capacitance can be smaller.

SiN_x diode arrays (Suzuki et al., 1986) share the advantage of a relatively simple structure with Ta_2O_5 MIM arrays. An ITO layer is deposited and patterned first. Then the 800–1500 Å thick SiN_x layer is deposited by PECVD. A top metal such as Cr is deposited and delineated to form the buslines and the top electrodes for the switch. The SiN_x film, which is partially transparent, depending on the nitrogen content, can be patterned using the Cr as a mask, or it can be patterned separately before the Cr deposition. The overlap area of ITO and Cr forms the device, which usually has an area of about 10×10 μm. As the dielectric constant of SiN_x is 7 and the SiN_x film thickness is about 1000 Å, the device capacitance is a less significant problem than in Ta_2O_5 MIM LCDs.

3.2.4 *pin* Diode Displays

pin diodes for displays are optimized to obtain the highest possible ratio between forward and reverse currents, and they have to be shielded from light to keep the reverse current low in the presence of backlighting and ambient light incident on the display.

Amorphous silicon *pin* diodes are fabricated by PECVD with silane (SiH_4) as a precursor for the undoped *i* layer and mixtures of SiH_4 with B_2H_6 and PH_3 for *p*-type and *n*-type layers, respectively. To avoid cross-contamination of doped and undoped layers, the films are deposited in a multichamber system with separate chambers for *p*, *i*, and *n* layers. The *i* layer has a thickness of 2000–6000 Å, and the doped layers are about 500 Å thick.

In Figure 3.14 the current–voltage characteristic of a typical optimized *pin* diode with an area of 20×20 μm is shown (Yaniv et al., 1986a, 1986b). The ratio between forward and reverse currents at $+3$ V and -3 V exceeds 8 orders of magnitude. The forward curve up to $+1$ V is described by the usual diode equation (Sze, 1969),

$$I = I_o [\exp(qV/nkT) - 1] \tag{3.9}$$

where I_o is the saturation current and n is the ideality factor. For optimized diodes the saturation current density is 10^{-13}–10^{-11} A/cm² and n is 1.3–1.5.

At forward voltages higher than 1 V the curve is no longer exponential and follows more closely a power law behavior (Cannella et al., 1985):

$$I = k(V - V_j)^m \tag{3.10}$$

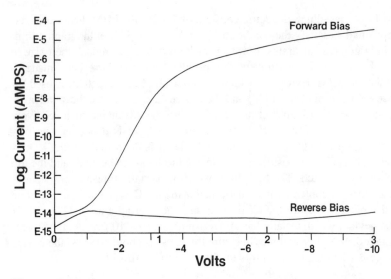

Figure 3.14 Current–voltage characteristic of 20 μm × 20 μm a-Si *pin* diode. [From Yaniv et al. (1986a), reprinted with permission from Society for Information Display.]

where V_j is the voltage drop across the *p-i* junction given by Eq. (3.9). The constant *m* is about 2. In this regime the current is limited by series resistance in the bulk of the a-Si layer.

At reverse bias the current is very low, as the *p* and *n* layers effectively block the injection of electrons and holes from the contacts and the *i* layer is depleted of charge carriers. At high reverse bias around 20 V, the current can increase sharply, and soft breakdown occurs. This soft breakdown is reversible, and the voltage at which it takes place depends on *i*-layer thickness and other deposition parameters.

The a-Si diode has asymmetric current–voltage characteristics with rectifying behavior. A single diode per pixel is therefore not compatible with the ac drive of LCDs. Any AMLCD with pin or Schottky diodes requires at least two devices per pixel. When two diodes are used in the back-to-back configuration of Figure 3.15a, the soft breakdown of the diodes can be used to obtain a symmetric switch with current–voltage curve very similar to that of MIM diodes (Szydlo et al., 1983). Since uniformity and reproducibility of soft breakdown are poor, displays based on this approach have not been successful.

By connecting two diodes in an antiparallel fashion (Figure 3.15b) a symmetric nonlinear device is obtained. This diode ring configuration was proposed by Togashi et al. (1985), and displays based on this principle were built by several organizations (Togashi et al., 1986; Urabe et al., 1989; Nicholas et al., 1990).

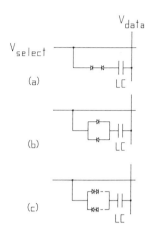

Figure 3.15 (a) Back-to-back diode configuration; (b) diode ring configuration; (c) multiple diode ring.

The I–V characteristic of the diode ring has a low threshold voltage of about 1V, incompatible with the higher threshold voltage of the TN LC cell of 2–3 V. To suppress leakage through the switch during the holding period, a holding voltage, sometimes also employed in MIM diode displays, is essential for the select line driving waveform in diode ring displays. The holding voltage is about equal to V_{50}, the voltage at which the transmission of the TN LC cell is 50%, and is required to obtain an acceptable contrast (Togashi et al., 1985).

The contrast ratio can also be improved by connecting two or more diodes in series in each branch of the ring, effectively increasing the threshold of the switch by a factor of 2 or more (Figure 3.15c). This has the additional advantage of reducing the effective overall device, capacitance of the diode ring.

The diode ring configuration utilizes only the forward characteristics of the diodes and hence does not take full advantage of the high ON/OFF current ratio of a-Si pin diodes. A configuration that uses both forward and reverse characteristics is the two-diode switch (Yaniv et al., 1986a, 1986b), depicted in the circuit diagram of Figure 3.16. Each pixel has two diodes, which are connected to different select lines. The anodes of diodes D_1 are connected to the select lines S_1, and the cathodes of diodes D_2 are connected to select lines S_2.

An example of a driving method for this configuration is also shown in Figure 3.16. The select lines S_1 are sequentially selected during the odd frames, while the select lines S_2 are at a high enough voltage to keep the diodes D_2 reverse-biased. The diodes D_1 are forward-biased during the select time, and the information on the pixel can be changed according to the voltage on the data lines.

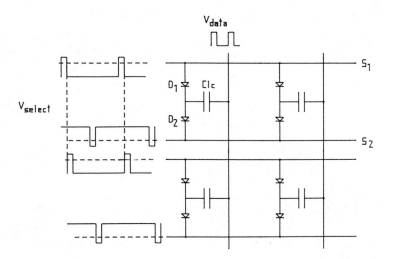

Figure 3.16 Two-diode switch configuration operated in the alternate scan mode.

During the even frames, the select lines S_2 are scanned while lines S_1 are kept at sufficiently low voltage that diodes D_1 are reverse-biased.

When the pixel is charged, the diode and LC pixel are in series, and a small voltage drop of about 1 V occurs across the diode. This voltage drop has to be uniform for all diodes across the display area to obtain a uniform gray scale. In practice, the forward characteristics are relatively easy to control within 0.1 V, and gray-scale uniformity is not a major problem. Gray scale can be obtained either by amplitude modulation of the video signal or by pulse-width modulation.

For the purpose of device redundancy, two or three diodes can be connected in series in each branch (Yaniv et al., 1988). When one diode is short-circuited, the other two diodes in the branch will prevent the occurrence of a pixel defect. The increased number of diodes per branch has the added benefit of reducing the effective overall device capacitance and of reducing the maximum reverse bias per diode. As mentioned before, a-Si *pin* diodes can suffer from soft breakdown at high reverse bias. The series connection of three diodes virtually eliminates this effect.

Attempts have been made to add a storage capacitor to displays with the two-diode switch (Baron et al., 1986), at the expense of increased complexity. In Figure 3.17 the circuit diagram for this configuration is shown. The pixel is split into two halves, with one half connected to the center point of the two-diode switch and the other half connected to the video line. A floating ITO layer buried under the two pixel halves and separated from them by the storage capacitor

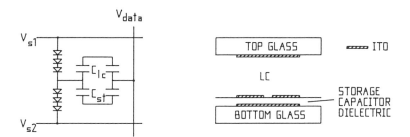

Figure 3.17 Two-diode switch circuit with added capacitance.

dielectric provides the added capacitance. On the top plate of the LCD another floating ITO layer forms the pixel counterelectrode. As discussed previously in connection with TFT LCDs, the added capacitance improves the uniformity of the display in gray-scale operation and at elevated temperatures.

Drawbacks of this approach are the increased complexity of fabricating the active matrix array, the recurrence of crossing lines on the same substrate, and the requirement to increase data voltages because the voltage is split over two pixel halves.

The two-diode switch approach requires two select lines for each row of pixels and therefore doubles the number of interconnections and driver circuits for rows. The D²R (double diode plus reset) circuit proposed by Kuijk (1990) does not suffer from this drawback. In Figure 3.18 the circuit diagram and driving waveforms of the D²R configuration are shown. Diodes D_1 are connected to their respective column lines, while diodes D_2 are connected to a common reference voltage. In this relatively complicated scheme the diode D_1 is used to charge the pixel in both the odd and even frames. Diode D_2 is forward-biased during the reset pulse prior to charging the pixel to the opposite polarity.

It can be shown (Kuijk, 1990) that with the proper choice of voltage levels (e.g., $V_d = \pm 1.5$ V, $V_{s1} = -4$ V, $V_{s2} = 2$ V, $V_{ns1} = 0$ V, $V_{ns2} = 6$ V, $V_{ref} = 4.5$ V, $V_{reset} = 10$ V), an ac voltage across the liquid crystal is obtained and the diodes are reverse-biased or at 0 V during the holding period.

Gray scale can be achieved either with modulation of the data voltage amplitude or by pulse-width modulation of the data voltage. As in diode ring and two-diode switch displays, the fabrication yield can be improved by connecting two or more diodes in each branch.

Arrays of *pin* diodes for displays require three to seven photolithography steps. The processes for diode ring, two-diode switch, and D²R displays can be basically the same. In one commonly used process (Togashi et al., 1985; Yaniv et al., 1986c), the ITO layer is patterned first, followed by deposition of the sandwich Cr-*pin*-Cr or W-*pin*-Cr (Figure 3.19). The top Cr is patterned and used

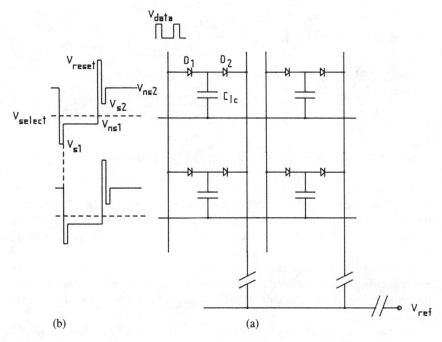

Figure 3.18 (a) Double diode plus reset (D²R) circuit; (b) addressing waveform for D²R circuit.

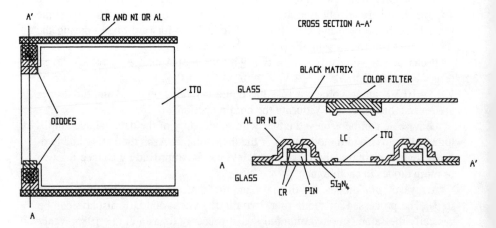

Figure 3.19 Layout and cross section of a pixel in a *pin* diode display with two-diode switch configuration.

as a mask to etch the a-Si layer in a dry etching process (McGill et al., 1986). The bottom metal is then patterned. A silicon nitride layer is deposited, and contact holes are opened on top of the diodes. Finally, a top metal such as Al, Mo, or Ni is deposited to connect the diodes and to form the buslines. The buslines consist of a double metal layer of the bottom metal and the top metal. This line redundancy increases the yield. The diodes are shielded from light because they are completely encapsulated by the metal electrodes.

A simpler, three-mask process, in which the diode is not completely shielded from light, has also been proposed (Togashi et al., 1986; Urabe et al., 1989).

3.2.5 Fabrication Issues

Many of the processes developed for the integrated circuit industry can be applied to the processing of AMLCDs. Standard mask aligners with some minor modifications to handle square glass substrates instead of silicon wafers can be used to fabricate displays with viewing areas up to 4 in. × 4 in. Projection aligners are preferred over contact and proximity aligners in order to improve yield. For large displays, special aligners have been built for the step-and-repeat exposure method illustrated in Figure 3.20.

The reticle has a standard size, and windows can be set to expose either the viewing area V; the edges with the interconnection leads A, B, C, D; or the corners E, F, G, H. The viewing area of the display is built up by multiple step-and-repeat exposure, usually without reduction. Stitching accuracy of the patterns to each other is better than about 1.5 μm. This versatile approach makes it

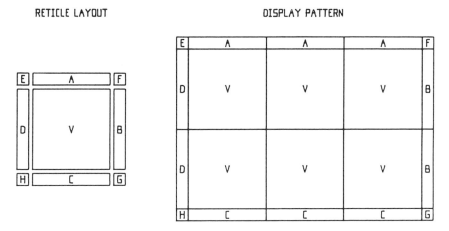

Figure 3.20 Reticle layout for composing a large-area display from a small reticle without reduction.

possible to fabricate displays with different viewing areas using the same set of reticles.

Metals and ITO are usually delineated by wet etching, whereas the a-Si layers and insulators are patterned by dry etching.

For large displays the busline resistances have to be low enough not to cause a distortion of the driving waveforms at the end of the line. This is particularly important for the row busline, through which, in the worst case, all pixels have to be discharged and charged to the opposite polarity at the same time. Low-resistivity metals such as Al and Ni are often employed to reduce the row line resistance.

The process for the top plate is different for TFT and diode displays. For black-and-white diode displays, ITO lines are patterned and a black matrix material, preferably an opaque insulator, is deposited and patterned in the interpixel area to improve the contrast ratio. For color displays, red, green, and blue color filters are formed prior to the ITO deposition. For TFT top plates, the process is simpler, because it is not necessary to pattern the ITO, and the black matrix can be a metallic conductor.

Several color filter technologies are used, including spin coating and patterning of dyed polyimides or dying of gelatins, electrodeposition of pigmented polymers, and printing. Depending on the application, the pixels on the active substrate and the color filters on the top plate have a certain arrangement. For alphanumeric displays a vertical stripe pattern is usually used, whereas for video applications a delta configuration is preferred.

The yield of the active matrix array fabrication and display assembly is important in determining the cost of the display. For small TFT LCDs up to 5 in. in diagonal size, yields in excess of 50% are obtained. To improve yield, redundancy schemes are often employed. The buslines can consist of double metallization to reduce the occurrence of open lines.

In TFT displays, two TFTs per pixel can be used to eliminate nonoperating pixels due to open TFTs. Another approach is to reduce the visibility of pixel defects by dividing the pixel into two, three, or four subpixels each with its own TFT and with equal data information (Nagayasu et al., 1988; Tomita et al., 1989). When a subpixel is not operating because of a defective switch, the other subpixel(s) will be functional and the pixel will be only partially defective.

Laser repair of short-circuited and open lines and pixels is also used and can be very effective for increasing yield, but it is time-consuming and not very compatible with mass production. It is very difficult to eliminate pixel defects completely, and for most applications one pixel failure per square inch can be tolerated.

Screens of more than 5 in. diagonal size still have a relatively low manufacturing yield, reportedly in the 10–20% range in 1990. Further improvement in yield by process optimization and automation is expected, however, and is crucial for the ultimate success of large-area AMLCDs.

3.3 DISPLAY PERFORMANCE OF ACTIVE MATRIX LCDs

Optimal performance of AMLCDs not only requires optimization of the active matrix array but also depends on the voltage levels and timing of the driving waveforms, the color coordinates of the color filters, the black matrix in the interpixel area, the spectrum and intensity of the backlight, and the LC material and cell gap. In addition, for display legibility under high ambient light conditions, the diffuse and specular reflectance from the display surface has to be minimized. It is difficult to compare the performance of different types of AMLCDs without taking all these factors into account.

Amorphous silicon TFT displays have been most successful to date. Pocket TVs with screens up to 5 in. in diagonal size have been on the market for several years. Many organizations, in particular in Japan, have developed larger screens as well, from 10.4 in. (Kanemori et al., 1990) to 14 in. (Nagayasu et al., 1988; Ichikawa et al., 1989).

In Figure 3.21 a photograph of a 10.4-in. diagonal color TFT LCD produced by OIS Optical Imaging Systems, Inc. is shown (reproduced in black and white). The $480 \times 640 \times 3$ (RGB) pixel display has a contrast ratio of about 100:1 in the normally white (crossed polarizer) mode at the optimum viewing angle.

By the end of 1990 the largest and highest resolution TFT LCD had a 15-in. diagonal size and 1600×1920 pixels (800×1280 color groups) (Wada et al., 1990). Figure 3.22 shows a typical example of the viewing angle dependence of contrast ratio in a TFT LCD (Katayama et al., 1988).

MIM diode displays have been commonly considered to have the lowest cost but also the lowest performance of all AMLCDs. In recent years, however, significant progress has been made with Ta_2O_5 MIM displays. After process optimization, a color MIM LCD TV with 240 rows was reported (Ono et al., 1990) with contrast ratio greater than 100 and color purity matching that of a-Si TFT displays. The maximum number of addressable rows in MIM displays has also recently increased. A 12 in., bilevel black-and-white display with 1152×900 pixels was developed for use in laptop engineering workstations (Morita et al., 1990).

SiN_x diode displays have been improved as well by optimization of processing. Bilevel black-and-white displays with 640×400 pixels and 9.8 in. (Toyama et al., 1987) and 9.3 in. (Ohira, 1990) diagonal size have been reported. A problem to be solved in SiN_x diode displays is the long-term stability of the SiN_x diode. Under prolonged operation the current in the diode decreases owing to charge trapping in the SiN_x film. This device degradation causes image retention

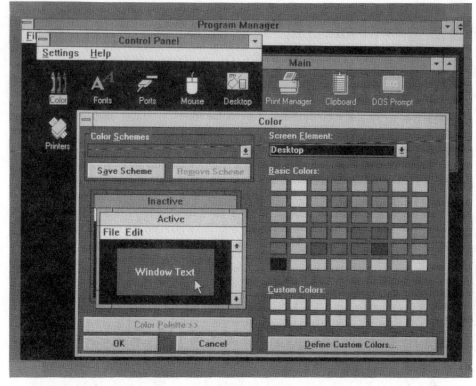

Figure 3.21 Photograph of 480 × 640 × 3 pixel 10.4 in. color TFT LCD produced by OIS Optical Imaging Systems, Inc. (reproduced in black and white).

in the display. For activated pixels, the ON current decreases gradually, so that the voltage across the LC pixel decreases with time (since LC capacitance and MIM diode are in series). A permanent afterimage is the result. This type of retained image does not fade away and has to be distinguished from the temporary afterimage due to dc voltage component over the liquid crystal.

pin diode ring displays were developed in the early 1980s (Togashi et al., 1985) and were gradually improved to medium resolution and medium size (Togashi et al., 1986; Nicholas et al., 1990). For higher resolution displays it is probably necessary to increase the number of diodes in the diode ring to obtain a higher threshold voltage for the switch. The resulting loss in active pixel area (pixel opening) and brightness might be unacceptable.

The highest resolution diode displays developed thus far use the two-diode switch configuration. An 8 in. × 8 in. display with 1296 × 1296 pixels was re-

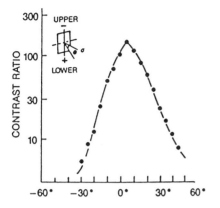 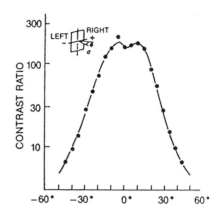

Figure 3.22 Viewing angle dependence of contrast ratio in a TFT LCD with crossed polarizers. [From Katayama et al. (1988), reprinted with permission from Society for Information Display.]

ported (Vijan et al., 1990) for avionic and military applications. This display used six diodes per pixel for redundancy purposes and the non-crossing line approach without storage capacitors. A maximum contrast ratio of about 50 was obtained, and surface reflectance was minimized to achieve a contrast ratio of 4 under ambient illumination of 10,000 foot-candles.

The D^2R configuration is a relatively novel approach, and 6-in. diagonal displays with 575 × 750 pixels have been built with a reported contrast ratio of more than 100 (Kuijk, 1990).

The overall performance of color displays is determined to a large degree by the color filters and the backlight. In Figure 3.23 the spectra of typical dyed polyimide color filters with a thickness of 1–2 μm are shown. When the thickness or the dye concentration is increased, more saturated colors can be obtained at the expense of a lower transmission. A trade-off between transmission and color purity is therefore necessary. The backlight is usually a cold cathode or hot cathode fluorescent tube or set of tubes. The lamp has specially blended phosphors to obtain a spectrum with distinct peaks for red, green, and blue wavelengths (Figure 3.24).

In Figure 3.25 the CIE color chromaticity diagram of a typical *pin* diode LCD that used the color filters and backlight of Figures 3.23 and 3.24, respectively, is shown. Good color performance was obtained, close to the color coordinates of a CRT, which are also shown for comparison.

When *pin* diode technology is chosen over TFT technology, it is usually justified by the absence of crossover short circuits in diode displays, because row and column lines are not on the same substrate. TFT display processing, how-

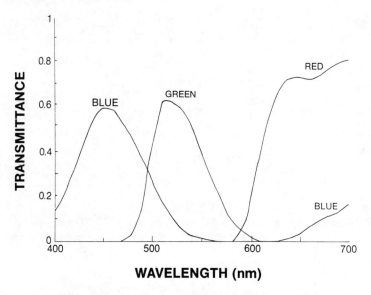

Figure 3.23 Spectra of dyed polyimide color filters.

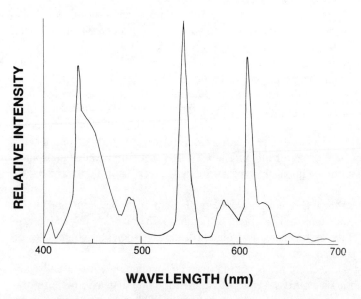

Figure 3.24 Spectral distribution of a fluorescent backlight for an RGB display.

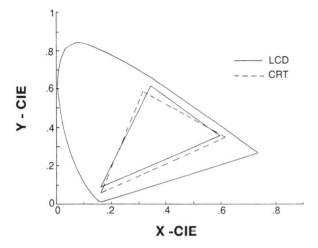

Figure 3.25 CIE color coordinates of a typical color AMLCD. For comparison, the color coordinates of a typical CRT display are also shown.

ever, has improved to the point that interlevel shorts are virtually eliminated by using double or triple insulation layers at crossover points. This merit of diode displays is therefore no longer a decisive factor in selecting the technology.

A drawback of all diode displays with columns and rows on opposite glass substrates is the relative complexity of the color plate fabrication. As mentioned earlier, the ITO lines on the color plate have to be deposited and patterned on top of the color filters and the black matrix to obtain the highest contrast ratio. The step coverage of the thin ITO film over the color filters is often insufficient to ensure line continuity, and planarization layers have to be inserted. In comparison, the color plate of TFT displays is less complicated because the ITO layer is continuous and not patterned. The black matrix material for a TFT display can be a metal, whereas in diode displays the black matrix in the interpixel area is preferably made from an opaque insulator to prevent short circuits between adjacent lines.

3.4 ACTIVE MATRIX LCDs: SUMMARY

In the past five years AMLCD technology has made progress to the point where large-scale mass production has become feasible. The question is no longer whether high-resolution, high-quality displays can be fabricated; both TFT and diode technology have demonstrated that capability. The remaining issue is cost. The glass part of the AMLCD module is the most expensive part, especially for large-area displays with a diagonal size of more than 5 in. With increasing yield and the continuing development of equipment for large-area photolithography, plasma deposition, dry etching, and metallization, the price will come down.

For intermediate complexity displays without gray scale, MIM LCDs appear very promising as they require the fewest process steps. TFT and *pin* diode displays are more costly to fabricate but can achieve the high resolution and gray scale required for high-definition TV and workstation monitors. AMLCDs are now penetrating the avionic and military display market, where cost is less critical. They outperform CRTs in size, weight, ruggedness, and, in particular, sunlight readability.

The next application will be in high-quality flat screens for portable computers. Another promising application is the use of small AMLCDs (about 3 in.) in projection systems. These systems usually employ a metal halide lamp with dichroic filters to split the spectrum into the three primary colors. Three LCDs modulate each color, and the three images are combined and projected onto a large screen. Projection TVs with NTSC resolution have already been put on the market, and prototype high-definition systems based on this principle have been developed (Kobayashi et al., 1989; Adachi et al., 1990).

The increasing number of applications will allow manufacturers to set up large-scale production plants. After mass production is established, a further increase in screen size may lead to hang-on-the-wall direct-view high-definition TV around the year 2000.

3.5 AMORPHOUS SILICON IMAGE SENSORS

The future of our society is based on information. The need for processing large quantities of information is growing continuously. Furthermore, the information age requires portability and miniaturization. Today, palm-size portable telephones, notebook computers, and portable fax machines are frequently used. As a result, the governing laws for sophisticated imaging electronics are (1) high reading speed, (2) high-quality image-reading capability, (3) small size, and (4) low price. To achieve small-size image sensors, one needs to use a contact-type image sensor that eliminates the need for a bulky reduction optical system.

The image sensor consists of a linear or two-dimensional array of individual light-sensitive elements. A document is exposed to the light-sensitive elements, which are integrated electronically by a peripheral multiplexing system. The collected information is then processed to build the document image.

Charge-coupled devices (CCDs) were at one time the only sensors available. These devices are fabricated on crystalline silicon substrates and as a result have some inherent disadvantages:

1. They are limited in length to a few centimeters.
2. Optical reduction of 1:10 is needed in order to read an A4 document that requires a long optical path (30–40 cm).

3. They require a high-quality lens and mechanical rigidity.
4. The substrate is opaque, complicating the illumination system.
5. Frequent adjustment is needed between the sensor and the optical system.
6. They give low yield because of the complexity of the CCD chip.
7. They require a high capital investment—parallel to one for a usual semiconductor foundry.

The thin-film sensor offers solutions to all of these issues. Although the cost is in the same range as for the CCD (considering the external chip drivers needed for the thin-film sensor), by offering a solution to the problems listed above, the thin-film sensor is becoming the main contender to dominate the image sensor field.

In the 1980s, amorphous silicon was recognized as the most suitable material for image sensors (Madan and Shaw, 1988). Image sensors using other alternatives such as CdS-CdSe chalcogenides are very rare. The main arguements in favor of amorphous silicon have to do with

1. Capability of large area deposition on usual glass substrates
2. High photoconductivity
3. Spectral response in the visible region
4. Simplicity of manufacture
5. Fast response time ($\sim$ μsec)
6. Efficient doping
7. The possibility of forming high-quality TFTs and *pin* diodes
8. Compatibility with conventional monocrystalline silicon technology

Contact image sensors using a-Si can be categorized into two groups: photoconductor and photodiode types. Furthermore, there are two kinds of photodiode-type sensors: direct-addressed and matrix-driven sensors (Rosan, 1991). Direct-addressed sensors are simpler to fabricate, but they have as many terminals as the number of picture elements, creating a very difficult interconnect problem. As they use diodes or TFTs fabricated on the same substrate, the matrix-driven sensors have a drastically reduced number of terminals. The next section describes the sensor structure, fabrication, and characteristics of these sensors.

3.5.1 Photoconductor-Based Sensor

The classic structure of a photoconductor shown in Figure 3.26 consists of a layer of amorphous silicon and two ohmic contacts. The structure can have a coplanar or a sandwich arrangement. In steady state the photocurrent is given by

$$I_{ph} = e \, \Phi \, g \qquad (3.11)$$

where e is electron charge, Φ is the generation rate of free electrons, and g is photoconductive gain, assuming that the electrons are the dominant charge car-

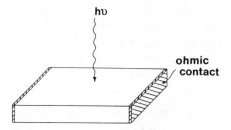

Figure 3.26 Structure of photoconductor for photoconductor-based image sensors.

riers (a good approximation for a-Si) and that the buildup of space charge is negligible.

The photoconductive gain g is defined as the number of free electrons created per absorbed photon and is equal to the ratio of recombination lifetime τ_r to the transit time τ_t required by the free electrons to move between the two electrodes (Rose, 1963):

$$g = \frac{\tau_r}{\tau_t} \qquad (3.12)$$

Equation (3.12) implies that the gain exceeds unity because of the slow photoresponse speed of a-Si photoconductors. At slow recombination rates of the photogenerated carriers ($\tau_t < \tau_r$), a secondary photocurrent is generated at the ohmic electrodes in order to preserve charge neutrality. As a result, the absorption of a single photon can generate a large number of carriers passing between the two electrodes.

The secondary photocurrent is sensitive to variations in the Fermi level and the density of the recombination centers. As a result, a-Si sensors based on the photoconductivity effect are strongly influenced by fabrication parameters, nonuniformities, changes in the density of recombination centers due to instability of a-Si (Staebler and Wronski, 1980), or other external factors.

The generation rate Φ is directly proportional to the photon flux. Both the gain and the generation rate are dependent on the geometry of the sensor and must be taken into account when designing or comparing sensor characteristics. The transit time is given by

$$\tau_t = \frac{l^2}{\mu V} \qquad (3.13)$$

where l is the electrode spacing, μ is the mobility, and V is the applied voltage. When the gap length is decreased, the gain will increase, but the area of illumination and consequently Φ will decrease. In practice, an increase in the photocurrent can be achieved by a change in geometry or by using a sandwich structure.

MICROELECTRONICS IN LCDS AND IMAGE SENSORS

The conclusion is that despite the simplicity of their structure the a-Si sensors based on the photoresistor effect cannot be used for high-performance (high-speed, high-resolution) applications because of their slow photoresponse, high dark current, and poor uniformity and reproducibility.

3.6 PHOTODIODE-BASED SENSORS

3.6.1 Amorphous Silicon Photodiodes

Hydrogenated amorphous silicon (a-Si:H) can be easily doped to form p- and n-type a-Si. As a result, various sandwich structures of photodiodes using a-Si can be achieved (Hayama, 1990). These photodiodes have a much faster response time than a-Si in the photoconductive mode and are more suitable for constructing contact-type image sensors. Some representative a-Si photodiode sandwich structures that were studied and compared extensively are p-i-n (Seki et al., 1983), p-i junctions (Hayama et al., 1986), heterojunctions (Kaneko et al., 1985), and Schottky junctions (Hamano et al., 1982). Their schematic structures are shown in Figure 3.27. The photocurrent of a reverse-biased diode is (neglecting the recombination of carriers during their transit) calculated as follows:

$$I_{ph} = eAd\,\phi$$

where A is the area of the photodiode, d is the thickness of the photodiode, and ϕ is the optical generation rate. Knowing the quantum efficiency η_{ph} for the op-

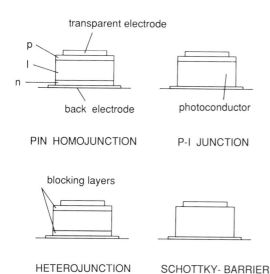

Figure 3.27 Structures of *pin* junction, *p-i* junction, heterojunction, and Schottky barrier junction used in photodiode-based image sensors.

tical generation of carriers and the incident photon flux F, we can obtain

$$\phi = \eta_{ph} F (1 - R)(1 - e^{-\alpha d})/d$$

where R is the reflection at the air–semiconductor interface and α is the absorption coefficient within the semiconductor.

The photodiode sandwich structure is characterized by blocking contacts that prevent injection of carriers from the electrode into the semiconductor. As a result, the speed of the photoresponse is due mainly to the transit carriers. Furthermore, the contribution of the trapped charges to the decay time, which can be up to 1 sec in photoconductors, is much reduced in photodiodes, as the photodiodes are working in depletion mode. Therefore, high-resolution, high-speed image sensors require the use of the sandwich photodiode structure.

As early as 1982, Silver et al. (1982) recognized the importance of the p-i-n amorphous silicon structure as a useful microelectronic device, in particular for large-area integrated circuits on glass. An uninterrupted p-i-n deposition sequence was developed, ensuring stable interfaces. Furthermore, a well-controlled VLSI process was implemented to preserve the high-quality diode characteristics after the completion of the mesa structure, including metallization, via contacts and etching (McGill et al., 1986).

A typical diode for sensor applications after processing has an active area of approximately 20 μm × 20 μm, an ideality factor of $n = 1.3$–1.5, and an extrapolated reverse leakage saturation current density of 10^{-13} A/cm². The diodes exhibit nearly 10 orders of magnitude rectification at ± 3 V, and the reverse-bias current density remains below 10^{-8} A/cm² for bias voltages of -15 V. The dc current–voltage curve for a typical device was shown earlier in Figure 3.14. Under pulsed forward bias, these diodes have been operated at current densities in excess of 300 A/cm². This is demonstrated in Figure 3.28 for a device with an area of 3.9×10^{-4} cm². This large diode size was used in order to be able to physically probe the device; however, it was fabricated using a process identical to that used for the 20 μm × 20 μm devices. In Figure 3.29, it is shown that the switching speed of these diodes is about 20 nsec.

Further, it has been demonstrated that by varying the intrinsic layer thickness of a p-i-n device, the current density can be varied in proportion to L^{-3}, where L is the intrinsic thickness (Cannella et al., 1986). By varying the device thickness, one can also control the capacitance per unit area and the reverse-bias current density.

The photodiode characteristics of the Schottky-type structure were extensively studied (Rosan and Brunst, 1986). It was found that the photocurrent is almost independent of the applied voltage, and in the transient mode the photocurrent settles within 2 nsec. The photoresponse slows down if contributions occur from photocarriers generated outside the sandwich structure. This process can be controlled by separating the photodiodes using the VLSI process de-

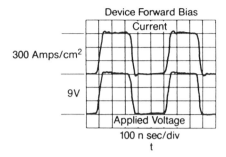

Figure 3.28 Pulsed operation characteristics of a-Si *pin* diodes illustrating high current capability of 300 A/cm² (load resistor = 50 Ω).

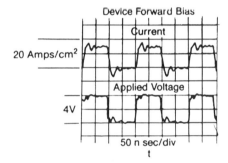

Figure 3.29 Forward-bias current pulsed in *pin* diodes demonstrating a switching speed of about 20 nsec.

scribed above. In such a way, higher resolution sensors can be obtained that limit the crosstalk between the sensor elements.

Although both the *p-i-n* diode structure and the Schottky structure show a very low dark current, further reduction and control can be obtained by using a-SiC$_{1-x}$ (*p*- or *n*-type) blocking layers, which have a larger bandgap than a-Si (Morozumi et al., 1984). Therefore, carriers emitted from the electrodes are better blocked by the n^+ and p^+ layers, resulting in a lower dark current.

In conclusion, the photodiode, because of its low current, fast photoresponse, and photocurrent largely independent of the applied voltage, is superior to photoconductors for image sensor applications. The *p-i-n* structure shows excellent reproducibility, uniformity, and stability. The Staebler–Wronski effect (Staebler and Wronski, 1980) does not affect the photodiode characteristics because the transit time of the majority carriers ($\sim$ 10–100 nsec) is much shorter than the recombination lifetime ($\sim$ 1 μsec to 10 msec).

3.6.2 Contact-Type Linear Image Sensors

The need for compact reading devices to scan large documents propagated the development of contact-type linear image sensors, where the document is scanned in direct contact with the image sensors or using a rod lens (self-focusing lens) array housing with only a few millimeters focal length and no reduction capability. There are two types of contact-type linear image sensors using a-Si photodiodes: direct-addressed sensors and matrix-driven sensors (Tsukada, 1984).

Direct-Addressed Sensors

Direct-addressed sensors use hybrid technology where the photodiodes are the only thin-film devices and are addressed by individual analog switches and shift registers integrated on crystalline silicon chips (Hamano et al., 1982; Ozawa et al., 1982). Generally, the LSIs are mounted on the substrate (chip-on-glass method) and electrically connected to the photodiode pads by wire bonding.

The usual structure of the photodiode is shown in Figure 3.30. The lower electrode metals, such as Al or Cr, can be evaporated or sputtered. The substrates can vary from 7059 glass to Pyrex, ceramics, or even stainless steel. The amorphous silicon is deposited by capacitive-coupled radio-frequency glow discharge (PECVD) decomposition of SiH_4. The film thickness varies between 0.5 and 1 μm, and very good thickness uniformity (better than 3%) can be obtained by good deposition control. The final step is the deposition of indium tin oxide (ITO), which forms the transparent electrode. The preferred method is dc sputtering, although reactive evaporation is also used. Direct-addressed sensors are already in use, in particular for G3 facsimile mode devices where a resolution of 200 dots per inch (dpi) and about 10 msec line scan time are required (Ozawa and Takenouchi, 1982).

Figure 3.31 shows the cross section of a sensor of this type (Saito et al.,

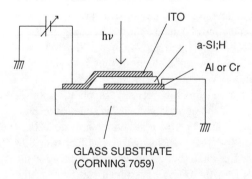

Figure 3.30 Structure of direct-addressed sensor.

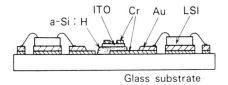

Figure 3.31 Cross section of direct-addressed sensor. [After Saito et al. (1984).]

1984). Data acquisition is performed in charge storage mode; the light-induced voltage drop across the diode capacitance (~ 1 pF) is evaluated by using the LSI readout chips. Typical values for the line scan time are 1–10 msec, and under illumination of 100 lux one can obtain a voltage signal of about 1 V across the diode.

Although the direct-addressed sensors have the advantage of using the most optimized signal processing (custom LSI), they suffer from both technical limitations and cost. As the resolution of the sensor is increased (to 400 or 600 dpi), the wire bonding method will not be feasible and the output signal will be reduced. Furthermore, the noise level will increase owing to the increased parasitic capacitance between relatively long neighboring circuitry lines.

From the economics point of view, the large number of drivers (14 drivers are needed for a 200 dpi 8.5 in. long sensor) are consuming valuable glass substrate area and production time, as 2000 wire bonds are required. The matrix-addressed sensor offers a solution to these problems by drastically reducing the number of readout chips.

Matrix-Addressed Sensors

To solve the problem of the high number of terminals used in direct-addressed sensors, a matrix-driven method must be implemented. Swartz et al. (1987) presented a page-width, high-speed, high-resolution, full contact line matrix-driven imager using a-Si alloy *p-i-n* diodes both as photosensing elements and as blocking diodes in the multiplexing scheme.

In this imager, an individual imaging element consists of a pixel diode and a blocking diode connected back to back. Figure 3.32 shows the equivalent circuit configuration, including the matrix configuration used to achieve faster reading and higher signal-to-noise (S/N) ratio. The matrix is constructed by connecting 16 elements in one block. These are simultaneously driven by a negative square pulse from the driver. The corresponding 16 current outputs are detected through current-to-voltage amplifiers in parallel. According to the resolution of the imager—either 200, 300, or 400 elements/in.—the imager has 100, 158, or 220 blocks, respectively.

The driving pulse is shifted in sequence to cover every block in the image (8.5

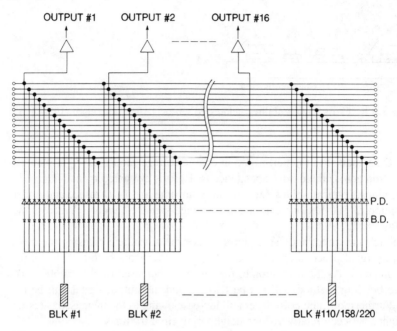

Figure 3.32 Circuit diagram of matrix-addressed sensor.

in.), and this cycle is synchronized with the paper transport. Figure 3.33 is the equivalent circuit to explain the readout operation, and Figure 3.34 shows the typical voltage changes of the pixel diode for both the dark and illuminated cases. In order to read out the pixel, a negative voltage pulse is applied to the blocking diode, which charges the diode down to the voltage of the negative pulse. The capacitance of the pixel diode retains this voltage while the pulse travels along the line to each of the subsequent blocks. Since the reverse current of the blocking diode is extremely small, the pixel diode is isolated from the other blocks, thus preventing crosstalk between elements.

In the absence of illumination (dark case), no change in the charge of the pixel diode will be detected during the next scanning cycle of the blocking diode, and therefore no current flow will be detected. If the pixel has been illuminated, however, the photocurrent induced in the pixel diode will discharge the capacitance of the pixel diode. This will create a voltage change on the pixel diode that will generate current flow in the next readout. The current is proportional to the illumination level and is converted to the voltage at the amplifier output.

The isolation diodes and photodiodes are deposited, as in the case of active matrix LCDs, using plasma-assisted CVD techniques. The structure of the p-i-n diodes include an a-Si:H:F intrinsic layer with an a-Si:H:F:P microcrystalline n^+

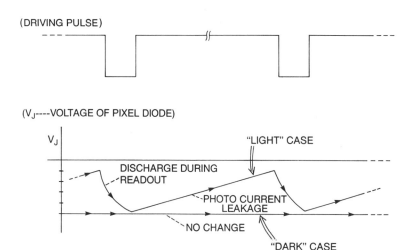

Figure 3.33 Equivalent circuit for one diode pair in matrix-addressed sensor. (B.D. = blocking diode, P.D. = photodiode.)

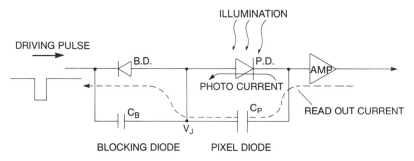

Figure 3.34 Typical photodiode voltage variation vs. time for dark and illuminated conditions.

layer and an a-Si:H:B p^+ layer. The layer thicknesses are approximately 6500, 500, and 500Å, respectively. The photodiodes are 105 μm × 125 μm for the 200 dpi resolution, and the isolation diodes are 32 μm × 32 μm. A cross section of the line imager is shown in Figure 3.35. The entire imager is coated with a silicon nitride passivation layer.

In Figure 3.36 a photomicrograph of the imager is shown. The I–V characteristic of an isolation diode is similar to that shown in Figure 3.14. Figure 3.37 shows the I–V characteristic of the photodiode under illumination, and the pho-

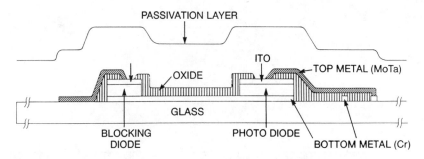

Figure 3.35 Cross section of line imager showing back-to-back diode pair.

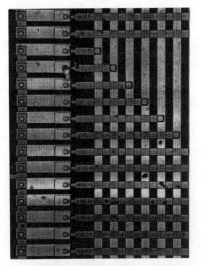 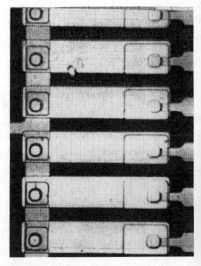

Figure 3.36 Photomicrograph of line imager sections showing multiplexed diode pairs.

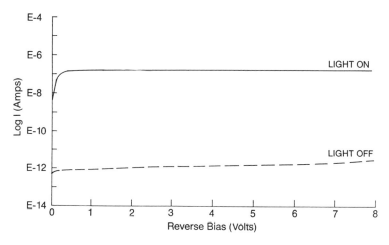

Figure 3.37 Current–voltage characteristics of a photodiode under dark and highly illuminated (5 mW/cm² white light) conditions.

tocurrent is seen to be 10^5 times the dark current for 5 mW/cm² white light illumination.

The major performance items of the 200 elements/in. imager are listed in Table 3.1. Typical output signals from the output amplifiers under both light and dark conditions of the imager are shown in Figure 3.38 for an illuminated light intensity slightly lower than that necessary to achieve the saturation exposure of 3.8 lx·sec (by fluorescent lamp). At this light level, the S/N ratio is greater than 40 dB. The imager can be used either in the true direct contact mode or with

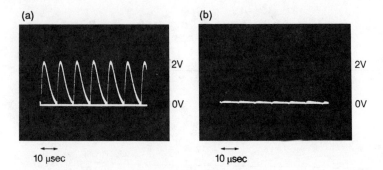

Figure 3.38 Oscillograms of the output amplified voltages for a line of photodiodes under (a) illuminated and (b) dark conditions.

Table 3.1 Performance of the 200 dpi Matrix-Driven Sensor

Items	Performance
Readout pulse	8 μsec (16 bits parallel)
Scan time	0.9 msec/line (8.5 in. wide)
Saturation exposure	≤ 3.8 px·sec
S/N ratio	≥ 40 dB
Disuniformity	≤ ± 5%

Source: Swartz et al. (1987).

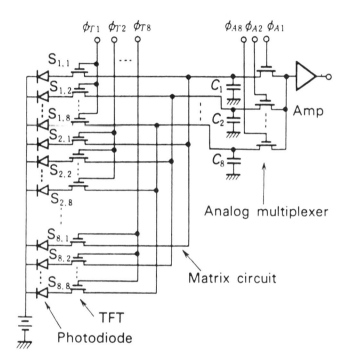

Figure 3.39 Equivalent circuit of a-Si linear image sensor driven by TFT matrix array. [After Okumura et al. (1983).]

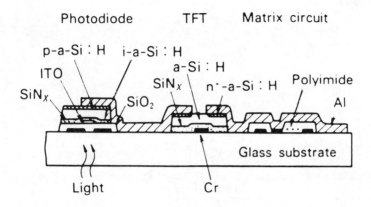

Figure 3.40 Cross section of linear imager operated by a-Si TFT array. [after Okumura et al. (1983).]

Selfoc-type lensing. The direct contact is achieved by back-projection of the light source through the glass substrate (McGill et al., 1989).

A typical readout pulse is 8 μsec (16 bits output parallel). For the 200 elements/in. imager, the typical scan time is about 0.9 msec/line (8.5 in. wide). This allows the reading of a letter size (8.5 in. × 11 in.) page in less than 2 sec. The use of a 10-W fluorescent lamp for the light source in the contact mode can achieve this speed easily. The photoresponse nonuniformity at this light condition is less than ±5% (McGill et al., 1989). Instead of an α-Si diode, an α-Si thin-film transistor (TFT) can be used for the multiplexing scheme (Okumura et al., 1983). This type of matrix-driven sensor, shown in Figure 3.39, consists of a photodiode array, TFT array for multiplexing, and external circuit. A cross section of the sensor is presented in Figure 3.40. The photodiode has a sandwich structure as presented above, whereas the TFT has a planar inverted staggered structure. The channel length can vary between 5 and 20 μm, depending on the photolithographic tool used and the required electrical characteristics. As a result, the TFT multiplexed sensor is expected to be slower than the diode multiplexed sensor. Furthermore, variations in the threshold voltage and TFT mobility

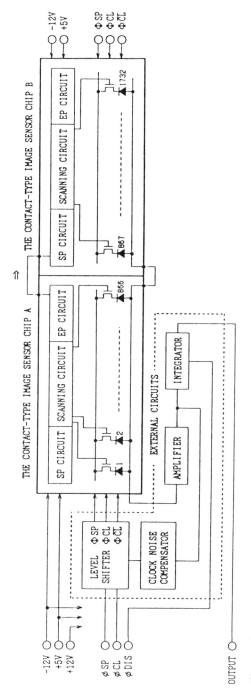

Figure 3.41 Circuit configuration of contact-type image sensor with integration of a-Si photodiodes, poly-Si TFTs, and poly-Si driving circuits. [from Takeshita et al. (1989), reprinted with permission from Society for Information Display.]

over the surface of the glass substrate can lower the S/N ratio and the overall quality of the sensor.

Further reduction of complexity was presented by Morozumi et al. (1984) and Takeshita et al. (1989) using the integration of a-Si photodiodes, polycrystalline silicon TFTs, analog switches, and poly-Si driving circuits on the same substrate. The equivalent circuit is shown in Figure 3.41.

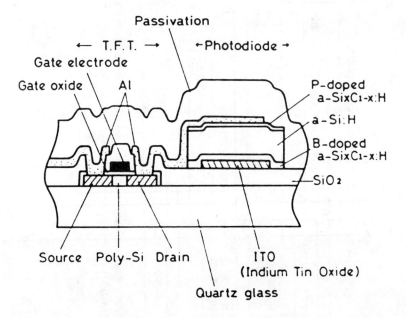

Figure 3.42 Cross-sectional view of integrated linear image sensor. [from Takeshita et al. (1989), reprinted with permission from Society for Information Display.]

Table 3.2 Performance of a Matrix-Driven Sensor with Integrated Poly-Si Thin-Film Transistors

Items	Performance
Supply voltage	0 V, −5 V, −16 V
Clock frequency	250 kHz (typical)
Readout time	2 μsec/bit (typical)
S/N ratio	≥ 40 dB
Saturation exposure[a]	≤ 0.89 Lx·sec
Uniformity	≤ ± 5%

[a]Light source is green LED.
Source: Takeshita et al. (1989).

The photodiode has a sandwichlike structure with B (boron)-doped a-SiC$_{1-x}$ and P (phosphorus)-doped a-SiC$_{1-x}$ on both sides of the photosensitive a-Si:H layer (see Figure 3.42). The poly-Si is deposited by LPCVD, and the gate insulator is formed by thermal oxidation of the poly-Si layer. High-performance linear sensors were obtained (see Table 3.2). The drawback of this approach is the high cost of the substrate. The poly-Si is deposited at approximately 650°C, and as a result quartz substrates must be used, adding a prohibitive cost to the sensor and limiting the substrate size. In this case, to scan A4 size documents, the combination of two image sensors is required, defeating the purpose of a full-length contact-type sensor and creating unnecessary alignment difficulties.

3.7 CONCLUSION

The revolution in electronic circuitry occurred with the development of active circuit devices. These devices are constructed of simple elements such as diodes and transistors to exploit the advancement in semiconductor technology from ICs through LSI and VLSI up to wafer-size LSIs. This advancement was possible because the semiconductor manufacturing technology increased our capacity to deal with microscopic circuits (less than 1 μm in line width!) through microfabrication of very high density integrated circuits.

We have discussed a new trend in the microelectronic industry: "giant microelectronics." This new trend, in contrast to the existing technology based on circuit miniaturization, has the goal of producing circuit elements on a very large "wafer" (14 in. × 14 in. and up to 40 in. diagonal).

To be able to create "giant microelectronics," one must produce and control the technology of depositing and processing high-quality semiconductor thin films on relatively large substrates. Although the semiconductor quality of these films is much lower than that of the usual crystalline semiconductors, they are preferred in many applications, such as solar cells, flat-panel displays, image sensors, and other simple operating circuits. Currently, the vast majority of display devices in use are based on CRT (Braun tube) technology, with the exception of a few application fields. Although excelling in cost and performance, the CRT cannot be made in any significantly larger format than currently available formats because a larger CRT would necessarily be heavier and deeper.

Although advances in the graphical form of information transmission are likely to require large display devices for receiving TV broadcasts, a CRT-based implementation of high-definition TV (HDTV) in 40-in. diagonal format would require a TV set weighing more than 200 kg and measuring more than 50 cm deep! The answer is the production of a thin-panel "picture on the wall" device.

Lightweight, thin-panel, lower-power-consumption display devices are needed as aircraft and automobile display devices as well as for use in homes and portable devices.

Currently, the most feasible method is considered to be the active matrix LCD, for which development of large-area circuit elements is needed.

The active matrix LCD and sensor technology presented in this chapter can be used alone or in combination for future exciting products such as two-dimensional flat scanners, ultrathin copying systems, and optical computers.

The current procedure for scanning color documents is to use a one-dimensional color scanner. The process is time-consuming (requiring about 10 min to process a sheet of an A4 size document), and the equipment is bulky. A more promising approach is the development of two-dimensional flat scanners consisting of a planar array of contact sensors that electronically read the document. Such a system would speed up the input process, eliminate the need for a lens system, and have no mechanical motion, so construction would be simpler.

The currently available copying systems are optical, drum-based systems that are bulky, noisy, and difficult to maintain (drum cleaning). Because of these problems, the realization of a copying system based on entirely new operating principles is desired.

An ultrathin copying system could be realized by combining large-area circuit elements like the two-dimensional imager and the active matrix LCD just described with a silver halide–based one-shot color-photosensitive material. This could lead to the development of an ultrathin copying system in which lenses, noise, bulkiness, and mechanical motion have been eliminated.

Parallel processing in two-dimensional space takes advantage of the ability of light to propagate in parallel to achieve an enormous gain in the operating speed of a computer. The main processing device for this computer can be designed to consist of two matrix LC display units and one two-dimensional image sensor, each made with large-area circuit elements.

To construct a computer the two active matrix display units are positioned to face each other, with the pixels aligned in direct correspondence. When planar light is directed from an external source perpendicular to the display plane, the light will pass only when the two opposite pixels are "on" at the same time; in other cases the light will be blocked. Thus, the light signal undergoes an AND spatial modulation.

In the above example, as the resolution of the display and two-dimensional sensor is increased, the greater the precision of the parallel computation will be. The ability to produce high-resolution large-area circuit elements holds the key to the realization of these exciting devices.

Many other new "ideas" can be formulated. The few presented above show the great potential of this new technology of "giant microelectronics." Numerous laboratories and companies in the United States, Europe, and Japan are working diligently to further develop the technology and establish manufacturing plants so that all of these ideas will soon be seen in the marketplace.

ACKNOWLEDGMENT

We gratefully acknowledge the help of Janet Beyer in typing the manuscript.

References

Adachi, M., T. Matsumoto, N. Nagashima, T. Hishida, H. Morimoto, S. Yasuda, M. Ishii, and K. Awane (1990). A high-resolution TFT-LCD for a high-definition projection TV, *SID 90 Dig.*, 338–341.

Alt, P. M., and P. Pleshko (1974). Scanning limitations of liquid-crystal displays, *IEEE Trans. Electron. Devices ED-21:* 146–155.

Baraff, D. R., J. R. Long, B. K. McLaurin, C. J. Miner, and R. W. Streater (1981). The optimization of metal-insulator-metal nonlinear devices for use in liquid crystal displays, *Proc. SID* 22(4): 310–313.

Baron, Y., A. Lien, V. Cannella, J. McGill, and Z. Yaniv (1986). Improved amorphous silicon alloy PIN diode active matrix LCD, *Japan Display '86*, 68–70.
Brody, T. P., J. A. Asars, and G. D. Dixon (1973). A 6 × 6 inch 20 lines-per-inch liquid-crystal display panel, *IEEE Trans. Electron. Devices ED-20:* 995–1001.
Cannella, V., J. McGill, Z. Yaniv, and M. Silver (1985). Experimental study of bulk limitation of the current in hydrogenated amorphous silicon diodes, *J. Non-Cryst. Solids 77/78:* 1421–1424.
Dohjo, M., T. Aoki, K. Suzuki, and M. Ikeda (1988). Low-resistance Mo-Ta gate-line material for large-area a-Si TFT-LCD's, *SID 88 Dig.,* 330–333.
Frenkel, J. (1938). On pre-breakdown phenomena in insulators and electronic semiconductors, *Phys. Rev. 54:* 647.
Hamano, T., H. Ito, T. Nakamura, T. Ozawa, M. Fuse, and M. Takenouchi (1982). An amorphous Si high speed linear image sensor, *Jpn. J. Appl. Phys. 21* (Suppl. 22–1): 245–249.
Hayama, M. (1990), Characteristics of *p-i* junction amorphous silicon stripe-type photodiode array and its application to contact image sensor, *IEEE Trans. Electron. Devices 37*(5): 1271–1275.
Hayama, M., K. Kobayashi, H. Miki, and Y. Ohnishi (1986). Influence of unsymmetrical electrode structure on a-Si photodiode characteristics, *Mat. Res. Symp. Proc. 70:* 689–694.
Ichikawa, M., S. Suzuki, H. Matino, T. Aoki, T. Higuchi, and Y. Oana (1989). 14.3-in.-diagonal 16-color TFT-LCD panel using a-Si:H TFTs, *SID '89 Dig.* 226–229.
Kaneko, S., Y. Kajiwara, F. Okumura, and T. Ohkubo (1985). Amorphous Si:H heterojunction photodiode and its application to a compact scanner, *Mat. Res. Soc. Symp. Proc.* 49: 423–427.
Kanemori, Y., M. Katayama, N. Nakazawa, H. Kato, K. Yano, Y. Fukuoka, Y. Kanatani, Y. Ito, and M. Hijikigawa (1990). 10.4 in.-diagonal color TFT-LCDs without residual images, *SID '90 Dig.* 408–411.
Katayama, M., H. Morimoto, S. Yasuda, T. Takamatu, H. Tanaka, and M. Hijikigawa (1988). High-resolution full-color LCDs addressed by double-layered gate-insulator a-Si TFTs, *SID 88 Dig.,* 310–313.
Kobayashi, I., M. Uno, S. Ishihara, K. Yokoyama, K. Adachi, H. Fujimoto, T. Tanaka, Y. Miyatake, and S. Hotta (1989). Rear-projection TV using high-resolution a-Si TFT-LCD, *SID 89 Dig.,* 114–117.
Kuijk, K. E. (1990). D^2R, a versatile diode matrix liquid crystal approach, *Proc. Eurodisplay '90,* 147–177.
LeComber, P. G., W. E. Spear, and A. Ghaith (1979). Amorphous silicon field-effect device and possible application, *Electron. Lett. 15:* 179–181.
McGill, J., V. Cannella, Z. Yaniv, P. Day, and M. Vijan (1986). VLSI processing of amorphous silicon alloy *p-i-n* diodes for active matrix applications, *Mat. Res. Soc. Symp. Proc. 70:* 637–641.
McGill, J., O. Prache, and Z. Yaniv (1989). Amorphous silicon alloy contact-type image sensors, *SID 89 Dig.* 251–253.
Madan, A., and M. P. Shaw (1988). *The Physics and Applications of Amorphous Semiconductors,* Academic, Orlando.

Maissel, L. I., and R. Glang (1970). *Handbook of Thin Film Technology,* McGraw-Hill, New York, pp. 5–17.

Matsushita, Y., T. Sunata, T. Yukawa, Y. Ugai, J. Tamamura, and S. Aoki (1987). A 5-in. active matrix full color LCD for displaying TV images, *Proc. SID 28*(2): 137–140.

Morita, H., K. Ishizawa, M. Shibusawa, Y. Tanaka, and K. Inoue (1990). Large-size TFD(MIM)-LCD for high grade lap-top computer terminals, *Proc. Eurodisplay '90,* 366–369.

Morozumi, S. (1984). 4.25-in. and 1.51-in. b/w and full-color LC video displays addressed by poly-Si TFT's, *SID '84 Dig.,* 316–317.

Morozumi, S., T. Ohta, R. Araki, T. Sonehara, K. Kubota, Y. Ono, T. Nakazawa, and H. Ohara (1983). A 250 × 240 element LCD addressed by lateral MIM, *Proc. Jpn. Display '83,* 404–407.

Morozumi, S., H. Kurihara, H. Ohshima, T. Takeshita, and K. Hasegawa (1984). Completely integrated a-Si:H linear image sensor with poly-Si TFT drivers, Ext. Abstracts 16th Conf. SSDM, Kobe, Japan, pp. 559–561.

Morozumi, S., R. Araki, H. Ohshima, M. Matsuo, T. Nakazawa, and T. Sato (1986). Low temperature processed poly-Si TFT and its application to large area LCD, *Proc. Jpn. Display '86,* 196–199.

Nagayasu, T., T. Oketani, T. Hirobe, H. Kato, S. Mizushima, H. Take, K. Yano, M. Hijikigawa, and I. Wasizuka (1988). A 14-in.-diagonal full-color a-Si TFT LCD, *Proc. 1988 Int. Display Res. Conf.,* 56–59.

Nicholas, K. H., A. G. Knapp, I. D. French, A. J. Guest, A. D. Pearson, J. R. Hughes, R. A. Ford, J. A. Chapman, H. C. J. Krekels, M. C. Hemings, A. J. van Roosmalen, and R. A. Hartman (1990). Diode network addressing of LC TV displays, *Proc. Eurodisplay '90,* 170–173.

Oana, Y. (1989). Technical developments and trends in a-Si TFT-LCDs, *J. Non-Cryst. Solids 115:* 27–32.

Ohira, K., Y. Hirai, E. Mizobata, T. Shiozawa, H. Uchida, S. Kaneko, O. Sukegawa, S. Hoshino, and C. Tani (1989). Analysis of high-resolution active-matrix LCD with SIN_x-diode array, *Proc. Jpn. Display '89,* 452–455.

Okumura, F., S. Kaneko, and M. Uchida (1983). Ext. Abstracts 15th Conf. SSDM, Tokyo, 201–205.

Ono, N., H. Aruga, T. Ushiki, K. Suzuki, T. Ushiyama, Y. Noda, T. Kamikawa, K. Kaneko, and S. Morozumi (1990). An MIM-LCD with improved TV performance, *SID '90 Dig.,* 518–521.

Ozawa, T., and M. Takenouchi (1986). Recent developments in amorphous silicon image sensors, *Proc. SPIE, 617:* 133–139.

Ozawa, T., M. Takenouchi, T. Hamano, H. Ito, M. Fuse, and T. Nakamura (1982). Design and evaluation of A4 amorphous Si hybrid image sensors, *Proc. Int. Microelectronics Conf., Tokyo,* 132–137.

Rosan, K. (1991). Amorphous semiconductor image sensors: physics, properties and performance, in *Amorphous and Microcrystalline Semiconductor Devices: Optoelectronic Devices* (J. Kanicki, Ed., Artech House, Boston.

Rosan, K., and G. Brunst (1986). a-Si:H image sensor: some aspects of physics and performance, *Mat. Res. Soc. Symp. Proc., 70:* 683–688.

Rose, A. (1963). *Concepts in Photoconductivity and Allied Problems,* Interscience, New York.
Saito, T., K. Suzuki, and Y. Suda (1984). Amorphous silicon contact image sensor, *Toshiba Rev. 149:* 33–36.
Seki, K., H. Yamamoto, A. Sasano, and T. Tsukada (1983). Hydrogenated amorphous silicon *pin* diodes with high rectification ratio, *J. Non-Cryst. Solids 59/60:* 1179–1182.
Silver, M., N. C. Giles, E. Snow, M. P. Shaw, V. Cannella, and D. Adler (1982). Study of the electronic structure of amorphous silicon using reverse-recovery techniques, *Appl. Phys. Lett. 41:* 935–938.
Staebler, D. L., and C. R. Wronski (1980). Optically-induced conductivity changes in discharge-produced hydrogenated amorphous silicon, *J. Appl. Phys. 51:* 3262–3266.
Suzuki, M., M. Toyama, T. Harajiri, T. Maeda, and T. Yamazaki (1986). A new active diode matrix LCD using off-stoichiometric SiN_x layer, *Proc. Jpn. Display '86,* 72–74.
Swartz, L., K. Kitamura, M. Vijan, J. McGill, V. Cannella, and Z. Yaniv (1987). A high-speed high-resolution contact line imager using amorphous silicon alloy *pin* diodes, *Mat. Res. Soc. Proc. 95:* 633–638.
Sze, S. M., (1969). *Physics of Semiconductor Devices,* Wiley, New York.
Szydlo, N., E. Chartier, J. N. Perbet, N. Proust, J. Magarino, and M. Hareng (1983). Integrated matrix addressed LCD using amorphous silicon back to back diodes, *Proc. Jpn. Display '83,* 416–418.
Takeshita, T., H. Kurihara, H. Ohshima, I. Yudasaka, and S. Morozumi (1989). Completely integrated a-Si/a-SiC heterojunction contact-type linear image sensor with poly-Si TFT drivers, *SID '89 Dig.,* 255–258.
Togashi, S., K. Sekiguchi, H. Tanabe, E. Yamamoto, K. Sorimachi, E. Tajima, H. Watanabe, and H. Shimizu (1985). An LC TV display controlled by a-Si diode rings, *Proc. SID* 26(1):9–15.
Togashi, S., K. Sekiguchi, H. Tanabe, T. Okigami, M. Okamoto, K. Sorimachi, E. Yamamoto, O. Sugiyama, N. Taguchi, S. Ishimori, M. Kikuchi, A. Suzuki, E. Tajima, and T. Aoyama (1986). A full color 6.7" diagonal LCTV addressed by a-Si diode rings, *Proc. Jpn. Display '86,* post-deadline paper PD4.
Tomita, O., K. Shimizu, Y. Asai, Y. Tanaka, N. Mukai, K. Shohara, N. Kokado, T. Yanagisawa, and K. Kasahara (1989). A 6.5-in. diagonal TFT-LCD module for liquid-crystal TV, *SID '89 Dig.,* 150–154.
Toyama, M., T. Harajiri, T. Maeda, Y. Kuroda, Y. Tsunoda, M. Suzuki, and T. Yamazaki (1987). A large-area diode-matrix LCD using SiN_x layer, *SID '87 Dig.,* 155–158.
Tsukada, T. (1984). Amorphous silicon linear image sensors, *JARECT* Vol. 16, *Amorphous Semiconductor Techniques and Devices* (Y. Hamakawa, Ed.), OHMSHA Ltd and North-Holland, Amsterdam, pp. 290–299.
Urabe, K., H. Fujisawa, M. Kamiyama, E. Tanabe, and T. Yoshida (1989). Application of a-Si thin film diodes to active matrix LCD, *J. Non-Cryst. Solids 115:* 33–35.
Vijan, M., A. Abileah, Y. Baron, V. Cannella, J. McGill, and Z. Yaniv (1990). A 1.7-Mpixel full-color diode driven AM-LCD, *SID '90 Dig.,* 530–533.
Wada, T., T. Masumori, Y. Takahashi, N. Kakuda, and T. Kawada (1990). 1280 × 800 color pixel 15 inch full color active matrix liquid crystal display, *Proc. Eurodisplay '90,* 370.

Yaniv, Z., Y. Baron, V. Cannella, J. McGill, and A. Lien (1986a). A new amorphous-silicon alloy *pin* liquid crystal TV display, *SID '86 Dig.*, 278–280.

Yaniv, Z., V. Cannella, A. Lien, J. McGill, and W. den Boer (1986b). Progress in two and three terminal amorphous silicon switching devices for matrix addressed LCDs, *Conf. Proc. SPIE 617:* 16–24.

Yaniv, Z., V. Cannella, Y. Baron, A. Lien, and J. McGill (1986c). Amorphous silicon alloy technology for active matrix displays, *Mat. Res. Soc. Symp. Proc. 70:* 625–635.

Yaniv, Z., V. Cannella, Y. Baron, J. McGill, and M. Vijan (1988): A 640/ × 480 pixel computer display using *pin* diodes with device redundancy, *Proc. 1988 Int. Display Res. Conf.*, 152–154.

4
Flat-Panel Displays

Alan Sobel

Consultant
Evanston, Illinois

4.1 INTRODUCTION

The cathode-ray tube (CRT) dominates the display field because of its performance, its cost, and the fact that it has been around for a long time and is thus a familiar, as well as a mature, technology. It is a durable and effective technology, but CRTs are bulky and clearly unsuitable for many applications. This is especially true where space is important. We do not carry CRTs on our wrists, and there are other portable applications, laptop and notebook computers being obvious examples, where the CRT must be replaced by some other technology. Another area in which the CRT is vulnerable is that of very large displays where space is at a premium. These include high-resolution, high-information-content desktop displays for such applications as desktop publishing and computer-aided drafting and design; even in roomy offices, the space demands of 19-in. and larger high-resolution CRTs are beginning to meet resistance. If flat panels of equivalent performance can be made available at competitive costs, flat panels will penetrate these markets. That time is approaching, though it has not yet arrived.

In general, then, flat panels are used where the end application or the product design make the CRT undesirable, unacceptable, or impossible. As flat-panel costs come down and performance improves, there will be more such areas.

The CRT has a number of technical advantages: simplicity of scanning, high energy density of the scanning mechanism (the electron beam), good luminous efficacy, and good to excellent resolution. Its disadvantages are its bulk, its thickness (for given panel dimensions), its requirement for high voltage, and the power and voltage demands of its scanning circuits.

The advantages of flat-panel displays are small bulk in relation to panel area, the discrete nature of the display pixels (though this is a disadvantage in some applications, such as the display of imagery), low operating voltages and currents, and the flatness of the display. Disadvantages include the number of drive circuits required, the interconnectedness of the display, which makes crosstalk reduction a major problem, and generally poor or marginal display performance (luminance, resolution, and color gamut). Another disadvantage is that there is a multiplicity of flat-panel technologies available but as yet no really dominant single technology, so research and development efforts are scattered among a number of attractive possibilities.

In this chapter I discuss the most important of the many technologies that have been used, concentrating on those with real market presence or potential. I first look at the problems besetting all flat-panel technologies and then discuss the individual approaches.

I have emphasized the economic bases of competition because display research is product-oriented. Most work gets done where managers see the possibility of a market payoff. Thus there is intense activity on liquid-crystal displays (LCDs), because this technology bids fair to become the dominant flat-panel technology in a number of market areas, including the very tempting consumer-TV areas. This chapter is largely confined to display technologies that are currently on the market or have good prospects of arriving at the marketplace. A number of devices that are technically interesting but have proved to be uneconomic are not discussed.

4.2 GENERAL PROBLEMS OF FLAT-PANEL DISPLAYS

There are three general problems associated with all kinds of flat-panel displays: selection nonlinearity, the number of devices and connections required, and power-handling. I discuss these in order in this section.

4.2.1 Selection Nonlinearity

Almost all flat-panel displays are arranged in matrix form; that is, each pixel or subpixel is located at the intersection of a row and a column electrode (Fig. 4.1). This has the advantage that n^2 elements can be accessed with only about $2n$ drive circuits. However, all of the pixels are interconnected, which leads to major problems.

In principal, one must be able to energize only one pixel while all others are unaffected. In practice, it is usually more difficult to energize all but one pixel and keep that one pixel unaffected. With a matrix connection, when one row and one column are driven, all of the pixels on the panel are excited to at least some

FLAT-PANEL DISPLAYS

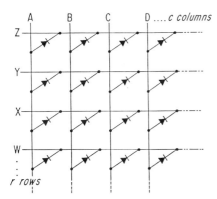

Figure 4.1 A matrix display. The matrix elements are shown as diodes, but they can equally well be symmetrical elements such as liquid-crystal pixels [From Sobel (1970); copyright IEEE.]

extent (Fig. 4.2). As the number of pixels increases, the *unselected* region grows larger, and essentially half of the applied voltage appears across each of the *hanging* regions. (It is possible to change this fraction by back-biasing, as is discussed below.)

Generally, more than one pixel is driven at a time (see Section 4.3). Figures 4.3 and 4.4 show how the circuit configuration changes as several columns are energized simultaneously. A potentially dangerous limit is reached when all col-

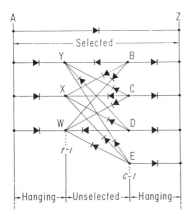

Figure 4.2 The matrix of Figure 4.1 redrawn to show different regions when a single element is driven. The element is at the intersection of column A and row Z. [From Sobel (1970); copyright IEEE.]

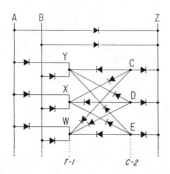

Figure 4.3 The matrix of Figure 4.1 with one row and two column electrodes driven simultaneously. [From Sobel (1970).]

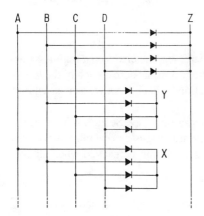

Figure 4.4 The matrix of Figure 4.1 with all columns driven simultaneously. There is now no sneak path between any column and the driven row Z. [From Sobel (1970).]

umns but one are driven (Fig. 4.5); then all the undesired current from the column-hanging region must flow through a single element in the row-hanging region (path *D-Z* in Fig. 4.5).

These drawings do not take into account the impedances of the driving circuits. If the driving circuits have high output impedances when they are not supplying current, then the situation is as depicted, and clearly trouble can result. However, if the driver output impedances are always low, then the potentials at points *W, X, Y,* and *D* in Figure 4.5 are constrained by the drive circuits; as long as these circuits can sink the currents that are flowing into them, excessive voltages will not appear across unselected or hanging pixels. This is a vital matter;

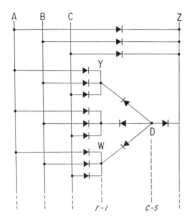

Figure 4.5 The matrix of Figure 4.1 with all but one column driven. Current from all the hanging elements concentrates on one hapless pixel. [From Sobel (1970).]

drive circuitry must be capable of both *sourcing* and *sinking* significant currents. In Section 4.2.3 we shall revisit this question.

In any event, each pixel must be unaffected by a half-amplitude voltage but produce its full output (maximum light or maximum opacity) when full voltage is applied. This defines *selection nonlinearity*. Figure 4.6 shows transfer curves for two types of devices—symmetrical and asymmetrical. For either, there is a knee—a driving voltage at which the output begins to rise sharply—and a maximum voltage, beyond which additional applied voltage either produces substantially less output or may actually damage the device. Real symmetrical devices usually have S-shaped curves; the knees are softer than desirable, and there is an upper limit to their outputs. For asymmetrical devices, such as light-emitting diodes (LEDs), the maximum voltage in the reverse direction typically is the voltage at which a dramatic increase in current results, albeit without the production of light.

If the drive circuits are of high output impedance, the voltage that can be applied to selected pixels without producing crosstalk (excitation of undesired pixels) is only slightly above the threshold voltage for displays with many rows and columns. With low-output-impedance drive circuits that have zero output voltage when they are not driving their rows or columns, the maximum voltage that can be applied to selected pixels is twice the threshold voltage (V_{th} in Fig. 4.6). If the drive circuits are returned to voltages in opposition to the applied voltage (called *back-biasing*), higher drive voltages can be applied. It can be shown (Livingston, 1956; Sobel, 1970) that with back bias of V_t applied (and assuming that the device is symmetrical or that $|V_a| \geq V_t$), the maximum select voltage is $3V_t$, a significant improvement. This relationship is not correct if the

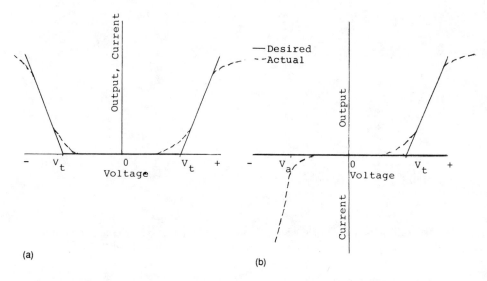

Figure 4.6 Desired and actual characteristics of matrix-display elements: *selection nonlinearity.* (a) Symmetrical device; (b) asymmetrical device.

device response is nonlinear; in particular, LCDs march to a different drummer (*cf.* Section 4.6.5).

To be able to produce gray scale, it is desirable that the slope above the knee, that is, the slope in the region between one-half and full applied-voltage amplitude, not be too steep (although for steep slopes, expedients are available to produce gray scale, as described later). It is also desirable that the knee be as sharp as possible, and this is a requirement that has led to the downfall of many proposed display devices. If the knee is soft, selection is poor; partially selected (sometimes called *half-selected*) elements will have some output, which degrades contrast. Finally, it is essential that all these characteristics be uniform across the entire display, since otherwise there will be undesirable shading or mottling. All of these requirements are difficult to meet; we shall discuss some of the ways in which they have been tackled later.

I do not know of any display devices that are current-controlled, that is, in which the independent variable is current (drive circuit output impedances much higher than display-device impedances). The nature of the problem, including the effects of parasitic shunt capacitances, militates against such devices.

4.2.2 The Tyranny of Numbers

A CRT requires deflection plus modulation: x and y synchronization plus a control input for each color—five inputs in all. In contrast, a matrix display needs a driver for every row and every column. A color VGA display (the current "standard" high-resolution computer display) has 480 rows, 640 columns, and 3 subpixels for each pixel. This is a total of 2400 drivers if there is a column driver for each color at each pixel.[1] (Other configurations can result in fewer drivers.) Furthermore, there are 921,600 subpixels on the panel (assuming three dots per pixel). This is an enormous number of devices and connections to the outside world. The sheer numbers of panel elements, interconnections, and drive circuits have presented major problems in the realization of high-performance matrix displays. We shall see later how these problems have been tackled.

4.2.3 Power Handling

All displays must produce enough output that the user can see the result, and this output must be distributed over a significant area. As a result, both the panel elements and the drive circuits must handle a substantial amount of power. For light-emitting devices, it is clear that the drive circuits must provide all of the power to be emitted, plus all the losses in the panel elements proper. Even for "passive" displays—displays that do not emit light but modulate it—the drive circuitry must provide enough power to change the state of the panel elements. For both active and passive displays, the drive circuitry must provide the power necessary to charge and discharge the capacitance of the matrix electrodes and the capacitance of the output elements.

It's instructive to look at some of the numbers. Consider a 10-in.-diagonal (254-mm) VGA display fully illuminated at 20 foot-lamberts (fL) (68.5 candela/m^2 or nits) and with a luminous efficacy (lumens out/power in) of 3 lumens/watt (lm/W). The display area is 48 in.^{2} or 0.333 ft^2, so the output luminous flux is 6.67 lm, and the input power is 2.22 W. Since there are 307,200 pixels, the power input per pixel is 7.23×10^{-6} W. If the operating voltage of the pixels is 100 V (a number that is of the right order of magnitude for gas-discharge and thin-film electroluminescent displays), then the average current per pixel is 113×10^{-9} A.

However, this is the *average* current per pixel. Since there are 480 rows, the duty factor for each row is 1/480, so the actual current during a row time is 480 times the average current, or 54.3×10^{-6} A. The row drivers must each sink the current from 640 columns, or 34.7×10^{-3} A! Note that these are "steady-state" currents; they do not include the substantial transient currents ($I = C\ dV/dt$) needed to charge and discharge the parasitic capacitances.

This is a major difference from image-pickup devices, where the problems are

---

[1] The requirements for the complete display of an NTSC TV image are similar to those for VGA. There are 480 active lines in the TV picture (the balance of the 525 lines are consumed by the vertical retrace time), and the 440 possible pixels of the TV picture need 640 matrix pixels when the Kell factor is taken into account. A major difference is that the NTSC picture is interlaced (*cf.* Section 4.3.1) while the VGA is a progressively scanned display, so that pixel times in VGA are one-half those of NTSC (less if the VGA frame rate is greater than 60 s^{-1}).

often sensitivity and signal-to-noise ratio but essentially no power need be handled. Furthermore, displays are almost always larger than pickup devices, so problems associated with capacitance are greatly magnified simply because of the size of the display.

4.3 SCANNING METHODS

Cathode-ray tubes are scanned point by point; that is, the electron beam delivers its energy to a single pixel at a time. (Multiple-beam CRTs are exceptions to this rule, but each beam in such a CRT is operating a single point at a time.) Flat-panel displays are almost never operated in this fashion, as we shall see.

4.3.1 Scanning and Flicker

Light from a pulsed light source appears to pulsate, or *flicker,* unless the luminance is low or the pulse frequency is high (Snyder, 1985). The brighter the light, the higher is the frequency at which the sensation of flicker disappears; this frequency is called the *critical fusion frequency* (*CFF*). The light-pulse duration affects the CFF; displays of longer persistence have fewer flicker problems than displays of short persistence (Farrell, 1987). Peripheral vision is more subject to flicker than is central or foveal vision; a pulsing light source seen at the edges of the visual field (especially at the top of the field) is more likely to be perceived as flickering than is one at which the viewer is gazing directly. A scanned display is a pulsed light source; the sensation of flicker can even be produced by an LCD viewed in steady ambient light if the reflectance of the LCD is modulated sufficiently slowly and to a sufficient depth. Flicker was recognized as a problem in television displays early in the development of TV, and *interlaced scanning* was introduced to combat the problem.

In an interlaced scan, the frame is divided into two or more *fields,* and these are displayed successively. The usual scheme is 2:1 interlace: All the odd lines of the frame are displayed, and then all the even lines are displayed. Thus 30-frame/sec 2:1 interlace produces 60 fields/sec, each field providing half the information. This expedient works well if the viewer is at some distance from the display, so that the smallest area that is resolved includes parts of at least two scan lines. This is typical for entertainment TV.

However, if the viewer is close to the display, as is typically the case when using a computer, interlaced scanning is disadvantageous. The viewer generally can resolve individual pixels, and these are pulsed at the frame rate, not the field rate. Furthermore, if text is being displayed on an interlace scan, different parts (adjacent rows) of individual letters are displayed in alternate fields, and this can be most annoying. Hence, computer displays usually employ *progressive* scan: The entire frame is scanned sequentially. In the United States, this is generally done at 60 frames/sec or more. This puts more of a burden on the drive circuitry than does interlaced TV scanning, but the difference in viewer preference is so marked that progressive scanning is virtually mandated by the market.

The interaction of a pulsed display with pulsating ambient light can cause annoying "beats" between the two light sources. It has been reported that this is especially a problem in offices with fluorescent lighting, which pulses at twice the power-line frequency. Although this is so fast (even at the European and Japanese line frequencies of 50 Hz) that the lighting itself does not flicker perceptibly, the interaction of the ambient light and light from the display may produce a problem.[2] For this reason, some display manufacturers use frame rates that are significantly different than the power-line frequency, usually going to higher frame rates such as 70 frames/sec (Sigel, 1989).

Because European and Japanese TV systems operate at frame rates of 25 sec^{-1} rather than the U.S. standard of 30 frames/sec, TV displays for use in these countries generally operate at lower luminance levels than do sets for the U.S. market. Presumably, high-luminance computer displays in Europe and Japan will have a similar problem and will be operated at lower luminances than their counterparts in the United States.

The perception of better image quality from progressive scanning, even for entertainment TV (as long as this has a great deal of detail) is such that some TV-set manufacturers have incorporated circuitry to convert the incoming interlaced picture to a progressively scanned display. Many of the proposed improved-definition TV and high-definition TV standards incorporate progressive scanning because a significant part of their *raisons d'être* is the provision of a more detailed and more pleasing picture.

4.3.2 Point-by-Point Scanning

In principle, a matrix display can be operated a single point at a time. There are a number of reasons why this generally doesn't work.

Pulse Distortion

The pulse width required for point-by-point scanning is inversely proportional to the number of pixels. As an example, consider the VGA standard again. For a monochrome display with a frame rate of 60 sec^{-1}, noninterlaced, and ignoring any retrace times, the pulse width for a single pixel is only 54 nsec. Transmitting this short pulse over the matrix electrodes, which typically have high series resistance and very high shunt capacitance, produces unacceptable distortion. In particular, pixels located far from the driver will receive drive pulses of drastically different shapes than those located closer.

---

[2] Electronic ballasts for fluorescent lamps generally drive the lamps at frequencies much higher than the line frequency. In environments where such ballasts are used, this beat problem should be negligible.

Energy Delivery

The CRT can deliver all the energy needed to excite a pixel for the entire frame time in a single pixel time, thanks to the high energy density in the electron beam. In contrast, the maximum voltage that can be applied to a row–column pair in a matrix display is limited by the transfer curve (*cf.* Section 4.2.1) and crosstalk considerations; it is only two or three times the voltage at the knee of that curve. (For most LCDs it is much lower than this.) There are no flat-panel display elements currently available that can produce the required outputs when driven with such short, low-amplitude pulses.

4.3.3 Line-at-a-Time Scanning

By far the most usual form of matrix-display scanning is line-at-a-time. Here a row electrode is energized for an entire line time, and the column electrodes are driven according to the data to be displayed. (Sometimes the row-driving voltage is called the *strobe* voltage and the column-drive voltages are called the *data* voltages.) Columns may be driven either on-off or with various forms of gray-scale modulation.

Line-at-a-time scanning has several advantages over point-at-a-time scanning. The time for which each pixel is energized is far greater, and this advantage is greater the more elements there are in the display. For the monochrome VGA display constants quoted above, the line time (again neglecting retrace times) is 34.7 μsec. This means that far more energy can be delivered to a pixel during each frame. Furthermore, since the pulse that delivers this energy is much longer, pulse distortion is a much less severe problem. (But for high-capacitance display devices, like thin-film electroluminescent and dc electroluminescent displays, pulse distortion can still cause difficulties.)

No penalty in numbers of electrode drivers is incurred; a driver is still required for each row and each column electrode. The fact that the drivers are handling longer pulses is an advantage, as there is not as much need to produce fast rise and fall times across large capacitances. Because the rise and fall times can be longer than for point-at-a-time, there will be less radiated electromagnetic interference.

The display controllers will be different for the two modes but not necessarily more expensive for one than for the other.

Some workers have divided the column electrodes in half and scanned the top and bottom halves of the panel simultaneously. This doubles the number of column drivers required but also doubles the time for which each element is excited. Furthermore, since each half of the display has a smaller number of rows, crosstalk problems related to the number of rows are reduced. This expedient has been particularly useful with LCDs, where the limitations on applied voltages are especially severe. It is also possible to divide the row electrodes vertically, so that the panel is cut into quadrants that can be scanned simultaneously. These ap-

FLAT-PANEL DISPLAYS

proaches are expensive and so have been resorted to only when other methods, such as improving the display-element performance, have not been successful.

Still another variant makes use of the particular properties of gas-electron-phosphor technology (described below) to scan a partial line at a time. (This approach has been given the acronym FLAT: fractional line at a time.) It is discussed in the section on gas-discharge displays.

4.3.4 "Three-Dimensional" Scanning; "Multidimensional" Scanning

Using a "third dimension" of selection could reduce the number of row and column drivers, albeit at the expense of panel complexity (Sherr, 1972). Several schemes have been devised along these lines. Some make use of specific properties of gas discharges and will be discussed later. First we shall look at "pure" 3D scanning.

Three-Dimensional Electron-Beam Scanning

A basic system is shown in Figure 4.7. This device, at first called Digisplay, was developed at Northrop Electronics and later at Texas Instruments in the 1970s (Goede, 1973, 1985; Scott *et al.*, 1978). In its conceptually simplest embodi-

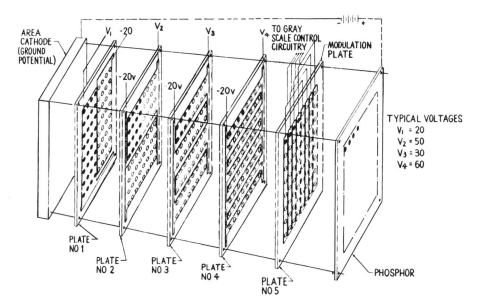

Figure 4.7 Exploded view of the *Digisplay* three-dimensional scanning system. The dark-shaded areas of the control plates are biased positive, allowing electrons to pass through. [From Goede (1973); copyright IEEE.]

ment, there is a succession of control plates, each of which allows half of the electrons from an area cathode to pass, but in different patterns. The figure shows six plates, each with two drivers, addressing an 8×8 array of phosphor dots. This requires only 12 drive circuits, compared to the 16 that would be required for a conventional matrix arrangement. Here the savings is trivial, but for a 16×16 array only 14 drive circuits would be needed, compared to 32 for the conventional matrix. As the number of elements increases, the saving in drive circuits becomes more substantial.

Alternative arrangements, using fewer switching plates and more drivers per plate, can be used. The one shown in Figure 4.8 uses the same number of drivers but fewer plates.

As shown, these schemes provide point-by-point scanning, with the advantage that no scanning order is required; random access is achieved. The scheme can be modified to allow for multiple beams on the phosphor, thus increasing the duty factor.

In practice, it has been desirable to use fewer plates and more drivers because the switching plates and their mounting are expensive. One scheme used four switching plates and 96 drive circuits to access a 512×512 array; the savings

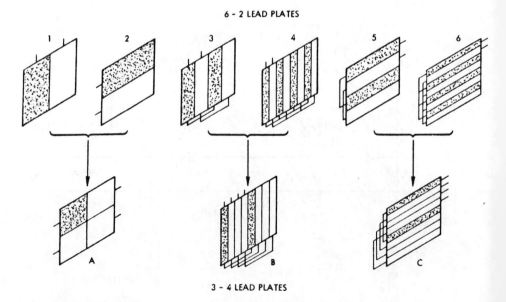

Figure 4.8 Alternative decoding techniques for *Digisplay* three-dimensional scanning. Both schemes use the same number of drivers but different numbers of plates. [From Goede (1973).]

FLAT-PANEL DISPLAYS

over the 1024 drivers required in a 2D matrix are evident. Texas Instruments took the approach as far as color TV displays, although with less than full NTSC resolution.

Both Northrop and Texas Instruments abandoned the work. As of early 1991 it was being pursued by Source Technology of Los Gatos, California. The device was of particular interest for laptop computer displays, but Source Technology's displays were considered too heavy and power-hungry for this application by computer manufacturers who studied it.

A Moving-Flap Multidimensional Addressing Scheme: *Micro Curl*

Micro-Curl is an ingenious scheme that uses electrostatic force to hold a coiled electrode in an uncoiled position (Simpson, 1985). The electrode can be a reflector or can be used to block light. Although this is called multidimensional, the physical components of the display are arranged in a two-dimensional plane, unlike the Digisplay, which is a "true" three-dimensional system.

Figure 4.9 shows the arrangement. The movable electrode A is separated from

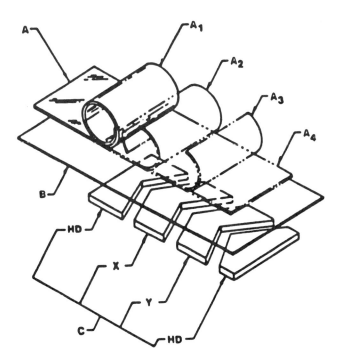

Figure 4.9 The Micro-Curl electrostatically deflected display element. In early units, electrode A was made of 0.00025-in. Mylar and dielectric B was 0.0005-in. acetate; the diameter of the curl was 0.03 in. (Simpson, 1985).

the various fixed electrodes by a thin insulating layer. When the electrode labeled HD (for hold-down) is energized with a voltage opposite to that on the movable electrode (typically 250–350 V), the coiled electrode moves from the off position, A_1, to position A_2. Energizing electrodes X and Y deflects electrode A further, until it comes within the field of the second HD electrode. Now if the voltages on the X and Y electrodes are removed, the movable electrode will remain in its extended position, held by the HD electrodes until they are deenergized.

The use of this in a two-dimensional scan arrangement is shown in Figure 4.10. This shows a four-pixel array. First the hold-down electrode, HD, which is common to all pixels, is energized. Then electrode X_2 is energized, deflecting electrodes A in the bottom two pixels. Electrode Y_1 is then energized, deflecting the A member in the lower left pixel but no others. The potential on HD keeps this electrode deflected when the potentials on X_2 and Y_1 are removed.

The scheme can be extended to three or more dimensions. Figure 4.11 shows a four-dimensional scheme; the circles in the X and Y electrodes indicate connections to different levels of conductors. In general, the number of drivers for a scheme like this is

$$S = dN^{1/d} + 2 \qquad (4.1)$$

where S is the number of drivers, d the number of "dimensions," and N the number of pixels. The factor 2 provides for drivers for the HD and A electrodes.

This arrangement does not require half-select nonlinearity; voltages on the various electrodes are not added to reach a threshold. This should make for wide operating windows and manufacturing tolerances.

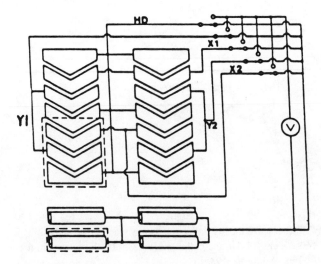

Figure 4.10 Two-dimensional scanning of the Micro-Curl device.

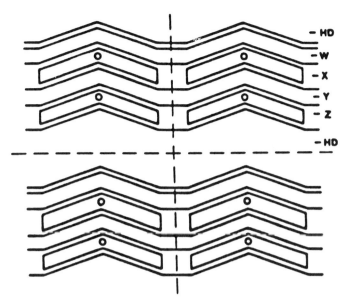

Figure 4.11 A "four-dimensional" scanning arrangement for the Micro-Curl device. In this arrangement, Micro-Curl claimed that 100 drive circuits could control 360,300 pixels (Simpson, 1985).

The scheme was disclosed by Micro-Curl Display Technology of Westport, Connecticut, in 1985. Development work was done under contract by Norden Systems. However, the device was never sold commercially, and there has been no public information since 1986.

What is interesting in the present context is the multiple-dimensional scanning. As usual, there is a trade-off between device complexity and the reduction in the number of drive circuits. For this technology, which requires the switching of voltages in excess of 250 V, reducing the number of drivers is crucial.

Multidimensional Selection in Gas-Discharge Displays

There are several schemes that use properties of the gas discharge, notably priming, to reduce the number of drive circuits required. Some of these are "true" multidimensional systems in that they require the coincidence of selection voltages on several electrodes; others are drive-circuit-reduction techniques using other means (*cf.* Section 4.6.6).

4.4 GRAY-SCALE METHODS

Except for the simplest text and line images, gray scale is necessary. For color displays, gray scale makes possible a palette of more than just simple combinations of the primary colors,[3] in addition to making possible more complex imagery.

Gray scale requires cooperation among the display device proper, its drive circuitry, and the *controller*—the circuit block that interprets information from the host system and delivers specific data and commands to the display peripheral circuitry. The controller must translate gray-level information to analog voltage levels or to the time intervals for which a pixel is on.

Some controllers for monochrome devices incorporate mapping algorithms to show different colors as varying gray levels. This has been of particular interest for monochrome flat panels for portable computers that use programs that generate color displays on desk-bound computers.

Many semiconductor manufacturers sell controllers; these are products that change rapidly to meet market requirements. Some controllers are capable of operating with a variety of display devices, while others are specialized to single types of displays. Manufacturers of some display devices, notably LEDs and LCDs, incorporate controllers into the device package, so that the user need only supply command signals and power. Controllers *per se* are not discussed in this chapter.

4.4.1 Analog Gray Scale

In the simplest form of gray-scale implementation, each dot is driven by a voltage corresponding to the desired output level. Although this is conceptually simple, it has a number of problems.

The voltage at the knee of the transfer curve and the slope of the transfer curve above that knee (*cf.* Fig. 4.6) must be uniform across the entire display. A gradual change in transfer-curve slope across the display may escape the viewer's notice, but short-range changes produce visible mottling, which can be annoying. Although variations of 2:1 in luminance level within short distances can be tolerated on a text display with no gray scale, such a large variation would be unacceptable for a display with any kind of imagery, even if the images contained large areas of uniform color, such as cartoons or some graphs.

This requirement on uniformity extends to the drivers. In general, an extended, correlated artifact, such as a single row or column performing very differently from its neighbors, is noticeable to almost all viewers, even if the performance difference is slight.

Despite the difficulty, analog modulation has been used successfully in many display devices.

---

[3] Without gray scale, eight colors are possible: red, green, blue, red plus green, red plus blue, green plus blue, red plus green plus blue, and no color (black, all subpixels off).

FLAT-PANEL DISPLAYS

4.4.2 Pulse-Width Modulation

If all the column drivers are pulsed to the maximum voltage allowable (determined by either crosstalk or saturation considerations) and the output is determined by the length of time each dot is driven, the concerns about uniformity are somewhat alleviated. The hope is that the maximum output will be more uniform across a panel than are the knee and slope of the transfer curve, a hope that is generally justified.

However, nothing comes free. The ratio of maximum output to that minimum output which is just noticeably different from no output becomes a function of the minimum column pulse width that can be handled. Using the monochrome VGA display parameters of Section 4.2.3 as an example, for 256 different output levels, the maximum column-pulse width is 34.7 µsec and the minimum pulse width will be $34.7/256 = 0.136$ µsec. This is a short pulse; one can anticipate problems with pulse distortion as it is propagated along the matrix electrodes. Because of this, pulse-width schemes have generally been restricted to a smaller number of gray levels. For example, 64 gray levels would require a minimum pulse length of 0.54 µsec. This is still a very short pulse, but it can be managed with some difficulty for at least some displays.

4.4.3 Combined Pulse-Amplitude and Pulse-Width Modulation

The combination of the two methods might alleviate the problems associated with either one. Such a scheme was described by Amano (1975) and is shown in Figure 4.12. Waveforms 1, 2, 4, and 8 are the basic pulse-width waveforms; three amplitude levels are possible. The result is a 16-level gray scale with a minimum pulse width of one-fourth the horizontal line time (16 µsec in this embodiment). Although this scheme appears to have worked well in the laboratory with dc gas-discharge displays, to my knowledge it has never been used in a commercial product.

Operation of this kind of scheme relies on the display output being a reasonably linear function of the energy supplied to each pixel, that is, the area under the curves of Figure 4.12. Not all displays operate this way; in particular, the output of an electroluminescent display is not simply proportional to the area under such a curve. However, this scheme should work with LEDs, LCDs, and dc gas-discharge displays.

4.4.4 Alternative Pulse-Width Methods

Several variants of the pulse-width method have been devised, often taking advantage of peculiarities of particular flat-panel devices. Although details of these methods will be found in the chapters covering the particular devices, I shall mention some of them here.

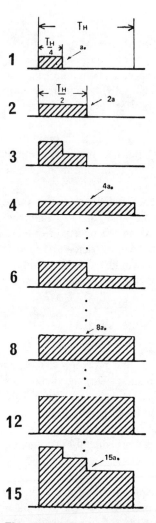

Figure 4.12 A combined pulse-width and pulse-amplitude (PWAM) gray-scale scheme. [From Amano (1975); copyright IEEE.]

Multiple Pulses Within One Frame

This technique resembles the one described in Section 4.4.2, except that instead of a single pulse, multiple pulses are used. These pulses may all be of the same length or they may be of differing lengths—for example, 1, 2, 4, and 8 times a basic pulse width. Frequently this approach is used with multiple line scans within a single frame time. Figure 4.13 (Michel, 1989) shows a scheme that has been used with ac gas-discharge displays. The frame is divided into N parts (four

FLAT-PANEL DISPLAYS 139

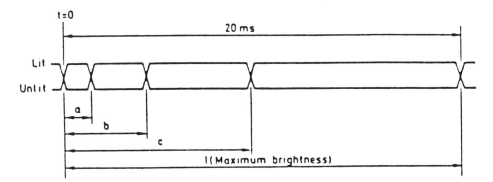

Figure 4.13 Gray scale by modulation within a frame time [From Michel (1989).]

in this illustration), and each pixel is on for from none to 2^N parts. Each pixel must be addressed N times during a frame to give 2^N gray levels. The duration of a frame is set by the persistence of the display and the tolerable flicker level. This French scheme has a frame rate of 50 sec^{-1} (the European power-line frequency), so the frame duration is 20 msec. Other, similar, systems have been reported (Anderson and Fowler, 1974). This approach is often referred to as *duty factor* or *duty cycle* modulation.

Multiple Frames

In this arrangement a number of frames may be considered as a kind of *super frame*, and each dot is turned on for one or more frame times within the super frame. This scheme can be implemented completely within the controller, and the drivers can be simple on-off devices. However, a consequence is that the effective frame time is increased, so fast-moving objects may be blurred or smeared.

The approach has been used with LCDs, especially supertwist devices. Its execution is similar to that described above for multiple pulses within a single frame. Supertwist LCDs tend to be slow in any event, and the resulting difficulty with moving images is often cited as a major disadvantage. It should be noted that slow devices also have difficulty with scrolled images and moving cursors.

4.4.5 Trading Spatial Resolution for Gray Scale

A group of elements of a single color can be treated as a single pixel. Turning these elements on or off in varying combinations provides different amounts of light and thus produces gray scale. This is similar to what is called in the hardcopy world *halftoning*; most printing processes provide gray scales by means of patterns of very small dots, since it is difficult to deposit ink in intermediate densities.

If the elements produce equal amounts of light (or equal changes in reflectance), the number of gray levels is then just equal to the number of elements per pixel. If the outputs of the elements are weighted, then a larger number of gray levels can be produced (Pleshko, 1990). In either case, since display elements on flat panels are relatively large, the regular breaking up of the display surface into large pixels can produce undesirable patterning. This has been mitigated by the use of *dithering*, as discussed below.

The approach is most attractive in displays with high resolution, as is possible with ac plasma panels.

Simple Halftoning Using Weighted Elements

Sarma *et al.* (1989) used a kind of halftone system in an active-matrix LCD. Rather than use pixels of different areas, their approach was to arrange the drive circuitry so that different subpixels would turn on at different drive voltages. This was done by tailoring the capacitances of the four control capacitors of each subpixel. In effect, then, this is a scheme for amplitude modulation of the LCD transmission, but it is done by dividing each pixel into four subpixels. This requires some increase in the size of each pixel to accommodate the subpixels, although only one control transistor is used for each pixel. Because the control characteristic is generated by controlling the geometry of passive elements (capacitors), it should be possible to achieve better uniformity than by relying on the control characteristics of the active elements (thin-film transistors) alone.

Halftoning Using Unweighted Elements—*Dither* and Other Methods

As mentioned above, use of a regular pattern of a group of display elements to constitute pixels will indeed give a gray scale, but the resulting patterning is generally unpleasant. Using a random pattern or a random threshold level to determine whether a particular subpixel is turned on or off can remove the regular pattern but substitutes an unpleasant graininess. A better solution is *ordered dither*, in which the randomization is confined to small areas (8×8 cells in Fig. 4.14) (Judice et al., 1974; Judice and Slusky, 1981).

Although this approach was pushed energetically for ac plasma displays, it has been largely supplanted by duty-factor methods, which impose costs in signal processing but do not give up any spatial resolution. Photonics Technology has demonstrated simple unweighted-element halftoning on their very large ac plasma panels; their 2048×2048-pixel, 1.5m-diagonal panel has its spatial resolution reduced to 512×512 to achieve a five-level gray scale (0, 1, 2, 3, or 4 elements turned on).

Other techniques for mitigating the noise effects of halftoning have been investigated (Baldwin, 1991). They include methods in which the difference between the possible output of a pixel (on or off) and the desired output is computed, and this difference is treated as an error to be spread among adjacent

FLAT-PANEL DISPLAYS 141

Figure 4.14 Examples of dither. (a) The effects of randomization over the entire panel; (b) the same image with the randomization done only in 8 × 8-element segments. Although the boundaries of the 8 × 8 blocks are visible, the overall appearance and intelligibility are improved. These pictures were made on a 512 × 512-pixel ac plasma panel. [From Judice et al. (1974); copyright IEEE.]

pixels. *Error propagation* and *dot diffusion* methods can give better-looking results than ordered dither.

4.5 COLOR

There are three different uses for color. The first is to *color-code* the display; that is, each color has a meaning. Examples are red for emergency, and yellow for warning. If color is used for coding, it is important that the set of colors be restricted; studies have determined that people can identify unambiguously only four to six different colors. If more than this are used, confusion may result. It is also important to pick a set of colors that will be so rendered on the display that all users can indeed identify them. The designer should bear in mind the possibility of color-blindness on the part of the user. For example, some 5% of males exhibit some difficulty in distinguishing red and green, while only 0.4% of females have the same problem (Farrell and Booth, 1984).

Note also the possibility of combining color with blinking for warning purposes.

The second use for color is to differentiate items on the display. The distinction between this and color-coding is that unambiguous identification is not required. It is sufficient that the different colors be reasonably distinct from each other. This kind of color enhancement of the information can make a complicated display much easier to interpret. An example is the use of color to differentiate layers in the diagram of a multilayer printed-circuit board. However, too many colors or the juxtaposition of colors that "fight" with each other can make a display unpleasant to use (Rice, 1991).

The third use for color is in rendering images. Here a large palette is necessary.[4] If there are only a few color (or gray-level) steps available, artificial contour lines may be introduced, especially in areas where the color changes gradually over an area. Although 16 or 32 colors may be quite satisfactory for differentiating elements of information, imagery is generally considered to require at least 256 possible colors, and more are desirable.

4.5.1 Color Emitters

The color CRT is a prime example of a display device carrying three different-color emitters on its surface. Gas-discharge and vacuum-fluorescent displays can also use different-color emitters on the display surface. LED displays have used

---

[4]*Gamut* is the range of colors that can be displayed. It is frequently shown as a triangle on the CIE chromaticity diagram, with the apices at the available primary colors. *Palette* describes the number of different colors that can be represented within the gamut. This is dependent, in part, on the number of different color codes that the display controller can handle.

FLAT-PANEL DISPLAYS

this approach, although blue LEDs have only recently become available, and only at very high prices compared to red, yellow, or green LEDs.

4.5.2 White Emitter with Superimposed Color Filters

Here each element comprises the display device proper and a superimposed filter. This approach has been proposed for electroluminescent displays, but it is most commonly seen with LCDs. The laying down and patterning of small filters precisely located with respect to the underlying display elements is an art that has advanced rapidly as display requirements have been increased (Goldowsky, 1990; Latham et al., 1987; Maurer et al., 1990).

Patterns of Color Dots

The pattern in which colors are placed on the display surface may be chosen for simplicity of construction, an apparent performance advantage, or best reproduction of the desired contents. This last consideration is tricky; a pattern that does well with continuous-tone imagery may introduce undesirable artifacts into line drawings or text if the dot structure is sufficiently coarse to be visible to the viewer (Silverstein et al., 1989).

The simplest arrangements to build use continuous stripes of color, either horizontal (which triples the number of rows required compared to a monochrome display) or vertical (which triples the number of columns). This arrangement is especially prone to introducing artifacts into the picture. Nevertheless, it has been used in commercial displays.

Triad arrangements, similar to those used on some CRTs, have been used in both true "triangular" geometries, corresponding to a hexagonal close-packed structure, and staggered "square" arrays.

Quad arrangements use four dots per pixel. Since some 70% of white light is green, these arrangements generally use two green dots, one red dot, and one blue dot for each pixel. IBM and Toshiba (Ichikawa et al., *1989*) have used an interesting variant with a red, green, blue, and white dot for each pixel of a large, high-resolution LCD (1440 × 1100 pixels monochrome, 200-μm dot pitch, 14.3-in. diagonal). This allows for a 16-color palette with only on-off devices. Furthermore, by energizing all three color dots to get white, an apparent doubling of the resolution is achieved.

4.5.3 Stacked, Switchable Color Filters

The color methods we have discussed so far are all *additive*; that is, the addition of the three primary colors produces white. The approach used in printing or painting is *subtractive*; the subtraction of all three primary colors—magenta, yellow, and cyan—produces black. This approach to light-emitting displays has been pioneered by In Focus, of Tualatin, Oregon. They use three layers of supertwist LCDs stacked on top of each other, with each layer tuned to modulate one of the three subtractive primaries. The advantage is in resolution; each layer has only one-third the number of dots required, and each of these dots can be about three times as large as would be required for a side-by-side arrangement. The disadvantage is that with three light modulators in sequence, the light losses are much higher than for a side-by-side arrangement; overall transmission of 12% is claimed (Conner, 1990; Gulick and Conner, 1991).

The approach was first presented in a "plate" for use with overhead projectors and has since been applied to direct-view LCDs as well. Parallax between color layers is avoided in the direct-view device by supplying the panel with collimated light and placing a diffuser between the panel and the user. Details on thickness penalties and light losses imposed by this arrangement have not been published as of this writing.

4.5.4 Field-Sequential Color

Rather than present the three primary colors simultaneously, they can be presented sequentially, as red, green, and blue fields. This arrangement requires a white emitter (or preferably an emitter with principal outputs at the three primary wavelengths) and switchable filters that can be switched fast enough. Since 30 frames of information per second is generally regarded as the minimum acceptable rate for electronic displays, this last requirement means that the display elements must be capable of switching at 180 frames/sec. This is a problem even with CRTs (Tektronix, 1990; Haven, 1991) and is more serious for LCDs, although work is under way to solve it.

Field-sequential color has been used with CRTs; in fact, the first broadcast color TV systems used field-sequential color, with rotating color wheels between the white-emitting CRT and the viewer. In flat-panel displays, the approach has been proposed for LCDs. Since the individual elements have significant persistence, the pixels at the top of the display will be brighter than the pixels at the bottom if the color-switched backlight illuminates the entire panel. Hence the backlight (or the overlay filter, if that is what is used) must be patterned into horizontal zones that can be switched at different times. It is also possible, with such an arrangement, to use different durations for the different colors, accomplishing (in part) what the use of two green dots in a four-dot quad arrangement does to increase the display luminance.

Field-sequential color has been described in connection with projection displays using LCDs (Lauer et al., 1990) but has not yet been implemented commercially with direct-view LCDs as of early 1991.

4.6 SPECIFIC FLAT-PANEL DEVICES

This section describes the various flat-panel-display technologies briefly; it is intended as a guide rather than an exhaustive treatise. For more detail, consult

FLAT-PANEL DISPLAYS 145

the excellent books by Bosman (1989) and Tannas (1985a). These arts are changing rapidly, and the changes are reported in trade publications and manufacturers' literature. The most current research and development results are reported at the annual symposia of the Society for Information Display (SID[5]) and at various meetings conducted by SPIE (The International Society for Optical Engineering[6]) and IS&T (The Society for Imaging Science and Technology[7]). Other sources of current information are the meetings held by SIGGRAPH, the Special Interest Group on Graphics of the Association for Computing Machinery (ACM), the National Computer Graphics Association, and the Electronic Imaging trade shows and meetings run twice a year by BISCAP, a commercial information provider.

4.6.1 Light-Emitting Diodes (LEDs)

Light-emitting diodes are single-crystal semiconductor devices from which light is emitted when a *pn* junction is forward-biased. Visible-light-emitting diodes use gallium arsenide (GaAs) or related ternary crystals such as GaAsP and GaAlAs. They are available in devices that produce red, yellow, or green light. Blue-emitting diodes, using silicon carbide or gallium nitride, are becoming available, but prices are currently (1991) some 10 times those for red emitters.

Operating voltage is in the range of 2 V, with currents for the typical small LED of the order of 10 mA. Typical reverse breakdown voltage is 5 V. The switching nonlinearity is excellent, so LEDs are very well suited to matrix displays. However, the manufacture of solid-state matrices, with many diodes fabricated as part of a single wafer, has not been feasible. Instead, LEDs are packaged either as single-emitter units or as single-character seven-segment or 5×7 dot arrays, often in a case with a lens or filter and a reflector to direct the emitted light into the lens or toward the viewer. Large arrays have been made by applying individual LEDs to printed-circuit or other mounting boards. Figure 4.15 shows three major types of LEDs.

Light-emitting diodes have proven convenient as single-element indicators or as alphanumeric displays of a few characters for a number of reasons. Their operating voltage is compatible with typical IC operating voltages, so special power

---

[5] 8055 West Manchester Ave, Playa del Ray, CA 90293; (213)305–1502. The symposia are managed by Palisades Institute for Research Services, Inc., 201 Varick St, New York, NY 10014; (212)620–3388. SID also cooperates in the annual International Display Research Conferences, which are held in the United States, Japan, and Europe in successive years. The technical paper digests of these meetings are the most current source of information on the display art.

[6] P. O. Box 10, Bellingham, WA 98227–0010; (206)676–3290. SPIE holds a large number of meetings on various topics, including display-related topics. Papers presented at these meetings are published, generally unrefereed, in *Proceedings,* which are available 5 months or more after the meeting has been held.

[7] 7003 Kilworth Lane, Springfield, VA 22151; (703)642–9090. IS&T is more concerned with image pickup and hard copy than with displays.

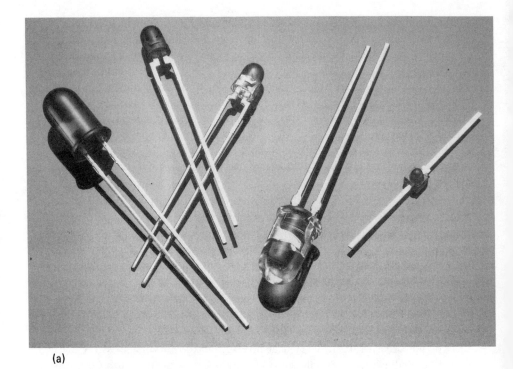

(a)

Figure 4.15 A selection of LED displays. (a) Single lamps. (Courtesy Hewlett-Packard Co.) (b) 16-segment "starburst" displays; control ICs are mounted on the other side of the board. (Courtesy Siemens Components Inc.) (c) Light bars. (Courtesy of Hewlett-Packard Co.)

supplies are not needed; they are compact and easily installed in equipment; and they are rugged and long-lived. They have not found much use in displays of high information content such as laptop computer displays because they are much more expensive than competitive technologies in such an application; the task of assembling over 300,000 discrete lamps into a small display is daunting. Also, although they operate at low voltages, this demands high currents; the VGA display of Section 4.2.3 would require 6.4 *amperes* per row if each pixel were operating at 10 mA. Note that although running the entire display at full luminance is unlikely, operating a single row at full luminance is not so abnormal, so each row driver must be capable of sinking the full current of all the pixels in a row.

Light-emitting diodes have found commercial application in physically large displays with small numbers of characters. "Sticks," typically one character high by some 18 characters wide, with the characters about 2 in. high, have proven to be relatively inexpensive and convenient to install and use. Such a display, using

FLAT-PANEL DISPLAYS

(b)

(c)

a 5×7 matrix for each character with one column of dots between characters, uses 756 LED lamps. The controller generally provides for moving text and simple block figures to attract viewers in such venues as airports, stores, and shopping malls.

A similar application on a larger scale is shown in Figure 4.16. This display, in the trading room of a major financial house, allows many traders to see important information. Not many of these very large displays have been sold. However, the stock "ticker" display at the top of the figure also uses LEDs—green in this case. Similar displays have been installed in many stock exchanges and brokerage offices. They replace electromechanical displays that, although effective, required more maintenance. This is an application in which reliability is considered very important; LEDs meet this requirement at reasonable cost.

Figure 4.16 The trading room at First Boston. The big board uses green LEDs, and the "ticker" display above it uses yellow LEDs. The board is 5 ft high and 8.6 ft wide, showing 20 lines of 60 characters/line. (Courtesy of Trans-Lux Corp.)

The last three years have seen a marked increase in the use of large LED displays of only a few (10–40) characters. They have been used at trade shows as attention-getting devices and in such more mundane applications as airport information panels, where they have proved to be a cost-effective way of displaying arrival and departure times and destinations at individual gates and baggage-claim carousels.

4.6.2 Electroluminescent (EL) Displays

In electroluminescent (EL) displays, an electric field applied to a phosphor (powder or thin film) produces light. Unlike LEDs (which are also sometimes referred to as electroluminescent devices), there are no fabricated *pn* junctions in EL displays. Instead, a high electric field inside the material excites carriers, which eventually decay with the emission of light.

Three types of EL devices are currently in use: ac powder, ac thin film, and dc powder. A fourth type, the dc thin-film EL (TFEL), has been the subject of some research but has not yet materialized as a display device.

AC Powder Electroluminescence

This is the oldest form of EL display device, dating back to the discovery by Destriau in 1936. As a display device, an ac EL has major disadvantages: its selection nonlinearity is poor; its luminance is low; and its life is poor.

The light output is a function of voltage and frequency; as either is raised, the luminance increases. However, life is essentially a function of the amount of light produced; as the luminance is increased, the life is shortened. A lamp running at a few foot-lamberts may have a lifetime of many thousands of hours, but at luminances high enough to be useful in display applications, lifetimes have been too short to be satisfactory. Work by Lehmann (1966) gave promise of overcoming this limitation, but Westinghouse management discontinued the project before its utility could be established.

AC powder EL is used in some lighting applications, notably as emergency lighting in passenger aircraft and as a backlight for LCDs. Its use in displays has been almost abandoned, although some work has been done in the People's Republic of China on very large character displays for assembly halls (Tongyu et al., 1986).

AC Thin-Film Electroluminescence

When ac thin-film electroluminescence (TFEL) was first introduced in 1974 (Inoguchi et al., 1974; Mito et al., 1974), it aroused great interest because of its extraordinarily long life—greater than 10^4 hr under continuous excitation. Since then TFEL has been the subject of intense effort; two companies are selling displays commercially.

The phosphor, generally ZnS doped with Mn, is a thin film sandwiched between layers of thin-film insulators such as Y_2O_3 (Fig. 4.17). All three layers are of the order of 300 nm thick. The layers are deposited on glass over a patterned transparent-conductor layer, usually indium tin oxide (ITO), which serves as the column electrodes. The row electrodes are generally thin aluminum, deposited on the back of the sandwich. The entire assembly is sealed, usually in a glass envelope, to prevent damage by atmospheric water vapor. Ultrahigh and ultraclean vacuum conditions are required to make defect-free, long-lived devices.

The mechanism of light emission involves trapping of carriers at localized sites at or near the phosphor–dielectric interface. These carriers tunnel to the conduction band when excited by a high electric field. There they become hot electrons that can excite localized centers by impact ionization, producing light (Gurman, 1989; Tannas, 1985b). Because the energy distribution of the hot electrons is ill-matched to the capture cross-section distribution of the traps, the process is inefficient. Furthermore, the indices of refraction of the phosphor and the dielectric are both high, so there is considerable trapping of light at their interfaces and at the dielectric–glass (or dielectric–ITO) interface, due to Fresnel reflection. Despite these difficulties, TFEL is an attractive display medium because of its excellent switching nonlinearity, good appearance, and long life.

Failure is generally due to dielectric breakdown at defects in the structure, and a good deal of effort has been devoted to developing structures in which this breakdown does not spread beyond the immediate vicinity of the defect.

The usual ZnS:Mn phosphor produces a yellow-orange light that is near the peak of the visual sensitivity curve and is comfortable to look at for long periods

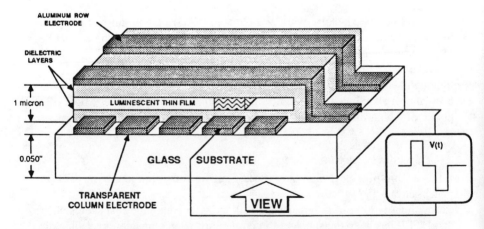

Figure 4.17 TFEL structure. The glass substrate must be capable of withstanding high processing temperatures. (Courtesy of Planar Systems, Inc.)

FLAT-PANEL DISPLAYS

of time. Contrast and viewing angle of TFEL displays can be excellent, although actual luminance is typically less than 30 fL in displays with many rows and columns. Color has been the subject of intense research activity, but the red and blue phosphors that have been reported to date are of limited luminance and efficacy.

The switching nonlinearity of these devices is excellent. Operating voltages of commercial displays are generally in the range of 175–200 V peak. Some TFEL material combinations show significant hysteresis, which might be used as a memory mechanism. However, this has not been sufficiently controllable to be used commercially so far.

Most of the energy that produces light is delivered during the rise and fall of the driving pulse, so simple pulse-width modulation is not very effective in producing gray scale. Combinations of pulse-width and pulse-amplitude modulation have been used (Steiner and Tsoi, 1988).

A typical mode of operation has been to drive the columns with positive-going pulses, scan down the entire display, and then drive the entire panel with a "refresh" pulse of the opposite polarity, to remove stored charge from all the pixels. Although the individual pixels behave like excellent capacitors, there is some loss of charge during a frame; as a result, different rows show differing lumi-

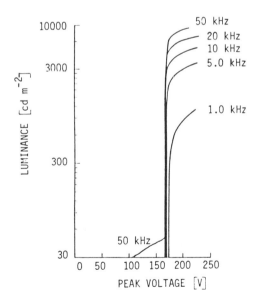

Figure 4.18 TFEL transfer curves. The curves are for continuous excitation (duty factor = 1); in a real display, the luminance is much lower because the duty factor is much less than unity. [From Gurman (1989).]

nances and there is some persistence, or "sticking," of still images. This can be largely cured by providing a refresh pulse at the end of every line time or every few line times.

Because of the high operating voltage, drive circuitry has presented a number of problems. This is compounded by the fact that the phosphor–dielectric sandwich has a very high effective dielectric constant, so the capacitance that the drivers must cope with and the resulting transient current during pulse rise and fall times are large. Texas Instruments and Supertex are the major manufacturers of ICs for this application. Another problem is reducing the required power; several schemes have been described for recovering the energy stored in the display's shunt capacitance and returning it to the power supply (Channing et al., 1989; Schmactenberg et al., 1989; Teggatz, 1989).

Planar Systems produces monochrome displays as large as 18 in. diagonal, with 1024 columns and 864 rows (Fig. 4.19). In this display, the column elec-

Figure 4.19 An 18-in. diagonal TFEL display. Resolution is 1024 × 800, luminance is 20 fL, and power consumption (including all driver circuitry) is 60 W. (Courtesy of Planar Systems.)

FLAT-PANEL DISPLAYS

trodes are divided along the horizontal midline of the display and are driven simultaneously from top and bottom; separate row drivers are provided for each half of the panel. The display is thus operated as two 1024 × 432-pixel displays; the increased duty factor makes possible a luminance of 18 fL at 60 frames/sec. Most TFEL displays are much smaller than this. The use of a circular polarizer in front of the display greatly improves the contrast in ambient light, but at the cost of half the luminance and at a significant monetary cost as well.

There are currently only two producers of TFEL displays: Planar Systems of Beaverton, Oregon, which acquired the TFEL business of Finlux (Espoo, Finland) in late 1990, and Sharp Corporation of Japan. The attractive appearance and mechanical durability of the technique make it a serious contender in flat-panel display markets, but it is still the most expensive monochrome display type commercially available in its size range.

DC Powder Electroluminescence

Another EL technology, dc powder electroluminescence (DCEL), uses powder phosphors and dc excitation. The phosphors are not the same as those used in ac powder EL. They can be deposited on the substrate by spraying or printing techniques that are far less expensive than the ultrahigh-vacuum methods needed for TFEL. The substrate is glass carrying ITO electrodes (Fig. 4.20); the rear electrodes are usually aluminum. The light is emitted from a thin active layer near the anodes which has been "formed" by applying continuous direct current for some time. Because the phosphor is conducting, it must be mechanically separated into columns to avoid crosstalk. Driving voltage is about half that required for TFEL, and pulse-width modulation works nicely to provide gray scale (Channing et al., 1989).

The life of DCEL, although much better than that of ac powder EL, is not as good as that of TFEL. Manufacturers have compensated for this by sensing the light output and varying the driving voltage appropriately, but differential aging (the decrease in luminance of often-used areas, such as the left side of text displays, compared to less-used areas of the display) is still a problem. Contrast is

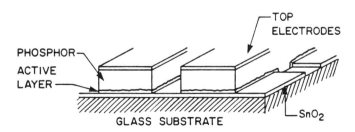

Figure 4.20 A dc powder EL display. [From Howard (1981).]

not as good as in TFEL; because the material is a powder, a circular polarizer does not work nearly as well as it does with the specular reflection from TFEL. Nevertheless, contrast is acceptable for many applications, especially since the cost should be comparable to that of LCDs.

There is currently only one commercial producer, Cherry Electrical Products. Much of the research and development work on this technology was done by Aron Vecht (1990) and his colleagues at Thames Polytechnic (London) and Phosphor Products.

Other Electroluminescent Technologies

There has been some work on organic EL materials, mostly in dc configurations (Adachi et al., 1989; Ishiko et al., 1989). These operate at much lower voltages than any of the technologies mentioned above, but so far life and luminance are severely limited.

There have also been attempts to reduce the operating voltage of EL materials by injecting carriers through junctions. It has so far been impracticable to fabricate such devices at any reasonable cost, but the possibility of low-voltage EL indicates that research along these lines will continue.

4.6.3 Vacuum Fluorescent Displays

Vacuum fluorescent displays [VFDs; sometimes called *fluorescent indicator panels* (FIPs)] are essentially very-low-voltage CRTs. A typical construction is shown in Figure 4.21. Low-voltage phosphor is coated on the anode electrodes; wire or mesh grids run orthogonal to these. Electrons are emitted from thin oxide-coated filaments operated at very low temperature; the filaments are invisible to the naked eye. The phosphor is viewed through the intervening structure of one or more grid layers plus the filaments.

The electron-accelerating voltage is generally well under 100 V. At these low voltages, conventional insulating cathodoluminescent phosphors would rapidly accumulate negative charges and repel additional electrons. Hence conductive phosphors, generally based on ZnO, are used.

These displays are used extensively as short-message displays, clock displays, and displays for the controls of home electronics devices. In these applications, in which the duty factors can be large, they produce adequate luminance and are clearly cost-effective, especially when produced in large volume.

The usual ZnO phosphor produces a wideband blue light that can be filtered to produce a variety of colors without too much loss of luminance. Recently, a variety of colored phosphors has become available, and VFDs with red, green, and yellow areas are now used.

Like other matrix displays, a driver is required for each row and each column. To reduce the number of connections to the outside world, manufacturers have worked on bonding driver chips to the glass substrate and have even put the drive

FLAT-PANEL DISPLAYS

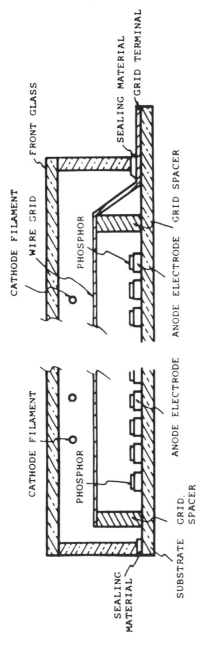

Figure 4.21 A vacuum fluorescent display. [From Morimoto (1982).]

circuitry inside the vacuum envelope. Taking this further, VFDs have been made with the substrate carrying a silicon IC, with phosphor coated directly onto the drains of FETs used as anode drivers (Yoshimura et al., 1986). This has made possible an extremely small display—216 × 246 pixels in an area of 6.9 × 9.1 mm (0.27 × 0.36 in.)—for use as a camcorder viewfinder or in similar applications.

Another tack has been to deposit the phosphor on the faceplate, like a conventional CRT. One such *front luminous* VFD was 4.4 × 7.1 in., showed 400 × 640 pixels, and was capable of displaying video with 16 gray levels using pulse-width modulation of the anode voltages (Nagasawa and Watanabe, 1988).

Although these devices have been very successful in their small-display niche, displays much larger than about 8 in. diagonal have not been seen commercially. The construction apparently becomes uneconomical for large information content. Making large displays requires increasing the glass thickness or curving (doming) the faceplate so that the envelope can withstand atmospheric pressure. These expedients have evidently not proven satisfactory.

The only producers are Japanese companies: Choa, Futaba, and Ise in particular.

4.6.4 Flat CRTs

There has been a vast amount of work in the area of flat CRTs over a period of many years, but most of it has fallen by the wayside. The major current contender is Matsushita (Nonomura et al., 1989). Early, small versions of this device, called *MDS* for *matrix drive and deflection system*, were demonstrated at the 1989 Consumer Electronics Show in Chicago. Flat CRTs are discussed in more detail along with other CRTs in Chapter 1.

The version most recently reported is shown in Figure 4.22. Electrons are emitted from the vertical thermionic cathodes and extracted from the cathode-region space charge by the vertical scanning electrodes. The resulting small beams of electrons are modulated by the potentials between grid-1 segments and the cathode. Beam-forming electrodes focus the modulated beams, which are then deflected in x and y to strike vertical color-phosphor stripes.

The construction is modular, with each module containing seven or eight 15.08—mm-wide sections. The electron beams in all the sections are deflected and modulated simultaneously. Thus the duty factor for each pixel is very much larger than for a conventional CRT. This has some of the desirable effects of multidimensional scanning discussed earlier. At 10 kV accelerating voltage, the luminance is 70 fL.

The aim of this development work is a 40-in.-diagonal display. To keep the weight reasonable and maintain a flat faceplate, internal support of that faceplate is necessary to withstand the atmospheric pressure of over a ton per square foot. The support is provided by thin "needle pillars" that transmit panel force to the

FLAT-PANEL DISPLAYS

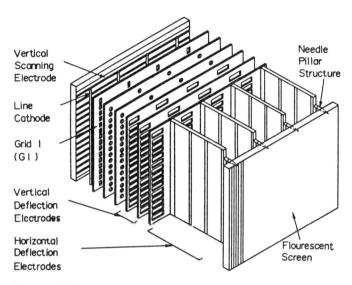

Figure 4.22 The Matsushita MDS flat CRT. [From Nonomura et al. (1989).]

vertical deflection plates, from which it is transmitted to the rear of the envelope through the electrode structure.

This support problem is faced by all large flat displays. Gas-discharge devices like those described below use various means to incorporate internal spacers. LCDs also require spacers to accurately maintain the spacing between opposite electrodes. Lack of a technique for handling this problem has been a major hindrance to making VFDs larger. Developers of the *Digisplay* (Section 4.3.4) and its successor devices never solved this problem, which is one of the reasons that the device is considered too heavy for a number of applications.

Channel Electron Multiplier–Flat Deflection CRT

This device, developed at Philips Research Laboratories in England, was an ingenious combination of a flat electron-beam scanning system followed by a channel electron multiplier to increase the current to the phosphor (Washington et al., 1985). Figure 4.23 shows the system. A low-voltage (about 400 V) electron beam is deflected electrostatically in the x direction. When it reaches the bottom of the panel, its direction is reversed, and it is steered up the other side of a dividing plate. At various locations along this second side, deflection plates steer the beam toward the screen, providing the y selection.

For this scheme to work, the beam voltage must be low so that deflection can be effected by reasonable voltages, and the beam current must be low to avoid excessive spot-size growth. This combination would result in an unacceptably low luminance. The cure is that the electron beam is steered, not to the screen,

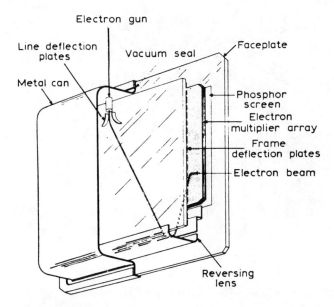

Figure 4.23 The Philips flat channel multiplier CRT. [From Washington et al. (1985).]

but to a channel electron multiplier, which increases the current by a factor of almost 800. The resulting beam, which in experiments was some 0.2 mA, is then further accelerated by about 10 kV before it strikes the phosphor. The resulting luminance is about 90 fL for a TV raster scan.

This scheme does not provide for internal support. An ingenious technique was devised that uses a metal back and a deformable seal to the glass faceplate. This allows the faceplate to deflect inwards under atmospheric pressure and made possible tubes up to about 12 in. diagonal. It is not clear how far this approach could be taken.

Color was also demonstrated, using two possible techniques to deflect the beam exiting from the channel multiplier.

The overall approach was abandoned around 1986. Although it was of considerable interest in military applications, commercial marketers were not convinced that this was a viable technology in their world. It did not appear that the approach could be extended to the 19-in. and larger sizes that are of principal consumer-TV interest, and the niche market of 12-in.-diagonal and smaller did not appeal to Phillips management.

Field-Emitter Cathodes

Microscopic points of metal, produced by IC-manufacturing techniques, will emit electrons if subjected to high electric fields. Because the points are submicrometer in radius, the required voltages are low. Hundreds of such emitters can

FLAT-PANEL DISPLAYS

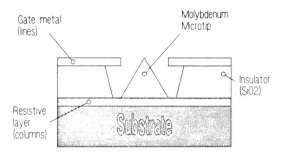

Figure 4.24 A cross-sectional schematic of a field-emitter cathode. [From Leroux et al., (1991).]

be contained within a single pixel area. Such a collection of electron emitters would be the cathodes of a flat CRT. The cross section of such a field-emitter cathode is shown schematically in Figure 4.24.

The approach has been worked on for years, especially at Stanford Research Institute (Brodie, 1979), and is beginning to generate serious results, notably at LETI in France (Meyer, 1990) and Coloray in Colorado. The work at LETI now incorporates a resistive layer to equalize the current among the many individual emitters that comprise a single pixel. The gate metal lines are part of the addressing system (Leroux et al., 1991). Many problems remain to be solved. Display devices would require row-and-column arrays, and it is not clear how these will best be fabricated. Crosstalk would presumably come from fringing fields at the intersections of the row and column arrays; avoiding this kind of problem may require significant device complication, although the most recent publication indicates that answers to many of these problems are at hand.

Either the device is limited to low electron-acceleration voltages, in which case the phosphor limitations of VFDs apply, or some clever way must be found to get the accelerating voltage up to many kilovolts without internal breakdown. The devices will presumably be driven a line at a time, so luminance should be adequate.

4.6.5 Liquid-Crystal Displays (LCDs)

The LCD is rapidly becoming the dominant form of flat-panel display. Early in their development, LCDs displayed LEDs in wristwatches and other applications requiring very small displays and very low power. They now dominate the laptop computer market, thanks to both their low power requirement and their thinness. Competitive technologies, although often boasting better appearance and wider viewing angles, appear to be losing ground, in significant part because of the

---

[8] I am indebted to Allan R. Kmetz for a careful reading and critique of this section. However, I must reserve for myself the responsibility for any errors or omissions.

enormous research, development, engineering, and marketing effort being put forth by LCD manufacturers.

The combination of LCD plus color filters plus (often) a backlight has dominated the market for small, hand-held TVs and will have a major impact on laptop computers (although costs are almost prohibitive at this writing). LCD panels as light modulators in projection systems are the subject of intense effort, and many early consumer HDTV projection systems will take this route.

The advantages of the technology are low power and the fact that LCDs are light modulators, so that the displays can be seen by ambient light, like print. Alternatively, in light-emitting displays, the tasks of generating light and imposing information upon it can each be carried out in optimized devices.

Disadvantages include marginal switching nonlinearity, restricted viewing angle, slow response speeds, and limited operating temperature range. All of these are being successfully attacked, with intense research and development activity under way, especially in Japan. More LCDs have been produced than any other flat panel type, in sizes ranging from wristwatches to wall-size displays.

Chapter 2 has discussed LCD fundamentals. In this section I shall describe some of the ways in which LCDs are used in displays.

LCDs Relying on their Intrinsic Switching Nonlinearity

The switching nonlinearity of LCDs is sufficient for multiplexing displays of several hundred rows. The sharpness of the knee in the transfer curve is far better for supertwist nematics (STN or SBE) than for simple twisted nematics (TN), but even TN displays have been used in medium-information-content displays.

Twisted and supertwisted nematic LCDs respond to the rms voltages impressed upon them—more specifically, to the square of the voltage integrated over a frame time (Kmetz, 1973; Alt and Pleshko, 1974). Operating waveforms are typically bidirectional rectangular pulses (Fig. 4.25); a dc component in the applied voltage leads to electrochemical reactions that drastically shorten the life of the display. The rms select and nonselect voltages are

$$V_{select} = \{[(V_1 + V_2)^2 + (N - 1)V_2^2]/N\}^{1/2} \qquad (4.2)$$

$$V_{nonselect} = \{[(V_1 - V_2)^2 + (N - 1)V_2^2]/N\}^{1/2} \qquad (4.3)$$

where the driving voltages are $\pm V_1$ for the rows and $\pm V_2$ for the columns and N is the number of rows. The optimum choice of

$$V_1 = V_2 N^{1/2} \qquad (4.4)$$

leads to a maximum ratio of the two voltages (Alt and Pleshko, 1974; Nehring and Kmetz, 1979):

$$\frac{V_{select}}{V_{nonselect}} = \left[\frac{N^{1/2} + 1}{N^{1/2} - 1}\right]^{1/2} \qquad (4.5)$$

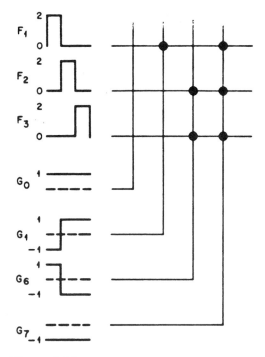

Figure 4.25 Voltages applied to an LCD using "3:1" addressing. The potential at any pixel is the difference between row and column voltages; hence, in the polarities shown, negative column (G_i) voltages turn selected pixels on. Both row and column polarities are reversed each frame, to avoid impressing a net dc voltage on the liquid crystal [From Nehring and Kmetz (1979); copyright IEEE.]

This ratio is small: only 1.387 for $N = 10$ and 1.106 for $N = 100$. Furthermore, both the turn-on speed and the viewing angle are proportional to the voltage above threshold, though not linearly so. If the number of rows is large, the permissible voltage above threshold is small, and rise time, viewing angle, and contrast are restricted.

The steep transfer curve of STN devices alleviates this problem, but even for STN displays it is generally desirable to divide the display horizontally, halving the number of rows, as described in Section 4.6.2, for displays of VGA or similar information content.

Ferroelectric LCDs promise improvements in multiplexing performance, but ferroelectric LCDs are still only developmental.[9] Hence these limitations have

---

[9] A session of six papers at the 1991 SID International Symposium (*SID Int: Symp. Dig. Tech. Papers XXII*: 383–407) described some current approaches to the problems of ferroelectric LC displays. The

driven workers to active-matrix technologies, in which the nonlinearity is provided by devices other than the liquid crystal.

LCDs with Additional Switching Devices: Active-Matrix LCDs

Active-matrix LCDs provide a switching element at each pixel or subpixel. For reflective displays, this has been done by using a silicon wafer as the substrate and providing a FET to drive each pixel. However, such an arrangement is limited to small, reflective displays. For large displays, the approach has generally been to make the switching elements from thin films of semiconductor, fabricating either *thin-film transistors* (TFTs) or thin-film diodes on one of the glass substrates.

Thin-film transistors can provide excellent nonlinearity. Driving voltage is now limited by the characteristics of the TFTs or diodes rather than by the stringent relationships cited above, leading to much better contrast and viewing angle. Furthermore, the combination of the high off-resistance of the TFT or diode and the high resistance and capacitance of the LC pixel can store a voltage on the pixel for a very large fraction of the frame time, increasing the effective duty factor and thus the contrast and viewing angle still further. (As noted earlier, driving voltage is reversed each frame so that there is no net dc voltage across the liquid crystal; this is essential for preventing undesirable electrochemical effects.) Twisted-nematic liquid crystals can be used; the additional complication of supertwist is unnecessary.

However, these gains come at substantial cost. This approach requires a large number of TFTs or diodes, and they must all work. Although the individual features are much coarser than those found in silicon ICs, the area over which the TFT array must be distributed is far larger than is found in any silicon IC. In effect, the requirement is for wafer-scale integration but without the ability to use discretionary wiring to replace faulty elements that marks most wafer-scale integration efforts.

Despite the difficulties, energetic work on active-matrix LCDs is going on, especially at Japanese companies, who dominate this market area. Small color-TV displays (typically some 200 × 200 pixels, and 2–4 in. diagonal) have been produced in large quantities, and active-matrix displays compete with STN displays in the laptop computer market. Most color displays for laptops are active-matrix devices. Although production yields of the order of 20% were reported in late 1990, these numbers will presumably be improved in the not too distant future.

One way to improve production yields is to incorporate redundancy. Figure 4.26 shows two ways in which this has been attacked, for TFTs and for diode

---

potential advantages of this technology are stimulating a great deal of research, and the field is moving so rapidly that any account here will be out of date by the time this book is printed.

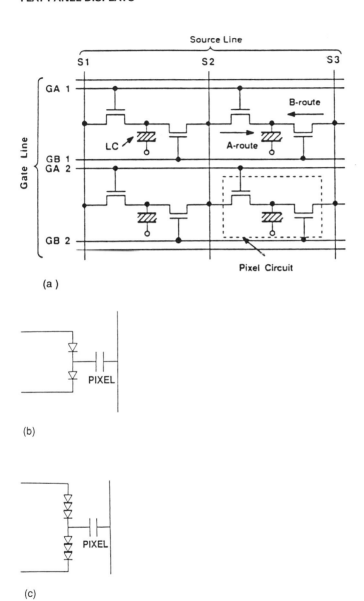

Figure 4.26 Redundant circuit elements can reduce the number of defects on an AM LCD. (a) Two gate transistors and two row electrodes for each pixel combat the effects of open elements (Takahashi et al., 1987). (b) Using *p-i-n* diodes, a simple configuration is vulnerable to open- or short-circuited diodes. (c) Multiple, series-connected diodes guard against the effects of short-circuited diodes. Opens are combatted by the use of double drive electrodes or by repair before assembly (Yaniv et al., 1988).

arrays. In addition, testing and repair before final assembly (by opening the connections to short-circuited active elements and short-circuiting open elements as appropriate) can raise yields.

The number of connections to the outside world is still a major concern. These can be reduced by mounting at least some of the driver chips directly on the display substrate, an approach being pursued vigorously by many workers (Kobayashi et al., 1990). It is also possible to fabricate these driving circuits as TFTs. Although it is difficult to make TFTs fast enough to distribute the video signals, both device and circuit improvements are making this possible (Emoto et al., 1989). Since the area occupied by the drive circuits is small compared to the area of the display, it is quite feasible to build substantial redundancy into the drive circuitry. The resulting decrease in connections to the display panel can produce major savings in cost and improvements in reliability, which can justify the cost of these additional on-panel components (Stewart, 1991).

Plasma Switching of LCDs

A new development from Tektronix, plasma switching of LCDs, represents a possible way of circumventing some of the problems of semiconductor active-matrix devices (Buzak, 1990a, 1990b). Here the switching element is the positive column of a glow discharge in a gas. As can be seen in Figure 4.27, each row of the display has a groove in the bottom glass sheet to contain the ionized

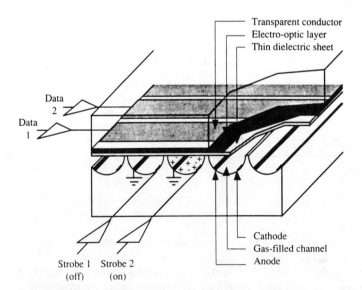

Figure 4.27 The plasma-addressed LCD.

gas. When the gas is ionized, it becomes conductive, and charge can move from the column electrodes through the liquid crystal to the ionized gas, via the capacitance of the thin glass sheet separating the liquid crystal from the gas. Since the ionized gas contains both electrons and positive ions, the voltage across the liquid crystal can be either positive or negative, and no net dc need appear across the liquid crystal. When the gas is not ionized, it is an excellent insulator, so that changes in the voltage of the column electrode do not affect the voltages across the unselected LC pixels. A single groove in glass, with its simple anode and cathode electrodes, thus replaces all the TFTs of a row.

This device has been demonstrated, using PDLCs (see below) in a full-color display about 7 in. diagonal with 300 × 300 dots. It is not yet clear what the limits of this technology are in terms of display size, speed, and resolution. It is potentially much less expensive than semiconductor AM LCDs.

Polymer-Dispersed LCs (PDLCs) or Nematic Curvilinear Aligned Phase (NCAP) Displays

In these devices (Fig. 4.28), the liquid crystal material is held in microscopic spheres in a plastic-film matrix. With no applied voltage, the film is strongly scattering; with an applied voltage, the LC material aligns with the electric field and the film becomes transparent. The director is the long axis of the LC molecule (Welsh and White, 1990). Because the LC material is embedded in plastic, it can be made in large sheets, it can be cut to any desired size or shape, and no tight-tolerance spacers between flat substrates are required. Polarizers and their attendant light loss are also not needed. The technique is still sufficiently new that a consistent name has not emerged.

There is essentially no switching nonlinearity, so external switches, such as TFTs or the plasma switch described above, must be used in a matrix display (Doane et al., 1990; Drzaic et al., 190; Welsh and White, 1990).

This is another area that is the subject of intense research and development. So far, applications to matrix displays have not emerged from the laboratory, but it is highly probable that they will in the near future.

4.6.6 Gas-Discharge Displays

There are two general types of gas-discharge displays: dc, in which the electrodes are in contact with the gas and are driven with unidirectional pulses, and ac, in which the electrodes are insulated from the gas and are driven bidirectionally. Both types have long histories, with a large variety of devices having been developed in each category. In this brief review I shall concentrate on current technologies and largely ignore types that are no longer being produced and that are unlikely to be revived.

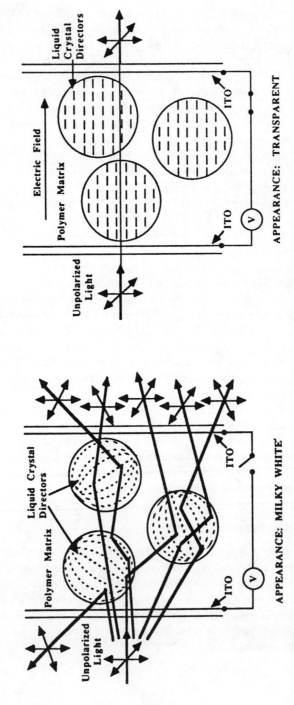

Figure 4.28 PDLC or NCAP configuration. [From Welsh and White (1990).]

Gas discharges are useful for several reasons. They can produce reasonable amounts of light; photoluminescent devices, in which the gas discharge produces ultraviolet light that excites phosphors (the fluorescent lamp is an example) can be efficient and can produce a variety of colors. Their switching nonlinearity is excellent. The gas discharge can also exhibit the property of *breakback*—the voltage when the discharge is on (operating or firing) is lower than the voltage needed to start it. This property can be used to provide *storage*: a pixel, once lit, stays on until it is deliberately turned off. (This is in contrast to *persistence*: a pixel, once excited, remains on for some time, but its output gradually decays to zero.) Gas discharges can make use of *priming*: a discharge can be induced to fire readily by the nearby presence of another operating discharge. This can be used to dramatically reduce the number of drive circuits required. This property is unique to gas discharges.

Both ac and dc discharges can produce color and gray scale, although these have been harder to achieve in ac than in dc devices.

DC Gas Discharges

The simplest form of dc device uses orthogonal cathode and anode electrodes on opposite substrates. The interior of the device is filled with neon gas, and insulating barriers prevent the discharge from spreading along the electrodes. What is observed is the *cathode glow*—an area of ionized gas in the immediate vicinity of the cathode. Operating voltages are of the order of 150 V.

A consequence of the breakback phenomenon described above is that the discharge in this usual *normal glow* regime exhibits *negative resistance*: The current increases even as the voltage falls. This can lead to destruction of the display. Hence a gas-discharge display requires a current-limiting resistance for every pixel. This can be a physical resistor, or it can be the current-limiting effect of a suitable electronic circuit. By adroit choice of driving voltage and resistance value, this breakback and negative resistance can be used to provide storage: A pixel, once fired, will stay lit until the voltage across it is reduced to a value below the *sustain* level and held at that value until enough ions and electrons have been lost from the gas (owing to recombination in the gas volume or at the pixel walls) that the pixel will not reignite.

The advantage of storage is that the display can be much brighter than without it, and continual refresh may be unnecessary. However, most attempts to achieve storage in this fashion have failed; the tolerances on pixel fabrication and especially on resistor values (a resistor in the range of 10 kilohms to 1 megohm is required for each pixel) have defeated attempts to make dc storage displays.

The usual gas is neon, which has the brightest visible output of any noble gas. (Reactive gases are not generally used because of the possibility of deterioration of the electrodes or the substrates.) Often about 0.1% of argon is added to take advantage of the Penning effect, which lowers the ionization voltage of the mixture and thus reduces the requirements on the drive circuits (Weber, 1985).

All gas-discharge devices require priming to ensure that they will ignite rapidly when pulsed. In most displays, the priming is provided by cells on the periphery of the display, which are continuously illuminated but are hidden from the viewer. Electrons, ions, metastable atoms, and photons from these peripheral priming cells diffuse into the main body of the display and promote the initiation of discharges.

Figure 4.29 shows a very large display made from dot-matrix modules. Each character is a 1-in.-high 5 × 7 dot matrix; each module contains two rows of 20 characters per row plus its own driving electronics. Regardless of the size of the display, each module is scanned independently, so that the duty factor per character is 1/20 (the inverse of the number of characters scanned). As a result, overall

Figure 4.29 A 1920-character modular dc gas-discharge dot-matrix display. Character height is 1 in. and luminance is 25–80 fL, requiring 24 W per module at the high luminance. Most gas-discharge displays are much smaller than this. (Courtesy of Quantum Electronics, Inc., Lewistown, PA.)

FLAT-PANEL DISPLAYS 169

luminance can be high, 25-80 fL, but at the cost of substantially more expensive drive circuitry than if the entire array were scanned as a simple row-column matrix.

The modular construction makes it convenient to build smaller or larger displays. Similar techniques are used by some vendors of LED displays.

Priming Using an Insulated Electrode Although simple orthogonal electrode dc displays have not made a great market impact, a variant on the approach, using priming provided by an insulated electrode, has been much more successful (Amano, 1975). The system is shown in Figure 4.30. Row and column electrodes are conventional; the barrier ribs prevent the discharge from spreading along the cathodes and also provide support against external atmospheric pressure (internal pressure is about 0.6 atm). The entire interior is fabricated by thick-film printing techniques, which were developed by DIXY to give extraordinarily high resolution.

The trigger electrode is insulated from the gas by a thin dielectric layer. When it is driven negative to about 100 V (an excursion of about 275 V), it produces a pulse of current that ionizes the gas. The resulting priming makes possible operation of the cathodes in the immediate vicinity of the pulsed trigger electrode with voltage excursions of 55 V on the cathodes and 65 V on the anodes (Awaji et al., 1986). No lit peripheral cells are required, so intrinsic contrast is excellent.

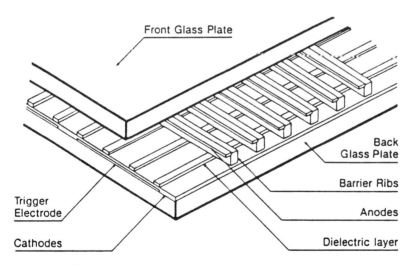

Figure 4.30 A buried-trigger-electrode dc gas-discharge display. Combining a trigger pulse with properly timed cathode and anode pulses reduces the number of cathode drivers required, since only those cathodes in the vicinity of the pulsed trigger will be sufficiently primed to fire (Awaji et al., 1986).

Note that the trigger electrode subtends only some of the cathodes. The priming it produces is confined to its immediate vicinity. More distant cathodes, pulsed to the same voltages as the nearby ones, will not fire their pixels because their gas is not primed. This can lead to a substantial reduction in the number of drive circuits required. Many cathodes can be connected to the same drivers, but only those cathodes that are primed by underlying trigger electrodes will fire. This is a kind of multidimensional scanning, as described in Section 4.3.4.

This device has been used to display real-time video. Gray-scale performance was good, but there is little enthusiasm for monochrome red in entertainment TV. However, the device is used extensively as a display for laptop computers.

This display was invented at Sony, was pursued at DIXY (a spinoff formed when Sony elected not to carry on further work), and is now produced by Matsushita. Using xenon (which has a strong UV output), Matsushita has demonstrated panels of one or two colors and even full-color displays. As of 1990 these were too dim to be salable, but improvements should be forthcoming.

Photoluminescent DC Gas Discharge Displays

This class of devices uses ultraviolet light from the discharge to excite phosphors, which then emit visible light. Some photoluminescent displays use UV light from the cathode glow, while others use UV from the positive column. The latter is potentially more efficient (this is the part of the discharge used in fluorescent lamps, for example), but these displays generally require more complex structures.

The working gas is usually xenon, which has the two advantages that it produces more UV than neon, and its visible light output is white. Although the white light may desaturate the output color (if the gas discharge itself, in addition to the phosphor, is visible), it can be less difficult to deal with than the red of neon.

A group at NHK Laboratories (the laboratories of the Japanese government broadcasting system) has been working along these lines for several years. A recent report (Murakami et al., 1990) describes a 33-in. multicolor panel with 800×1024 display cells, using a He-Xe gas mixture. Maximum white luminance is 25 fL (85 cd/m^2), intrinsic contrast ratio is 820:1, but luminous efficacy is only 0.15 lm/W. This implies that if the panel were fully illuminated at full output, it would draw 616 W.

This display has an interesting form of storage. The anode voltage is applied in short pulses, so that the discharge never really comes to a steady state but once fired is left primed for another firing at reduced voltage. As a result, successive short pulses produce light from a cell once fired, and this succession of short light pulses can be prolonged for a substantial fraction of a frame time.

A somewhat similar approach has been taken at Hitachi. Again, the device uses the positive column, but in this case in either pure xenon or a Ne-Xe mixture. The difference is that the anode pulses are very short, so the device never breaks from the Townsend region into a positive column. Not only does this provide storage, as in the NHK device, but the result of this short pulse is that the electrons produced in the discharge are very energetic and produce an intense burst of UV light to excite the phosphors (Mikoshiba et al., 1990; Suzuki et al., 1990). This approach has achieved higher luminance (146 fL) and higher luminous efficacy (2.2. lm/W) than the NHK work, but Hitachi has not built panels as large as NHK's.

Gas-Electron-Phosphor (GEP) Displays

This is a hybrid technology: A gas discharge is used to produce electrons, which are then accelerated to excite cathodoluminescent phosphors. Hence it can be considered a gas-discharge technology or a flat-CRT technology.

An early form was developed at Siemens (Schauer, 1982). Row and column electrodes were deposited on opposite sides of a glass-ceramic plate that had a hole for each pixel. The distance between this plate and the phosphor was so short that the gas did not break down into a discharge, although some 10 kV was applied. In this space, the product of the gas pressure and the distance, pd, was well to the left of the Paschen curve, which governs breakdown.

On the other side of the plate, where the cathode was located, there was ample space for a conventional glow discharge. The electrodes on the plate served as anodes for this discharge; electrons were drawn through the holes in the plate and accelerated to strike the phosphors. Full-color displays and real-time TV were demonstrated. However, there were sufficient difficulties that the program was terminated.

A somewhat similar device, invented at Zenith Electronics Corporation and carried forward at Lucitron Inc., overcame many of the problems encountered by Siemens (DeJule et al., 1986). The device is shown in Figures 4.31 and 4.32. It incorporates internal supports, so that the display could be very large without requiring excessively thick glass for the faceplate. (The internal pressure is only $\frac{1}{500}$ atm, so this is a vacuum tube as far as mechanical considerations are concerned.) It is divided into vertical "regions" 16 columns wide, all of which are scanned simultaneously. Thanks to priming, only a limited number of scan drivers are required: 10 in the vertical direction and 16 in the horizontal direction.

The cathodes are "hollow" cathodes, which are more effective in producing electrons than planar cathodes. In each region, the discharge at the anode is drawn into a small region of intense ionization—the *plasma sac*. Electrons are extracted from the sac and accelerated through a distance of about $\frac{1}{8}$ in. to strike the phosphors.

The sac is scanned in the vertical and horizontal directions by manipulating the voltages on the row and column grids. The volume in the immediate vicinity of the sac is primed by particles and photons from the sac, so when a row electrode is turned off and its neighbor turned on, the sac will be reestablished only

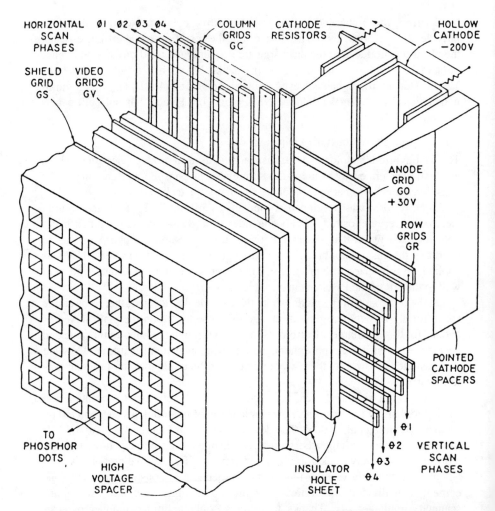

Figure 4.31 A simplified perspective view of the Lucitron® Flatscreen® panel. The supports for the hollow cathodes are also the spacers that sustain the flat faceplate against atmospheric pressure. The reset electrodes are omitted. The drawing is simplified for clarity (DeJule et al., 1986).

at the nearest row electrode and not at more distant electrodes that have undergone the same voltage change. As a result of this approach, a 256 × 352-element monochrome display was scanned with only 26 scan drivers. The process is initiated by setting up a "reset" plasma sac at the top of the display in each region, outside the view of the observer. This requires two additional electrodes; a panel this wide of any height could be scanned with a total of 28 scan drivers.

FLAT-PANEL DISPLAYS 173

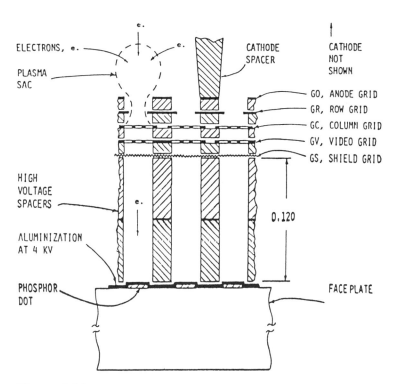

Figure 4.32 A cross-sectional view of the Flatscreen® GEP device. The anode grid keeps the discharge alive during switching times; the shield grid, which is fine mesh, helps to prevent the high voltage from striking through to the cathode space and causing a destructive arc (DeJule et al., 1986).

Note that since each row of the display is scanned in 16-pixel regions, the duty factor per row is 1/16—lower than full line-at-a-time scanning but much higher than point-at-a-time scanning. Once again, this is a kind of multidimensional scanning. This scheme was dubbed *fractional line-at-a-time scanning* (FLAT).

Since there is only one plasma sac in each cathode region, a single video grid, common to all the pixels in the region, suffices to provide gray scale. The overall economy of drive circuitry is unusual; a total of only 50 pulse circuits were required for the 256 × 352-pixel monochrome display (90,112 pixels). A conventional matrix of the same size would require 608 drivers.

In color versions, the phosphor was laid down in vertical stripes. Scanning could be done at three times the monochrome rate, or the video grid could be divided into three grids and the plasma sac made to subtend three subpixels. The latter technique was used in Lucitron's experimental color panels.

Monochrome devices with 35-in. diagonals and 8.5-in. color devices were demonstrated. Although the technology is promising, Lucitron met with financial reverses and closed its doors in December 1987. No further work has been done.

AC Gas Discharges

In ac gas-discharge devices the electrodes are isolated from the gas, typically by a layer of glass about 0.001 in. thick. This insulating layer is further coated with a thin (a few hundred nanometers thick) layer of a durable, low-work-function material such as MgO. The overcoat reduces the operating voltage and also protects the underlying insulator, which is generally a low-melting-point glass containing a great deal of lead, from sputtering and browning due to ion bombardment. The gas is usually a Penning mixture of neon and argon at about 400 torr.

In the usual operating arrangement (Fig. 4.33), the entire panel is supplied with alternating-polarity rectangular-pulse voltage, the *sustain voltage*, of an amplitude just too low to cause firing in any pixel. The capacitance of the gas layer is much less than that between the electrodes and the gas, C_{wall}, so most of the applied voltage appears across the gas. When the gas fires, charge moves through the circuit and results in a voltage, the *wall voltage*, appearing on both walls of the pixel. The discharge is terminated in a few microseconds because the peak current is high, resulting in a substantial wall voltage that is in the opposite polarity to the applied voltage. The *breakback* phenomenon, described earlier, contributes to this process. The series capacitance thus performs the current-limiting function for which resistors are used in most dc devices (Pleshko, 1980).

When the next half-cycle of sustain voltage is applied, the wall voltage adds to it, so the cell fires again. This process repeats indefinitely, until the wall charge is reduced by an appropriately timed pulse. The ac panel operated in this fashion has *storage*. Typical sustain frequencies are 25–100 kHz; although the amount of light produced for each pulse is small, the light averaged over a millisecond or so can be tens of foot-lamberts. Furthermore, it is flicker-free; the sustain frequency is far higher than the critical fusion frequency. As a result, large panels of high information content can be built.

To turn on a pixel, a write pulse of greater amplitude than the ignition voltage, is applied, so a discharge takes place, producing a wall voltage (dotted waveform). The wall voltage, in conjunction with the applied voltage, continues to fire the selected pixel until erasure. An erase pulse occurs just before a sustain pulse of the same polarity; it reduces the wall voltage sufficiently that the following sustain pulses when added to the remaining voltage are insufficient to cause the pixel to fire. A selected cell continues to fire because the sum of the sustain and wall voltages is greater than the ignition voltage. An unselected cell has essentially zero wall voltage, so it never fires. Note that a selected cell primes itself very effectively; unselected cells are primed by photons and other particles

FLAT-PANEL DISPLAYS

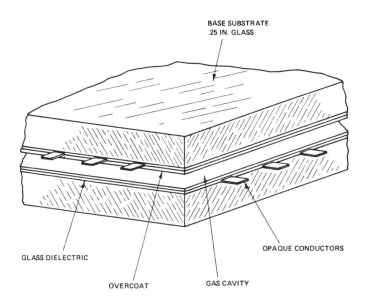

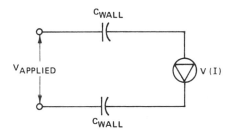

Figure 4.33 Construction and equivalent circuit of an ac gas-discharge display.

by "pilot" cells located along the boundary of the panel and hidden from the viewer (Michel, 1989).

Operating voltages must be carefully controlled. There is a distribution of firing and erase voltages across a panel, due to mechanical tolerances. An important operating parameter is the *memory margin*: the range of sustain voltages below which some cells cannot be fired and above which some cells will fire without being selected. Memory margins are currently of the order of 20 V.

Since the gas is below atmospheric pressure, the substrates must be held apart by spacers. For high-resolution panels, positioning of the spacers is a difficult problem; their presence distorts the electric field distribution and so alters firing

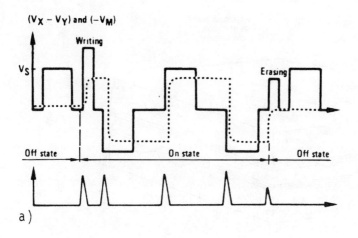
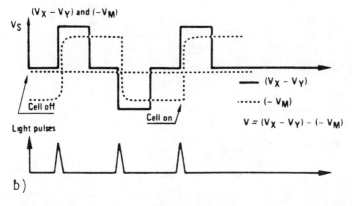

Figure 4.34 AC plasma-panel waveforms. (a) Write and erase voltages. (b) Steady-state voltages.

and erase voltages in their vicinity. This is one of a number of problems that have been attacked successfully over the years of development of this device. Spacers are placed carefully in locations at which the electrode patterns are altered to compensate for the presence of the spacers.

Performance of these devices is sensitive to extremely small levels of contamination; the manufacturing environment must be at the same level of cleanliness as is used in IC manufacturing (although electrode linewidths are measured in mils rather than in micrometers).

Currently, ac panels are used primarily for displays of high information content such as computer displays; they are no longer used for single-line alphanumeric displays. Sizes range from some 6 in. diagonal up to 59 in. diagonal. The largest monolithic flat-panel display now available is an ac plasma panel from Photonics Technology (Northwood, Ohio) with 2048 × 2048 resolution and dimensions of 41.8 in. square (1.5-m diagonal) and an area of 12.1 ft^2 (Wedding et al., 1987).

In addition to high panel-manufacturing costs, ac panels have been hurt by the high cost of drive circuitry. Only two United States-based semiconductor manufacturers, Supertex and Texas Instruments, make driver ICs, and their prices have remained higher than the display manufacturers would like. Plasmaco, a spinoff from IBM, has pushed on, mounting driver circuitry on the panel to reduce costs, and has also pioneered a circuit technique to reduce the number of drivers, at a slight cost in resolution (Warren and Weber, 1990). There has also been a continuing effort to reduce the power requirements of the display to make it more competitive with LCDs for portable and laptop applications (Weber and Wood, 1987). At present, the power consumption of an ac panel displaying text is comparable to that of a backlit LCD. However, when the display must show imagery or graphics, power consumption increases.

A variant form of operation, called *ac refresh*, runs the panel at much higher sustain frequencies—several hundred kilohertz. Memory margin is reduced to zero, but the higher sustain frequency produces correspondingly higher light output, even though the panel must now be scanned like any non-storage device, with resulting lowered duty factors. NEC of Japan used this approach; they scored a major success in selling over 100 large displays for a new trading floor at the New York Stock Exchange. However, they recently announced that they are withdrawing from this product activity, for reasons undisclosed in the press.

Gray Scale Because this device has storage, it has been difficult to provide gray scale. Two general approaches have been used. One is to trade spatial resolution for amplitude resolution, that is, for gray scale. This can be done simply by considering a group of pixels as a unit and turning members of this unit on and off in various combinations. The large Photonics panel mentioned above uses this technique to get a five-level gray scale with resolution reduced to 512 × 512. Methods of reducing the resulting patterning have been discussed above (Judice et al., 1974; Judice and Slusky, 1981; Baldwin, 1991).

Another approach is to turn pixels on and off for varying durations. This requires that the information be organized in frame times, so there is a maximum possible on duration. Several workers have done this at video rates, although with limited gray scale (Criscimagna et al., 1986; Lee et al., 1986).

Color Color is provided by using a xenon-rich gas to excite phosphors (Friedman et al., 1991). A continuing problem has been that bombardment by ions damages the phosphors; the conventional ac-panel construction does not provide a good way of separating the phosphor from ions produced in the discharge. Countermeasures have included coating the phosphor with a layer that transmits the far UV that xenon emits while resisting the passage of ions. Other approaches include laying down the phosphor in patterns that keep it away from the most intense ion bombardment.

A promising approach has been to use a *single-substrate* construction, in which both row and column electrodes are fabricated on one substrate by thick-film techniques. The phosphor can then be deposited on the other substrate, where few ions will be accelerated to strike it and cause damage (Gay et al., 1990; Uchiike et al., 1990; Doyeux et al., 1991). Thomson Tubes Electroniques displayed a 17-in. diagonal panel using this approach at the 1990 SID International Symposium; its luminance was about 50 fL, but its luminous efficacy was only 0.21 lm/W. Improvements in this performance, plus the addition of gray scale to increase the color gamut, have been forecast by the company; some were shown at the 1991 symposium. The single-substrate construction is still more expensive than the two-substrate method, and the capacitances that must be driven are also larger.

4.6.7 Miscellaneous Display Types

Electrochromic displays use changes in the color or reflectance of certain chemicals as their chemical state is changed. Essentially these are batteries that are charged or discharged by external sources. They have the potential of excellent viewing angle and contrast but limited color (Ando et al., 1986). Some work is going on, but most has been dropped because lifetime problems could not be solved. The Alpine Group of Hackensack, New Jersey, has announced work on what is apparently a variant of this technology, but few technical details have been published.

Electrophoretic displays use suspensions of microscopic particles of one color in fluid of a contrasting color. The particles are naturally charged. When exposed to an electric field, they migrate to the surface of the display, changing its color. Under an opposite charge, they move to the rear surface and the color of the liquid is revealed. As originally shown, they had excellent contrast and viewing angle but no switching nonlinearity. Switching nonlinearity could be added by incorporating a grid structure (Beilin et al., 1986). The problems have been difficult enough that work on this technology seems to have been abandoned everywhere.

Arrays of *magnetic balls or particles* could be addressed by the coincidence of row and column currents. Typically the balls were magnetized and black on one hemisphere and white or colored on the other. They could be rotated by the magnetic field and would remain in their last position until disturbed. It is not clear why this technique has not been pursued more energetically, but at present it seems to be quite dead.

Magnetic vanes or *flaps* are single-pixel devices carrying a small permanent magnet that are rotated into one of two positions by an associated coil and held in that position by a mechanical or magnetic latch. They have wide viewing angle, are generally direct-driven, and typically cost as much as $1/pixel. They are used on buses and trains and in large arrays at transportation terminals and stock exchanges.

Magnetic flaps carry complete messages, such as city names. They are very expensive but have found use in airports and train stations. Typically each line of the display will have an array of such flaps, only one side of which is visible at a time. They are accessed by what is essentially a small motor. They are not used for moving messages, but since they carry silk-screened characters, a choice of colors and styles is available, and the viewing angle is as good as that of a printed poster.

Very Large Panels

Notice boards in transportation terminals, group-viewing displays for command and control applications, and displays at sports stadia are examples of applications for displays measuring many feet in the diagonal. Although projection techniques can be used, especially if the ambient illumination can be controlled (as in theaters), the requirements for many of these applications mandate the use of flat panels. The panels are usually modular in construction, for reasonable cost in fabrication and transportation.

Liquid-crystal displays have been used for very large panels, generally with driving arrangements such that the duty factor for each pixel is close to unity. By this means reasonable viewing angle and contrast can be obtained; cost is high, but this is often not a principal limitation in these applications.

Matsushita, Mitsubishi, and Sony have used special CRTs for this application (Kamogawa et al., 1991; Sakaguchi et al., 1991). These are nonscanned, flood-gun devices, typically with a simple electron-emitting and control arrangement to excite a single dot of phosphor of one color at close to unity duty factor. Depending on the overall display size, 1–12 pixels may be housed within a single vacuum envelope. Displays as large as 1860 ft^2 have been constructed. The control circuitry takes video and distributes it to the drive circuitry, which is typically modular and mounted directly behind the individual display modules. Because high luminances and large sizes are required, power demands of tens of kilowatts are normal.

An alternative approach is to use vacuum-fluorescent technology for the modules (Xi et al., 1991). The technology is simpler because the accelerating voltage is lower (less than 1 kV compared to the 8–12 kV used by the CRT producers), but the luminance is also substantially lower.

Large modular EL displays have also been fabricated, but with substantially poorer performance.

4.7 CONCLUDING REMARKS

The display is the interface between human and machine. As electronic devices of all types become more pervasive, more competent, and more complex, improved display performance will be required. Displays will become smaller and less power-hungry for many applications, and larger in physical size and information-handling capability for others. It is unlikely that any single technology will dominate the display world, because there are so many different applications and requirements.

A major impetus to the development of display technology is work on high-definition TV (HDTV) and various methods of image storage and image compression. These techniques make sense only if the high-definition images being produced can be displayed. The development of high-performance displays will be spurred by the prospect of large-volume consumer markets. Indeed, the pervasiveness of color CRTs is due in large part to the spread of consumer color TV. We can expect a similar impetus from the development of HDTV, with the fruits of this development available for more technical applications.

The CRT will not be completely supplanted for many years; it is too capable and too mature a technology. However, there are areas, like laptop computers, where it cannot compete, and improved performance and lower cost in these areas will lead to increasing competition in market areas such as desktop computers where the CRT is now dominant.

Displays have been a fertile field for inventors. We can anticipate that new display devices will appear. The only question about the eventual obsolescence of this chapter is how soon it will occur.

References

Adachi, C., T. Tsutsui, and S. Saito (1989). Organic thin-film electroluminescent device, *Proc. 9th Int. Display Res. Conf. (Jpn. Display '89)*, 708–711.

Alt, P. M., and P. Pleshko (1974). Scanning limitations of liquid-crystal displays, *IEEE Trans. Electron Devices ED-21*: 146–155.

Amano, Y. (1975). A flat-panel TV display system in monochrome and color, *IEEE Trans. Electron Devices ED-22*: 1–7.

Anderson, B. C., and V. J. Fowler (1974). AC plasma panel TV display with 64 discrete intensity levels, *SID Int. Symp. Dig. Tech. Papers V*: 28–29.

Ando, E., K. Matsuhiro, and Y. Masuda (1986). Large-area dot format electrochromic display, *SID Int. Symp. Dig. Tech. Papers XVII*: 132–135.

Awaji, N., J. Endo, Y. Amano, H. Yamamoto, and H. Uchiike (1986). Characteristics of discharge in a dc-plasma display panel with trigger electrodes, *SID Int. Symp. Dig. Tech. Papers XVII*: 391–394.

Baldwin, W. A. (1991). A comparison of halftone techniques, *SID Int. Symp. Dig. Tech. Papers XXII*: 517–520.

Beilin, S., D. Zwemer, and R. Kulkarni (1986). 2000-Character electrophoretic display, *SID Int. Symp. Dig. Tech. Papers XVII*: 136–140.

Bosman, D., Ed. (1989). *Display Engineering: Conditioning, Technologies, Applications*, North-Holland, Amsterdam.

Brodie, I. (1979). Microcathode field emission arrays for CRT application, *SRI Res. Brief*, No. 26R.

Buzak, T. S. (1990a). A new active-matrix technique using plasma addressing, *SID Int. Symp. Dig. Tech. Papers XXI*: 420–423.

Buzak, T. S. (1990b). Switching pixels with gas, *Inf. Display 6*: 7–9, 14 (Oct. 1990).

Channing, D., T. Theroux, and A. Wolf (1989). EL addressing technology, *SID Int. Symp. Dig. Tech. Papers XX*: 54–57.

Conner, A. R. (1990). Subtractive color STN-LCD display, *Proc. 10th Int. Display Res. Conf. (Eurodisplay '90)*, pp. 362–365.

Criscimagna, T. N., H. S. Hoffman, and W. R. Knecht (1986). Enhancement of write/erase speeds for ac-plasma panels, *SID Int. Symp. Dig. Tech. Papers XVII*: 395–398.

DeJule, M., C. S. Stone, A. Sobel, and J. Markin (1986). A four-square-foot monochrome flatscreen display, *SID Int. Symp. Dig. Tech. Papers XVII*: 410–413.

Doane, J. W., J. L. West, J. B. Whitehead, Jr., and D. S. Fredley (1990). Wide-angle-view PDLC displays, *SID Int. Symp. Dig. Tech. Papers XXI*: 224–226.

Doyeux, H., G. Baret, J. Deschamps, O. Hamon, S. Salavin, and P. Zorzan (1991). A 23-in. color ac plasma display, *SID Int. Symp. Dig. Tech. Papers XXII*: 721–723.

Drzaic, P. S., R. Wiley, J. McCoy, and A. Guillaume (1990). High-brightness reflective displays using nematic droplet/polymer films, *SID Int. Symp. Dig. Tech. Papers XXI*: 210–213.

Emoto, F., K. Senda, E. Fujii, A. Nakamura, A. Yamamoto, Y. Uemoto, K. Kobayashi, M. Kyougoku, T. Kamimura, and G. Kano (1989). A 0.92-in. active-matrix LCD with fully integrated poly-Si TFT drivers of new circuit configuration, *Proc. 9th Int. Display Res. Conf. (Jpn. Display '89)*, pp. 152–154.

Farrell, J. E. (1987). Predicting flicker thresholds for visual displays, *SID Int. Symp. Dig. Tech. Papers XVIII*: 18–21.

Farrell, R. J., and J. M. Booth (1984). *Design Handbook for Imagery Interpretation Equipment*, Boeing Aerospace Company, Seattle, WA, Section 5.2, Color, especially pp. 5.2–22–5.2–23.

Friedman, P. S., A. Rahman, R. A. Stoller, and D. K. Wedding (1991). A 17-in.-diagonal full-color ac plasma video monitor with 64 gray levels, *SID Int. Symp. Dig. Tech. Papers XXII*: 717–720.

Gay, M., S. Salavin, and J. Deschamps (1990). A 17-in. 8-color ac plasma display panel with simplified structure, *SID Int. Symp. Dig. Tech. Papers XXI*: 477–480.

Goede, W. F. (1973). A digitally addressed flat-panel CRT, *IEEE Trans. Electron Devices ED-20*: 1052–1062.

Goede, W. F. (1985). Flat cathode-ray tube displays, in *Flat-Panel Displays and CRTs* (L. E. Tannas, Jr., Ed.), Van Nostrand Reinhold, New York, pp. 202–207.

Goldowsky, M. (1990). Economical color filter fabrication for LCDs by electro-mist deposition, *SID Int. Symp. Dig. Tech. Papers XXI*: 80–83.
Gulick, P. E., and A. R. Conner (1991). Stacked STN LCD's for true color projection systems, *Proc. SPIE Conf. Large Screen Projection, Avionic, and Helmet-Mounted Displays*, 1456.
Gurman, B. (1989). Electroluminescent displays, in *Display Engineering* (D. Bosman, Ed.), North-Holland, Amsterdam, pp. 221–224.
Haven, T. J. (1991). Reinventing the color wheel, *Inf. Display* 7: 11–15.
Howard, W. E. (1981). Electroluminescent display technologies and their characteristics, *Proc. SID 22*: 47–56.
Hughes, A. J. (1989). Liquid crystal displays, in *Display Engineering: Conditioning, Technologies, Applications* (D. Bosman, Ed.), North-Holland, Amsterdam, p. 167.
Ichikawa, K., S. Suzuki, H. Matino, T. Aoki, T. Higuchi, and Y. Oana (1989). 14.3-in.-diagonal 16-color TFT-LCD panel using a-Si:H TFTs, *SID Int. Symp. Dig. Tech. Papers XX*: 226–229.
Inoguchi, T., M. Takeda, Y. Kakihara, Y. Nakata, and M. Yoshida (1974). Stable high-brightness thin-film electroluminescent panels, *SID Int. Symp. Dig. Tech. Papers V*: 84–85.
Ishiko, M., K. Utsugi, K. Nunomura, S. Takano, and C. Tani (1989). Matrix-addressed organic thin film EL display panel, *Proc. 9th Int. Display Res. Conf. (Jpn. Display '89)*, pp. 704–707.
Judice, C. N., and R. D. Slusky (1981). Processing images for bilevel digital displays, *Adv. Image Pickup Display 4*: (B. Kazan, ed), Academic Press, New York, 157–229.
Judice, C. N., J. F. Jarvis, and W. H. Ninke (1974). Bi-level rendition of continuous-tone pictures on an ac plasma panel, *Conf. Record, 1974 Conf. Display Dev. and Syst.*, pp. 89–98.
Kamogawa, H., K. Tatsuda, Y. Seko, S. Uemura, T. Shimogo, K. Shibayama, Z. Hara, Y. Kani, and S. Iwata (1991). A lighting element for high-resolution large-screen video displays, *SID Int. Symp. Dig. Tech. Papers XXII*: 573–576.
Kmetz, A. R. (1973). Liquid-crystal display prospects in perspective. *IEEE Trans. Electron Devices ED-20*: 954–961.
Kmetz, A. R. (1975). Matrix addressing of non-emissive displays, in *Nonemissive Electrooptic Displays* (A. R. Kmetz and F. K. von Willisen, Eds.), Plenum, New York, p. 272.
Kobayashi, I., T. Tamura, M. Uno, K. Adachi, Y. Bessho, S. Nakamura, M. Takeda, and S. Hotta (1990). 2.8″ defect-free a-Si TFT-LCD module using stud-bump-bonding COG for projection ED-TV, *Proc. 10th Int. Display Res. Conf. (Eurodisplay '90)*, pp. 48–51.
Latham, W. J., T. L. Brewer, D. W. Hawley, J. E. Lamb III, and L. K. Stichnote (1987). A new class of color filters for liquid-crystal displays, *SID Int. Symp. Dig. Tech. Papers XVIII*: 379–382.
Lauer, H.-U., E. Lueder, M. Dobler, K. Schleupen, J. Spachmann, T. Kallfass, P. Jones, and B. Macknick (1990). A frame-sequential color-TV projection display, *SID Int. Symp. Dig. Tech. Papers XXI*: 534–537.
Lee, J. Y., M. J. Marentic, J. R. Moore, and L. F. Weber (1986). High-speed asynchron-

ous video addressing of ac-plasma display incorporating brightness control, *SID Int. Symp. Dig. Tech. Papers XVII*: 399–402.

Lehmann, W. (1966). Hyper-maintenance of electroluminescence, *J. Electrochem. Soc. 113*: 40.

Leroux, T., A. Ghis, R. Meyer, and D. Sarrasin (1991). Microtips displays addressing, *SID Int. Symp. Dig. Tech. Papers XXII*: 437–439.

Livingston, D. C. (1956). Electroluminescent television panel, U.S. Patent 2,774,813 (Dec. 18, 1956).

Maurer, R., D. Andrejewski, F.-H. Kreuzer, and A. Miller (1990). Polarizing color filters made from cholesteric LC silicones, *SID Int. Symp. Dig. Tech. Papers XXI*: 110–113.

Meyer, R. (1990). 6″ diagonal microtips fluorescent display for T.V. applications, *Proc. 10th Int. Display Res. Conf. (Eurodisplay '90)*, pp. 374–377.

Michel, J. P. (1989). Large area gas discharge displays or plasma displays, in *Display Engineering: Conditioning, Technologies, Applications* (D. Bosman, Ed.), North-Holland, Amsterdam, p. 203.

Mikoshiba, S., S. Shinada, A. Kohgami, M. Suzuki, and F. L. Curzon (1990). High-speed addressing of a Townsend-discharge panel TV display using predischarges, *SID Int. Symp. Dig. Tech. Papers XXI*: 474–476.

Mito, S., C. Suzuki, Y. Kanatani, and M. Ise (1974). TV imaging system using electroluminescent panels, *SID Int. Symp. Dig. Tech. Papers V*: 86–87.

Morimoto, K. (1982). A high resolution graphic display, *SID Int. Symp. Dig. Tech. Papers XIII*: 218–219.

Murakami, H., R. Kaneko, M. Seki, T. Yamamoto, T. Kuriyama, T. Katoh, T. Takahata, H. Ohnishi, and M. Tsuji (1990). Multi-color picture display with a 33-inch gas-discharge pulse memory panel, *Proc. 10th Int. Display Res. Conf.*, pp. 30–33.

Nagasawa, S. and H. Watanabe (1988). A 640 × 400 graphic VFD with 16 gray levels, *SID Int. Symp. Dig. Tech. Papers XIX*: 301–304.

Nehring, J., and A. R. Kmetz (1979). Ultimate limits for matrix addressing of rms-responding liquid-crystal displays, *IEEE Trans. Electron Devices ED-26*: 795–802.

Nonomura, K., F. Yamazaki, J. Hashiguchi, M. Takahashi, K. Hamada, T. Nakatani, S. Kitao, T. Shiratori, T. Kataoka, K. Tomii, H. Miyama, K. Yoshikazu, and J. Nishida (1989). A 40-in. matrix-driven high-definition flat-panel CRT, *SID Int. Symp. Dig. Tech. Papers XX*: 106–109.

Pleshko, P. (1980). AC plasma display device technology: an overview, *Proc. SID 21*(2): 93–100.

Pleshko, P. (1990). Halftone gray scale for matrix-addressed displays, *Inf. Display 6*: 10–11 (Oct. 1990).

Rice, J. F. (1991). Ten rules for color coding, *Inf. Display 7*: 12–14 (Mar. 1991).

Sakaguchi, Y., T. Ohki, Y. Namikoshi, T. Ozone, and W. Ogawa (1991). Large-area color display "Skypix," *SID Int. Symp. Dig. Tech. Papers XXII*: 577–579.

Sarma, K. R., H. Franklin, M. Johnson, K. Frost, and A. Bernot (1989). Active-matrix LCDs using gray-scale in halftone methods, *SID Int. Symp. Dig. Tech. Papers XX*: 148–150.

Schauer, A. (1982). Plasma panel lights up 14-in. flat-panel display, *Electronics 55*: 128–130 (Dec. 15, 1982).

Schmachtenberg, R., T. Jenness, M. Ziuchkovski, and T. Flegal (1989). A large-area 1024 × 864 line ACTFEL display, *SID Int. Symp. Dig. Tech. Papers XX*: 58–60.

Scott, W. C., W. C. Holton, W. G. Manns, D. F. Weirauch, M. R. Namordi, F. Doerbeck, and J. E. Gunther (1978). Flat cathode-ray-tube display, *SID Int. Symp. Dig. Tech. Papers IX*: 88–89.

Sherr, S. (1972). Three axis matrix display addressing techniques, *IEEE Conf. Rec. 1972 Conf. Display Devices*, pp. 32–45.

Sigel, C. (1989). CRT refresh rate and perceived flicker, *SID Int. Symp. Dig. Tech. Papers XX*: 300–302.

Silverstein, L. D., R. W. Monty, F. E. Gomer, and Y.-Y. Yeh (1989). A psychophysical evaluation of pixel mosaics and gray-scale requirements for color matrix displays, *SID Int. Symp. Dig. Tech. Papers XX*: 128–131.

Simpson, G. S. (1985). Micro-electro-mechanical displays with capacitance latching switches at each pixel, Paper distributed by Micro-Curl Display Technology, Inc.

Snyder, H. L. (1985). The visual system: capabilities and limitations, in *Flat-Panel Displays and CRTs* (L. E. Tannas, Jr., Ed.), Van Nostrand Reinhold, New York, pp. 54–69, especially pp. 61–64.

Sobel, A. (1970). Selection limits in matrix displays, *Conf. Rec., 1970 IEEE Conf. on Display Devices*, New York, pp. 74–84.

Sobel, A. (1971). Some constraints on the operation of matrix displays, *IEEE Trans. Electron Devices ED-18*: 797–798.

Steiner, S. A., and H. Y. Tsoi (1988). High-performance column driver for gray-scale TFEL displays, *SID Int. Symp. Dig. Tech. Papers XIX*: 31–34.

Stewart, R. G. (1991). Self-scanned active-matrix liquid-crystal displays, *SID Int. Symp. Dig. Tech. Papers XXII*: 530–534.

Suzuki, M., S. Mikoshiba, S. Shinada, A. Kohgami, and F. L. Curzon (1990). Discharge cross-talk in a Townsend discharge panel TV, *Proc. 10th Int. Display Res. Conf.*, pp. 80–83.

Takahashi, Y., T. Nomura, S. Kohda, and T. Kawada (1987). A new overhead projection system using a defectless 640 × 400 pixel active-matrix LCD together with an input pad, *SID Int. Symp. Dig. Tech. Papers XVII*: 79–81.

Tannas, L. E., Jr., Ed. (1985a). *Flat-Panel Displays and CRTs*, Van Nostrand Reinhold, New York.

Tannas, L. E., Jr., (1985b). Electroluminescent displays, in *Flat-Panel Displays and CRTs* (L. E. Tannas, Jr., Ed.), Van Nostrand Reinhold, New York, pp. 237–288.

Teggatz, R. E. (1989). A power-efficient 32-bit electroluminescent display column driver, *SID Int. Symp. Dig. Tech. Papers XX*: 68–70.

Tektronix, Inc. (1990). RGB liquid crystal shutter display, Prelim. Spec. (5/9/90), Tektronix Display Products, Beaverton, OR.

Tongyu, L., M. Xianxin, and Z. Zhonghou (1986). A large-area matrix-addressed ACEL display device, *SID Int. Symp. Dig. Tech. Papers XVII*: 254–256.

Uchiike, H., A. Kubo, S. Harada, T. Kanehara, S. Hirata, and Y. Fukushima (1990). Very-fine-resolution 170-lines/in. color ac surface-discharge plasma displays, *SID Int. Symp. Dig. Tech. Papers XXI*: 481–484.

Vecht, A. (1990). AC and dc electroluminescent displays, *SID Seminar Lecture Notes II*: F-2.2–F-2.48.

Warren, K. W., and L. F. Weber (1990). Increased address rate for the independent sustain and address ac plasma display, *SID Int. Symp. Dig. Tech. Papers XXI*: 489–492.

Washington, D., J. R. Mansell, D. L. Lamport, A. G. Knapp, and A. W. Woodhead (1985). Progress of the flat channel multiplier CRT, *SID Int. Symp. Dig. Tech. Papers XVI*: 166–169.

Weber, L. F. (1985). Plasma displays, in *Flat-Panel Displays and CRTs* (L. E. Tannas, Jr., Ed.), Van Nostrand Reinhold, New York, pp. 332–414.

Weber, L. F., and M. B. Wood (1987). Energy recovery sustain circuit for the ac plasma display, *SID Int. Symp. Dig. Tech. Papers XVIII*: 92–95.

Wedding, D. K., Sr., P. S. Friedman, T. J. Soper, T. D. Holloway, and C. D. Reuter (1987). A 1.5-m-diagonal ac gas discharge display, *SID Int. Symp. Dig. Tech. Papers XVIII*: 96–99.

Welsh, L., and L. White (1990). NCAP displays: optical switching and dielectric properties, *SID Int. Symp. Dig. Tech. Papers XXI*: 220–223.

Xi, H., G. Shichao, J. Weichen, and L. Huabin (1991). Flat full-color pixel display panel and ultra-large-screen video display, *SID Int. Symp. Dig. Tech. Papers XXII*: 571–572.

Yoshimura, M., K. Fujii, S. Tanake, S. Vemura, and M. Horic (1986). High-resolution VFD on-a-chip, *SID Int. Symp. Dig. Tech. Papers XVII:*$ 403–406.

Yaniv, Z., *et al.* (1988). *Proc. 1988 Int. Display Res. Conf.*, 152.

5
Color Image Display with Black-and-White Film

Guo-Guang Mu, Zhi-Liang Fang, and Xu-Ming Wang

*Nankai University,
Tianjin, China*

Yu-Guang Kuang

*Academia Sinica,
Changehun, China*

5.1 INTRODUCTION

A large amount of color information must be stored for display in the future. As a permanent storage material, the multilayered color film made of organic dyes is not an ideal candidate. Many valuable color films suffer fading, for the recording organic dye material is unstable and color information fades gradually as time passes. On the other hand, color information recorded on black-and-white silver halide films can be stored for years without fading. Historically, many efforts have been devoted to the recording and storage of color information with black-and-white film. An attempt was made to record the color object on a black-and-white film through a close-contact transparent colored screen. To display the color image, the same screen is placed in perfect unison with the recorded film. Another common technique involves the repetitive application of primary color filters so that the color information can be preserved on three separate rolls of black-and-white film. Three projectors with proper primary color filters project the primary color images, simultaneously and coinciding with each other perfectly, onto a screen for color image display, or on a fresh color film. A color image that has been stored for a long time on black-and-white film can thus be displayed in its original colors. But these techniques encounter certain drawbacks, one of which is that the color display system is too complicated and its operation is rather elaborate. Highly precise adjustment is needed.

White-light image processing provides a simple and economical way to store and display color images. Mueller (1969) proposed a method for encoding a color scene on a black-and-white film with a single exposure by using a three-color grid screen. For optical demonstration, he used a sequential technique to simulate and encode a stationary scene on a black-and-white film. Three monochromatic sources were used for color decoding and display.

To solve the problem of archival storage of color film, Yu (1980) described a technique for spatial color encoding and white-light color image retrieval. He recorded color images on black-and-white film with sequential encodings by a Ronchi ruling. Similar works have been presented by Macovski (1972) and Grousson and Kinany (1978). Fang et al. (1984) presented a technique for archival storage of color films through a three-primary-color grating formed by interference between diffraction orders of a black-and-white orthogonal Ronchi ruling in a white-light processor.

Color image storage and display can also be accomplished by other modulators, among them the oriented speckle screen (Mu et al., 1985, 1987), image hologram (Mu et al., 1983b), and contact screen (Chiang et al., 1984).

Mu et al. (1983a) designed and fabricated a special modulator (tricolor grating encoder). With this total optical color modulator, color encoding is fulfilled in one photographic step. There is no need for post-photo-taking sequential encoding. During exposure, the tricolor grating is contacted with a panchromatic black-and-white film, and the color scene is dispersed and encoded onto the film simultaneously in a single exposure. The photochemical development process yields an encoded black-and-white transparency. When the transparency is placed in the input plane of a white-light processor, and its three first orders of spectra are filtered through the corresponding three primary color filters at the Fourier plane, the original color image is displayed at the output plane. This technique has two major advantages:

1. A single exposure takes the place of sequential encoding, and real-time color photography with black-and-white film becomes possible.
2. The positive encoded transparency can be directly obtained; a two-step encoding process (first negative, then positive) may be unnecessary.

This technique may be used not only for storage of color images on black-and-white films, but also for recording color scenes with black-and-white film with an ordinary camera; the color image is then displayed in a color image display system. A white-light process can also provide a technique to accomplish color enhancement for faded color film (Yu et al., 1981) if the fading is not severe. In this chapter, we place emphasis on color image display with black-and-white film utilizing a tricolor grating.

5.2 PROPERTIES OF A WHITE-LIGHT OPTICAL INFORMATION PROCESSOR

Let us first introduce some basic principles concerning Fourier optics. A Fourier transform can be performed optically by the use of converging lenses with coherent light illumination. Information processing is carried out by frequency domain filtering. This concept can be extended to collimated white-light illumination.

5.2.1 Optical Fourier Transform Performed by a Converging Lens

A converging lens is able to perform two-dimensional Fourier transform (parallel processing). The phase transformation of a converging lens (Goodman, 1968) can be simply expressed as

$$t_l(u,v) = \exp\left[-\frac{ik}{2f}(u^2 + v^2)\right] \tag{5.1}$$

where k is the wavenumber of the illumination, f is the focal length of the converging lens, and (u,v) is the coordinate system of the lens.

Let a transparency with amplitude transmittance $t(x,y)$ be placed in the front focal plane P_1 of the converging lens L_1 as shown in Figure 5.1, where a normal monochromatic plane wave of unit amplitude is applied to illuminate the transparency. The field distribution at the back focal plane P_2 of the lens is given according to the Fresnel diffraction integration by

$$\begin{aligned}T(\alpha,\beta) &= K \int \left[\int t(x,y) \exp\left\{\frac{ik}{2f}[(\alpha-x)^2 + (\beta-y)^2]\right\} dx\, dy\right] \\ &\quad \times t_l(u,v) \exp\left\{\frac{ik}{2f}[(\alpha-u)^2 + (\beta-v)^2]\right\} du\, dv \\ &= K' \int t(x,y) \exp\left[\frac{-ik}{f}(x\alpha + y\beta)\right] dx\, dy \end{aligned} \tag{5.2}$$

where (α,β) denotes the spatial coordinate system of P_2, and K and K' are proportional constants. Equation (5.2) is the two-dimensional Fourier transform of $t(x,y)$. Denoting $p = \alpha k/f$ and $q = \beta k/f$ as the angular frequency coordinates, the above equation can be rewritten as

$$T(p,q) = \int t(x,y) \exp[-i(xp + yq)]\, dx\, dy \tag{5.3}$$

where the proportional constant is ignored for simplicity. When the input transparency is placed in any position in front of the transforming lens, the Fourier transform property is kept, but an additional quadratic phase factor is added (Thompson, 1978).

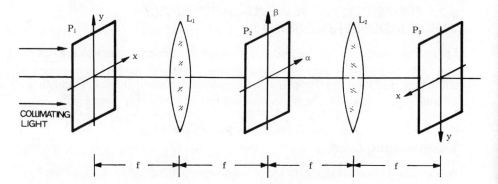

Figure 5.1 Schematic representation of an optical information processor.

5.2.2 Frequency Domain Filtering

Suppose a filter function $F(p,q)$ is inserted at the frequency plane P_2. The output distribution at plane P_3 is given by

$$E(x,y) = t(x,y) * f(x,y) \tag{5.4}$$

where * denotes convolution, and $f(x,y)$ is the original reference object, whose Fourier transform is $F(p,q)$.

Assume that a Ronchi ruling is superimposed on the input object; the Fourier spectrum would then be

$$\begin{aligned} T_M(p,q) &= \int t(x,y) \left[\frac{1}{2} + \frac{1}{2}\mathrm{sgn}(\cos p_0 x)\right] \exp[-i(px+qy)] \, dx \, dy \\ &= \frac{1}{2} T(p,q) + \frac{1}{4} \sum_{n=1}^{\infty} a_n T(p \pm np_0, q) \end{aligned} \tag{5.5}$$

where

$$\mathrm{sgn}(\cos p_0 x) = \begin{bmatrix} 1, & \cos(p_0 x) \geq 0 \\ -1, & \cos(p_0 x) < 0 \end{bmatrix}$$

p_0 is the spatial frequency of the ruling, and a_ns are Fourier coefficients. This expression describes a multiorder diffraction. Each order contains the Fourier spectrum of the input object. Multiple filtering is capably carried out on different orders for multipurpose information processing. Furthermore, parallel multiobject processing is possible, if a prior multiexposure recording is fulfilled with each exposure being made through a different object contact with a Ronchi ruling at a different azimuth position. The Fourier spectrum of this encoded transparency is

COLOR DISPLAY WITH BLACK-AND-WHITE FILM

$$T_M(p,q) = \frac{1}{2}\sum_{i=1}^{N} T(p^i,q^i) + \frac{1}{4}\sum_{i=1}^{N}\sum_{n=1}^{\infty} a_n T(p^i \pm np_0, q^i) \quad (5.6)$$

where N is the number of encoded objects and (p^i, q^i) denotes the angular frequency coordinate system for the ith encoded object. The ith object's diffraction patterns are arranged in p^i orientation.

5.2.3 Optical Information Processor with White-Light Illumination

If the illuminating monochromatic light is substituted by a collimated white light, the field for light of wavelength λ at the Fourier plane P_2 is

$$T_M(p,q;\lambda) = \frac{1}{2}\sum_{i=1}^{N} T(p^i,q^i;\lambda) + \frac{1}{4}\sum_{i=1}^{N}\sum_{n=1}^{\infty} a_n T(p^i \pm np_0, q^i;\lambda) \quad (5.7)$$

Each diffraction order contains the Fourier spectrum of the input object modulated on all the visible wavelengths. So the processor is inherently suitable for color image processing. The different objects described above may be the color components of a color image.

To perform frequency domain synthesis, the exact Fourier transform is needed. But that is not the case for color filtering. For the purpose of color filtering is to prevent the passage of all colors other than those of the filter's permitted wavelengths. No phase match or compensation is applied. The input-encoded transparency can be placed anywhere in front of the transform lens provided the output plane is adjusted accordingly.

5.3 COLOR IMAGE ENCODING WITH A TRICOLOR GRATING

First in this section, let us describe the encoding technique by which a color scene is spatially encoded onto a black-and-white transparency in a single exposure. To record a color image on a black-and-white film, color components should be distinguished by their representations in black-and-white forms. The colors corresponding to these representations are projected by the use of a tricolor grating, a totol optical color modulator (TOCM).

5.3.1 Total Optical Color Modulator

Figure 5.2 schematically shows a typical total optical color modulator (TOCM), the tricolor grating, which combines the three additive primary color gratings. These gratings are red with black rulings at the x azimuth position, green with black at the x' azimuth position, and blue with black at the x'' position. The intensity transmittance of a TOCM is given by

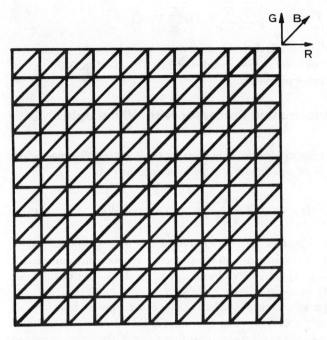

Figure 5.2 Schematic drawing of the total optical color modulator.

$$T_T(x,y) = \left[\frac{1}{2} + \frac{1}{2}\text{sgn}(\cos p_0 x)\right]_R + \left[\frac{1}{2} + \frac{1}{2}\text{sgn}(\cos p_0 x')\right]_G \qquad (5.8)$$
$$+ \left[\frac{1}{2} + \frac{1}{2}\text{sgn}(\cos p_0 x'')\right]_B$$

where the subscripts R, G, and B denote the three primary colors the transmittances stand for; (x,y), (x',y'), and (x'',y'') are the coordinate systems with respect to the orientations of the different primary color gratings; and (x',y') and (x'',y'') can be obtained by applying linear coordinate transforms on (x,y).

A tricolor grating can be fabricated in the laboratory as follows. A fresh color slide film is superimposed with a close-contact Ronchi ruling of angular spatial frequency p_0. The recording is made with sequential exposures of the fresh slide film to three primary collimated light illuminations (or white-light illumination of color temperature about 5500° K with sequential primary color filtering). Between exposures, the Ronchi ruling is rotated a certain angle, say 45°, as shown in the example pictures of this chapter. That is, expose the fresh film with the superimposed Ronchi ruling to red illumination while the grating is at the 0° azimuth position for the first time. Then orient the Ronchi ruling to a new azimuth position of 45° and expose the slide film to blue collimated illumination.

COLOR DISPLAY WITH BLACK-AND-WHITE FILM

The third exposure, to green illumination, takes place after the ruling has been rotated by another 45° to a position of 90°. The developing and corresponding fixing processes, in which the gamma of the slide film is maintained at unity, yield a color transparency whose intensity transmittance can be described by Eq. (5.8). Proper design of the tricolor grating will greatly eliminate the appearance of moiré fringes.

5.3.2 Encoding of Color Information

The encoding of a color image is performed by a single recording on a black-and-white photographic film that has been superimposed with a tricolor grating. This process can be carried out using an ordinary camera or other instrument to take the photograph. The optical scheme of color encoding is shown in Figure 5.3. A tricolor grating is placed at the imaging plane of the imaging lens L, in contact with a panchromatic black-and-white film. The color scene is imaged by the lens onto the black-and-white film through the modulation of the tricolor grating. Denote the irradiance of the color image as

$$T(x,y) = [T_r(x,y)]_R + [T_g(x,y)]_G + [T_b(x,y)]_B \tag{5.9}$$

where T_r, T_g, and T_b are the red, green, and blue primary color components of the color image irradiance, and subscripts R, G, and B denote the display colors. For simplicity of illustration, we express the green and blue components in coordinate systems (x',y') and (x'',y''), respectively. That is,

$$T(x,y) = [T_r(x,y)]_R + [T_g(x',y')]_G + [T_b(x'',y'')]_B \tag{5.10}$$

The intensity distribution on the panchromatic black-and-white film is expressed by the product of $T(x,y)$ and $T_T(x,y)$. For each primary color component of the color scene, the other primary color gratings seem to contain opaque parts only; the green and blue transparent parts do not allow the passage of the red component, for example. For the red component of the input color scene, the tricolor

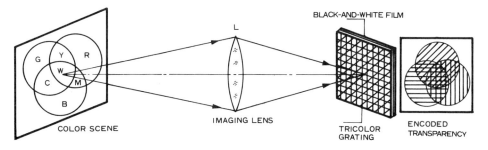

Figure 5.3 Optical configuration for the encoding of a color scene.

grating behaves as a Ronchi-type grating at an azimuth position corresponding to the coordinate system (x,y). For the green component, the tricolor grating acts as a Ronchi-type grating at an azimuth position corresponding to the coordinate system (x',y'). And the blue component reaches the photographic film as if it passed through a grating at a position corresponding to the (x'',y'') coordinate system. Properly designing the relationship of the three primary gratings of the TOCM modulator will ease the encoding process. There is no need to balance the color exposures as in the multiexposure encoding process. After exposure and development, a negative encoded transparency is obtained whose intensity transmittance is expressed as

$$\begin{aligned} T_n(x,y) &= [T(x,y)T_T(x,y)]^{-\gamma_{n1}} \\ &= \left\{ T_r(x,y) \left[\frac{1}{2} + \frac{1}{2} \mathrm{sgn}(\cos p_0 x) \right] \right. \\ &\quad + T_g(x',y') \left[\frac{1}{2} + \frac{1}{2} \mathrm{sgn}(\cos p_0 x') \right] \\ &\quad \left. + T_b(x'',y'') \left[\frac{1}{2} + \frac{1}{2} \mathrm{sgn}(\cos p_0 x'') \right] \right\}^{-\gamma_{n1}} \end{aligned} \qquad (5.11)$$

where γ_{n1} is the gamma of this negative film. For simplicity of illustration, we omit the proportional constant hereafter. This is the same expression reported by Yu (1980, 1983) with a triple exposure encoding process. This tricolor grating enables the color encoding to be performed in a single exposure.

By contact printing on another black-and-white film with a gamma of γ_{n2}, a positive encoding transparency is obtained whose intensity transmittance is given by

$$\begin{aligned} T_p(x,y) &= [T_n(x,y)]^{-\lambda_{n2}} \\ &= \left\{ T_r(x,y) \left[\frac{1}{2} + \frac{1}{2} \mathrm{sgn}(\cos p_0 x) \right] \right. \\ &\quad + T_g(x',y') \left[\frac{1}{2} + \frac{1}{2} \mathrm{sgn}(\cos p_0 x') \right] \\ &\quad \left. + T_b(x'',y'') \left[\frac{1}{2} + \frac{1}{2} \mathrm{sgn}(\cos p_0 x'') \right] \right\}^{\gamma_{n1}\gamma_{n2}} \end{aligned} \qquad (5.12)$$

The films used in negative encoding and positive printing are chosen such that the product of their gammas is equal to 2. Then the intensity transmittance of the positive encoded transparency is given by

$$\begin{aligned} T_p(x,y) &= \left\{ T_r(x,y) \left[\frac{1}{2} + \frac{1}{2} \mathrm{sgn}(\cos p_0 x) \right] \right. \\ &\quad + T_g(x',y') \left[\frac{1}{2} + \frac{1}{2} \mathrm{sgn}(\cos p_0 x') \right] \end{aligned}$$

COLOR DISPLAY WITH BLACK-AND-WHITE FILM

$$+ \left. T_b(x'',y'')\left[\frac{1}{2} + \frac{1}{2}\text{sgn}(\cos p_0 x'')\right]\right\}^2 \tag{5.13}$$

We have the amplitude transmittance of this encoded transparency expressed as

$$t_p(x,y) = T_r(x,y)\left[\frac{1}{2} + \frac{1}{2}\text{sgn}(\cos p_0 x)\right]$$
$$+ T_g(x',y')\left[\frac{1}{2} + \frac{1}{2}\text{sgn}(\cos p_0 x'')\right]$$
$$+ T_b(x'',y'')\left[\frac{1}{2} + \frac{1}{2}\text{sgn}(\cos p_0 x'')\right] \tag{5.14}$$

This is the encoded positive transparency obtained with a single exposure with a TOCM modulator. Primary color components are modulated by gratings of different azimuth orientations. The rightmost part of Figure 5.3 shows a schematic example.

This two-step color encoding process can be combined and replaced by a one-step encoding. Applying a different developing process to the encoded film recorded with the use of a tricolor grating, a positive black-and-white encoded transparency is obtained whose intensity transmittance is described as

$$T_p(x,y) = \left\{T_r(x,y)\left[\frac{1}{2} + \frac{1}{2}\text{sgn}(\cos p_0 x)\right]\right.$$
$$+ T_g(x',y')\left[\frac{1}{2} + \frac{1}{2}\text{sgn}(\cos p_0 x')\right]$$
$$+ \left.T_b(x'',y'')\left[\frac{1}{2} + \frac{1}{2}\text{sgn}(\cos p_0 x'')\right]\right\}^\gamma \tag{5.15}$$

where γ is determined by the particular film used and the photochemical process. If we control development conditions so that γ is equal to 2, an encoded positive transparency is obtained directly. The positive linear amplitude transmittance of this encoded black-and-white transparency is given by

$$t_p(x,y) = T_r(x,y)\left[\frac{1}{2} + \frac{1}{2}\text{sgn}(\cos p_0 x)\right]$$
$$+ T_g(x',y')\left[\frac{1}{2} + \frac{1}{2}\text{sgn}(\cos p_0 x'')\right]$$
$$+ T_b(x'',y'')\left[\frac{1}{2} + \frac{1}{2}\text{sgn}(\cos p_0 x'')\right] \tag{5.16}$$

This is the same expression as Eq. (5.14). The second contact printing of the two-step encoding process is omitted.

Another way to have one-step color encoding is to have the recording in the linear region of the T–E curve (Chao et al., 1980) of the used fresh photographic film, where the developed film's amplitude transmittance can be described as

$$t(x,y) = a - bE(x,y) \tag{5.17}$$

where a and b denote proper constants determined by the film used, and $E(x,y)$ is the exposure. Then we have the amplitude transmittance

$$\begin{aligned} t(x,y) = a - b\bigg\{ & T_r(x,y)\bigg[\frac{1}{2} + \frac{1}{2}\text{sgn}(\cos p_0 x)\bigg] \\ + & T_g(x',y')\bigg[\frac{1}{2} + \frac{1}{2}\text{sgn}(\cos p_0 x')\bigg] \\ + & T_b(x'',y'')\bigg[\frac{1}{2} + \frac{1}{2}\text{sgn}(\cos p_0 x'')\bigg]\bigg\} \end{aligned} \tag{5.18}$$

From the view of Fourier filtering, this is equivalent to Eq. (5.14).

Whether a positive or a negative encoding transparency is used, the displayed output image formed by the first diffraction order of its Fourier spectrum is always positive.

5.4 COLOR IMAGE DISPLAY WITH BLACK-AND-WHITE FILM

Now let us describe the decoding process that will give a faithful color reproduction of the original color scene on the output plane of the decoder with an encoded black-and-white transparency.

5.4.1 Color Image Decoding

The decoder is a white-light optical image processor, a typical 4-f system. The principle of color image display is based on Fourier spectrum color filtering by the Fourier transform technique. The stored color image can be totally optically displayed by the use of primary color filters. As shown in Figure 5.1, the encoded positive transparency $t_p(x,y)$ is placed at input plane P_1. Under white-light illumination, its multichromatic Fourier spectra are formed on the Fourier plane P_2 by the Fourier transform of the first achromatic Fourier lens L_1. Red, green, and blue primary color filters are inserted at the corresponding first-order positions to modulate the first Fourier transform orders of the transparency in the proper primary colors, as shown in Figure 5.4.

The complex light distribution for the wavelength λ at the Fourier plane P_2 is

$$E(p,q;\lambda) = \iint t_p(x,y) \exp[-i(px+qy)] \, dx \, dy \tag{5.19}$$

Evaluating this equation, we have

$$E(p,q;\lambda) = T_r(p,q) + \frac{1}{2} \sum_{n=1}^{\infty} a_n T_r(p \pm np_0, q)$$

$$+ T_g(p',q') + \frac{1}{2} \sum_{n=1}^{\infty} a_n T_g(p' \pm np_0, q')$$

$$+ T_b(p'',q'') + \frac{1}{2} \sum_{n=1}^{\infty} a_n T_b(p \pm np_0, q'') \quad (5.20)$$

where (p,q), (p',q'), and (p'',q'') are the angular spatial frequency coordinate systems and $T_r(p,q)$, $T_g(p',q')$, and $T_b(p'',q'')$ are the Fourier transforms of $T_r(x,y)$, $T_g(x',y')$, and $T_b(x'',y'')$, respectively. Equation (5.20) can be written in terms of the linear spatial coordinates of the Fourier plane as

$$E(\alpha,\beta;\lambda) = T_r(\alpha,\beta) + \frac{1}{2} \sum_{n=1}^{\infty} a_n T_r\left(\alpha \pm \frac{n\lambda f}{2\pi} p_0, \beta\right)$$

$$T_g(\alpha',\beta') + \frac{1}{2} \sum_{n=1}^{\infty} a_n T_g\left(\alpha' \pm \frac{n\lambda f}{2\pi} p_0, \beta'\right)$$

$$T_b(\alpha'',\beta'') + \frac{1}{2} \sum_{n=1}^{\infty} a_n T_b\left(\alpha'' \pm \frac{n\lambda f}{2\pi} p_0, \beta''\right) \quad (5.21)$$

where $\alpha = (\lambda f/2\pi)p$, $\beta = (\lambda f/2\pi)q$, $\alpha' = (\lambda f/2\pi)p'$, $\beta' = (\lambda f/2\pi)q'$, $\alpha'' = (\lambda f/2\pi)p''$, $\beta'' = (\lambda f/2\pi)q''$, and f is the focal length of the transform lens. By this we know that the Fourier diffraction orders of different components of the encoded transparency at the Fourier plane are spatially separated except for the zero orders. If the spatial frequency of the tricolor grating is greater than twice the highest frequency of the encoded scene, there will be no overlap between different

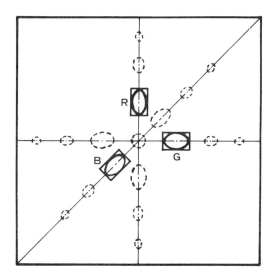

Figure 5.4 Primary color filters.

Fourier orders. Color filtering is done at the three first orders of the diffraction patterns at the Fourier plane. That is, the three first orders are filtered by a red, a green, and a blue transparent film, respectively, while the other orders are blocked. This filtering function can be described as

$$F_T(\alpha,\beta) = \left[F\left(\alpha - \frac{\lambda_R f}{2\pi}p_0, \beta\right)\right]_R + \left[F\left(\alpha' - \frac{\lambda_G f}{2\pi}p_0, \beta'\right)\right]_G$$
$$+ \left[F\left(\alpha'' - \frac{\lambda_B f}{2\pi}p_0, \beta''\right)\right]_B \tag{5.22}$$

where

$$F(\alpha - \alpha_0, \beta) = \begin{cases} 1, & |\alpha - \alpha_0| \leq \alpha_0/2;\ |\beta| \leq \alpha_0/2 \\ 0, & \text{otherwise} \end{cases}$$

λ_R, λ_G, and λ_B are the wavelengths for red, green, and blue light, respectively, and the terms in brackets $[\ ]_R$, $[\ ]_G$, and $[\ ]_B$, denote the color components for which the transmittances act. As discussed before, the transmittance will become zero (opaque) for color components other than the one denoted by the subscript. Then the field at the Fourier plane immediately behind the color filters is given by

$$E'(\alpha,\beta) = \frac{1}{2}a_1\left\{\left[T_r\left(\alpha - \frac{\lambda_R f}{2\pi}p_0, \beta\right)\right]_R + \left[T_g\left(\alpha' - \frac{\lambda_G f}{2\pi}p_0, \beta'\right)\right]_G\right.$$
$$\left. + \left[T_b\left(\alpha'' - \frac{\lambda_B f}{2\pi}p_0, \beta''\right)\right]_B\right\} \tag{5.23}$$

We know that the Fourier spectrum for the red component of the original color scene is represented by red light, the original green component is represented by green light, and the blue component is given by blue light. By means of an inverse Fourier transform made by the second achromatic lens L_2, the retrieved color image with the original colors is displayed on the output plane P_3. The multiple color field distribution is described as

$$E''(x,y) = [T_r(x,y)\ \exp(ixp_0)]_R + [T_g(x',y')\ \exp(ix'p_0)]_G$$
$$+ [T_b(x'',y'')\ \exp(x''p_0)]_B \tag{5.24}$$

As the primary colors are mutually incoherent, interference between different primary color components will not occur at the output plane of the decoder. The output image intensity distribution is expressed as

$$I(x,y) = [T_r^2(x,y)]_R + [T_g^2(x',y')]_G + [T_b^2(x'',y'')]_B \tag{5.25}$$

This retrieved color image can be viewed directly by placing a sheet of ground glass at the output plane of the decoder or fed to a color monitor by a portable video camera as we did. Then the color image is displayed directly on a monitor

COLOR DISPLAY WITH BLACK-AND-WHITE FILM

by the whole display system. It can be fed to a computer by an interface card for further digital processing and display, or recorded on color slide film for display on instruments other than the above-described display system.

Figure 5.5 shows an encoded black-and-white transparency of a color scene. We can see that the red component of the original color scene is encoded on the black-and-white film by horizontal modulation, the green by vertical modulation, and the blue by oblique modulation. Figure 5.6 shows the black-and-white picture of the corresponding retrieved color image produced by the display system with the encoded transparency.

5.4.2 Complementary Color Image Display

To directly produce a faithful color image on color print paper, a complementary color image should be formed at the output plane of the decoder. The principle of complementary color production is based on the implication that the tricolor grating has a minimum transmittance greater than zero.

Let t_{min} and t_{max} denote the minimum and maximum amplitude transmittance,

Figure 5.5 Picture of an encoded black-and-white transparency.

Figure 5.6 Black-and-white picture of the corresponding color image.

respectively, of the exposed film after developing and fixing. The amplitude transmittance of a recorded grating can be expressed as

$$t(x) = t_{min} + (t_{max} - t_{min})\left[\frac{1}{2} + \frac{1}{2}\operatorname{sgn}(\cos p_0 x)\right] \tag{5.26}$$

If this transparency is inserted into the input plane of an optical processor, its Fourier transform appearing at the Fourier plane can be written as

$$t(p) = \frac{1}{2}(t_{max} + t_{min})\delta(p) + \frac{1}{4}(t_{max} - t_{min})\sum_{n=1}^{\infty} a_n \delta(p \pm np_0) \tag{5.27}$$

The first-order diffraction spectrum is expressed as

$$t_1(p) = (a_1/4)(t_{max} - t_{min})\,\delta(p - p_0) \tag{5.28}$$

The output field re-formed by the first-order Fourier spectrum through the second Fourier transform lens is described by

$$t'(x) = (a_1/4)(t_{max} - t_{min})\exp(ip_0 x) \tag{5.29}$$

COLOR DISPLAY WITH BLACK-AND-WHITE FILM

So the output irradiance is

$$I(x) = |t'(x)|^2 = (a_1 2/16)(t_{max} - t_{min})^2 \qquad (5.30)$$

That is, the output irradiance when all the Fourier spectrum orders except for the first are blocked is proportional to the squared difference of the maximum and minimum local transmittance, or the degree of modulation, $(t_{max} - t_{min})^2$.

It has been proved experimentally that when a grating is printed on a negative black-and-white film, there exists a certain exposure threshold. When the exposure is below that threshold level, the degree of modulation of the recorded grating increases as the exposure increases. In contrast, when the exposure is above the threshold, the degree of modulation decreases as the exposure increases. Figure 5.7 shows the relationship between the degree of modulation and the optical density of the most exposed part of the film (corresponding to the transparent part of the superimposed grating) for a grating of 81 p/mm and another of 201 p/mm. The threshold values for these two gratings are noted in the figure as F_{Q1} for the grating of 81 p/mm and E_{Q2} for the grating of 201 p/mm.

This can be explained as follows: Both t_{max} and t_{min} are functions of the exposure. If the exposure is small enough, the rate of decrease of t_{min} is higher than that of t_{max} as the exposure increases. When the exposure reaches the threshold level, the rate of decrease of t_{min} slows down. For when the optical density is high enough, there is less relaxing room left for t_{min} to decrease further. On the other hand, the rate of decrease of t_{max} is maintained. So the degree of modulation decreases as the exposure increases if the optical density D, relative to t_{min}, is higher than E_Q.

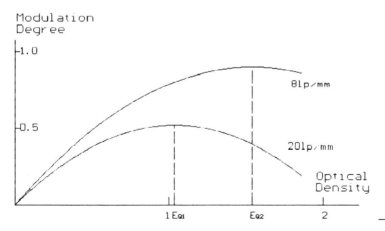

Figure 5.7 Relationship between the modulation degree of the recorded grating and the optical density of the most exposed part.

When a photograph is taken with a tricolor grating and the minimum exposure is above the threshold level, a complementary image can be obtained. The displayed color image, with the three complementary primary color filters of cyan, magenta, and yellow placed at the corresponding first-order positions of the Fourier plane of the decoder, will be the complementary color image of the original scene. Recording this output on color print paper yields a color image in the original colors.

The color image display system can be constructed with simple optical elements. An alternative white-light processing scheme used in the optical demonstration consists of a single lens. The input and output planes are arranged to be conjugate planes. The color filters are placed at the rear focal plane of this imaging lens. A similar derivation of this system shows its compatibility to the 4-f system consisting of two Fourier transform lenses. A photograph of such an experimental setup is shown in Figure 5.8. A schematic drawing of the color image decoder is given in Figure 5.9.

5.5 TECHNICAL NOTATIONS

We have described the principle of color image storage and display with black-and-white film in detail. In this section, we will deal with some technical prob-

Figure 5.8 An experimental setup of the color image display system.

COLOR DISPLAY WITH BLACK-AND-WHITE FILM

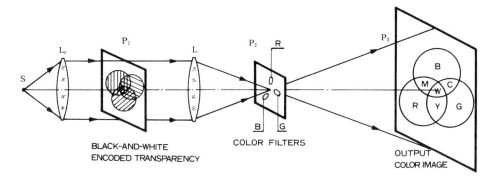

Figure 5.9 Color image decoder.

lems encountered in applications of this technique. To demonstrate optical color information encoding and decoding and to seek other possible applications of this technique, the tricolor grating is designed and fabricated first, then the color-encoding camera and the color image display system.

5.5.1 The Tricolor Grating

The tricolor grating is the key optical element for color image encoding, and its quality will influence the application of this technique. The design of this encoding element enables a color image to be dispersed into three primary color components and simultaneously encoded on black-and-white film. The recorded resolution of the color information and the color saturation and image quality of the displayed color image rely mainly on the quality and specifications of the tricolor grating.

Spatial Frequency of the Tricolor Grating

The spatial frequency of the tricolor grating determines the resolution of the encoded image. The tricolor grating acts as a sampling element, encoding the sampled image components. It is obvious that the higher the spatial frequency of the tricolor grating, the more sampling points there are and the higher the encoded image resolution is. According to the sampling theorem, the sampling rate should be greater than $2M$ provided the highest resolution requirement of the photo image is M. For ordinary color photography, a spatial frequency range of 301–501 p/mm is satisfactory for the resolution requirement.

Choice of Primary Colors

The three primary colors of the tricolor grating are red, green, and blue. The choice of primary colors in fabrication of the tricolor grating depends on exposure conditions. The color temperature of the illuminating light and the spectral response of the black-and-white film used determine the transmittance–wavelength relationship of each additive primary color grating. The original color information is properly encoded and recorded on the black-and-white film only if the primary colors are properly chosen. The tricolor grating used for photographing under daylight illumination differs from that used under artificial light, for example.

Diffraction Efficiency and Intensity Transmittance

Encoding color information with a tricolor grating can be understood as selectively printing the gratings on the black-and-white film. The quality and other specifications of the duplicated grating rely on the print operation and the contrast, shape, and transparent/opaque ratio of the grating used. Mathematical derivation shows that the diffraction efficiency of the printed grating is maximized for an absorption grating with a transparent/opaque ratio of 1:1. The higher the contrast of the grating is, the higher the diffraction efficiency of the printed grating is.

Because, in the color image decoding step, the color filtering is performed at the positions of the diffraction orders of the encoded black-and-white transparency, the diffraction efficiency of the encoded black-and-white film directly influences the brightness of the displayed color image and other specifications. Experiments provide proof that increases in the diffraction efficiency of the encoded black-and-white film improve the brightness of the displayed color image as well as the output signal-to-noise ratio.

To eliminate the exposure loss caused by the introduction of the tricolor grating, the average intensity transmittance of the tricolor grating should be as high as possible. The optimal value for the average transmittance ranges from 40% to 50%.

Orientations of Color Gratings

Color filtering is performed to the first orders of the Fourier spectrum of an encoded black-and-white transparency. Generally this spectrum has three sets of diffraction orders of 0, ± 1, ± 2, etc. If the gammas of the black-and-white films do not exactly coincide with the mathematical prediction, cross orders will appear. If the gratings are not properly oriented, the first orders will overlap with other cross orders, and first-order filtering is impossible. The passing of the cross orders brings moiré fringes to the output color image. To determine the optimal orientations of the color gratings, mathematical calculations are used. The positions of diffraction orders are mathematically determined. It is given that the grating orientations should differ by 45°, that is, the optimal orientations are 0°, 45°, and 90°. Then the diffraction orders can be spatially separated.

5.5.2 Color-Encoding Camera

Figure 5.10 is a photograph of a color-encoding camera that can encode a color scene, including stationary and moving objects, on a black-and-white film in a

COLOR DISPLAY WITH BLACK-AND-WHITE FILM

Figure 5.10 The color-encoding camera.

single exposure. The performance and operation are described in the following paragraphs.

Structure and Performance

There is not much difference between the encoding camera shown in Figure 5.10 and an ordinary camera. The imaging lens, scene-recording system, and shutter are the same as those of an ordinary camera, except for the tricolor grating on the imaging plane. The tricolor grating is mechanically connected to the film advance system. When the film advance is in action, this tricolor grating is pulled into the camera in order to prevent friction with the advancing film. When the next frame is in position, the grating is pushed to its duty position by a spring and is in close contact with the film. The light energy loss due to the introduction of the tricolor grating is about 50%, so exposure time is doubled.

The operation of this encoding camera is similar to that of an ordinary camera. The tricolor grating is removable. Without the load of the tricolor grating, the encoding camera can be used as an ordinary camera; it is a dual-function instrument.

Choice of Black-and-White Film

The black-and-white film chosen for the color image encoding should be a high-resolution panchromatic film. Its spectral response in the visible range of light is approximately constant in order to faithfully record the color information.

To obtain an encoded transparency with high quality, high contrast is needed. This requires a high resolution of the black-and-white film, say 3 or 4 times that of the spatial frequency of the tricolor grating. The developing and fixing processes are the usual ones for the chosen film. The result is a satisfactory black-and-white encoded transparency.

5.5.3 Color Image Decoder

The color image decoder is the main body of the color image display system, whose function is to retrieve the original color image from a black-and-white encoded transparency. Figure 5.11 presents a view of the color decoder. Figure 5.12 shows an overall view of the display system, which consists of a color image decoder, a portable TV camera, and a color video monitor.

Structure and Function

The optical system of this instrument is schematically shown in Figure 5.13. Figure 5.14 is a photograph of the structure of an actual decoder designed and manufactured by the Academia Sinica.

The white-light source, an indium arc, is imaged by the achromatic converg-

Figure 5.11 The total optical color decoder.

COLOR DISPLAY WITH BLACK-AND-WHITE FILM

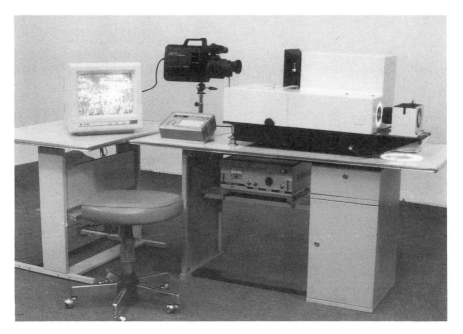

Figure 5.12 The color image display system.

ing lens onto the pinhole filter. A natural spot of the indium arc is formed. The following collimating lens is a high aperture ratio (1:3) achromatic lens. Therefore a large parallel white-light beam is available to illuminate the input encoded transparency. The white-light Fourier transform lens is a complex achromatic objective (semisymmetric). The input transparency is Fourier transformed by this achromatic Fourier lens, and a satisfactory Fourier spectrum is obtained. The three first orders of the Fourier spectrum are filtered by a red, a green, and a blue filter, respectively. At the output plane of the decoder, a faithful color image is formed.

The optical elements in the optical system are professionally and precisely designed. Coaxial conditions must be satisfied for the whole system. To increase the utility of power energy, the reflecting loss of the converging lens is minimized. This color image decoder and the display system can be further reduced in size to be portable.

Operation of the System

The operation is simple. Either a single piece of encoded transparency or a roll of transparencies may be mounted in the input plane. Encoded images are shown in the input aperture one by one by a rolling system. The spectral color filters can be individually adjusted in three dimensions to have the colored low-pass filters

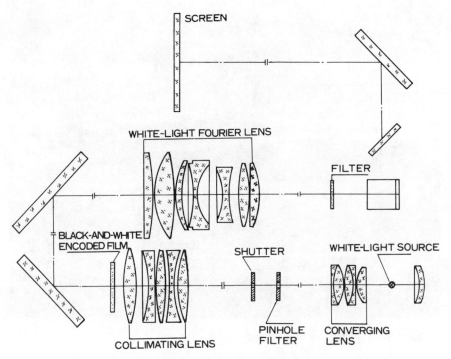

Figure 5.13 Schematic drawing of the inner structure of the color image decoder.

coincide with the corresponding first-order diffractions. The output color image is taken by a portable video camera and displayed on a color monitor. It can be viewed directly by placing a ground glass at the output plane or fed into a computer by an interface card. To record the color image on photographic paper or slide film, an optional magazine is available. The inside shutter of this decoder is used to control the exposure.

5.6 SOME APPLICATIONS

The color information recorded on black-and-white films is not subject to fading. The chemical treatment of black-and-white film is simple. This technique of color information recording can be applied to remote sensing, aerial photography, multispectra photography, civilian photography, and optical experimental training. And it can be applied for cinema photography. As an educational instrument, the color image display system can be used to perform the following experiments: optical information processing based on Fourier optics, including the

Figure 5.14 A photograph of the structure of a color image decoder.

addition and subtraction of images; optical correlation and convolution; image differentiation; pattern recognition; and pseudocolor encoding or other processing utilizing spatial or color filtering.

5.7 CONCLUSIONS

Color image storage and display using black-and-white film have been described. A special total optical color modulator is placed in an ordinary camera. With a single exposure, a color scene can be encoded on black-and-white film. After filtering in a white-light image processor, the original color image is retrieved from the encoded black-and-white transparency. The use of black-and-white film to record color images and its potential applications have been discussed. If this technique is applied to the making of cinemascope pictures, color film can be directly recorded on black-and-white films for storage for a prolonged period of time. Expanding the technique to ordinary color photography will produce great economic benefits.

This technique takes advantage of the color filtering of white-light image processing and provides a new way to take, preserve, and display color photographs.

Acknowledgment

We thank Mr. Fu-Lai Liu for making the drawings and photographs used in this chapter.

References

Chao, T. H., S. L. Zhuang, and F. T. S. Yu (1980). White-light pseudocolor density encoding through contrast reversal, *Opt. Lett.* 5:230.

Chiang, C. K., G. G. Mu, and H. K. Liu (1984). Contact-screen pulse-width encoding of a polychromatic image on a single piece of black-and-white film, *Acta Opt. Sinica* 4:706.

Fang, Z.-L., J.-Q. Wang, and G.-G. Mu (1984). An encoding technique for archival storage of color films by diffraction orders interference, *Acta Opt. Sinica* 4:701.

Goodman, J. W. (1968). *Introduction to Fourier Optics*, McGraw-Hill, New York p.77.

Grousson, R., and R. S. Kinany (1978). Multi-color image storage on black and white film using a crossed grating, *J. Opt.* 9:333.

Macovski, A. (1972). Encoding and decoding of color information, *Appl. Opt.* 11:416.

Mu, G.-G., J.-Q. Wang, Z.-L. Fang, and X.-Y. Li (1983a). A white-light processing technique for color photography with a black-and-white film and a tricolor grating, *Chin. J. Sci. Instrum.* 4:124.

Mu, G. G., F. X. Wu, and Z. Q. Wang (1983b). Holographic image encoding and white-light image processing, *Dig. Int. Conf. Lasers*, Guangzhou, China, p. 376.

Mu, G. G., Z. Q. Wang, Q. Gong, Q. W. Song, and F. X. Wu (1985). White-light image processing using oriented speckle-screen encoding, *Opt. Lett.* 10: 376.

Mu, G. G., X. M. Wang, and Z. Q. Wang, (1987). The application of computer-generated oriented speckle screen to image processing, *Proc. SPIE 673*: 508.

Mueller, P. F. (1969). Color image retrieval from monochromatic transparencies, *Appl. Opt.* 8: 2051.

Thompson, B. J. (1978). Optical transforms and coherent processing systems—with insight from crystallography, in *Optical Data Processing* (David Casasent, Ed.), Springer-Verlag, New York, p.17.

Yu, F. T. S. (1980). White-light processing technique for archival storage of color films, *Appl. Opt.* 19: 2457.

Yu, F. T. S. (1983). *Optical Information Processing*, Wiley, New York, p.313.

Yu, F. T. S., G. G. Mu, and S. L. Zhuang (1981). Color restoration of faded color films, *Optik 58*: 389.

6
Projection Display Technologies

Karen E. Jachimowicz*

*Honeywell, Inc.,
Phoenix, Arizona*

6.1 OVERVIEW

6.1.1 What Is a Projection Display?

A projection display uses projection optics to relay an image, either real or virtual, to the viewer. This chapter discusses those projection systems that create real images viewed with the use of a screen. Virtual image projection displays, which include head-up displays and helmet-mounted displays, are covered in other chapters.

Real image projection displays comprise a light source, an image source, the projection optics, and a screen, as shown in Figure 6.1. The image source and the light source can be one and the same, as in the case of CRT projection displays, or they can be separate, as with light valve systems.

Projection displays are configured with the imaging source in either the front or the rear of the screen. Front projection displays (Fig. 6.2) project the image onto the screen from the viewing side. The screen, which is separate from the projection unit, reflects the image light back into the viewer's eyes. Rear projection displays (Fig. 6.3) project the image onto the back of the screen. The screen can be separate from the projector, as with front projection systems, or the screen and image projector can be included in a single enclosure (a self-contained sys-

---

Current affiliation: Motorola, Inc., Tempe, Arizona.

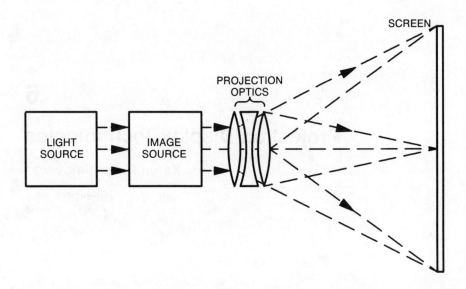

Figure 6.1 Projection display components.

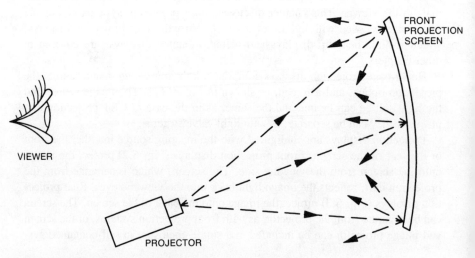

Figure 6.2 Front projection display.

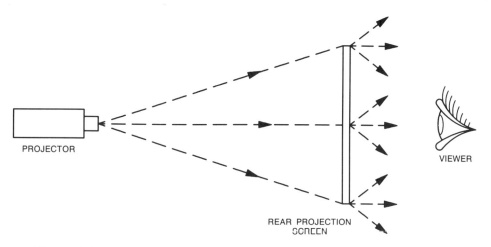

Figure 6.3 Rear projection display.

tem). When the screen and projector are mounted in the same enclosure, the optical path can be folded to create a more compact system. Self-contained systems are less flexible, because screen size or placement cannot be adjusted. In their favor, however, they require less maintenance because operator adjustments for defocus and misalignment are minimal due to the fixed position of the components.

6.1.2 Why Use a Projection Display?

Projections displays are used in applications where the screen is very large or where the screen size is not fixed. Although direct-view displays are getting larger all the time, for screen sizes greater than 30–40 in. diagonally, the practical way to create the image is to generate a small image and project it onto a large screen. Theaters, conference centers, exhibition halls, simulators, education centers, teleconferences, and military command and control centers typically utilize large-screen projection displays.

Projection displays are well suited to applications where the image size is not fixed. A projector with variable focus allows the screen size to be changed to suit the particular situation, unlike direct-view displays, which have a fixed image size.

6.1.3 Characteristics of Projection Displays

Choosing a projection display for a specific application involves reviewing and comparing the capabilities of different projection displays. Display characteristics such as resolution and luminance must be capable of providing the image quality desired. Unfortunately, the techniques used to achieve reported performance characteristics vary widely, so these values cannot be easily compared. The recently published ANSI Standard IT7.215 has been developed to set forth standards for projection display test procedures. This will hopefully result in reported performance data that can be reliably compared. Data comparison is complicated by the fact that the performance of self-enclosed projection displays will include the effects of the screen, whereas the performance of projectors not supplied with a particular screen will not include screen effects. The amount of ambient illumination is critical to performance parameters such as contrast and gray scale, yet a standard environment for taking measurements has not yet been accepted by manufacturers. Consequently, data are taken in varied conditions.

Determining the resolution of a display is therefore not as easy as gathering data from the manufacturer. The information published on displays, including projection displays, is usually the addressability of a system, not its resolution. The information given in this chapter (1) is that reported by the manufacturer, (2) is usually the addressability of the display (a 1280 × 1024 system has an addressability of 1280 × 1024, but if a viewer cannot see this many discrete elements the resolution is a lower value), and (3) should be used for relative comparison very carefully. The issue of resolution and addressability and how to measure them has been handled in depth in display literature (Snyder, 1988; Tannas, 1985), which should be referred to if more information is desired.

It is becoming popular to use the modulation transfer function (MTF) measurement as a measure of display resolution. The MTF is a measure of the image modulation present at a particular spatial frequency of line pairs (one line pair equals one white "on" line and one black "off" line). It is useful to remember that the MTF of a projection display system can be either measured directly at a particular screen or calculated from the MTF measurements of the individual components. Several recent works give the methods for determining the MTF of the different components and the system as a whole (Barten, 1986, 1991; Veron, 1989; Fendley, 1983; Banbury and Whitfield, 1981). The MTF of the display system is the product of the MTFs of the image source, optics, and screen:

$$\text{MTF}_{display} = \text{MTF}_{image\ source} \times \text{MTF}_{optics} \times \text{MTF}_{screen} \qquad (6.1)$$

The MTF of the image source is itself a product of its individual component MTFs, such as the MTF of the electronics and the MTF of the phosphor screen if the image source is a CRT. Although the resolution capability of a display system is usually not given in MTF, sometimes it is possible to obtain these values for the individual components. Optics themselves are usually specified in terms of MTF, and it is common to specify the resolution capability of a projection display screen in terms of its MTF.

PROJECTION DISPLAY TECHNOLOGIES

The luminance of a particular projection display system depends on the flux out of the projection optics and the size and gain of the projection screen used:

$$B_s = \Phi \, G/A \tag{6.2}$$

where B_s is the screen luminance in foot-lamberts; Φ is the flux out of the projection optics, in lumens; G is the screen gain; and A is the area of the image, in square feet. Gain is actually a function of angle (see Section 6.6), and so the screen luminance is a function of angle also. However, it is common to specify gain and luminance at the on-axis (0°) angle, and therefore the angle dependence is usually left out of the equation.

Luminance values for a self-enclosed projection display will usually be measured at the screen, as the system will always be used with the same screen. A separate unit projector will specify the lumens out of the projection optical system, so the user can then calculate the screen luminance when a screen of a particular size and gain is used.

Reported luminance values are most often obtained when all image sources are full on, which is a situation that rarely occurs in actual use. Luminance can also be specified as peak line luminance, which is a measure of the maximum luminance of a line that is full on, sometimes called "highlight brightness." This is measured by scanning a single "on" line with a photometer, obtaining a luminance profile. The maximum value is the peak line luminance. The numbers given in this chapter are those reported by the manufacturers. Luminance measurement techniques and conventions are available (Csaszar, 1991; Tannas, 1985; RCA, 1974) for obtaining more exact values.

Both monochrome and color projection displays are available. A monochrome projection display uses a single monochrome image source. Most color projection displays use multiple image sources—red, green, and blue—and combine the colors to create a full color image. The color gamut of a particular projection display depends on the chromaticity of the image sources used, which can vary considerably among display types. In order to display a white image, the luminances of the image sources are not equal. The exact mix depends upon the chromaticity of the image sources, but in general, to achieve a standard white the mix is approximately 70% green, 20% red, and 10% blue.

Characteristics such as contrast, modulation, and gray scale are initially determined by the image source itself. The optical system and screen are designed to preserve these characteristics as much as possible. The final characteristics are very dependent upon the situation in which the display is used. The screen affects these characteristics by allowing ambient illumination into the viewing volume. In a situation where there is little ambient illumination, the screen can be designed such that it will preserve the image quality of the image source. As the amount of ambient illumination increases, the final image quality depends largely on how well the screen minimizes ambient illumination reflected into the viewing volume. Reflected ambient illumination degrades the contrast, modulation, and gray scale of a projected image. This phenomenon is covered in more detail in the section on projection screens (Section 6.6).

Convergence is a measure of how well the three individual images of a color projection display are spatially aligned with each other on the screen. Misconvergence in a projected image can cause a pixel that is supposed to be white seem to be a different color because the red, green, and blue images are not fully superimposed. Misconvergence has several causes, including the individual images not being of the correct geometry, the images not being properly positioned at the screen, or image distortions being introduced by the optical system. An example of misconvergence of images is shown in Figure 6.4.

Different types of projection displays have different sources of convergence errors and different strategies for dealing with them. Some systems require considerable time for converging before they can be used, which is a factor in how well the system will meet the display need. Convergence techniques are covered in more detail in Section 6.7.

Other considerations involved in the choice of a projection display include the physical dimensions of the projector, the power consumption, and the speed of the display. Although most display applications require the system to work in real time, some extremely high resolution projection displays are available, spe-

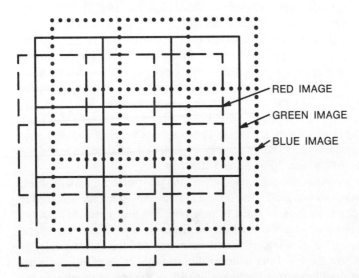

Figure 6.4 An example of misconvergence of projected images. The red, green, and blue images should superimpose one another.

cifically for displaying CAD images or maps, that are not required to operate in real time.

The size, weight, and power consumption of the different systems depend on the particular type of system. Active matrix LCLV video projection displays are probably the smallest color video projectors, with screen sizes as small as 20 in. diagonally. There is almost no practical limit to the maximum size of a projection display. Laser projection displays routinely provide backdrops for concert hall performances and large outdoor crowd shows, which can cover thousands of square feet.

The rest of this chapter provides an overview of the major projection display types, how they operate, and their general characteristics.

6.2 CRT PROJECTION DISPLAYS

The most common type of projection displays in use today are those that use CRTs as the image sources. CRT projection displays provide a high-quality dependable image, relying on mature CRT technology. These systems are very flexible, many being capable of use in either front or rear projection implementations, with multisync capability allowing a wide range of signals to be accepted, and with variable-focus optical systems allowing use with a variety of screen sizes. Consumer CRT projection TVs have proliferated in recent years, illustrating the strengths of CRT projections: high-quality image, low cost, reasonable size, and reliability.

The major drawback of CRT projection displays is the inherent inverse relationship between CRT luminance and CRT resolution. Raising the luminance of a CRT requires more electrons, resulting in a larger electron beam, which lowers the resolution capability of the CRT. Although many new techniques for improving CRT luminance are becoming available and will be discussed, in general to raise both the luminance and resolution of a CRT, the CRT size also must increase.

6.2.1 Operating Principles of CRT Projection Displays

In a CRT projection display, one or more monochrome CRT images are projected onto a viewing screen. By overlaying the red, green, and blue images on the screen, a full color display is formed.

Projection CRTs are very similar to monochrome direct-view CRTs. The basic construction and operation of CRTs are covered in Chapter 1. Flat CRTs are introduced in Chapter 4. Projection CRT differences stem from the fact that projection CRTs must be very bright so the image can be magnified. Projection CRTs are specifically designed to achieve maximum brightness and resolution simultaneously. They are operated at high anode voltages (30–50 kV is common); use special cathodes such as dispenser cathodes, which provide high electron beam current; and may use liquid-cooled faceplates, as operation at high beam current raises the temperature, which lowers the phosphor efficiency. A typical projection CRT, which operates at 35 kV and several milliamperes of beam current, is shown in Figure 6.5.

CRT projection display optical systems can be either refractive or reflective. The first high-performance CRT projection displays used reflective (Schmidt) optics. Reflective optical systems (Fig. 6.6) can be designed to collect a large portion of the light emitting from a CRT. A high-quality Schmidt system will have a collection efficiency of about 33%, compared to a collection efficiency of 20% for a high-quality $f/1.0$ refractive lens system (Todd and Sherr, 1986). This high performance is achieved in Schmidt optical systems without introducing serious aberrations (Patrick, 1972). Reflective systems are still used, and one system includes the reflective optics within the CRT (Forrester, 1990).

Refractive CRT projection optics (Fig. 6.7) have become the more popular. Their advantages are small size and low weight, especially when plastic optics are used. Refractive optics are also easily implemented into a variable-focus design. The major disadvantage of refractive projection optics is that high collection efficiency is at odds with the requirements for small size, minimum aberra-

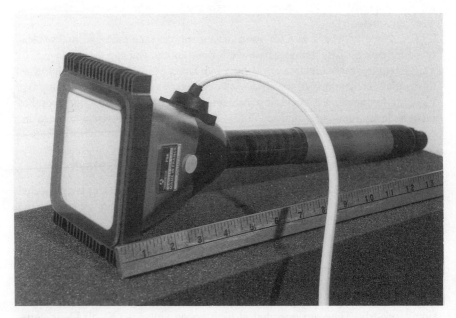

Figure 6.5 A Thomson-CSF projection CRT with a liquid-cooled faceplate.

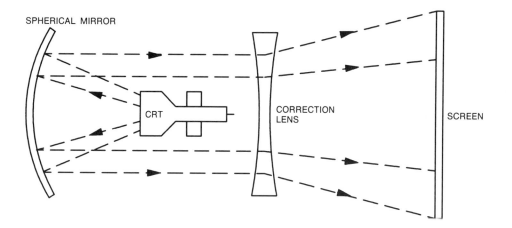

Figure 6.6 CRT projection display with reflective (Schmidt) optical system.

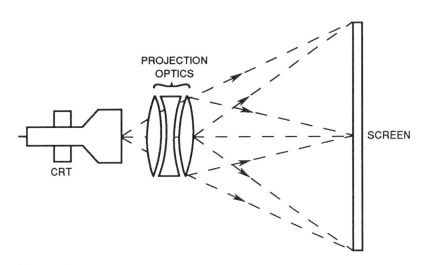

Figure 6.7 CRT projection display with refractive optical system.

tions, and low cost. This trade-off has been mitigated by the increased use of computer optical design programs, aspheric optics, and plastic optics. A large amount of work has gone into optimizing refractive projection optical systems, resulting in high-performance systems becoming available that provide a relatively high collection angle simultaneously with the resolution and costs consistent with the application.

In a color CRT projection display, the images from red, green, and blue monochrome CRTs are combined to create a single full-color image. Image combination is performed using either an off-axis or an on-axis technique. In the off-axis method (Fig. 6.8), the three CRTs direct their images to the screen from different angles, off-axis from each other, and the images are made to superimpose at the screen. The on-axis technique (Fig. 6.9) uses a beam combiner to merge the images before transmission through the projection optics. The three images are then coaxial and form a single full-color image, which is then projected onto the screen.

The advantages of the off-axis system are that packaging is simpler and the optics can be specifically designed for the appropriate color of CRT being used. The off-axis system is the least expensive system and is the most common image-combining technique among CRT projection displays.

Disadvantages of the off-axis technique are the need for trapezoidal distortion correction and the fact that the three images must be reconverged whenever the screen size is changed. Trapezoidal distortions occur because the two outer CRTs

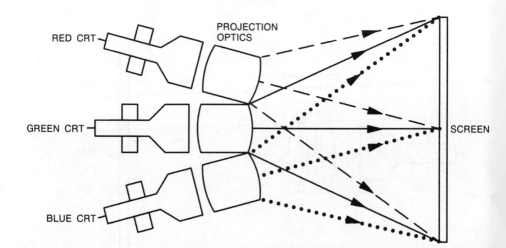

Figure 6.8 Off-axis technique for combining monochrome images in a full color CRT projection display.

PROJECTION DISPLAY TECHNOLOGIES

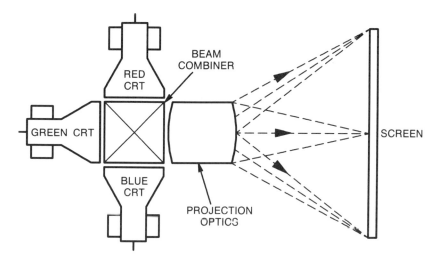

Figure 6.9 On-axis technique for combining monochrome images in a full color CRT projection display.

are at an angle to the screen, creating distortion in the images and resulting in misconvergence at the screen (Fig. 6.10).

Trapezoidal distortion can be corrected optically at the CRT faceplate (Hockenbrock and Rowe, 1982) or by electronically predistorting the image on the CRT so the image will be correct when it reaches the screen. An example of predistorted CRT images for the correction of trapezoidal distortions is shown in Figure 6.11. A particular distortion correction is valid only for one particular throw distance, however. Each time the screen size or placement is changed, the system must be reconverged.

The on-axis optical technique uses a beam combiner to merge the three CRT images before the projection optics. The image is then projected onto the screen with a single set of projection optics. The beam combiner is an optical device that uses dichroic coatings to selectively reflect or transmit the different wavelengths of light, resulting in the images being coaxial. The operation of a cube beam combiner as used in a CRT projection display is shown in Figure 6.12. The beam combiner can be constructed of individual prisms coated and cemented to form a cube or individual coated glass plates. There are a number of beam-combining implementations besides the cube format (Scholl, 1987), but for CRT projection displays the most common technique is the cube beam combiner.

Beam combiners operate such that green is allowed to pass straight through the cube, red is reflected off a dichroic coating designed to reflect red light incident at 45° and pass all other wavelengths (Fig. 6.13), and blue is reflected off a

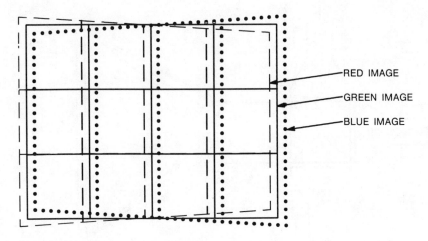

Figure 6.10 Trapezoidal distortion in images caused by off-axis projection.

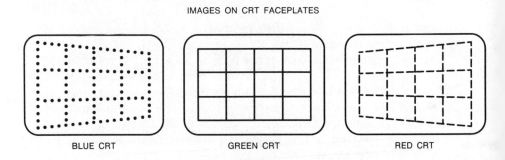

Figure 6.11 Predistorted CRT images for correction of trapezoidal distortion.

coating that reflects blue light incident at 45° and passes all other wavelengths (Fig. 6.14), resulting in all three images exiting the same side of the cube, into the projection optics.

The advantages of the on-axis approach are that trapezoidal correction is not necessary because the images are coaxial before arriving at the screen, only one projection lens is required, and changes in screen size do not require reconvergence of the images.

The disadvantages of an on-axis approach are that the packaging can become large because of the beam combiner and resulting optics and that the design and fabrication of a beam-combining system can be expensive when used with CRTs.

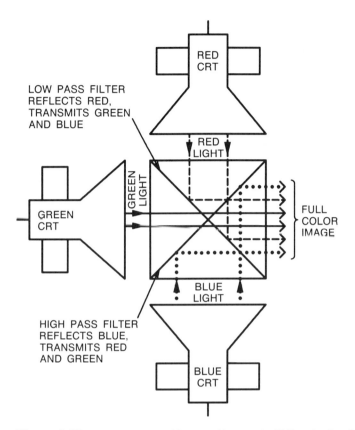

Figure 6.12 Cube beam combiner used in on-axis CRT projection display to combine red, green, and blue images.

Dichroic coatings perform optimally with polarized, collimated, single-wavelength illumination, characteristics that CRTs do not possess. CRTs approximate Lambertian radiators, emitting unpolarized light in all directions, so the cone of light being transmitted through the beam combiner covers a large range of incidence angles and a range of wavelengths. This leaves the design of the projection optics and the beam combiner dichroic coatings in conflict, because it is desirable to collect as much light as possible with the projection optics, requiring a small f number (i.e., a large cone angle), whereas the transmission and color characteristics of the beam combiner work best with a small cone angle. This trade-off usually results in the beam combiner becoming a large, expensive optical component, and so its use in CRT projection displays is limited to higher quality, higher cost systems.

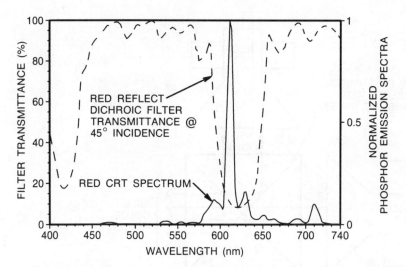

Figure 6.13 Action of red-reflecting dichroic coatings.

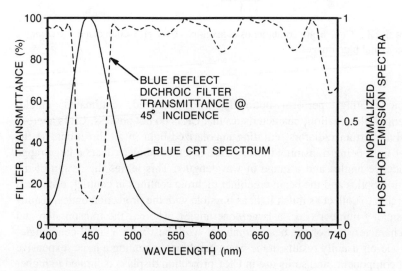

Figure 6.14 Action of blue-reflecting dichroic coating.

6.2.2 Characteristics of CRT Projection Displays

The color gamut of a CRT projection display is determined by the emission spectra of the particular CRT phosphors being used. Typical high-performance projection CRT phosphors are P53 green, P55 blue, and P56 red. These phosphors are specifically designed to withstand the high electron beam currents necessary in a projection CRT. The emission spectrum of the P53, P55, and P56 phosphors is shown in Figure 6.15, and one possible color gamut resulting from the use of these phosphors in Figure 6.16. The color gamut of any particular projection display can be manipulated easily with filters, or different phosphors, so the particular color gamuts shown in this chapter should be considered as examples only.

The screen luminance of a CRT projection display system is given by (Kingslake, 1983)

$$B_s = B_{CRT}TG/4F^2 (1+m)^2 \qquad (6.3)$$

where B_{CRT} is the luminance of the CRT in foot-lamberts, T is the transmission of the optical system, m is the system magnification, and F is the f number of the optics. In the design of a CRT projection display, these values are adjusted for maximum screen luminance.

One method for maximizing the screen luminance is to maximize B_{CRT}, the luminance of the CRT itself. This is feasible only to the extent that the increase in luminance does not degrade the resolution characteristics of the CRT to be below the system requirements. CRT design improvements aimed at improving

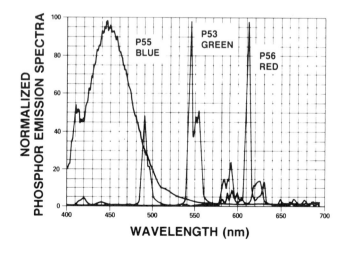

Figure 6.15 Emission spectra of projection CRT phosphors P53, P55, and P56.

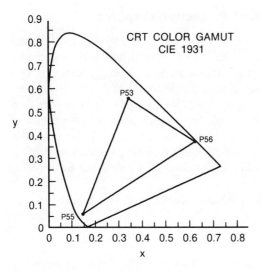

Figure 6.16 Color gamut of CRT projection display using P53, P55, and P56 phosphors.

luminance without compromising resolution include special phosphors and electron guns (Chevalier and Deon, 1985), liquid cooling, interference filters (Vriens et al., 1988), and curved CRT faceplates (Asano et al., 1989; Malang, 1989). Projection CRT phosphors are designed to withstand high beam currents while maintaining adequate lifetimes. Interference filters and curved faceplates both change the luminance distribution so that it is not Lambertian; instead, the luminance is directed more on-axis, and therefore the optics collect more light than they would with a Lambertian source. Liquid cooling between the faceplate and the first optical element not only cools the CRT phosphor, making it more efficient, but also acts as liquid coupling, reducing the losses at glass–air interfaces.

Reducing the magnification and/or the f number of the projection optical system is very helpful in increasing screen luminance, as the luminance varies as the inverse square of these terms. Optical systems have been optimized until refractive systems with f numbers between $f/1.0$ and $f/1.4$ are the most common (Clarke, 1988). The use of aspherical and plastic optics has helped designers create low f number systems that maintain low weight and cost.

The trade-off involved in reducing magnification is that of the physical size of the unit. System magnification is the ratio of the screen size to the image source size. Assuming that a certain screen size is desired, reducing magnification means using larger CRTs, which will improve resolution capability as well as luminance, but at the cost of a larger unit.

Screen gain and optics transmission are the last two parameters that can be manipulated to increase screen luminance. Optical transmission is maximized by good antireflection coating design and a minimum of glass–air interfaces. Screen gains have increased considerably in the last several years. Screens with a gain of 3–10 are common these days. Increased gain is achieved only by decreasing the available viewing zone, however. Factors involved in the choice of screen gain are discussed in Section 6.6.

Convergence is a major consideration for CRT projection systems. Individual CRTs have their own nonlinearities, which must be corrected. The three images must be overlaid, and they must be corrected until they are all the same shape. Some CRT projection displays have automatic convergence systems, while others have to be manually converged. The type of convergence system used has an effect on how easy the display is to use. Convergence techniques are discussed in more detail in Section 6.7.

There are a wide variety of CRT projection displays available on the market today. The applications for CRT projection systems are numerous, and the cost and performance of systems span a broad range.

Figure 6.17 shows a consumer CRT projection television with a 40-in. screen diagonal. Like the one shown, most home CRT projection TVs are off-axis systems, which are self-contained so a minimum of convergence and focus adjustments are required by the consumer. These home TVs accept standard NTSC video signals. Home CRT projection TVs are available in a range of sizes from small units with 40-in. diagonal screens, a screen luminance of several hundred foot-lamberts, and weighing about 150 lb, to large units with 70-in. diagonal screens, a screen luminance of about 150 fL, and weighing several hundred pounds.

For these large-screen consumer TVs, CRT projection techniques have provided the proper blend of performance, size, and cost, resulting in steadily increasing popularity. Several companies, including Hitachi (Ando et al., 1989), Toshiba (Murakami et al., 1989), and Mitsubishi (Toide et al., 1991), have developed CRT projection systems for next-generation high-definition television applications.

Cathode-ray tube projection displays are also used to supply large-screen displays for industrial applications. These systems display both video and computer data, with screen sizes ranging from about 60 in. to over 250 in. diagonally. Video/data projectors accept a wide range of input signals, from 525-line video to 1500-line computer-generated information. These CRT projection displays are usually off-axis systems in which the screen is separate from the projector. This results in a projector that can be used with different screens and screen sizes and in both front and rear projection implementations. Figure 6.18 pictures a typical industrial video/data CRT projection display. Typical unit sizes are about 20–25 in. wide, 10–15 in. high, and 30–40 in. long, with weights of about 100–200 lb. The luminance of these units is specified as lumens out of the projector be-

Figure 6.17 A consumer off-axis CRT projection TV with a 40-in. diagonal screen. This system is made by Pioneer.

cause the screen is supplied separately. A large range of luminance output capabilities are available, depending on the performance and cost of a system. The range covers 300–1500 lumens. Video/data CRT projection displays are used for conferences, exhibitions, education, teleconferencing, CAD/CAM, and other audiovisual presentations where the audience is usually large and/or a large screen is necessary.

Another common application of CRT projection displays is in high-performance military and aerospace applications such as flight simulators (El-

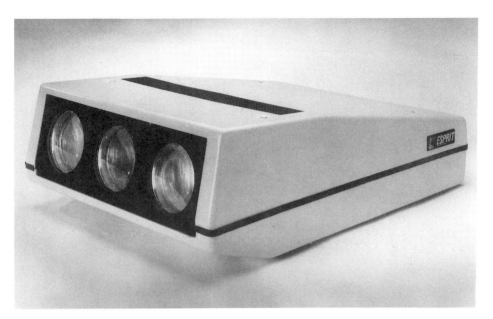

Figure 6.18 Off-axis industrial CRT video/data projection display. This system, by Ampro Corp., uses a reflective CRT. (Photo courtesy of Ampro Corporation, Woburn, MA)

mer, 1982; Holmes, 1987a) and command and control centers. Figure 6.19 shows the internal construction of a very high performance CRT projection display used in instances where the resolution and image quality requirements are very demanding. This is an on-axis system, which gives better performance but costs more than an off-axis system and is usually larger. Most of these high-performance CRT projection displays are built as custom units, so the performance varies with the application, but typical requirements are for resolutions above 1000 lines and light outputs of 800 lumens.

6.2.3 CRT Projection Displays: Summary

CRT projection is a very popular way to achieve large screen sizes and high-resolution images. The technology is well developed, resulting in systems with high performance for a relatively low cost. Applications range from consumer TVs to industrial and military high-performance video and data projectors. As the requirements for these systems expand, however, the shortcomings of CRT

Figure 6.19 Internal construction of a high performance on-axis CRT projection display. (Photo courtesy of TDS Development Corp., Canoga Park, Calif.)

projection become evident: The inverse relationship between luminance and resolution of a CRT forces high-performance systems to use large CRTs, increasing the volume and power consumption of the unit.

6.3   OIL-FILM LIGHT VALVE PROJECTION DISPLAYS

A light valve differs from a CRT in that it does not create light; instead, it controls its transmission. Light valve projection displays consist of a separate light source and image source and the projection optics and screen (Figure 6.20). The light valve is used to modulate the red, green, and blue light from the source into an image.

A very important advantage of a light valve display is that resolution and luminance are no longer interrelated as they are in CRT systems. This allows luminance to be increased without affecting resolution.

The major drawback of light valve systems has historically been their large size, high cost, and high power consumption. They have been able to provide high performance in terms of luminance and resolution, but the size has been large also. This is changing, however, with the recently developed active-matrix

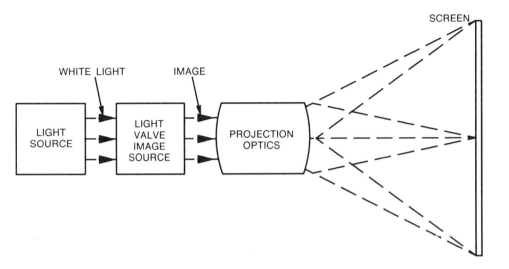

Figure 6.20 Light valve projection display components.

liquid-crystal light valve projection displays, which have the potential to fill a niche for a projection display in a package smaller, less expensive, and lighter than a CRT projection display.

There are many types of light valve display systems. Light valves (LVs) can be made from oil films, deformable mirrors, and liquid or solid crystals. Images are formed by refraction, diffraction, birefringence, or absorption, to name a few examples.

This section and the next review the operating principles of the most common types of light valve projection displays: oil-film light valve (OFLV) projection displays and liquid-crystal light valve (LCLV) projection displays.

6.3.1 Operating Principles of Oil-Film Light Valve Projection Displays

Oil-film light valve (OFLV) projection displays are used when it is necessary to display high-luminance, full-color, real-time video and graphics and the large size and higher cost and power consumption of these systems can be tolerated. Applications of oil-film light valve projection displays include displays for large-audience educational and sporting events, command and control centers, dome simulators, and conference and symposium events.

Oil-film light valve projection displays have been around a long time and represent some of the earliest image projectors. The first OFLV was developed by Fischer in Switzerland around the year 1944 (Baumann, 1953; Johannes, 1989).

A derivative of this system is the Gretag Eidophor projection display, still successfully marketed today. General Electric has developed a projection display using an OFLV, the Talaria, also commonly used today. Both of these displays use a dark-field schlieren system with an oil-film modulator. The advantage of a schlieren optical system is that the system throughput can be very high. There are no absorbers, such as polarizers or dyes, in the system. If a very efficient diffractor is used, almost no light is lost, resulting in a high light throughput system.

In a dark-field schlieren system, a combination of bars and slots is used to block or transmit light, as illustrated in Figure 6.21. The lens images the slots of the object plane onto the bars of the image plane. The input and output spatial filters (bars and slots) form the first set of conjugate surfaces in a schlieren system. The bars are positioned to block the image of the slots, and no light is transmitted.

A transparent deformable film, the oil film, is placed between the bars and slots and a lens to image the film surface onto a projection screen (Fig. 6.22). The deformable film and the projection screen form the second set of conjugate surfaces within the schlieren system. A disturbance or thickness variation in the deformable layer diffracts light from its original path. The light bypasses the bars and is imaged onto the projection screen. Each pixel on the screen corresponds to a pixel on the deformable layer. An unmodulated pixel on the deformable layer

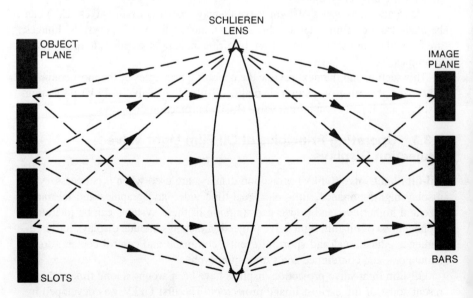

Figure 6.21 Bars and slots of dark-field schlieren system.

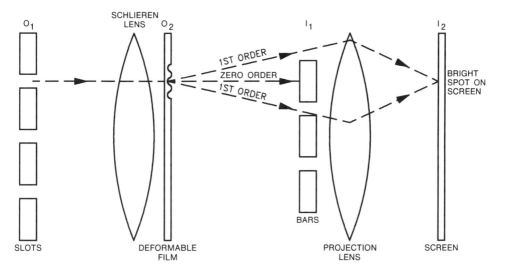

Figure 6.22 Light transmission through a schlieren system using a deformable oil film.

corresponds to a dark pixel on the screen (hence the name dark-field schlieren system), and vice versa.

An electron beam is used to deposit a charge onto the oil film, which provides the modulation. The electron beam moves over the oil film in a raster format. The velocity of the electron beam determines the amount of charge deposited, which determines the depth of the sinusoidal diffraction grating formed in the oil film. When no charge is deposited, the oil is not deformed, and adjacent oil-film raster lines overlap, creating a smooth oil-film surface. There is no diffraction, and no light transmission occurs. When charge is deposited, sinusoidal raster grooves are formed in the oil film, creating the phase grating. Diffraction angle and amount of light diffracted are determined by the basic diffraction equations (Hutley, 1982). Maximum grating depth results in maximum pixel intensity. Gray scale is provided by levels of charge between minimum and maximum. Both the Gretag Eidophor and the GE Talaria use this basic system, with some color implementation differences.

The Gretag Eidophor uses three schlieren light valves, one for each color, with a high-intensity light source providing white light. The white light is separated into its constituent red, green, and blue components with dichroic filters, in a process that is the reverse of the beam combining used in the CRT on-axis projection display. The separate red, green, and blue light is then sent to the corresponding light valves. The basic configuration of an Eidophor light valve is shown in Figure 6.23. The oil-film control layer resides on a spherical mirror. Light reflects off the mirror, and the set of bars act as both bars and slots.

The Talaria OFLV projector takes advantage of the separation of colors that occurs in diffraction to provide all three color images with a single light valve (Glenn, 1958; Good, 1968). By using the angle of diffraction dependence on wavelength (Fig. 6.24) and carefully controlling the spacing of slots and bars, a single light valve unit can provide full color control.

The GE OFLV uses two sets of slots and bars, one to control green light and another to control red and blue light combined (magenta light). The slots consist of two sets of dichroic filter slots overlaid one on top of another (Fig. 6.25). The vertical slots, bars, and control layer pattern are used to modulate the red and blue light. The wavelengths are far enough apart that both can be controlled with the same slots/bars, and the two diffraction frequencies are overlaid at the control layer. Modulation of the green light by the magenta grating does not affect it because diffraction occurs along the green bars, and vice versa.

Green light modulation, occurring orthogonal to the red and blue, takes place just as in a monochrome system. Color selection in the GE light valve is illustrated in Figure 6.26.

The GE single-electron-beam, single-control-layer, full-color light valve (Fig. 6.27) was introduced in 1968. It has advantages of size and registration because it uses only one light valve.

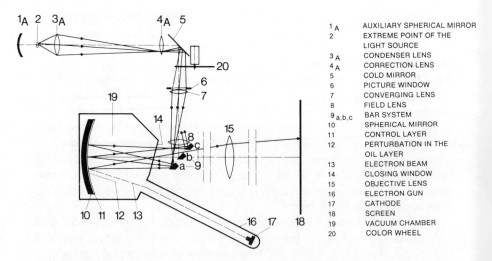

Figure 6.23 Gretag Eidophor oil-film light valve configuration. (Courtesy SAIC, McLean, Va.)

PROJECTION DISPLAY TECHNOLOGIES

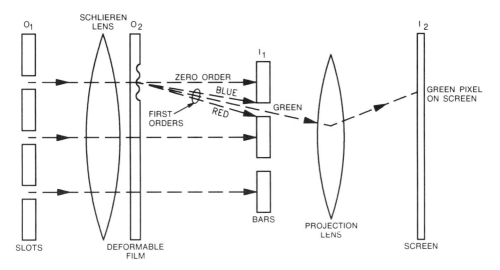

Figure 6.24 Color separation in oil-film light valve by diffraction.

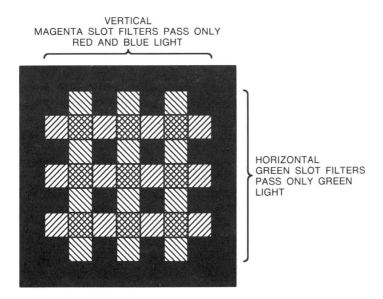

Figure 6.25 Green and magenta color filter slots are superimposed on one another in GE schlieren light valve.

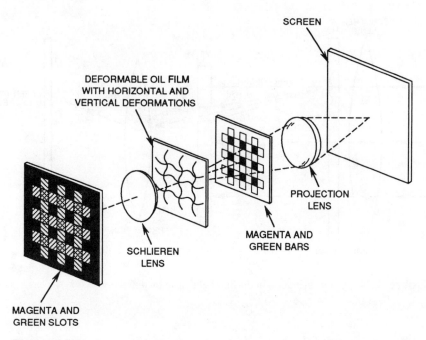

Figure 6.26 Green and magenta color selection occurs orthogonally in GE schlieren light valve.

6.3.2 Characteristics of Oil-Film Light Valve Projection Displays

Both the Talaria and Eidophor systems provide full color, although their respective color gamuts are determined differently. In the GE system, color is determined by the dichroic filter slots and the filtering action of the magenta slots on the red and blue colors individually. The University of Dayton (Howard, 1989) has characterized the color performance of the Talaria display. In the Gretag display, color is determined by the dichroic filters used to divide the white light into red, green, and blue components.

Oil-film LV projection displays typically operate at 60 Hz refresh rates and are available in both monochrome and full-color systems. There are a range of systems available, with varying capabilities. Lumens out of the systems run from 2000 to 8000 white lumens, and addressabilities of over 1000 scan lines are available in both types. These systems are quite large and heavy, but they can provide very large bright video images. Figure 6.28 is a picture of the Gretag Eidophor projector, and Figure 6.29 shows the GE Talaria projector.

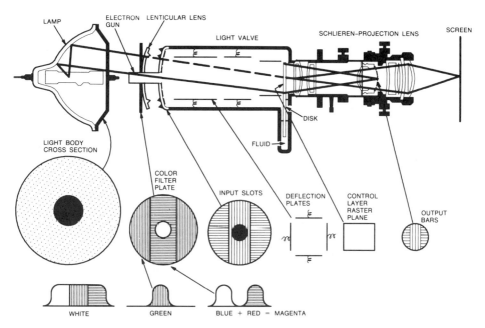

Figure 6.27 GE Talaria full-color light valve diagram. [From Good (1976), courtesy of the author.]

The configuration of the GE system has size advantages because it contains only one light valve, whereas the Eidophor has three. It also does not need to be converged like a separate light valve system does. The use of a single light valve does tend to limit luminance capability, however. General Electric has also introduced a two-light-valve system (True, 1987) in which one light valve modulates green and the other modulates red and blue. This has increased the total light output capability of the Talaria.

6.3.3 Oil-Film Light Valve Projection Displays: Summary

An OFLV projection display uses a schlieren optical system with an oil-film modulator to create an image. One advantage of these displays is their high light output, which results from a high-output lamp and an efficient schlieren optical system. Another advantage is that the resolution and luminance of the display are not related as they are with CRTs. Consequently, OFLV projection displays are capable of projecting images with both high luminance and high resolution. Their large size and relatively high cost limit their use to applications where this high performance is necessary and affordable. Oil-film LV projection displays

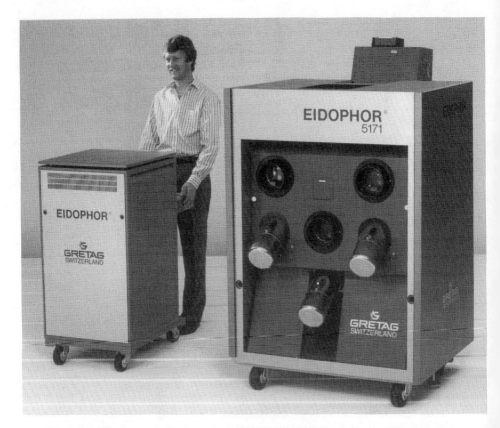

Figure 6.28 Gretag Eidophor oil-film light valve projection display. (Photo courtesy of SAIC, McLean, Va.)

are used in military and aerospace applications for dome and helmet-mounted simulators and command and control centers. Non-aerospace applications include large coliseum and stadium displays.

6.4 LIQUID-CRYSTAL LIGHT VALVE PROJECTION DISPLAYS

Liquid-crystal light valve (LCLV) projection display technology has been growing extremely rapidly in recent years, building upon the results of research activity in liquid crystals and liquid-crystal displays. In this type of display, liquid-crystal light valves are used to modulate white light from the light source into an image, which is then projected onto the screen. The LCLVs can be used in either

Figure 6.29 General Electric Talaria MP oil-film light valve projection display. (Photo courtesy of General Electric, Syracuse, N.Y.)

a transmissive or reflective configuration. These are illustrated in Figures 6.30 and 6.31. Reflective LCLV projection displays usually use a polarizing beam splitter to reflect light toward the light valve on the first pass, then transmit the modulated image to the projection optics on the return path.

The biggest advantage of LCLV image projection displays is the separation of light generation from image generation, allowing the two to vary independently.

A disadvantage of many LCLV projection displays is that only light of a single polarization can be used by the LCLV, resulting in an immediate loss of half the system light. This disadvantage can be eliminated, however, by using recent developments in techniques to repolarize light of the wrong polarization and render

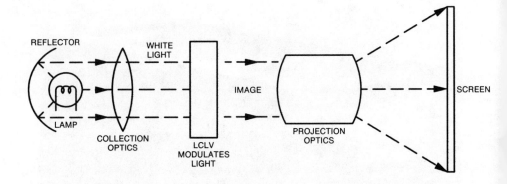

Figure 6.30 Transmissive LCLV projection display components.

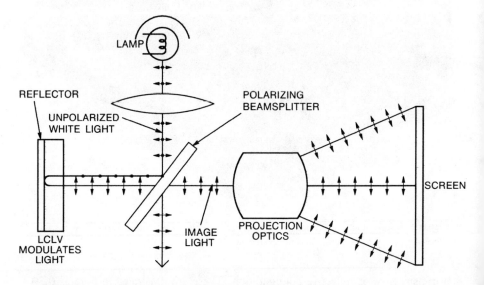

Figure 6.31 Reflective LCLV projection display components.

it usable, or by using new liquid-crystal types that do not operate on the polarization of the light.

There are a variety of ways in which light valves can be addressed. These include optical addressing, which can be done with a laser or a CRT—for example, thermal addressing, which is accomplished with an IR laser—or, more recently, active-matrix addressing. The active-matrix-addressed LCLV uses a matrix of active elements, such as thin-film transistors (TFTs), to provide individual addressing of each pixel.

6.4.1 Active-Matrix-Addressed LCLV Projection Displays

The results of intense research and development that has occurred recently in the area of active-matrix-addressed LC direct-view displays have been used in the area of projection displays. In 1986 Seiko-Epson introduced the first full-color projection display using active-matrix-addressed LCLVs (Morozumi et al., 1986). Other systems quickly followed. Active-matrix LCLV (AMLCLV) projectors have added another class of projection displays: small, portable, and inexpensive consumer and low-end industrial systems.

Operating Principles of AMLCLV Projection Displays

An active-matrix light valve used in a projection display is very similar to the ones used for direct-view active-matrix liquid-crystal flat-panel displays. Active elements are used to provide individual light transmission control of each pixel. In projection display applications the diffuse backlight used for direct-view flat panels is replaced with a pseudocollimated light source. Light passing through the light valve is modulated with image information, which is projected onto the diffuse screen, where the image is formed. This type of system is illustrated in Figure 6.32.

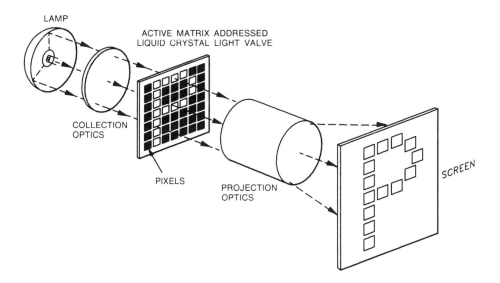

Figure 6.32 AMLCLV projection display components.

The majority of LC projection displays being developed use a twisted nematic liquid-crystal configuration similar to direct-view LCDs. The light valve blocks or transmits light by operating on its polarization state. Since the light valve uses only one of the polarization states, 50% of the light is lost. This is being remedied with techniques for converting light of the "wrong" polarization so that it is usable, minimizing polarization losses (Toide and Kugo, 1991; Schadt and Funfschilling, 1990).

A new type of AMLV being developed for projection displays uses polymer-dispersed liquid crystals (PDLC). In a PDLC light valve, small liquid-crystal spheres are suspended in a polymer matrix. The liquid crystal is designed such that in one orientation its index of refraction matches that of the polymer and the light valve is clear. In the other state the refractive indices do not match, and light is scattered by the light valve (Doane et al., 1988; Fergason, 1985). Therefore, in one state—the on state—the LCLV transmits all light through to the projection lens; and in the off state, light is scattered and does not make it to the projection lens. Several systems of this type are under development (Jones et al., 1991; Hirai et al., 1991; Takizawa et al., 1991).

One difference between a direct-view and a projection AMLCLV display is that the projection display uses three different light valves—one each for red, green, and blue—as opposed to the direct-view technique of using color filters over the pixels to provide the different colors. This eliminates the color filter layer of the liquid-crystal cell and provides a resolution and light throughput advantage over using one light valve to provide all three colors.

White-light sources typically used are tungsten halogen, metal halide, and xenon. Dichroic mirrors are used to divide the light into its component colors. The red, green, and blue light then passes through the respective light valves, which are coded with video information. The three color images can then be combined, again with dichroic coatings, and projected onto the screen or combined at the screen as in the off-axis CRT system. The layout of the Seiko-Epson display is shown in Figure 6.33. This system is pictured in Figure 6.34.

The beam-combining and beam-splitting operations in the active-matrix LCLV display are more efficient in terms of light throughput than in a CRT system. As discussed in Section 6.2, dichroic coatings operate most efficiently with polarized, collimated, single-wavelength light. The light transmitted through an LV system is not of a single wavelength, but it is usually polarized and semicollimated. This greatly simplifies the design of the beam-combining system. For this reason, most active-matrix LCLV projection systems use an on-axis optical system, which does not add greatly to the size or cost of the system as it does in the CRT projection display.

Characteristics of AMLCLV Projection Displays

The color gamut of an active-matrix LCLV projection display is determined by the spectral characteristics of the dichroic coatings used on the beam splitters and

PROJECTION DISPLAY TECHNOLOGIES 243

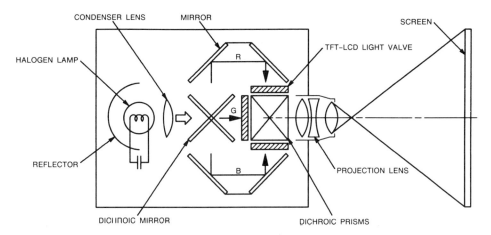

Figure 6.33 Seiko-Epson full-color AMLCLV projection display layout (Courtesy Morozumi et al. 1986)

Figure 6.34 Picture of Seiko-Epson AMLCLV projection display. (The light valves are made by Seiko-Epson, but the system was originally marketed by Kodak.)

combiners. Figure 6.35 gives an example of the color gamut of the Seiko-Epson system (Morozumi et al., 1986).

The resolution of a particular AMLCLV projection display depends on the number of pixels contained in the light valve. In order to increase the resolution of the display, either the pixel density or the size of the light valve must be increased. Several systems are commercially available, including the Seiko-Epson unit with a pixel resolution of 320 × 220 and a Sharp display with a resolution of 382 × 234. Many more systems have been reported (Sakamoto et al. 1991; Takeuchi et al., 1991; Fukuta et al., 1991; Kobayashi et al., 1989; Kunigata et al., 1990; Noda et al., 1989; Timmers et al., 1989) with resolutions available up to 1422 × 960 pixels (Takubo et al., 1989). These resolution numbers refer to the number of pixels in the light valve, which is usually the resolution-limiting element in an active-matrix LCLV projection display. These resolution values do not directly compare with values reported for other systems such as CRT or OFLV projectors (Barten, 1991; Stroomer, 1989).

The image luminance provided by a liquid-crystal light valve projector is based on the lumens out of the lamp and the efficiency of the optical system and light valve, given by the equation

$$B_S = W_{lamp} \eta_{lamp} \Omega T_{lv} T_{optics} G/A \tag{6.4}$$

where W_{lamp} is the lamp power in watts, η_{lamp} is the lamp efficacy in lumens per watt, Ω is the collection efficiency of the collection optics, T_{lv} is the transmission of the light valve, and T_{optics} is the projection optical system transmission. Major

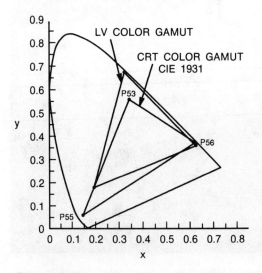

Figure 6.35 Color gamut of Seiko-Epson AMLCLV projection display (Morozumi et al., 1986) and CRT projection display.

lossy areas are the collection efficiency of the lamp (which is very good if it is 50%) and the transmission of the light valve. The active matrix transmits about 60%, and the light valve and substrate reduce transmission further, until total transmission through a light valve is on the order of 5–30%. Optics transmission of the light valve system includes the efficiency of the dichroic beam-splitting and -combining components as well as the projection optics transmission. Typical systems use 150–300-W tungsten or metal halide lamps, achieving between 100 and 300 lumens out of the projection optics.

One major advantage of an active-matrix liquid-crystal light valve display is the ease of convergence. The image geometry is controlled by the pixel geometry, which is fixed. The individual pixels do not move with respect to each other, and become nonlinear. Two light valve pixel geometries that are manufactured identically need only initial mechanical alignment and will stay converged thereafter.

Active-matrix LCLV projectors operate at video rates with sufficient modulation and contrast to provide video images. These systems have expanded the applications for projection displays, as they have provided a small-size projection display suitable for video and data presentation on screen sizes in between those of direct-view and projection CRTs. The smallest screen size available with CRT projection systems is about 40 in. diagonal, whereas AMLCLV projectors can project onto screens as small as 20 in. and are much smaller, lighter, and less expensive than CRT projection displays. Figure 6.36 compares the size of an active-matrix LCLV video projector with that of a CRT video projector. The light

Figure 6.36 Comparison of Sony CRT projection display (left) and Kodak/Seiko-Epson AMLCLV projection display (right).

valve systems do not yet reach into the higher performance end of the CRT systems but have considerably expanded the applications for projection systems in smaller screen sizes and have the potential to increase their capabilities substantially.

6.4.2 Optically Addressed LCLV Projection Displays

Optically addressed LCLVs combine the attributes of a high-luminance white-light source and a high-resolution, low-luminance image source. A high-resolution image generator, such as a CRT, is used to modulate high-intensity white light, achieving higher luminance than that possible by magnifying the CRT image. The larger sizes and higher cost of these displays limit their use to high-performance applications.

Operating Principles of Optically Addressed LCLV Projection Displays

Optically addressed liquid-crystal light valves are used in a reflective mode, as shown in Figure 6.31. In this configuration an optical image is written onto one side of the light valve. The image source is most commonly a CRT but can be a scanned laser image or other image source. A high-luminance white-light source is used to supply polarized light to the side opposite the writing side of the light valve. The light valve varies the polarization of the reflected light in proportion to the luminance of the written image. A polarizer/analyzer pair turns this polarization modulation into a gray-scale image that is projected onto a screen.

Hughes Aircraft Company (HAC) designed, developed, and marketed a reflective liquid-crystal light valve that works in this manner (Efron et al., 1981; Grinberg et al., 1975). They also developed projection displays that use the liquid-crystal light valve (Bleha et al., 1977; Ledebuhr, 1986; Fritz, 1990).

The structure of the Hughes LCLV is shown in Figure 6.37. The light valve combines an ac-driven photoconductor/dielectric mirror substrate with a nematic liquid crystal operated in the voltage-controlled birefringence mode. Between two transparent conductive electrodes of indium tin oxide are a photoconductor, a light-blocking layer, a dielectric mirror, and a liquid-crystal layer. The image is written on the photoconductor side. An ac bias voltage is applied between the transparent electrodes. Where there is no image light impinging on the photoconductor, the ac bias voltage is primarily across the photoconductor and not the liquid crystal. When an image pixel is on, light impinges on the photoconductor, the impedance at that point drops, and the voltage is across the liquid crystal. This voltage across the liquid-crystal layer is used to vary the birefringence of the layer.

Polarized illumination generated by a xenon arc lamp enters the projection side of the light valve. It passes through the liquid-crystal layer, reflecting off the dielectric mirror and back through the liquid-crystal layer before exiting the light valve. Voltage across the liquid-crystal layer varies the birefringence of the layer,

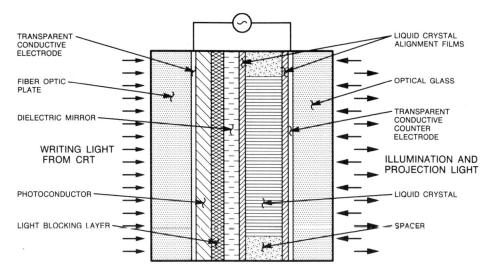

Figure 6.37 Structure of Hughes Aircraft Company optically addressed LCLV (Gold and Ledebuhr, 1985).

which alters the plane of polarization of the reflected light as it passes through the liquid crystal.

No image light impinging on the photoconductor results in projection illumination exiting the LCLV in the same polarization state as it entered and being blocked by the polarizer/analyzer (Fig. 6.38).

Image light impinging on the photoconductor results in a voltage drop across the liquid crystal. Projection illumination emerges with a polarization that is rotated 90° to the incident light and passes through the polarizer/analyzer. This full "on" condition of the light valve is illustrated in Figure 6.39.

Light levels and resulting voltage gradations between full on and full off create the gray scale of the image.

The projection display systems that Hughes has marketed use a CRT to write the image onto the light valve, but other sources are feasible. A display system has been fabricated that uses a scanning laser beam to write the image (Trias et al., 1988).

Characteristics of Optically Addressed LCLV Projection Displays

Full color is created by using three different light valves, one each for red, green, and blue image light, as in the active-matrix-addressed LCLV display. Figure 6.40 illustrates the layout of a full-color Hughes projection display using three optically addressed LCLVs and CRTs to provide the writing image.

The luminance of a display using the Hughes LCLV is dependent on the lamp

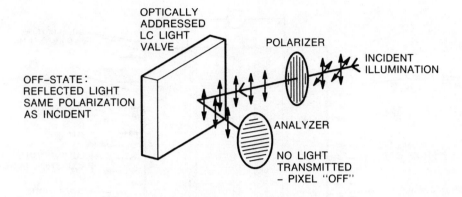

Figure 6.38 Optically addressed LCLV in "off" state.

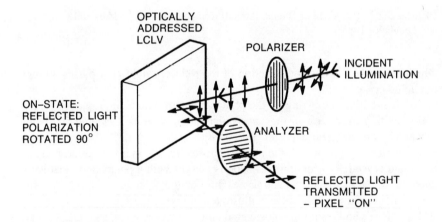

Figure 6.39 Optically addressed LCLV in "on" state.

efficacy, collection efficiency, optical system transmission, and light valve transmission [Eq. (6.4)]. A range of systems are available, with varying xenon lamp sizes and performance characteristics. Lamp sizes run from 500 to 2000 W, with resulting outputs up to 2500 white lumens.

The resolution of these systems is 1024 visible scan lines × 1400 pixels per line, with contrast ratios reported to be 50:1 (Fritz, 1990).

The systems presently being produced use a cadmium sulfide photoconductor, which is operated at 30 Hz interlaced video rates but is not as fast as desired.

PROJECTION DISPLAY TECHNOLOGIES

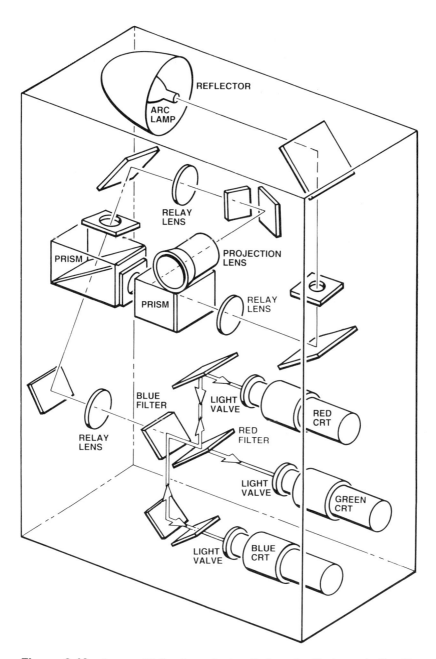

Figure 6.40 Layout of full-color projection display using Hughes optically addressed LCLV (Ledebuhr, 1986).

Recently, a new light valve design was reported (Sterling et al., 1990) that uses an amorphous silicon photoconductor, promising faster operating speeds.

The geometry and linearity of the image are determined by the writing CRT, so the convergence and registration techniques are the same as for a CRT system (Section 6.7). The Hughes displays use an active convergence feedback technique to provide misconvergence detection and correction.

Optically addressed LCLV systems are used to provide high-performance large-screen displays for commercial and military/aerospace environments. The systems are used with screen sizes ranging from 1 to 5 square meters. Ruggedized versions of these systems are being used in naval shipboard applications (Fritz, 1990; Gold, 1980). Other applications include large-screen command and control (Gold and Ledebuhr, 1985) and simulation applications (Sterling et al., 1990). Figure 6.41 pictures a projection display system using the Hughes optically addressed LCLV. This system is 24 in. wide × 72 in. high × 44 in. deep.

Figure 6.41 Picture of Hughes Aircraft projection display using CRT-addressed LCLV. This system is 24 in. wide, 72 in. high, and 44 in. deep. (Photo courtesy of Hughes Aircraft Co., Fullerton, Calif.)

6.4.3 Thermally Addressed LCLV Projection Displays

Thermally addressed LCLVs are used to create full-color systems capable of displaying very high density alphanumeric and graphic data, such as maps or large CAD and engineering drawings. This extremely high resolution capability is the principal advantage of thermally addressed LCLV projection displays. In a situation similar to optically addressed LCLV displays, a high-resolution image source—in this case a deflected IR laser beam—is used to create the image and modulate a high-luminance white-light source. Thermally addressed LCLVs do not operate at video rates, however, and this low speed is the main disadvantage of such displays, limiting their use to applications where real-time speeds are not needed.

Operating Principles of Thermally Addressed LCLV Projection Displays

A smectic liquid-crystal light valve can be designed to have two different states: a transparent state and a scattering state. Smectic liquid crystals exhibit a hysteresis effect, and the present state depends on its history and temperature (Kahn, 1973). In a display application the temperature is controlled to be slightly below the transition temperature where the cell turns from clear to scattering. Local heating causes local areas of scattering, which has little effect on neighboring areas, and so will have very sharp edges. This is illustrated in Figure 6.42.

The thermally addressed LCLV is used in a display application in a reflective mode, with an IR laser providing the writing illumination and a white-light source providing the projection illumination (Fig. 6.43). The IR laser (usually a

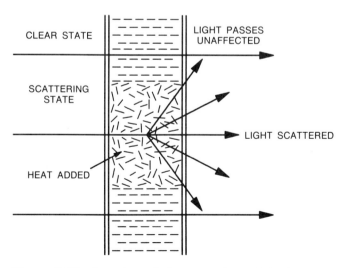

Figure 6.42 Scattering and clear states of smectic liquid-crystal light valve.

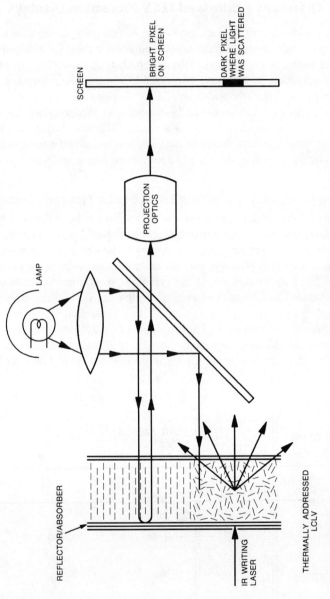

Figure 6.43 Layout of projection display using thermally addressed LCLV.

laser diode) is modulated with video information and deflected across the light valve to write the image (laser modulation and deflection techniques are covered in Section 6.5). Coatings on the light valve reflector/absorber are designed to absorb IR laser light, which heats the liquid-crystal material, while reflecting projection light.

Wherever the IR laser beam writes onto the light valve, the region undergoes local heating, which turns that region (pixel) into a scattering state. Illumination light hitting the liquid crystal is highly scattered, with little light being accepted by the projection lens. This creates a dark pixel on the screen where the laser writes on the light valve. In unwritten areas, the liquid crystal is transparent. Projection illumination reflects off the light valve unaltered and is collected by the projection optics and imaged onto the screen, creating a bright pixel.

The light valve has a semipermanent memory. After being written, an image remains until erased or updated. A local change requires only local erasing and rewriting, as opposed to rewriting the entire image.

Characteristics of Thermally Addressed LCLV Projection Displays

Just as with other LCLV projection displays systems, color is implemented by using multiple monochrome light valves, and the color gamut is largely determined by the spectral characteristics of the dichroic filters used to separate the white light into components.

The resolution capability of thermally addressed LCLV projection displays is unmatched by other types of projection displays. These systems have been under investigation by several companies (Dewey, 1984; Tsai, 1981). Several systems are offered as products: the Hitachi liquid-crystal large-screen display and the Greyhawk Softplot and LAD systems. The Hitachi system (Nagae et al., 1986) projects an image with an addressable resolution of 2000 × 2000 pixels onto a 6.5 ft × 6.5 ft screen, with a resulting image luminance of 40 fL. Greyhawk has several systems (Stepner and Kahn, 1986; Kahn et al., 1987), including the Softplot, a 40-in. diagonal display with a resolution of 3400 × 2200, and the LAD (large-area display) system with a 7 ft × 10 ft screen and 5000 × 7500 pixels. Figure 6.44 pictures the Greyhawk LAD display.

These systems are used for static images, with writing times for a whole screen ranging from 30 secs to 30 min. Particular areas can be erased and rewritten, allowing interactive changes and updates.

Thermally addressed LCLV displays are used for displaying and checking circuit diagrams and engineering drawings; for displaying maps, networks, and command and control information; for displaying and monitoring plant operations; and for displaying other high-information-content static images, which alternatively must be plotted out on hard copy to view. Their resolution is higher than other projection display systems considered in this chapter, but video rate operation is not possible at this time.

Figure 6.44 Picture of Greyhawk LAD projection display using thermally addressed LCLVs. (Photo courtesy of Greyhawk Systems, Inc., Milpitas, Calif.)

6.4.4 Liquid-Crystal Light Valve Projection Displays: Summary

Liquid-crystal light valves are implemented into display applications using a variety of addressing techniques, including active-matrix addressing, optical addressing, and thermal addressing. These displays have in common the advantage that their resolution and luminance are not interrelated. Besides this common advantage, each type has its own distinguishing characteristics. Active-matrix-addressed LCLV projection displays have added a new class of small, lightweight, low-cost video projectors competing with CRT projection displays. Optically addressed LCLV projection displays are high-luminance, high-resolution systems but are also large and relatively expensive. Thermally addressed LCLV projection displays possess resolution characteristics unmatched by other projection displays but are not currently capable of operation at video rates.

6.5 LASER PROJECTION DISPLAYS

A laser projection display creates an image by writing directly onto the projection screen with a laser beam. The laser light is diffused by the screen, making the real image visible to the viewer. Laser displays possess several inherently high-quality aspects: the fully saturated colors, the high-resolution capability of a focused laser beam, and the high luminance and contrast capability of lasers. Laser displays have not achieved a large amount of commercial success, primarily because of the size and inefficiency of lasers themselves. Recent progress in small visible lasers is creating practical alternatives to the larger lasers.

6.5.1 Operating Principles of Laser Projection Displays

In a laser projection display the image is written on the viewing screen with a scanning laser beam, very much like an image is written on a CRT with an electron beam. The screen does not luminesce, however; instead, the laser light is diffused by the screen, which is placed at the real-image location. The basic components of a laser projection display, shown in Figure 6.45, are the laser light sources; the modulators, which encode the laser beam with intensity variations corresponding to the video information; the deflectors, which provide movement of the laser beam to trace out the image on the screen; and the screen itself.

Monochrome laser projection displays use a single laser. To create a full-color laser display, one laser for each color is used. It is possible to obtain "white" lasers, which combine several lasing sources into a single package.

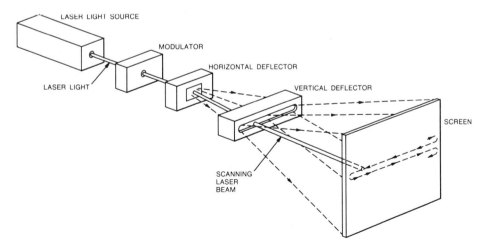

Figure 6.45 Laser projection display components.

Historically, laser projection displays have used argon-ion lasers for the green and blue colors and either a krypton laser or an argon-ion pumped dye laser for red. These lasers are inefficient, and to obtain the watts of output power required, water-cooled lasers requiring kilowatts of input power are necessary. This has limited the use of laser displays to extremely large image size systems, such as concert hall and laserium displays.

Recent developments in diode lasers and diode-pumped solid-state lasers hold the potential to open up the range of applications for laser projection displays. Diode lasers have seen their introduction into visible wavelengths and have been steadily increasing in power and decreasing in wavelength. Moreover, diode-pumped solid-state lasers are providing visible light with efficiencies that are orders of magnitude greater than that of argon-ion and dye lasers. These small, efficient, visible lasers have the potential to make small laser displays a reality. Figure 6.46 is a picture of a small 532-nm diode-pumped solid-state laser.

Video information is encoded into the laser beam with a modulator synchronized with the scanning mechanism. There are numerous methods that can be used to modulate a laser beam (O'Shea, 1985), the most common being an acousto-optic modulator.

An acousto-optic (A-O) modulator uses an acoustic signal to create a bulk diffraction grating in a crystal (Yariv and Yeh, 1984). This is accomplished by using the acoustic wave as a pressure wave applied to the crystal, which creates

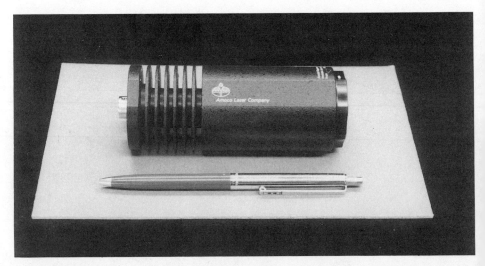

Figure 6.46 Picture of Amoco laser diode-pumped solid-state laser with 532-nm output.

a periodic variation of the index of refraction within the crystal. This is seen by a laser beam passing through the crystal as a bulk diffraction grating.

Acousto-optic modulators operate in the Bragg regime (Lekavich, 1986), where, at the particular Bragg incidence angle, most of the light is diffracted into the first order. The diffraction efficiency (percentage of light diffracted into the first order) depends on the amplitude of the acoustic driving signal. Laser beam modulation is implemented by varying the amplitude of the acoustic drive signal, which in turn varies the amplitude of the light passed to the first order (Fig. 6.47). The zero order is blocked, while the modulated first order travels through the rest of the system and onto the screen.

Laser beam deflection can be accomplished by one of several means, depending on the speed, size, and accuracy desired. Deflection techniques include mechanical mirror deflection (specifically rotating polygon, galvanometer mirror, and hologon deflection) and acousto-optic deflection.

A rotating polygon with mirror facets is used when repetitive scans at a fixed frequency are desired. A motor rotates the polygon, and the laser beam reflects off the mirrored facets. As the polygon rotates, the angle of incidence of the laser beam is changed, which in turn changes the angle of reflection, and the laser beam traces out one line for each facet. As the polygon continually rotates, the laser will continuously trace out a horizontal line (Fig. 6.48). The polygon is

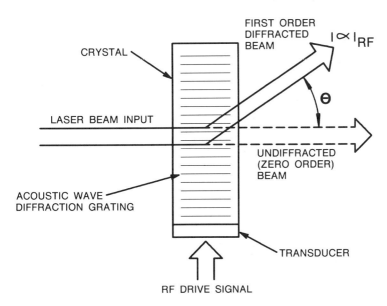

Figure 6.47   Principles of acousto-optic laser modulation.

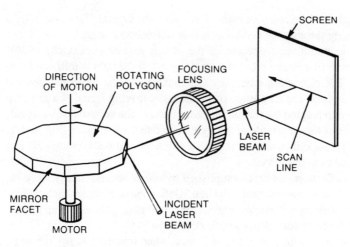

Figure 6.48 Rotating polygon laser deflection.

commonly used in laser raster projection displays to provide horizontal deflection.

Polygon deflection frequency depends on the revolutions per minute of the polygon and the number of facets on the polygon. In general, very high speeds, and very high resolutions, can be achieved with the rotating polygon. In display applications, a polygon mirror is used for the horizontal deflector in raster displays with addressabilities ranging from 525 scan line TV systems to HDTV systems running at 1125 scan lines. The performance trade-off is that of speed versus size. To provide many scan lines, the number of polygon facets can be increased, but this results in an increase in polygon size and in power required. Alternatively, the polygon rotational speed can be increased, providing a smaller polygon, but there is a limit to how small and how fast a polygon can be operated.

A new twist on the rotating polygon is the rotating hologon, which uses holographic segments to reflect the laser beam as the hologon turns (Fig. 6.49). The holograms are quite easy to replicate, making their production costs low. An example of a popular use of hologons is in supermarket scanners.

A galvanometer deflector uses a moving coil principle to provide single-axis rotation of a mirror, which in turn deflects the laser beam, as shown in Figure 6.50. Galvanometer mirrors are commonly used in a random access mode for large outdoor laser displays in stadiums, amusement parks, or concert halls. They are also used in a resonant mode to provide a raster deflection pattern. The speed of deflection is limited to below about 25 kHz, so this type of deflector is commonly used as the vertical deflection device in raster laser projection displays.

PROJECTION DISPLAY TECHNOLOGIES

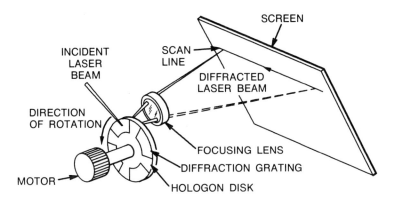

Figure 6.49 Rotating hologon laser deflection.

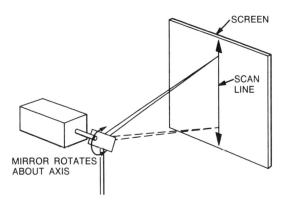

Figure 6.50 Galvanometer mirror laser deflection.

Acoustic-optic deflectors work very similarly to acousto-optic modulators. A refractive index grating is created in a crystal with an acoustic wave. Laser light incident at the Bragg angle diffracts into a strong first order. For deflection to occur, the frequency of the acousto-optic wave is changed, which varies the angle of diffraction of the first order (Fig. 6.51), causing the laser beam to trace out a line. This type of deflection is commonly used for the horizontal deflection in a raster laser display.

The acousto-optic modulator and deflector are very similar in operation, the difference being that the A-O modulator varies acoustic drive *amplitude* to modulate the first order, and the A-O deflector varies acoustic drive *frequency* to deflect the first-order beam.

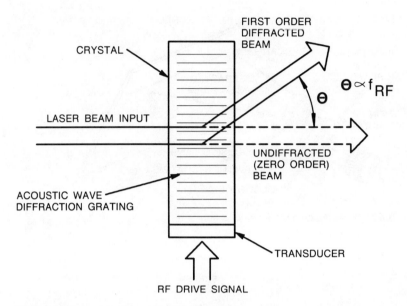

Figure 6.51 Acousto-optic laser deflection.

There are numerous methods for implementing these scanning and modulating components in a laser projection display (Hubin, 1991; Johnson and Montgomery, 1976; Merry and Bademian, 1979) and design rules that help determine the best method for a particular application (Beiser, 1974, 1986; Zook, 1974; O'Shea, 1985).

Laser video displays typically consist of an acousto-optic modulator, a galvanometer mirror for vertical deflection, and either a polygon mirror or an acousto-optic deflector providing the horizontal deflection. Figure 6.52 shows the configuration of a system using a polygon mirror, and Figure 6.53 gives the configuration for a system using an acousto-optic deflector.

Speckle is a phenomenon unique to laser systems, occurring because of the coherent nature of laser light. Speckle is the sparkling/granularity effect visible in laser images, which comes about from the interference of the coherent laser beam with itself after passing through or reflecting off a diffuse screen. The coherent laser beam is redirected by the screen, and then different parts of the beam interfere with each other to set up an interference pattern in space. The positive and negative interference regions cause light and dark spots to appear in the image, which move as the viewer moves within the viewing volume (because the pattern is not on the screen, it is in space). This movement of the speckle pattern causes the sparkling effect.

Speckle is present in most laser displays, including those that are rastered and/

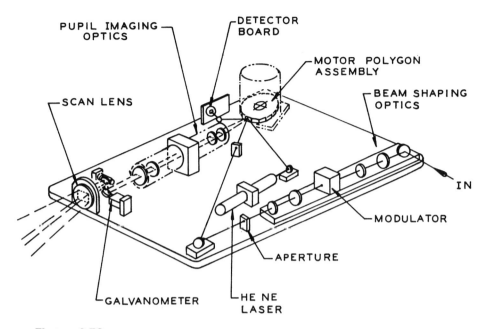

Figure 6.52 Layout of the Lincoln Laser RS-3A laser raster video display utilizing polygon horizontal deflection. The He-Ne laser is used to derive sync signals for the system. (Courtesy of Lincoln Laser Corp., Phoenix, Ariz.)

or full-color or multicolor laser displays. Its presence is noticed less at lower luminance, and it can become quite brilliant at higher luminances. The principle behind speckle-removal techniques is to either remove the coherency of the light or to overlay many different speckle patterns in space so that they average out to be a smooth image (Welford and Winston, 1989). The most common speckle-removal technique is to place a moving diffuser at an intermediate image plane, which overlays multiple interference patterns at the viewing plane image.

6.5.2 Characteristics of Laser Projection Displays

The color gamut of a particular laser display depends on the wavelengths of the lasers used in the system. The color gamut of laser displays is typically larger than that of other display types because the colors are fully saturated, lying on the outside of the chromaticity diagram, as shown in Figure 6.54. Changing the laser wavelengths slides the triangle corners along the outside of the CIE diagram.

The resolution of a laser display depends on the electrical bandwidth of the laser modulator, the speed of the scanning device, and the smallest spot that the

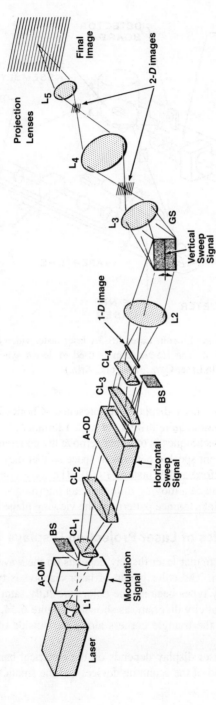

Figure 6.53 Components of laser projection display using acousto-optic horizontal deflection (A-OD) and acousto-optic modulation (A-OM) (O'Shea, 1985). BS = beam stop, CL = cylindrical lens, and L = lens.

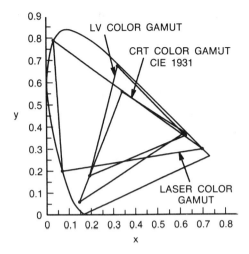

Figure 6.54 Color gamut of laser projection display, using green at 514 nm, red at 647 nm, and blue at 488 nm.

laser display can be focused to. Owing to its coherent nature, laser light can be focused to a very small spot. This focused spot can be so small that in practice the laser beam may have to be defocused because scan lines are visible. The laser beam can also be collimated so the distance to the screen does not affect focus. This characteristic can be very desirable and is unique to laser displays. The speed of the scanning device and resulting system resolution depends on the particular scanning device and implementation.

The luminance of a laser display depends on the laser power used and the light throughput of the display system.

$$B_s = W_{laser} \eta_{laser} T_{system} G/A \qquad (6.5)$$

where W_{laser} is the power out of the laser in watts; η_{laser} is the luminous efficacy of the laser, in lumens per watt; and T_{system} is the system transmission. The system transmission depends very much on the deflection system used, varying anywhere from less than 5% to over 50%. Acousto-optical components are not as efficient as mirrors, so the more A-O components in a system, the lower the transmission. However, A-O components are compact and have no moving parts, a desirable feature in many applications.

Laser displays can operate at video rates, with modulation, contrast, and speed consistent with these requirements. Unlike many other display systems, the laser display has no persistence or memory.

Convergence must be addressed in color laser displays, as with other projection displays. This is very often handled with photosensor devices that pick off a portion of light to determine the amount of misconvergence, providing feedback to a mirror or deflector that performs correction. Convergence techniques are discussed further in Section 6.7.

The size, weight, and power consumption of laser displays vary widely. Since the most common application of laser displays is to create images larger than any other projector is capable of handling, the systems tend to be large and power-hungry. Entertainment establishments such as Disneyland and SeaWorld use laser displays to provide large-scale visual effects unobtainable with other light and image sources.

Although the most common application of laser displays is for very large custom displays, several companies have marketed laser displays for applications where the screen size is less than 25 ft. The Naval Training Equipment Center has been very successful in implementation of a laser display for a flight simulator (Barber, 1984). Several small 525-line systems are available, such as the IntraAction system shown in Figure 6.55. This system is a monochrome projector using an A-O modulator, an A-O horizontal deflector, and a galvanometer mirror vertical deflector.

Figure 6.55 Monochrome laser video projection display (IntraAction Corp., Bellwood, Ill.).

6.5.3 Laser Projection Displays: Summary

Laser projection displays write an image directly onto the projection screen with one or more laser beams. Laser projection displays have several high-quality aspects due to the coherent, monochromatic nature of laser light, such as high resolution and a large color gamut. The large size and inefficiency of the laser light sources have kept these displays from becoming very popular, but recent advances in small, efficient visible lasers promise new territory for laser projection displays.

6.6 PROJECTION DISPLAY SCREENS

6.6.1 Introduction

The projection screen is a very important part of the projection display. A high-quality, high-resolution image source and projection optical system can be degraded by a projection screen that does not preserve the image quality of the display.

This chapter covers screens placed at the real image plane of the projection system as diffusers to view the image. Virtual image displays and the "screens" used with them do not fall into this category.

The characteristics of the projection screen can determine the final image quality of the displayed image, including luminance, resolution, contrast, and color. The goal of a projection screen design is to present the projected image to the viewer with little to no image quality degradation within a specified viewing volume. This is accomplished by using various types of diffusion and lens action, including refraction and reflection provided by lenslets, diffusion by scattering centers, and absorption from dyes.

Projection screens are usually designed to be used as either a front projection screen or a rear projection screen, usually not both. These two types of screens are shown in Figures 6.2 and 6.3. Front projection screens reflect the projection light into the viewing volume. Rear projection screens transmit the projected light through the screen to the viewer. Two advantages of front projection are that the screen can be curved to provide gain, and no projection space is needed behind the screen. The advantages of rear projection screens are that less ambient illumination is directed into the viewing volume and the display can be made more compact by folding.

Front projection used to be the more popular and still is for large auditorium presentations where the room is relatively dark. Rear projection has become the more popular implementation for home TV systems and other applications where ambient illumination may be a problem.

Front and rear projection screens are characterized by the same parameters: gain, reflectance, colorimetry, and contrast under ambient illumination. These parameters are measured and given in terms of bend angle. The bend angle is the angle through which a principal ray of light from the projector must bend, as it hits a point on the screen, to get directed into the viewer's eyes. The viewing angle, sometimes confused with the bend angle, is the angle between the viewer's line of sight and the normal to the screen. If the viewer is looking at the center of the screen, then the viewing angle is equivalent to the bend angle. Figures 6.56 and 6.57 illustrate these concepts.

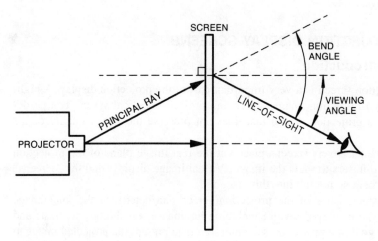

Figure 6.56 Bend angle vs. viewing angle.

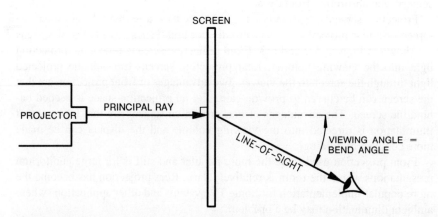

Figure 6.57 The bend angle is equivalent to the viewing angle when the viewer looks at the center of the screen.

The viewing zone of a particular projector–screen combination is that range of angles wherein the image quality, including luminance, resolution, and contrast, falls within specified parameters.

6.6.2 Screen Types

The number of types of screens available for use with projection displays has expanded greatly in recent years. In addition to standard diffusion techniques to create a real image, optical elements have been added to the screens to direct the light and tailor the viewing zone in such a way that little light is wasted. Screens are now available with lenticular lenslets, Fresnel lenses, and contrast-enhancing black stripes in addition to the diffusing element. These elements combine to make the screen a highly developed optical system.

Diffuse screens, those that use only diffusion to create an image and a viewing zone, include front diffuse screens, rear diffuse screens, flexible diffuse screens, and rigid diffuse screens. This screen category includes ground glass and opal glass. There is an extremely wide range of types of diffusion and the resulting characteristics.

Flexible diffuse screens have a diffusing coating applied to a vinyl substrate. They are low-cost and lightweight and can be made in very large sizes. Flexible diffuse screens can be rolled up when not in use. Some flexible diffuse screens can be used for both front projection and rear projection applications. Flexible screens are sometimes dyed to add contrast or have holes perforated in them to let sound through. A major drawback to flexible diffuse screens is that they are subject to motion during use as a result of air currents or pressure differentials within the environment, causing unwanted image distortions.

Rigid diffuse screens can be fabricated from glass or plastic, with either bulk or surface diffusion added. Surface diffusion is created by grinding, acid etching, or coating the surface of a glass or acrylic substrate. Bulk diffusion is created by adding particles within the substrate, which cause progressive diffusion as the light travels through the screen (Goldenberg and McKechnie, 1985). Rigid diffuse screens include ground glass, opal glass, and marata plates, among others.

Fresnel lenses are used with rear projection screens to direct the rays falling on the screen's outer edges toward the viewer. This helps create an image with even luminance across the screen, avoiding image luminance rolloff at the edges. The operation of a Fresnel lens used with a diffuse rear projection screen is illustrated in Figure 6.58.

Lenticular lenslets are used with both front and rear projection screens to tailor the light distribution. Von Rolf Moller first discussed the use of lenticular screen elements in 1939 (Moller, 1939). The lenslets are usually cylindrical, although both spherical and toroidal lenslets have been discussed and demonstrated (Henkes, 1982; Mihalakis, 1987; Takatsuka et al., 1982). The lenslets use both reflection and refraction to distribute image light into a particular viewing zone.

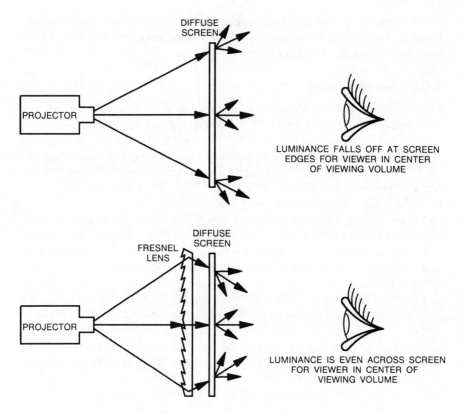

Figure 6.58 Action of Fresnel lens with rear projection screen.

Their most common implementation is to widen the image viewing zone in the horizontal direction while not affecting the vertical (Fig. 6.59). Recent designs have also implemented lenslets that account for color separation caused by off-axis CRTs.

Lenticular structures in a rear projection screen may include black stripes, which are used to absorb ambient illumination and lower the overall reflectance of the screen (Bradley et al., 1985). Lenslets are used to direct image luminance away from the black stripes but allow ambient illumination to be absorbed.

High-performance projection screens have become complex and detailed systems. The screens used with consumer projection TVs consist of a Fresnel lens, a diffusion layer, at least one set of lenticular lenslets, and black stripes. Figure 6.60 shows an example of a complete screen structure using all of these elements.

PROJECTION DISPLAY TECHNOLOGIES

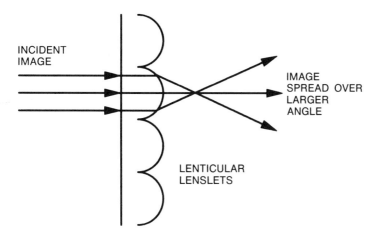

Figure 6.59 Action of lenticular lenslets.

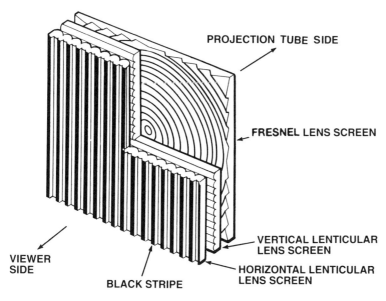

Figure 6.60 Structure of rear projection screen using Fresnel lens, horizontal and vertical lenticular lenslets, and black stripes (Murakami et al., 1989).

6.6.3 Screen Characteristics

Gain is a measure of the relative luminance of an image provided by a particular screen and is probably the most important and most commonly used measure of comparison between screens. Screen gain curves describe the relative image luminance versus bend angle provided by a particular screen. The *gain* of a screen is defined to be its luminance at a given angle relative to the luminance that would be achieved if a Lambertian screen were used:

$$G(\theta) = B_S(\theta)/B_L \tag{6.6}$$

where $G(\theta)$ is the screen gain as a function of bend angle, $B_S(\theta)$ is the screen luminance as a function of angle, and B_L is the luminance that would be achieved if the screen were Lambertian. A Lambertian screen is considered to be perfectly diffuse, diffusing light into all angles with equal luminance, and its luminance is therefore not angle-dependent. Figure 6.61 illustrates the viewing zone of a perfectly diffuse Lambertian screen, defined to have a gain of 1 at all angles.

The gain of a screen is often referred to and used in calculations without its angle dependence, which implies that the on-axis (0°) angle is being used.

Gain versus bend angle curves show how a screen distributes the image luminance. A screen with a gain greater than 1 at a particular angle directs more light to that direction than a Lambertian screen would. Conservation of energy cannot be violated, of course, and so a screen with high gain at some angles must have lower gain at other angles. Figure 6.62 shows the viewing volume of a screen with a gain greater than 1 at many angles.

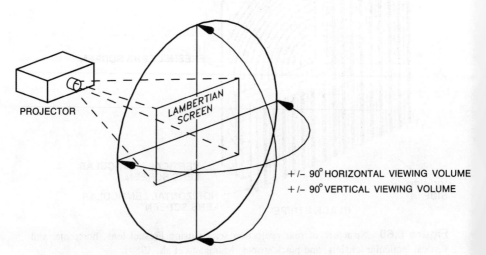

Figure 6.61 Lambertian screen viewing volume (gain equals 1 at all angles).

PROJECTION DISPLAY TECHNOLOGIES

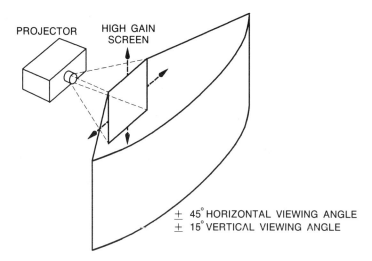

Figure 6.62 High-gain screen viewing volume.

Figure 6.63 shows the gain curves for a set of six different rear projection screens: two rigid diffuse, two flexible diffuse, and two composite screens. The composite screens each consist of a Fresnel lens, a diffuser, and a set of lenticular lenslets. The high-gain screens provide high luminance on-axis, but the luminance falls off rapidly, resulting in a smaller viewing volume. The lower gain screens have a lower luminance on-axis, with a correspondingly larger viewing volume.

Composite screens do not have the same gain curves in the horizontal and vertical directions, owing to the action of the lenticular lenslets. Figure 6.64 compares the horizontal and vertical gain of the two composite screens of Figure 6.63, which both have the lenticules oriented vertically. The widening of the viewing zone in the horizontal direction is evident. The other four screens of Figure 6.63, which do not use lenticular lenslets, have circularly symmetric gain curves.

Reflectance is a useful parameter for the characterization of rear projection screens. A front projection screen is designed to reflect, and the gain curve illustrates how well and in what form the screen does this. For rear projection screens, however, the gain curve shows how well the screen transmits and tailors the light distribution. It is still necessary to determine what portion of the ambient illumination will be reflected into the viewing volume. Reflectance curves are useful in determining this. Projection screens have two reflectance terms: diffuse reflectance and specular reflectance. Specular reflectance is mirror reflectance, where little scattering occurs and the angle of reflection is equal to the

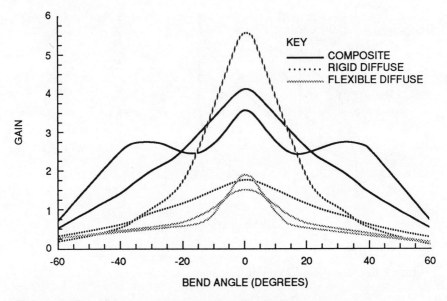

Figure 6.63 Gain vs. bend angle for six rear projection screens.

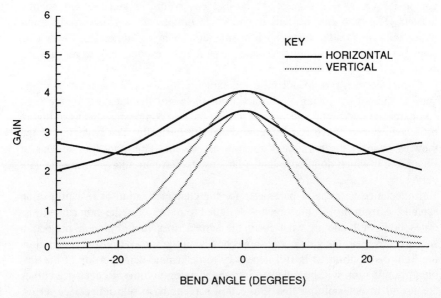

Figure 6.64 Horizontal and vertical gain for two rear projection screens with lenticular lenslets running vertically, widening horizontal viewing zone.

PROJECTION DISPLAY TECHNOLOGIES

angle of incidence. Diffuse reflectance is scattered reflection, which occurs within a large range of angles. Diffuse reflectance causes a general image washout (low contrast), whereas specular reflectance may not be noticeable at most angles but can render the image unviewable at the particular specular angle.

Figure 6.65 shows reflectance versus angle for the six rear projection screens of Figure 6.63. Both specular and diffuse reflection is evident. The illumination was incident from the $+60°$ angle, leading to the greatest amount of reflection (the specular reflection) occurring at $-60°$. The specular reflection falls off rather quickly, and the amount of diffuse reflection can be read from the positive angle readings. The composite screen that does not exhibit strong specular reflection contains black reflection-inhibiting stripes within its structure for just this purpose.

Direct measure of image contrast provides information on how the gain and reflectance characteristics of the screen combine to present an image on the screen. Figure 6.66 shows image contrast versus bend angle for four rear projection screens under room ambient illumination (30–60 fc) conditions. All of the screens shown provide contrast sufficient for comfortable viewing over a wide range of angles. Figures 6.67 and 6.68 illustrate what happens to the image contrast as the ambient illumination incident on these screens is increased to 1000 fL and then to 2500 fL. The effect of both specular and diffuse reflection can be

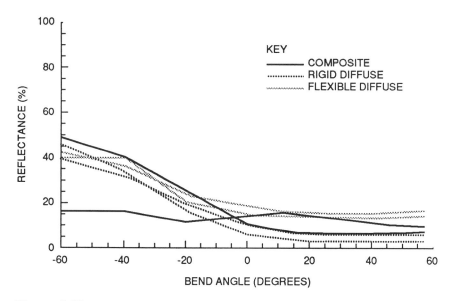

Figure 6.65 Reflection vs. bend angle for six rear projection screens from Figure 6.63.

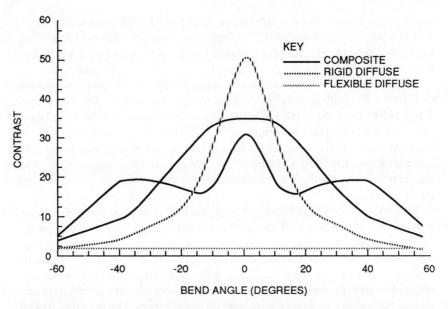

Figure 6.66 Contrast vs. bend angle in room ambient illumination for four rear projection screens.

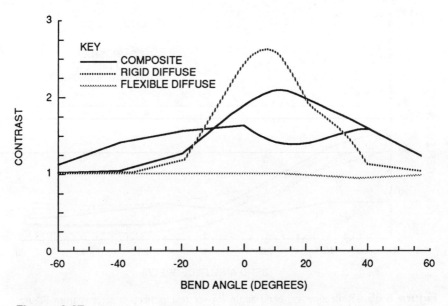

Figure 6.67 Contrast vs. bend angle in 1000-fc ambient illumination for the four screens from Figure 6.66.

PROJECTION DISPLAY TECHNOLOGIES

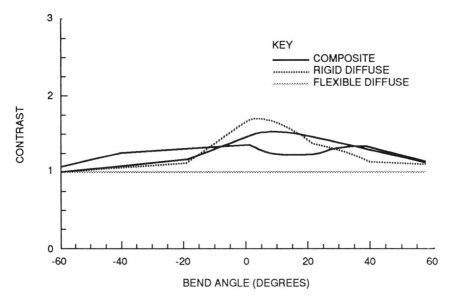

Figure 6.68 Contrast vs. bend angle in 2500-fc ambient illumination for the four screens from Figure 6.66.

seen. The ambient illumination, incident from +60°, lowers contrast the most where specular reflection occurs.

In addition to image contrast, the projection screen also can affect the colorimetry and resolution of the image, as well as introduce other artifacts that may be disturbing. The color balance of off-axis projection systems may not be even across the screen because of the differing incidence angles. Some lenticular lenslets are designed to correct for this effect. The lenticular lenslets can degrade the image resolution if they are not designed and implemented properly. The lenslet should be several times the size of the image pixel, or disturbing moiré effects occur. This depends on the exact lenslet design, however, and the absolute pixel size changes with changes in screen size, so each case must be evaluated separately. Whether effects such as these are significant depends on the specific application and implementation of the projection system. Recent works (Jenkins, 1981; Bradley et al., 1985) have characterized some of these phenomena.

6.6.4 Projection Screen Summary

Projection screen choices have expanded greatly in recent years, and choosing a screen for a particular application should take into account the specific image characteristics desired. Screen gain is the most useful screen parameter, describing the image luminance versus bend angle achieved from a particular screen.

High-gain screens have higher luminance at particular angles, but the luminance tends to fall off rather sharply, giving a smaller viewing angle. Low-gain screens supply lower image luminance, but the luminance remains constant over a larger range of angles, providing a larger viewing volume. Screen optical systems such as Fresnel lenses and lenticular lenslets have improved the light-tailoring ability of projection screens, providing even image luminance and a tailored, nonsymmetric gain curve. Ambient illumination degrades image contrast very rapidly. In high ambient illumination environments, rear projection screens with absorbing black stripes are helpful in maintaining image contrast.

6.7 CONVERGENCE AND IMAGE BLENDING

In most full-color projection displays the final image on the screen is generated by projecting and superimposing the images from separate monochrome image sources. Converging the display is the process of aligning the monochrome images on the screen so they overlap to create a full-color image. The individual images must be corrected for nonlinearities and geometric distortions and be focused across the screen. These functions are all included in the convergence process.

Mismatch between projected images is also a problem when the images from more than one projection display are tiled (i.e., mosaicked) to create a single picture. Image blending is the matching of two or more tiled projected images so edge seams and/or nonuniformities are not visible to the eye. The separate projected images must be aligned with each other for linearity and overlap at the edges, and their focus, luminance, and colorimetry must be matched.

6.7.1 Convergence

Converging a color projection display is necessary not only to align the monochrome images, but also to account and correct for the effects of individual optics and the unique nonlinearities associated with each image source. It has been shown (Mitsuhashi, 1990) that the individual red, green, and blue images must be converged to within ½ pixel to prevent compromising the display resolution. Techniques for converging projection displays include provisions for correcting linearity and focus errors in the image. These controls must be dynamic and adjust for drifting of the focus and geometry of the images over time.

The convergence techniques discussed in this chapter were developed primarily for CRT projection displays but can be applied to all types of projection displays, including laser and light valve displays. However, there are at least two types of display technology that do not need extensive converging, as their design inherently creates converged images. The GE single-light-valve Talaria display requires little convergence, as the colors are created and controlled by the same light valve. Fixed-matrix displays, such as AMLCLV projection displays,

PROJECTION DISPLAY TECHNOLOGIES

are relatively easy to converge because their geometry is fixed. The geometry of the red, green, and blue images with respect to one another does not vary, so initial mechanical alignment of the images with respect to each other is usually all that is required over the life of the display.

Cathode-ray tube geometry and convergence errors are corrected by inputting a correction signal to the deflection current. This can be done by adding a signal to the main deflection yoke or, more commonly, by adding a separate coil to the deflection yoke to accept convergence signals (Fig. 6.69). The correction signal is a polynomial that is a function of the x and y deflection signals. The image is corrected by adjusting the coefficients of the polynomial terms. Each of the terms in the polynomial controls a different type of correction, as shown in Figure 6.70.

Focus across the screen is adjusted in much the same way, by adding a current and/or voltage signal to the focus circuit. When magnetic focusing is used, a separate focus coil is used to provide dynamic focus correction.

Correction signals historically have been adjusted by analog means, using a potentiometer for each of the polynomial terms. Analog circuits drift, however, which means that the display must be reconverged after a period of use. This has led to the development of convergence and focus correction circuitry that is either completely or partly digital to minimize drift, improve accuracy, and lower the difficulty of converging (Holmes, 1987a). Correction values corresponding to the different parts of the image are stored in memory and read out as needed (Fig.

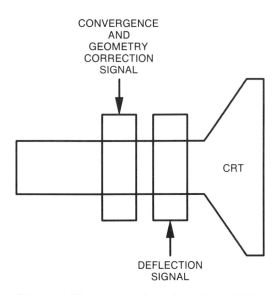

Figure 6.69 Separate deflection yoke on CRT used to provide convergence corrections.

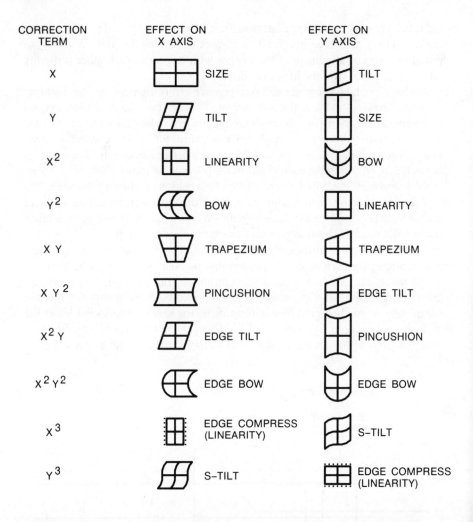

Figure 6.70 CRT convergence and geometry correction functions (Elmer, 1982).

6.71). Storing the correction function in digital memory eliminates that portion of drift caused by the potentiometers.

Each monochrome image source will have at least one correction matrix for convergence errors (sometimes more than one is used) and a correction matrix for focus errors. Several methods have been developed that simultaneously minimize memory requirements and maximize correction accuracy and resolution (Holmes, 1987a; Lyon and Black, 1984). Correction values are stored for spe-

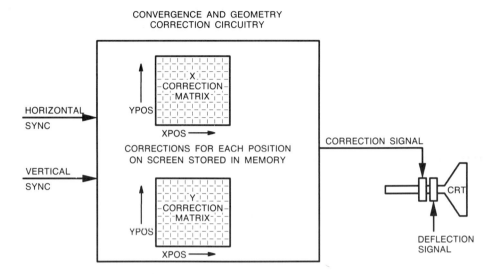

Figure 6.71 Correction matrices store correction values according to position on screen.

cific points, and interim values are interpolated, saving memory space while providing high-resolution correction.

Initial values for convergence and focus correction matrices are determined at the first display setup. A test pattern is projected, usually a grid pattern. An interactive program is used with which the operator enters correction values and immediately sees the effect on the image. The green image is corrected first for nonlinearities and geometric distortions. After the green image is geometrically correct, the red and blue images are projected, and correction values are entered until these images superimpose on the green image. These convergence correction values are stored in corresponding memory chips for each CRT.

Focus correction values are entered in the same manner. The individual images are projected, and focus correction values are manually entered for specific positions across the screen. Again, interpolation is used to fill in the interim points. These values are stored in the focus correction memory for each CRT.

Once the image is initially converged and focused, provisions must be implemented for periodic adjustments for drift over time and temperature. Some display systems permit periodic manual resetting of focus and convergence, thus correcting for changes over time. This can be time-consuming and does not permit continuous system use, so automatic convergence adjustments are becoming standard.

Automatic focus and convergence adjustment systems incorporate feedback from the image to provide data to reconverge and refocus the image. These systems are of many configurations, including optical sensors at the screen (Lyon and Black, 1984) or on a mirror for detecting errors. Some systems use data generated from the actual image to provide feedback, while others project special alignment images outside the field of view or during flyback. One system uses a CCD camera to look at the entire image and detect errors (Kanazawa and Mitsuhashi, 1989); this permits precise error detection and correction.

6.7.2 Image Blending

In applications where the required image luminance and/or resolution cannot be furnished by a single projector, multiple-projector systems can be used to create a single image. Large multisegment images can be assembled in a tiled configuration or an area-of-interest configuration. In either approach the separate projected images must be matched in linearity, geometry, color, and luminance, so the viewer sees a continuous seamless image.

In the tiled approach, the images from several projectors are lined up in a matrix, creating one large image that is a mosaic of smaller images (Fig. 6.72). The segments of the tiled image have similar resolution, creating one large image with constant resolution throughout. This technique is most useful where high resolution is needed in all parts of the display, such as in multiviewer large-screen entertainment systems, simulator displays, or command and control centers.

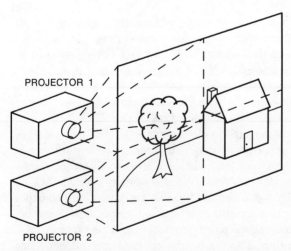

Figure 6.72 Tiled image approach to multiprojector scenes, using four projectors (projectors 3 and 4 hidden).

PROJECTION DISPLAY TECHNOLOGIES

The area-of-interest approach was developed for single-viewer simulator applications (Cowdry, 1985; Spooner, 1982). Projection displays are used in flight simulators to present a wide field-of-view image to the pilot. The projected imagery fills the pilot's visual field of view, giving the sensation that the projected imagery is the outside world. It has been shown that the human eye sees high resolution only in the forward foveal view and not peripherally (Bunker and Fisher, 1984). The area-of-interest technique was developed to take advantage of this fact, providing high resolution only in the direction in which the viewer is looking. In this way display hardware and processing power are not wasted displaying resolution and detail that will not be used. In most cases two projectors are used, one to project a low-resolution background, the other to project a high-resolution inset. These two images can be optically combined and projected onto the screen with a single projection system as shown in Figure 6.73. The high-resolution inset, typically with a field of view of about 25°, tracks the viewer's eyes. The low-resolution background fills the remainder of the field of view.

The area-of-interest technique has the advantage of requiring less projection display hardware and image-processing power but can be used in single-viewer applications only, and adds the requirement of head-tracking the high-resolution inset.

Regardless of which multiple-image technique is used, the images must be aligned and blended at the edges for geometry, convergence, and luminance, and overall for color (in addition to convergence of each individual channel). Edge

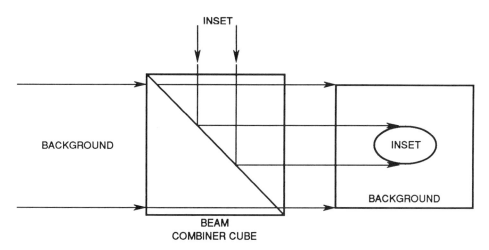

Figure 6.73 The background and inset in an area-of-interest multiple-projector image provide a high-resolution image where the viewer is looking, with a low-resolution background filling the field of view.

matching is particularly critical, because the eye is very sensitive to discontinuities. Attempts to butt edges together without blending are rarely successful, resulting in either a luminance line or a gap, as well as geometric mismatches (Figs. 6.74 and 6.75).

Image-blending techniques are similar to those used for convergence and fo-

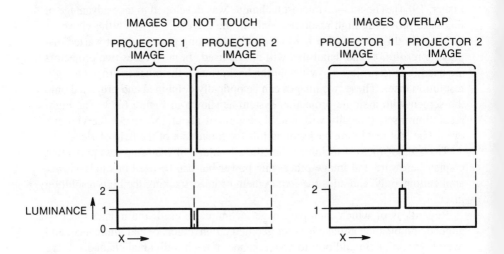

Figure 6.74 Luminance nonlinearities result when edges are not blended.

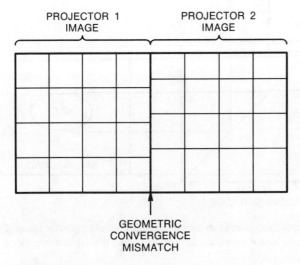

Figure 6.75 Line discontinuities result when edges are not blended.

cus correction, consisting of control circuitry for matching color, intensity, and linearity of separate projected images (Holmes, 1987b, 1989; Green and Lyon, 1988). The color hue of each individual projector is adjusted until all projected images are matched for white. This can be done by eye or with a spectrophotometer. Systems have been developed that perform these adjustments automatically by using sensors at the screen (Lyon and Black, 1984).

Linearity and geometry matches at the edges are accomplished by using the individual convergence circuitry of each projector. A grid pattern is again used, and interactive convergence circuitry is used to match the geometry of the edges until there are no linearity mismatches. Recent techniques have increased the accuracy and ease of use of this process. These techniques include adding a grid of lights to the screen to give the operator a pattern to converge to and using a finer grid at the edges to allow better edge linearity matching (Green and Lyon, 1988).

Intensity nonlinearities are eliminated by permitting the images to overlap and attenuating the luminance at the edges, resulting in a smooth transition. With a perfect blend the two image luminances meet at the 50% luminance points, resulting in even luminance across the screen (Figs. 6.76 and 6.77). If the edges do not match up exactly, the luminance variation caused by the slight mismatch would not create as large a discontinuity as if the image luminance were to fall off quickly.

In the tiled approach the edge luminance is gradually attenuated, using either a fixed attenuation function or an operator-adjustable attenuation. In the area-of-interest technique the background is fully attenuated where the inset is to be, with an edge gradient leading to no attenuation throughout most of the image. The

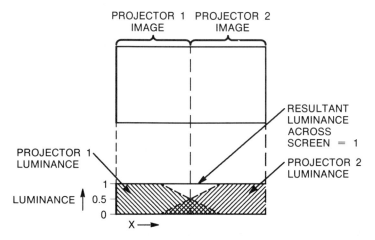

Figure 6.76 Luminance blending of tiled images.

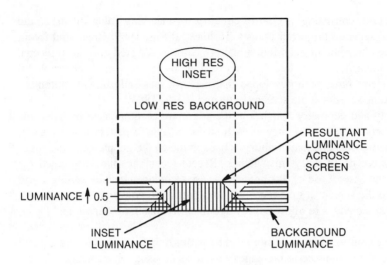

Figure 6.77 Luminance blending of area-of-interest images.

inset is attenuated at its outer edges. The attenuation of the inset and background is adjusted until a smooth transition occurs between the two.

Automatic feedback systems are used in the image-blending process to correct for errors over time, just as with the convergence process. Sensors are used to detect discontinuities in linearity, convergence, color, or luminance and make the proper corrections.

6.7.3 Convergence and Image Blending: Summary

Convergence is the process of correcting projected image geometry and focus errors and aligning the monochrome images with each other. This can be a difficult and time-consuming process. If multiple projectors are used to create a single image, as in the tiled or area-of-interest techniques, the process is expanded to include matching the luminance, geometry, and chromaticity of the separate images. Convergence and image-blending circuitry has evolved from analog to digital systems where correction values are stored in memory according to their location on the image. Techniques for automatic error detection and correction provide stable convergence and blending once the system is initially set.

ACKNOWLEDGMENT

I would like to thank all of the people in Honeywell's Advanced Display group, with whom I worked while writing this chapter. They provided needed assistance and guidance in the compiling of the manuscript. In particular I thank Bob Trimmier for his boundless information, Doug Harding for taking many of the photographs in the chapter, and Dan Schott for his thorough review of the manuscript. I also thank Ron Gold of Hughes, Tony Busquets of NASA Langley, and Manou Akhavi and A. W. Malang of TDS Development for their helpful comments on the text, and all the companies and individuals who provided information and/or pictures of their display systems.

References

Ando, K., M. Ohki, M. Ohta, M. Ogino, and E. Yamazaki (1989). Projection display series for high definition TV, in *Projection Display Technology, Systems, and Applications* (F. J. Kahn, Ed.), Proc. SPIE 1081, pp. 38–45.

Asano, T., Y. Morita, E. Yamazaki, Y. Tanaka, and M. Ohsawa (1989). High-performance aspherical phosphor screen projection CRT, *Proc. Jpn. Display '89*, pp. 546–549.

Banbury, J. R., and F. B. Whitfield (1981). Measurement of modulation transfer function for cathode ray tubes, *Displays*, January, 189–198.

Barber, B. (1984). The use of lasers in wide-angle visual displays, *Proc. Image III*, May 30–June 1, pp.223–237.

Barten, P. G. (1986). Resolution of projection TV systems, *SID 1986 Dig.*, pp. 455–458.

Barten, P. G. J. (1991). Resolution of liquid-crystal displays, *SID 1991 Dig.*, pp. 772–775.

Baumann, E. (1953). The Fischer large-screen projection system, *J. SMPTE 60*: 344–356.

Beiser, L. (1974). Laser scanning systems, in *Laser Applications* (M. Ross, Ed.), Academic, New York, pp.53–159.

Beiser, L. (1986). Imaging with laser scanners, *Opt. News*, November: 10–16.

Bleha, W. P., J. Grinberg, A. D. Jacobson, and G. D. Myer (1977). The use of the hybrid field effect mode liquid crystal light valve with visible spectrum projection light, *SID 1977 Dig.*, pp. 104–15.

Bradley, R., J. F. Goldenberg, and T. S. McKechnie (1985). Ultra-wide viewing angle rear projection television screen, *IEEE Trans. Consumer Electronic. CE-31*(3): 185–193.

Bunker, R., and R. Fisher (1984). Considerations in an optical variable acuity display system, *Proc. Image III*, May 30–June 1, 1984, pp. 240–251.

Chevalier, J., and J. M. Deon (1985). Projection CRTs for advanced display systems, *SID 1985 Dig.*, pp. 54–57.

Clarke, J. A. (1988). Current trends in optics for projection TV, *Opt. Eng.* 27(1): 16–22.

Cook, A. M. (1988). The helmet-mounted visual system in flight simulation, *Proc. Flight Simulation: Recent Developments in Technology and Use*, London, England, Apr. 13–14, pp. 214–232.

Cowdry, D. A. (1985). Advanced visuals in mission simulators, *Proc. Advisory Group for Aerospace Research and Development Conference Proceedings* No. 408, Cambridge, U.K., Sept. 30–Oct. 3, 1985, pp. 3-1—3-10.

Csaszar, I. A. (1991). Data projection equipment and large screen data displays, test, and performance measurements, *SID 1991 Dig.*, pp. 265–267.

Dewey, A. G. (1984). Laser-addressed liquid crystal displays, *Opt. Eng.* 23(3): 230–240.

Doane, J. W., A. Golemme, J. L. West, J. B. Whitehead, and B.-G. Wu (1988). Polymer dispersed liquid crystals for display application, *Mol. Crystal Liq. Crystal 165*: 511–532.

Efron, U., J. Grinberg, P. O. Braatz, and M. J. Little (1981). A silicon photoconductor-based liquid-crystal light valve, *SID 1981 Dig.*, pp. 142–143.

Elmer, S. J. (1982). A color calligraphic CRT projector for flight simulation, *Proc. SID* 23(3): 151–157.

Fendley, J. R. (1983). Resolution of projection TV lenses, *Proc. SID* 24(1): 49–51.

Fergason, J. L. (1985). Polymer encapsulated nematic liquid crystals for display and light control applications, *SID 1985 Dig.*, pp. 68–70.

Forrester, H. (1990). CRT video projection systems, *Inf. Display*, June: 6–9.

Fritz, V. J. (1990). Full-color, liquid crystal light valve projector for shipboard use, in *Large Screen Projection Displays II*, SPIE Vol. 1255, pp. 59–68.

Fukuda, K., S. Mori, K. Sato, N. Kabuto, and K. Ando (1991). 100-in. extra-slim liquid crystal rear-projection display, *SID 1991 Dig.*, pp. 423–426.

Glenn, W. E. (1958). New color projection system, *J. Opt. Soc. Am.* 48(11): 841–843.

Gold, R. S. (1980). Liquid crystal light valve projector for shipboard use, in *Optomechanical Systems Design*, SPIE Vol. 250, pp. 59–68.

Gold, R. S., and A. G. Ledebuhr (1985). Full color liquid crystal light valve projector, *Advances in Display Technology V*, Proc. SPIE 526, pp. 51–58.

Goldenberg, J. F., and T. S. McKechnie (1985). Diffraction analysis of bulk diffusers for projection-screen applications, *J. Opt. Soc. Am.* 2(12): 2337–2348.

Good, W. E. (1968). A new approach to color television display and color selection using a sealed light valve, *Proc. Natil. Electron. Conf.*, XXIV: 771–774.

Good, W. E. (1976). Projection television, *Proc. SID* 17(1): 3–7.

Green, M., and P. Lyon (1988). A new computer–human interface for aligning and edge-matching multichannel projector systems, *SID 1988 Dig.*, pp. 109–112.

Grinberg, J., W. P. Bleha, A. D. Jacobson, A. M. Lackner, G. D. Myer, L. J. Miller, J. D. Margerum, L. M. Fraas, and D. D. Boswell (1975). Photoactivated birefringent liquid-crystal light valve for color symbology display, *IEEE Trans. Electron Devices* ED-22(9): 775–783.

Henkes, J. L. (1982). Development of a rear-projection screen for projection television, *SID Proc.* 23(3): 141–145.

Hirai, Y., S. Niiyama, Y. Ooi, M. Kunigata, H. Kumai, M. Yuki, and T. Gunjima (1991). Liquid crystal/polymer composite devices for active-matrix projection displays, *SID 1991 Dig.*, pp. 594–597.

Hockenbrock, R., and W. Rowe (1982). Self-converged, three-CRT projection TV system, *SID 1982 Dig.*, pp. 108–109.

Holmes, R. E. (1987a). Large screen color CRT projection system with digital correction, in *Large Screen Projection Displays II* SPIE Vol. 760, pp. 16–21.

Holmes, R. E. (1987b). Digital remote control for matrixed simulator visual displays, *Proc. IMAGE TV*, 23–26 June, 1987, Phoenix, AZ.

Holmes, R. E. (1989). Videorama—a new concept in juxtaposed large screen displays, in *Projection Display Technology, Systems, and Applications* (F. J. Kahn, Ed.), Proc. SPIE 1081, pp. 15–20.

Howard, C. M. (1989). Color performance of light-valve projectors, in *Projection Display Technology, Systems, and Applications* (F. J. Kahn, Ed.), Proc. SPIE 1081, pp. 107–114.

Hubin, T. (1991). Acousto-optic color projection system, presented at SPIE Electronic Imaging Conf. February 1991.

Hutley, M. C. (1982). *Diffraction Gratings*, Academic, London, Chapter 2.

Jenkins, A. J. (1981). Photometry and colorimetry of high gain projection screens, in *Light Measurement '81*, SPIE Vol. 262, pp.110–116.

Johannes, H. (1989). *The History of the EIDOPHOR Large Screen Television Projector*, Gretag Aktiengesellschaft, Zurich, Switzerland.

Johnson, R. H., and R. M. Montgomery (1976). Optical beam deflection using acoustic-traveling-wave technology, in *Acousto-Optics*, SPIE Vol. 90, pp.40–48.

Jones, P., A. Tomita, and M. Wartenberg (1991). Performance of NCAP projection displays, SPIE/SPSE Conf., San Jose, February 1991.

Kahn, F. J. (1973). IR-laser-addressed thermo-optic smectic liquid-crystal storage displays, *Appl. Phys. Lett.* 22(3): 111–113.

Kahn, F. J., P. N. Kendrick, J. Leff, L. J. Livoni, B. E. Loucks, and D. Stepner (1987). A paperless plotter display system using a laser smectic liquid-crystal light valve, *1987 SID Dig*.

Kanazawa, M., and T. Mitsuhasi (1989). Automatic convergence correction systems for projection displays, *SID 1989 Dig.*, pp. 264–267.

Kingslake, R. (1983). *Optical System Design*, Academic, Orlando, FL, p. 131.

Kobayashi, I., M. Uno, S. Ishihara, K. Yokoyama, K. Adachi, H. Fujimoto, T. Tanaka, Y. Miyatake, and S. Hoota (1989). *SID 1989 Dig.*, pp.114–117.

Kunigita, M., Y. Hirai, Y. Ooi, S. Niiyama, T. Asakawa, K. Masumo, H. Kumai, M. Yuki, and T. Gunjima, (1990). A full-color projection TV using LC/polymer composite light valves, *SID 1990 Dig.*, pp. 227–230.

Ledebuhr, A. G. (1986). Full-color single-projection-lens liquid-crystal light-valve projector, *SID 1986 Dig.*, pp. 379–382.

Lekavich, J. (1986). Basics of acousto-optic devices, *Lasers Appl.* April: 59–64.

Lyon, P., and S. Black (1984). A self-aligning CRT projection system with digital correction, *SID 1984 Dig.*, pp. 108–111.

Malang, A. W. (1989). High briteness projection video display with concave phosphor surface, in *Projection Display Technology, Systems, and Applications* (F. J. Kahn, Ed.), Proc. SPIE 1081, pp. 101–106.

Merry, J. B., and L. Bademian (1979). Acousto-optic laser scanning, in *Laser Printing*, SPIE Vol. 169, pp.56–59.

Mihalakis, G. (1987). New technology for improvement of lenticular high gain screens, in *Large Screen Projection Displays* SPIE Vol. 760, pp. 29–35.

Mitsuhashi, T. (1990). HDTV and large screen display, in *Large Screen Projection Displays II*, SPIE Vol. 1255, pp.2–12.

Moller, R. (1939). *The Lenticular Screen*, Fernseh, A. G., Hausmitteilungen, Vol. I (in German).

Morozumi, S., T. Sonehara, and H. Kamakura, (1986). LCD full-color video projector, *SID 1986 Dig. XVII*: 375–378.

Murakami, S., T. Tanaka, and S. Arai (1989). 55 inch high definition projection display, in *Projection Display Technology, Systems, and Applications* (F. J. Kahn, Ed.), Proc. SPIE 1081, pp. 46–52.

Nagae, Y., E. Kaneko, Y. Mori, and H. Kawakami (1986). Full-color laser-addressed smectic liquid-crystal projection display, *SID 1986 Dig.*, pp.386–371.

Noda, H., T. Muraji, Y. Gohara, Y. Miki, K. Tsuda, I. Kikuchi, C. Kawasaki, T. Murao, and Y. Miyatake (1989). High definition liquid crystal projection TV, *Proc. Jpn. Display '89*, pp. 256–259.

O'Shea, D. C. (1985). *Elements of Modern Optical Design*, Wiley, New York, p. 310.

Patrick, N. W. (1972). Developments in cathode-ray tubes for the Schmidt projector, *1972 IEEE Conf. Display Devices*, pp. 109–115.

RCA (1974). *RCA Electro-Optics Handbook*, Tech. Ser. EOH-11, RCA Corporation, Harrison, N.J.

Sakamoto, M., M. Imai, H. Moriyami, S. Tsujikawa, H. Ichinose, and S. Kaneko (1991). High-quality-image EDTV liquid-crystal-projector, *SID 1991 Dig.*, pp. 419–422.

Schadt, M., and J. Funfschilling (1990). Novel polarized liquid-crystal color projection and new TN-LCD operating modes, *SID 1990 Dig.*, pp. 324–326.

Scholl, M. S. (1987). Three beam-combining schemes in a color projection display, in *Current Developments in Optical Engineering II*, SPIE Vol. 818, pp.196–205.

Shikama, S., E. Toide, and M. Kondo (1990). A polarization transforming optics for high luminance LCD projectors, 1990 SID Int. Display Res. Conf., pp. 64–67.

Snyder, H. L. (1988). Toward the determination of electronic display image quality, in *Advances in Man–Machine Systems Research*, Vol.4, JAI Press, pp. 1–68.

Spooner, A. M. (1982). The trend towards area of interest in visual simulation technology, *Proc. 4th Int. Service/Industry Training Equipment Conf.*, p. 205.

Stepner, D. E., and F. J. Kahn (1986). Liquid crystal light valves display large plots in real time, *Comput. Technol. Rev.*, Fall: 143–147.

Sterling, R. D., R. D. Te Kolste, J. M. Haggerty, T. C. Borah, and W. P. Bleha (1990). Video-rate liquid-crystal light-valve using an amorphous silicon photoconductor, *SID 1990 Dig.*, pp. 327–329.

Stroomer, M. V. C. (1989). CRT and LCD projection TV: a comparison, in *Projection Display Technology, Systems, and Applications*, SPIE Vol. 1081, pp. 136–143.

Takatsuka, T., T. Haranou, S. Takakusa, M. Takehara, and R. Tamamura (1982). Anisotropic front projection screen with high and uniform brightness, *Proc. SID* 23(3):147–150.

Takeuchi, K., Y. Funazo, M. Matsudairs, S. Kishimoto, and K. Kanatani (1991). A 750-TV-line-resolution projector using 1.5-megapixel a-Si TFT LC modules, *SID 1991 Dig.*, pp. 415–418.

Takizawa K., H. Kikuchi, and H. Fijikake (1991). Polymer-dispersed liquid-crystal light valves for projection displays, *SID 1991 Dig.*, pp. 250–253.

Takubo, Y., M. Takeda, T. Tamura, H. Iwai, K. Ukita, Y. Bessho, H. Takahara, K. Komori, and I. Yamashita (1989). High density reflective type TFT array for high definition liquid crystal projection TV system, *Proc. Jpn. Display '89*, pp. 584–587.

Tannas, L. E. (1985). *Flat-Panel Displays and CRTs*, Van Nostrand Reinhold, New York, Chapters 2–4.

Timmers, W. A. G., E. Stupp, R. van den Plas, and M. V. C. Stroomer (1989). A full resolution LCD front-projection TV with high brightness, *Proc. Jpn. Display '89*, pp. 260–263.

Todd, L. T., and S. Sherr (1986). Projection display devices, *Proc. SID 27*(4): 261–268.

Toide, E., S. Shikama, and M. Kondo (1991). High-definition projector with optimized interference filter, *SID 1991 Dig.*, pp. 163–166.

Trias, J., W. Robinson, T. Phillips, B. Merry, and T. Hubin (1988). A 1075-line video-rate laser-addressed liquid-crystal light-valve projection display, *SID Proc. 29*(4): 275–277.

True, T. T. (1973). Color television light valve projection system, *Proc. IEEE Int. Convention and Exposition*, Mar. 26–30, 1973.

True, T. T. (1987). High-performance video projector using two oil-film light valves, *SID 1987 Dig.*, pp.68–71.

Tsai, R. C. (1981). High data density 4-color LCD system, *Inf. Display*, May: 3–6.

Veron, H. (1989). A resolution measurement technique for large screen displays, in *Projection Display Technology, Systems, and Application* (F. J. Kahn, Ed.), Proc. SPIE 1081, pp. 21–28.

Vriens, L., J. H. M. Spruit, J. C. N. Rijpers, and M. R. T. Smits (1988). The interference filter projection TV CRT, *SID 1988 Dig.*, pp. 214–217.

Welford, W. T., and R. Winston (1989). *High Collection Nonimaging Optics*, Academic, San Diego, CA.

Yariv, A., and P. Yeh (1984). *Optical Waves in Crystals*, Wiley, New York.

Zook, J. D. (1974). Light beam deflector performance: a comparative analysis, *Appl. Opt. 13*(4): 875–887.

7
Stereoscopic Display

Larry F. Hodges

*Georgia Institute of Technology,
Atlanta, Georgia*

7.1 INTRODUCTION

The term *three-dimensional display* often refers to the display of images that are created based on a three-dimensional coordinate system and then displayed as a parallel or perspective projection onto a flat CRT (cathode-ray tube) screen. An observer's perception of depth in the image is conveyed by cues such as shading, shadowing, occlusion, motion, and linear perspective, as well as structure and size of familiar objects. *Stereoscopic display* adds the additional depth cue of *stereopsis*. When an observer looks at a three-dimensional scene, because the eyes are horizontally separated the images formed at the back of each eye for any particular point in the scene differ in their horizontal position. This effect is referred to as *binocular disparity* or *binocular parallax*. Stereopsis involves the merging of these two slightly different two-dimensional images by the brain into a single three-dimensional image. In a stereoscopic display system we generate two views of a scene and display them so that only the left eye sees the left-eye view and only the right eye sees the right-eye view of the scene. The result is that the observer sees an image that appears truly three-dimensional. A striking example of the importance of this effect on depth perception is provided by the random-dot stereograms devised by Julesz (1971). A random-dot stereogram provides an observer with binocular parallax cues to the depth in an image while eliminating all other cues. If only the left- or right-eye view of the stereogram is observed, no pattern or depth is apparent. However, when viewed stereoptically, a three-dimensional pattern emerges. Figure 7.1 illustrates this effect.

Recent research indicates that time-multiplexed CRT-based stereoscopic displays provide better user performance at many three-dimensional visual tasks than two-dimensional displays. Recent studies have compared user performance with 2-D perspective display versus time-multiplexed stereoscopic display for both accuracy and reaction times. Stereoscopic displays have been judged superior for visual search and interactive cursor-positioning tasks (Beaton, 1990; Beaton et al., 1987), for spatial judgement tasks (Yeh and Silverstein, 1990a), and for communication of design information (McWhorter et al., 1990). A recent study indicates that, in 3-D task environments, worker acceptance of display formats is higher for stereoscopic displays than for 2-D perspective displays (Reinhart et al., 1990).

Improperly constructed and displayed stereoscopic images, however, can adversely affect both user performance and user acceptance. Images produced with poorly designed algorithms can result in vertical parallax and spatial distortions (Hodges and McAllister, 1990). Ghosting in an image can strongly affect subjective ratings of display quality even though its effects on actual task performance may be minimal (Yeh and Silverstein, 1990b). My goal in this chapter is to provide an introduction to the current technology for computer-based stereoscopic display and to also provide an overview of factors that are important to the proper construction and display of computer images in a time-multiplexed environment.

7.2 EARLY STEREOSCOPIC DISPLAYS

The history of stereoscopic display goes back to the early 1800s. Early devices include the Wheatstone stereoscope and the Brewster stereoscope. With the

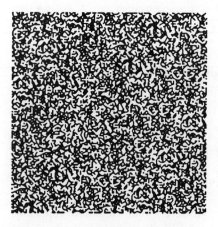 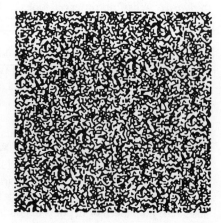

Figure 7.1   Random-dot stereogram.

STEREOSCOPIC DISPLAY

Wheatstone stereoscope, two photographs or stereo drawings were displayed on a viewer that used mirrors to deliver the correct perspective view to each eye. The Brewster stereoscope replaced the mirrors with prisms (and later with convex lenses) and became popularly known as the parlor stereoscope.

In the early 1900s people began to consider the idea of three-dimensional images without viewing apparatus. F. E. Ives proposed a method known as the parallax stereogram. A parallax stereogram consists of a fine vertical slit plate and an image with left- and right-eye perspectives printed in narrow stripes and placed behind the slit plate. The slits force each eye to see only the correct perspective view of the image. A second early autostereoscopic display technique replaced the parallax barrier with a linear array of cylindrical lenses known as a lenticular sheet. This sheet is usually made in such a thickness so that its rear surface coincides with the focal plane of the lenses, so that the correct perspective view is directed to each eye.

7.3 STEREOSCOPIC CRT DISPLAYS

Stereoscopic CRT display systems are usually divided into two basic categories: time-parallel and time-multiplexed (see Fig. 7.2).

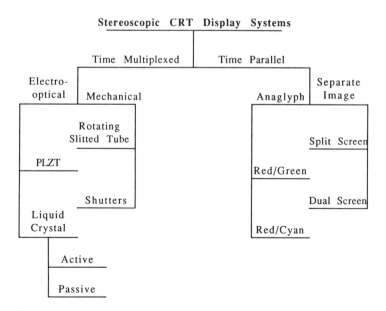

Figure 7.2 Types of stereoscopic CRT display systems.

7.3.1 Time-Parallel Stereoscopic Display

Time-parallel systems present left- and right-eye views simultaneously on separate CRT screens or on a single CRT screen and require a special optical apparatus to deliver the correct perspective view to each eye. Examples of time-parallel systems include anaglyph displays, viewer displays, certain types of polarized displays, and head-mounted displays.

With anaglyph displays, the right- and left-eye perspectives are filtered with complementary or near-complementary colors and are superimposed on a CRT screen. The observer wears glasses with filters that match the projection filters. For black-and-white images, red and green filters are usually used. For color images, red and cyan or green and magenta filters are used. A problem with this technique is that it distorts the colors of an image (Lane, 1982).

Viewer displays usually display the right- and left-eye views side by side on a single CRT screen. The observer looks at the image through a viewer that directs each image to the correct eye. The viewers used for these types of displays are primarily updated versions of the Wheatstone and Brewster stereoscopes.

Polarized time-parallel displays use two CRTs arranged at right angles to each other and a partially silvered mirror. Polarizers are placed on each monitor, with the filters arranged at right angles to each other. The half-silvered mirror is placed between the monitors so that it reflects one perspective view and transmits the other. The observer wears correspondingly polarized glasses so that each eye sees the correct view.

Head-mounted stereoscopic displays present many unique design issues compared to the other types of time-parallel displays or to the time-multiplexed stereoscopic displays. Many of these issues are discussed in Chapter 9, 10, and 17. The majority of this chapter deals with issues primarily associated with time-multiplexed stereo.

7.3.2 Time-Multiplexed Stereoscopic Display

Time-multiplexed stereoscopic systems present the stereoscopic image by alternating right- and left-eye views of an object on a CRT. (See Chapter 1 for fundamentals of CRTs). Early implementations used an alternation rate of 60 Hz (30 left-eye views and 30 right-eye views per second). This is easily done on a 30-Hz interlaced monitor by writing one perspective to the even scan lines and the other perspective to the odd scan lines. On a 60-Hz noninterlaced monitor, two complete views must be stored in frame buffer memory so that the refreshed image switches between views at the beginning of each vertical refresh cycle. Both of these approaches suffer from flicker because each eye sees an image that is being updated only 30 times a second (Hodges and McAllister, 1985).

In order for an observer to see the stereo image, the alternating left- and right-eye perspectives must be synchronized with a shutter system that occludes the left eye when the right-eye view is displayed and occludes the right eye when the left-eye is displayed. Early systems used a rotating cylinder with right- and left-eye slits or other types of mechanical shutters. In the late 1970s and early 1980s the mechanical shutters were replaced by electro-optical shutters based on lead lanthanum zirconate titanate (PLZT) ceramic wafers. Shutter assemblies consisted of front and rear polarizers with a PLZT ceramic wafer in between. The axes of polarization of the polarizers were rotated 90° with respect to each other and were oriented at 45° with respect to an electric field applied to the ceramic wafer. In the off state of the shutter, light traversing the front polarizer was blocked by the rear polarizer. With the application of a sufficient voltage potential, the PLZT ceramic wafer acted as a half-wave retarder, and light passing through the front polarizer was rotated 90° so that it passed through the rear polarizer. The shutters were designed to resemble eyeglasses in size and weight. The major disadvantages were that a cable was attached to the shutter glasses and the shutters transmitted only about 15–17% of the light energy (Roese and McCleary, 1979).

Beginning around 1985, PLZT shutters have been replaced by shutter systems based on liquid-crystal modulator (LCM) polarization. (See Chapter 2 for the fundamentals of liquid crystals.) LCM-based shutters transmit approximately twice as much light energy as PLZT shutters, resulting in a much brighter image. Time-multiplexed stereoscopic systems with LCM polarization encoding are available in two basic configurations. These two implementations contain the LCM either as part of the glasses worn by the viewer or as a large panel covering the entire CRT display area. These are called, respectively, active and passive systems and are shown in Figures 7.3 and 7.4.

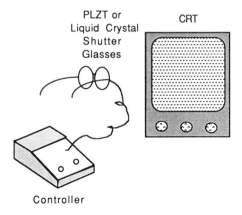

Figure 7.3 Active shutters worn by observer.

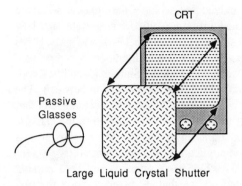

Figure 7.4   Passive glasses worn by observer.

In the active polarization system, the shutter system is entirely contained in goggles or glasses worn by the viewer. A cable or an infrared signal conveys the synchronization signals from the CRT display to the LCM. This system can also be implemented with glasses designed to minimize ghosting.

In a passive system, a large shutter is mounted on the front of a CRT screen. The shutter in this case is designed using LCM cells and one circular polarizer so that the left- and right-eye perspectives displayed on the screen are circularly polarized in opposite directions. The observer wears passive glasses with the left and right lenses circularly polarized to allow the proper views to be seen. Passive stereoscopic systems have been available since 1986.

7.3.3   Liquid-Crystal Modulators

Fast liquid-crystal modulators (LCMs) are the key components in all of the current time-multiplexed stereoscopic 3-D displays. Three basic types exist for use in glasses, whereas only one technique is applicable for the passive, large-screen switching LCMs (Bos et al., 1988; Hodges et al., 1988).

The first type involves a scattering LCM that switches between a transmissive and light-scattering state. This LCM does not require the use of polarizers and therefore has high transmission. Switching speeds are adequate, but light leakage in the scattering or blocking state allows excessive ghosting. In addition, the viewer will integrate the transmission state with the scattering state and thus always sees a fuzzy picture.

The second type is a standard twisted nematic liquid-crystal cell with crossed polarizers. This device transmits light in the electrically unexcited state and blocks light in the excited state. Switching speeds are slow, and viewing angles are poor. The blocking ability of these devices can be very good when the device is fully switched, but switching speeds are so slow that the devices cannot switch fully within a video frame, and thus system ghosting is severe.

The fastest switching and lowest ghosting LCMs are based on a variable retarder principle and switch between 0 and ½ wave states. When placed between polarizers, these devices block or transmit light with the maximum transmission similar to a twisted nematic device—about 30%. Ghosting is low but still noticeable. These devices can be encoded with either circular or linearly polarized light. Circular polarization encoding is more expensive but results in the best performance in passive systems for two reasons. First, when the viewer's head tilts, ghosting does not increase. Second, compensation for wavelength-dependent ghosting can be made with circular polarization.

7.4 STEREOSCOPIC DISPLAY ALGORITHMS

There are two basic approaches to computing the left- and right-eye perspective views of a stereoscopic image. The two approaches are mathematically identical but produce slightly different images in practice. We shall refer to the two approaches as the *off-axis projection* and the *on-axis projection* (Hodges, 1991). Similar concepts have been described by others using different terminology (Baker, 1987; Hodges et al., 1988; Rogers and Adams, 1990; Tessman, 1990; Williams and Parrish, 1990).

7.4.1 Off-Axis Projection

For an off-axis projection we assume two different, horizontally aligned centers of projection. The right-eye view will be produced by projection to a right center of projection (RCoP). The left-eye view will be produced by projection to a left center of projection (LCoP). We assume that the projections are implemented in conjunction with a standard viewing transformation such as is described in the textbooks by Hearn and Baker (1986) or Foley et al. (1990). After the viewing transformation has been implemented, the view plane is located parallel to the xy plane and passes through the origin. We are in a left-handed coordinate system, and a standard center of projection for a nonstereoscopic perspective view of the scene is located on the z axis at $(0,0,-\mathbf{d})$ as shown in Figure 7.5. The resulting projection of a point $P = (x,y,z)$ onto the projection plane has coordinates (x_p,y_p) with

$$x_p = x\mathbf{d}/(\mathbf{d}+z), \qquad y_p = y\mathbf{d}/(\mathbf{d}+z) \tag{7.1}$$

To produce the two views necessary for a stereoscopic image, we move the RCoP off the z axis to $(\mathbf{e}/2, 0, -\mathbf{d})$ and the LCoP to $(-\mathbf{e}/2, 0, -\mathbf{d})$. $\mathbf{e}$ represents the total horizontal separation between the centers of projection (Fig. 7.6). Choosing an appropriate value for $\mathbf{e}$ is an important issue and will be discussed in detail in a later section.

For an arbitrary point $P = (x,y,z)$, the projected value of P for the left-eye view, $P_l = (x_l,y_l)$, has projection plane coordinates

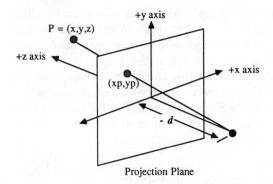

Figure 7.5 Perspective projection.

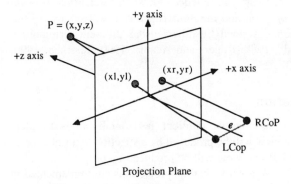

Figure 7.6 Off-axis projection.

$$x_l = (x\mathbf{d} - z\mathbf{e}/2)/(\mathbf{d}+z), \qquad y_l = y\mathbf{d}/(\mathbf{d}+z) \qquad (7.2)$$

The projected value of P for the right-eye view, $P_r = (x_r, y_r)$, has projection plane coordinates

$$x_r = (x\mathbf{d} + z\mathbf{e}/2)/(\mathbf{d}+z), \qquad y_r = y\mathbf{d}/(\mathbf{d}+z) \qquad (7.3)$$

Scrutiny of Eqs. (7.2) and (7.3) results in two straightforward but valuable observations. First, y_l and y_r are identical. Therefore we need only compute the projected y value once for both views. Identically projected y coordinates also guarantee that we have not introduced vertical parallax artifacts into our image. Our second observation is that x_l and x_r are produced by different combinations of the same basic terms. From the terms computed for x_l we can compute x_r with only two additional operations. By taking advantage of the coherence between

STEREOSCOPIC DISPLAY

the two projections, we can reduce the number of operations needed to compute both of them. Depending on the algebraic format of the equations, this approach results in a computational savings ranging from 30 to 36%.

The field of view (FOV) for the stereoscopic image computed with the off-axis projection consists of three regions: a stereoscopic region that is seen by both eyes plus two monoscopic regions that are seen by only the left eye or the right eye (Fig. 7.7). Together these three regions presented a wider FOV than is achieved from a single-perspective image computed with the same parameters. In addition to this increased FOV due to the nonoverlapping regions, we also encode more information about a scene because the left- and right-eye views of the overlapping region show different perspectives of the scene.

7.4.2 On-Axis Projection

Whereas an off-axis projection is accomplished by moving the center of projection, the on-axis projection produces the same set of projected points by horizontally moving the data. If we begin with Eqs. (7.2) and (7.3) for x_l and x_r, but apply the partial fraction technique to put them into a slightly different algebraic form, we get

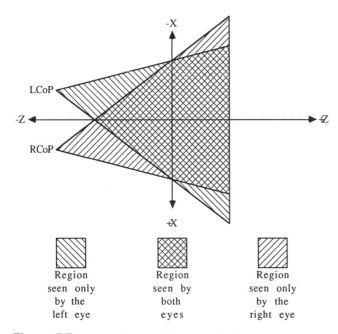

Figure 7.7 Field of view of off-axis projection.

$$x_l = \frac{d(x + e/2)}{d+z} - \frac{e}{2} \tag{7.4}$$

and

$$x_r = \frac{d(x - e/2)}{d+z} + \frac{e}{2} \tag{7.5}$$

Comparing Eq. (7.1) for the projected x value of a standard projection to Eq. (7.4) for the x value of a left-eye projection, we observe that x_l can be obtained by (1) translation of x by $e/2$, (2) standard perspective projection, and (3) translation of the projected x by $-e/2$. Furthermore, since step 3 is independent of the value of the projected value of x, we can compute the complete left-eye view by the algorithm

1. For every point $P = (x,y,z)$ to be projected
   1.1. Translate the x to $x + e/2$
   1.2. Project P using the standard perspective projection
2. Pan the entire image by $-e/2$.

The algorithm for the right-eye view is similar but uses a translation of x to $x - e/2$ and a pan of the entire image by $e/2$.

Although this algorithm has been previously described in the literature, its relationship to the off-axis projection, the effect of the pan on the view volume, and the motivation for using this approach are incompletely described (Baker, 1987; Hodges et al., 1988; Tessman, 1990; Williams and Parrish, 1990). As can be seen from the derivation, the off-axis algorithm and the on-axis algorithm are mathematically equivalent. However, their implementation results in different advantages and disadvantages from each approach.

First we consider the FOV of the on-axis projection. Referring to Figure 7.8, let the rectangular box represent the volume of interest (Fig. 7.8a). For the right-eye view we translate the box to the left, project onto the projection plane (Fig. 7.8b), and pan the image back to the right (Fig. 7.8c). The pan results in a loss of projected data on the right side of the screen and an empty field on the left side of the screen. For the left-eye view, we translate the box to the right, project onto the projection plane (Fig. 7.8d), then pan the image back to the left (Fig. 7.8e). The pan results in a loss of projected data on the left side of the screen and an empty field on the right side of the screen. Although the FOV still consists of three regions, as before, it is no larger than that of a single-perspective projection (Fig. 7.8f).

In spite of this limitation as compared to the off-axis projection, the on-axis projection has one significant advantage. Few graphics workstations provide hardware graphics routines that allow programmer control of the center of projection. The off-axis projection, therefore, must be implemented in software. The

STEREOSCOPIC DISPLAY

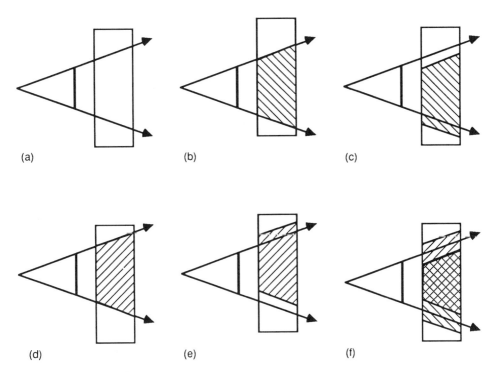

Figure 7.8 On-axis projection. (a) Standard position and projection; (b) translated and projected for right-eye view; (c) panned for right-eye view; (d) translated and projected for left-eye view; (e) panned for left-eye view; (f) zones seen by right eye, left eye, and both eyes.

on-axis projection, however, can be implemented using the standard perspective projection in conjunction with translations and pans. As these operations are usually available in hardware, the on-axis projection results in better overall performance.

7.4.3 Rotations

In some applications, it may be desirable to compute stereoscopic images based on parallel projections of a scene. In this case the left- and right-eye views are computed by rotations about a vertical axis through the center of the scene. The left-eye view is rotated from left to right, and the right-eye view is rotated from right to left. Total rotation should be approximately 4°. An alternative to this technique that is useful when front and back clipping planes are also desired is to replace the rotations with a z shear of the data. The effect of the shear is to preserve the horizontal movement of the data produced by rotation while eliminating the front-to-back motion. Mathematical details of this approach are described by Lipscomb (1979). In some scenes, rotation or shearing in conjunction with parallel projections results in a reverse parallax effect so that objects appear to get larger as they move away from the observer in depth.

Because rotations in conjunction with perspective projection have been widely used in the literature, there is a note on their use for generating stereoscopic views. There is conclusive evidence that stereoscopic images of perspective views produced by rotating a scene relative to the viewpoint contain artifacts that adversely affect image quality and content. These artifacts include vertical parallax and spatial distortions (Baker, 1987; Butts and McAllister, 1988; (Hodges and McAllister, 1990; Saunders, 1968).

7.5 HORIZONTAL PARALLAX

The choice of **e**, the horizontal distance between the LCoP and RCoP, is critical in order for an observer to merge the left- and right-eye views of an object into a single stereoscopic image. A proper value for **e** depends on a surprising number of variables, including the dominant color of the viewed object, its location on the screen, whether the object employs negative (crossed) or positive (uncrossed) horizontal parallax, the observer's own visual system and experience, distance of the observer from the screen, size of the display screen, and linear distance of the object from the projection plane.

The amount of horizontal parallax on the display screen resulting from a particular value of **e** depends on the orthogonal distance from the projection plane of the objects in the scene and can be computed for any point from the difference of the projected x coordinates (x_r, x_l). From Eqs. (7.4) and (7.5), horizontal parallax **p** is computed as

$$\begin{aligned} \mathbf{p} &= x_r - x_l \\ &= \frac{\mathbf{d}(x - \mathbf{e}/2)}{\mathbf{d} + z} + \frac{\mathbf{e}}{2} - \left[\frac{\mathbf{d}(x + \mathbf{e}/2)}{\mathbf{d} + z} - \frac{\mathbf{e}}{2}\right] \\ &= \mathbf{e}[1 - \mathbf{d}/(\mathbf{d} + z)] \end{aligned} \quad (7.6)$$

For example, an object in a plane $z = \mathbf{d}$ units on the opposite side of the projection plane from the two centers of projection would have parallax equal to **e**/2. An object in a plane 2**d** units on the opposite side of the projection plane from the two centers of projection would have parallax equal to 2**e**/3. For an object at an infinite distance from the projection plane, $\mathbf{p} = \mathbf{e}$. In general, the value of **e** is related to the amount of horizontal parallax **p** by

$$\mathbf{e} = \mathbf{p}/[1 - \mathbf{d}/(\mathbf{d} + z)] = \mathbf{p}(i + 1)i \quad (7.7)$$

where $i*\mathbf{d}$ is the maximum depth of the area containing all objects in the scene.

7.5.1 e in Terms of Visual Angle

Display screens for stereoscopic images vary from small raster graphics screens to large projection screens. We choose, therefore, to measure **e** not in absolute terms but in terms of horizontal visual angle (HVA). For a horizontal parallax on a screen of **p** units, the HVA would be defined as the angle β necessary to subtend **p** units from a point orthogonal to the screen from a distance **d**. From the geometry (Fig. 7.9),

$$\mathbf{p} = 2\mathbf{d}\tan(\beta/2). \tag{7.8}$$

Substituting Eq. (7.8) for **p** in terms of HVA into Eq. (7.7), we can compute an appropriate value for **e** given a maximum allowable HVA as

$$\mathbf{e} = 2\mathbf{d}\tan(\beta/2)(i+1)/i \tag{7.9}$$

where **e** is the distance between LCoP and RCoP, **d** is the orthogonal distance of the centers of projection from the projection plane, β is the maximum allowed horizontal visual angle (HVA), and i is the depth of scene in units of **d**.

7.5.2 Appropriate Values for the Maximum Horizontal Visual Angle

Choosing appropriate values for the maximum HVA is a critical part of composing stereoscopic images. If HVA is too large, all or parts of the stereoscopic image will be difficult to fuse. If HVA is too small, the stereoscopic effect is lost. It would seem natural, when determining a value for maximum HVA, that we should base it on factors such as the interocular distance between an observer's eyes and her distance from the display screen. For example, if we assume average interocular distance to be approximately 6.35 cm and that the observer is sitting 61 cm from the display screen, the maximum disparity for an object at infinity would subtend a visual angle in the horizontal plane of approximately 6°, and the HVA for an object a distance of 61 cm behind the screen would subtend

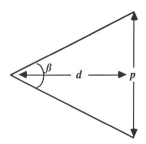

Figure 7.9 Horizontal visual angle.

an angle of 3°. It is my experience that parallax of this magnitude results in an image that cannot be fused by many observers and that a smaller value should be used. Valyus (1962) recommends that maximum allowed HVA be limited to 1.6°. Lipton (1982) judges the 1.6° limit to be "fairly accurate" but points out that "negative parallax values several times the rule given by Valyus are permissible for objects moving rapidly out of the frame or held on the screen only briefly." Levonian (1954) argues that a value one-third the size of the maximum parallax should be used to reduce a visual fatigue.

A controlled experiment on the amount of allowable parallax within a time-multiplexed computer graphics display environment was recently published by Yeh and Silverstein (1990b). They point out that the maximum HVA that can be fused is affected by the interocular crosstalk between the left- and right-eye perspective views of a scene. In addition, the stimulus duration of the image and whether the horizontal parallax was crossed or uncrossed affects the HVA. Their mean value (1.57°) for uncrossed HVA with a stimulus duration of 2 sec agrees with the value suggested by Valyus, while their mean crossed parallax value (4.93° s) with a stimulus duration of 2 sec exceeds his recommendation, as noted by Lipton. As a rule of thumb, therefore, we suggest that maximum HVA be limited to 1.5°, which can be accomplished by setting $e = 0.028d$. In their study, Yeh and Silverstein suggest that more conservative values of between 27° of arc ($e = 0.008d$) for crossed disparity and 24' of arc ($e = 0.007d$) for uncrossed disparity based on a stimulus duration of 200 msec.

7.6 FACTORS AFFECTING IMAGE QUALITY

7.6.1 Interocular Crosstalk and Ghosting

In a perfect time-multiplexed system, when the right-eye image is on the screen, the left-eye image should be completely extinguished, and when the left-eye image is on the screen, the right-eye image should be completely extinguished. In a real system, interocular crosstalk between the images occurs. Factors that affect the amount of interocular crosstalk include the dynamic range of the shutter, phosphor persistence, and vertical screen position of the image. Factors affecting the perceived ghosting between the left- and right-eye images as a result of crosstalk are image brightness, contrast, textural complexity, and horizontal parallax.

Crosstalk is caused by a combination of the effects of the leakage of the shutter in its closed state and the phosphorescence from the opposite eye's image when the shutter is in its open state. The amount of leakage in the shutter's closed state is quantified in terms of its dynamic range. The dynamic range of the shutter is defined as the ratio of the transmission of the shutter in its open state to the transmission of the shutter in its closed state. The amount of phosphorescence from the opposite eye's image is defined by phosphor persistence, or the length of time it takes for the phosphorescence to decay to 10% of its initial light output. Persistence is related to color of the object in that the P22 red, green, and blue phosphors used in most CRTs do not decay at the same rate. The red and blue phosphors have substantially faster decay rates than the green phosphor. The phosphorescence contribution also increases in relationship to distance to the bottom of the CRT screen. This increase is an artifact of the top-to-bottom scanning of an image on a raster graphics display screen. Scan lines near the top of the screen have had longer to decay when the shutter switches than scan lines at the bottom. Yeh and Silverstein (1990b) have quantified this effect as a function of image color and screen position based on P22 phosphors and Tektronix 808-012 four-segment liquid-crystal (LC) stereo goggles in terms of extinction ratio (the luminance of the correct eye image divided by the luminance of the opposite eye ghost image).

Lipton (1987) pointed out that the ghosting effect produced by crosstalk between the left- and right-eye images is mitigated by a number of factors such as image brightness, contrast, textural complexity, and horizontal parallax. The effect of these factors on perceived ghosting between the images is not as easy to quantify in terms of exact measurements but does provide useful "rules of thumb" for improving image quality. In general, ghosting in images is directly proportional to brightness, amount of horizontal parallax, and high contrast. Ghosting is inversely proportional to the textural complexity or amount of detail in the image.

7.6.2 CRT Refresh Rate

The CRT refresh rate contributes to the perceived flicker in an image. Currently 60- and 120-Hz refresh rates predominate in the high-end time-multiplexed stereoscopic display market. Shutters and monitors that can be driven at approximately 60 Hz provide the observer with a 30-Hz refresh rate to each eye. Switching is usually done by a hardware double buffer, with each perspective view shown at full resolution. Interactive displays are difficult to achieve because all image updates must be done in $\leq 1/60$ sec to maintain correspondence between perspective views and the observer's eyes. Systems at 120 Hz provide a 60-Hz refresh rate to each eye but at reduced vertical resolution because each frame buffer is divided into logical left- and right-eye buffers. Interactive displays for hardware double-buffered systems are more practical because the programmer is working with the logical equivalent of four buffers: two to display the current image and two in which to draw the next stereo image (Baker, 1987). I am aware of no studies that compare the actual performance of 60-Hz full-resolution systems to 120-Hz reduced resolution systems. The accepted assumption is that a faster refresh rate is more advantageous than full resolution.

7.6.3 Interactive Devices

Another important factor is the interface for interaction with and input to a stereoscopic image. There are currently many contenders but no universally accepted paradigm for interaction with a 3-D space. Studies in 1987 and 1988 concluded that planar thumbwheels and slider devices were most accurate for 3-D cursor positioning for both time-multiplexed stereoscopic and perspective displays of three-dimensional data (Beaton et al., 1987; Beaton and Weiman, 1988). Since that study, several new devices for interaction with 3-D space have become available, including the SpaceBall, Bird, and various types of data gloves (Zimmerman et al., 1987). The optimal device is still to be determined.

7.6.4 Image Scaling

The apparent depth of points or objects within a stereoscopic display depends on the observer's position relative to the display screen. This relationship is suggested in Figure 7.10. As the observer moves further away from the center of the display, the horizontal parallax for a fixed point decreases, and as the observer moves closer, the horizontal parallax increases.

Since the observer's distance from the CRT is usually unspecified or unknown, the left- and right-eye views are normally computed assuming that an observer is located in the center of and at a fixed distance from the projection plane, and (except in head sensing displays) the views are not dynamically changed as the observer changes his or her head position. This assumption results in a fixed horizontal parallax for points at a particular distance from the projection plane regardless of the position of the observer. The fixed horizontal parallax causes distortion of the image along the viewing axis normal to the plane of the viewer as shown in Figure 7.11. As the observer moves away from the display, the image is elongated. As the observer moves closer, the image contracts. As the observer shifts her head from side to side, the same view of the image shifts in conjunction with her head movement. The practical implication of these fac-

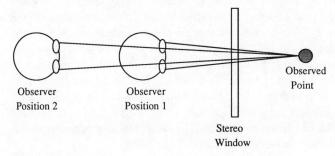

Figure 7.10   Parallax change with head movement.

STEREOSCOPIC DISPLAY

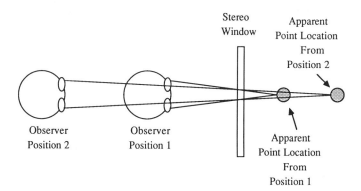

Figure 7.11 Distortion caused by head movement and no parallax change.

tors is that there is an optimal position at which to observe the stereoscopic image so that the front-to-back scaling of the image is in correct proportion to the up–down and left–right scaling.

Up to this point I have not made any distinction between the image modeling geometry and the image display geometry. The description of the objects in a scene are modeled, and the values of **e** and **d** are specified relative to a unitless image modeling geometry that is independent of the actual device on which the image is to be displayed. When the stereoscopic image is displayed on a particular device, the image size and the corresponding horizontal parallax are scaled relative to the width and height of the display screen. The optimal viewing distance is d_s, the value of **d** from the modeling geometry after being adjusted by the same scale. Observing the image from this distance also ensures that the maximum horizontal visual angle is not violated.

7.7 SUMMARY

The technology for stereoscopic display of computer image has progressed remarkably since the mid-1980s. Increasingly, research results have shown that stereoscopic displays provide many advantages over monoscopic displays in a variety of application areas. As scientists and engineers become aware of those advantages, we can expect stereoscopic display to become a firmly entrenched area of electro-optical display technology.

References

Baker, J. (1987). Generating images for a time-multiplexed stereoscopic computer graphics system, in *True 3D Imaging Techniques and Display Technologies*, SPIE Proc. 761, pp. 44–52.

Beaton, R. J. (1990). Displaying information in depth, *SID Dig. XXI*: 355–358.
Beaton, R. J., and N. Weiman (1988). User evaluation of cursor positioning devices for 3-D display workstations, in *Three-Dimensional Imaging and Remote Sensing Imaging*, SPIE Proc. 902, pp. 53–58.
Beaton, R. J., R. J. DeHoff, N. Weiman, and P. W. Hildebrandt (1987). An evaluation of input devices for 3-D computer display workstations, in *True 3D Imaging Techniques and Display Technologies*, SPIE Proc. 761, pp. 94–101.
Bos P., et al. (1988). High performance 3D viewing systems using passive glasses, *SID Dig. Tech. Papers XIX*: p. 19.
Butts, D. R. W., and D. F. McAllister (1988). Implementation of true 3-D cursors in computer graphics, in *Three-Dimensional Imaging and Remote Sensing Imaging*, SPIE Proc. 902, pp. 74–84.
Foley, J. D., A. vanDam, S. K. Feiner, and J. K. Hughes (1990). *Computer Graphics: Principles and Practice*, 2nd ed., Addison-Wesley, New York.
Hearn, D., and M. P. Baker (1986). *Computer Graphics*, Prentice-Hall, Englewood Cliffs, NJ.
Hodges, L. F., and D. F. McAllister (1990). Rotation algorithm artifacts in stereoscopic images, *Opt. Eng.* 29(8): 973–976.
Hodges, L. F., and D. F. McAllister (1985). Stereo and alternating pair techniques for display of computer-generated images, *IEEE Comput. Graphics Appl.* 5(9): 38–45.
Hodges, L. F. (1991). Basic principles of stereographic software development, *Stereoscopic Displays and Applications II*, Proc. SPIE 1457, pp. 9–17.
Hodges, L. F., P. Johnson, and R. J. DeHoff (1988). Stereoscopic computer graphics, *J. Theoret. Graphics Comput.* 1(1): 1–12.
Julesz, B. (1971). *Foundations of Cyclopean Perception*, Univ. Chicago Press, Chicago.
Lane, B. (1982). Stereoscopic displays, *Processing and Display of Three-Dimensional Data*, Proc. SPIE 367, pp. 20–32.
Levonian, E. (1954). Stereoscopic cinematography: its analysis with respect to the transmission of the visual image, M.A. Thesis, Univ. Southern California.
Lipscomb, J. (1979). Three-dimensional cues for a molecular computer graphics system, Ph.D. Dissertation, Dept. of Computer Science, Univ. North Carolina, Chapel Hill.
Lipton, L. (1982). *Foundations of the Stereoscopic Cinema*, Van Nostrand Reinhold, New York.
Lipton, L. (1987). Factors affecting "ghosting" in time-multiplexed plano-stereoscopic CRT display systems, *True 3D Imaging Techniques and Display Technologies*, Proc. SPIE 761, pp. 75–78.
McWhorter, S., L. F. Hodges, and W. Rodriguez (1990). Evaluation of 3-D display techniques for engineering design visualization, Proc. ASEE Engineering Design Graphics Annual Mid-Year Meeting, pp. 121–130.
Reinhart, W. F., R. J. Beaton, and H. Snyder (1990). Comparison of depth cues for relative depth judgements, in *Stereoscopic Displays and Applications*, Proc. SPIE 1256, pp. 12–21.
Roese, J. A., and L. E. McCleary (1979). Stereoscopic computer graphics for simulation and modeling, *Proc. SIGGRAPH* 13(2): 41–47.
Rogers, D. F., and J. A. Adams (1990). *Mathematical Elements for Computer Graphics*, 2nd ed., McGraw-Hill, New York.

Saunders, B. G. (1968). Stereoscopic drawing by computer—is it orthoscopic? *Appl. Opt.* 7(8): 1499–1503.
Tessman, T. (1990). Perspectives on stereo, in *Stereo Displays and Applications*, Proc. SPIE 1256, pp. 22–27.
Valyus, N. A. (1962). *Stereoscopy*, Focal Press, New York; quoted in Lipton (1982).
Wickens, C. D. (1990). Three-dimensional stereoscopic display implementation: guidelines derived from human visual capabilities. *Stereo Displays and Applications*, Proc. SPIE 1256, pp. 2–11.
Williams, S. P., and R. V. Parrish (1990). New computational control techniques and increased understanding of the depth-viewing volume of stereo 3-D displays, in *Stereo Displays and Applications*, Proc. SPIE 1256, pp. 73–82.
Yeh, Y.-Y., and L. D. Silverstein (1990a). Visual performance with monoscopic and stereoscopic presentation of identical three-dimensional visual tasks, *SID Dig.* XXI:359–362.
Yeh, Y.-Y., and L. D. Silverstein (1990b). Limits of fusion and depth judgement in stereoscopic color displays, *Human Factors* 32(1): 45–60.
Zimmerman, T. G., J. Lanier, C. Blanchard, S. Bryson, and Y. Harvill (1987). A hand gesture interface device, *Proc. SIGCHI 87*, 189–192.

8
Peripheral Vision Displays

Harry M. Assenheim

Consultant
Georgetown, Ontario, Canada

8.1 INTRODUCTION

The requirement for a peripheral vision display in aircraft stems from the increase in the amount of information presented to the pilot. This results in the problem of display cluttering.

In the past the presentation of information has taken the form of multiple instruments crowded in a restricted space around the center of the pilot's field of view. One proposed solution to this problem has been the superimposing of data on cathode-ray tube (CRT) displays. This results in a dense symbology requiring visual switching between several information sources, and the processes of eye movement, accommodation, and convergence inhibit the rate of information acquisition (Wulfeck et al., 1958). In addition to this, the increase in clutter in integrated CRT displays gives rise to error and increased workload.

One solution to these problems is to use peripheral vision to access some of the information otherwise competing for space in foveal vision. This not only frees the pilot from concentration on those parts otherwise displayed in the foveal vision, but also gives a powerful and efficient stimulus to those parts of the visual process involved in the prevention of disorientation and vertigo.

8.1.1 Historical Background

The idea that peripheral vision could be used in aircraft as an instrument landing aid was probably conceived in the late 1950s by Majendie and Lowe, who suggested the use of a peripheral vision rate-field display (Vallerie, 1968). The first attempt to use this concept was made at Smith's Industries in the United Kingdom, was known as a Para-Visual Indicator, and consisted of a servo-driven black-and-white "barber pole," the rotation of which gave the illusion of movement along the pole. This was tested as an approach and landing aid in a DC-8 by KLM, the Dutch airline (Reede, 1965). A similar device developed in the United States by Collins Radio used two concentric "barber poles," creating a moiré pattern and giving apparent motion in any direction. It was known as a Peripheral Command Indicator, had little commercial success, but gave rise to a considerable amount of research in the studies of visual and perceptive processes, which led to the development of the only commercial peripheral vision display, the Malcolm Horizon, the concept of which will be described later.

A review of earlier systems has been given by Stokes et al. (1990), and a brief description of early tests will be given here for completeness.

The first serious studies of the performance of the Para-Visual Indicator were carried out by Brown et al. (1961), who compared four displays intended to aid pilots in the final approach phase of flight. They were the Para-Visual Indicator, an instrument landing system (ILS), and two peripheral vision displays with streaming or flashing lights. Results of the tests showed that all three peripheral vision displays were less useful than the ILS for providing information on the size of errors, and that head movements could change the apparent rate of movement in the "barber poles" and cause subjects to react to nonexistent indications or miss real ones.

Many of the rate-field peripheral display techniques demonstrated considerable potential, particularly with the advent of other electronic display technologies. However, more recently, most attention has been given to the Malcolm Horizon, which was developed by Varian Canada and subsequently sold to Garrett Canada, where it was redeveloped and marketed.

The first commercial Malcolm Horizon was the Model B, produced by Varian Canada. This used a xenon arc discharge lamp as the light source, with large optics incorporated in a rather cumbersome projector, through which roll and pitch were obtained by analog movement of the optics. This was redesigned to use a laser source and two optical scanners to produce the projected bar of light. This will be discussed in Section 8.2.4.

Some mention should be made of telepresence systems, in which the operator becomes part of the display. These systems are still in the early research stages and are designed to give the operator stereoscopic visual display with an advanced interactive panoramic electro-optical system and components, which have been developed for use in multisensory telepresence and simulation systems. Two such systems are described by Webster (1989) and Ritchey (1989), and at this time probably represent the state of the art.

8.2 THE MALCOLM HORIZON

The concept of the Malcolm Horizon was invented in the mid-1960s with the research of Richard Malcolm at the Defence and Civil Institute of Environmental Medicine (DCIEM; Downsview, Ont., Canada) in a study of disorientation in the aircraft cockpit. His studies involved visual and vestibular proprioception, and an excellent summary of his work is available (Malcolm, 1984).

8.2.1 The Physiology of the Malcolm Horizon

Recent statistical analyses have shown that about 15% of all aircraft crashes are due directly to the pilot suffering disorientation or vertigo (Money, 1982). In about 25% of aircraft crashes, disorientation and vertigo are considered to be major contributing factors (Benson, 1978). It is necessary at this stage to consider the reasons why disorientation and vertigo are induced, and to analyze the factors that cause these physiological effects.

Disorientation occurs when the pilot does not know where he is or his attitude with respect to the real world. Vertigo exists when he thinks he knows his orientation but is wrong, or perceives that he is wrong. Typical examples may be considered with reference to a pilot in a blackened cockpit: Disorientation effects are clear; however, if the pilot feels and is convinced that he is diving steeply, even although the plane is on a horizontal course, then he is suffering from vertigo. He will pull back on his control stick to avert the dive; he will climb, and the plane may stall.

Disorientation and vertigo are caused by conflicting signals of balance that are sent to the cerebral cortex in the brain. About 90% of balance information is received through the eye (visual), and about 10% via the ear (vestibular). Some balance information may be tactile. The balance signals received by the eye and sent to the brain may be considered instantaneous, whereas the balance signals from the fluid level in the inner ear experience some small delay before being received by the brain. It is the conflict between these two signals at any moment in time that causes disorientation and vertigo.

Balance information detected by the eye is almost entirely in the peripheral vision (90%), and only a very small part (10%) is detected in foveal vision. Hence peripheral vision accounts for about 80% of our balance (Fig.8.1). Foveal vision encompasses an angle of about 2–10° directly in front of the eyes, and peripheral vision refers to the areas to the side, up to an angle of 190° overall (Fig.8.2). Information received in the foveal vision is geared to alphanumerics and the resolution of detail. It is softwired into the brain, to brain cells that are very much general-purpose cells; the information is received, processed, and compared to memory. The logic circuits commence to work, and information is sent to other parts of the body to respond to the received image. In other words,

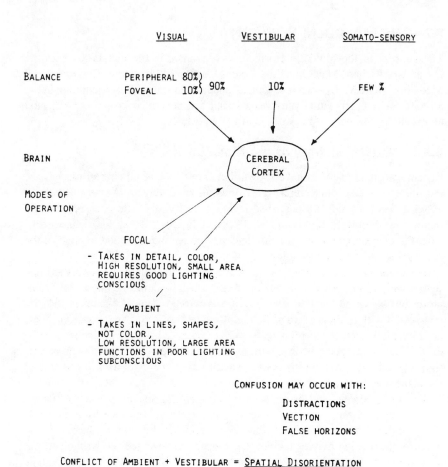

Figure 8.1 Proprioception at the onset of spatial disorientation.

considerable brain activity is involved in foveal vision, even though it is subconscious. Peripheral vision is effectively hardwired into the brain. If an object is viewed in peripheral vision, each element of information that is sent to the brain goes to a different brain cell. Thus, a different brain cell exists for interpreting each piece of information with regard to angle, attitude, and so on. This means that there is effectively no brain activity during viewing in peripheral vision. No logic is required.

Consider a pilot in a cockpit who does not have a good visual horizon, for reasons of weather, clouds, night flying, or whatever. This is termed IFR or Instrument Flying Regulations flying. He sits in an environment where he

PERIPHERAL VISION DISPLAYS

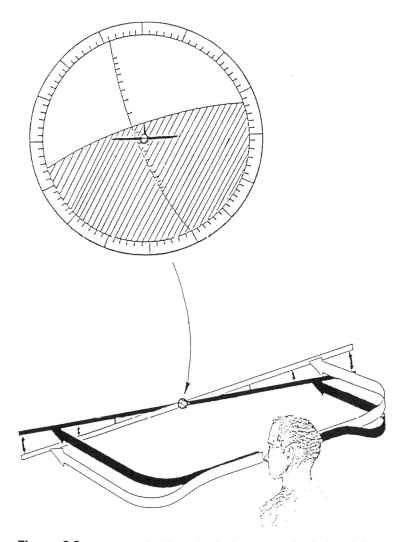

Figure 8.2 Example of wide-angle displacement and velocity relative to standard artificial horizon display.

uses his foveal vision to read meters that give him balance information. In addition, his peripheral vision sees a fixed cockpit and fixed canopy; in other words, his major balance organ tells him that he is straight and level, whatever his orientation in space. If he is rolling or pitching, his vestibular proprioception (supplying only 10% of balance information) is clearly in conflict with the visual proprioception. Hence, there is a strong probability of disorientation occurring.

It has been found that under difficult flying conditions, the pilot looks at his small artificial horizon meter (8 ball) about 80% of the time, leaving about 20% for other problems. Thus with the peripheral vision device the pilot's brain activity is freed to spend 100% of his time interpreting other data, or five times the time normally available under conditions of stress (Leibowitz and Dichgans, 1980; Ninow et al., 1971).

In addition, when a pilot looks at an instrument, most of the time he is not interested in the actual value of the reading it shows but only wishes to know whether it has changed significantly since he last looked at it.

Tests that have been carried out by placing a pilot in a flight simulator and adding increasing stress (Money at al., 1984) have shown that after about five stress factors, the pilot breaks down. As an example of how this may occur (Steele-Perkins and Evans, 1978; Tormes and Guedry, 1974), consider a pilot who is placed in a flight simulator under good conditions; storm conditions are simulated; darkness is simulated; one engine fails; then perhaps a malfunction of the rudder mechanism is simulated; and so on. Thus, the stress is increased until the pilot breaks (i.e., "crashes" the simulator). The same tests have been repeated using a peripheral vision device, and it has been found that the pilot can tolerate about twice as much stress before breakdown.

Thus it can be concluded that the peripheral vision system does the following things:

1. It reduces stress on the pilot, enabling him to cope with other emergencies.
2. It effectively gives the pilot VFR (Visual Flying Regulations) flying under IFR conditions.
3. The device substantially decreases the risk of the aircraft crashing.

Analysis of the brain's interpretation of peripheral vision shows that five functions can be translated and fed directly to the brain in peripheral vision before confusion or overloading occurs. The functions, as applied to a pilot, are pitch, roll, vertical speed, airspeed, and heading.

So far, we have split the visual input into foveal and peripheral vision. The proprioceptors for foveal vision are the cones of the retina, which are frequency-sensitive and require a reasonable illumination to activate. Peripheral vision stimulates the rods of the retina, which are not frequency-sensitive and require considerably less illumination to activate. Figure 8.3 gives the relative numbers of rods and cones in the retina versus angle away from central and also versus visual acuity. The cones, which constitute the central area for peripheral vision, encompass the angle from about $+2½°$ to $-2½°$ and gradually decrease as the rods (foveal vision) increase to a maximum of 18–20° on either side of the central axis (the fovea). Relative visual acuity closely follows the angular variation of the cones (peripheral vision).

PERIPHERAL VISION DISPLAYS

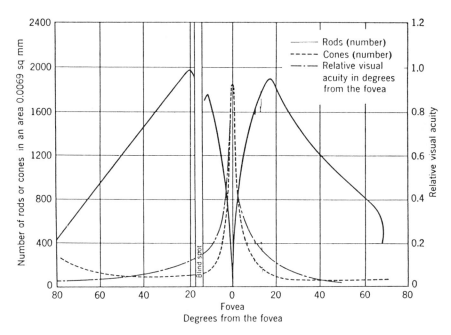

Figure 8.3 Distribution of rods and cones along a horizontal meridian. Parallel vertical lines represent the blind spot. Visual acuity for a high luminance as a function of retinal location is included for comparison. [From Woodson (1954); data from Osterberg (1935) and Wertheim (1894).]

As the brain is the basic interpreter of visual and vestibular input, we can now consider two modes of vision as interpreted by the brain, the focal and ambient modes, rather than foveal and peripheral, which describe two forms of retinal activation.

8.2.2 The Visual System in Spatial Disorientation

Thus far, it has been thought that spatial disorientation is due to conflicting inputs between the vestibular system and the visual system. It is now known that this is only part of the story. Other sources of orientation information, hence potential conflict, include the somatosensory system and the visual system. Of these, a major source of conflict can arise within the visual system itself, between its two modes of processing. These two modes are

1. The *focal mode*, which focuses and is used for tasks requiring acuity or resolution. The focal mode is exclusively visual, requires good lighting and good optics, and typically involves conscious attention. This is equivalent to what I have called foveal vision.

2. The *ambient mode*, which attempts to orient the viewer to the "ambient" environment, tells where he/she is, and determines relative motion. The ambient mode supplies its inputs to the same orientation centers in the brain into which information is fed from the other orientation senses.

Rather than possessing an isolated ambient visual system, humans actually possess an ambient orientation system into which vision and the other senses each contribute a share of the inputs, veridical or otherwise. This is what we call peripheral vision.

There are several fundamental differences between the two modes. The ambient mode functions quite well at very low light levels (see Fig. 8.4), which shows the variation of resolution efficiency against the logarithm of luminance level in foot-candles (fc). The ambient system (peripheral) is considerably more efficient at low luminance levels than the focal system (foveal). The rod–cone discontinuity (i.e., foveal–peripheral discontinuity) occurs at about 10^{-2} fc log luminance level.

Though one cannot read in a dark room, one can orient oneself, provided that there is some light. The ambient mode requires minimal, if any, resolution. Whereas the focal mode actively discriminates and records fine detail, the uncritical ambient mode passively takes in the quality of the surroundings—for ex-

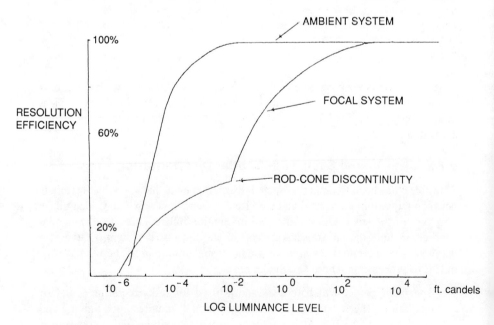

Figure 8.4 Visual processing: two-mode system.

ample, the "surfaceness" of a surface, the "wallness" of a wall, or "horizonness" of a horizon. The ambient mode functions as a reflex rather than at a conscious level, and, provided the stimulus is visible, orientation responses appear to occur on an "all-or-none" basis. The ambient mode acts in concert with the other senses to subserve spatial orientation, balance, posture, and gaze-stability.

An important aspect of these two modes of processing is that they can be dissociated, as demonstrated by the fact that one can read while walking. This dissociation has some impact on night driving, for example. One steers the vehicle by the ambient mode, which is relatively unaffected at night. As long as one can steer, one maintains a high level of confidence in one's ability to drive and therefore tends to drive at the same speed as during daytime, or commonly a little faster. The problem is that the focal mode, hence hazard recognition, has been selectively degraded. That, coupled with the fact that reactions are slower at lower light levels, contributes to the disproportionately higher accident rate at night.

Object recognition by the focal mode spans the full range of spatial frequencies, whereas the ambient mode is adequately activated by low spatial frequencies, typically stimulating large areas of the visual field. As would be expected in terms of spatial frequency, focal vision is less efficient in the periphery. Although ambient functions are less efficient if restricted to a small area of the periphery as compared with central vision, unlike focal vision, ambient functions are typically optimized the larger the area of the visual field stimulated.

With regard to locus of the two modes, whereas the focal mode is concentrated toward the optical center of the eye, the ambient mode encompasses the entire retina, including central vision. There is much more peripheral field (over 3000 times as much), hence much more receptor area available for confusing inputs peripherally. Ambient mode inputs with the potential for confusion include the following.

Distractions Any visible motion attracts the eye by a reflex mechanism that is difficult to suppress. For example, if the hood of one's car jiggles while driving, the altered pattern of motion will attract the eye nearly every time it occurs. In an aircraft, moving canopy reflections can do the same.

The Vection Illusion The false sensation of self-motion processed via the ambient mode is known as "vection." Vection is familiar to anyone who, while waiting for the traffic lights to change, notices the next car moving backwards but misinterprets himself as moving forward and slams on the brakes of his motionless car.

False Horizons Of interest is the discovery that the visual cortex subserving the ambient mode appears to contain receptors specifically responsive to lines and is quite ready to accept uncritically any line with the quality of "horizonness" as a horizon line. Hence the commanding nature of sloping cloud decks or terrain, of a lighted shoreline or highway, or of other false or misplaced horizons that can subtly misorient the pilot. A particularly lethal combination can occur during a departure across a false horizon in reduced visibility—such as crossing a lighted shoreline at night. The balance organ cannot distinguish between acceleration and climb, and as what appears to be the horizon passes beneath his wing line, the pilot becomes convinced that he is doing a loop. The natural tendency is to dump the nose and fly back into the water.

The foregoing (distractions, vection illusion, and false horizons) illustrate how inputs to the ambient mode may confuse, disorient, or misorient the pilot. On the other side of the coin, lack of inputs to the ambient mode can also produce misinterpretation, confusion, or misorientation. In keeping with its overall orientation functions, another important role of the ambient mode is to provide a general reference for the scaling of viewed objects in terms of size, distance, and position in space—a baseline for perspective, so to speak. In the absence of ambient inputs, the accuracy of such judgments is totally random.

Spatial orientation is a common problem to be expected in situations in which one's visual system is either bombarded with distracting, conflicting, or disorienting cues or denied valid orienting cues—namely, the true horizon or surface—thus setting one up for the equally disorienting vestibular illusions. The best course is to prevent spatial disorientation by maintaining visual dominance of valid orientation cues, (gauges). A well-disciplined composite instrument cross-check is requisite to averting not only spatial disorientation but also the more subtle forms of misorientation that are likely to occur in deceptive visual conditions.

Should spatial disorientation occur, effective coping requires the timely establishment of visual dominance on valid orientation cues. Under IFR, the only valid orientation cues are instruments, and, of those, primarily the attitude indicator. In the presence of confusing and conflicting ambient mode inputs, it appears necessary to increase the ratio of valid to invalid cues. Many pilots instinctively reduce the ratio of invalid inputs by adjusting their lights, lowering their head, or lowering their seat while simultaneously leaning forward to concentrate their full attention on the attitude indicator, thus expanding the "ratio" of this valid cue. They should then force the attitude indicator to indicate straight and level for 30–60 sec, to allow vestibular inputs to subside.

8.2.3 The Vestibular System in Spatial Disorientation

Vestibular input plays a larger part in visual orientation than would otherwise be expected. Focal vision, for "locking on" to a moving object is less efficient than one might think. Tracking a moving object is made much easier if the head or body moves, thus giving vestibular input. Ablative studies on cats and primates have shown that the conventional tracking of a moving object by moving the head depends directly on a vestibular "locking on" rather than a visual lock.

An analysis performed by Money (1982) poses questions as to why peripheral vision is preferred to foveal vision and whether there is convincing evidence that peripheral vision is particularly well suited to the processing of orientation information. With regard to the "why" part of the question, there are four obvious advantages for supplying orientation information by peripheral vision rather than foveal vision:

1. Peripheral vision is the kind of vision normally used for orientation and posture (Leibowitz and Dichgans, 1980) and is therefore well suited to the effortless and correct processing of orientation information.
2. Peripheral vision still works well when the retinal image is blurred, as it often is by severe turbulence or vibration. Foveal vision, on the other hand, fails rapidly as the clarity of the retinal image is degraded.
3. Having provided attitude to peripheral vision, foveal vision then needs to be used for checking the standard artificial horizon much less frequently. This means that foveal vision can be used for other things, and other things should then be done better.
4. With attitude information provided by peripheral vision, the pilot is continuously receiving "artificial horizon information" no matter what else he/she is looking at. The constant provision of orientation information will, in all likelihood, reduce the frequency of the kinds of information that are precipitated by unperceived changes in the attitude of the aircraft.

With regard to the nature of the evidence that peripheral vision is particularly well suited to processing orientation, there are five different kinds of evidence indicating that peripheral vision is much more involved in orientation functions than is foveal vision.

1. Studies of humans with discrete brain lesions have shown that people without focal vision can retain good ambient vision and good visual orientation and bodily equilibrium. These observations in humans have been confirmed by experiments with animals (Leibowitz and Dichgans, 1980; Schneider 1969).
2. Postural tests have shown that ambient vision makes a much greater contribution to bodily equilibrium than does focal vision. Artificially imposed movement of the peripheral visual field can cause people to experience self-motion and to fall down, whereas movement of central visual fields has no such effects (Henn et al., 1980).
3. Ambient vision has been found to be much more important than focal vision in a variety of orientation/equilibrium phenomena, including circular vection, linear vection, and optokinetic nystagmus (Berthoz et al., 1975; Brandt et al., 1973; Dichgans, 1977; Dichgans et al., 1972; Henn et al., 1980). In some experiments, conflicting information inputs have been provided to the

ambient and focal systems, and the ambient system has always determined the orientational responses.
4. There are single neurons in visual areas of the brain that are responsive only to lines or edges that are oriented at particular angles and located to stimulate certain discrete parts of the retina. For some such single neurons (although possibly not most), the effective lines must stimulate a specific peripheral area of the retina in order to provoke a response from the neuron (Hubel and Wiesel, 1962).
5. Rotation of the peripheral visual field can actually cause systematic alteration of activity in certain semicircular canal units (neurons) in the vestibular nuclei in the brain stem. The vestibular nuclei are areas of the brain known to be largely concerned with orientation and self-motion; the fact that peripheral retinal areas are physically connected to these particular nuclei is good evidence that ambient vision is involved in orientation and self-motion (Henn et al., 1980).

The basic differences between focal and ambient vision have been summarized by Leibowitz and Dichgans (1980) and are shown in Table 8.1.

Money (1982) concludes that because of the abundance of evidence, the dominant role of ambient vision (as opposed to focal vision) in orientation is now generally accepted by scientists working in this area. It is therefore reasonable to expect that an instrument for providing information about orientation will be more effective if it presents the information to peripheral retinal areas.

Thus we see that orientation is a complex interplay between visual and vestibular inputs, and the Malcolm Horizon offers a strong veridical visual cue to this end.

Table 8.1 Basic Differences Between Focal and Ambient Vision

Focal vision	Ambient vision
Answers the question "what."	Answers the question "where."
Small stimulus pattern, fine detail.	Large stimulus patterns.
Optical image quality and light intensity are important.	Optical image quality and light intensity are relatively unimportant.
Central retinal areas only.	Peripheral (and central) retinal areas.
Well represented in consciousness.	Not well represented in consciousness.
Serves object recognition and identification.	Serves spatial localization and orientation.

Source: Leibowitz and Dichgans (1980).

8.2.4 Basic Operation of the Malcolm Horizon Peripheral Vision Display

The basic operation of the Malcolm Horizon peripheral vision display system, which is produced and marketed by Garrett Canada, a division of Allied Signal, is to display a bar of light on the instrument panel of an aircraft, via input from the aircraft's gyroscope, such that it remains parallel to the real horizon at all times. The bar of light is, in principle, stationary in space and brings the external horizon into the cockpit. This provides sufficient spatial cues via the peripheral vision to eliminate, or at least to reduce, disorientation and vertigo.

The source of light is a He-Ne laser (this will be discussed in Section 8.3), and the scanning is achieved by two optical scanners, one x-axis and one y-axis. Roll and pitch information is converted via a microprocessor into the voltages required to feed the x and y scanners. Figure 8.5 shows a typical functional block diagram, and Figure 8.6 shows the complete system comprising the projector, the processor, and the control panel.

8.3 VISUAL PERCEPTION FOR A PERIPHERAL VISION DISPLAY

The following analysis attempts to look into the basic requirements of a peripheral vision display of the Malcolm Horizon type, for avionic use, with regard to color vision and visual perception. In many cases, the results obtained are those ideally required rather than readily available. It remains to be seen whether the design objectives are practical, and whether feasibility, weight, size, cost, and so on are compatible with these objectives.

First, a note on the units used. The units of illuminance are

1 lumen/m^2 = 1 candela/m^2 = 1 lux (lx)
1 foot-candle = 1 candela/ft^2 = 10.76 lx

Retinal illuminance is given in units of the troland, where 1 troland = illuminance in lux $\times$ pupil area in mm^2, often used in the form $\log_{10}$ trolands. The eye pupil diameter may vary from about 7 mm when fully dilated (after 20 min of total dark adaptation), to about 1 mm under bright light. Under normal conditions, the diameter is between 2 and 4 mm, although it should be noted that the pupil dilates abnormally when the subject has taken drugs or alcohol or under any other medically abnormal conditions.

Luminance is measured or specified in foot-lamberts (fL) where an illumination of 1 fc on a perfectly diffusing surface of luminous reflectance R will produce R fL luminance. Internal lighting in the cockpit, for example, HUD, CRT, or edge-lit panels, is specified in foot-lamberts.

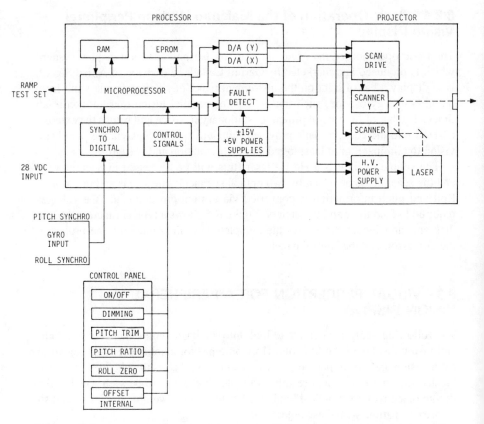

Figure 8.5 Block diagram of the Garrett Canada production model peripheral vision display.

8.3.1 Ambient Lighting

The spectral irradiance of the sun reaches a maximum at a wavelength of 480 nm as measured above the atmosphere and 500 nm at sea level.

Typical luminance levels are given in Table 8.2.

8.3.2 Retinal Sensitivity to Color Vision

Spectral vision of various colors varies widely for foveal and peripheral vision. Foveal vision stimulates the cones in the retina, and peripheral vision stimulates the rods. Figure 8.7 gives the relative radiance in logarithmic units required for foveal and peripheral vision at different wavelengths (Hecht and Hsia, 1945) and

Figure 8.6 The production model peripheral vision display produced by Garrett Canada. The system comprises the projector, the processor, and the control panel.

shows that retinal sensitivity can vary by two orders of magnitude at the lower wavelengths, while being essentially identical above about 650 nm.

Foveal vision is predominantly photopic vision and makes use of the retinal cones that give color vision. Photopic vision requires a relatively high level of illuminance.

Peripheral vision is predominantly scotopic or dark-adapted vision and uses the retinal rods, which are critical to the dark-adapted eye. For scotopic vision the level of illuminance may be so low that the retinal cones are not stimulated, and there may be no color vision.

Extracting the sensitivity to foveal and peripheral vision, for the same colors and on the same arbitrary scale [taken from graphs similar but not identical to Fig. 8.7; see Wald (1945)], gives the comparison of values in Table 8.3. The relevant points to note here are that in peripheral vision the eye is 660 times more sensitive to green than to red. Yellow/green and blue are 330 and 500 times better, respectively. For foveal vision, green gives no substantial advantage, and blue gives a slight disadvantage.

It is important to note, however, that the retinal sensitivity in peripheral vision is meaningful only under scotopic viewing, which is well below foveal threshold levels. Also to be considered is the color contrast factor, in which one must con-

Table 8.2 Typical Luminance Levels

Mil Std. 884C, 10,000 fc	107,000
Brightest white cumulus cloud	30,000
Exteriors in daylight, typical	20,000
Clear sky, average	10,000
Overcast sky, bright	5,000
Interiors in daylight, typical	100
Overcast sky, dark	25
Panel lighting (0.02 fl)	6×10^{-2}
Exteriors at night, typical	10^{-2}

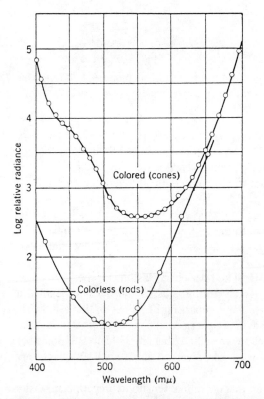

Figure 8.7 Relative radiance required for rod and cone vision at different wavelengths. The curves here may be considered to apply to conditions that give minimum thresholds for each type of receptor. [From Hecht and Hsia (1945).]

Table 8.3 Sensitivity to Foveal and Peripheral Vision

		Sensitivity							
		Foveal				Peripheral			
Color	λ (nm)	Bels	Antilog	Recip.	Norm to red peripheral	Bels	Antilog	Recip.	Norm to red peripheral
He-Ne, red	630	0.7	5	0.2	0.7	0.4	3½	0.3	1
orange	600	0.2	1½	0.7	2.3	−0.7	1/5	5	16
yellow	580	0.1	1¼	0.8	2.6	−1.2	1/15	15	50
yellow/green	555	0.0	1	1.0	3.3	−2.0	1/100	100	330
He-Se, green	520	0.3	2	0.5	1.7	−2.3	1/200	200	660
He-Cd, blue	470	1.0	10	0.1	0.3	−2.2	1/150	150	500

Source: After Wald (1945).

sider projection color, sunlight (ambient), power densities, and cockpit lighting. An integration of these factors, wavelength for wavelength, according to the Munsell chromaticity diagram (CIE tristimulus values) will give the effective signal-to-noise contrast. These values must be interpreted in the context of the MacAdam just-perceptible color differences, which take into account the considerable visual nonuniformity of the retina.

8.3.3 Color Discrimination

The color discrimination sensitivity has been shown to be most sensitive in the yellow from about 570 to 580 nm.

Scotopic Luminosity
With failing light, the brightness of different colored objects alters, the colors toward the red end of the spectrum becoming relatively darker, those toward the violet end brighter, so that finally the reds appear almost black and the blues bright. This is known as *Purkinje's phenomenon* and applies only under scotopic conditions in the peripheral vision; it is absent in the fovea.

In the dark-adapted eye, red light is recognized as red over an angle of 7–9° from the central position (overall angle 14–18°), a factor of 3 or 4 times that observed with green light (1.5–2.5° from central). Yet the red light is not seen outside this larger angle (14–18°), whereas green (or blue) light, while recognizable as green only over the much more restricted angle, is seen as a bright white light over an extensive angle (much greater than 18°).

As we are concerned with peripheral vision displays, it is clear that we must use the region of maximum rod sensors, which, from Figure 8.3, extends from $-20°$ to $+20°$, which is the scan angle on the commercial Malcolm Horizon peripheral vision display. Thus, it is clear that under scotopic conditions, green (or blue) will extend the retinal sensitivity well into the peripheral vision, whereas red light will cut out beyond $-9°$ to $+9°$.

Figure 8.3 shows that equality of visual acuity between rods and cones occurs at about $+2.5°$ to $-2.5°$, so at any angle beyond this, peripheral vision should predominate, and only in peripheral vision may one use the increased sensitivity of the retinal rode.

8.3.4 Critical Flicker Fusion Frequency

The critical fusion frequency (CFF) for flicker is the frequency at which a flickering light appears to be steady, assuming 100% modulation of the stimulus.

For flicker frequencies less than 10 Hz, luminance is proportional to the reciprocal of the duration of flash. With reference to Figure 8.8, the critical flicker frequency is low at low luminances and rises fairly rapidly as luminance increases until it achieves a constant value of about 17 Hz (for 19° diameter of test

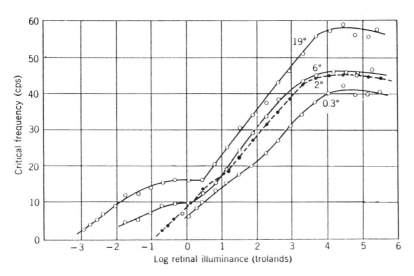

Figure 8.8 Influence of area of centrally fixated test field on the relation between critical frequency and log retinal illuminance. [From Hecht and Smith (1936).]

field with an illuminated nonflickering surround extending to 35°). This is maintained up to about 1 troland or 10^{-2} fL, which is equivalent to the lowest level of panel illumination.

Above this value, critical frequency rises again and finally reaches a level of about 60 Hz at high luminances. Since the flickering area subtends a visual angle of 19° (half that of the commercial Malcolm Horizon) we may suppose that both foveal cones and peripheral rods are stimulated and that each type of receptor makes its contribution to the curve. Analysis indicates that the low-luminance branch of the curve is due to the activity of rods, the high-luminance branch, to cones.

Under photopic conditions, where the cones are active, the CFF is critically dependent on luminance level and rises linearly with log retinal luminance up to the CFF of around 60 Hz.

At a flicker rate of between 2 and 20 Hz, the brightness of a flickering light may appear to be greater than that of the continuous sensation of complete fusion for the same luminance. Maximum brightness enhancement occurs at about 17 Hz. It would appear unlikely that we can take advantage of this effect, as 17 Hz still produces a very noticeable flicker for angular fields above about 0.3°, or with an illuminance above about 1 fL (dark and scotopic).

It is difficult to arrive at hard and fast conclusions for the best flicker frequency at which to operate, 15–20 Hz for best sensitivity, or 45–55 Hz for absence of flicker effect. One conclusion that arises from this anaysis is that when a display is available that can compete with 10,000 fc, the flicker will then be more noticeable and should perhaps be increased to 45–50 Hz, with the ensuing loss of sensitivity.

8.3.5 Color and Brightness Requirements for Illumination of a Peripheral Vision Display

Tests that I carried out and that were published by Ziegler (1983) used lasers for the line illumination in the Malcolm Horizon system marketed by Garrett Canada, the reference wavelengths being as follows:

He-Ne red	632 nm
Argon green	514 nm
He-Cd blue	442 nm

The conclusions were

1. At high ambient light levels the eye is most sensitive to yellow light (550–575 nm), where retinal sensitivity is approximately 6 db better than for red light (633 nm). Retinal sensitivity to green (514 nm) is approximately 4 db better than to red (633 nm), whereas for blue (442 nm) the eye is 3 db less sensitive than to red.
2. At low ambient light levels the eye is most sensitive to green (475–525 nm), where it is approximately 22 db better than for red. Sensitivity to blue light (442 nm) is about 11 db better than sensitivity to red.
3. There was a distinct psychological preference in the eight test subjects for the green line and a definite repulsion against the blue line. The blue line was noted as being difficult to look at. The red line was considered satisfactory as a line, but was not preferred, as red represents danger in many situations.

The Malcolm Horizon peripheral vision display produced by Garrett Canada uses a red He-Ne laser (1–3 mW illumination). It has been known for some time that a red He-Ne display is not the optimum color for a peripheral vision display, but it does have the advantages of being inexpensive, reliable, and available.

The disadvantages of red, other than that of retinal sensitivity, are that (1) red signifies danger, and (2) red light striking a red enunciator light will give a momentary red flash to the pilot, which adds stress to his flying.

In order to choose the optimum color for display, a research program was undertaken to measure retinal sensitivity against varying ambient conditions for the three lasers mentioned above (He-Ne, argon, and He-Cd).

Figure 8.9 gives the plot of the measured values, up to about 600 lx ambient illumination, with the curves extrapolated to 1000 lx. The primary conclusion is that a green laser gives about threefold improvement in retinal sensitivity for

PERIPHERAL VISION DISPLAYS   331

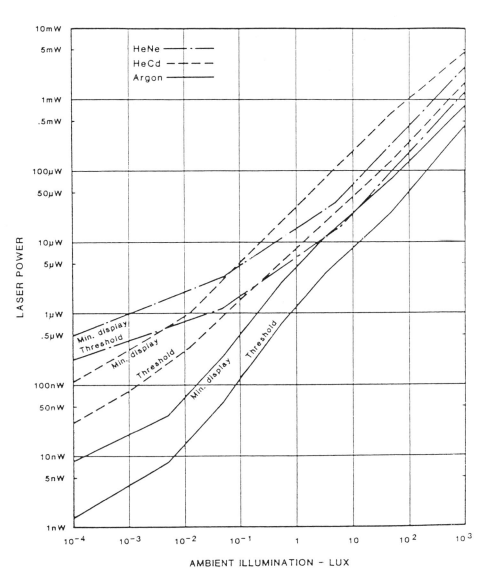

Figure 8.9 Retinal sensitivity. Plotted for laser power versus ambient illumination. The colors are He-Ne, red; He-Cd, blue; and argon, green. Threshold measurements were minimum for perception; Minimum display were minimum for comfortable viewing.

photopic vision. Or, put another way, the existing 3-mW red line could be replaced by a 1-mW green line and compete with the same ambient conditions.

Unfortunately, the large increase in retinal sensitivity under scotopic conditions cannot be used, as the system is well dimmed for dark adaptation when used under low ambient light conditions.

The next question that arises is how bright to make the display if it is to be used with the Military Specification requirement of 10,000 fc.

Measurements made by the author (unpublished) give approximate power requirements for threshold detection for each color. These were extrapolated from plots similar to those of Figure 8.9 to a background illumination of 10,000 fc (107,000 lx) and gave the following power requirements for optimum viewing:

He-Ne	red	70 mW
Argon	green	20 mW
He-Cd	blue	150 mW

8.4 FLIGHT TESTING OF THE MALCOLM HORIZON PERIPHERAL VISION DISPLAY

Early airborne testing of the Malcolm Horizon was in the CH-135 (Twin Huey UH-1N) and CH-124A (Sea King, S-61, H-3) helicopters of the Canadian forces, the UH-18 (Twin Otter) of the U.S. Air Force (USAF) Test Pilot School at Edwards Air Force Base (McNaughton, 1983), and the Jetstar of the Lockheed California Company.

Further testing was performed on an F-4 at the USAF Test Pilot School, a CP-140 (P-3) of the Canadian forces, and an NT-33 aircraft operated by Arvin/Calspan of Buffalo, New York, for the USAF and U.S. Navy (USN).

Lessons learned were that the peripheral vision display was useful for many different kinds of aircraft, and, for each aircraft, of more use in some functions than in others. An example of this that, when refueling, the pilot of the aircraft being refueled is essentially flying formation and cannot read his own instrumentation except for rapid glances. At the same time the tanker may be performing turns or other maneuvers that the pilot would like to know about. This is more important in some aircraft than in others, such as a B-52 rather than an F-16.

Testing up until about 1982 had produced an airworthy practical system, but it was realized that the hardware design had not taken full consideration of the military requirements for form factor, modularity, self-test, and other characteristics necessary for widespread adoption as distinct from special application.

A redesign program was initiated by Garrett Canada in 1982 in which many improvements were incorporated. These improvements were built into systems that were installed in the USAF SR-71 Blackbird surveillance aircraft (*Aviation & Aerospace*, 1990), where they were used in a secondary flight instrument system to provide the pilot with attitude and orientation information through peripheral vision, allowing him to concentrate on other cockpit duties.

Flight test results on early tests were generally positive (Kennedy 1989), although somewhat inconclusive. Pilot comments varied from "significantly improved pilot situational awareness" in a USN SH-2F Helicopter (U.S. Navy, 1988) and "helped to reduce pilot workload . . . may reduce the probability of disorientation" in a Royal Aircraft Establishment Sea King helicopter (Royal Air Force, 1986), to "failed to meet any of the test objectives" in a USAF F-16.

The shortcomings given by the pilots generally included the brightness and color of the display and in some cases included the small irregularities or discontinuities of the projected line over the complex instrument panels used, that is, irregularities caused by instrument bezels, protruding or set-back CRTs, etc. The unfavorable initial report from the USAF on the F-16 tests were attributed to restricted pitch movement and restricted display brightness.

The conclusions given in a recent survey (Kennedy, 1989) were;

1. Peripheral vision display is a pioneer.
2. It's a long way from prototype to operational system.
3. Operational personnel have a natural resistance to the use of information acquired subliminally.
4. Peripheral vision displays will make a large contribution to future cockpits.

So we see that the peripheral vision display is many things to many people. It is doubtful whether further tests will be more conclusive. Perhaps only its operational use in life-threatening situations will convince aircraft designers that it is a powerful tool in the prevention of many aircraft crashes.

8.5 THE FUTURE

The next generation of peripheral vision displays will most probably be based on the Malcolm Horizon, as this is the only viable system to date. This may possibly be incorporated in a helmet-mounted display.

In order to reduce the size of the projector unit in the cockpit, it is often necessary to place the laser source in a more remote location. This requires the use of a fiber-optic link to take the laser light to the projector and project onto the instrument panel. In order to do this, it is necessary to recollimate the laser light from the fiber such that it has similar characteristics to the laser beam before it is focused into the fiber. The characteristics of the light beam, which is to be collimated and with small diameter (for example, about 1 mm for a He-Ne laser), may be retained. This is difficult to achieve unless full use is made of the Lagrange invariant, which states, briefly, that in any closed optical system the product of image size, angle of field, and refractive index of the medium is invariant whenever a ray path crosses the optical axis. I have successfully achieved the design for such a fiber-optic link (Assenheim, 1989).

The display color should be green, and a suitable source for the future could be a frequency-doubled Nd-YAG laser, pumped by a Ga-Al-As laser. The doubled frequency would give a line of 532 nm wavelength and could be achieved by using a crystal of potassium titanyl phosphate (KTP) as the nonlinear element. Such a laser source has been built by GTE Electro-Optics Division in Palo Alto, California. It is lightweight and compact and has low power consumption (a few tens of watts), giving a 10-mW output at 532 nm.

The next forecast for the future relates to the problems of projecting a bar of light on the changing contours of modern-day instrument panels. In addition, small instrument panels, as in some small helicopters, preclude the possibility of using a true peripheral display. A holographic display of the peripheral stimulus would eliminate these problems. Holography offers the possibility of giving the pilot a horizon bar in space, either inside or outside the cockpit. It would enable the horizon to be wound around him, so giving true peripheral vision. It could also allow us to forget the concept of a horizon "line," and present to the pilot a view similar, if not identical, to the true horizon—an interface between two areas, sky or ground, blue or brown—or even to present the complete picture of an airport runway, regardless of whether the airport was in fact visible. This could be achieved using the holographic techniques of optical phase conjugation.

We are still in the early stages of this powerful technique, but the potential of peripheral vision displays is extremely impressive and stimulating.

References

Assenheim, H. M. (1989). U.S. Patent 4,818,049 (April 4.).
Aviation & Aerospace (1990). *Aviation & Aerospace*, November 7, p. 14.
Berthoz, A., B. Pavard, and L. R. Young (1975). Perception of linear horizontal self-motion induced by peripheral vision (linear vection). Basic characteristics and visual–vestibular interactions, *Exp.Brain Res.23*: 471–489.
Brandt, T., J. Dichgans, and E. Koenig (1973). Differential effects of central versus peripheral vision on egocentric and exocentric motion perception, *Exp. Brain Res.16*: 476–91.
Brown, I., S. Holmquist, and M. Woodhouse (1961). A laboratory comparison of tracking with four flight director displays, *Ergonomics 4*:229–251.
Dichgans, J. (1977). Optically induced self-motion perception, in *Life-Sciences Research in Space*, Proceedings of a symposium held at Cologn/Porz, Germany, ESA SP-130, pp. 109–112.
Dichgans, J., R. Held, L. R. Young, and T. Brandt (1972). Moving visual scenes influence the apparent direction of gravity, *Science 178*: 1217–1219.
Hecht, S., and Y. Hsia (1945). Dark adaption following light adaption to red and white lights, *J. Opt. Soc. Am. 35*: 261–267.

Hecht, S., and E. L. Smith (1936). Intermittent stimulation by light. VI. Area and the relation between critical frequency and intensity, *J.Gen.Physiol.* *19*: 979–991.
Henn, V., B. Cohen, and L. R. Young (1980). Visual-vestibular interaction in motion perception and the generation of nystagmus, *Neurosci. Res. Program Bull.* *18*(4): 459–651.
Hubel, D. H., and T. N. Wiesel (1962). Receptive fields, binocular interaction and functional architecture in the cat's visual cortex, *J.Physiol.160*: 106–154.
Kennedy, A. E. (1989). Flight testing and application of a peripheral vision display, *Proc. SPIE Conf. Display Syst. Optics 2*, Vol.1117, Orlando,FL.
Leibowitz, H. W., and J. Dichgans (1980). The ambient visual system and spatial disorientation, NATO/AGARD-CP-287, pp. B4-1—B4-4.
McNaughton, G. B. (1983). The role of the visual system in spatial orientation/disorientation, U.S. Air Force, Norton AFB.
Malcolm, R. M. (1984). Pilot disorientation and the use of a peripheral vision display, *Aviation, Space Environ. Med.*, March: 231.
Money, K. E. (1982). Theory underlying the peripheral horizon device, DCIEM Tech. Commun. 82-C-57, Defence and Civil Institute of Environmental Medicine, Downsview, Ontario, Canada.
Money, K. E., B. S. Cheung, J. P. Landolt, and J. C. Pellow (1984). Quantitative influence of the peripheral vision device on instrument flying in a simulator, DCIEM Tech. Commun. 84-c-02, Defence and Civil Institute of Environmental Medicine, Downsview, Ontario, Canada.
Ninow, E. H., W. F. Cunningham, and F. A. Radcliffe (1971). Psychophysiological and environmental factors affecting disorientation in naval aircraft accidents, NATO/AGARD-CPP-95-71, pp. A5-1—A5-4.
Osterberg, G. A. (1935). Topography of the layer of rods and cones in the human retina, *Acta Ophthalmol. Suppl.6*.
Reede, C. H. (1965). KLM research on the lowering of weather minima for landing of aircraft, *Ingenieur 77*(.11): L1–L13; Royal Aircraft Establishment, Library Transl. No.1250.
Ritchey, K. J. (1989). Telepresence systems—display system optics, *Proc. SPIE Conf. Display Syst. Optics 2*,Vol.1117, Orlando, FL.
Royal Air Force (1986). Flight assessment of a peripheral vision display in a Sea King Mk1 helicopter, Royal Air Force (U.K.) Institute of Aviation, Report No.533, October.
Schneider, G. E. (1969). Two visual systems, *Science 163*: 895–902.
Steele-Perkins, A. P., and D. A. Evans (1978). Disorientation in Royal Naval helicopter pilots, NATO/AGARD-CP-225, p.p. 48-1–48-5.
Stokes, A. F., C. Wickens, and K. Kite (1990). *Display Technology: Human Factors Concept*, The Society of Automotive Engineering, Warrendale, PA.
Tormes, F. R., and F. E. Guedry (1974). Disorientation phenomena in naval helicopter pilots, Bureau of Medicine and Surgery MF51.524.005-7026; BAIJ, Naval Aerospace Medical Research Laboratory, Pensacola, FL, 29 July.
U.S. Navy (1988). ASW Helo Night Landing Assist 2, U.S. Navy Report No.SY-OOR-XX, Naval Air Test Center.

Vallerie, L. L. (1968). Peripheral vision displays, phase II report. NASA Rep. CR-1239, National Aeronautics and Space Administration, Washington, DC.

Wald, G. (1945). Human vision and the spectrum, *Science 101*: 635–638.

Webster, J. A. (1989). Stereoscopic full field of vision display system to produce total visual telepresence, *Proc. SPIE Conf. on Display System Optics 2*, Vol.1117, Orlando, FL.

Wertheim, T. (1894). Uber die indirekte Sehscharfe, *Z. Psychol. 7*: 172–187.

Woodson, W. E. (1954). *Human Engineering Guide for Equipment Designers*, Univ. California Press, Los Angeles.

Wulfeck, J. W., A. Weisz, and M. W. Raben (1958). Vision in military aviation, USAF:WADC TR 58-399, Wright-Patterson Air Force Base, OH.

Ziegler, D. A. (1983). Vision colour testing for a peripheral vision display, Garrett Canada Rep. 83E1496, Garrett Canada, Rexdale, Ontario, Canada.

9
Holographic Head-Up Displays

Robert B. Wood

*Hughes Flight Dynamics, Inc.,
Portland, Oregon*

9.1 INTRODUCTION

Head-up display (HUD) systems are electro-optical devices that present attitude, navigation, guidance, targeting, and other information into the pilot's forward field of regard through the aircraft windshield. The information viewed by the pilot is projected at optical infinity and is conformal with and overlays his real-world view. Thus, the pilot flying with a head-up display is able to keep his attention focused on the real-world view while simultaneously viewing and monitoring critical flight parameters and data during all phases of flight. A collimated HUD display, or a display projected at optical infinity, minimizes the need for the pilot to come into the cockpit and refocus his eyes to view conventional panel-mounted displays. Eliminating vision transitions minimizes reaction time, enhances safety, and contributes to reducing pilot workload.

Head-up display systems have been used extensively in high-performance aircraft for the past 25 years for targeting and navigation purposes. In recent years, fighter aircraft HUD technology has been adapted for use in commercial and transport aircraft. In these cockpits, HUDs provide enhanced attitude awareness, low-visibility guidance information, navigation information, and autopilot monitoring information (Johnson, 1990; Long, 1990; Edelman, 1990; Boucek et al., 1983). Recently HUDs have been identified as the best means of displaying conformal weather-penetrating raster imaging sensor data to the flight crew (Rioux, 1990). Finally, simple head-up displays are beginning to be designed into automobiles. Although automotive HUDs are fundamentally different from high-performance aircraft systems, the optical principles remain the same.

All head-up displays require an image source, generally a high-brightness cathode-ray tube, and an optical device that projects the display information at optical infinity. The display is projected onto a semitransparent element referred to as the HUD combiner. Combiners are generally characterized as highly transparent, partially reflective optical devices positioned between the pilot's eyes and the aircraft windshield. The combiner reflects the HUD information to the pilot while simultaneously transmitting the real-world view of the outside world. The pilot is thus able to see both the real world and the display information focused at infinity. In essentially all HUD systems, the use of wavelength-selective holographically manufactured coatings in place of traditional partially reflective dielectric or metallic coatings improves the combiner reflectivity while simultaneously improving the real-world see-through transmission. These characteristics significantly enhance the photometric performance of HUDs and allow raster displays with stroke overlay to have adequate contrast ratios in moderate lighting conditions. It is for these reasons that most modern HUDs are designed with holographic combiner elements, and many older HUDs are being upgraded with replacement holographic combiner assemblies.

Head-up display systems generally comprise two major subsystems: the pilot display unit (PDU) and the HUD processor. The PDU interfaces electrically and mechanically with the aircraft and provides the optical interface to the pilot. The HUD processor interfaces electronically with aircraft avionics and sensors, runs a variety of algorithms in real time related to data formatting and verification, and generates the characters and symbols making up the display. More sophisticated HUD processors are capable of generating high-integrity guidance commands and cues for precision low-visibility landings and takeoffs, driving multiple independent HUD displays, and performing raster video processing. The output of the HUD processor is generally X and Y deflection and video bright-up signals for controlling the intensity of the cathode-ray tube in the PDU.

This chapter reviews the technology performance characteristics of both refractive and wide field-of-view head-up display optics. Included is a discussion of the performance advantages available when wavelength-selective holographic elements are used as the HUD combining elements. This chapter also briefly discusses HUD electronics and high-integrity HUD information processing and provides examples of several HUD systems.

9.2 HEAD-UP DISPLAY OPTICS

The purpose of this section is to describe and compare conventional refractive and wide field of view HUD optical systems in terms of parameters that are of

HOLOGRAPHIC HEAD-UP DISPLAYS 339

particular importance in modern commercial aircraft. Before the technologies can be compared, basic HUD performance characteristics are defined and described.

The HUD optical system provides the visual interface between the pilot and the aircraft electronics and sensors. As a result, the characteristics of the HUD optics must not adversely affect the pilot's performance or restrict the operational capabilities of the aircraft.

An example of a HUD optical system characteristic that could limit the capabilities of an aircraft is the instantaneous horizontal field of view (FOV) discussed in detail below. A high-integrity HUD system must have a horizontal instantaneous FOV large enough to provide a conformal display in high cross-wind conditions. Generally, this requires a horizontal display FOV of at least ± 5°. If the display has a smaller instantaneous FOV, or if large amounts of head motion are required to see the limits of the total FOV, the regulatory agencies involved with technical and operational certification might limit the crosswind landing conditions of the aircraft when the display is being used. The design of the HUD optics and electronics must be consistent with the system and aircraft operational objectives.

A simplified block diagram of a HUD optical system is presented in Figure 9.1.

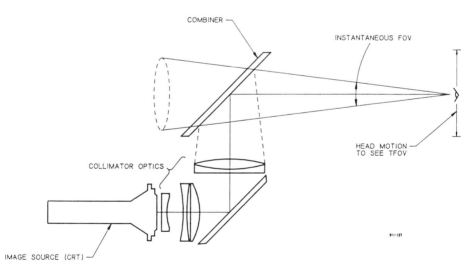

Figure 9.1 A simplified refractive HUD schematic illustrating the CRT image source, the collimating optics, and the combiner.

9.2.1 Symbology Collimation

A fundamental characteristic of aircraft HUDs is that essential flight data and information are projected at optical infinity onto the forward field of view seen through the HUD combiner and aircraft windshield. The process of projecting information at optical infinity is called *collimation*.

A collimated display has several important characteristics and advantages for aircraft pilots. First, it eliminates the need for the pilot's eyes to change focus when viewing either the real world or the HUD symbology through the HUD combiner. The pilot stays in constant visual contact with the real world, and both the real world and symbology are in focus simultaneously. Second, a collimated display minimizes head-dependent visual changes in the position of symbology relative to the real world as seen through the combiner. The collimated HUD image precisely overlays the real world and remains conformal with the real world independent of the pilot's head position within the head motion box. Finally, because a HUD display minimizes the need for the pilot to refocus into the cockpit for essential flight information, head-up to head-down to head-up transitions are eliminated. Transition time includes the time needed to refocus onto head-down displays and the time needed to accommodate for the brightness differences inside the cockpit. Eliminating or minimizing transition times improves reaction times, reduces workload, and enhances safety.

There are two basic forms of HUD collimation systems in widespread use today. They are refractive optics systems using lenses and flat-plate combiners, and wide field-of-view optical systems using combiners with optical power and holographic coatings. Both the refractive and the wide field-of-view (WFOV) approaches perform the same basic collimation functions described above and present the pilot with conformal information superimposed at infinity on a portion of his or her forward vision through the aircraft windshield. In both optical approaches, the source of the displayed information is a high-brightness cathode-ray tube (CRT) capable of generating fine linewidths (0.005 in. typical). However, the way in which the collimation function is performed and the resulting HUD optics characteristics are quite different. In particular, the holographic WFOV offers significantly larger instantaneous and binocular overlapping fields of view and significantly improved photometric performance compared with conventional refractive optics systems.

The sections that follow discuss the optical performance parameters that are important for HUD systems and compare the performance characteristics available with the two collimation methods.

9.2.2 HUD Optical Performance Parameters Defined

This section defines and describes the most important head-up display optical characteristics for transport aircraft. The following characteristics are discussed:

HOLOGRAPHIC HEAD-UP DISPLAYS

1. Total field of view (TFOV)
2. Instantaneous FOV (IFOV)
3. Binocular overlapping FOV
4. Monocular FOV
5. Head motion box
6. Display linewidth
7. Binocular parallax errors
8. Display accuracy
9. Display brightness and contrast ratio

Characteristics and advantages of using holographic elements for combiners in both refractive and WFOV systems are discussed in Section 9.3.

Display Fields of View

The display fields of view (FOV) are angular regions over which the collimated display is visible by the pilot. There are four distinct characteristics needed to fully characterize the FOV performance. These characteristics are the total field of view, the instantaneous FOV, the binocular overlapping FOV, and the monocular instantaneous FOV.

Figure 9.2 illustrates the FOV characteristics of a refractive HUD optical system. Figure 9.3 illustrates the FOV characteristics of a WFOV system.

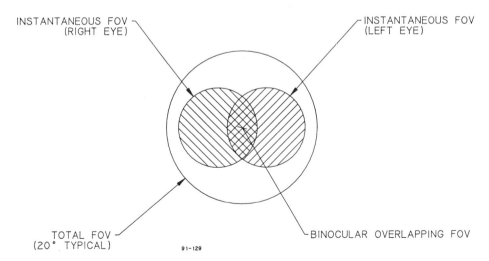

Figure 9.2 The display fields of view of a refractive HUD. The instantaneous FOV is less than the total display FOV, and head motion is needed to see the entire display. The overlapping binocular FOV is generally small in refractive HUDs.

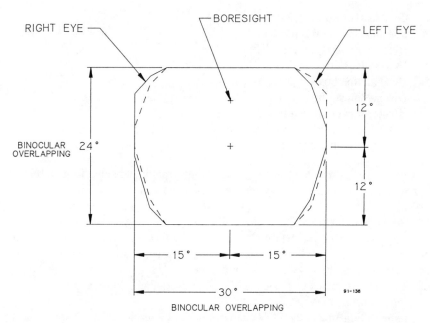

Figure 9.3 The display fields of view of a wide FOV HUD. The instantaneous and total displays FOVs are essentially the same in WFOV systems, and no head motion is needed to see the entire display. In some WFOV systems, the overlapping binocular FOV is the same as the total FOV. (Flight Dynamics, Inc.).

Total Field of View The total display field of view (TFOV) is the maximum horizontal and vertical angular extent of the collimated display allowing head motion within the head motion box. It is specified in horizontal and vertical degrees. The total display FOV can be significantly larger than the portion of the display that is visible from a fixed head position. The TFOV is often specified from the desired operational requirements of the HUD system and aircraft. It can be restricted by the cockpit geometry, physical packaging constraints associated with the cockpit interior, optical constraints due to the packaging envelope, and weight and cost constraints.

High-performance WFOV HUD systems in use today have total display fields of view approximately 24° vertically by 30° horizontally. It is often desirable to have several degrees of vertical field of view at the limits of the horizontal FOV.

Instantaneous FOV The instantaneous FOV (IFOV) is the angular extent of the HUD display as seen from either eye from a fixed head position. Thus, the IFOV is composed of what the left eye sees plus what the right eye sees from a fixed head position. The horizontal eye spacing provides a larger horizontal IFOV than the vertical IFOV because each eye can see a different portion of the total display. The horizontal eye spacing is generally assumed to be 2.5 in.

In refractive HUD systems, the IFOV depends on the physical dimensions of the collimator aperture, the distance between the viewer's eyes and the collimator aperture, and the interpupillary distance. If large-aperture collimators are used and the pilot sits close to the combiner, the IFOV can be as large as 17° vertically by 30° horizontally.

In most WFOV systems in use today, the IFOV is nearly the same as the TFOV. Thus, essentially all of the display information is visible to the pilot from a fixed head position without head movement. The minimum desirable instantaneous FOV for high-performance WFOV HUD systems is about 24° vertically by 30° horizontally.

The size and shape of the IFOV often change with head motion within the head motion box, especially in the vertical direction. In most WFOV systems, the combiner vertical dimension is designed and sized to pass the vertical FOV from the cockpit eye reference point (ERP). Thus, if the pilot moves her head above the ERP location, the top of the vertical FOV is vignetted at the combiner because there is no combiner glass to cover these angles. Generally, the instantaneous horizontal FOV remains nearly unchanged with horizontal head movement.

Binocular Overlapping FOV The binocular overlapping FOV is the angular extent of the HUD display that is common to both eyes simultaneously. From the standpoint of viewer comfort, it is desirable to have the binocular overlapping FOV as large as possible. Generally, the horizontal overlapping binocular FOV is less than the total horizontal FOV, and the edges of the display are seen with a single eye (i.e., left eye looking right or right eye looking left). It is always true that the larger the binocular overlapping FOV, the more complex the optical system.

In refractive HUDs, the binocular overlapping FOV is generally only a portion of the total FOV. In particular, a refractive HUD with an instantaneous FOV of 16° vertically by 30° horizontally may have only an irregularly shaped overlapping FOV of 16° vertically by 12° horizontally. In high-performance WFOV HUD systems, on the other hand, the binocular overlapping FOV can be equal to the total display FOV, or about 24° vertically by 30° horizontally. Thus, the entire display FOV is visible with either eye when the head is centered in the head motion box.

Monocular Instantaneous FOV The monocular instantaneous FOV is the angular extent of the HUD display as seen by a single eye.

In refractive HUD optical systems, the angular size of the collimator aperture as seen from an eye location defines the monocular instantaneous FOV. The shape and size of the monocular IFOV do not change much with head motion owing to the optics geometry as long as the angle between the eye and any point on the collimator aperture is less than the TFOV. However, only a portion of the TFOV is seen at a time. Typical monocular instantaneous fields of view for conventional HUDs are between 10° and 18° in diameter, depending on the collimator aperture and viewing distance.

In WFOV optical systems, the monocular instantaneous FOV from the center of the head motion box can be limited by the combiner size in the vertical direction (if less than the total vertical FOV) and can be equal to the total FOV in the horizontal direction. Thus, in many WFOV systems, the monocular instantaneous FOV can be equal to the total FOV from the center of the head motion box. As the eye approaches the edge of the head motion box, however, the monocular instantaneous FOV decreases rapidly because of combiner and relay lens aperture vignetting.

Head Motion Box

The HUD head motion box, or eyebox, is a three-dimensional region in space surrounding the cockpit eye reference point (ERP) in which the display can be viewed with at least one eye. The center of the head motion box can be displayed forwards or aft, or upward or downward, with respect to the cockpit ERP to better accommodate the actual sitting position of the pilot. The positioning of the cockpit eye reference point (Stone, 1987) is dependent on a number of ergonomically related cockpit issues such as head-down display visibility, the over-the-nose down look angle, and the physical location of various controls such as the control yoke and the landing gear handle. In many cases, the cockpit eye reference point and where the pilot actually sits can vary by several inches.

The head-up display head motion box should be as large as possible to allow maximum head motion without losing the display information. The minimum desired size of the head motion box, independent of the optical system technology, is as follows:

Horizontal	4.5 in.
Vertical	2.5 in.
Depth	6.0 in.

The optical performance requirements must be met throughout the eye motion box volume, including display brightness, parallax errors, and display accuracy.

There are basic differences between the head motion box of a refractive HUD system and that of a WFOV system. To begin with, in refractive HUD systems, the pilot must move his or her head vertically and laterally to view the extremes of the total FOV (the "porthole" effect). In some refractive HUD systems, a 5.0 in. wide head motion box is required to view all the display information. This effect is discussed by Chorley (1974) and Vallance (1983). In many WFOV systems, on the other hand, the total display is visible without any head motion.

(Horizontal head movement within a 5.0 in. wide head motion box alters the monocular FOV from each eye but does not change the horizontal IFOV.) Even though head motion is not needed to view the edges of the display, a large head motion box is desirable to provide head movement freedom without losing binocular overlapping FOV.

In WFOV optical systems, the head motion box is a well-defined exit pupil. (WFOV HUDs are often referred to as "pupil-forming HUDs" for this reason.) From near the center of the pupil, the total display FOV can be visible instantaneously. Outside of the pupil, no display is visible. In a high-performance WFOV system, where the overlapping binocular FOV is equal to the total FOV, the total horizontal display FOV is visible instantaneously with up to $\pm$ 2.5 in. of horizontal head movement.

In all HUDs, the monocular instantaneous FOV vignettes (cuts off) with lateral or vertical eye displacement from the center of the head motion box. Because the size of the monocular IFOV changes with head motion, a definition of the head motion box size should include a minimum monocular IFOV definition from the edge of the region. This ensures that even when the pilot's head is decentered such that one eye is at the edge of the head motion box, there is still a useful monocular FOV from both eyes. Typical minimum monocular FOV limits are dependent on the system operational requirements, but 10° horizontally by 10° vertically is generally a very usable display field. Note that in a WFOV system, when one eye is at the horizontal edge of the eyebox, the other eye is near the eyebox center and can therefore see the total display FOV.

As shown in Figure 9.2 for refractive HUDs, significant head motion is required to see the edges of the total FOV, and the instantaneous FOV is less than the total FOV. In Figure 9.3, however, the entire display is seen instantaneously.

Display Linewidth

The HUD display linewidth is dependent on two principal factors, the effective focal length of the optical system and the physical linewidth on the CRT. The angular display linewidth is related to the optical system focal length (F) and the physical CRT linewidth (W) (generally specified at the 50% intensity points) as follows:

$$\text{Linewidth (milliradians)} = \arctan (W/F) \tag{9.1}$$

Thus, if the focal length of an optical system is 5.0 in. and the CRT linewidth at the 50% intensity points is 0.005 in., the display linewidth will be about 1 milliradian (mrad).

Optical aberrations will adversely affect the apparent display linewidth. These aberrations include chromatic aberrations (lateral color) and residual uncompensated coma and astigmatism.

Acceptable HUD display linewidths are between 0.8 and 1.2 mrad at the 50% intensity points. The linewidth should not vary outside of these limits across the full brightness range of the HUD system, generally between about 0.1 and 2400 foot-lamberts (fL) measured at the pilot's eye. In HUDs that require short CRTs for packaging reasons, dynamic focus may be required to maintain the desired CRT linewidth across the usable screen area.

Display Binocular Parallax Errors

The object of the HUD display is to provide a virtual image of the CRT at optical infinity. In the binocular overlapping portion of the IFOV, the left and right eyes view the same information on the CRT. If the optical system is not perfect (and they never are), there will be slight angular errors between two horizontally displaced eyes. These errors are displayed binocular parallax errors or collimation errors. The binocular parallax error for a fixed field point within the total FOV is a measure of the angular difference in rays entering two eyes separated by the interpupillary distance, assumed to be 2.5 in. If the projected virtual display image were perfectly collimated at infinity from all head motion box positions, the two ray directions would be identical, and the parallax errors would be zero.

Parallax errors are generally divided into horizontal and vertical components. Horizontal errors are divided into convergence, or eyes looking inward (i.e., objects inside of infinity), and divergence, or eyes looking outward. Vertical errors, or dipvergence, represent eyes skewed out of the horizontal eye plane. These errors, which describe how two eyes are oriented with respect to each other to binocularly merge symbols, are illustrated in Figure 9.4.

Normal eyes cannot readily compensate for divergent or dipvergent parallax errors. For this reason, these errors must be made as small as possible, generally 1.5 mrad or less. Convergent parallax errors can be easily compensated for, but cause the apparent display depth to vary from infinity (0 mrad) to about 83 ft (2.5 mrad). To maximize the display accuracy, the convergent parallax errors should also be minimized.

Convergence, divergence, and dipvergence errors originate because the light rays from the CRT to the pilot's eyes travel different nonsymmetrical paths through the optical system. This happens whenever the pilot's head is laterally displaced from the bilateral symmetry plane of the optical system (which generally passes through the design eye location) or when the pilot views the display at nonzero azimuth angles.

To characterize the HUD display for parallax errors, the instantaneous FOV is sampled at regular horizontal and vertical display increments from a variety of head positions within the head motion box. Typical FOV increments are about 1° vertically by 2° horizontally, resulting in about 300 grid intersections in a 24° × 30° display FOV. Similarly, the head motion box can be divided into 0.5 in. vertical and 0.75 in. horizontal increments in the plane perpendicular to the op-

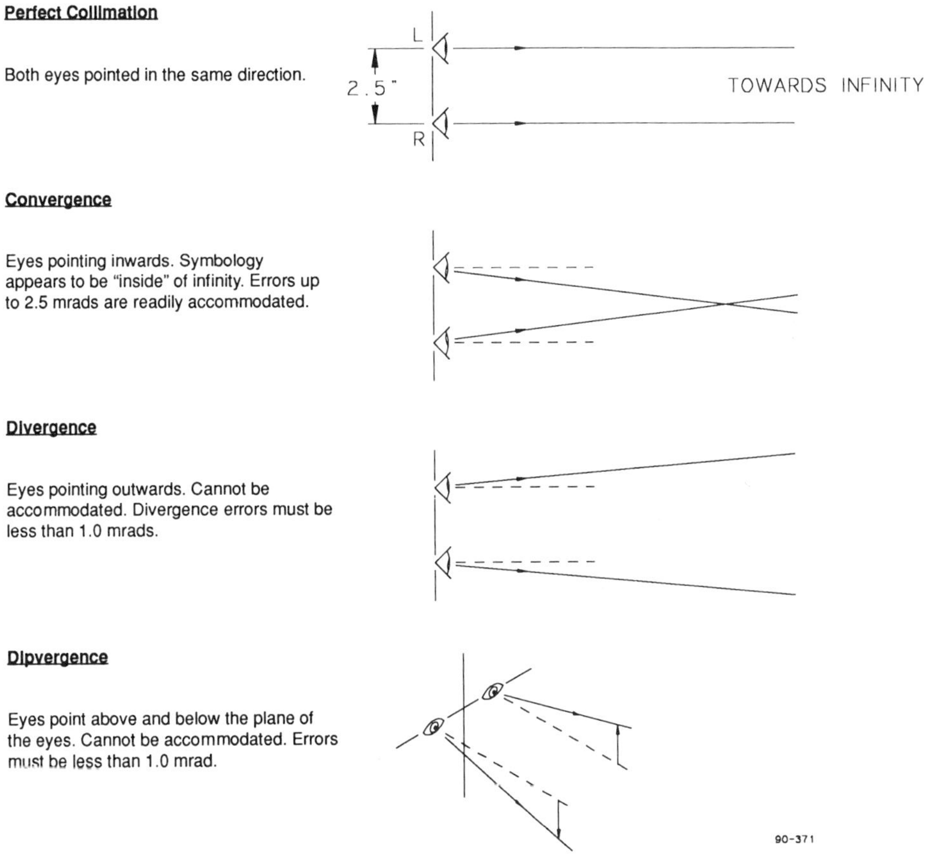

HOLOGRAPHIC HEAD-UP DISPLAYS

Perfect Collimation

Both eyes pointed in the same direction.

Convergence

Eyes pointing inwards. Symbology appears to be "inside" of infinity. Errors up to 2.5 mrads are readily accommodated.

Divergence

Eyes pointing outwards. Cannot be accommodated. Divergence errors must be less than 1.0 mrads.

Dipvergence

Eyes point above and below the plane of the eyes. Cannot be accommodated. Errors must be less than 1.0 mrad.

Figure 9.4 Binocular parallax errors. Convergent errors are readily accommodated but cause the apparent display image distance to change. Dipvergence and divergence cannot be visually accommodated if the errors are greater than about 1.5 mrad.

tical chief ray. Two or more planes are often selected to characterize the head motion box in the longitudinal direction. (Since the HUD display is collimated, the binocular parallax does not change rapidly with longitudinal distance from the collimator. The instantaneous FOV, however, is a strong function of longitudinal spacing to the collimator.) This eyebox sampling can result in about 100 evaluation positions in a 3 in. × 5 in. head motion box.

Thus, the total number of evaluation samples used to fully characterize the parallax performance across the total FOV and head motion box can be in the thousands. (Field-of-view vignetting near the edge of the head motion box results in fewer field samples compared with the head centered numbers.) Of those total data points, a percentage will meet a specified performance criterion.

The following parallax error values are representative of what can be achieved throughout the head motion box and across the total display FOV in a refractive HUD optical system with a binocular overlapping FOV of 16° vertically by 6° horizontally.

Convergence	100% of data points < 0.8 mrad
Divergence	100% of data points < 0.5 mrad
Dipvergence	100% of data points < 0.5 mrad

Thus, all of the field points when viewed from any head position have display parallax errors of less than 1 mrad. This represents a very well optically corrected system.

Because WFOV systems are much more complex than refractive systems (they are not rotationally symmetric), and because more of the display FOV is seen binocularly, the parallax errors are larger than those of refractive systems. The following parallax error values are representative of what can be achieved in a well-corrected WFOV display with a 24° × 30° overlapping display FOV:

Convergence	95% of data points < 2.5 mrad
Divergence	95% of data points < 1.0 mrad
Dipvergence	95% of data points < 1.5 mrad

Thus, approximately 95% of the data points sampled from across the display FOV and head motion box have errors less than 3 mrad.

In essentially every case, the sampled points that do not meet the desired performance occur at the edges of the overlapping FOV (near the edges of the total FOV) and from the vertical and horizontal limits of the head motion box. The performance is generally very good near the center of the head motion box.

Display Accuracy

Display accuracy is a measure of the precision with which the projected HUD image overlays the real-world view seen through the combiner and windshield

HOLOGRAPHIC HEAD-UP DISPLAYS 349

from any eye position within the head motion box. Display accuracy is generally a monocular measurement. For a fixed field point, the accuracy is numerically equal to the angular difference between the position of a real-world feature as seen through the combiner and windshield and the HUD projected symbology.

The HUD system total display accuracy error budget includes sensor input conversion errors, computation and roundoff errors, electronic gain and offset errors, optical errors, CRT- and yoke-related errors, installation variations, windshield variations, environmental conditions, assembly tolerances, and thermally induced errors. Optical errors are dependent on both head position and field angle. Errors associated with the basic sensor input data are not generally included in HUD display accuracy errors but are accounted for in the overall system design.

The contributions from the HUD optics to the total display accuracy error budget are made up of four sources: uncompensated pupil and field errors originating in the optical system, distortion map curve-fitting inaccuracies, manufacturing variations, and the tolerances associated with combiner positioning.

The following display accuracy values are achievable in wide FOV HUDs with a stowable combiner when all HUD system error sources are included:

Boresight (optical reference axis)	± 3.0 mrad
Total display accuracy	± 7.0 mrad

The error contribution from the optics alone in the total display accuracy value, not including the stowable combiner alignment tolerances, is about 3.0 mrad. The optics contribution to the error budget drops to about 1.5 mrad in refractive HUD optical systems.

The boresight direction is the single reference direction (somewhat arbitrary) used to verify the angular alignment accuracy of the HUD hard points. It generally corresponds to the aircraft longitudinal axis (where the aircraft is pointed). The boresight direction is also used as the calibration direction for zeroing all electronic errors. The boresight errors include mechanical installation tolerances, electronic drifts due to thermal variations, and optical errors at a fixed point. These errors also include manufacturing tolerances for positioning the combiner element into the mechanical assembly.

Display Contrast Ratio

Head-up display contrast is the ratio of the brightness of the virtual display relative to the real-world background brightness. HUD contrast ratio is defined as

$$\text{Contrast ratio} = \frac{\text{display brightness} + \text{apparent real-world brightness}}{\text{apparent real-world brightness}} \tag{9.2}$$

or

$$\text{Contrast ratio} = 1 + \frac{\text{display brightness}}{\text{apparent real-world brightness}} \quad (9.3)$$

The apparent display brightness is dependent on the luminance of the CRT source, the losses associated with the lens assembly, and the combiner phosphor reflectivity. Generally, the most inefficient component of the HUD optical system is the combiner. The use of holographic coatings to enhance the combiner reflectivity is discussed in Section 9.3. The visual display brightness is determined by cascading the CRT phosphor output with the relay lens transmission, the CRT sidelobe suppression filter, and the combiner phosphor reflection efficiency.

The apparent real-world brightness is a product of the actual real-world brightness and the combiner photopic transmission. (The transmission of the aircraft windshield is generally left out of the apparent real-world brightness computation, even though the losses through many cockpit windshields can be significant.) The worst-case apparent real-world brightness as far as HUD contrast is concerned is assumed to be a cloud illuminated by direct sunlight at an altitude of 10,000 ft. It is generally assumed that this condition results in a background brightness of about 10,000 fL.

The minimum acceptable HUD contrast ratio when operating in the stroke only display mode is specified against a 10,000 fL background. (When a HUD system is operating in the stroke only display mode, the displayed information is written on the CRT as it would be drawn by hand. That is, each character and line segment is individually drawn as a separate discrete line. See the section on raster brightness, below.) A stroke only display contrast ratio of 1.2 is adequate for comfortable display viewing.

Simple refractive HUD systems utilizing dielectric partially reflective combiner coatings with 20% reflectivity can achieve a 1.2 contrast ratio with a CRT light output of about 8800 fL. If additional solar rejection filters are included, the CRT must be driven even harder to achieve an acceptable contrast ratio.

One important advantage of using holographic wavelength-selective combiners in HUD systems is the increased phosphor reflectivity. Increased phosphor reflectivity allows a fixed contrast ratio to be achieved with lower CRT light output. Reducing the CRT drive levels extends the life of the CRT, thereby lowering the life-cycle cost of the HUD system.

Holographic HUDs in commercial aircraft operation are capable of providing more than 2500 fL of display brightness, thus allowing a display contrast ratio greater than 1.3 against a 10,000 fL background. A 1.2 contrast ratio can be achieved in HUDs with holographic combiners with a CRT light output of about 5200 fL.

Raster Brightness and Contrast Ratio

With the development of improved performance, cost-effective millimeter wavelength radar and infrared imaging sensors, there is growing interest in the commercial aviation community in providing a video image to the pilot on a HUD with stroke guidance information symbology overlay. The combination of weather-penetrating sensor video with a guidance information overlay on a conformal WFOV HUD may allow pilots to fly safely into low-visibility conditions without the need for specially monitored ILS equipment and special runway lighting. The video image is written on the HUD CRT in a raster format, much like standard television.

This capability requires the use of high-speed deflection circuitry and a high-speed vector and character generator. The sensor provides a standard RS-170 (or equivalent) raster video to the HUD system. During the raster vertical retrace interval at the end of the frame, the stroke guidance information is written.

Most HUD displays are refreshed at a rate of 60 sec^{-1} to prevent visible display flicker. For all the display information to be written on the CRT at this rate, or in 16.6 msec, the stroke writing rate is generally in the range of 4000–8000 in./sec. This represents an angular writing rate of between 46,000 and 92,000°/sec. A typical stroke only display brightness measured at the design eye location for a writing rate of about 7000 in./sec. (81,000°/sec) is about 2400 fL.

The light output from an element of phosphor screen is dependent on how much beam energy is delivered during each refresh interval. Thus, for a given beam current, the faster the electron beam sweeps across the faceplate (or the faster the writing rate), the lower the light output from that particular phosphor element.

When operating in the raster mode, roughly half of the (RS-170) 488 actual video lines are displayed in about 15.3 msec. This allows approximately 1.3 msec for the vertical retrace interval. (In interleaved video, odd lines are displayed in the first 16.6 msec, followed by the even lines in the next frame. Thus, the total video image is updated every 1/30 sec.) Each line in a standard RS-170 video signal is allocated about 58.5 μsec. If the horizontal dimension of the CRT is 2.6 in., the raster line writing rate is about 45,000 in./sec.

Assuming that the light output from the phosphor is inversely proportional to the writing rate, the maximum raster display brightness using a CRT capable of emitting 20,000 fL at a stroke writing rate of 8000 in./sec is approximately 700 fL at the pilot's eye. This brightness level assumes a holographic combiner.

A raster gray level or shade is often defined as an increase in brightness of 1.41 (the square root of 2) over the next lowest level. The number of shades of gray visible on a raster HUD display depends on the ambient background brightness. The approximate background brightness limits are given in Table 9.1 as a function of the desired number of shades of gray. This analysis assumes that the maximum raster brightness is 700 fL.

Thus, the number of gray shades visible to the pilot during the raster or stroke/raster display mode is strongly dependent on the brightness of the real-world ambient background and on the light output capabilities of the CRT. For ex-

Table 9.1

Ambient background brightness (fL)	Number of visible raster gray shades
100	8
150	7
230	6
365	5
625	4
1255	3
4400	2

ample, if the real-world background brightness is 1250 fL, a raster display brightness of 700 fL will provide three raster shades of gray.

Fortunately, the raster image on the HUD is required primarily when the visibility is poor, typically below 100 ft of RVR (runway visual range). The natural filtering of heavy cloud cover and fog will generally reduce the ambient brightness to less than 100 fL in this runway environment.

For the stroke display to be visible against the raster image, the stroke information should be about two tones brighter than the brightest raster shade. This is not difficult to achieve in practice because the stroke writing rate, even in the stroke raster mode, is slower than the raster writing rate. In some cases, there is inadequate time to write the full stroke guidance display in the vertical retrace interval (about 1.3 msec). In this case, extra time can be "borrowed" from the top or bottom of the raster image. If, for example, an additional 2.3 msec (making a total of 3.9 msec) is needed to draw the complete stroke display, approximately 15% of the raster image will be borrowed and not displayed. In a WFOV HUD with a 24 degree vertical TFOV, the vertical FOV of the raster image will be approximately 20 degrees.

9.2.3 HUD Optics Comparisons

In this section, refractive optics systems are compared with wide field-of-view holographic optical systems.

Refractive Optical Systems Behavior

A refractive HUD optical system is generally composed of a rotationally symmetric lens assembly designed to collimate the CRT image and a single flat-plate beamsplitter or beam combiner that redirects the collimated ray bundles into the pilot's direct line-of-sight view through the aircraft windshield. The partially reflective combiner coating can be either a wavelength-insensitive dielectric coating or a wavelength-selective holographically manufactured coating (discussed in Section 9.3).

HOLOGRAPHIC HEAD-UP DISPLAYS 353

The rotationally symmetric collimator lens assembly is almost always a variant of the Petzval lens form. In many cases, it is configured with a large air gap located between the exit elements (closest to the combiners) and the field flattener elements (closest to the CRT). The air gap is designed to allow a front surface folding mirror to be inserted between the two lens groups, thus allowing the optical system to be folded at approximately 90°. Folding the lens assembly allows the vertical height of the HUD chassis to be minimized, thereby facilitating packaging the lens and CRT combination in highly constrained cockpits. Note that folding the optical system with a front surface mirror does not change the rotational symmetry of the optical system. Figure 9.5 illustrates a typical folded Petzval form collimator with internal solar rejection filters and dual flat-plate combiners. This generic type of HUD system has been described in the literature by Chorley (1974), Banbury (1982), and Vallance (1983). Because the collimating lens has rotational symmetry and the combiners have no optical

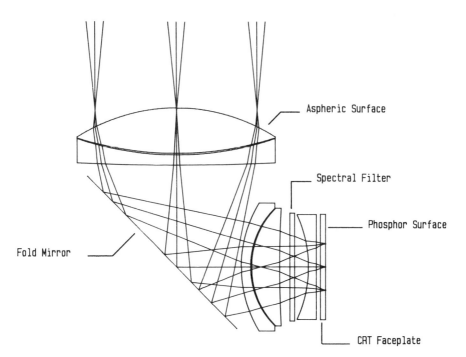

Figure 9.5 A typical folded refractive HUD optical system. This collimator design is based on a Petzval lens form, with a large air gap between the exit group and field flattener elements. This lens was designed for use with dual holographic combiners.

power (i.e., the combiners are flat), the optical performance of refractive HUDs is generally excellent both in terms of parallax errors and display accuracy. Table 9.2 summarizes the optical performance of a typical refractive HUD optical system. Because of the rotational symmetry and the small number of individual lens elements making up the assembly, refractive HUD optical assemblies are generally reasonably priced and straightforward to assemble and test.

In some critical HUD applications, aspheric surfaces may be required to control spherical aberrations. In other cases, the collimating lens aperture may not be round but truncated to improve the shape of the instantaneous FOV and to optimize the HUD system packaging. In both of these cases, the cost of the collimating assembly will increase commensurate with the additional complexity.

Limited Instantaneous Field of View   The fundamental drawback of refractive optics systems is that the IFOV is smaller than the total display FOV and a limited portion of the display is viewed binocularly. Thus, not all of the available display is visible to the pilot instantaneously, and head motion is required to see the edges of the display.

There are two geometric parameters needed to determine the instantaneous

Table 9.2 Typical Refractive HUD Optical System Characteristics

Characteristic	Performance
Field of view	
Total[a]	25° circular typical
Instantaneous[b]	21° horiz × 18° vert
Overlapping	7° horiz × 18° vert
Aperture size[c]	5.0 in. diameter truncated vertically to 4.0 in.
Head motion box size	4.5 in. diameter typical
Head motion needed to see TFOV[a]	±0.7–2.0 in.
Parallax performance[d]	
Convergence	100% < 0.8 mrad
Divergence	100% < 0.5 mrad
Dipvergence	100% < 0.8 mrad
Elements in collimator assembly	5 or 6 (typical)
Combiner photopic transmission	Coating-dependent
Phosphor type	P1, P53
Contrast ratio	< 1.2: 1

[a]Dependent on geometry and collimator aperture.
[b]Dual combiners.
[c]Design variable, dependent on IFOV requirements, geometry, etc.
[d]Sampled from all pupil and field positions.

display FOV characteristics of a refractive HUD collimator: the physical collimator aperture diameter D and the spacing between the pilot's eyes and the center of the collimator aperture, d. In single-combiner refractive HUD optics, the vertical and horizontal display fields of view are approximately as follows:

$$\text{Vertical instantaneous FOV} = 2 \arctan [(D/2)/d] \quad (9.4)$$
$$\text{Horizontal instantaneous FOV} = 2 \arctan [D/2 + 2.5/2)/d]$$

Note that in the horizontal FOV calculation, the pilot's eye separation (assumed to be 2.5 in.) increases the horizontal FOV. The horizontal binocular overlapping display FOV is computed as

$$\text{Overlapping binocular FOV} = 2 \arctan [(D/2 - 1.25)/d] \quad (9.5)$$

In many modern refractive HUDs, the collimator aperture is not circular but is cropped vertically to optimize lens packaging. In this case, the vertical instantaneous FOV is computed from the vertical aperture dimension that is different from the horizontal aperture.

Typical values for the display fields of view for a refractive HUD using a single combiner are as follows:

Total FOV	25° circular
Vertical instantaneous FOV	14.3°
Horizontal instantaneous FOV	21.2°
Overlapping binocular FOV	7.2°

This example is for a round 5 in. collimator lens diameter located 20 in. from the viewer. The IFOV can be improved by increasing the size of the collimator aperture or by physically positioning the pilot closer to the aperture. In most cockpit geometries, a variety of direct and indirect restrictions are placed on the HUD geometry. These restrictions include the physical chassis width, clearance between the combiners and the inner surface of the windshield, the angle of the over-the-nose vision line, the distance between the pilot and the combiner, and, if applicable, the ejection plane position and angle. It is a never-ending challenge for the designers to increase the display IFOV while meeting the increasingly more restrictive envelopes of both fighter and commercial cockpits.

One proven method of increasing the vertical field of view of refractive optical systems is to position a second flat combiner element parallel to and above the lower combiner, as illustrated in Figure 9.6. The upper combiner element forms a second, vertically displaced virtual image of the collimator aperture. The effect of this is to increase the instantaneous look-up angle from a fixed head position. This is a common combiner configuration on many military HUDs in use today.

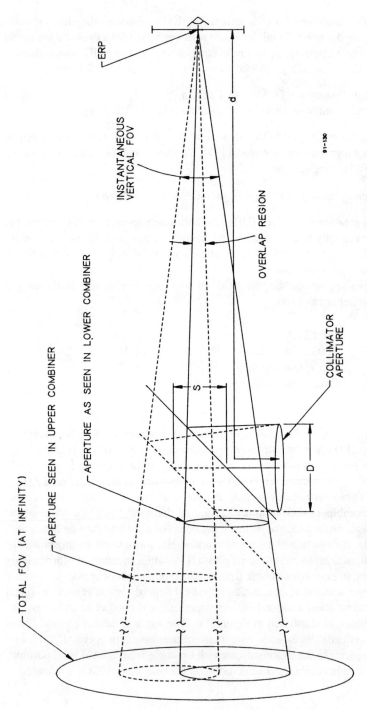

Figure 9.6 The vertical instantaneous FOV can be increased in refractive HUDs if a second parallel combiner plate is positioned above the lower combiner. The increase in vertical instantaneous FOV is about 5°.

The vertical instantaneous FOV for this case is given approximately as

Vertical instantaneous FOV = up FOV (from upper combiner)
+ down FOV (from lower combiner)
= arctan $[(D/2 + s/2)/(d+s)]$ + arctan $[(D/2 + s/2)/d]$ (9.6)

where s is the vertical spacing between the upper and lower combiners, illustrated in Figure 9.6. Note that s must be less than the vertical diameter of the collimating aperture, d, to prevent gaps from appearing in the instantaneous display FOV from near the top of the head motion box. There is no increase in the instantaneous horizontal display FOV when dual combiners are used.

Typical values for the display FOV for refractive HUDs using dual combiners are as follows:

Total FOV	25° circular
Vertical instantaneous FOV	19°
Horizontal instantaneous FOV	21.2°
Overlapping binocular FOV	7.2°

The fact that the instantaneous display fields of view are smaller than the total FOV implies that head motion is required to see the total FOV.

Refractive Optics Summary Table 9.2 summarizes the optical performance achievable with refractive HUD optics. The specific performance characteristics will vary depending on the geometry of the system. The instantaneous FOVs are dependent on the physical size of the collimating aperture and the total distance between the design eye and the aperture.

Wide Field-of-View HUD Optics

Wide FOV optical systems differ from conventional systems in that the collimation and combining functions, performed separately in refractive optics systems by the collimating lens assembly and the flat combiners, are optically merged into a single large-area combiner/collimator. This combination allows the collimator aperture to be as large as the physical size of the combiner. This approach allows a collimator aperture larger than what could be physically packaged into the constrained HUD optics envelope. The increase in collimator aperture available in WFOV systems permits significantly larger vertical and horizontal instantaneous display fields of view and allows the binocular overlapping FOV to approach the instantaneous FOV.

Merging the collimation function together with the combiner function necessarily means that there is optical power in the combiner. Incorporating optical power generally implies that the active combiner surface is physically curved. Typically, the active combiner surface of a reflective WFOV HUD combiner is a spherical section. In some holographic combiners, a complex holographic grating structure introduces aspheric terms into the combiner power, causing the element to behave as a wavelength-selective aspheric reflector. The most common form of WFOV combiner in use today is a spherical substrate with a simple holographic coating.

If a CRT faceplate were to be placed at the focal surface formed by the curved combiner, a virtual image of the CRT would be formed, assuming all optical aberrations were ignored. Although straightforward to visualize, this configuration does not lend itself to packaging in real cockpit geometries because of the close proximity of the CRT faceplate and the combiner/collimator and physical interference with the pilot.

The packaging problem is solved by placing a relay lens assembly between the CRT and the combiner. The relay lens forms a real aerial image of the CRT faceplate at one focal distance from the combiner/collimator. Figure 9.7 illustrates a simplified WFOV overhead-mounted HUD geometry. The introduction of a relay lens into the optical system allows a great deal of flexibility in terms of CRT size, aberration compensation, and optical system packaging and configuration.

In refractive HUD optics, the flat-plate combiners do not change the curvature characteristics of the reflected collimated ray bundles. Thus, the refractive HUD combiner bend angle, or the angle through which the display information is bent between the collimating lens and the pilot, can be optimized strictly for packaging reasons. Typical refractive HUD combiner bend angles vary between 50° and 130°.

Because of the restricted cockpit volume, the WFOV combiner/collimator is also used off-axis in order to fold the optical package into the available cockpit envelope. Unlike the refractive HUD case, however, the off-axis combiner/collimator used in conjunction with a large head motion box and large display FOV results in severe off-axis optical aberrations. These aberrations, if not reduced, will result in poor optical performance in terms of parallax and collimation errors, display accuracy, and symbology clarity. Compensating for off-axis aberrations is one of the principal functions of the relay lens assembly and the principal reason for using aspheric power in the combiner element itself. The bilaterally symmetric relay lens assembly is designed to form a preaberrated intermediate aerial image at one focal length from the combiner element. The preaberration of the intermediate image is designed to compensate for the aberrations introduced by the optically fast off-axis combiner/collimator. The combiner reimages the preaberrated intermediate image at optical infinity while redirecting the rays toward the pilot. Thus, the relay lens and combiner/collimator are optically integrated to make up the HUD system. Any change to the spacing between the combiner and relay lens or any change to the combiner off-axis angle will require an optical reoptimization.

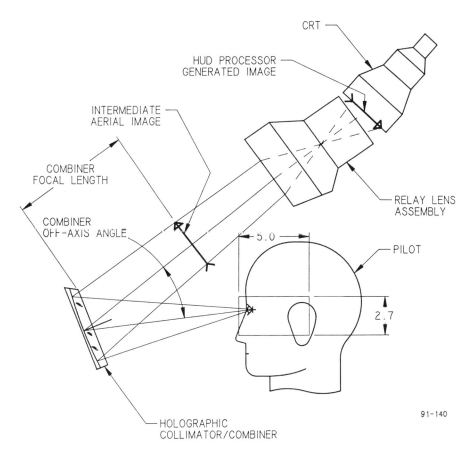

Figure 9.7 A schematic view of an overhead-mounted WFOV HUD. In this case, the relay lens and cathode ray tube are mounted over the pilot's head. The combiner is positioned about 12 in. forward of the pilot.

WFOV Performance Advantages—Display Fields of View The fields of view in WFOV HUDs are not as straightforward to calculate as refractive collimator HUDs because of the formation of the intermediate image. The relay lens aperture, the spacing between the collimator and the relay lens, the collimator focal length, and the distance between the pilot and the collimator all affect the display fields of view, especially in the horizontal plane. If the relay lens aperture is large enough to not vignette the display from any head position near the head centered position, then the dimensions of the combiner can be designed to be the limiting aperture in terms of the instantaneous display FOV.

Typical values for the display fields of view for an overhead-mounted WFOV system are as follows:

Total FOV	24° vert × 30° horiz
Instantaneous FOV	24° vert × 30° horiz
Overlapping binocular FOV	24° vert × 30° horiz

Drawbacks of WFOV HUD Optical Systems The performance drawbacks of WFOV optical systems are related to the uncompensated residual aberrations resulting from the complex system geometry. In particular, the parallax errors are generally two to five times poorer in the best WFOV systems than well-corrected refractive HUD systems. (Note, however, that the binocular overlapping field of view of the WFOV display may be between 5 and 20 times as large as the overlapping FOV of a refractive HUD.) In addition to parallax errors, the display accuracy over the TFOV and the head motion volume is generally poorer than in refractive systems by a factor of between 2 and 4.

A second disadvantage of WFOV systems is that the eye motion box surrounding the cockpit design eye location is generally smaller than the viewing region associated with conventional HUD optics. However, unlike in refractive HUDs, where head motion is required to see the total display FOV, very little if any head motion is needed to see the total display FOV in WFOV systems. It is therefore difficult to directly compare a refractive HUD eyebox with its limited FOV to a WFOV HUD eyebox where the total FOV can be seen with up to ± 2.50 in. of horizontal head motion.

Finally, most WFOV relay lens assemblies contain elements that are decentered, tilted, cylindrically shaped, or aspheric. These lens complexities are required to compensate for the combiner-induced optical aberrations and to achieve the best possible optical performance. These relay lens characteristics and complexities cause the assemblies to be several times more expensive compared with refractive HUD systems.

Even with the drawbacks discussed here, WFOV HUD systems are beginning to dominate the military and commercial HUD markets. The overwhelming advantages of larger area fields of view more than overcome the parallax, accuracy, and cost drawbacks.

Wide Field-of-View Optics Performance Summary Table 9.3 summarizes the optical performance achievable with a typical WFOV HUD. The performance of specific parameters will vary depending on the geometry of the system.

9.3 CHARACTERISTICS OF HOLOGRAPHIC ELEMENTS USED AS HUD COMBINERS

The design of refractive HUD systems using spectrally insensitive dielectric combiners entails a trade-off between the real-world photopic transmission

Table 9.3 Performance Characteristics of a Wide FOV Holographic HUD (FDI Model HGS–1000WS)

Characteristic	Performance
Field of view	
Total	30° horiz × 24° vert
Instantaneous	30° horiz × 24° vert
Overlapping binocular	30° horiz × 24° vert
Head motion box size	
Horizontal	4.7 in.
Vertical	2.7 in.
Depth	6.0 in.
Head motion needed to see TFOV	Not required
Parallax performance[a]	
Convergence	95% < 2.0 mrad
Divergence	98% < 1.0 mrad
Dipvergence	94% < 1.5 mrad
Combiner to design eye distance	12.0 in.
Combiner/collimator off-axis angle	32°
Combiner line-of-sight errors	Zero
Combiner photopic transmission	> 80%
Phosphor type	P53
Contrast ratio[b]	1.3 : 1

[a]Sample size is approximately 3100 points over 15 pupil positions and about 200 field positions.
[b]Against 10,000 fL background.

through, and the CRT light reflection from, the combiner. These photometric characteristics affect the display contrast ratio for a given image source brightness and therefore affect the pilot's ability to view HUD information against high-brightness ambient backgrounds and to pick out real-world features or targets (Ginsburg, 1983).

When using conventional non-wavelength-selective combiner coatings, the trade-off between combiner reflectivity and real-world transmissivity can be expressed as

$$\text{Reflectivity} + \text{transmissivity} + (\text{absorption and scatter}) = 1 \quad (9.7)$$

In most dielectric HUD combiner elements, the absorption and scatter losses are small. If the combiner reflectivity is high, the display appears bright even under the brightest real-world conditions, and the CRT drive requirements are modest to achieve a given display contrast ratio. Under such conditions, however, the apparent brightness of the real world is low, which limits the acquisition distance to important ground features.

A typical spectrally insensitive dielectric combiner has a photopic transmissivity of about 75% and a reflectivity of about 22%. To overcome the poor combiner reflectivity and to produce an acceptable display contrast ratio, the CRT must be driven with a high beam current. This can have the effect of shortening the CRT lifetime. Therefore, some HUD systems designed with conventional combiners often have nonoptimal display contrast ratios against high-brightness backgrounds and poor CRT lifetimes.

During the last decade HUD combiners have been designed that use wavelength-selective coatings to improve the photometric efficiency of the display. Although thin-film rugate coatings have been investigated for use on HUD combiners (Norris, 1991), the most common wavelength-selective coatings in production today are volume phase reflection holograms manufactured in dichromated gelatin. There are several advantages of holographic coatings:

The peak efficiency of the hologram can be very high, typically greater than 85%.
The spectral bandwidth can be precisely controlled during manufacturing.
The peak diffraction efficiency wavelength can be spectrally matched to the CRT phosphor peak emission wavelengths.
The hologram construction geometry can be optimized for a specific end-use cockpit geometry, resulting in a controlled wavelength shift across the combiner.

By tuning the spectral response of the holographic coating to the specific CRT emission wavelength, the total eye-weighted phosphor reflectivity can be increased to greater than 35%, a 50% increase compared to wavelength-insensitive combiners. At the same time, the photopic transmission through the holographic coating can be greater than 80%, a significant improvement compared with conventional approaches. Thus, the use of holographic combiners can result in an increase in the absolute brightness of the display, an increase in the display contrast ratio, an increase in the combiner transmissivity, and reduced CRT beam currents to achieve a given contrast ratio.

Holographically manufactured gratings with complex aspheric diffraction characteristics can also be manufactured in dichromated gelatin. Although holographic combiners with complex fringe structures are more difficult to manufacture and have lower yields than simple conformal fringe holograms, the aspheric optical characteristics can offer significant performance enhancements and advantages in HUDs required to work in difficult geometries.

9.3.1 Hologram Diffraction Theory

The spectral and angular diffraction efficiency of a reflection hologram used as a HUD combiner can be designed and analyzed using the coupled-wave analysis described by Kogelnik (1969). This general treatment describes the diffraction characteristics of transmission and reflection gratings in lossy and lossless dielectric media. Of particular interest to holographic HUD combiner designers is the special case of conformal reflection holograms with unslanted fringe planes in a lossless dielectric medium. Dichromated gelatin as well as recently introduced photopolymers (Smothers et al., 1989; Ingwall and Fielding, 1985; Ingwall and Troll, 1988; Owen, 1985; Samollovich et al., 1980; Evans, 1985; Hay and Guenther, 1988) meet the simplifying assumptions and conditions made by Kogelnik very well. Kogelnik's analysis accurately predicts the performance of reflection holograms made in these emulsion types. A rigorous coupled-wave analysis that includes high-order diffraction has been performed by Moharam and Gaylord (1981).

The performance of holographic HUD combiners is determined from the hologram spectral and angular bandwidths and the peak diffraction efficiency. These characteristics depend on the hologram bulk parameters: the emulsion thickness and the index of refraction modulation within the emulsion due to exposure. According to Kogelnik's coupled-wave theory for unslanted lossless dielectric gratings, the diffraction (reflection) efficiency, Δn, can be computed as follows:

$$n = 1/[1 + (1 - \xi^2/\nu^2)/\sinh^2(\nu^2 - \eta\xi^2)^{1/2}] \tag{9.8}$$

where

$$\nu = \pi \frac{\Delta n d}{\lambda[(\cos \theta)(\kappa\lambda/2\pi n - \cos \theta)]^{1/2}} \tag{9.9}$$

and

$$\xi = \frac{d/2(\kappa \cos \theta - \kappa^2\lambda/4\pi n)}{2 \cos \theta_0 - \cos \theta} \tag{9.10}$$

For a conformal grating,

$$\kappa = 4\pi n \cos \theta_0/\lambda_0 \tag{9.11}$$

In these equations, λ_0 is the construction wavelength, λ is the evaluation wavelength, θ_0 is the construction angle, θ is the evaluation angle in the emulsion, n is the bulk index of refraction of the emulsion, and Δn is the index of refraction modulation within the emulsion. All angles are measured internal to the emulsion. The index of refraction modulation is the amplitude of the spatial modulation of the refractive index due to exposure. The term ν is a complex parameter describing the coupling constant between the forward and backward traveling wave fronts due to the grating. The term ξ represents the dephasing of the hologram with respect to the Bragg condition.

For light polarized in the plane of incidence, the equation for ξ is modified as follows:

$$\xi(//) = \xi(\perp) \cos 2\theta \tag{9.12}$$

The diffraction efficiency for randomly polarized light, n(total), is

$$n(\text{total}) = (1/2)[n(//) + n(\perp)] \tag{9.13}$$

It can be seem from the equation for ξ, a measure of dephasing between the wave fronts, that changes in the playback angle or playback wavelength can have similar effects. From a practical standpoint, this implies that changes in the playback angle or wavelength will have similar effects on the diffraction efficiency characteristics of the hologram. In HUDs using holographic combiners and fixed-wavelength CRT phosphors, detuning due to angle is generally of critical importance.

If the dephasing measure is set to zero, a new continuum of Bragg conditions is available, and the effects of changes in the incidence angle on the peak wavelength can be determined. Thus,

$$\cos(\theta) = (\kappa/4\pi n)\lambda \tag{9.14}$$

where κ, the grating vector, is a constant after the exposure and processing of the hologram emulsion. This equation can be written for two different playback conditions as follows:

$$\cos\theta_1/\lambda_1 = \kappa/4\pi n = \cos\theta_2/\lambda_2 \tag{9.15}$$

Note that both θ_1 and θ_2 are measured internal to the emulsion. Thus, if the Bragg condition after processing is known, the shift in wavelength for any playback angle can be determined.

9.3.2 Hologram Spectral and Angular Characteristics

Figure 9.8 illustrates the diffraction efficiency characteristic as a function of playback wavelength for a reflection hologram recorded at 45° with respect to the surface normal (in air). The exposure wavelength is assumed to be 550 nm, and the reconstruction angle is 45°. Note that as the playback wavelength approaches the exposure wavelength, the Bragg condition is satisfied, and the diffraction efficiency approaches 100%.

A similar plot can be generated for the diffraction efficiency characteristics of the same hologram but assuming a playback angle of 35°. In this case, also illustrated in Figure 9.8, as the playback wavelength approaches the exposure wavelength the original Bragg condition is not satisfied, and the diffraction efficiency remains low. As the wavelength continues to increase to about 580 nm, the Bragg condition is once again satisfied, and the diffraction efficiency approaches 100%. Note that for equal dephasing, the diffraction characteristics appear to be nearly identical. The wavelength shift resulting from the changes in playback angle is predicted by the equations above.

The diffraction efficiency as a function of angle is illustrated in Figure 9.9. In

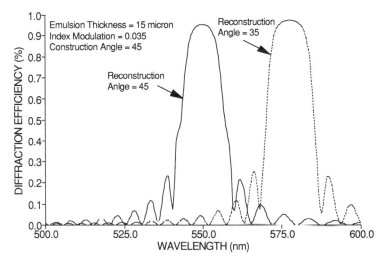

Figure 9.8 The diffraction efficiency characteristic as a function of playback wavelength for a reflection hologram recorded at 45° at 550 nm. At a reconstruction angle of 45°, the Bragg condition is satisfied at 550 nm. If the reconstrution angle is changed to 35°, the Bragg condition is satisfied at a playback wavelength of about 576 nm.

this case, the playback wavelength is fixed at 550 nm, and the playback angle is varied. As above, the Bragg condition is satisfied at about 45°.

In summary, the peak diffraction efficiency for the hologram occurs at the Bragg condition (wavelength and angle) initially set up as a result of the hologram exposure and processing. As discussed above, new Bragg conditions can be satisfied if the reconstruction wavelength and angle are appropriately adjusted.

Hologram Bulk Parameters

The two most important controllable bulk parameters in reflection holograms for HUD combiners are the emulsion thickness and the index of refraction modulation. The emulsion thickness is controlled during substrate coating by controlling the solution concentrations and the thickness of the gel layers before drying. The index of refraction modulation is dependent on the exposure energy density (mW/cm^2), the relative coherence of the exposing beams, the exposure time, and emulsion processing. Figure 9.10 illustrates the theoretical effects of different emulsion thicknesses and index of refraction modulations on the hologram diffraction efficiency characteristics as a function of wavelength. The spectral bandwidth and, to some extent, the peak diffraction efficiency and shape of a reflection hologram can be effectively altered by controlling the emulsion thickness and the refractive index modulation. The flexibility of dichromated gelatin al-

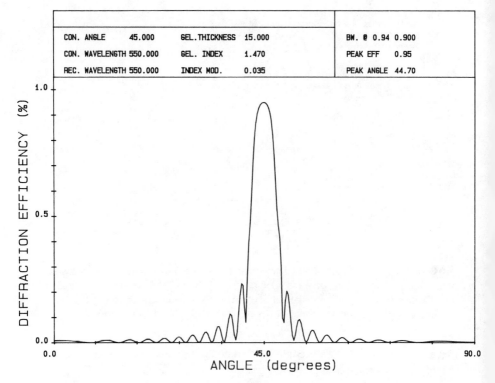

Figure 9.9 The diffraction efficiency as a function of angle for the hologram in Figure 9.8. The playback wavelength is fixed at 550 nm, and the playback angle is varied.

lows the holographic combiner to be fully optimized in terms of peak diffraction efficiency and spectral bandwidth for a specific HUD application.

Hologram Formation in Dichromated Gelatin

The holographic emulsion used in essentially all holographic HUD combiner elements in production today is dichromated gelatin (DCG). There are several reasons why it is currently the best material for manufacturing high-quality holographic elements. First, DCG can be uniformly coated onto both curved and flat substrates. Second, its diffraction characteristics are predictable, controllable, and repeatable. Finally, DCG can be reprocessed to chemically alter its peak wavelength and bandwidth characteristics. An additional advantage of DCG is that wide-bandwidth holograms can exhibit very low scatter, lower than any other available material. Numerous processes for preparing emulsions and making DCG holograms are documented in the literature (McCauley et al., 1973; Shankoff, 1968; Chang, 1980; McGrew, 1980). Although the processes are sim-

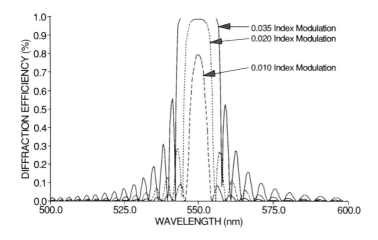

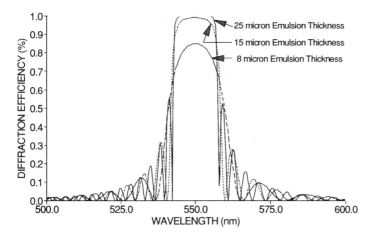

Figure 9.10 The theoretical effects of (a) index of refraction modulation and (b) emulsion thickness on the hologram diffraction efficiency characteristics as a function of wavelength.

ilar, modifications are needed for specific applications. The drawbacks of DCG are that it is sensitive only to the blue end of the visible spectrum without dye sensitization, and it is a photographically slow material, requiring about 250 mJ/cm^2 of exposure at 514.5 nm. A further drawback is that DCG is sensitive to ambient humidity in all phases of manufacturing, making laboratory environmental control a necessity.

Many theories of hologram formation in DCG and their applications have been documented in the literature by Chang (1980), Margarinos and Coleman (1985), and others. The basic process for forming a hologram in DCG consists of three steps: exposure, wet processing, and dehydration. The first step in the exposure process is to rigidly mount the sensitized emulsion (assumed to be attached to a substrate) in an exposure apparatus such that two spatially coherent beams interfere across the surface area and throughout the bulk of the material. A standing-wave intensity pattern is formed within the emulsion. During the exposure time, photons from the laser source photo induce the reduction of the $+6$ chromate ion to a $+3$ state, thereby cross-linking the carboxylate group between and within gelatin molecular strands. In this way, the regions of peak standing-wave intensity have an increased cross-link density compared with the less exposed regions. Typical exposures vary from 50 to about 400 mJ/cm^2.

After exposure, the emulsion is swelled in a wet process designed to maximize the density differential between the highly exposed and cross-linked regions and the less exposed regions. After swelling, the emulsion is dehydrated and dried in a high-temperature oven for several hours. The refractive index of the emulsion is, to first order, proportional to the molecular density. Thus, areas that are heavily cross-linked due to exposure are more dense than the lightly exposed or weakly cross-linked areas and thus have a higher refractive index. These density variations result in parallel iso-index of refraction planes throughout the bulk of the emulsion. The alternating high–low index matrix results in highly peaked spectral and angular diffraction characteristics near the Bragg condition as illustrated in Figures 9.8–9.10. Note that only emulsions with resolution on the molecular level can record bulk interference fringes with periods on the order of 0.2 μm corresponding to about 5600 lines mm. In a 25 μm thick DCG emulsion, there are about 140 iso-index fringe planes formed within the hologram at 514.5 nm.

Hologram Exposure for Optimum Bragg Condition

Figure 9.11 illustrates a simplified refractive HUD collimator optics geometry using a single holographic combiner. The design eye location, an ergonomically determined position within a given cockpit that is the nominal head position when viewing the HUD, is illustrated on the right side of the figure. Rays have been traced from the design eye location showing the limits of the IFOV and from the vertical limits of the head motion volume to the optical center of the combiner.

To minimize the angular mismatch between the exposure angle and wavelength and the playback angle and wavelength, the exposure point source is selected to be located at the center of the head motion volume. In this case, the playback angles from the design eye location exactly match the exposure angles with no angular detuning.

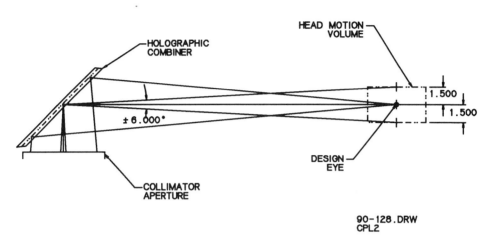

Figure 9.11 A simplified refractive HUD optics geometry illustrating the collimating optics, the aperture, and a single holographic combiner. Rays are traced from the center of the combiner to the limits of the head motion volume.

If the pilot were to move his head vertically with respect to the design eye and view the display, an entirely new set of playback angles at each position on the combiner would result. Figure 9.11 illustrates the playback angles for a point near the center of the combiner when the display is viewed from the vertical limits of the head motion volume. From eye positions above the exposure or construction point, the playback angles on the combiner are increased, resulting in a blue shift of the diffraction efficiency curve. Below the exposure point, the hologram is red-shifted with respect to the viewer. (The effective angular detuning with horizontal motion away from the design eye is very small and can, in general, be ignored.) The hologram angular bandwidth (and hence the spectral bandwidth) in conjunction with the phosphor emission must be adequate to cover the entire range of head movement in order to achieve acceptable display brightness uniformity.

When using dichromated gelatin as the hologram emulsion, it is often necessary to expose the gelatin at one wavelength (most commonly at 514.5 nm) and to swell it to the desired end-use wavelength. When designing the holographic combiner for use at playback wavelengths typical of CRT phosphors (i.e., about 548 nm), the emulsion is swelled to the desired wavelength at the desired reconstruction angle. A wavelength shift between the exposure and playback conditions of a hologram generally leads to nonoptimal imaging and photometric performance unless the construction geometry is specifically designed to compensate for the wavelength shift [see, e.g., Hayford (1981)].

In the case of conformal holograms, however, the change in exposure angle and point source location needed to compensate for the wavelength shift between the construction and end-use wavelengths is very small. For example, when the air gate (Wood and Cannata, 1986) exposure technique is used, the deviation between the playback and construction geometries for a wavelength shift of 35 nm is less than 0.3% of the distance to the point source and requires less than a 1% change in construction angle. These corrections are small enough in conformal holograms that they can generally be ignored. In complex holograms manufactured using aberrated wave fronts, the wavelength shift must be accounted for in the construction optics design.

9.3.3 Image-Forming Characteristics of Holographic Elements

In refractive optics HUD applications, flat wavelength-selective holographic combiners are used to enhance the photometric characteristics of the display. In these cases, the holographically manufactured coating contains no optical power, and the image-forming characteristics of the collimating lens are not altered by the combiner.

WFOV Combiners with Conformal Holographic Coatings

In WFOV HUD optical systems where the combiner off-axis angle is relatively small (generally under about 45°) and the physical combiner aperture is under 10 in., combiners using spherical substrates can often be used. Even though the combiner is operating over an extended field and aperture and at a fairly large combiner off-axis angle, the aberrations introduced by the combiner can be compensated for in the relay lens assembly and by predistorting the image on the CRT. Thus, as in the refractive HUD case, the holographically manufactured combiner coatings can be constructed to provide a reflection grating with essentially no slant (i.e., the grating κ vector is parallel with the local substrate surface normal). This is referred to as a *conformal hologram* because the holographic fringes are conformal with the substrate curvature. Thus, the holographic coating does not alter the image-forming characteristics of the combiner substrate. It does, however, provide wavelength selectivity, which enhances the photometric performance of the display.

WFOV Combiners Using Holographic Optical Elements

In WFOV optical systems required to operate in highly constrained cockpit geometries, the combiner/collimator is often required to operate at very large off-axis angles (greater than 60°) over large fields of view and over large collimator apertures. Under these conditions the aberrations introduced by a spherical combiner element often cannot be economically overcome using a conventional refractive relay lens assembly. It is for these reasons that some WFOV HUD systems use combiners incorporating complex holographic optical elements (HOEs) with diffractive power. [The first discussion of the use of diffractive power in

HUD combiners for aberration correction that I know of is that of Close (1973 and 1974).]

The operating geometries of the complex combiner are such that the image-forming characteristics and aberration properties of the HOE must be taken into account (Close et al., 1974). Under these conditions, HOE combiners exhibit many of the same aberrations as conventional refractive and reflective optics. Sweatt (1977) has shown that an HOE exhibits the same aberrations as conventional thin refractive elements with very high (e.g., $n \cong 10,000$) indices of refraction. Thus, a HOE used as a collimator/combiner in a HUD system will exhibit varying amounts of spherical aberration, coma, astigmatism, distortion, and color.

There are several ways in which the residual aberrations of the HOE collimator can be reduced. A basic requirement is to use the HOE in a symmetric configuration about the surface normal. Thus, the coherent construction wave fronts are incident at equal and opposite angles at the center of the HOE. Although it might seem attractive to use asymmetric angles for packaging reasons, asymmetric construction results in a constant grating slant that introduces substantial chromatic aberrations due to grating dispersion, even when narrowband phosphors (e.g., P43) are used. The use of slanted gratings also can have negative ramifications on the see-through performance of the combiner. Finally, variations of monochromatic aberrations are unfavorable for the asymmetric HOE case.

If symmetric construction is assumed, the next major issue is the minimization of the off-axis angle at which the HOE is used. This is desirable because aberrations such as coma, astigmatism, and distortion vary with the first, second, and third power, respectively, of the off-axis angle. These aberrations can be minimized (or even driven to zero) at one point in the field of view. However, the need for high diffraction efficiency and a well-corrected system over a large aperture and field drives the system away from this essentially zero pupil special condition.

An additional degree of freedom is provided by the use of a curved substrate for the HOE combiner. Although HOE first-order properties (such as focal length) are independent of substrate shape, that is not the case for HOE aberrations. The use of a spherical substrate allows better aberration correction over aperture and field to be achieved (Hayford, 1981).

Finally, the HOE can be constructed using aspheric wave fronts. Because of the location of the HOE combiner in the HUD system, an aspheric wave front HOE will have a positive effect on aberrations over both the aperture and field, reducing but not eliminating them. This is similar to what can be achieved using aspheric surfaces on conventional optical elements but with the potential for lower production costs compared with the fabrication of large aspheric surfaces. The use of diffractive power in the combiner can have a negative impact on the see-through characteristics of the HUD combiner, which in many cases are not acceptable. However, most HUD vendors are able to construct complex HOE combiners with minimal amounts of undesirable residual transmission grating see-through behavior.

9.3.4 Photometric Characteristics of Holographic Combiners

The most important attributes of holographically manufactured combiner coatings in HUDs have to do with the photometric performance characteristics. As shown above, a reflection hologram can be manufactured and tuned to operate over a narrow-wavelength band with very high peak diffraction efficiency. These characteristics allow the photopic transmission of the combiner to be above 80% while simultaneously providing high phosphor reflection efficiency. Unlike conventional wavelength-insensitive combiners for which the sum of transmission and reflection is essentially 1; in holographic combiners the sum of photopic transmission and phosphor reflection efficiency is greater than 1. This section examines the photometric performance attributes of holographic wavelength-selective coatings in terms of performance parameters important in HUD systems.

Hologram Transmission Characteristics
The hologram photopic transmission is the eye-weighted transmission of the real world as seen through the combiner. The combiner spectral transmission efficiency, $T(x,y,\theta,\lambda)$, is a function of position on the combiner surface, the construction geometry and wavelength, and the reconstruction angle and wavelength. This characteristic is determined by ray tracing the combiner construction geometry and the HUD playback or reconstruction geometry and applying Kogelnik's coupled-wave analysis for points on the combiner. The total photopic transmission, $T_p(x,y,\theta)$, is the normalized area under the curve of the product of the spectral sensitivity curve of the eye, $E(\lambda)$, the background spectral output characteristic assumed to be a blackbody source, $B(\lambda)$, the spectral transmissivity of the holographic combiner, $T(x,y,\theta,\lambda)$, and the transmission characteristics of the antireflection coatings, $T_{ar}(\lambda)$. Thus,

$$T_p(x,y,\theta) = \frac{\int E(\lambda)B(\lambda)T(x,y,\theta,\lambda)[T_{ar}(\lambda)]^2 \, d\lambda}{\int E(\lambda)B(\lambda) \, d\lambda} \qquad (9.16)$$

The numerical integration is carried out across the visible spectrum (approximately from 400 to 700 nm). Since the hologram is wavelength-selective, the photopic transmission is dependent on the effective real-world color temperature.

Hologram-Induced Real-World Color Shift
Holographic combiners are designed and tuned to diffract specific incident CRT phosphor wavelengths into the head motion box. The phosphor wavelengths that enter the eye are also removed from the real-world view seen through the com-

biner. This notch filter effect reduces the photopic transmission and introduces a slight discoloration or tinting of the real-world view seen through the combiner. The real-world discoloration is minimized by designing the holographic notch with an angular bandwidth just adequate to cover the angular bandwidth requirements derived from the vertical dimensions of the HUD eye motion box.

Real-world discoloration can be computed by determining the 1976 CIE color coordinates u' and v' with and without the filtering effects of the holographic combiner. The u' and v' color coordinates are computed from the X, Y, and Z tristimulus values as follows:

$$u' = 4X/(X + 15Y + 3Z) \tag{9.17}$$

and

$$v' = 9Y/(X + 15Y + 3Z) \tag{9.18}$$

The X, Y, Z tristimulus values are computed as follows:

$$X(x,y,\theta) = \int x(\lambda)B(\lambda)T(x,y,\theta,\lambda)\, d\lambda \tag{9.19}$$
$$Y(x,y,\theta) = \int y(\lambda)B(\lambda)T(x,y,\theta,\lambda)\, d\lambda \tag{9.20}$$
$$Z(x,y,\theta) = \int z(\lambda)B(\lambda)T(x,y,\theta,\lambda)\, d\lambda \tag{9.21}$$

where $x(\lambda)$, $y(\lambda)$, $z(\lambda)$ are the CIE 1931 color matching functions, $B(\lambda)$ is the background source, often assumed to be a blackbody source, and $T(x,y,\theta,\lambda)$ is the transmission spectrum looking through the wavelength-selective combiner. The transmission spectrum of the hologram is a function of position on the hologram, the playback angle of incidence on the combiner, and the construction parameters.

The real-world color shift is determined by computing the difference in color coordinates between the real world views with and without the wavelength-selective filter in position. Thus,

$$\text{Color shift} = [(u - u')^2 + (v - v')^2]^{1/2} \tag{9.22}$$

where the primed and unprimed coordinate systems represent the real-world chromaticity coordinates with and without the combiner in position, respectively. The vector direction of the real-world color shift is directly away from the color coordinates of the color of the display at the cockpit eye reference point.

If minimizing the real-world coloration is an important requirement, a color-compensating filter, such as a red hologram, could be used to balance the real-world color shift at the expense of photopic transmission.

Hologram Phosphor Reflection Efficiency

The brightness of the HUD symbology at the pilot's design eye location depends on the light output from the CRT, the light transmission of the CRT energy through a sidelobe suppression or solar rejection filters (if present), the light transmission through the relay lens assembly or collimating optics, and the hologram phosphor reflection efficiency. The hologram phosphor reflection efficiency is the net phosphor energy reflected (or more properly, diffracted) by the hologram to a position with the HUD head motion box.

In general, the hologram diffraction efficiency characteristic is a function of the construction angle and wavelength and the playback angle, as described above. The light transmission through the optical system (relay lens or collimator) is usually a relatively weak function of wavelength (except for the sidelobe suppression filter), whereas the light output from the CRT and the combiner reflectivity are highly wavelength-dependent. Thus, to determine the hologram phosphor efficiency, a numerical integration must be carried out over the visible spectrum.

The net hologram phosphor reflection efficiency, $n(x,y,\theta)$, is the normalized area under the curve of the product of the eye-weighted phosphor spectral output and the hologram spectral reflectivity. Because of the exposure geometry, the hologram spectral reflectivity is a function of the position where a particular analysis ray strikes the hologram surface, the exposure angle, and the playback angle. The angles are determined by ray tracing the exposure and playback geometries. Thus,

$$n(x,y,\theta) = \frac{\int E(\lambda)P(\lambda)H(x,y,\theta,\lambda)\,d\lambda}{\int E(\lambda)P(\lambda)\,d\lambda} \qquad (9.23)$$

where $E(\lambda)$ is the spectral content of the photopic eye response, $P(\lambda)$ is the spectral content of the phosphor, and $H(x,y,\theta,\lambda)$ is the spectral reflection characteristic of the hologram. Note that the sum of the hologram transmission and the hologram reflection is equal to 1 at each specific wavelength.

The display brightness variations due to the combiner are determined by numerically computing the hologram phosphor efficiency values for each combination of head position and field angle of interest. Variations among the computed phosphor efficiencies represent the variations in brightness of the display due to the combiner. Since the phosphor emission characteristics are spectrally fixed, the phosphor efficiency characteristics for any combination of head position or look angle can be optimized by adjusting the hologram construction geometry as well as the emulsion bulk parameters. For HUD geometries where the pilot sits close to the combiner, the angular bandwidths on the combiner needed to cover the head motion box are large, an acceptable brightness uniformity cannot be achieved without compromising photopic transmissivity. In this case, combiners with multiple exposed holograms (Wood and Thomas, 1990; Wood et al., 1987) can be used to achieve both high reflectivity and high transmissivity.

CRT Phosphors

There are three types of CRT phosphors that are used predominantly in modern HUD systems: P1, P43, and P53. All three types have been used successfully in

HUDs with holographic combiners. Each type of phosphor has a unique set of attributes that affects its selection for use in HUDs. Of particular importance when using holographic wavelength-selective combiners are the phosphor spectral characteristics. Other phosphor parameters of interest to HUD designers are the phosphor peak luminance, persistence, resolution, graininess, burn characteristics, and screen luminance maintenance. These parameters are discussed by Seymour (1988).

P1 Phosphor   Figure 9.12 illustrates the eye-weighted spectral energy distribution of the P1 phosphor. This phosphor is commonly used in refractive HUDs both with and without wavelength-selective holographic combiners. The P1 phosphor is characterized by a wide spectral emission bandwidth, spanning approximately 90 nm at the 10% intensity points. Because of this wide emission bandwidth, the P1 phosphor can be used only with conformal fringe dispersion-free holographic combiners.

Figure 9.13 illustrates the P1 phosphor energy diffracted from a hologram designed to operate at 45° and to have a photopic transmission of about 83%. Also shown in the figure is the same hologram reconstructed at 39° representing a head-down position in the head motion box of Figure 9.11. Note that the hologram diffraction characteristic shifts red from this head position, and the total

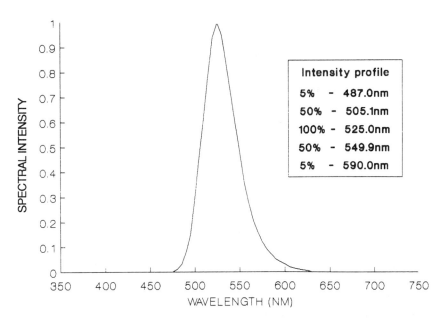

Figure 9.12   The spectral energy distribution of the P1 phosphor having a spectrally wide emission bandwidth, spanning approximately 90 nm at the 10% intensity points.

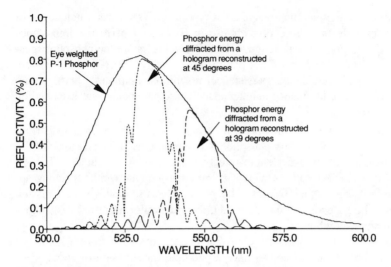

Figure 9.13 The P1 phosphor energy diffracted from a hologram with a photopic transmission of 83% designed to operate at 45°. From the center of the head motion box, the hologram diffracts the central wavelengths of the phosphor emission. From the bottom portion of the head motion box, the hologram spectral response shifts red, and a different portion of the phosphor spectral output is diffracted.

area under the hologram phosphor curve is reduced slightly. The combination of a spectrally wide phosphor with a narrow-bandwidth holographic element provides an excellent combination of real-world transmission, phosphor reflectivity, and display brightness uniformity.

P53 Phosphor Figure 9.14 illustrates the spectral energy emission distribution of the P53 phosphor, a rare earth phosphor containing multiple narrowband emission peaks. Of particular interest in HUD systems is the primary central emission peak spanning about 18 nm at the 10% emission levels. This peak contains about 65% of the photopic light output from the phosphor. One of the important advantages of the P53 phosphor is its ability to be driven with high beam currents without saturation. P53 phosphor screens with brightness levels as high as 50,000 fL have been reported.

This phosphor is used in both refractive HUD combiners and WFOV HUD combiners containing no hologram diffractive power (i.e., conformal fringe holograms). Figure 9.15 illustrates the P53 phosphor energy diffracted from a hologram designed to operate at 45° and to have a photopic transmission of about 83%. Also shown is the same hologram reconstructed at 39° (representing a head-down position in the head motion box in Figure 9.11). Note that the hologram diffraction characteristic shifts red, yet the total area under the hologram

HOLOGRAPHIC HEAD-UP DISPLAYS

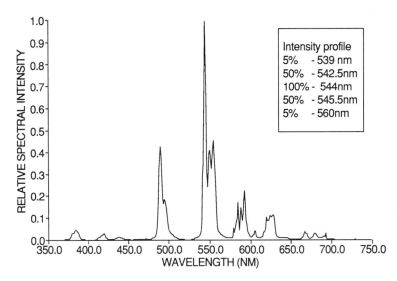

Figure 9.14 The spectral energy emission distribution of the P53 phosphor showing the many sidelobes. Only the central green peak, spanning from 539 to 560 nm is of interest in HUD systems.

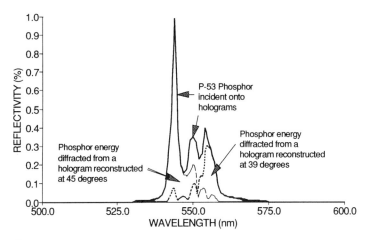

Figure 9.15 The P53 phosphor energy diffracted by a holographic combiner with 85% photopic transmission from the center and top of the eyebox.

phosphor curve remains fairly high. The hologram phosphor efficiency for this case is approximately 35% with a variation of about ± 25% with respect to the mean. Note that in this case the hologram spectral characteristic is less efficient and has a large spectral bandwidth than the P1 hologram discussed above. These characteristics are needed to keep the display brightness variations to less than 25%.

P43 Phosphor   Figure 9.16 illustrates the spectral energy emission distribution of the P43 phosphor, an efficient rare earth phosphor containing multiple narrowband emission peaks. The primary central peak in the P43 phosphor is even narrower than in the P53 phosphor, spanning about 14 nm at the 10% emission levels. This peak contains about 65% of the photopic light output from the phosphor. One of the disadvantages of the P43 phosphor is that it saturates severely at moderate screen loading and therefore does not provide as much brightness as the P53.

This phosphor is used primarily in WFOV HUD combiners containing complex holographic fringe structures. In these systems, the slanted gratings introduce significant chromatic dispersion in reflection. To minimize line blur due to the grating dispersion, a narrowband phosphor is needed. P43 is often selected for this reason. Conformal holograms designed for use with the P53 phosphor

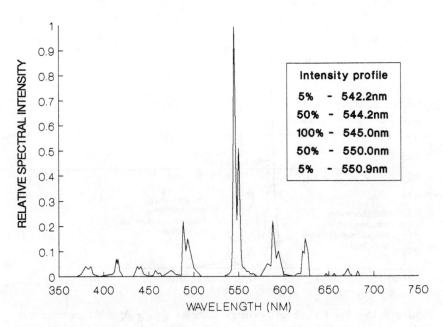

Figure 9.16   The spectral emission charactersitics of the P43 phosphor.

will often work with the P43 phosphor with a slight reduction in the display brightness uniformity.

9.3.5 Holographic Combiner Construction Geometries

In this section, several holographic element exposure geometries are discussed. Exposure geometries including a simple method of hologram exposure for flat and curved holographic combiners, dual aberrated wave front exposure geometries for WFOV combiners with diffractive power, and a single-beam exposure geometry of an aspheric holographic element.

Flat Combiner Hologram Construction Geometry

Figure 9.17 illustrates a hologram exposure technique for manufacturing high-quality conformal reflection holograms (Wood and Cannata, 1986). As shown, the "air gate" is a single-beam construction method where the second exposure beam (object beam) is generated from the primary beam (reference beam) at the air–gelatin emulsion surface (hence "air gate"). The index of refraction difference between air and the gelatin emulsion (refractive index of about 1.5) is enough to generate a backward-traveling wave front that interferes with the primary beam wave front, thereby forming a standing wave that is recorded in the sensitized emulsion. No additional reflection enhancement devices (e.g., mirrors with index-matching fluids) are required. This method significantly simplifies the exposure setup complexity and allows extremely high quality holograms to be manufactured with high yields. This technique can be applied to many different reflection hologram geometries.

Dual Aberrated Wave Front Hologram Construction Geometry

Figure 9.18 illustrates a very complex exposure geometry used to manufacture a dual aberrated wave front off-axis holographic combiner element at Flight Dynamics, Inc. This method was first attempted by Close et al. (1974), then Witherington (1976), and later perfected by Close (1985). Because of the large path lengths that both the object and reference beams traverse, and because of the extreme stability required over an exposure of several minutes, active interferometric fringe locking methods are often required.

Single-Beam Aberrated Wave Front Exposure Geometry

Figure 9.19 illustrates a single-beam exposure geometry for manufacturing an aspheric holographic element for use as a turning mirror in a WFOV HUD system (Wood and Hayford, 1988). As shown, a single reference beam is used. The object beam is derived from the reference beam after reflecting from a front surface tenth-order aspheric mirror. In this geometry, the holographic emulsion is immersed in an index-matching fluid, which eliminates the generation of unwanted parasitic holograms. The advantages of single-beam geometries are that there is inherent stability between the two interfering construction beams and

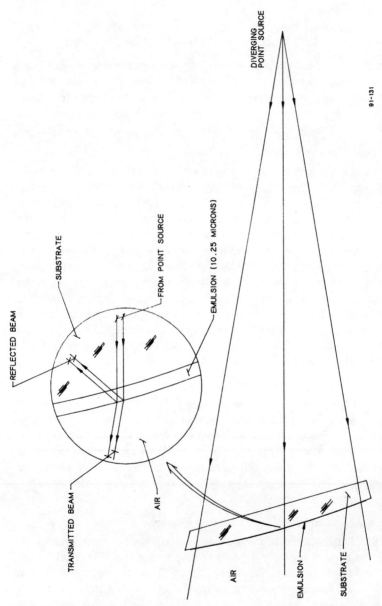

Figure 9.17 The "air gate" exposure technique (Wood and Dannata, 1986). A simple holographic combiner exposure method using a single exposure beam. The second exposure wave front is generated from the first at the gelatin–air interface. This method can be used on flat-plate or curved combiners.

HOLOGRAPHIC HEAD-UP DISPLAYS

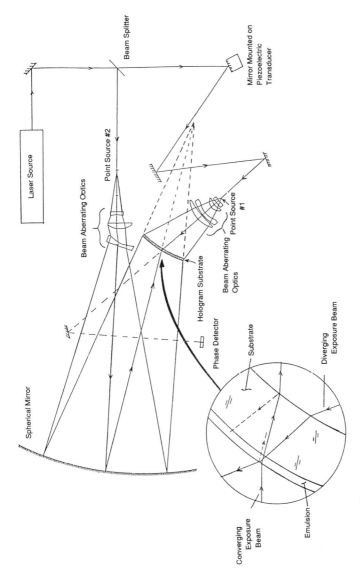

Figure 9.18 A dual-beam aberrated wave front exposure setup. (Figure courtesy of Flight Dynamics, Inc.) This exposure setup was used to manufacture a HUD combiner with diffractive power for a prototype commercial HUD application in 1981. The two aberrated wave fronts interfere at the hologram substrate. Because of undesirable transmission hologram behavior, this exposure approach was abandoned by Flight Dynamics in favor of the air gate method.

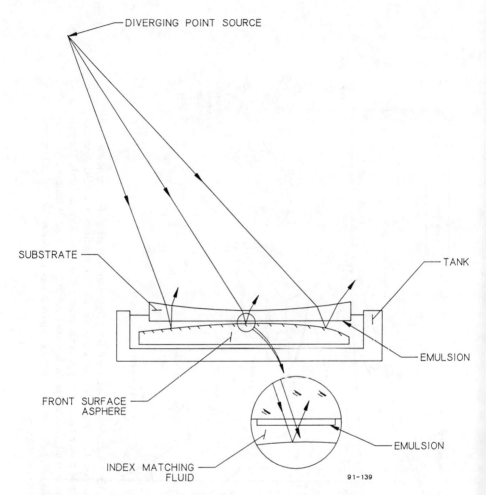

Figure 9.19 A single-beam aberrated wave front exposure setup. This setup was used to manufacture a holographic fold element used in a WFOV glareshield-mounted HUD. A single aberrated wave front was generated after reflection from an aspheric front surface.

fewer parasitic holograms are generated. This method has been used for manufacturing WFOV HOE HUD combiners (Arns et al., 1984a, 1984b; Weihrauch, 1983).

9.4 HIGH-INTEGRITY HUD PROCESSING AND PDU ELECTRONICS

Regardless of the mission the HUD is performing, HUD systems generally comprise two major subsystems: the pilot display unit and the HUD computer or processor. The HUD processor interfaces electronically with the aircraft sensors and avionics, derives and processes symbol-positioning algorithms, and generates deflection signals. The pilot display unit (PDU) receives the deflection signals from the HUD processor, converts the signals into an image on a high-brightness CRT, collimates the image, and combines it with the real-world scene for viewing by the pilot. This section briefly discusses the HUD processor and the pilot display unit electronics.

The HUD system requirements generally determine the HUD processor architecture and configuration. One of the most sophisticated and demanding HUD applications today is a fail-passive head-up guidance system (HGS) that computes and displays precision guidance commands, allowing a pilot to manually fly low-visibility approaches and landings to Category IIIa limits. Other guidance applications include detecting windshears and providing windshear recovery guidance. A system with these capabilities is described by Desmond and Ford (1984), Desmond (1986), and Johnson (1990).

9.4.1 HUD Processor

The functions of a high-integrity HUD processor include providing an electronic interface with aircraft sensors and avionics, validating the incoming data, performing symbol forming and positioning calculations, and generating X, Y, and Z deflection and video signals for driving the HUD pilot display unit.

In 1985, a stand-alone HUD guidance system, the Flight Dynamics HGS-1000WS (Desmond and Ford, 1984), was certified by the FAA. This system generates control commands for ILS beam tracking and flare guidance for low-visibility landings and operations. For manually flown low-visibility approaches and landings, the FAA required, high-integrity, fully monitored processing. The performance of the system was shown to meet the stringent statistical touchdown performance requirements defined in the FAA advisory circulars (AC 20-57A). In 1987, windshear detection and escape guidance was added to the system. Finally, in early 1991 the HGS-1000WS system was certified for low-visibility takeoffs down to 300 ft of runway visual range, the lowest takeoff minimum allowed in the United States.

As HUD systems become common in commercial cockpits, they will be fully integrated with the aircraft core avionics. This will inevitably lead to the HUD becoming the primary flight instrument on the aircraft. The HUD as the primary flight instrument is discussed by Taylor (1990). In this case, the HUD system, in conjunction with the autopilot functions and the core avionics, will provide full-time flight director guidance head-up. Depending on the HUD system and core avionics architectures, a stand-alone HUD processor may contain multiple display generators capable of providing high-integrity guidance to both the pilot and copilot. Finally, as weather-penetrating imaging sensor technology evolves, the HUD processor may also perform real-time raster video processing.

Flight Dynamics, Inc. (FDI) was the first company to apply advanced holographic WFOV HUD technology in a manner suitable for a manually flown fail-passive low-visibility landing system. Since no FAA regulations or guidelines existed for HUD guidance systems, FDI designed the system to meet the performance and safety requirements already established for automatic landing systems. There are three principal FAA performance requirements that must be demonstrated in certifying a HUD guidance system as a low-visibility fail-passive landing system. These are as follows (AC 20-57A, AC 25.1309-1A, AC 120-28C):

1. The probability of landing outside of the "landing area" must be less than 1 in a million.
2. The probability of an undetected HUD failure that could lead to a hazardous event must be near 1 in a billion.
3. The landing function must be available at least 99% of the time from 1500 feet to touchdown.

These basic requirements drive the HUD processor architecture, design, and monitoring techniques for both hardware and software. The following paragraphs summarize the fail-passive HUD processor's capabilities.

Input Data Processing and Verification

The required high levels of hardware and software integrity cannot be achieved with a single data path, even if the HUD processor could be shown to be always error- and fault-free. This is because single string sensors providing the principal aircraft information required by the HUD generally do not have the reliability and integrity necessary to ensure the function. To ensure high-integrity landing guidance, dual independent sources are used for all of the primary sensor information. This includes two sources of attitude (two inertial systems, or one inertial and an alternate source of attitude), two sources of radio altitude, two sources of air data providing barometric altitude and airspeed, and two ILS navigation receivers. In computing the probability of an undetected error leading to a haz-

HOLOGRAPHIC HEAD-UP DISPLAYS 385

ardous event, it is assumed that the dual sensors are independent of each other, and the wiring is physically separated. This physical signal separation common to dual-channel architectures is carried through to the input section of the HUD processor, which contains dual interface processing subsystems.

The first operation in the high-integrity fail-passive HUD processor is to validate the input data. All data that are used in any way to position symbols on the HUD display are validated prior to further processing. Some common input-validating techniques are as follows:

Parity Checks The parity of each serial digital data stream word is verified. This determines whether a complete transmission of each digital word has been received.
Status Checks The portion of each serial data stream digital word that indicates the validity of the corresponding data is verified.
Range The range of operation of each input parameter is verified.
Activity The required update or transmission rate for parameters on each data bus is checked. Any parameter not received within its minimum update rate is considered to be invalid until the next transmission of that parameter is received.
Input Subsystem Built-In Test (BIT) The input subsystem that receives the input signals is extensively monitored with hardware and software built-in tests.

All sources of display information are validated before further processing. Once all sources are shown to be valid, the HUD processor compares the redundant inputs to determine if the independent values agree within predetermined tolerances. If two sources of redundant data are available, and the compared values do not agree within a specified tolerance of each other, then the corresponding parameter value used by the HUD processor for flight guidance and/or symbology presentation calculations is invalidated. Any data marked as invalid are removed from the display, and the capabilities of the system are degraded accordingly. If a third source of redundant data is available to the HUD, then the processor can selectively remove the bad data source and continue on with the two agreeing redundant sources. Note that only the pilot-side sensor data are actually displayed to the pilot. The copilot sensors are used as comparators only and are generally not displayed. This ensures that what the pilot sees head-up and what he sees head-down are consistent.

Figure 9.20 illustrates the sensor interface block diagram of a HUD processor capable of computing flight guidance designed for manually flown low-visibility approaches. Figure 9.21 is a block diagram of the data flow through the fail-passive HUD processor. Figure 9.22 illustrates a symbology set certified by the FAA for manually flown CAT IIIa approaches and landings.

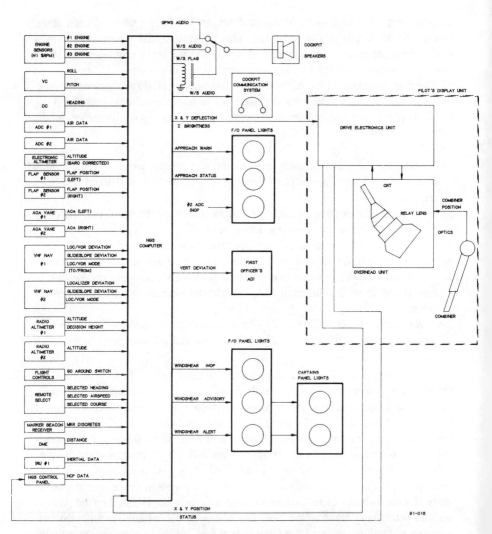

Figure 9.20 The sensor interface block diagram of a HUD processor capable of computing flight guidance designed for manually flown low-visibility approaches. (Courtesy of Flight Dynamics, Inc.)

HOLOGRAPHIC HEAD-UP DISPLAYS

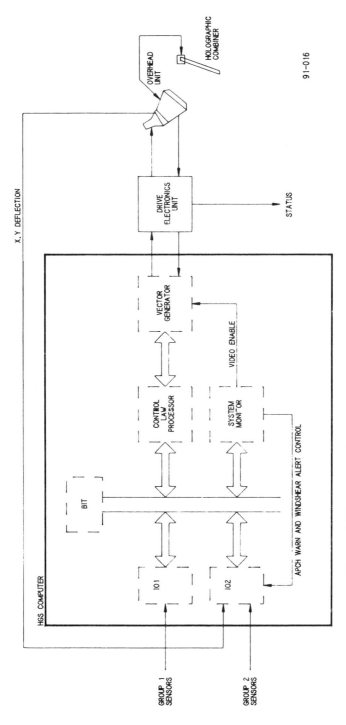

Figure 9.21 A simplified block diagram of the data flow through a fail-passive HUD processor used for low-visibility operations. (Courtesy of Flight Dynamics, Inc.)

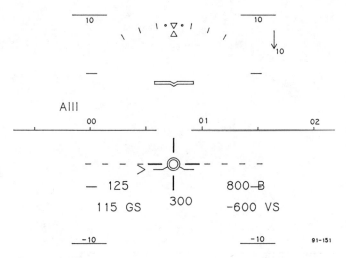

Figure 9.22 The symbology set for a certified CAT IIIa HUD.

Control Law Processing

Once the sensor input data are received and validated, the data are available for further processing. The control law processor (CLP) performs the computations and runs the algorithms required to position and form the symbology for display. The CLP also executes sensor status logic, mode select logic, and the overall system status logic.

Display Monitoring

To provide the high degree of safety necessary for low-visibility HUD operations, the HUD processor continuously evaluates the position of critical HUD symbols on the CRT and monitors the progress of the aircraft during an approach. If a fault occurs anywhere in the system, including the aircraft sensors, the pilot is apprised of the situation on the combiner. The pilot is thus provided with sufficient information to evaluate the status and health of the HUD system. The monitoring functions may include the following.

HUD Fault Monitor The HUD system contains extensive self-test hardware and software that continuously determines the operational status of each Line Replaceable Unit. If a detected fault can be isolated, then the unaffected portions of the HUD system can continue to operate.

Critical Symbol Monitor The critical symbol monitor (Desmond et al., 1987) on the HGS-1000WS (Flight Dynamics, Portland, OR) uses an inverse function method to monitor and verify the positional accuracy of several critical symbols on the HUD display in real time. In this scheme, the measured positions on the CRT of key display elements are processed with an inverse function to determine the aircraft sensor input values needed to position a specific display element in the specific location. The inverse functions consist of inverse optical distortion computations and inverse symbol-positioning computations. The state of the aircraft derived by inverse calculations is compared with the current state of the aircraft equipment and sensors received by the HUD. Discrepancies result in the blanking of the unverified symbol and the downgrading of HUD capabilities.

Approach Monitor The approach monitor evaluates the progress of the approach to the runway. Any out-of-tolerance condition (i.e., unacceptable ILS beam deviations) causes a warning indication to be presented on the HUD display. The approach monitor is active during the HUD approach mode and is functionally divided into a tracking monitor and a flare monitor. The tracking monitor evaluates the state of the aircraft's approach relative to airspeed, localizer and glideslope deviations, and other parameters. The flare monitor determines the possibility of an unsafe landing by evaluating the aircraft's sink rate, lateral displacement, airspeed relative to the selected approach airspeed, and a variety of other parameters.

Rollout Monitor The rollout monitor evaluates the progress of the aircraft's rollout down the runway. An out-of-tolerance condition causes a warning indication on the HUD. The rollout monitor is active from touchdown until the aircraft reaches a ground speed suitable for taxiing. The rollout monitor indicates the possibility of an unsafe rollout condition by evaluating the aircraft's lateral displacement from the centerline of the runway, distance traveled along the runway, and other parameters.

Takeoff Monitor The takeoff monitor evaluates the progress of the aircraft's takeoff roll down the runway. An out-of-tolerance condition causes a warning indication on the HUD. The takeoff monitor is active through the takeoff rotation and indicate the possibility of an unsafe takeoff roll condition by evaluating the aircraft's lateral displacement from the centerline of the runway, distance traveled along the runway, and other parameters.

Using performance-monitoring techniques, the HUD system has been shown to meet the safety requirements specified by the FAA (AC 120-28C). In addition, the HUD, when considered separately and in relation to other aircraft systems, is designed so that any failure that would prevent the continued safe flight and landing of the aircraft (such as misleading data that could lead to a hazardous event) is extremely improbable (on the order of 1 event in a billion approaches). Warning information is designed to alert the crew to any unsafe system operating conditions and enable them to take appropriate corrective actions.

Display Generation

The control law processor computes the position of Earth-referenced symbols from the input data, computes the position and movement of the flight director symbology, and builds and executes the display list representing the complete display. Display lists are completely updated every 50 msec. The symbol generator receives the display list, verifies the integrity of the list, and converts it into X, Y deflection and Z-axis modulation signals representing the display characters and symbols for display by the PDU. The display is refreshed 60 times per second to eliminate display flicker. In some HUD models, the symbol generators are capable of prioritizing symbology and blanking lower priority information in the event of physical symbology overlap.

With the development of improved performance cost-effective millimeter-wave, infrared, and imaging radar sensors, there is much interest in providing a conformal video image to the pilot with guidance information overlay. The combination of weather-penetrating sensor video with guidance information overlay on a conformal WFOV HUD may allow pilots to fly into low-visibility conditions without the need for specifically monitored ground equipment and special runway lighting.

To achieve this capability, the HUD processor display generator subsystem must be capable of synchronizing with the raster source and providing high-speed stroke information during the vertical retrace interval of the imaging source. This can be accomplished using high-speed deflection circuitry and a high-speed vector and character generator.

In this case, the HUD processor provides an input port for the sensor video source. The video signal can be manipulated in the HUD processor to generate appropriate signals for the PDU to produce a stroke only, a stroke on raster, or a reference level restoration, timing generation, video buffering, and level conversion. More exotic image processing can be performed for feature or edge extraction, if required.

9.4.2 The HUD Pilot Display Unit

The primary function of the pilot display unit (PDU) is to convert deflection and video signals from the HUD processor into a high-brightness image on a CRT. The CRT image is positioned with respect to the HUD optical system such that a collimated virtual CRT image is formed at infinity for viewing by the pilot. The PDU optics provides a virtual display over a display area, or field of view, that is commensurate with the operational requirements of the aircraft. Further, the PDU provides a high-brightness display capable of being viewed against high-brightness real-world backgrounds. Finally, the PDU combiner transmits the real-world view with minimal attenuation.

The PDU receives the X, Y position and the Z bright-up signals from the HUD processor and generates deflection and drive signals for the CRT to produce the display. The PDU consists of four major electrical subsystems: predistortion mapping, deflection, video, and BIT. Descriptions of these subsystems follow. A simplified block diagram of typical PDU electronics is presented in Figure 9.23.

Distortion Mapping and Deflection

The deflection circuit converts X and Y output voltages from the HUD processor into X and Y deflection currents used to drive the CRT yoke. Processing of the X and Y position signals includes distortion correction and deflection amplification. Distortion correction includes compensation for both CRT pincushion and optical distortions.

The CRT pincushion distortion originates when a flat-face CRT is used as the display image source. The magnitude of the pincushion distortion depends on the physical size of the CRT image and the distance between the effective beam deflection position and the phosphor surface.

Optical distortions originate in both refractive and wide FOV optical systems. Although it is possible to minimized the distortions introduced by the optical systems in some cases, the added optical complexity is often not worth the benefit. This is true in refractive optical systems because it is straightforward to electronically compensate for the rotationally symmetric distortions. These distortions typically represent a few percent of the full-scale deflection.

In wide FOV systems, the combiner/collimator off-axis angle causes the geometric distortions to be very large and anamorphic. (Geometric distortions generally increase with the cube of the combiner off-axis angle. As in conventional optical systems, it is, in principle, possible to compensate for the distortions originating in the optical design. In practice, however, the optical system is already overconstrained, and any additional image distortion constraints would reduce the optical performance and increase complexity.

One approach to distortion compensation is to approximate the theoretical optical distortion characteristics (determined by ray tracing) with a two-dimensional image predistortion polynomial. The polynominal approximation to the actual data can be computed in the HUD processor prior to the digital-to-analog conversion or incorporated in the PDU by using precision analog circuitry. Figure 9.24 illustrates three predistortion maps: a benign characteristic for a refractive HUD with a 25° diameter FOV, a moderately complex characteristic for an overhead-mounted WFOV system, and a complex distortion map for a WFOV HUD system with a large off-axis angle.

In highly asymmetric WFOV systems, manufacturing variations in the relay lens assemblies can cause the actual distortion mapping characteristic to vary from the nominal characteristics by amounts unacceptable for meeting display

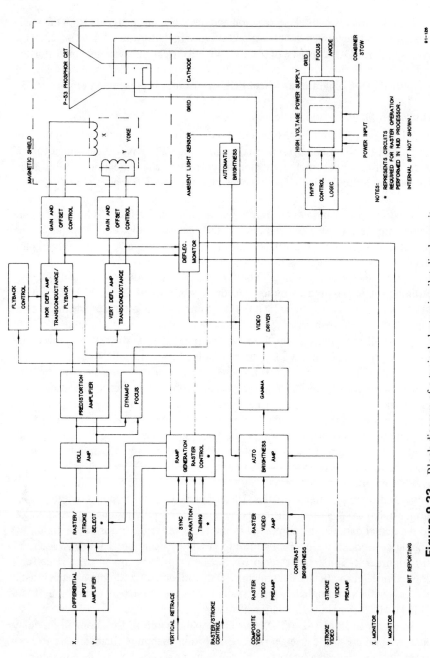

Figure 9.23 Block diagram of a typical electronic pilot display unit.

HOLOGRAPHIC HEAD-UP DISPLAYS

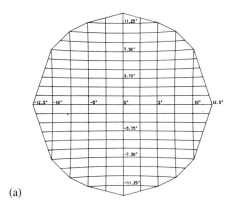

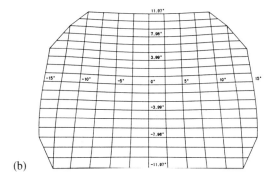

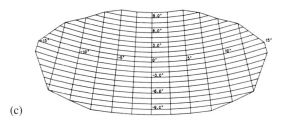

Figure 9.24 Distortion mapping characteristics for (a) a refractive optics (25° TFOV) system, (b) an overhead-mounted WFOV HUD, and (c) a glareshield-mounted WFOV system.

accuracy requirements. In this case, the actual distortion "residuals" can be characterized and used to modify the nominal polynomial constants.

The output from the distortion mapping electronics is fed into a pair of precision transconductance amplifiers that drive the CRT yoke windings. The resulting magnetic field deflects the electron beam to the desired location on the CRT phosphor faceplate. The deflection amplifiers include feedback to stabilize the deflection loop and provide high deflection accuracy.

Video

The video circuitry amplifies the Z-axis output signals from the HUD processor to a level necessary to drive the CRT grid. Included in the video path is an automatic brightness control circuit that uses the output from an ambient light sensor detector to control the CRT grid voltage in order to maintain a constant display contrast ratio as a function of the ambient light level. Manual brightness control is also available. Systems capable of raster operation include a linear video amplifier needed to provide continuous shades or gray and CRT gamma correction.

Built-In Test

The built-in test (BIT) functions of a PDU are designed to monitor the status of all electronic circuitry within the unit. Some of the built-in test functions include high- and low-voltage power supplying monitoring, CRT protection, deflection and video activity monitors, and a variety of combiner position sensors. In addition, an output is provided from the PDU that monitors the X, Y position of the electron beam. This output is fed back into the HUD processor and is used to determine whether the beam is being properly positioned at the CRT faceplate.

In HUDs with stowable combiners, it is important to monitor the mechanical position of the combiner element to ensure proper display overlay on the real-world scene. This monitoring is accomplished in the Flight Dynamics, Inc. HUD by using a modulated infrared light-emitting diode and a linear detector assembly (Fig. 9.25). If desired, the error signal is fed back into the deflection amplifier to electronically offset the display in proportion to the error signal. Thus, the Earth-referenced display elements remain locked onto the real world, independent of combiner position. The detector assembly is designed to not saturate when exposed to direct sunlight.

9.5 EXAMPLES OF HUD OPTICAL SYSTEMS AND FUTURE TRENDS

In this section, several HUD optical systems are discussed. Each system uses a different configuration holographic combiner. The HUD optical systems discussed here include a refractive optics collimator with a single wavelength-selective holographic combiner; a refractive optics collimator with dual holo-

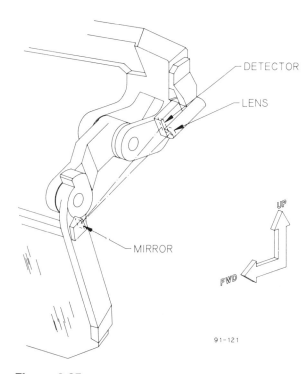

Figure 9.25 A combiner position sensor used to continuously monitor and verify the mechanical position of the combiner. An IR beam and detector assembly is mounted to the fixed position of the combiner, and a small mirror is mounted to the movable portion. The return beam is incident on the detector assembly, where the signal is processed to determine alignment.

graphic combiners, a WFOV optical system designed for a commercial transport cockpit using a conformal holographic combiner, and finally, two WFOV systems designed for military fighter aircraft.

Also in this section is a description of an automotive HUD system using the windshield, with or without an embedded holographic element, as the combiner element.

9.5.1 Refractive HUD with Single Holographic Combiner

Figure 9.26 is a raytrace of the collimator and combiner for a compact overhead-mounted HUD optical system configured for commercial transport applications. This optical system consists of three basic components: a CRT image source, a five-element rotationally symmetric collimator, and a flat wavelength-selective

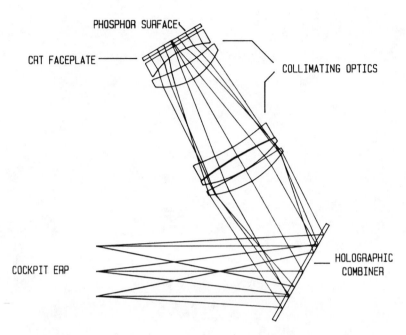

Figure 9.26 A raytrace of a collimator lens for a compact overhead-mounted optical system. This lens assembly was designed to be used with the P1 phosphor and with a single flat holographic combiner. This lens provides a total display field of view of 30° diameter.

holographic combiner. Figure 9.27 is a photograph of the integrated pilot display unit showing the cockpit speaker and map light.

The flat holographic combiner is constructed using the air-gate technique illustrated in Figure 9.19. To maximize the brightness and brightness uniformity with vertical head motion, this PDU uses the broadband P1 phosphor (Fig. 9.13) in combination with a narrowband holographic combiner (Fig. 9.14). The performance of the optical system is summarized in Table 9.4.

Figure 9.28 is a raytrace of dual holographic combiners designed for use with a P1 phosphor. The collimating lens aperture is shown. The dual holographic combiners are constructed using the air-gate technique illustrated in Figure 9.19.

In refractive HUDs with dual combiners, there is a horizontal display zone located near the center of the instantaneous FOV where the display is viewed after reflecting from both the upper and lower combiner elements simultaneously. This region is known as the *overlap* region. To prevent a double image of symbols in the overlap region, the optics must be well corrected across the full collimating aperture, and the combiners must be parallel. Refractive optical systems

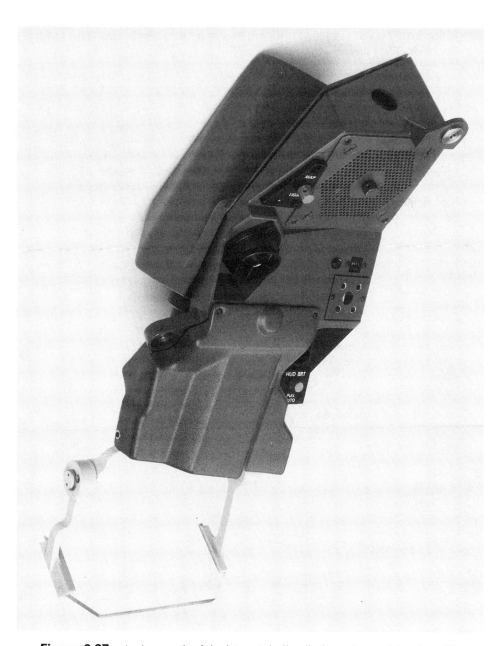

Figure 9.27 A photograph of the integrated pilot display unit containing the collimating lens shown in Figure 9.26.

Table 9.4 Optical Performance Characteristics of an Overhead-Mounted Refractive HUD with a Flat Holographic Combiner

Characteristic	Performance
Field of view	
Total	30° horiz × 24° vert
Instantaneous	25° horiz × 16° vert
Overlapping	4° horiz × 16° vert
Eyebox size	5.0 in. Horiz, 3.0 in. vert
Head motion needed to see TFOV	0.5 inc. horiz, 1.0 inc. vert
Parallax performance	
Convergence	100% < 1.0 mrad
Divergence	100% < 1.0 mrad
Dipvergence	100% < 1.0 mrad
Optical display accuracy	100% < 2.5 mrad
Combiner/collimator off-axis angle	55°
Combiner line-of-sight errors	0 mrad
Elements in collimator	Total of five
Combiner photopic transmission	82%
Combiner phosphor reflection	35%
Phosphor type	P1
Contrast ratio[a]	1.3 : 1

[a] Against 10,000 fL background.

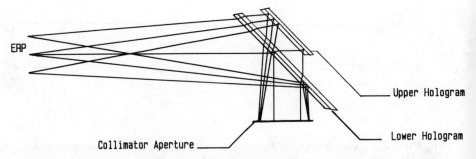

Figure 9.28 A raytrace of the dual holographic combiners used with the lens shown in Figure 9.5. These combiners are constructed using the air-gate technique.

Table 9.5 Optical Performance Characteristics of a Glareshield-Mounted Refractive HUD with Dual Holographic Combiners

Characteristic	Performance
Field of view	
Total	25° diameter
Instantaneous	19° horiz × 16° vert
Overlapping	6° horiz × 16° vert
Eyebox size	6.0 in. horiz × 3.0 in. vert
Head motion needed to see TFOV	0.7 in. horiz, 1.2 in. vert
Parallax performance	
Convergence	100% < 0.9 mrad
Divergence	100% < 0.0 mrad
Dipvergence	100% < 0.5 mrad
Optical display accuracy	100% < 1.5 mrad
Combiner/collimator off-axis angle	95°
Combiner line-of-sight errors	0 mrad
Elements in collimator	
Spherical	4
Aspheric	1 surface
Combiner photopic transmission	75%
Combiner phosphor reflection	35%
Phosphor type	P53 or P1
Contrast ratio	1.3 : 1

[a]Against 10,000 fL background.

with dual combiners must therefore have optical correction adequate to eliminate image doubling in the overlap region.

As seen by the pilot, there is a transmission zone within the overlap region due to the overlapping coatings. This zone can be objectionable because of excessive real-world attenuation. When holographic dual combiners are used, the hologram edges can be tapered in the overlap region, thus minimizing or eliminating transmission zones. The performance characteristics of an optical system designed for use with dual holographic combiners (this lens is illustrated in Figure 9.5) are summarized in Table 9.5.

9.5.2 WFOV Overhead-Mounted Holographic HUD

Figure 9.29 is a raytrace of a holographic WFOV overhead-mounted optical system showing the eye reference point, combiner, relay lens, and CRT faceplate. Figure 9.30 is a photograph of the system in a commercial transport cockpit. This WFOV system uses a Mangin mirror combiner design, where the spherical ho-

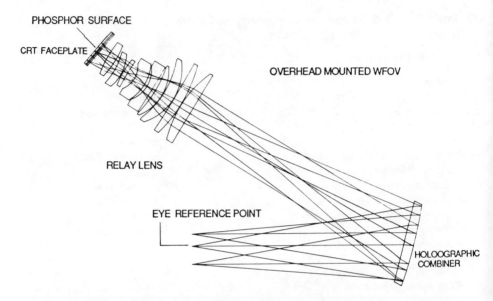

Figure 9.29 A raytrace of a holographic WFOV overhead-mounted optical system showing the cockpit ERP, the holographic combiner element (without cover), the relay lens, and CRT faceplate. (Figure courtesy of Flight Dynamics, Inc.)

logram surface is embedded between two planar surfaces. This design approach eliminates any combiner-induced real-world distortions, line-of-sight errors, and parallax errors. Nondistorting combiners are mandatory in commercial aircraft with flat windshields.

The optical system operates over a 30° display FOV with a 4.75 in. horizontal eyebox. The combiner/collimator off-axis angle is about 31°. The nine-element relay lens assembly is designed to compensate for aberrations introduced by the low f/number combiner, the large display field of view, and the off-axis combiner/collimator aberrations. The hologram construction geometry is based on the airgate technique illustrated in Figure 9.19. This optical system has been certified by the FAA for CAT IIIa operations as part of the Flight Dynamics, Inc. HGS-1000 HUD guidance system. Figure 9.31 is a photograph taken through the HGS combiner showing symbology superimposed on the real world. This HUD system has the largest display field of view of any production HUD flying today. The optical performance of this system is summarized in Table 9.3.

9.5.3 Holographic HUD Designs for Fighter Cockpits

In modern tactical aircraft, head-up displays play an important role as both the primary fight display and the primary weapons-aiming system. Unfortunately,

HOLOGRAPHIC HEAD-UP DISPLAYS

Figure 9.30 A photograph of the Flight Dynamics Head-Up Guidance System (1983) installed in a Boeing 727–200 cockpit.

fighter HUD PDU envelopes are often heavily constrained by the basic cockpit geometry, forcing the HUD optical system into nonoptimal configurations. This often leads to HUD optics geometries with large combiner/collimator off-axis angles.

The combiner envelope in fighter WFOV systems is restricted by the over-the-nose vision plane (the maximum down look angle from the cockpit eye reference point), the ejection clearance plane, and the clearance between the top of the combiner element and the inner surface of the aircraft windshield. The ejection clearance defines the maximum aft position of any part of the HUD PDU including the combiner if the pilot is to safely eject from the aircraft. Finally, the combiner clearance to the inner surface of the windshield must be adequate to allow easy removal of the combiner assembly and to prevent the windshield from contacting the top of the combiner in the event of a bird striking the canopy.

In addition to the general HUD performance requirements discussed in this chapter, fighter HUDs are often designed to operate in conjunction with conical

Figure 9.31 A photograph of commercial HUD symbology taken through the HGS holographic combiner.

aircraft canopies. The optical system is therefore designed to provide a display that is conformal with the real world as seen through the combiner and canopy.

Single-Combiner WFOV Glareshield-Mounted HUD

Figure 9.32 is a simplified raytrace of a glareshield-mounted WFOV optical system for use in a high-performance fighter aircraft. This HUD is designed to fit within the highly constrained cockpit of the F-15 aircraft and to compensate for real-world parallax errors and line-of-sight errors introduced by the aircraft canopy. Illustrated in the figure are the HUD system cockpit constraints discussed above: the ejection line, the clearance to the aircraft canopy, the over-the-nose vision line, and the chassis packaging envelope. For this optical system to provide acceptable levels of optical performance, the holographic combiner/collimator is manufactured with a dual-beam aberrated wave front construction technique (Fisher, 1989), resulting in a wavelength-selective combiner with highly aspheric diffraction characteristics. In addition, some lens elements within the

Table 9.6 Performance Characteristics of a Two-Hologram WFOV Glareshield-Mounted Prototype HUD

Characteristic	Performance
Field of view	
Total	30° horiz × 21° vert
Instantaneous	30° horiz × 18° vert
Overlapping binocular	25° horiz × 18° vert
Pupil size	5.0 in. horiz × 3.0 in. vert
Head motion needed to see TFOV	±0.5 in.
Parallax performance[a]	
Convergence	80% < 2.0 mrad
Divergence	80% < 1.0 mrad
Dipvergence	80% < 1.0 mrad
Combiner/collimator off-axis angle	56°
Combiner line-of-sight errors	2.0 mrad max
Elements in relay lens	8 total
Spherical	5
Cylindrical	1 surface
Aspheric	2 surfaces
Combiner photopic transmission	85% maximum
Phosphor type	P43
Contrast ratio	1.3 : 1

[a]Demonstrator system (Flight Dynamics, Inc., 1984). Parallax measured with respect to the real world as seen through the windshield. Represents in excess of 2500 combined field and pupil positions.

relay lens assembly are decentered or aspheric. More than 200 of these systems have been delivered and are in operation today.

Two-Hologram WFOV Glareshield-Mounted HUD

Figure 9.33 is a raytrace of a glareshield-mounted WFOV optical system using a simple conformal holographic combiner collimator designed with no aberration compensation diffractive power (Wood, 1988). (The construction approach is illustrated in Figure 9.19.) The hologram is optimized to enhance the photometric characteristics of the display only and has a photopic transmission at the installation angle of 85%. Figure 9.34 illustrates the completed HUD unit.

This glareshield WFOV HUD configuration also uses a holographically manufactured wavelength-selective aspheric fold element designed both to compensate off-axis aberrations and to minimize stray light from sunlight entering the relay lens assembly. Although there is a significant amount of diffractive aspheric optical power in the fold element, arbitrary amounts of power cannot be used

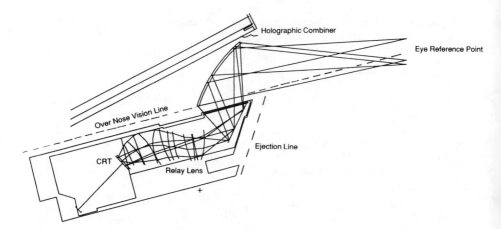

Figure 9.32 A simplified raytrace of the glareshield-mounted F-15E aircraft WFOV holographic head-up display designed and manufactured by Kaiser Electronics and Kaiser Optical illustrating the various optical system subsystems. The large combiner/collimator off-axis angle requires both a complex relay lens assembly and an aspheric wave front holographic combiner construction process to achieve the desired optical performance. (Figure courtesy of Kaiser Optical Systems, Inc., Ann Arbor, MI.)

because of chromatic dispersions introduced by the holographic grating. Using the narrowband P43 phosphor reduces the grating dispersion to an acceptable level. The fold element construction geometry is illustrated in Figure 9.22.

Finally, this system is designed and optimized with the optical effects of the canopy fully integrated into the relay lens and fold element designs. This occurs during the raytrace and optimization of the optical system, resulting in better overall display accuracy and display conformality. The optical performance of this system is summarized in Table 9.6.

Multiple-Element WFOV Glareshield-Mounted HUD

Figure 9.35 is a schematic raytrace of a glareshield WFOV optical system designed for use in the F-16D aircraft (Anon 1981; Vallance, 1983; Warwick, 1986). This system uses a unique (Ellis, 1981; Woodcock, 1983) configuration of multiple wavelength-selective conformal holographic elements as the combiner assembly. As seen in the figure, two flat holographic elements are used to fold the light path between the relay lens and the collimating combiner element. The collimating combiner reimages the intermediate image to infinity for viewing by the pilot. The two flat holograms fold the optical path to minimize the collimator off-axis angle, thereby reducing relay lens complexity and improving performance. The two flat folding holograms have high photopic transmission as

HOLOGRAPHIC HEAD-UP DISPLAYS

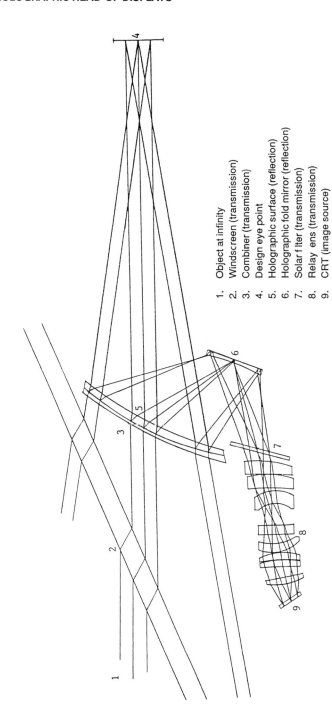

Figure 9.33 A raytrace of a glareshield-mounted WFOV optical system using a simple conformal holographic combiner/collimator, a holographically manufactured aspheric fold element, a nine-element relay lens assembly, and the CRT interface.

1. Object at infinity
2. Windscreen (transmission)
3. Combiner (transmission)
4. Design eye point
5. Holographic surface (reflection)
6. Holographic fold mirror (reflection)
7. Solar filter (transmission)
8. Relay lens (transmission)
9. CRT (image source)

Figure 9.34 A photograph of the completed HUD unit illustrated in Figure 9.33. (Courtesy of Flight Dynamics, Inc., 1985.)

measured from the head motion box, even though they are highly efficient reflectors at the CRT wavelengths and at the incident angles.

This optical configuration takes advantage of the angle–wavelength characteristics of reflection holograms. As seen in Figure 9.35, the P43 phosphor light leaving the upper hologram transmits through the forward hologram (the collimator) and diffracts from the flat reflecting rear hologram. This element redirects the light back toward the forward hologram. When the light is incident on the forward hologram from the rear hologram side, the incident angles are such that the Bragg condition is satisfied, and the diffraction efficiency for the central P43 phosphor wavelengths is high. Similarly, the light leaving the forward hologram

Table 9.7 Performance Characteristics of an Automotive HUD Using a Projection Lens

Display fields of view	
Total	9.4° horiz × 10.4° vert
Instantaneous[a]	5.4° horiz × 3° vert
Eye ellipse	5.0 in. horiz × 3.4 in. vert
Head motion needed to see TFOV[b]	None
Parallax performance	
Convergence	95% < 24 ± 2 mrad
Dipvergence	95% < 3.0 mrad
Windshield off-axis angle	134° (total)
Elements in projection lens	3
Windshield photopic transmission	70% max
Contrast ratio[d]	1.3 : 1
Image source	Back-illuminated LCD

[a]The display image source can be positioned such that the full display can be seen from any head position within the eye ellipse.
[b]The usable display is sized to fit within the instantaneous FOV.
[c]Display distance is about 9 ft.
[d]Against a 5000 fL background.

will transit through the rear hologram because the rear hologram is detuned from the Bragg condition. This combiner design approach fully exploits the spectral and angular characteristics of holographic reflection elements.

One drawback of this configuration is that some portions of the real-world field of view are seen through three independent holographic elements and glass substrates. This leads to a photopic transmission of approximately 70%. A further drawback is the mechanical packaging of the upper hologram. The structure required to support and align the upper hologram must be adequate to withstand airblast at several hundred knots. This robust structure tends to increase forward field-of-view obscuration more than competing systems. However, hundreds of these systems have been manufactured and are in operation, on the F-16D aircraft.

9.5.4 Automotive HUD Systems

Automotive HUD systems are similar to aircraft HUDs in that a virtual display can be seen without having to look down into the vehicle. There are, however, several fundamental differences in the function of automotive HUD systems. Automotive HUDs (AHUDs) are not used to provide the viewer with vital informa-

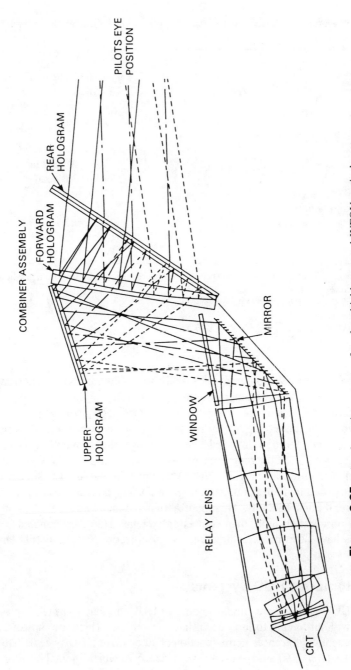

Figure 9.35 A schematic raytrace of a glareshield mounted WFOV optical system designed for use in the F-16D aircraft as described by Vallance (1983). This design features a three-hologram combiner assembly as described in the text.

tion needed to enhance the operation of the vehicle but are used to enhance safety by eliminating head-down transitions. There are experimental data on the safety benefits and reaction time benefits of HUDs (Dellis, 1988; Sakata et al., 1987; Okabayashi et al., 1989; Weihrauch et al., 1989; Evans et al., 1989).

Some of the basic differences between aircraft WFOV conformal HUDs and AHUDs are as follows:

1. The AHUD is not collimated at infinity but is generally projected at a distance corresponding to the front bumper of the automobile. This is a compromise distance designed to eliminate driver confusion potentially originating from having a display distance greater than real-world obstacles (i.e., traffic). Drivers are already experienced at focusing from outside of the car to inside to view conventional dashboard displays. Therefore, "focusing in" slightly to see the HUD display is a practiced and natural action. A display at the front bumper minimizes the focus change, yet still keeps the driver's attention outside.

2. The AHUD is not conformal with the real world and must be positioned such that it does not interfere with the normal operation of the vehicle. Unlike aircraft HUD systems where the projected display is conformal with the real-world view, AHUD images are positioned either below or to one side of the primary driving FOV seen through the windshield. One proven location is below the primary driving FOV.

3. The AHUD has a very limited display FOV, generally restricted to a few square degrees. The limited FOV is due to the severe restrictions on size, location, and cost of the projection system.

4. The AHUD uses a fixed-format display source. Thus, display information is altered by switching discrete elements that make up the display. High-brightness CRTs have not been used in production AHUDS to date because of the high cost of stroke-writing electronics and because of the low display brightness inherent with raster displays. Vacuum fluorescent displays are the predominant AHUD image source.

5. The AHUD uses the vehicle windshield as the combiner element. Because of the aerodynamic considerations in modern automobiles, the installation angle of any windshields is large, and the associated reflectivity is therefore high. Although windshield combiners are feasible in aircraft HUDs, the optical performance characteristics, especially the display FOV and parallax performance, and packaging options are severely compromised when using the windshield as the combiner.

6. Finally, the AHUD system cost is several orders of magnitude less than that of a high-performance aircraft system. In addition, the potential market for AHUDs is several orders of magnitude greater per year than for aircraft HUDs.

A block diagram of a demonstrated AHUD system is presented in Figure 9.36. The display system consists of an all-plastic optical projection system, a standard or reflection enhanced windshield combiner, a display source (i.e., vacuum fluorescent or LCD), the associated display drive electronics, and a computer interface to the automobile sensors. Figure 9.37 is a photograph of the AHUD display taken through the automobile windshield.

The AHUD Optical System

There are several optical methods for projecting information from the windshield for viewing by the driver described in the literature (Weihrauch et al., 1989; Okabayashi et al., 1989; Wood and Thomas, 1988). Most of the optical systems described are designed to project a virtual image of the display source at about 9

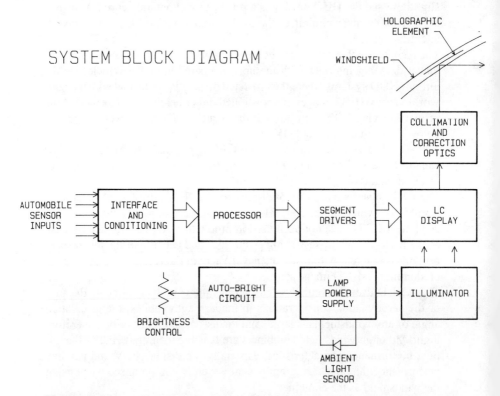

Figure 9.36 A block diagram of an automotive HUD system. The display system consists of an all-plastic optical projection system designed to correct for the aspheric curvature characteristics of the windshield, a microprocessor interface to the automobile, and a high-brightness image source. This system uses a back-illuminated five-color liquid-crystal module for the image source.

Figure 9.37 A photograph of the AHUD system display illustrated in Figure 9.36 taken through the automobile windshield.

ft forward of the automobile eye ellipse or head motion box. The projection optics, acting as a magnifier, form a virtual erect image of the display source forward of the eye ellipse after reflecting from the windshield. Since the windshields have aspheric curvatures, the projection systems are designed to compensate for the windshield-induced aberrations, resulting in clear, well-corrected virtual displays. The windshield shape can be described as generally anamorphic and aspheric, with different base radii in the vertical and horizontal planes. To optically compensate for complicated optical effects of the anamorphic windshield, at least one of the surfaces within the projection optics assembly must be aspheric and/or decentered. Dipvergence is the principal binocular parallax error requiring compensation. The projection assemblies are generally designed to be replicated or molded in plastic.

Table 9.7 summarizes the performance of an AHUD projection system.

The Windshield Combiner

The reflectivity of a standard two-ply automobile windshield depends on the installation angle. Each surface reflection can be determined from Fresnel equations for randomly polarized light assuming a glass index of refraction of about

1.52. The results show that the reflectivity of a conventional windshield at an incidence angle of 67° is about 22%.

To increase the display brightness and improve the contrast ratio, the windshield reflectivity can be increased by introducing a holographic element into the windshield (Hartman, 1986; Wood et al., 1989) or by adding a wavelength-selective conventional coating on the inner surface of the windshield (Okabayashi et al., 1989). Although these methods can enhance the windshield reflectivity, the total transmission through the windshield must not fall below the federal requirement of 70%. If a holographic coating is used in the windshield (Wood and Thomas, 1988), the total combiner reflectivity can be increased to about 33% while maintaining the 70% transmission requirement.

Display Image Source

The major design considerations in selecting an image source for an AHUD are output brightness, resolution limitations, packaging considerations, power dissipation, and the recurring costs associated with the display source and the drive electronics. The display technology selected for most of the AHUD systems in use today is the vacuum fluorescent display (VFD). The VFD uses a high-brightness blue-green phosphor with a peak near the eye response peak.

References

AC 20-57A. DOT FAA Advisory Circular AC No. 2057A, Automatic Landing Systems.
AC 25.1309-1A. DOT FAA Advisory Circular AC No. 25.1309-1A, System Design and Analysis.
AC 120-28C. DOT FAA Advisory Circular AC No. 120-28C, Criteria for Approval of Category III Landing Weather Minima.
Anon. (1981). F-16's headup display developed for Lantirn, *Aviation Week Space Technol.*, June 8: 141.
Arns, J., T. Edwards, G. Moss, and J. Wreede (1984a). System for forming improved reflection holograms with a single beam, U.S. Patent 4,458,977.
Arns, J. T. Edwards, G. Moss, and J. Wreede (1984b). Double beam systems for forming improved holograms, U.S. Patent 4,458,978.
Banbury, J. (1982). Wide field of view head-up displays, AGARD Panel Conf., Advanced Avionics and the Military Aircraft Man/Machine Interface, Blackpool, U.K., pp. 89–96.
Boucek, G. T. Pfaff, and W. Smith (1983). The use of holographic head-up display of flight path symbology in varying weather conditions, SAE 2nd Aerospace Behavioral Engineering Technol. Conf. Proc. Long Beach, CA, pp. 103–109.
Chang, B. (1980). Dichromated gelatin holograms and their applications, *Opt. Eng. 19*: 642
Chorley, R. (1974). Head-up display optics, in *Optoelectronics*, AGARD Lecture Ser. No. 71.
Close, D. (1973). The use of ray intercept curves for evaluating holographic optical elements, SPIE Applications of Geometrical Optics, San Diego, CA, pp. 101–106.

Close, D. (1974). Hologram optics in head-up displays, SID Int. Symp., San Diego, CA, pp. 58–59.
Close, D. (1985). Method and apparatus for production of holographic optical elements, U.S. Patent 4,530,564.
Close, D., A. Au, and A. Graube (1974). Holographic lens for pilot's head-up display, Hughes Research Laboratories, Malibu, CA., April 1974, N62269-73-C-0388; Phase II, April 1975, N62269-74-C-0642; Phase III, February 1976, N62269-75-C-0299; Phase IV, August 1978, N62269-76-C-018.
Dellis, E. (1988). Automotive head-up displays: just around the corner, *Automot. Eng.* 96(2):107.
Desmond, J. (1986). Improvements in aircraft safety and operational dependability from a projected flight path guidance display, SAE Aerospace Technol. Conf. and Exposition, Long Beach, CA.
Desmond, J., and D. Ford (1984). Certification of a holographic head-up display for low visibility landings, AIAA/IEEE 6th Digital Avionics Systems Conf., Baltimore, MD, p. 441.
Desmond, J., D. Ford, M. Fossey, M. Stanbro, and K. Zimmerman (1987). Method and apparatus for detecting control system data processing efforts, U.S. Patent 4,698,785.
Edelman, D. (1990). Heads up display applications for wide body transport aircraft—an operational point of view, SAE Aerospace Technol. Conf. Exposition, Long Beach, CA.
Ellis, S. M. (1981). Head up displays, U.S. Patent 4,261,647.
Evans, R. (1985). The development of dichromated gelatin for holographic optical element applications, *Applications of Holography*, SPIE Vol. 523, pp. 302–304.
Evans, R., A. Ramsbottom, and D. Sheel (1989). Head-up displays in motor cars, Second International Conf. on Holographic Systems, Components and Applications, September 1989, Bath, U.K. pp. 56–62.
Fisher, R. L. (1989). Design methods for a holographic head-up display curved combiner, *Opt. Eng.* 28(6):616.
Ginsburg, A. (1983). Direct performance assessment of HUD display systems using contrast sensitivity, NAECON 83 Mini Course Notes, pp. 33–44.
Hartman, N. (1986). Heads-up display system with holographic dispersion correcting, U.S. Patent 4,613,200.
Hay, W., and B. Guenther (1988). Characterization of Polaroid's DMP-128 holographic recording photopolymer, *Holographic Optics: Design and Applications*, SPIE Vol. 883, pp. 102–105.
Hayford, M. (1981). Holographic optical elements—a guide for use in optimal system design, Optical Research Associates, Pasadena, CA. See references.
Ingwall, R., and H. Fielding (1985). Hologram recording with a new Polaroid photopolymer system, *Applications of Holography*, SPIE Vol. 523, pp. 306–312.
Ingwall, R., and N. Troll (1988). The mechanism of hologram formation in DM photopolymer, *Holographic Optics: Design and Applications*, SPIE Vol. 883, pp. 94–101.
Johnson, T. (1990). Alaska Airlines experience with the HGS-100 head-up guidance system, SAE Aerospace Technol. Conf. and Exposition, Long Beach, CA.
Kogelnik, H. (1969). Coupled wave theory for thick hologram gratings, *BSTJ* 48(9):2909.
Long, H. (1990). HUD potential for narrow-bodied air carrier aircraft, SAE Aerospace Technol. Conf. and Exposition, Long Beach, CA.

McCauley, D., C. Simpson, and W. Murbach (1973). Holographic optical elements for visual display applications, *Appl. Opt. 12*: 232.
McGrew, S. (1980). Color control in dichromated gelatin reflection holograms, *Recent Advances in Holography*, SPIE Vol. 215.
Moharam, M., and T. Gaylord (1981). Rigourous coupled-wave analysis of planar-grating diffraction, *J. Opt. Soc. Am. 71*(7):811.
Norris, G. (1991). UK tests thin-film combiner HUDs, *Flight Int.*, April 24–30:17.
Okabayashi, S., M. Sakata, J. Fukano, S. Daidoji, C. Hashimoto, and T. Ishikawa (1989). Development of practical heads-up display for production vehicle application, SAE SP-770, Detroit, MI, pp. 69–78.
Owen, H. (1985). Holographic optical elements in dichromated gelatin, *Applications of Holography*, SPIE Vol. 523, pp. 296–301.
Rioux, M. (1990). Advanced Sensing Applications Seminar, AEEC General Session, 1990.
Sakata, M., S. Okabayashi, J. Fukano, S. Hirose, and M. Ozono (1987). Contributions of head-up displays (HUDs) to safe driving, 11th Int. Tech. Conf. on Experimental Safety Vehicles, Washington, DC.
Samollovich, D., A. Zeichner, and A. Friesem (1980). The mechanism of volume hologram formation in dichromated gelatin, *Photogr. Sci. Eng. 24*(3):161.
Seymour, R. (1988). Phosphors for head-up display cathode ray tubes, *Displays*, July 1988: 123–130.
Shankoff, T. (1968). Phase holograms in dichromated gelatin, *Appl. Opt. 7*: 2101.
Smothers, W., T. Trout, A. Weber, and D. Mickish (1989). Hologram recording in Du Pont's new photopolymer material, IEE 2nd Int. Conf. on Holographic Systems, Components and Applications, Bath, U.K.
Stone, G. (1987). The design eye reference point, SAE 6th Aerospace Behavioral Eng. Technol. Conf. Proc., *Human/Computer Technology: Who's in Charge?*, pp. 51–57.
Sweatt, W. (1977). Describing holographic optical elements as lenses, *J. Opt. Soc. Am. 67*: 803.
Taylor, C. (1990). The HUD as primary flight instrument, SAE Aerospace Technol. Conf. and Exposition, Long Beach, CA.
Vallance, C. H. (1983). The approach to optical system design for aircraft head up display, *Proc. SPIE 399*: 15–25.
Warwick, (1986). Headup holograms show the way, *Flight Int.*, March 15, 1986.
Weihrauch, M. (1983). Manufacturing technology for head-up displays using diffraction optics, Hughes Aircraft Co. Final Report for Period September 1978 to May 1981; Materials Laboratory, Wright Patterson AFB, OH, Rep. AFWAL-TR-83-4027, Vol 1.
Weihrauch, M., G. Meleony, and T. Goesch (1989). The first head up display introduced by General Motors, SAE Int. Congress and Exposition, Spec. Proc. SP 770, Detroit, MI, pp. 55–62.
Witherington, R. (1976). Optical display systems utilizing holographic lenses, U.S. Patent 3,940,204.
Wood, R. (1988). Head up display system, U.S. Patent 4,763,900.
Wood, R., and R. Cannata (1986). Holographic device, U.S. Patent 4,582,389.
Wood, R., and M. Hayford (1988). Holographic and classical head-up display technology for commercial and fighter aircraft, SPIE Applications of Holography, Vol. 883, pp. 36–52.

Wood, R., and M. Thomas (1988). A holographic head-up display for automotive applications, *Automotive Displays and Industrial Illumination*, SPIE Vol. 958.

Wood, R., and M. Thomas (1990). Holographic head-up display combiners with optimal photometric efficiency and uniformity, *Cockpit Displays and Visual Simulation*, SPIE Vol. 1289.

Wood, R., J. Hung, and N. Jee (1987). Holographic optical display system with optimum brightness uniformity, U.S. Patent 4,655,540.

Wood, R., M. Thomas, J. Valimont, E. Littel, and G. Freeman (1989). Vehicle display system using a holographic windshield prepared to withstand lamination process, U.S. Patent 4,842,389.

Woodcock, B. (1983). Volume phase holograms and their applications to avionic display, *Proc. SPIE 399*: 333–338.

10
Biocular Display Optics

Philip J. Rogers

*Pilkington Optronics,
Clwyd, United Kingdom*

Michael H. Freeman

*Optics and Vision Ltd.,
Clwyd, United Kingdom*

10.1 INTRODUCTION

Biocular display optics refers to those optical systems that provide two-eye vision of an object seen from only one viewpoint. More specifically, biocular viewers can be defined as lenses that have a single axis of symmetry but are of sufficient size to present a magnified image of an object to both eyes simultaneously.

Most human observers have full use of two eyes. Each eye has a separate lens system and retina to form and receive an image of the scene. The brain behind the eyes forms a single impression of the scene, but the use of two eyes allows a better perception of depth and an improvement in contrast sensitivity, a measure of the signal-to-noise ratio of the visual system. A larger part of depth perception comes from *stereopsis*, the ability of the visual system to use the small differences between the images in the two eyes because they are viewing the scene from slightly different positions.

The eyes are accustomed to operating as a pair and exhibit fusional lock, the ability to point accurately at the same part of the scene so that the brain receives images that are closely similar. If this fusional lock is broken, discomfort and fatigue are experienced. The body of scientific knowledge associated with all aspects of two-eye viewing is called *binocular vision*.

Optical systems for visual use are more acceptable to observers if they make full use of both eyes. A monocular telescope is not as visually satisfactory as a "pair" of binoculars. Modern microscopes are now predominantly fitted with two eyepieces so that both eyes can be used. If the eyepieces view through two objectives or different parts of a large objective, the instrument is a binocular microscope, and stereoscopic depth can be seen in the image by the normal observer. If the eyepieces are linked, by semireflecting prisms, for example, to view through the center of a single objective, no stereopsis is available, but the eyes can operate their fusional lock mechanism and so experience minimum discomfort and fatigue.

The basic microscope (and telescope) is an optical system comprising two lenses or lens systems separated by a space in which an intermediate image is formed. To view through an eyepiece, the eye must be placed within the exit pupil (eye ring) formed by the combination of eyepiece and objective. When an eyepiece or similar lens system of short positive focal length is used to view directly (and magnify) an object, it is conventionally called a *magnifier* to distinguish it from an eyepiece that operates in combination with an objective.

With the simple magnifier there is no constraining exit pupil, and if the magnifier is made large enough, both eyes may see some or all of the single magnified image. The eyes of the average observer are about 64 mm apart, and magnifiers with similar diameters are called *biocular magnifiers,* where the absence of the *n* of the word binocular is meant to signify a single ocular used by two eyes. Freeman (1973) attempted to define a consistent usage of the word in different contexts. For instance, the microscope described above, which uses two eyepieces viewing through two objectives or different parts of a single objective, is rightly called a *binocular* microscope, while the microscope that uses semireflecting mirrors to divide the light from a single objective into two eyepieces could be called a *split biocular* microscope

The main use of the word *biocular*, however, has been to describe biocular magnifiers of reasonable magnifying power ($\times 5$) for applications such as viewing the output phosphors of small cathode-ray or image intensifier tubes. These present a difficult problem for the optical designer. To be useful, the designs must have a combination of first-order properties—notably f-number and field of view—which, in most cases, rules out the possibility of complete correction of aberrations, even those of third order. The lens design problem is, therefore, that of correcting the aberrations of a biocular magnifier to a balance of residuals that both eyes in combination will deem to be acceptable.

This assessment of visual acceptability is outside the scope of standard optical design computer programs (and standard optical designers!). Section 10.3 describes the difficult interface between optical design technology and binocular visual science. Section 10.2 covers the relevant characteristics of the eyes, singly and in combination, and the later sections describe actual optical designs.

10.2 RELEVANT ASPECTS OF VISION

10.2.1 One-Eye and Two-Eye Vision

The eye as a separate entity has an optical imaging system of some 17 mm equivalent focal length. In vision terms this is 60 diopters of optical power, where the unit has the dimensions of reciprocal length so that 1 diopter is the power of an optical system of 1 m focal length and 2 diopters is the power of a system with 500 mm focal length. In the eye most of the optical power is provided by the cornea, the internal lens having an average power of 10–15 diopters. Figure 10.1 shows the optical system of the eye. In observers up to about 50 years of age, the internal lens can increase its power: by up to 10 diopters at age 20 but by only 4 diopters at age 40. This mechanism of *accommodation* allows near objects to be sharply focused on the retina, as close as 100 mm at age 20 but only 250 mm at age 40. By the age of 50 most observers have only a depth-of-focus effect depending on the aperture of the eye and require the assistance of reading spectacles for near-vision tasks. While the accommodation response can be voluntarily controlled, it is largely automatic.

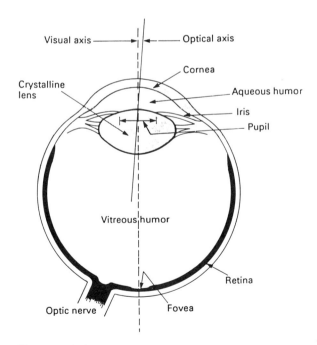

Figure 10.1 Horizontal section of right eye viewed from above. [From Freeman (1990), by permission of Butterworths, Publishers.]

The aperture of the optical system of the eye is controlled by an iris that forms a pupil of 7–8 mm diameter in dark conditions and reduces with increasing illumination to a minimum of about 2 mm. The extent of this change becomes less with older observers. The pupil response is entirely involuntary, being governed mainly by the luminance of the observed scene. Both accommodation and pupil responses can be modified by drugs.

The optical quality of the eye is acceptably good on-axis, being nearly diffraction-limited at 2.5 mm pupil, although in most observers the optical performance is reduced at larger pupil sizes. The eye is not corrected for chromatic aberration, showing over 1 diopter power difference between light of 650 nm and light of 450 nm. The optical performance is also poor off-axis. This poor optical performance is matched by the retina, which has a very dense array of receptors only in a central region subtending between 1° and 2° in the outside world. In order to see clearly, the eye must *fixate,* that is, point accurately in the direction of interest. The six external muscles holding and controlling each eyeball are therefore as important a part of good vision as are the optical surfaces and retinal detectors.

For good binocular vision the eyes must operate together and both fixate the same point in the same scene. There is an extensive feedback mechanism involving the six muscles and the image received by the brain from each eye. When this is operating correctly, the brain *fuses* the two images into one image while retaining the stereoscopic information on depth provided by the differences between the two images. When observing a distant object the eyes must fixate so that their axes are substantially parallel. For a distant object to the right or left, the two eyes rotate as required but remain substantially parallel. If the object moves closer to the observer, the eyes must now turn inwards to direct their axes at this new position. This is called *convergence* and is a natural and easy function for most people.

The accuracy of the convergence is very high and involves the fusion of two images into one—a function known as *fusional lock.* Convergence is generally automatic but can be voluntarily controlled. By imagining an object just in front of the nose, an incorrect convergence can be established when actually viewing a more distant scene. When an incorrect convergence is adopted or imposed, the brain sees two separated images, a condition known as *diplopia.* Under these conditions it is often found that one image will predominate and the other image appear to fade away due to the brain suppressing the image from one eye. This suppression usually shows a slow alternation between the eyes.

The action of the fusional lock can be shown by the fact that it is easy to maintain an incorrect convergence when it is substantially incorrect, but if the angle of error is reduced, a point is reached where the two images jump together. Maintaining an incorrect convergence just outside this fusion effect is very fatiguing and much more stressful than a large error where the brain invokes the suppression mechanism.

For relatively near objects the eyes must accommodate as well as converge in order to see an in-focus and fused image. The muscles performing both these tasks are linked together and, together with the pupil changes, make up what is called the *triad* response. In the normal observer, this mechanism ranges over the requirements for objects in a normal scene, even expecting to find close objects in a downward direction and distant objects at the horizontal or above. Optical instruments for visual use can distort these normal conditions, and even the prescribing of corrective spectacles has to take these aspects of binocular vision into consideration. Of particular concern is the inability of the normal triad response to diverge the eyes. An optical system that requires this, even in small measure, will be found to be stressful and uncomfortable.

Other uncomfortable conditions relate to the vertical movements of the eyes. In normal conditions the same angle is required for both eyes. Any difference in vertical angle is known as *dipvergence,* and stress can again result from relatively small values.

10.2.2 Anthropometrics of Binocular Vision

The human skull provides a socket for the globe of each eye so that the eyes are free to rotate but are protected as much as possible. The brow protrudes above the eyes, and the cheekbones below. The socket or orbit of the skull allows room for six external muscles to rotate the eye about a center of rotation that wanders by about 1 mm between up-and-down movements and side-to-side movements of differing extent. For the purposes of optical instrument design, the single eye has the dimensions and ranges given in Figure 10.2.

The prime dimension associated with two-eye vision is the distance between the eyes. This is defined by the interpupillary distance (IPD or, sometimes, PD). This has an average value of 60–64 mm but ranges more than 10 mm to each side. The range for females is greater than that for males, and differences also exist between racial types. Table 10.1 provides the results of some studies. When eyes converge to view a near scene, the IPD decreases as the eyes rotate about their centers. The formula for the reduction (R) in IPD is

$$R = 20 \sin (\theta/2) \tag{10.1}$$

where θ is the convergence angle (between both eyes). A display viewed at 2/3 m rather than at infinity therefore shows a reduction of about 1 mm.

The eyes can rotate horizontally to angles greater than 40°, but a comfort limit is experienced at a value between 30° and 40° during unaided viewing. Observers tend to rotate their head for angles greater than these. Figures 10.3 and 10.4 give

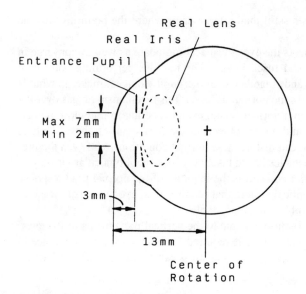

Figure 10.2 Principle optical dimensions of the eye.

Table 10.1 Interpupillary Distance (IPD) Table (IPD in millimeters)

	Percentile			
	2.5th	50th	97.5th	Ref.
Females	53.3	63.5	71.1	Dreyfuss (1967)
U.S. Air Force personnel	56.6	63.2	70.7	Hertzberg (1954)
U.K. Air Force personnel	58.3	64.2	70.7	Hobbs (1975)
U.S. Army drivers Black	57.9[a]	62.0	71.1[b]	Damon et al. (1962)
White	54.1[a]	58.9	64.0[b]	Damon et al. (1962)
U.K. Army personnel	56.5	62.4	68.2	Dudley (1978)

[a] Value for 5th percentile.
[b] Value for 95th percentile.

BIOCULAR DISPLAY OPTICS **423**

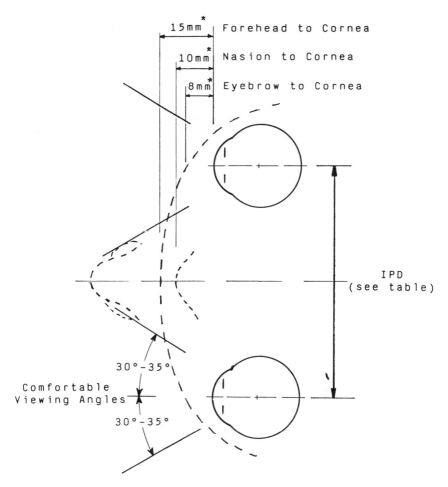

Figure 10.3 Binocular viewing and head shape. The values with asterisk (*) are the approximate mean values for an unstructured survey of 40 adult white males in the United Kingdom.

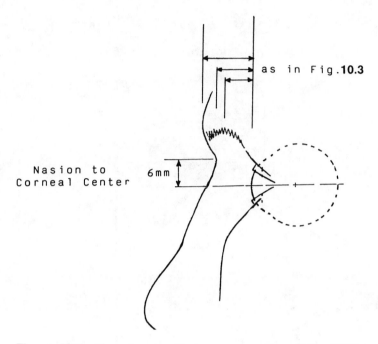

Figure 10.4 Vertical section. Nasion to cornea. [From Hobbs (1975).]

some salient dimensions for two-eye vision that are of value in the design of biocular and binocular instruments.

10.2.3 Binocular Visual Performance

In Section 10.2.1 the optical quality of the human eye was described as being very good on-axis with small pupils but not good at larger field angles and pupil sizes. This section reviews this in more detail, but for a fuller treatment of the eye as an optical instrument see Freeman (1990).

The variation of optical quality with pupil size is indicated by Figure 10.5a, which shows the reduction in MTF of an eye with increasing pupil size calculated from wave-front aberrations. Each curve is labeled with the pupil diameter in millimeters. The curve for 2 mm is essentially diffraction-limited, while the dashed line gives the diffraction-limited response for a 3-mm pupil (Walsh and Charman, 1985). There are, however, considerable differences between subjects at the higher pupil size. From the same reference, Figure 10.5b shows the MTF at 5-mm pupil diameter for 10 subjects.

The wave-front aberrations associated with these curves have some overall spherical aberration but mainly consist of irregular aberrations with wide varia-

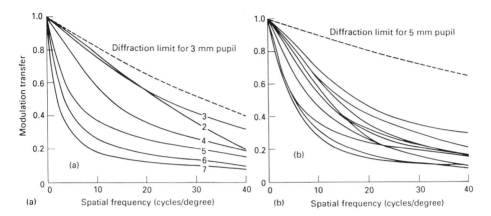

Figure 10.5 (a) Modulation transfer functions for the eye of a typical subject at different pupil sizes. (b) MTF for 10 subjects all at 5 mm pupil size. Deduced from the corresponding wave-front aberration data at 590 nm wavelength for vertical sinusoidal gratings. [After Walsh and Charman (1985).]

tions from eye to eye and from subject to subject. This gives the use of two eyes a probability of improvement in visual acuity over one eye. However, because the retina is often the limiting factor in resolving small detail, the improvement is generally accepted to be 5–10% in the spatial frequency of just-resolved high-contrast targets.

For low-contrast targets, the eye has a maximum sensitivity at a spatial frequency around 2–3 cycles/degree. At this frequency the aberrations of the eye have little effect, the limit being determined by the signal-to-noise ratio at the retina. Because the retinal noise is random, the summation of signals from two eyes provides a $\sqrt{2}'$ improvement over one-eye viewing. Visual experiments confirm this to be the case. An improvement of some 40% in contrast sensitivity over a wide range of frequencies was reported by Campbell and Green (1965). Although contrast sensitivity decreases with decreasing illumination, the benefit of binocular viewing is maintained at some 20% even if the luminance is reduced by the use of beam splitters to send light to both eyes, provided the resulting luminance is in the region of 40 cd/m².

Figure 10.6 gives a general contrast sensitivity curve under optimum viewing conditions. The dashed lines give an estimate for the best and worst of 90% of people, again under optimum viewing conditions using one-eye viewing. The contrast modulation (Cm) is here defined for a cyclical target, where

$$\text{Cm} = \frac{|Bt - Bb|}{Bt + Bb} \tag{10.2}$$

where Bt is luminance of the bright bars and Bb that of the dark bars.

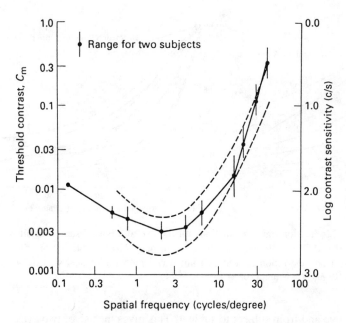

Figure 10.6 Contrast threshold under optimum conditions. [After Van Meeteren and Vos (1972).] The dashed lines indicate an estimate of contrast sensitivity for the best and worst of 90% of people under optimum viewing conditions. (Reproduced from *Design Handbook for Imagery Interpretation Equipment,* Farrell and Booth, by permission of Boeing Aerospace Company.)

A further, but undefinable, advantage of binocular viewing concerns the coherent interaction of the aberrations of a display with the random aberrations of the eyes. The binocular visual response will be determined by the "best" of the two eyes compared to the average value with one eye.

10.3 BIOCULAR VIEWER OPTICS

10.3.1 Introduction

The optical design requirements for noncollimated biocular viewers differ considerably from those where near-collimation is a prerequisite, for example, head-up display optics. Aberration correction requirements are less stringent and are defined in many cases only by the ability of the eyes to accommodate and fuse the separate images received by each eye. Against this must be weighed the more extreme optical parameters required of biocular magnifiers (a convenient term to describe noncollimated viewers used with eyes closeup to the optics), a very high numerical aperture being necessary to give adequate apparent fields of view from small-format cathode-ray and image intensifier tubes.

Single-lens magnifiers of sufficient diameter to be used by both eyes simultaneously are not new. Sherlock Holmes used one (according to Conan Doyle), though it is unlikely that he would have used both eyes, for reasons that will become obvious in the following.

A single-lens monocular magnifier of low magnifying power will almost inevitably give adequate monochromatic axial image quality simply because of the low aperture, and hence low spherical aberration, defined by the eye pupil diameter. At finite field angles the situation is different, and aberrations with a lower dependence on aperture—such as coma and, particularly, astigmatism—can be significant. In this centered eye case it is not difficult, however, to achieve an aberration balance that will give adequate imagery over a moderate field of view. This is because it is not necessary to correct spherical aberration fully provided that large decentrations of the eye pupil are not required. In the case of a biocular magnifier, large eye pupil decentrations with respect to the magnifier optical axis are not only required but are indeed inherent given the typical interpupillary distance (IPD) of 64 mm. The latter means that the eyes are nominally decentered symmetrically by 32 mm either side of the optical axis. They will thus sample parts of the lens aperture where all aberrations are likely to be large; in this case even adequate axial imagery cannot be guaranteed. For example, uncorrected spherical aberration will give rise to axial astigmatism even for small eye pupil diameters. This astigmatism is due to the difference in the local radial and tangential curvatures of the spherically aberrated emergent wave front at its intersections with the eyes. The fact that decentered spherical aberration can give rise to astigmatism (and coma) may be implied by traditional stop-shift equations. It does suggest, however, at least partial replacement of conventional Seidel aberrations by more meaningful visual aberrations related to the eye–optics interface (Sands, 1971; Haig, 1972; Freeman, 1975a; Palmer and Freeman, 1983).

Another major difference between biocular magnifiers and most other forms of optical design is the nonscalability of the IPD. A biocular that operates quite satisfactorily at a large diameter and low magnification will almost certainly perform badly if scaled down in diameter and up in magnification. In this case the IPD will represent a much greater proportion of the lens aperture, and the eyes will look through a completely different, and uncorrected, optical path.

The IPD also influences the smallest diameter biocular magnifier that can reasonably be envisaged. In close-up viewing—eyes between, say, 50 and 80 mm from the optics—quite a wide field of view can usually be perceived, as each eye can see up to a considerable angle off-axis at the side of the field of view opposite itself. For magnifier diameters only slightly in excess of the maximum IPD (see Section 10.2.2), the majority of the image may be perceived only monocularly;

although this has some benefit in terms of easing the optical design, a study (Landau, 1990) has indicated that a small biocular overlap (35%) relative to the total field of view is objectionable to some observers. A 79 mm diameter (the metric equivalent of 3.1 in.) has been determined empirically as a compromise between biocular overlap and difficulty of optical correction.

10.3.2 Visual Surfaces and Aberrations

When a pair of eyes observe an image through a biocular magnifier (Fig. 10.7), the brain attempts to fuse the separate pictures seen by each eye. Given that noncollimated bioculars are set to give a virtual image that is a finite distance in front of the eyes, fusion will usually involve convergence of the optical axes of the eyes. For a given object point, there will be a position in image space where the two lines of sight connecting each eye to the corresponding image point approach each other most closely (or, in some cases, intersect). At this position, the line between the two closest points represents the line of closest approach. The locus of the center points of the lines of closest approach generated over the biocularly viewed field of view defines the *convergence surface,* the surface over which the two eyes' lines of sight are most likely to converge.

The eyes do not necessarily focus over the convergence surface. In fact, there will be separate (pseudo-Coddington) sagittal and tangential foci along every line

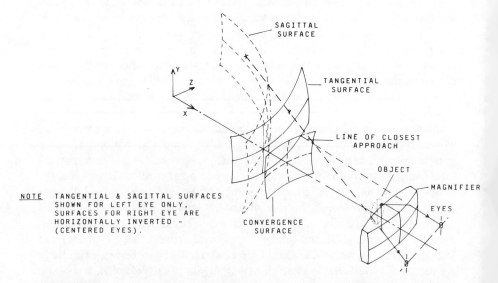

Figure 10.7 Convergence and astigmatic surfaces due to a biocular magnifier. [After Freeman, (1975b).]

BIOCULAR DISPLAY OPTICS

of sight of each eye. The locus of the midpoint between the S and T foci corresponding to each point on the image defines, for each eye separately, the *accommodation surface,* the surface over which each eye is independently most likely to focus.

Using the surfaces defined in Figure 10.7, seven visual aberrations can be identified (Fig. 10.8) that together define the performance of a biocular magnifier. These aberrations are the following.

V_1, *Astigmatic difference* Separation between S and T surfaces (either eye), positive if the S focus is nearer the observer.

V_2, *Image curvature* Departure of the accommodation surface from a plane through its intersection with the optical axis (either eye), positive if the edge of the accommodation surface is further away from the observer.

V_3, *Convergence defocus* Separation between the convergence and accommodation surfaces (either eye), positive if the accommodation surface is nearer the observer than the convergence surface.

V_4, *Binocular distortion (horizontal)* Horizontal component of the line of closest approach; less important than V_5 as it represents a small horizontal vergence error.

V_5, *Biocular distortion (vertical)* Vertical component of the line of closest approach (unsigned); important as it represents dipvergence (and therefore requires that the optical axes of the eyes be vertically misaligned for fusion to occur).

Image blur Angular image blur (of a point object), including effects of chromatic aberration and corresponding to a given eye pupil diameter (either eye).

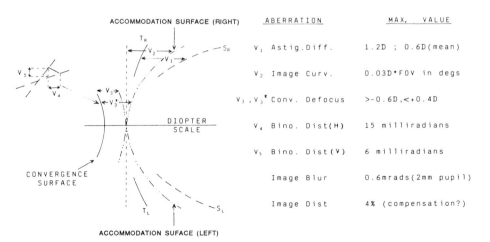

ABERRATION	MAX. VALUE
V_1 Astig.Diff.	1.2D ; 0.6D(mean)
V_2 Image Curv.	0.03D*FOV in degs
V_3, V_3^* Conv. Defocus	>-0.6D,<+0.4D
V_4 Bino. Dist(H)	15 milliradians
V_5 Bino. Dist(V)	6 milliradians
Image Blur	0.6mrads(2mm pupil)
Image Dist	4% (compensation?)

Figure 10.8 Visual aberrations and tolerances. [After Freeman (1975b).]

Image distortion A measure of the straightness of either top or bottom of the image of a rectangular object (usual aspect ratio 4:3, horizontal to vertical). Defined as the percentage change in the height of the format relative to its top center value; positive implies pincushion shape.

The above visual aberrations, preferably evaluated for a number of head positions, are sufficient to define fully the optical performance of a biocular magnifier as perceived by both eyes.

10.3.3 Visual Aberration Tolerances

The coherent link between visual optics and the eye, and the nonlinear behavior of the latter, mean that there is no direct analytical method for calculating tolerances for visual aberrations; these must instead be estimated from empirical data. The maximum values suggested in Figure 10.8 are based on extensive trials carried out by Freeman (1975b) with a number of observers; in most cases the values are in reasonable agreement with those obtained by other authorities (Giles, 1977; Charman and Whitefoot, 1978; Burton and Home, 1980; Mouroulis, 1982; Burton and Haig, 1984), although some of these were investigating monocular optics.

10.4 DESIGN OF BIOCULAR VIEWERS

10.4.1 Simple Biocular Magnifiers

Some years ago, details were published of an updated version of the Sherlock Holmes type of simple biocular magnifier (Coulman and Petrie, 1949), the design in this case being an acrylic singlet having one nonspherical (conic) surface. The authors made the point that it was important to achieve close coincidence of the accommodation and convergence surfaces—achieved by use of the conic surface—even to the detriment of distortion. Scaling this design down to a magnifier diameter of 79 mm and reoptimizing employing a 6th-order aspheric surface rather than a conic leads to a maximum reasonable magnification of $\times 2.3$, defining magnification as the angle subtended by the image relative to that of the object as seen by the unaided eye from a distance of 250 mm.

10.4.2 Compound Biocular Magnifiers

The previous paragraph described a simple biocular magnifier having a magnification of $\times 2.3$. That this magnification is inadequate for viewing small-format image intensifier or cathode-ray tubes can be demonstrated by consideration of the contrast sensitivity curve shown earlier in Figure 10.6; note that this "one-eye" curve is used because the majority of the field of view of a wide-angle biocular magnifier is viewed monocularly. Arbitrarily, assume that the viewer

BIOCULAR DISPLAY OPTICS

needs to detect a minimum contrast difference of 0.01 between adjacent CRT raster lines. The intercept of the 0.01 contrast line with the solid curve gives a corresponding spatial frequency of about 12 cycles/degree. The requirement, therefore, is that the raster lines must be magnified such that they represent an apparent spatial frequency of 12 cycles/degree to the eye.*

For a typical raster pattern of 625 TV lines (312.5 line pairs), the apparent vertical field of view that is required of the biocular will therefore be 312.5/12 = 26°. Figure 10.9 shows the thin-lens paraxial relationship between vertical field of view and magnification (plus associated focal lengths and object space f numbers) for three typical CRT-usable diagonals. Clearly the smallest CRT is inadequate, requiring a magnification in excess of $\times 7$; a 41 mm CRT solution is the most extreme that is possible, requiring a magnification of $\times 4.7$ and an associated f number of 0.54.

The above parameters are well beyond those that can be achieved with a simple magnifier such as the one described in Section 10.4.1. Consequently, a number of compound designs achieving the required magnification have been described in the literature (Rogers, 1972a, 1972b; Shenker, 1973; Anderson and Moyers, 1973; Roberts and Rogers, 1975; Faggiano et al., 1983). Figure 10.10 shows one example designed to give good imagery over a reasonable spectral

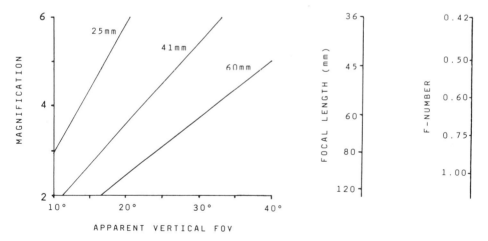

Figure 10.9 Relationship between magnification and apparent vertical field of view and associated focal lengths and object space F. Numbers for various CRT diagonals (thin-lens eye relief 100 mm; dioptric setting -2.0; FOV aspect ratio 4:3, horizontal to vertical).

---

*Approximate, as a raster is really a square wave.

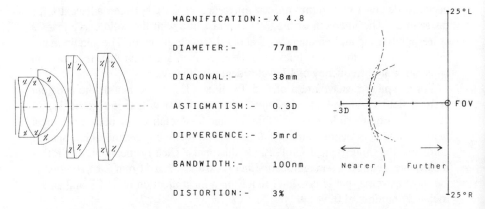

Figure 10.10 Performance analysis for a "straight-through" biocular magnifier design, and criteria and terminology used in analysis. (-··-) Accommodation curve, right eye; (---) accommodation curve, left eye; (···) convergence curve. Accommodation and convergence diopter curves are for the diagonal section (4:3 format) of the field of view. Zero diopter is equivalent to an image at infinity. Diagonal is the maximum object diagonal (4:3 format) as defined by the maximum field of view.

bandwidth, with consequent increase in optical complexity relative to an unachromatized design. The figure also illustrates the criteria and terminology used for this and all subsequent biocular magnifier design analysis.

Magnification is defined as the increase of angular subtense when viewing from the nominal eye relief (50 mm for magnifiers, 500 mm for relaxed view bioculars) relative to that of the object viewed by the unaided eyes (from 250 mm for magnifiers, from 500 mm for relaxed viewers).

Biocular overlap is given by extent of convergence curve relative to total field of view. The maximum diagonal field of view is defined either by the accommodation surface beginning to move rapidly toward infinity or by physical cutoff.

Astigmatism and dipvergence are both maximum values, the former for the diagonal section of the field of view only. Bandwidth is the spectral band over which maximum angular color reaches 1.5 mrad (5mm eye pupil) for the "better" eye where the viewing is biocular.

The biocular design in Figure 10.10 is a "straight-through" optical system. The most extreme application for a biocular magnifier arises from the periscopic configuration of, for example, an image-intensified passive driving sight; here an internal fold must be incorporated into the biocular design. The combination of a diagonal field of view of at least 45° (the minimum for night driving), a consequent f number of almost $f/0.5$ (for a 40–45 mm object diagonal), plus an internal fold represents a major design challenge. In general this leads to either re-

BIOCULAR DISPLAY OPTICS

sorting to an expanding fiber optic to increase the effective object size (Baird, Inc., private communication, 1985) or abandoning any pretense to correction of chromatic aberration or image curvature (Rogers, 1972b). Figure 10.11 illustrates the latter approach (Rogers, 1972a); it is fairly clear that this example represents the limit in terms of performance adequacy using the aberration criteria defined in Section 10.3.3. Other examples exist (Trowbridge and Ness, 1972; Walker, 1973) of biocular magnifiers that combine an internal fold with moderate to high magnification. A compendium of straight-through and folded designs appears in Rogers (1985a).

10.4.3 Relaxed View Bioculars

The biocular viewers described thus far have been designed nominally for close-up viewing, that, is with the eyes positioned some 50–80 mm from the magnifier and with the head most probably restrained by a brow pad. This is typical of many applications, but in some situations a much greater eye relief is required. In this case the biocular is generally larger in diameter than the typical 79 mm of the close-up type and is usually referred to as a relaxed view biocular.

Typical parameters for a relaxed view biocular are eye reliefs of up to a meter and a low dioptric setting (for example, -0.75 diopter at the eye point) in order to maintain a reasonable magnification at the greater eye relief.

A point should be made about the magnification of a relaxed view biocular. Magnification defined relative to the object viewed by the unaided eye at 250 mm distance is not particularly useful when the eye relief distance itself is in excess of 250 mm. A better definition of magnification in this case relates the angular subtense of the image to that of the object when viewed from the eye relief dis-

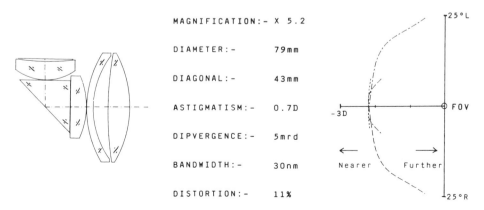

Figure 10.11 Performance analysis for a folded biocular magnifier design.

tance. This is the definition that will be used in all further discussions of relaxed view bioculars.

Parameters of Relaxed View Bioculars

In many situations where relaxed view bioculars are employed, it is desirable that all of the image be seen when the eyes are used in combination, if only from one head position. This requirement gives rise to a value for maximum magnification that can be employed to view an object of given dimensions while retaining visibility of all of the image using both eyes (Rogers, 1982). The following method can be used to determine the value of the maximum magnification (all lengths in millimeters).

Object distance:

$$L = \{1/F - D/[1000(B+1)]\}^{-1} \tag{10.3}$$

Magnification:

$$M = ER\,(B+1)/L \tag{10.4}$$

Total field of view:

$$TFOV = 2\arctan(MH/ER) \tag{10.5}$$

where ϕ is the biocular diameter, F is focal length (minimum realistic value 0.64ϕ), D is the dioptric setting (negative), ER is eye relief, $B = ER.D/1000$, and H is the half-diagonal of the object.

The total field of view (TFOV) defined above results only from object size and lens magnifying power; the maximum diagonal field of view that is visible instantaneously (IFOV) is defined by biocular diameter and eye relief. Given that A represents the ratio of the width of the object format to its diagonal, then the biocular diameter that corresponds to a given IFOV can be found from

Image semi diagonal:

$$H' = -[ER\,\tan[(IFOV/2)]/B \tag{10.6}$$

Biocular diameter*:

$$\phi = 2[(H'B)^2 + 0.25\,I^2\,(B+1)^2 + ABH'I(B+1)]^{1/2} \tag{10.7}$$

where I is the interpupillary distance (assume 64 mm) and A is the aspect ratio defined above.

The maximum magnification M and the corresponding minimum biocular diameter ϕ at which IFOV = TFOV can be found using Eqs. (10.3)–(10.7). Figure 10.12 gives a typical example for a dioptric setting of -0.75 and an eye relief of 500 mm for the same CRT diagonals (and aspect ratios) as those in

---

*Valid provided that IFOV > 2 arctan $(I/2A\,ER)]$

BIOCULAR DISPLAY OPTICS

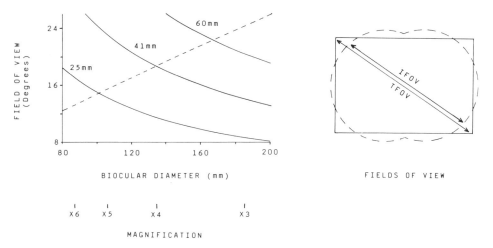

Figure 10.12 Instantaneous and total fields of view (IFOV and IFOV) versus relaxed view biocular diameter. Various CRT diagonals; focal length of biocular 0.64 ϕ; thin-lens eye relief 500 mm; dioptric setting -0.75.

Figure 10.9. The intersection of the IFOV curve with each TFOV curve represents the minimum biocular diameter (and therefore maximum magnification) that can be employed for that object diagonal.

Note that the above is restricted to what is essentially a thin-lens paraxial approximation on the basis that in a real system, lens construction and image distortion will affect the answer. It is useful for establishing approximate magnification and dimensions of a relaxed view biocular, but precise parameters should be determined for a real design.

Relaxed View Biocular Examples

In the earliest reference, Jeffree (1949) described a biocular originally designed in World War II as a relaxed viewer "with an eye distance of about two feet," the design being an acrylic singlet with one conic surface. A trace of this design indicates good imagery over a small field of view, somewhat smaller than that shown in Figure 10.13, where the ray trace has been carried out up to a field angle where the accommodation surface begins to head rapidly for infinity. Distortion over a field of view of 5° (probably the author's intended value) is less than 1%. The relaxed view magnifying power of only $\times$ 1.8 is, however, inadequate for many applications.

Figure 10.14 shows an enhancement of the Jeffree biocular where the singlet (now with one sixth-order aspheric surface) has been augmented by another plastic (spherical) lens, the two being disposed in simple Petzval form. The improve-

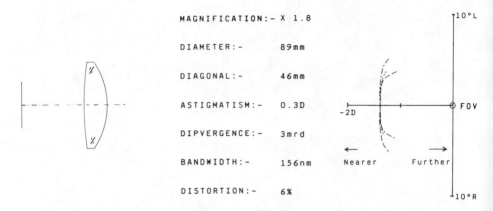

Figure 10.13 Performance analysis of Jeffree single acrylic lens relaxed view biocular design.

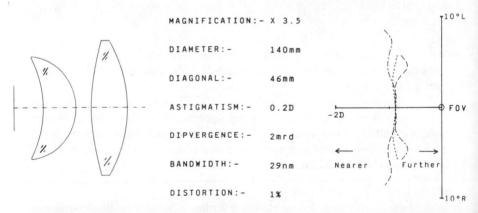

Figure 10.14 Performance analysis of two acrylic lens relaxed view biocular design.

ment is considerable: The diameter is larger (140 mm vs. 89 mm), but both the magnification and field of view are much higher. Note that this design, like the Jeffree example, is unachromatized and therefore suitable for use only with a filtered narrowband phosphor. It is possible to achromatize a relaxed view biocular for use with a wideband source, but the increase in complexity is considerable. In situations where a color display is essential, some means of combining separate colored narrowband sources within the biocular design makes most sense.

BIOCULAR DISPLAY OPTICS

10.4.4 Optimization of Biocular Viewers

Any attempt to optimize a biocular viewer in the traditional manner employed for optics where the integrated effect of the full aperture of the lens is considered (camera lenses, etc.) is doomed to failure. The aberrations of an optic having the combination of aperture and field angle required for a biocular viewer will always be fairly large judged by the usual standards. Indeed, a system optimized to have the smallest possible aberrations, but with rapid aberration slopes, will probably be found to be unacceptable. Any optical software for use in biocular design must consider the real situation—that is, a pair of small focusable apertures of fixed separation that are mobile over the pupil of the optic—and use the aberration criteria discussed in Section 10.3.

Approaches that can be used are either to adapt existing software or to devise software that is specific to the problem. A method described by Rogers (1972b) is probably the earliest reference to the latter. In this, a computer program was described that employed a standard pattern of 37 field points (Fig. 10.15) covering three head positions in a plane perpendicular to the optical axis. The program tested each field point for monocular/biocular vision and then employed the appropriate criteria for optimization. Final analysis, not described in the paper, was then carried out to define the convergence, sagittal, and tangential surfaces detailed in Section 10.3.2.

More recently, a method of biocular optimization was described (Wickholm, 1985) that used a standard commercial optical design program, in this case CODE V. Wickholm used the multiconfiguration option in the program to define separate 20 mm circular apertures centered on each eye separately (configurations 1 and 2) to control astigmatic/accommodation surface shapes. A pair of approximately square apertures were then defined with their centers 65 mm apart

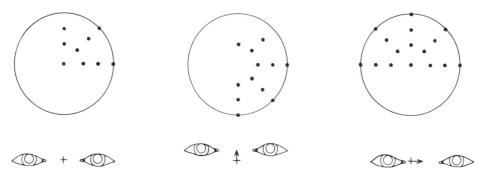

Figure 10.15 Eye positions and corresponding field points for the optimization of biocular magnifiers. [After Rogers (1972b).]

and disposed symmetrically about the optical axis (configuration 3) to control the convergence surface. The three configurations were optimized using common parameters and image plane.

Other authorities (Coulman and Petrie,1949; Jeffree, 1949; Shenker, 1973; Walker, 1973; Faggiano et al., 1983)—some of them pre-computer-optimization era—have also inferred that conventional optimization methods can be suitably modified. The use of sets of rays based on two small decentered pupils with simultaneous control over the quality and geometry of the image due to each pupil is indeed a feasible way of optimizing biocular viewers. This method does not, however, elicit the same warm feeling and confidence as direct control over visual aberrations.

10.5 SPECIAL BIOCULAR VIEWERS

10.5.1 Panoramic Systems

Conventional bioculars can give moderately wide fields of view of around 50°, but in some situations this may be insufficient. A "continuous" biocular that would appear as an optical window of restricted height but considerable width, and through which a panoramic field of view could be seen, would be an attractive proposition. Such a system can be constructed by putting together a series of rectangularly truncated bioculars with their optical axes converging on a point midway between the observer's eyes (Fig. 10.16), each biocular being fed by a separate CRT (Tuck, 1986). To give true perspective, the CRTs need to be supplied with images from objective lenses that are angled away from or toward a common point (equivalent to the midpoint between the eyes at the display end).

A system of this type in which the horizontal fields of view of the individual bioculars have no overlap is not comfortable to use, as dead areas appear to at least one eye at certain head positions. A compromise must therefore be reached between total horizontal field of view and permissible head movement.

10.5.2 Flight Simulator Optics

The only case of a collimated biocular viewer that is being considered in this chapter is that of flight simulator optics. The simplest version of the latter involves the use of a large spherically surfaced concave mirror that displays a wide-angle collimated image to the observer, the plane through the observer's eyes being that which contains the effective aperture stop of the system. Simple equations have been given (Smith, 1966) for the angular aberrations of a spherical concave mirror due to spherical aberration, coma, and astigmatism. These equations show, for the simplest and preferred case, that where the aperture stop is at the center of curvature of the mirror, coma and astigmatism are identically zero. In fact, given an object convex to the mirror and with a radius of curvature half that of the mirror, spherical aberration is the only aberration anywhere in the field

BIOCULAR DISPLAY OPTICS

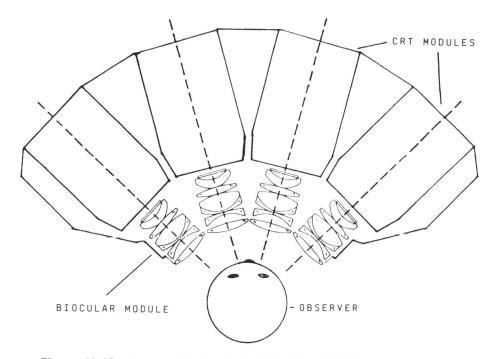

Figure 10.16 Panoramic biocular display. [After Tuck (1986).]

of view. Its value is given by Smith's equation as $0.0078/(f\#)^3$ radians where $f\#$ is the f number of the mirror focused for minimum angular aberration.

Assume that the collimating mirror is slightly decollimated (Shenker, 1972) such that eyes disposed symmetrically 32 mm either side of the optical axis have to converge by C mrad to achieve fusion. The spherical aberration equation can then be transposed to give the minimum mirror focal length F that corresponds to 2 mrad of divergence (a sensible maximum) at the limit of a lateral head movement of Δd.

$$F = 28.9 \left[\Delta d^2/(C+2) \right]^{1/3} \tag{10.8}$$

where F and Δd are in millimeters and C is in milliradians.

A graph of F versus Δd is given in Figure 10.17, which also illustrates a twin mirror (pilot and copilot) simulator setup. Note that for large values of F, Δd can be sufficiently large to allow both pilots to observe via a single mirror optic (Shenker, 1972) and that the graph is approximate for small values of F as the relationship is only of third-order accuracy.

The principal problem with the simple mirror simulator optic is the limited

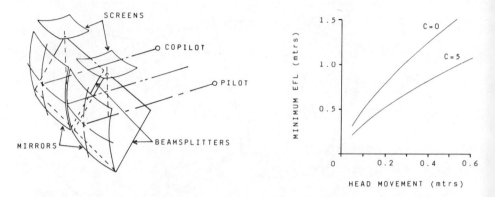

Figure 10.17 Concave spherical mirror flight simulator optics [from Shenker (1972)] and minimum focal length of mirror versus head movement for two values of axial convergence (in mrad).

vertical field of view caused by the beamsplitter (Shenker, 1972). This particular problem is removed in a novel system devised by La Russa (1966) and called a "pancake window" because of the relatively small thickness of the whole assembly. In this system the concave mirror is in a generally vertical position, and light from the CRT is reflected onto it by a vertical beamsplitter in close proximity (Fig. 10.18). A straight-through view of the CRT is eliminated by the use of (effectively) crossed linear polarizers. Transmission efficiency is rather low, typically 1.2%, but image luminance can be relatively high by comparison with a conventional system due to the proximity of the whole assembly to the observer's eyes. A more recent development of the pancake window by La Russa (La Russa and Gill, 1978) replaces the concave mirror with a plate containing a powered holographic element; basic operation is similar to the original, but the overall "pancake" is conveniently reduced in thickness.

10.5.3 Catadioptric Bioculars

The previous section made it clear that large biocular optics need not necessarily be refracting: this is equally true of smaller, high optical power biocular magnifiers. An unobscured, folded catadioptric magnifier was devised by Freeman (1972) and developed into a system that achieved a magnification of ×5.4 and a field of view of 58°.

The catadioptric magnifier was based on a beamsplitter coating that, by transmitting and reflecting the light beam in sequence, allowed the optical path to be folded within a beamsplitter prism (Fig. 10.19). The usual problem of low transmission due to a 50% loss at each intersection of the beam with the beamsplitter was overcome by the use of a polarizing beamsplitter coating that selectively

BIOCULAR DISPLAY OPTICS

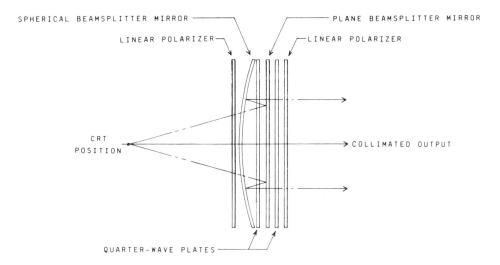

Figure 10.18 Farrand "pancake window" biocular display. [From La Russa and Gill (1978).]

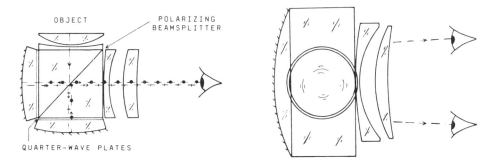

Figure 10.19 Catadioptric biocular magnifier. [After Freeman (1972).]

transmitted (P) or reflected (S) the plane-polarized light incident upon it. A double pass of a quarter-wave retarder plate was used to rotate the plane of polarization between successive intersections with the beamsplitter. The magnification was provided by a large catadioptric concave spherical mirror, which, given that the eyes were nominally situated in a plane containing the effective center of curvature of the mirror (as in the previous section), suffered only from spherical aberration and field curvature. These two aberrations were corrected by simple optics placed at the eye and object sides of the prism, respectively. The overall transmission of this biocular was optimized by the addition of another magnifying mirror and quarter-wave plate at the base of the beamsplitting prism.

10.5.4 Fresnel Bioculars

Because they are powerful optics of reasonably large diameter, biocular viewers tend to contain thick lens elements with consequences in terms of the bulk and mass of the system. The classic way to reduce the "fat" nature of a powerful lens is to make it into a thin plano-parallel plate, the requisite power being provided by a Fresnel surface; an obvious corollary to this is a considerable reduction in weight.

Several multielement Fresnel designs have been described in the literature (Cox, 1978; Hilbert, 1978). Figure 10.20 shows a biocular magnifier (Rogers, 1982) that uses Fresnel lenses to replace the acrylic lenses in the relaxed view biocular described in Section 10.4.3. A single Fresnel lens replaced the aspheric lens in the latter design, but it was found necessary to use two identical Fresnel lenses and one smaller spherically surfaced conventional lens to replace the acrylic lens nearest the object. More Fresnel lenses seem to be required to achieve a level of performance similar to that attained with the two acrylic lenses, probably because, although it can simulate the effect of an aspheric lens, a Fresnel lens is not usually given the freedom to bend. The latter has, however, been suggested (Egger, 1979). There are risk areas associated with using Fresnel lenses in this way, in particular, stray light, moiré patterns, and resolution loss due to the structured surface.

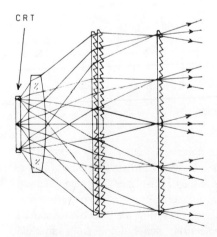

Figure 10.20 Fresnel lens relaxed view biocular. [After Rogers (1982).]

10.5.5 Stereoscopic Bioculars

Early in this chapter biocular viewers were defined as optics that have sufficient diameter to allow the image of a single object to be seen by both eyes simultaneously. How, then, can a biocular be made to provide a stereoscopic image? Answer: only by devious, even slightly dishonest, means. One such solution (Rogers, 1985b) is illustrated in Figure 10.21, where separate image intensifier tubes (or CRTs) and collimators provide right- and left-eye images—similar to conventional binocular arrangements. However, a solid prismatic block is positioned between the collimators and the eyes in order to present an apparently continuous optical window to the observer. The block contains two pairs of angled semireflective and totally reflective faces that serve to spread the exit pupils of the collimators inwards until they overlap. A negative dioptric setting is achieved by a plano-concave lens cemented onto the block.

From the nominal, preferably close-up, eye positions, the left eye sees a right-hand-side biased field of view via the left half of the biocular, and vice versa. In the biocular overlap region, a stereoscopic image is visible. Toward the edge of the field of view, the image seen by a given eye flips over to the other view; at that point there is a small jump in picture perspective. Figure 10.21 shows some of the ray paths for one eye; those for the other eye are mirror images (for centrally positioned eyes).

The drawbacks of this method are the above-mentioned perspective jump plus

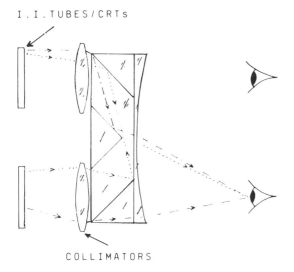

Figure 10.21 Stereoscopic biocular magnifier. [After Rogers (1985b).]

a cigar-shaped field of view due to the long path lengths in the prism block at the horizontal edges of the field of view.

References

Anderson, D. K., and T. E. Moyers. (1973). Modular, biocular eyepiece display for thermal systems, *Proceedings of Electro-Optics—Principles and Applications*, SPIE Vol. 38, Boston, MA, pp. 171–176.

Burton, G. J., and N. D. Haig. (1984). Effects of Seidel aberrations on visual target discrimination, *J. Opt. Soc. Am. 1A*: 373.

Burton, G. J., and R. Home. (1980). Binocular summation, fusion and tolerances, *Opt. Acta 27*: 809.

Campbell, F. W., and D. G. Green. (1965). Monocular versus binocular visual acuity, *Nature 208*: 191.

Charman, W. N., and H. Whitefoot. (1978). Astigmatism, accommodation, and visual instrumentation, *Appl. Opt. 17*: 3903.

Coulman, C. E., and G. R. Petrie (1949). Some notes on the designing of aspherical magnifiers for binocular vision, *J. Opt. Soc. Am. 39*: 612.

Cox, A. (1978). Application of Fresnel lenses to virtual image display, *Proceeding of Visual Simulation and Image Realism*, SPIE Vol. 162, San Diego, CA, pp. 130–137.

Damon, A., H. K. Bleibtreu, O. Elliott, and E. Giles. (1962). Predicting somatotype from body measurements, *Am. J. Phys. Anthropol. 20*: 461–473.

Dreyfuss, H. (1967). *The Measure of Man in Human Factors in Design*, Whitney Library of Design, New York.

Dudley, R. C. (1978). The distribution of eye interpupillary distance in a survey of 430 soldiers, Memorandum No. 7/78, Army Personnel Research Establishment, Farnborough, U.K.

Egger, J. R. (1979). Use of Fresnel lenses in optical systems: some advantages and disadvantages, *Proceedings of Optical Systems Engineering*, SPIE Vol. 193, San Diego, CA, pp. 63–69.

Faggiano, A., C. Gadda, and P. Moro. (1983). A lens design program, *Proceedings of Optical Systems Design, Analysis and Production*, SPIE Vol. 399, Geneva, Switzerland, pp. 136–141.

Freeman, M. H. (1972). Reflecting eyepieces, *Proceedings of Electro-Optics International*, Brighton, U.K., pp. 234–240.

Freeman, M. H. (1973). On the use of the term "biocular," *Opt. Laser Technol. 5*: 266.

Freeman, M. H. (1975a). Visual performance with biocular magnifiers, *Opt. Laser Technol. 7*: 266.

Freeman, M. H. (1975b). Ph.D. Thesis, City Univ., London, U.K.

Freeman, M. H. (1990). Optical performance of the eye, *Optics*, 10th ed., Butterworths, Keene, N.J., Chapter 15.

Giles, M. K. (1977). Aberration tolerances for visual optical systems, *J. Opt. Soc. Am. 67*: 634.

Haig, G. Y. (1972). Visual aberrations of large pupil systems, *Opt. Acta 19*: 643.

Hertzberg, H. T. E. (1954). Anthropometry of flying personnel 1950, WADC-TR-53-3210, Wright-Patterson AFB, OH.

Hilbert, R. S. (1978). U.S. Patent 4,293,196 (date is U.S. priority).
Hobbs, P. C. (1975). An anthropometric survey of 500 Royal Air Force Aircrew Heads, Tech. Rep. 73137, RAE, Farnborough, U.K.
Jeffree, J. H. (1949). Use of aspherical surfaces in the design of magnifiers for binocular vision, *Nature 164*: 1006.
Landau, F. (1990). The effect on visual recognition performance of misregistration and overlap for a biocular helmet mounted display, *Proceedings of Helmet-Mounted Displays II*, SPIE Vol. 1290, Orlando, FL, pp. 173–184.
La Russa, J. A. (1966). U.S. Patent 3,443,858 (date is U.S. priority).
La Russa, J. A., and A. T. Gill. (1978). The holographic pancake window TM, *Proceedings of Visual Simulation and Image Realism*, SPIE Vol. 162, San Diego, CA, pp. 120–129.
Mouroulis, P. (1982). On the correction of astigmatism and field curvature in telescopic systems, *Opt. Acta 29*: 1133.
Palmer, J. M., and M. H. Freeman. (1983). Biocular magnifiers for electro-optic displays—assessment of visual comfort, *Proceedings of Image Assessment—Infrared and Visible*, SPIE Vol. 467, Oxford, England, pp. 8–15.
Roberts, M., and P. J. Rogers. (1975). U.S. Patent 4,183,624 (date is U.K. priority).
Rogers, P. J. (1972a). U.S. Patent 3,604,787 (date is U.K. priority).
Rogers, P. J. (1972b). Monocular and biocular magnifiers for night vision equipment, *Proceedings of Electro-Optics International*, Brighton, U.K., pp. 37–43.
Rogers, P. J. (1982). Biocular magnifiers for use with cathode ray tube (CRT) displays, *Proceedings of Max Born Centenary Conference*, SPIE Vol. 369, Edinburgh, Scotland, pp. 90–95.
Rogers, P. J. (1985a). Biocular magnifiers—a review, *Proceedings of International Lens Design Conference*, SPIE Vol. 554, Cherry Hill, NJ, pp. 362–370.
Rogers, P. J. (1985b). U.K. Patent 2,199,154 (date is U.K. priority).
Sands, P. J. (1971). Visual aberrations of afocal systems, *Opt. Acta 18*: 627.
Shenker, M. (1972). Biocular optical systems, *Tech. Dig. OSA Annual Meeting* (summary only), San Francisco, CA.
Shenker, M. (1973). Optical systems for direct view night vision devices, *Proceedings of Image Intensifiers: Technology, Performance, Requirements and Applications*, SPIE Vol. 42, San Diego, CA, pp. 45–48.
Smith, W. J. (1966). The rapid estimation of blur sizes for simple optical systems, *Modern Optical Engineering*, McGraw-Hill, New York, Section 13.6.
Trowbridge, B. D., and K. J. Ness. (1972). U.S. Patent 3,764,194 (date is U.S. priority).
Tuck, M. J. (1986). U.S. Patent 4,772,942 (date is U.K. priority).
Van Meeteren, A., and J. J. Vos. (1972). Resolution and contrast sensitivity at low luminances, *Vision Res. 12*: 825.
Walker, B. H. (1973). Eyes in the night—LLL imaging devices. Part 2, *Photonics Spectra 7*: 36.
Walsh, G., and W. N. Charman. (1985). Measurement of the axial wavefront aberration of the human eye, *Ophthalmol. Phys. Opt. 5*: 23.
Wickholm, D. R. (1985). Merit function for biocular magnifiers, M.S. Thesis, Optical Sciences Center, Univ. Arizona.

11
Standardization of Nondiscrete Displays

Abdul Ahad S. Awwal

*Wright State University,
Dayton, Ohio*

11.1 INTRODUCTION

The main purpose of a display system is to faithfully reproduce the visual information as perceived by its image-gathering end. A display system is merely the end interface of a complete information transmission system that transfers visual information from a distant location acquired by optical, opto-electronic, or some other means to the eye of an observer. The transmittal of two-dimensional information from a distant space to the vicinity of the observer eliminates the need for an observer to be present at the scene. The quality of a display system is the measure of its ability to reproduce the original image as acquired by the acquisition end of the imaging system.

The most basic elements of an image are points (samples) of infinitesimally small spatial dimension and infinitesimally short duration (for displaying dynamic images, temporal aspects of the delta function become important). If one can reproduce a single point with infinite precision, one can reproduce any information completely. The completeness derives from the fact that the information can be completely reconstructed if the group of single elementary points that compose the image can be reproduced, a result that follows directly from Whittaker (1915) and Shannon's (1949) sampling theorem, when the sampling interval goes to zero. The theorem states that a band-limited signal/image can be recovered completely from its samples if the sampling frequency (Nyquist frequency) is more than twice the highest frequency component of the image (Gaskill, 1978). As simple as it may sound, reproduction of just a single point challenges an engineer with an unmanageable, impenetrable task. A single point is composed of all possible frequencies from zero to infinity of equal strength, and any physical system, however perfect it may be, can never possess an infinite passband; therefore, the generation of a point is an impossible achievement. High-frequency information is always lost in the transmission process, and therefore a point appears to have an extended dimension.

A physicists' (Klein and Furtak, 1986) point of view of an imaging system (Hopkins, 1981) is that a cone of light diverging from a point source of light of an aberration-free system should converge to a point. However, owing to aberration the ray does not converge to a single point. For an aberration-free system, geometrical optics predicts that a single point will be imaged into a single point. However, it turns out that geometrical optics is no longer able to describe the actual physical phenomenon of the imaging of point sources, and one has to apply the diffraction theory of light. According to this theory, the image of a single point through a finite-aperture lens will be the Fraunhofer diffraction pattern of the lens aperture. But the diffraction pattern is nothing but the Fourier transform of the aperture. We can also utilize the linear system theory to reach the same conclusion. The output is the product of the Fourier transform of the input and the transform of the lens aperture. The lens makes the output appear in the form of a Fourier transform. A system engineer (be it electrical, mechanical, or optical) has the ability to simplify the view of the world by using an input-output relationship. The mathematical relationship, simplified in the transform domain, expresses the output transform by a simple multiplication of the input transform with the system characteristics.

To characterize display systems, both physical and psychological measures and combinations thereof have been developed. Figure 11.1 shows the flow of visual information from the scene to different parts of the display system. The physical measure is concerned with what transformation takes place between the actual scene and the scene displayed on the screen, whereas the psychological measure determines how it is being perceived by the human eye. However, many of the psychological measures are empirically developed and are heavily biased by observer performance, and as a result, they are not always reliable and mathematically treatable.

11.2 LINEAR SYSTEM: IMPULSE RESPONSE

An image can be expressed as a two-dimensional spatial sum of infinitesimally small individual points. In order to construct an image from these individual points, the most important assumption is that of linearity, which allows one to apply the principle of superposition to superimpose discrete point images of the

STANDARDIZATION OF NONDISCRETE DISPLAYS

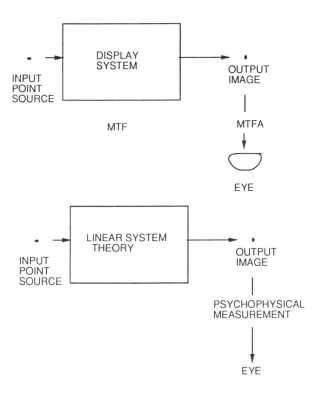

Figure 11.1 Transmission of visual information.

imaging system to construct the original image. More appropriately, if a system is linear, one can construct an image from the images of individual point responses; this is known as a point-spread function (PSF). In other words, one can apply the superposition sum to the point spread function and the input image function to estimate the output. If the PSF is independent of the space on the display on which it appears, then this space invariance can be applied to simplify the superposition sum into the convolution sum (Jong, 1982). Applying further the mathematical tools of Fourier transform, one can convert the convolution sum into multiplication of the input transform and the system transform. This is the nicety of the mathematical tools that linear system theory offers for analyzing a physical system (Gaskill, 1978).

Now we attempt to express mathematically the above description. First, we express the linearity by stating that a system is linear only if it satisfies the principle of superposition. Later, by assuming shift invariance, the convolution integral is formulated.

11.2.1 Superposition

The linearity of a system can be verified by applying the law of superposition. If two inputs $I_1(x,y,t)$ and $I_2(x,y,t)$ when applied to a general system produce outputs $y_1(x,y,t)$ and $y_2(x,y,t)$, then the system will be linear, if for two arbitrary constants a_1 and a_2 the output produced by input $[a_1 I_1(x,y,t) + a_1 I_2(x,y,t)]$ is given by $a_1 y_1(x,y,t) + a_2 y_2(x,y,t)$. Thus, using the principle of superposition, the response due to a linear combination of inputs can be computed by combining the individual responses due to each input with appropriate weights. A direct consequence of this is that, expressing the input as a sum of weighted delta functions, one can calculate the output by summing the responses of those delta functions. Any input can be expressed as an integral sum of delta functions as follows:

$$f(x) = \int_{-\infty}^{\infty} f(\alpha) \delta(x - \alpha) \, d\alpha \qquad (11.1)$$

However, for an image-recording medium, the input-output relationship is given by

$$y = c I_i^\gamma \qquad (11.2)$$

Here, I_i represents the input intensity, c is an arbitrary constant, γ is the characteristic of the recording media, and y is the output intensity. An electro-optical imaging system has a gamma (Goodman, 1968) of unity, which makes the output image intensity proportional to the input object intensity. Thus the superposition principle can be applied to such systems.

11.2.2 Shift Invariance

A system is said to be *shift-invariant* (isoplanatic, time-invariant, space-invariant) provided that when the input is shifted by some amount the output is shifted only by the same amount without a change in shape or magnitude. In other words, for a shift-invariant system, if output y_1 is produced by a linear system,

$$y_I(x,y,t) = \mathcal{O}\{I_I(x,y,t)\} \qquad (11.3)$$

where $\mathcal{O}$ is the linear system operator, then a shifted version of the input will produce a shifted version of the original output y_1, that is,

$$y_I(x - x_I, y - y_I, t - t_I) = \mathcal{O}\{I_I(x - x_I, y - y_I, t - t_I)\} \qquad (11.4)$$

where x_1, y_1, t_1 are constants relating to the amount of shift in the three axes. Thus in such a system the behavior of the system does not vary with the translation of the independent variables.

STANDARDIZATION OF NONDISCRETE DISPLAYS

11.2.3 Three-Dimensional Impulse Response

An electrical system is traditionally characterized by its response to an impulse input that is an infinite height and zero base temporal function. Likewise, optical systems are characterized by the point spread function, which is defined as the response to a two-dimensional spatial impulse input. An electro-optical imaging system, on the other hand, combines an electrical temporal signal with optical temporal signals, where two-dimensional spatial information varying in time is being displayed. To combine the spatial and temporal merit functions, we define a three-dimensional impulse function $\delta(x,y,t)$, which has infinitesimally small spatial dimension and also exists for an infinitesimally small duration. Then the response of an electro-optical system to such an input, appropriately termed the *point impulse input*, is defined as the *impulse point-spread function* (IPSF) denoted by $p(x,y,t)$. The wording was chosen so that it matches with the point-spread function (as the PSF already has two dimensions out of three, we are obliged to keep it unaltered); the word *impulse*, which usually goes with the word *response*, is placed in front to remind us of the dynamic flavor of the proposed specification. By definition, IPSF is a three-dimensional function, two dimensions of space and the dimension of time. Inherent in the wording one can visualize that the IPSF of a physical optical system describes the way it spreads the point impulse input.

For a general system, the three-dimensional impulse response will be a function of space and the time at which it is applied. For an input located at x',y',t', the output at (x,y,t) is denoted by $p(x;x',y;y',t;t')$. Thus,

$$p(x;x',y;y',t;t') = \mathcal{O}\{\delta(x - x', y - y'; t - t')\} \tag{11.5}$$

Assuming shift invariance in both space (isoplanatic) and time (time-invariant), the output due to $\delta(x - x', y - y'; t - t')$ is nothing but $p(x - x', y - y', t - t')$. Using the assumption of linearity, the output of an electro-optical display to any input is expressed by the superposition integral

$$y(x,y,t) = \int_{-\infty}^{\infty}\int_{-\infty}^{\infty}\int_{-\infty}^{\infty} I(x',y',t')p(x;x',y;y',t;t')\, dx'\, dy'\, dt' \tag{11.6}$$

Here, $I(x',y',t')$ is the input; x',y',t' are the variables of integration; and the limits of the integration are usually set by the input object or the impulse point-spread function. From now on we shall omit the limits of the convolution integrals. They are usually $-\infty$ to ∞, and for all physical systems they will be determined by the physical extent of the input or the IPSF. Applying the assumption of shift invariance,

$$p(x - x', y - y', t - t') = p(x;x',y;y',t;t') \tag{11.7}$$

Accordingly, the superposition sum expressed by Eq. (11.6) reduces to the famous convolution integral given by

$$y(x,y,t) = \iiint I(x',y',t')p(x - x'; y - y', t, - t')\, dx'\, dy'\, dt' \qquad (11.8)$$

The output of a linear shift-invariant system is given by the aforementioned three-dimensional convolution integral. The evaluation of this integral can become very cumbersome even for simple functions. From the Fourier transform properties of the convolution integral, the output image transform is the product of the input transform and the system characteristics. For a display system, then, the system characteristic also known as the optical transfer function is the complex Fourier transform (Lavin and Quick, 1974) of the impulse point-spread function,

$$H(u,v,f) = \iiint p(x,y,t)e^{-i2\pi(ux+vy+ft)}\, dx\, dy\, dt \qquad (11.9)$$

Here u and v are, respectively, the horizontal and vertical spatial frequencies, and f is the temporal frequency. The magnitude of the optical transform function $H(u,v,f)$ is known as the *modulation transfer function* (MTF). The three-dimensional MTF essentially describes the spatial frequency and temporal frequency response of the display system. If the object to be displayed is known, then knowing the three-dimensional MTF (Levi, 1970) allows one to correctly predict the displayed image. The frequency spectrum of the output is, as will be shown later for both static and dynamic imageries, the product of a 3-D MTF and the input spectrum. Note that the MTF as used in the literature refers to MTF only under static imaging conditions.

In practice, the point impulse input is a difficult thing to come by. To overcome the problem of generating a point impulse input, one uses an extremely narrow slit, which behaves like a delta function in one dimension. Thus the MTF is often defined as the modulus of the line-spread function, which is the response to a line input. Generally an assumption is made that the display exhibits a rotationally symmetric response. Consequently, the line-spread response is used to construct a radially symmetric point spread function. Therefore, the one-dimensional and two-dimensional MTFs are the same, which validates this later definition of MTF as related to line response. However, the responses of a display system may be different along different directions and at different places (Simmonds, 1981). This may happen because of the nonuniform thickness of the phosphor screen, different point-spread functions at different locations, and type of scanning of the display. For example, in a scanning type of display, a horizontal line and a vertical line are not displayed with the same precision (Hearn and Baker, 1986). Next we concentrate on the measurement of the static MTF.

11.3 MEASUREMENT OF THE STATIC MTF

11.3.1 Periodic Cosine Grating

The modulation transfer function of a display system is the magnitude frequency response to a sinusoidally varying signal input. Therefore the MTF can be estimated by measuring the magnitude of the response to the sinusoid of each frequency. If a system is linear, the output to a sinusoid is also a sinusoid but with a different magnitude and some phase shift (Gaskill, 1978).

To determine the response experimentally, sinusoids of unity amplitude but different frequencies can be displayed on the screen. This can be achieved by supplying a 1 V peak-to-peak voltage signal to the video terminal of a TV or CRT display (Verona, 1968). Then the maximum and minimum values of the displayed pattern are measured. Accordingly, the normalized amplitude of the output signal can be used directly to get the ordinates of the MTF curve corresponding to different frequencies (abscissa). Thus from Figure 11.2 one can observe that the output is

$$y(t) = \frac{L_{max} + L_{min}}{2} + \frac{L_{max} - L_{min}}{2} \cos(2\pi u x + \theta) \qquad (11.10)$$

where L_{max} and L_{min} are, respectively, the maximum and minimum luminance of the display. The MTF, m, is the normalized output amplitude at the particular frequency u,

$$m = \frac{L_{max} - L_{min}}{L_{max} + L_{min}} \qquad (11.11)$$

The luminance values L_{max} and L_{min} can be measured by a scanning aperture with a microphotometer. For a phosphor screen, the accuracy of luminance measurements is affected by several factors. The granular nature and nonuniformity of

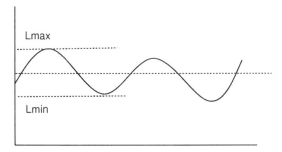

Figure 11.2 A cosine wave output.

the phosphor screen may cause spatial variations of luminous intensity, so the apparent MTF may vary at different places. Also, since a very small area is viewed in order to accurately determine the maxima and minima, little light energy is collected, and the signal-to-noise ratio is rather low. This is overcome by moving the sine wave bar pattern across the screen at a constant speed, leaving the optical detector stationary (Dore and Anstey, 1981). However, when the phosphor decay time is too large, stationary targets are preferred (Williams, 1981), so that the residue of the previous scan does not degrade the image MTF (will be shown in next section). Among other approaches, triangular rasters are preferred over sinusoids (Leunen, 1981). The veiling glare (Banbury, 1981) and dynamic noise also interfere with the measuring process, particularly in the low-frequency region of the MTF curve (K. Taylor, 1981).

11.3.2 Periodic Square Wave Pattern

Although sinusoidal patterns yield direct measurement of the MTF, they are difficult to generate and measure accurately. To overcome this difficulty, square wave patterns are generated instead, and then the square wave modulation is measured. A review of display specifications will reveal that square waves were more popular in the early days and sine wave responses have gained popularity only recently, owing to the advances in linear systems and communications theory and its interaction with optics. From the measurement point of view, when the maximum and minimum values of a sinusoid are measured, the intensity of the finite measuring aperture varies. For a square wave, the maximum and minimum values remain constant for some portion of the period, and therefore use of square waves minimizes the error in measurement.

To obtain the sinusoidal frequency response, the mathematical relationship between the square wave and the sinusoidal wave is exploited along with the assumption of linearity (Coltman, 1954). Let $r(u)$ represent the square wave response; then the sinusoidal wave response is given by

$$S(u) = \frac{\pi}{4}\left[r(u) + \frac{r(3u)}{3} - \frac{r(5u)}{5} + \frac{r(7u)}{7} + \frac{r(11u)}{11} - \frac{r(13u)}{13} + \ldots + a_k\frac{r(ku)}{k}\right] \quad (11.12)$$

where k is an odd number and a_k is given by $a_k = (-1)^{(m+1)} (k-1)/2$ for $r = m$, $a_k = 0$ if $r < m$. Here m = total number of prime factors of k, r = number of different prime factors in k.

When noise is present in the display system, it becomes difficult to estimate the image amplitude. In a noisy situation, variance can help in reducing the effect of noise. MTF measurement based on variance of the periodic rectangular image utilizes this fact to make the use of square waves more pragmatic (Droege and

Rzeszotarski, 1984). Using approaches similar to the one described in this section, it may be possible to use triangular input for estimating the display MTF (Nijhawan et al., 1975).

11.3.3 Edge Response

Note that the earlier approaches to MTF determination are based on the fact that the MTF is a response to periodic sinusoidal patterns. In this section, we utilize an aperiodic function that contains all possible frequencies for determining the MTF. A point impulse input, if Fourier transformed, will reveal that it contains all possible frequencies of equal strength. Also it follows directly from the definition of MTF that it can be measured by calculating the magnitude of the Fourier transform of the point-spread function. However, as it is difficult to generate an impulse input of three dimensions (since if exposed for an extremely short period, it will not have enough energy for a detector to detect the PSF) or, as a matter of fact, even two dimensions (excluding time), we simplify the situation by exposing the system by a line input. A line input is nothing but a one-dimensional impulse function. Sometimes the line input does not guarantee the exciting input to be a perfect almost-zero-width function, so the approach is further simplified by exposing the system to an edge input. Differentiation of an edge results in a one-dimensional delta function. Assuming the system to be linear, the differentiation may be done at the input or at the output as both lead to the same result. The derivative of the edge response is the line response, which can later be Fourier transformed to obtain the MTF. As we shall see later, the dynamic part of the dynamic modulation transfer function (Awwal et al., 1989) is also calculated from the temporal edge response.

$$\frac{d}{dx}e_i(x) = l_i(x) \tag{11.13}$$

Here $e_i(x)$ is the edge response and $l_i(x)$ is the line response, and the differentiation process is illustrated in Figure 11.3. The MTF is given by

$$\text{MTF} = |F\{l_i(x)\}| \tag{11.14}$$

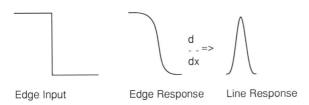

Edge Input   Edge Response   Line Response

Figure 11.3 Edge and line response.

Assuming the point-spread function to be circularly symmetric, the edge response in any one direction is sufficient to create the two-dimensional MTF. Sophisticated instruments have been developed to measure the edge response with micrometer resolution, which automatically scan the edge response and then determine the MTF from it by using software. Usually the scanned edge response data are discretely differentiated, and then fast Fourier transform (FFT) algorithms (Jong, 1982) or Winograds (Heshmaty-Manesh and Tam, 1982). FFT algorithm are are employed to obtain the MTF. However, the differentiation is usually a very noisy process, and the data may have to be smoothed first before performing the discrete differentiation. Smoothing algorithms such as three-point running average or least squares smoothing may be applied to smooth the data before processing.

11.4 STATIC IMAGING IN PHOSPHOR DISPLAYS

Consider the case when the image being displayed is a static one. This assumption is also applicable for a dynamic image with a very short time constant phosphor. The other assumption that needs to be made is that the three-dimensional point spread function is separable. The separability issues are discussed in detail by Levi (1983). The three-dimensional IPSF may be expressed, then, as a product of the static PSF, denoted by $p(x,y)$, and a dynamic impulse response symbolized as $d(t)$, given by

$$p(x,y,t) = p(x,y)\, d(t) \tag{11.15}$$

The output image transform is then given by the Fourier transform of the convolution integral between the static image input $f(x,y)$ and the IPSF $p(x,y,t)$:

$$\begin{aligned}
G(u,v) &= \int_{-\infty}^{\infty}\int_{-\infty}^{\infty}\int_{-\infty}^{\infty} f(x,y)*p(x,y,t) e^{[-i2\pi(ux+vy+ft)]} dx\, dy\, dt \\
&= \iiint\iiint f(x',y')p(x-x',y-y',t-t') \\
&\quad dx'\, dy'\, dt'\ e^{[-i2\pi(ux+vy+ft)]}\, dx\, dy\, dt \\
&= \iiint\iiint [f(x',y')p(x-x',y-y',t-t') \\
&\quad e^{[-i2\pi(ux+vy+ft)]}\, dx\, dy\, dt]\, dx'\, dy'\, dt' \\
&= \iiint\iiint [f(x',y')p(x-x',y-y')d(t-t') \\
&\quad e^{[-i2\pi(ux+vy+ft)]}\, dx\, dy\, dt]\, dx'\, dy'\, dt' \\
&= \iiint f(x,y)S(u,v)D(f)e^{[-i2\pi(ux'+vy'+ft')]}\, dx'\, dy'\, dt'
\end{aligned}$$

STANDARDIZATION OF NONDISCRETE DISPLAYS

$$= F(u,v)S(u,v)D(f) \int e^{-i(2\pi ft')} dt'$$
$$= F(u,v)S(u,v)D(f)\delta(f)$$
$$= F(u,v)S(u,v) \quad (11.16)$$

where $*$ denotes a convolution operation, and $F(u,v), S(u,v), D(f)$, and $\delta(f)$ are, respectively, the input spatial spectrum, static two-dimensional transfer function, dynamic transfer function, and Dirac delta function. The output spectrum is the product of the input spectrum and the two-dimensional optical transfer function. The physical significance of the delta function in Eq. (11.16) can be understood by noting that the output spectrum exists only at a dc temporal frequency. This implies that the output spectrum exists only as a static spectrum and no effect of the dynamic transfer function is manifested in it. Accordingly, in the absence of relative motion, the output spectrum is completely characterized by the static MTF of the imaging system. However, when the object being displayed is not static, specifying only the static MTF does not characterize the imaging completely. If a long-persistence phosphor is used in a dynamic environment, severe deterioration of the image takes place (Rash and Verona, 1987). To account for this degradation the concept of dynamic MTF (Awwal et al., 1991) is introduced in the next section.

11.5 CHARACTERISTICS OF A DYNAMIC IMAGING SYSTEM

When the object is in motion, the input function can be expressed as

$$f(x,y,t) = f(x + \mathbf{m}t, y + \mathbf{n}t) \quad (11.17)$$

where, for simplicity, we have assumed that the velocity is constant and is equal to $\mathbf{m}$ along the x direction and to $\mathbf{n}$ in the y direction. It should be pointed out that m and n are velocity vectors that are added as vectors in two orthogonal directions. For nonuniform velocity, the term $\mathbf{m}t$ should be replaced by the integral of $\mathbf{m}$ over time. Further, we assume that the light energy emitted by the moving body is not varying with time. Accordingly, the luminance intensity emitted from the object, which is a function of both space and time, is reduced to a function of space alone, where the position of the object is changing with time.

$$G(u,v,f) = \int_{-\infty}^{\infty}\int_{-\infty}^{\infty}\int_{-\infty}^{\infty} f(x,y,t)*p(x,y,t)$$
$$e^{[-i2\pi(ux+vy+ft)]} dx\, dy\, dt$$
$$= \iiint\iiint f(x',y',t')p(x-x',y-y',t-t')$$
$$dx'\, dy'\, dt'\, e^{[-i2\pi(ux+vy+ft)]} dx\, dy\, dt$$

$$
\begin{aligned}
&= \iiint\!\!\iiint [f(x',y',t')p(x-x',y-y',t-t') \\
&\quad e^{[-i2\pi(ux+vy+ft)]}\,dx\,dy\,dt]\,dx'\,dy'\,dt' \\
&= \iiint\!\!\iiint [f(x',y',t')p(x-x',y-y')d(t-t') \\
&\quad e^{[-i2\pi(ux+vy+ft)]}\,dx\,dy\,dt]\,dx'\,dy'\,dt' \\
&= \iiint f(x'+\mathbf{m}t',y'+\mathbf{n}t')S(u,v)D(f)e^{[-i2\pi(ux'+vy'+ft')]}\,dx'\,dy'\,dt' \\
&= F(u,v)S(u,v)D(f)\int \exp[-i2\pi(ft'+u\mathbf{m}t'+v\mathbf{n}t')]\,dt' \\
&= F(u,v)S(u,v)D(f)\delta(f - u\mathbf{m} - v\mathbf{n}) \\
&= F(u,v)S(u,v)D(u,v) \quad\quad\quad\quad\quad\quad\quad\quad\quad\quad (11.18)
\end{aligned}
$$

The dynamic $D(u,v)$ term becomes a function of u and v when the known values of $\mathbf{m}$ and $\mathbf{n}$ are substituted. The output spectrum is now related to the input spectrum by an overall dynamic transfer function $H(u,v,f)$, which is composed of two parts:

$$H(u,v,f) = S(u,v)D(u,v) \quad\quad\quad\quad (11.19)$$

The first part, $S(u,v)$, is the static MTF, and the second part is the dynamic part of the MTF, which has a nonunity value when there is relative motion between the object and the sensor. Equation (11.19) can be used to predict the effect of both object speed and the static MTF on the quality of the displayed image. The overall product is called the dynamic modulation transfer function (DMTF).

The measurement technique for the static MTF was discussed earlier. To evaluate the dynamic part of the DMTF, one notes that for a nonscanning type of display (Karim, 1990) such as image tubes (Csorba, 1985), if a stimulus is applied for a certain amount of time and then removed, it is equivalent to applying a negative step input. The response to the step input for a phosphor type of display is an exponentially decaying function (Torr, 1985). Performing a time derivative of which one can obtain the line response and then the Fourier transform yields the dynamic part of the DMTF. Accordingly,

$$d(t) = -\frac{d}{dt}[I_0 e - t/\tau] = \frac{I_0}{\tau}e - t/\tau \quad\quad\quad\quad (11.20)$$

where τ is the exponential decay time constant of the phosphor screen. The dynamic part of the DMTF is then

$$D(f) = -\frac{1}{1 + i2\pi f\tau} \quad\quad\quad\quad (11.21)$$

Replacing the temporal frequency f with $\mathbf{m}u + \mathbf{n}v$, the corresponding spatial frequency equivalent of the dynamic part of the DMTF can be obtained. For a nonscanning display screen, the DMTF is inversely proportional to the phosphor time constant and also the relative velocity of the moving target. Obviously, if the phosphor time constant is higher, severe image degradation will result when it is used in a dynamic environment. For some phosphors, due to its composition, the decay may be associated with several time constants (Sandel et al., 1986) and may also vary with temperature (Torr, 1985; Horn and McCutcheon, 1970). The dynamic MTF of P20 and P1052 phosphors with 10% time constants of 0.2 msec and 20 msec, respectively, are shown in Figure 11.4 (Awwal et al., 1991). The image degradation is readily apparent from this figure. However, the characteristic of the DMTF is such that the degradation MTF $[D(u,v)]$ has no zeros in the frequency plane. Therefore, one can easily restore such degradation by using an inverse filter (Cherri et al., 1989). The MTF analysis in the previous sections assumes that the display is continuous and nonscanning. In the next section, we derive the MTF relationship of scanning type displays.

11.6 THE MODULATION TRANSFER FUNCTION FOR CRT DISPLAY

A raster scan system is continuous in the horizontal direction and discrete in the vertical direction. Consequently, CRTs are not isotropic, which implies that they do not respond uniformly in all directions. However, the system satisfies the

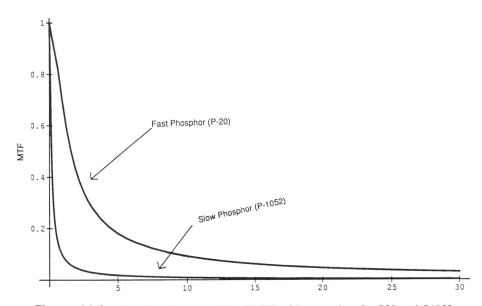

Figure 11.4 The dynamic part of the DMTF of image tubes for P20 and P1052 phosphors.

linearity or superposition principle, and hence the concept of the MTF can still be applied. The effect of anisotropic behavior is that instead of a circularly symmetric PSF, one obtains a PSF that is a function of both x and y, and thus the convolution must be done in two dimensions; but anisotropy does not imply that it is shift-variant. To specify the anisotropic behavior, both the vertical and horizontal MTFs are required to estimate the output spectrum from the input spectrum information. The scanning electron beam on the photoconductive surface acts as a scanning aperture with a transmittance distribution $a(x,y)$. The output on the display is then an average of the input luminance over the aperture and is given by

$$f_a(x,y) = \int_{-\infty}^{\infty} \int_{-\infty}^{\infty} f(\alpha,\beta) a(\alpha - x, \beta - j\Delta y) \, d\alpha \, d\beta \qquad (11.22)$$

where $(x, j\Delta y)$ corresponds to the position of the aperture, the (α,β) are the coordinates of the aperture plane, and Δy is the vertical spacing between the scan lines. The position coordinates of the pixel x and $j\Delta y$ are changing with time. If Δt is the period of a single line scan, then the integer division of total time t by Δt yields the integer j that corresponds to the vertical position of the scanning electron beam at a time t. Again if v is the scanning speed and t_h represents the time elapsed since the starting of a scan line, then vt_h represents the horizontal position of the scanning beam. Accordingly, the following equations are obtained.

$$j = \text{int}[t/\Delta t] \qquad (11.23)$$
$$t_h = t - j\Delta t \qquad (11.24)$$
$$x = vt_h \qquad (11.25)$$

The signal expressed as a function of time alone is given by

$$f(t) = \int\int f(\alpha,\beta) a\left[\alpha - vt_h, \beta - \left(\frac{t}{\Delta t}\right)\Delta y\right] d\alpha \, d\beta \qquad (11.26)$$

It is not a coincidence that the spatial TV signal is a function of time alone. The transmitted TV signal was originally a purely temporal signal. Now if the processing electronics have an impulse response given by g_t, then the output is given by $f'(t) = f(t) * g_t(t)$. The resultant intensity at any point in the displayed image is

$$f(x,y) = \sum_{j} \int_{j\Delta t}^{(j+1)\Delta t} f'(t) p(x - vt_h, y - j\Delta y) \, dt \qquad (11.27)$$

where $j\Delta t$ and $(j + 1)\Delta t$ represent the start and end times respectively, for the scan; vt_h represents the x coordinate of the position of the scanning spot at time t; and $p(x,y)$ represents the point-spread function of the image scan spot. For a CRT, the point-spread function is often approximated by a Gaussian function

STANDARDIZATION OF NONDISCRETE DISPLAYS

(Clodfelter, 1986), as illustrated in Figure 11.5. Note that the Fourier transform of the Gaussian spot is also another Gaussian (Gaskill, 1978), which is the static two-dimensional MTF of CRT displays.

However, display screen phosphors have finite memory, which makes it necessary to account for the residue of the luminance from the previous scan period. If τ is the time constant, then the intensity $I(t)$ can be expressed in terms of an exponential sum,

$$I(t) = \sum_{n=0}^{5} I_o \exp\left[\frac{-1}{\tau}(t - t_o + nT)\right] \quad (11.28)$$

Assuming that the residual luminance is negligible after five periods, only a finite number of terms, as shown in the above equation, have to be considered. Here T is the vertical frame period and t_0 is the time elapsed between the latest scan and the time of observation. The $n = 0$ term represents the effect of the most recent

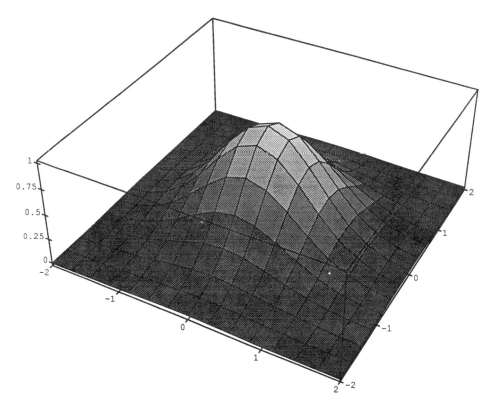

Figure 11.5 Point-spread function of CRT display.

excitation, and the other terms represent the residuals from five previous excitations (Rash and Becher, 1982).

11.7 DYNAMIC IMAGING OF CRT DISPLAY

A dynamic imaging condition exists when there is relative motion between the object and the sensor. To facilitate the MTF measurement process, the moving target can be modeled by a sinusoidal spatial variation whose phase is a function of time. Suppose the static sinusoidal signal displayed on a CRT screen has a spatial frequency u. Then the intensity function is given by

$$I(t) = \frac{I_0}{2}[1 + \sin(2\pi ux + \theta)] \sum_{n=0} I_0 \exp\left[\frac{-1}{\tau}(t - t_0 + nT)\right] \tag{11.29}$$

where x represents horizontal distance and θ represents the offset at space $x = 0$. With the introduction of a time-varying phase function, the sine wave will appear to drift on the screen. The rate of drift is proportional to the relative motion between the target and the display sensor. If ω represents the relative angular speed of the target, then the intensity is given by

$$I(t) = \frac{I_0}{2} \sum_{n=0} [1 + \sin(2\pi ux - \omega nT + 0)] \exp\left[\frac{-1}{\tau}(t - t_0 + nT)\right] \tag{11.30}$$

Note that the sinusoidal variation is also a function of time, which brings it inside the summation sign. If the object is stationary, then an object point is repeatedly mapped to the same image point. Now a dynamic target can be treated as a stationary target undergoing a periodic variation of intensity, due to the fact that the same point is not mapped onto the same point. The minimum intensity point of the dynamic sine wave pattern is obtained by setting $2\pi ux + \theta = 3\pi/2$ at time $t = t_0$, when the $n = 0$ term goes to zero. Thus,

$$I_{min} = \frac{I_0}{2} \sum_{n=1} \left[1 + \sin\left(\frac{3\pi}{2} - \omega nT\right)\right] e^{-nT/\tau} \tag{11.31}$$

$$= \frac{I_0}{2} \sum_{n=1} [1 - \cos(\omega nT)] I_0 e^{-nT/\tau} \tag{11.32}$$

The maximum intensity point of the dynamic sine wave target is obtained by setting $2\pi ux + \theta = \pi/2$ at time $t = t_0$. Accordingly, the maximum intensity is given by

$$I_{max} = I_0 + \frac{I_0}{2} \sum_{n=1} \left[1 + \sin\left(\frac{\pi}{2} - \omega NT\right)\right] e^{-nT/\tau} \tag{11.33}$$

STANDARDIZATION OF NONDISCRETE DISPLAYS

$$I_{max} = I_0 + \frac{I_0}{2}\sum_{n=1}[1 + \cos(\omega NT)]I_0 e^{-nT/\tau} \quad (11.34)$$

A pixel with spatial frequency u moving across the screen at a speed **m** is driven at a temporal frequency $f_t = \mathbf{m}u$. Therefore, $\omega = 2\pi f_t = 2\pi mu$.

Now, using Eq. (11.11), the MTF is obtained to give

$$D = \frac{1 + \sum_{n=1}\cos(\omega nT)e^{[-nT/\tau]}}{1 + \sum_{n=1}e' - nT/\tau} \quad (11.35)$$

D represents the dynamic MTF, which is expressed as a function of velocity of the object. In the above question $\omega = 0$ corresponds to a static target. When plotted for different velocities ωT, this equation represents the deterioration of the dynamic MTF with different angular velocities. Interestingly, when $\omega T = 2\pi$, the image appears as a static one. Usually, considerable degradation effects are observed for $\omega T < \pi$ as seen from Figure 11.6. It can be noted that ωT represents the ratio of the angular velocity of the object to the frame frequency. The physical significance of $\omega T = 2\pi$ is that when the new shifted image is formed, the new frame scanning starts at exactly the same time. One way to counter the effect of velocity could be to change the frame rate such that the image appears as a static one on the screen and does not exhibit any smearing.

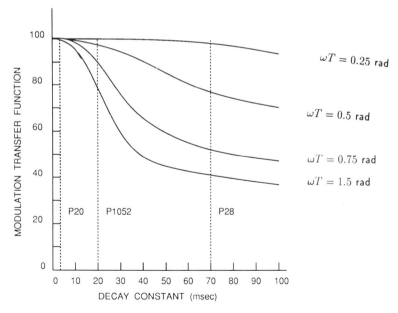

Figure 11.6 Dynamic MTF of CRT display.

11.8 MTF DEGRADATION DUE TO VIBRATION

As shown earlier, the dynamic part of the MTF in the absence of any relative motion is unity. However, when the input is under vibration, a $\delta(x)$ input creates a new point-spread function input as seen by the system. On the other hand, when the display on which the image is being displayed is in constant oscillatory motion, the situation is similar to that of the previous case because the relative motion involved is the same in both cases. Let the input $\delta(x)$ of amplitude a be vibrating at a frequency f. Since the vibration can be modeled as a harmonic motion, the displacement of the input can be expressed as

$$x = a \sin(2\pi f t) \tag{11.36}$$

where t represents time. The linear velocity of the input is given by (Shi-ming, 1982)

$$\dot{x} = 2a\pi f \cos(2\pi f t) \tag{11.37}$$

Let $\theta = 2\pi f t$ represent the angular displacement with time t; θ varies between 0 and π. Thus the angular velocity differs at different points of the travel. Therefore the length of time the input remains at a space δx will also be a variable. If the luminance of the light L_0 falling on a space Δx remains constant for a time Δt and then repeats with a frequency f, then the average light energy collected is given by

$$\overline{L(x)} = L_0 f \, \Delta t \tag{11.38}$$

But the time Δt is related to the instantaneous velocity by the following relation:

$$\Delta t = \frac{\Delta x}{\dot{x}} = \frac{\Delta x}{2a\pi f \cos \theta} \tag{11.39}$$

Then the average luminance is given by

$$\overline{L(x)} = L_0 f \frac{\Delta x}{2a\pi f \cos \theta} = \frac{1}{a\pi \cos \theta} = \frac{1}{\pi a \sqrt{1 - (x/a)^2}} \tag{11.40}$$

where $L_0 \Delta x/2$ is assumed to be unity for the sake of normalization. Since the dynamic MTF is $\delta(t)$ in the absence of motion, $\overline{L(x)}$ is the line-spread function for the vibrating image itself. For an aerial photography system, $\overline{L(x)}$ is the point-spread function of the film for an image under vibration. The degradation MTF due to the relative vibratory motion is then the Fourier transform of this function, which is

$$M_r(f) = J_0(2a\pi u) \tag{11.41}$$

Figure 11.7 illustrates the effect of amplitude of vibration a on MTF deterioration. Now, owing to the finite exponential decay characteristics, there will be an

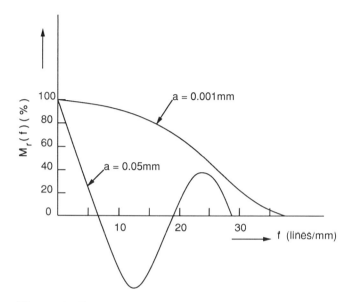

Figure 11.7 $M_r(f)$ due to vibration.

additional dynamic effect from the residue of the previous exposure, as in the case of scanning type displays. Given that the input at any point $x = x_1$ is $L(x_1)$, if T is the period of vibration, then in each period the same point is illuminated twice. The signal passes through the same point after at time $2(T + t)$, where t represents the time from the origin. The luminance during the first period is given by

$$I_m = L(x_1)\left\{1 + \exp\left[-\frac{2}{\tau}\left(\frac{T}{4}+t\right)\right]\right\} \tag{11.42}$$

Then the residue signal that was left one time period ago is $I_m \exp(-t/\tau)$. Summing all residues, one gets the overall luminance:

$$\begin{aligned}
I(x,t) &= I_m + I_{m-1} + I_{m-2} + I_{m-3} + \cdots \\
&= I_m[1 + e^{-T/\tau} + e^{-2T/\tau} + e^{-3T/\tau} + \cdots] \\
&= I_m \sum_{n=0}^{\infty} e^{-nT/\tau} \\
&= \left\{1 + \exp\left[-\frac{2}{\tau}\left(\frac{T}{4}+t\right)\right]\right\} \sum_{n=0}^{\infty} e^{-nT/\tau} \\
&= [1 + be^{-2t/\tau}]c
\end{aligned} \tag{11.43}$$

where

$$b = e^{-T/2\tau} \tag{11.44}$$

and

$$c = \sum_{n=0}^{\infty} e^{-nT/\tau} \tag{11.45}$$

Here, $I(x,t)$ is similar to a negative step response. Therefore, the dynamic part of the DMTF due to phosphor persistence is

$$h_v(r) = -\frac{d}{dt}I(t) = \frac{2\alpha bc}{\tau}e^{-2t/\tau} = Ae^{-2t/\tau} \tag{11.46}$$

where

$$A = 2\alpha bc/\tau \tag{11.47}$$

is a constant. Equation (11.46) has the same form as the dynamic impulse response for nonscanning displays. Now taking its Fourier transform, one obtains the normalized MTF as a function of temporal frequency:

$$M_v(t) = \frac{1}{1 + j\pi f_t \tau} \tag{11.48}$$

For harmonic vibration of the image point, the temporal frequency f_t and the spatial frequency u are related by

$$f_t = u\dot{x} = 2\pi ufa \cos\theta \tag{11.49}$$

Then the MTF is given by

$$M_v(f,\theta) = \frac{1}{[1 + (\pi f_t \tau)^2]^{1/2}} = \frac{1}{[1 + (2\pi^2 fau\tau \cos\theta)^2]^{1/2}}$$
$$= \frac{1}{P(1 - Q^2 \sin^2\theta)^{1/2}} \tag{11.50}$$

where

$$P = [1 + (2\pi^2 \tau afu)^2]^{1/2} \tag{11.51}$$

and

$$Q = \frac{2\pi^2 \tau afu}{[1 + (2\pi^2 \tau afu)^2]^{1/2}} < 1 \tag{11.52}$$

Equation (11.50) expresses the MTF due to vibration as a function of the instantaneous angular displacement θ. A weighting average over 0 to π yields M_v as a function of frequency of vibration alone given by

STANDARDIZATION OF NONDISCRETE DISPLAYS

$$M_v(f) = \frac{2}{\pi} \int_0^{\pi/2} M_v(f,\theta) \, d\theta$$

$$= \frac{2}{P\pi} \int_0^{\pi/2} \frac{d\theta}{[1 - Q^2 \sin^2\theta]^{1/2}} = \frac{2}{\pi P} I_e \quad (11.53)$$

where I_e represents elliptic integration, which can be expressed as a series. If $Q \ll 1$, then

$$M_v(f) = \frac{1}{[1 + (2\pi^2 \tau a f u)^2]^{1/2}} \quad (11.54)$$

The overall MTF of the imaging display is given by the product of the three MTFs: the static MTF, the radial degradation function, and the dynamic MTF due to vibration. The overall MTF, therefore, is found to be

$$H_v(u) = H_s(u) J_0(2\pi a u) \frac{1}{[1 + (2\pi^2 \tau a f u)^2]^{1/2}} \quad (11.55)$$

Note from the above equation that the spectrum is modified only along the u direction, which is true only for motion along the x direction. If the vibration is present in any other direction, then in the two-dimensional expression of the MTF, u should be replaced by a combination of velocity vectors in both the u and v directions. Figures 11.8 and 11.9 show the effect of the vibration MTF and the

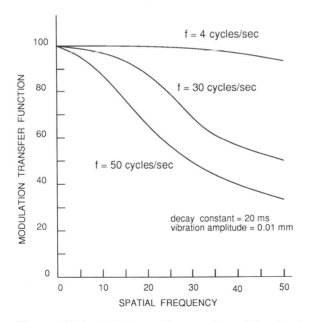

Figure 11.8 MTF due to vibration: effect of phosphor decay.

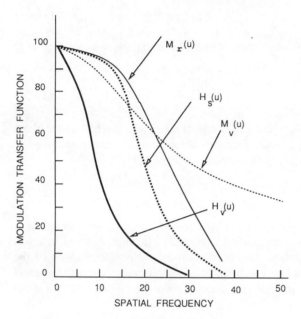

Figure 11.9 Overall MTF due to vibration.

overall MTF. If the imaging system is not an electro-optic one such as a photographic film, then the effect of vibration is a wider point-spread function. In that case, one of the approaches for recovery of the undegraded image concerns the number of photographic shots that must be taken in order to capture at least one image when the velocity of vibration is zero, that is, at the end point (Wulich and Kopeika, 1987). For low-frequency sinusoidal vibrations, experiments have been performed recently to verify the theory of image recovery (Rudoler et al., 1991).

The mathematical and analytical tools provided by the MTF simplify the input-output relationship. One question that still remains unanswered is, How good is the MTF measurement when it comes to be judged by a human observer? Ultimately, the human eye has to gather the final information. A visual display may have excellent high-frequency response, but human perception is limited to a certain maximum frequency. Therefore, ultimately, the information pertaining to the eye response must be incorporated into the display response characteristics to make it a meaningful quality criterion.

11.9 MODULATION TRANSFER FUNCTION AREA

Charman and Olin (1965) coupled the purely observer-independent physical parameter MTF with the observer response commonly known as the contrast threshold function. For a viewer, a minimum level of modulation is required before one is able to detect the presence of a spatially varying sinusoidal pattern. The level of contrast (modulation) required for an observer to detect the presence of a sine wave pattern with a 50% probability level as a function of spatial frequency is known as the *contrast threshold function* (CTF). The MTFA is the integrated difference between the system's ability to transmit modulation and the observer's contrast threshold function (Mouroulis, 1981) for the spatial frequencies zero to the point beyond which the observer's modulation threshold exceeds the system capability as shown in Figure 11.10. Mathematically,

$$\text{MTFA} = \int_0^{u_0} [S(u) - T(u)] \, du \qquad (11.56)$$

where $S(u)$ is a static MTF factor at spatial frequency u, $T(u)$ is the 0.5 probability threshold modulation at frequency u for the system display, and u_0 is the limiting resolution. The MTFA (Keesee, 1976) includes the effect of sensor and display bandwidth, linewidth, signal-to-noise ratio, gamma, contrast, mean luminance, edge enhancement, and spatial filtering by the eye. To verify the usefulness (Snyder, 1973) of MTFA measurement (Borough et al., 1967), Snyder (1974) con-

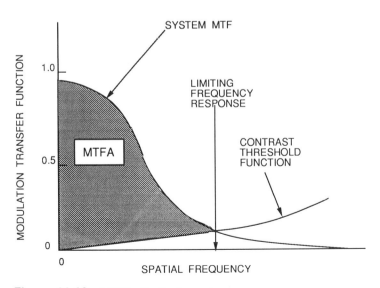

Figure 11.10 MTFA of a display system.

ducted experiments to determine the correlation between correct target recognition and the MTFA. A correlation value of 0.965 was determined between the MTFA and correctly identified targets. One disadvantage of the MTFA, however, is the empirical procedure to determine it for every individual system, which may well vary from observer to observer.

The MTF value below the corresponding CTF value is not usable to the observer. Thus, the MTFA is a measure of usable contrast above the threshold value. Once the amplitude of the MTF has exceeded the threshold level, the extended portion of the MTF well above the CTF does not provide any additional information for the observer. However, Eq. (11.56) still predicts a higher MTFA value. The MTF just above the CTF detection level is more important than those much above it. Task and Verona (1976) introduced a log transformation to weigh the MTFs above and near the CTF more heavily than those farther above the CTF curve. The transformation converts the modulation (m) axis into a "shades of gray (G)" or "just noticeable difference" axis by using the formula

$$G = 1 + \frac{\log[(1 + m)/(1 - m)]}{\log 1.414} \tag{11.57}$$

Thus this formula proides a compression of the higher levels and an expansion of lower levels, a technique usually adopted for communications systems to reduce quantization error at lower amplitude levels and compress the variation at higher levels.

11.10 MODELING AN ELECTRO-OPTICAL IMAGING SYSTEM

One of the attractions of the MTF is its use in linear systems modeling. When a number of linear systems are used in cascade form, then the overall MTF of the complete system is obtained by multiplying the individual MTFs (Stark et al., 1969). If all the components that comprise a display system are cascaded as shown in Figure 11.11, such as the input sensor, display device, and its internal electronics having transfer functions H_1, H_2, H_3, respectively, then the overall transfer function is given by

$$H = H_1 H_2 H_3 \tag{11.58}$$

Before the complete model of the imaging system is derived (Datta et al., 1984), the transfer function for each of the sections must be evaluated (Clarke and Yeadon, 1973). The output luminance of the image under certain conditions must be proportional to the input current (Guest, 1971). Moreover, the isoplanatic (shift-invariant) and time-invariant conditions must also be satisfied. Then one can predict the displayed image quality by means of digital image processing techniques (Gonzalez and Wintz, 1987).

STANDARDIZATION OF NONDISCRETE DISPLAYS

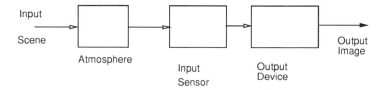

Figure 11.11 Composite MTF of an electro-optical imaging system.

In the systems modeling approach, small nonlinearities may be linearized or a discrete simulation model of the whole visual system may be developed. For large nonlinearity, spatial domain processing can be applied after the nonlinearity has been modeled. Each part of the visual communication system (Patterson, 1986) may also have its own noise sources that can be modeled by the Monte Carlo simulation technique (Cheney and Kincaid, 1985). In addition to modeling the input scene (D. J. Taylor, 1981), the atmosphere must also be modeled. All of these can then be incorporated into a huge computer model to simulate a real electro-optic display system. The output screen of the night vision goggles also has an additional thermal noise source (Karim, 1990) that also can be estimated or simulated. This model can then be used in testing manufactured devices. For example, by changing one design parameter at a time while keeping all others constant, the device performance can be optimized with respect to a particular parameter value. Next, another parameter can be varied to attain optimized performance; the end result will be an optimization with respect to all variables that are controllable. Using the model one can then compare the field performance with the simulated performance and further refine the model or the device.

11.11 PIXEL ERROR

For digital image processing, it is very convenient to compare the pixel gray levels with the original gray-level images. The average standard deviation of the pixel intensity levels from the intensity levels of the original image across several pixels is taken as the pixel error measure. Normalized mean square error, point squared error, perceptual MSE, image fidelity, structural content, and correlation quality are examples of such quality criteria (Snyder, 1985). However, these error concepts are simply discrete representations that are related to MTF-based metrics (Task, 1979).

11.12 SUMMARY

The characterization of nondiscrete displays discussed in this chapter mainly focuses on the displays in both static and dynamic modes of operation. It may be noted that there are more characteristics that are derivatives of the MTF. Snyder (1985) summarizes many such parameters related to the MTF specification. However, I have concentrated on only MTF-based specification, recognizing its tremendous advantage in systems modeling. In Chapter 12, using the formulas developed here, image degradation simulation results are observed, and then some restoration schemes with simulation results and possible future image recovery architectures are proposed. At the end of this chapter I list some special SPIE publications on display systems that contain more articles on display characterization issues that may be of interest to the reader.

References

Awwal, A. A. S., A. K. Cherri, M. A. Karim, and D. L. Moon (1989). Dynamic response of an electro-optical imaging system, *Helmet Mounted Displays*, SPIE *1116*, pp. 185–196.

Awwal, A. A. S., A. K. Cherri, M. A. Karim, and D. L. Moon (1991). Dynamic modulation transfer function of a display system, *Appl. Opt. 30*: 201.

Banbury, J. R. (1981). Evaluation of modulation transfer function (MTF) and veiling glare characteristics for cathode ray tube displays, in *Assessment of Imaging Systems: Visible and Infrared* SPIE Proc. 274, pp. 130–138.

Borough, H. C., R. F. Fallis, R. H. Warnock, and J. H. Pritt (1967). Qualitative determination of image quality, Boeing Company Tech. Rep. D2-114058-1.

Charman, W. N., and A. Olin (1965). Tutorial: image quality criteria for aerial camera systems, *Photogr. Sci. Eng. 9*: 385.

Cheney, W., and D. Kincaid (1985). *Numerical Mathematics and Computing*, Brooks/Cole, Monterey, CA.

Cherri, A. K., A. A. S. Awwal, M. A. Karim, and D. L. Moon (1989). Restoration of motion degraded images in electro-optical displays, *Proc. Soc. Photo-Opt. Instrum. Eng. 1116*: 198.

Clarke, J. A., and E. C. Yeadon (1973). Measurement of the modulation transfer function of channel image intensifiers, *Acta Electron. 16*: 33.

Clodfelter, R. M. (1986). Modulation transfer function for the display engineer, *Advances in Display Technology VI*, SPIE 624, pp. 113–118.

Coltman, J. W. (1954). The specification of imaging properties by response to a sine wave input, *J. Opt. Soc. Am. 44*: 468.

Csorba, I. P. (1985). *Image Tubes*, Howard W. Sams, Indianapolis, IN.

Datta, P. K., P. D. Gupta, O. P. Nijhawan, and R. Hradaynath (1984). Influence of temperature on the MTF of cascaded image-intensifier tubes, *Appl. Opt. 23*: 1967.

Dore, M. J., and G. Anstey (1981). Automatic measurement of cathode ray tube modulation transfer functions (MTFs), *Assessment of Imaging Systems: Visible and Infrared* SPIE Proc. 274, pp.139–145.

Droege, R. T., and M. S. Rzeszotarski (1984). Modulation transfer function from the variance of cyclic bar images, *Opt. Eng. 23*: 68.

Gaskill, J. D. (1978). *Linear Systems, Fourier Transform and Optics*, Wiley, New York.

Gonzalez, R. C., and P. Wintz (1987). *Digital Image Processing*, 2nd ed., Addison-Wesley, Reading, MA.
Goodman, J. W. (1968). *Introduction to Fourier Optics,* McGraw-Hill, NY.
Guest, A. J. (1971). A computer model for channel multiplier plate performance, *Acta Electron. 14*: 79.
Heshmaty-Manesh, D., and S. C. Tam (1982). Optical transfer functions calculations by Winograd's fast Fourier transform, *Appl. Opt. 21*: 3273.
Hearn, D., and M. P. Baker (1986). *Computer Graphics*, Prentice-Hall, Englewood Cliffs, NJ.
Hopkins, H. H. (1981). Modern methods of image assessment, in *Assessment of Imaging Systems: Visible and Infrared* SPIE Proc. 274, pp. 2–11.
Horn, J. E., and M. J. McCutcheon (1970). Decay time of some image tube phosphors as a function of excitation time, *Proc. IEEE 58*: 592.
Jong, M. T. (1982). *Methods of Discrete Signal and System Analysis*, McGraw-Hill, New York.
Karim, M. A. (1990). *Electro-Optical Devices and Systems*, PWS-Kent, Boston, MA.
Keesee, R. L. (1976). Prediction of modulation detectability thresholds for line-scan displays, Tech. Rep., Aerospace Medical Res. Lab., AMRL-TR-76-38.
Klein, M. V., and T. E. Furtak (1986). *Optics*, 2nd ed., Wiley, New York.
Lavin, H., and M. Quick (1974). The OTF in electro-optical imaging systems, *Image Assessment Specification*, SPIE Proc. 46, Rochester, NY, pp. 279–286.
Leunen, H. A. J. V. (1981). An instrument for the measurement of the modulation transfer factor of image transfer devices, *Assessment of Imaging Systems: Visible and Infrared*, SPIE 274, pp. 99–110.
Levi, L. (1970). On combined spatial and temporal characteristics of optical systems, *Appl. Opt. 17*: 869.
Levi, L. (1983). Spatiotemporal transfer function: recent developments, *Appl. Opt. 22*: 4038.
Mouroulis, P. (1981). Contrast sensitivity in the assessment of visual instruments, *Assessment of Imaging systems: Visible and Infrared*, SPIE 274, pp. 202–210.
Nijhawan, O. P., P. K. Datta, and J. Bhusan (1975). On the measurement of MTF using periodic patterns of rectangular and triangular waveforms, *Nouv. Rev. Opt. 6*: 33.
Patterson, M. L. (1986). An analysis of visual communication based on electrical signal theory, *Proc. SID 27*: 289.
Rash, C. E., and J. Becher (1982). Analysis of image smear in CRT displays due to scan rate and phosphor persistence, U.S. Army Aeromed. Res. Lab., USAARL Rep. 83-5, Fort Rucker, AL.
Rash, C. E., and R. W. Verona (1987). Temporal aspects of electro-optical imaging systems, *Proc. SPIE 765*: 22–25.
Rudoler, S., O. Hadar, M. Fisher, and N. S. Kopeika (1991). Image resolution limits resulting from mechanical vibrations. Part 2: Experiment, *Opt. Eng. 30*: 577.
Sandel, B. R., D. F. Collins, and A. L. Broadfoot (1986). Effect of phosphor persistence on photometry with image intensifiers and integrating readout devices, *Appl. Opt. 24*: 3697.
Shannon, C. E. (1949). Communication in the presence of noise, *Proc. IRE 37*: 10.

Shi-ming, X. (1982). MTF deterioration by image motion in electro-optical imaging system, *Proceedings of Electro-Optics/Laser International UK Conference*, March 1982, Brighton, England, pp. 62–73.

Simmonds, R. M. (1981). Two dimensional modulation transfer function of image scanning system, *Appl. Opt. 20*: 619.

Snyder, H. L. (1973). Image quality and observer performance, in *Perception of Displayed Information* (L. M. Biberman, Ed.), Plenum, New York.

Snyder, H. L. (1974). Image quality and face recognition on a television display, *Human Factors 16*: 300.

Snyder, H. L. (1985). Image quality: measures and visual performance, in *Flat Panel Displays and CRTs* (L. E. Tannas, Jr., Ed.), Van Nostrand Reinhold, New York, pp. 70–90.

Stark, A. M., D. L. Lamport, and A. W. Woodhead (1969). Calculation of the modulation transfer function of an image tube, *Advances in Electronics and Electron Physics 28* (part B): 567.

Task, H. L. (1979). An evaluation and comparison of several measures of image quality for television displays, Air Force Aerospace Med. Res. Lab., Tech. Rep. AMRL-TR 79-7, 1979.

Task, H. L., and R. W. Verona (1976). A new measure of television display quality relatable to observer performance, Air Force Aerospace Med. Red. Lab., Tech. Rep. AMRL-TR 76-73, 1976.

Taylor, D. J. (1981). Systems engineering approach to imaging systems assessment, *Assessment of Imaging System: Visible and Infrared*, SPIE 274, p. 264.

Taylor, K. (1981). Measurement of veiling glare in 2nd generation image intensifiers, *Assessment of Imaging System: Visible and Infrared*, SPIE 274, p. 315.

Torr, M. R. (1985). Persistence of phosphor glow in microchannel plate image intensifiers, *Appl. Opt. 24*: 793.

Verona, R. W. (1968). Sine wave response of TV displays, MS Thesis, Electrical Engineering, Univ. Dayton.

Whittaker, E. T. (1915). On the functions which are represented by the expansions of interpolation theory, *Proc. Roy. Soc. Edinburgh, Sect. A 35*: 181.

Williams, T. L. (1981). Modulation transfer function system for image intensifier units, *Assessment of Imaging System: Visible and Infrared* SPIE 274, 148.

Wulich, D., and N. S. Kopcika (1987). Image resolution limits resulting from mechanical vibrations, *Opt. Eng. 26*: 529.

Recent Related SPIE Proceedings

Wilson and Balk Eds. (1987). *Scanning Imaging Technology*, SPIE Proc. 809.

Cox and Hartman, Eds. (1987). *Display System Optics*, SPIE 778.

Freeman, Ed. (1987). *Imaging Sensors and Displays*, SPIE 765.

Granger and Baker, Eds. (1985). *Image Quality—An Overview*, SPIE 549.

Schlam, Ed. (1986). *Advances in Display Technology VI*, SPIE 624.

Wight, Ed. (1987). *Electro-Optical Imaging Systems Integration*, Crit. Rev. SPIE 762.

Wilson, Ed. (1988). *Scanning Imaging*, SPIE 1028.

12
Restoration of Dynamically Degraded Images in Displays

Abdul Ahad S. Awwal

*Wright State University,
Dayton, Ohio*

12.1 INTRODUCTION

Imaging faithfully in a dynamic environment is an important task for an electro-optical imaging system. An environment is referred to as dynamic when there exists a finite relative motion between the object and the image-gathering system. Computer vision, medical imaging of live organisms through electro-optic displays (Ohyama *et al.*, 1987), remote sensing from a space probe orbiting a planet (King *et al.*, 1988), and reconnaissance from an aircraft or real-time pilotage using night imaging optics are illustrations of dynamic imaging situations. In dynamic environments, the image degradation becomes severe enough that it may be necessary to take measures to restore the original undegraded image. In extreme cases, the targeted object may lose its contrast with the background, and as a result the observer may fail to discriminate the targets from their background, or overlap of several frames may blur the image so much that it may become impossible to identify the object and its true location. As the degree of dynamicity increases, the degradation effect may become a hundred times more drastic than that which can be explained by the modulation transfer function (MTF) alone.

Proper knowledge of the mechanism of degradation is essential to carry out effective image restoration (Bates and McDonnel, 1986). In Chapter 11, on characterization of displays, we investigated several such models of dynamic degradation, in particular, degradation due to linear and vibratory motion in nonscanning displays an linear motion in scanning type displays. It has already been pointed out that phosphor persistence affects the temporal response of a CRT display (Rash and Becher, 1982; Verona et al., 1979). In a dynamic environment, a more persistent phosphor is able to drastically reduce the modulation contrast and eventually cause the loss of gray levels as shown in Chapter 11. In particular, for a phosphor-based display system, the dynamic modulation transfer function (DMTF) has been shown to be a product of the classical MTF and a dynamic function dependent on phosphor persistence and object motion.

In this chapter, the smearing of the displayed image is demonstrated by means of simulation by considering the appropriate temporal response for the imaging condition under analysis. It is shown that the phosphor type as well as the magnitude and direction of the relative motion between the target and the sensor play significant roles in image degradation. Following the identification of the proper degradation function, real-time techniques are identified for the restoration of motion-degraded images.

12.2 DMTF REVIEW

Assume that an input object $f(x,y,t)$ is operated on by the impulse point-spread function (IPSF) $p(x,y,t)$ of an imaging system to produce an output image $g(x,y,t)$. Mathematical simplification of the output-input relationship occurs when the IPSF $p(x,y,t)$ is separable, which can be obtained from the product of a purely spatial function $p(x,y)$ and a purely temporal function $d(t)$. Accordingly, when

$$p(x,y,t) = p(x,y)d(t) \qquad (12.1)$$

one may evaluate the Fourier transform of $g(x,y,t)$ as

$$G(u,v,f) = D(f)S(u,v)F(u,v)\delta(f - u\mathbf{m} - v\mathbf{n}) \qquad (12.2)$$

where m and n denote velocity components in the x and y directions, respectively; u and v represent the corresponding spatial frequencies; f is the temporal frequency; $D(f)$ is the dynamic transfer function; $S(u,v)$ is the static transfer function; and $F(u,v)$ is the spectrum of the static image. The presence of the factor $\delta(f - u\mathbf{m} - v\mathbf{n})$ implies that during the evaluation of $G(u,v,f)$, the variable f in $D(f)$ must be replaced with $u\mathbf{m} + v\mathbf{n}$. The overall system block diagram of the electro-optical display is shown in Figure 12.1, which includes both static and dynamic transfer functions. The product $D(f)S(u,v)\delta(f - u\mathbf{m} - v\mathbf{n})$ is referred to as the DMTF. This new form for the spatiotemporal transfer function of the imaging system can be used effectively to characterize the motion-degraded images. It is important to note that Eq. (12.2) is correct only as long as the separability condition of Eq. (12.1) is valid. Fortunately, this separability condition remains valid as long as phosphor-based electro-optical display systems are con-

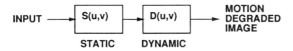

Figure 12.1 Image degradation model of an electro-optical display.

sidered (Levi, 1983; Awwal et al., 1991). Also for imaging using a time-varying blur function (such as imaging through random media), the spatial and temporal variations are separable in space and time (Ward and Saleh, 1987). Note that this type of spatiotemporal impulse response has a lot of similarity with our imaging situation. When the function is nonseparable, then numerical techniques can be used to evaluate the convolution integral.

In the case of linear motion of a scanning type of display such as cathode-ray tube (CRT), the above equation still applies; the only difference is in the form of $D(f)$. The dynamic transfer function for a CRT type of screen is given by

$$D = \frac{1 + \Sigma_{n=1} \cos(\omega nT) e^{-nT/\tau}}{1 + \Sigma_{n=1} e^{-nT/\tau}} \tag{12.3}$$

which is expressed as a function of angular frequency ω, which in turn is a function of spatial frequency u of the object, vertical frame rate T, and time constant τ of the phosphor. The terms in the summation are taken as long as the magnitude of the terms contribute significantly to the sum. For image under vibratory motion, the overall spectrum is given by

$$G(u,v,f) = M(f)J_0(2\pi au)S(u,v)\delta(f - u\mathbf{m} - v\mathbf{n}) \tag{12.4}$$

where $M(f)$ is the dynamic transfer function corresponding to vibration resulting from phosphor decay, and $J_0(.)$ is the zeroth-order Bessel function, which is the Fourier transform of the vibration impulse response.

Next, we investigate the particular form of the dynamic component of DMTF. For a phosphor-based display, the impulse response can be estimated from the knowledge of its decay characteristics. When a phosphor screen is excited from a certain period of time and then the excitation is removed, it is equivalent to applying a negative step input. Thus the corresponding decay is nothing but that due to a negative step response, which upon differentiation yields the impulse response. In general, phosphor emission is an incoherent phenomenon; hence the decay is associated with more than one decay constant. Rather, the exponential decay characteristics involve a broad spectrum of emission, governed by several decay constants. One can approximate the phosphor decay as (Sandel et al., 1986)

$$I(t) = a_1 e^{-t/\tau_1} + a_2 e^{-t/\tau_2} + a_3 e^{-t/\tau_3} \tag{12.5}$$

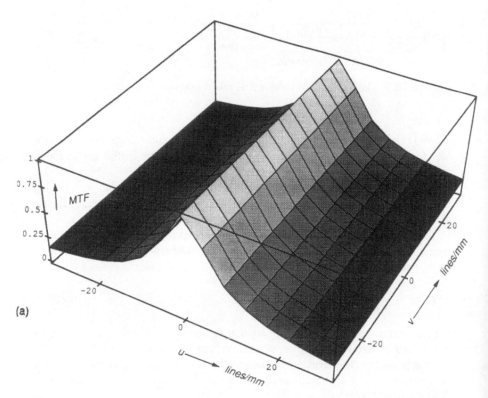

Figure 12.2 A 3-D representation of the 2-D dynamic component of DMTF for phosphor display. (a) **m** = 4cm/sec; (b) **m** = 2cm/sec, **n** = 3 cm/sec. (τ = 8.68 msec in both cases.)

where $I(t)$ is the luminance of the phosphor screen, a_i is a factor representing the proportion of the phosphor in the ith state, and τ_i is the corresponding time constant defined as the length of time during which the phosphorescence decays to $1/e$ of its maximum value a_i. In general, the last two terms of Eq. (12.5) represent fluorescence modes of a phosphor light output. When compared to the first term, which represents the phosphorescence mode of light output, the contributions of the last two terms are insignificant (Sandel *et al.*, 1986). Accordingly, differentiation of Eq. (12.5) can be approximated as

$$|d(t)| = a_I/\tau \exp(-t/\tau) \qquad (12.6)$$

where, for simplicity, the constant term can be set to unity. Taking the Fourier transform of Eq. (12.6), the dynamic component of the DMTF for the phosphor screen can be obtained as

RESTORATION OF DYNAMICALLY DEGRADED IMAGES

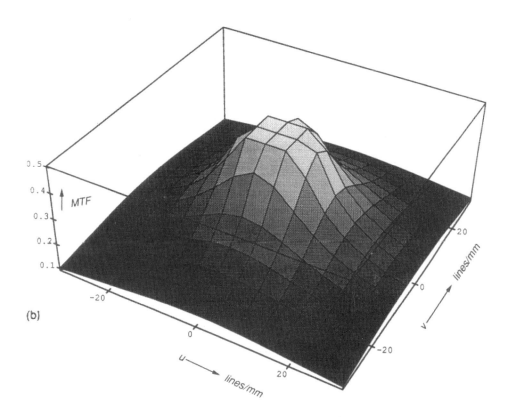

(b)

$$D(f) = \frac{1}{1 + j2f\pi\tau} \qquad (12.7)$$

Equation (12.7), representing the temporal transfer function of a phosphor screen, can be introduced into Eq. (12.2) to predict the overall degradation of motion involved in the displayed images. However, $D(f)$ is a function of the time constant of the phosphor used in the electro-optic display under consideration. For instance, the 10% time constant of P1052 phosphor is 20 msec (Csorba, 1986). Accordingly, we may set $\tau = 8.68$ msec for the exponential decay constant of type P1052 phosphor. In order to predict the effect of the temporal response with Eq. (12.7), however, one needs to convert the temporal information into a spatial format. This is accomplished simply by replacing f with $u\mathbf{m} + v\mathbf{n}$ when the object is moving relative to the sensor in either the x or y direction or a combination thereof. Figure 12.2a shows, for example, the three-dimensional image of the two-dimensional dynamic MTF for the 1052 phosphor when the horizontal velocity is 4 cm/sec, and Figure 12.2b shows the same when the velocity is 2 cm/sec in the x direction and 3 cm/sec in the y direction.

The dynamic component of DMTF for a CRT screen is given by Eq. (12.3), where the angular frequency ω can be expressed as a function of the spatial frequency u moving across the screen at a speed **m**, driven at a temporal frequency $f_t = \mathbf{m}u$. Therefore, $\omega = 2\pi f_t = 2\pi \mathbf{m}u$. The equation, when plotted for a different velocity ωT, represents the deterioration of the dynamic MTF with a different angular velocity. As stated earlier, when $\omega T = 2\pi$, the image appears as a static one. Usually, considerable degradation effects are observed for $\omega T < \pi$, as can be seen from Figure 11.6. For $T = 17$ msec, $\tau = 20$ msec, and a velocity of 10 cm/sec, Figure 12.3 shows the corresponding two-dimensional MTF for a CRT display. It may be noted that ωT represents the ratio of the angular velocity of the object to the frame frequency $1/T$. The physical significance of $\omega T = 2\pi$ is that when the new shifted image is formed, the new frame scanning starts at exactly the same time. One way to counter the effect of velocity could be to change the

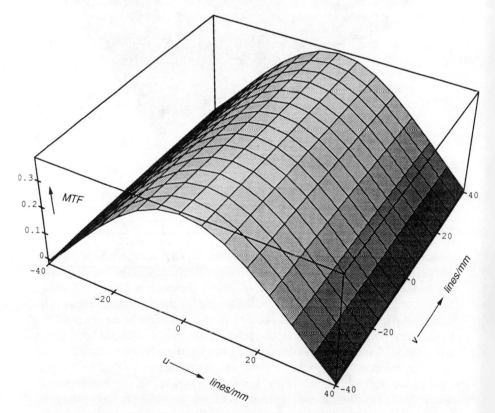

Figure 12.3 A 3-D representation of the 2-D dynamic component of DMTF for CRT for $T = 17$ msec, $\tau = 20$ msec, and $m = 10$ cm/sec.

frame rate such that the image appears as a static one on the screen and does not exhibit any smearing. Considerations such as these are used to restore dynamic images acquired with a charge-coupled device (CCD) when the CCD camera is in rotary motion, such as in a spacecraft orbiting a planet (King *et al.*, 1988).

Next, for electro-optical imaging under vibration, the overall MTF of the imaging display is given by the product of the three MTFs: the static MTF, the vibration MTF resulting from vibration impulse response, and the dynamic transfer function resulting from phosphor persistence. The overall MTF, therefore, is found to be

$$H_v(u) = H_s(u)J_0(2\pi au)\frac{1}{[1 + (2\pi^2\tau afu)^2]^{1/2}} \qquad (12.8)$$

where $H_s(u)$ is the static MTF and a is the amplitude of vibration, while f is the frequency of the corresponding vibration. Note from Eq. (12.8) that the spectrum along the u direction is modified. This corresponds to the fact that the vibration is present only along the x direction. If the vibration is present in any other direction, then in the two-dimensional expression of MTF, u should be replaced by a combination of velocity vectors in both the u and v directions. Experiments to restore vibratory images by taking "lucky shots" and to estimate degradation amplitude have been performed recently (Rudoler *et al.*, 1991). The three-dimensional plots of an electro-optic display due to vibration are shown in Figure 12.4. Figure 12.4a corresponds to $a = 0.01$ mm and P1052 phosphor, while Figure 12.4b corresponds to $a = 0.1$ mm and the same phosphor. From Figure 12.4b one notes that as the vibration amplitude increases, the frequency domain degradation function introduces zeros in the spectrum of the output image.

12.3 SIMULATION OF IMAGE DEGRADATION

To comprehend the extent of image degradation in an electro-optical display, a particular image intensifier tube, such as that used in night vision goggles, a CRT display under motion, and display under vibration are considered. In this application, it is estimated that the maximum angular movement of the head (with the goggles) is 120°/sec. By assuming that the distance between the phosphor screen and the object lens is 3 in., for example, the spatial speed **m** is calculated to be $[120°/180°](3 \text{ in.})/\text{sec} = 15.958$ cm/sec.

The calculated value of **m** and the experimentally determined value for the time constant of the type P1052 phosphor, for example, can be used in Eq. (12.7) to calculate the corresponding MTF of the dynamic response. A computer program can then be used to simulate the effect of the DMTF on images. The simulation algorithm involves the following steps:

Step 1. Obtain the fast Fourier transform (FFT) of the static image.
Step 2. Obtain the FFT of $d(t)$ for a particular motion.

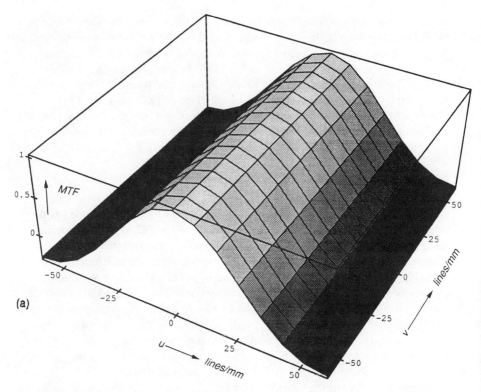

Figure 12.4 A 3-D representation of the 2-D dynamic component of DMTF due to vibration. (a) $a = 0.01$ mm; (b) $a = 0.1$ mm. ($\tau = 8.68$ msec for both.)

Step 3. Multiply (point by point) the FFTs obtained in steps 1 and 2.
Step 4. Take the inverse FFT of the product to estimate degradation.

A 64 × 64 pixel binary image, as shown in Figure 12.5, is used as the input image in the simulation. Figure 12.6 shows the corresponding degraded images when the head movement speed along the x direction approaches (a) 10 cm/sec and (b) 15 cm/sec. It is very evident that with increasing speed the image degrades more and more. Since motion is directed only along the x axis, the degradation of vertical edges becomes more significant. For comparison, we next consider a movement in both the x and y directions, that is, at an angular orientation. Figures 12.7a and 12b show motion-degraded images for the case of object moving along the x axis with spatial speeds of 7 and 15 cm/sec and along the y axis for spatial speeds of 4 and 10 cm/sec, respectively. In Figure 12.7, one finds that both horizontal and vertical edges have been degraded.

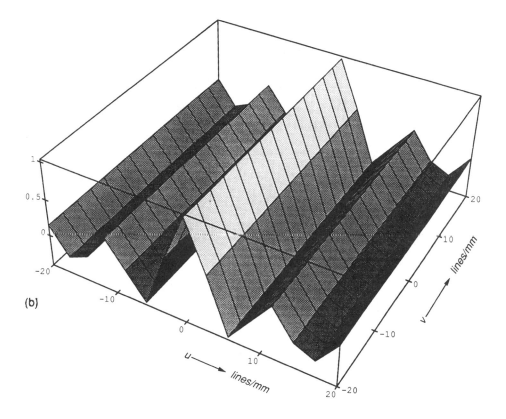
(b)

12.4 IMAGE RESTORATION

In this section, some of the images degraded in the previous section will be restored using our noise-free restoration model. It is shown that approximate estimation of the restoring filter provides acceptable restoration of images. Then some possible restoration schemes are identified with reference to other restoration models.

12.4.1 Simulation: Noise-Free Inverse Filtering

In Section 12.3 I showed that the DMTF can degrade input images (in the absence of noise, for simplicity) to such a degree that the object is scarcely recognizable. When noise is present the restoration filter becomes a Weiner filter (Weiner, 1949) provided the image and noise statistics are known to some extent. In such cases, a real-time optical Weiner filter may be implemented (Ikeda et al., 1986). Coherent optical and digital techniques based on least squares image res-

Figure 12.5 A 64 × 64 undegraded binary image.

toration as an extension of the Weiner type restoration algorithm was also performed earlier by Helstrom (1967), Rino (1969), and Horner (1969). To restore the degraded images, consider the Fourier domain degradation model shown in Figure 12.1. In this model, the object image $f(x,y,t)$ is operated on by a degradation function $H(u,v,f)$, which produces a degraded output image $g(x,y,t)$. Our main objective is to obtain an approximation of $f(x,y,t)$ given $g(x,y,t)$ and some knowledge of the degradation function $H(u,v,f)$. However, the degradation function as given by Eq. (12.2) consists of two separable functions, $S(u,v)$ and $D(f)$, where $S(u,v)$ is a purely static (spatial) function and $D(f)$ is a purely dynamic (temporal) function. Therefore, by eliminating the effects of $D(f)$, a reasonably good displayed image can be obtained. This is accomplished simply by dividing $H(u,v,f)$ by $D(f)$. Such a restoration technique, based on the concept of inverse filtering (Cherri *et al.*, 1991), can be applied in a straightforward fashion because the modulus of $D(f)$ is nonzero except at infinity. If this were not true, such a condition would have posed a major restriction to inverse filtering. The simulation of the restoration process consists of the following steps:

RESTORATION OF DYNAMICALLY DEGRADED IMAGES

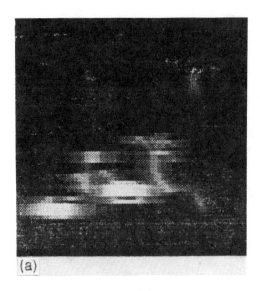

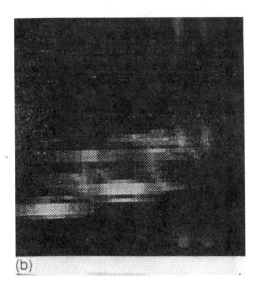

Figure 12.6 Image degraded due to horizontal speeds of (a) 10 cm/sec; (b) 15 cm/sec.

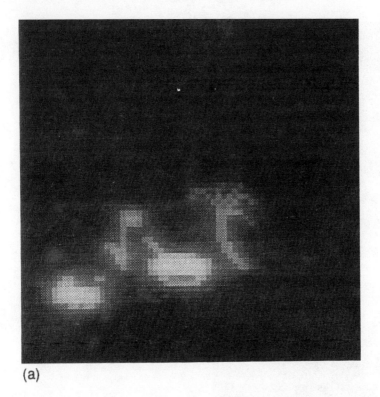

(a)

Figure 12.7 Images degraded due to (a) horizontal speed of 7 cm/sec and vertical 4 cm/sec; (b) horizontal 15 cm/sec and vertical 10 cm/sec.

Step 1: An FFT of the degraded image $g(x,y)$ is obtained. This gives $S(u,v) D(f)$.
Step 2: Given the velocity of the object and its direction, the corresponding inverse filter, $[D(f)]^{-1}$, is calculated.
Step 3: The results obtained in steps 1 and 2 are multiplied (point by point) in the frequency domain.
Step 4: An inverse FFT of the result of step 3 yields the restored image.

To restore the degraded image, one needs to know the phosphor decay time constant(s) and the magnitude and direction of relative velocity between the sensor and the target. This information is necessary to realize $1/[D(f)]$ as discussed earlier. Although there is no problem in meeting the first requirement, often it may be difficult to exactly estimate the value of the relative motion.

12.4.2 Restoration Models

A real-time restoration model is shown in Figure 12.8a. In this model, both the magnitude and direction of the relative motion are determined first. This infor-

RESTORATION OF DYNAMICALLY DEGRADED IMAGES

(b)

mation is used in the design of the restoration filter (Ambs *et al.*, 1988). A simple hybrid setup to perform the restoration process is illustrated in Figure 12.8b, where a spatial light modulator (SLM) can be used to imprint the inverse filter. Note that this real-time restoration model is suitable in applications where both speed and direction of the relative motion can be reasonably estimated. Nonlinear crystals have also been illustrated to record holograms for restoring a linearly smeared image (Fainman, 1989) with arbitrary magnitude and direction of the degradation function (Vachss and Hesselink, 1988). Real-time space domain deblurring filters based on Weiner filters using CCD image sensors and a video frame integrator have also been proposed (Ikuta, 1985). Ohyama et al. (1987) also proposed restoring image degradation in CRT display by using phase matching and resampling.

For the particular applications where the relative motion between the target and the sensor is not known exactly, a simpler restoration model like the one shown in Figure 12.9 can be used. In this simple system, a fixed Fourier hologram can be used to store the inverse filter for a single specific speed, or an optimum composite inverse filter for a specific range of speeds (Gao *et al.*, 1990). The image of Figure 12.6, for example, can be restored to some degree

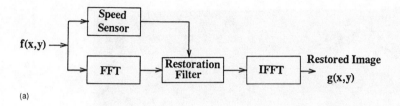

(a)

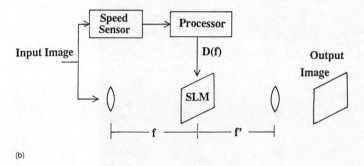

(b)

Figure 12.8 Adaptive restoration (a) model and (b) system.

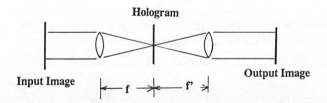

Figure 12.9 Nonadaptive restoration using Fourier hologram as an inverse filter.

as in Figure 12.10, using an inverse filter designed for a speed value of 7 cm/sec. Similarly, Figure 12.11 shows a restoration of a motion-degraded image due to horizontal motion of 10 cm/sec when the image is subjected to a restoration filter designed for a speed of 15 cm/sec. Figure 12.12 shows an example of restoration of the smeared image of Figure 12.7 (a velocity of 10 cm/sec at an angle of 30°) obtained by processing with a filter corresponding to 14.14 cm/sec at an angle of 45°. Nevertheless the restored images show great improvement in that they are quite similar to the original images, observations suggest that it is not always

RESTORATION OF DYNAMICALLY DEGRADED IMAGES

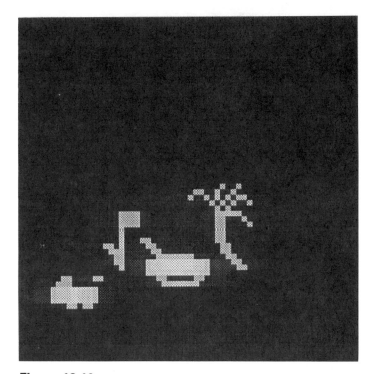

Figure 12.10 Restoration of the image of Figure 12.6a using an inverse filter corresponding to 7 cm/sec.

necessary to know the exact value of the relative velocity. A value close to the relative speed of the target is more likely to give a better output image of an electro-optical display than one obtained without using a restoration filter at all. A linear holographic filter with approximate velocity has been used for this purpose (Jiang and Xu, 1983; Jiang et al., 1985). It is worth mentioning that a high-frequency filter can also be used as a restoration filter because the degradation function behaves like a low-pass filter (Gonzalez and Wintz, 1987).

The use of a holographic filter may be avoided (Celaya and Mallick, 1978) by adopting a joint Fourier transform setup, where the problem of accurate alignment and prefabrication of filter hologram do not arise. Instead of the Fourier plane degradation filter, the degradation function is placed side by side with the degraded image in the input plane (Javidi et al., 1988, 1989).

12.5 SUMMARY

In this chapter, I have used the dynamic transfer function to predict the image quality and the performance of a phosphor-based electro-optical display. By in-

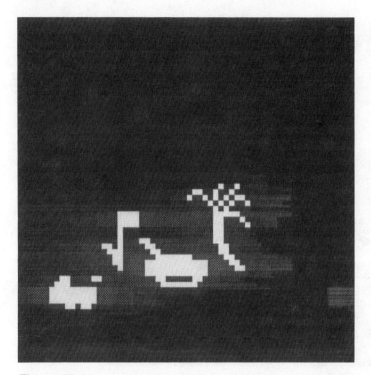

Figure 12.11 Restoration of the image degraded due to a horizontal speed of 10 cm/sec using an inverse filter corresponding to a horizontal speed of 15 cm/sec.

corporating the concept of temporal response, which is otherwise missing from the traditional imaging systems analysis, one is able to quantify degradation of the output image of an electro-optical display. This quantification is then used to successfully propose efficient and real-time techniques for restoration of motion-degraded images. I have also shown that for the restoration of motion-degraded displayed images it is necessary only to have a reasonable knowledge of relative velocity. The three-dimensional dynamic component of DMTF plots for three different types of degradation show that the nature of the degradation function is the same in each case. Accordingly, I illustrated results for only one of the three. A complete display system may have more than one component connected in cascade. Computer modeling of the whole system (Guest, 1971) provides the advantage of devising a way to restore the effect of all the degradations, even including degradation due to dynamic weather conditions (Dinstein *et al.*, 1988), in a single step. With improvements in image processing electronics, it may become practical to consider computer-based restoration schemes implementing real-time digital signal processing techniques. Smears caused by time-varying

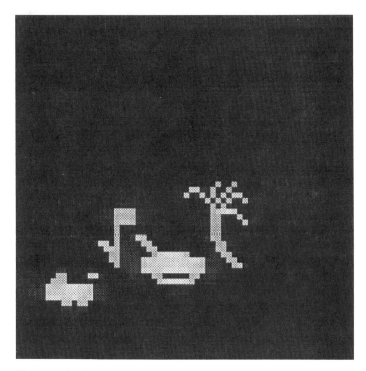

Figure 12.12 Restoration of the image degraded due to horizontal speed of 15 cm/sec and vertical speed of 10 cm/sec (Fig. 12.7b) using filter corresponding to horizontal speed of 10 cm/sec and vertical speed of 10 cm/sec.

(Ward and Saleh, 1986, 1987), spatially varying (Angel and Jain, 1978), or randomly varying (Ward and Saleh, 1985) degradation functions can be restored by microprocessor-based electronic hardware. The added flexibility may enable one to explore iterative techniques (Mammone and Rothacker, 1987; Maeda, 1985) that may even be extended to realize nonlinear restoration schemes (Meinel, 1986).

References

Ambs, P., Y. Fainman, S. H. Lee, and J. Gresser (1988). Computerized design and generation of space-variant holographic filters. 1. System design considerations and applications of space variant filters to image processing, *Appl. Opt. 27*: 4753.

Angel, E. S., and A. K. Jain (1978). Restoration of images degraded by spatially varying pointspread functions by a conjugate gradient method, *Appl. Opt. 17*: 2186.

Awwal, A. A. S., A. K. Cherri, M. A. Karim, and D. L. Moon (1991). Dynamic modulation transfer function of a display system, *Appl. Opt. 30*: 201

Bates, R. H. T., and M. J. McDonnell (1986). *Image Restoration and Reconstruction*, Clarendon Press, Oxford.

Celaya, L., and S. Mallick (1978). Incoherent processor for restoring images degraded by a linear smear, *Appl. Opt. 17*: 2191.

Cherri, A. K., A. A. S. Awwal, M. A. Karim, and D. L. Moon (1991). Restoration of moving binary images degraded owing to phosphor persistence, *Appl. Opt. 30*: 3734.

Csorba, I. P. (1986). Image intensifiers in low light level and high speed imaging, Electronic Imaging '86, Boston, MA, pp. 3–6.

Dinstein, I., H. Zoabi, and K. S. Kopeika (1988). Prediction of effects of weather on image quality: preliminary results of model validation, *Appl. Opt. 27*: 2539.

Fainman, Y. (1989). Applications of photorefractive devices for optical computing, *SPIE Proc. 1150*: 120.

Gao, M. L., S. H. Zheng, M. A. Karim, and D. L. Moon (1990). Restoration of dynamically degraded gray level images in phosphor based display, *Opt. Eng. 29*: 878.

Gonzalez, R. C., and P. Wintz (1987). *Digital Image Processing*, 2nd ed., Addison-Wesley, Reading, MA.

Guest, A. J. (1971). A computer model for channel multiplier plate performance, *Acta Electron. 14*: 79.

Helstrom, C. W. (1967). Image restoration by the method of least squares, *J. Opt. Soc. Am. A 57*: 297.

Horner, J. L. (1969). Optical spatial filtering with the least-mean-square-error filter, *J. Opt. Soc. Am. A 59*: 553.

Ikeda, O., T. Sato, and H. Kojima (1986). Construction of a Weiner filter using a phase-conjugate filter, *J. Opt. Soc. Am. A 3*: 645.

Ikuta, T. (1985). Active image processing, *Appl. Opt. 24*: 2907.

Javidi B., H. J. Caulfield, and J. L. Horner (1988). Real-time deconvolution by nonlinear image processing, *Tech. Dig. Annual Meeting of Opt. Soc. Am.*, Santa Clara, CA, p.89.

Javidi B., H. J. Caulfield, and J. L. Horner (1989). Image deconvolution by nonlinear image processing, *Appl. Opt. 28*: 3106.

Jiang, Y., and Y. Xu (1983). Simple method for image deblurring, *Appl. Opt. 22*: 784.

Jiang, Y., C. Zhang, and C. Zheng (1985). Restoration of blurred images by convolution of arbitrary point spread function, *J. Opt. Soc. Am. A 2*: 73.

King, R. A., A. L. Broadfoot, B. R. Sandel, and V. A. Jones (1988). Correcting image distortion with fiber-optic tapers, *Appl. Opt. 27*: 2048.

Levi, L. (1983). Spatiotemporal transfer function: recent developments, *Appl. Opt. 22*: 4038.

Maeda, J. (1958). Image restoration by an iterative damped least-squares method with non-negativity constraint, *Appl. Opt. 24*: 751.

Mammone, R. J., and R. J. Rothacker (1987). General iterative method of restoring linearly degraded images, *J. Opt. Soc. Am. A 4*: 208.

Meinel, E. S. (1986). Origins of linear and nonlinear recursive restoration algorithms, *J. Opt. Soc. Am. A 3*: 787.

Ohyama, N., E. Badiqué, M. Yachida, J. Tsujiuchi, and T. Honda (1987). Compensation of motion blur in CCD color endoscope images, *Appl. Opt. 26*: 909.

Rash, C. E., and J. Becher (1982). Analysis of image smear in CRT displays due to scan rate and phosphor persistence, U.S. Army Aeromedical Res. Lab., USAARL Rep. 83-5, Fort Rucker, AL.

Rino, C. L. (1969). Band limited image restoration by linear mean-square estimator, *J. Opt. Soc. Am. A* 59: 547.

Rudoler, S., O. Hadar, M. Fisher, and K. S. Kopeika (1991). Image resolution limits resulting from mechanical vibrations. Part 2. Experiment, *Opt. Eng. 30*: 577.

Sandel, B. R., D. F. Collins, and A. L. Broadfoot (1986) Effect of phosphor persistence of photometry with image intensifiers and integrating readout devices, *Appl. Opt.* 24: 3697.

Vachss, F., and L. Hesselink (1988). Synthesis of a holographic image velocity filter using the nonlinear photorefractive effect, *Appl. Opt.* 27: 2887.

Verona, R. W., H. L. Task, V. Arnold, and J. H. Brindle (1979). A direct measure of CRT image quality, U.S. Army Aeromed. Res. Lab., USAARL Rep. No. 79-14, Fort Rucker, AL.

Ward, R. K., and B. E. A. Saleh (1985). Restoration of images distorted by systems of random impulse response, *J. Opt. Soc. Am. A* 2: 1254.

Ward, R. K., and B. E. A. Saleh (1986). Restoration of images distorted by systems of random time-varying impulse response, *J. Opt. Soc. Am. A* 3: 800.

Ward, R. K., and B. E. A. Saleh (1987). Image restoration under random time-varying blur, *Appl. Opt.* 26: 4407.

Weiner, N. (1949). *The Extrapolation, Interpolation and Smoothing of Stationary Time Series*, Wiley, New York, Chapter III, p. 81.

13
Discrete Display Devices and Analysis Techniques

John C. Feltz

*Systems Research and Applications Corporation,
Arlington, Virginia*

13.1 INTRODUCTION

A display device is classed as *discrete* when it presents a set of individual picture elements, or pixels, as contrasted with *continuous display* devices, which present a continuously addressable area for displaying information. Discrete display devices play a very important role in both commercial applications and display research. Because of their widespread use and the variety of display types and applications, it is useful to have a general discussion of these devices. A common understanding of how all discrete displays behave, how they are characterized, and how they are used is beneficial for any professional who may need to use discrete displays. This chapter, rather than examining a single technology or issue of image displays, examines the similarities and common characteristics of all discrete display devices and presents general analysis techniques for them.

Discrete displays are like any other displays in that their primary use is to present dynamic information to human observers. Some discrete displays are now finding use in electro-optic information processing systems, but these applications are still in the research and development phases. Most of this chapter focuses on the much more common situations where the display device is the final component in a system, presenting information to human observers. This section briefly introduces the applications of discrete displays, explains why they must be treated separately from continuous devices, and presents an outline of the chapter.

13.1.1 Applications

Discrete displays are very useful devices, with applications found throughout industry, business, research, and consumer products, especially in the display of computer-generated information. Discrete displays are especially well suited to the many applications that involve transmitting time-sampled information serially over some distance, such as commercial video and television. The development of discrete display technologies has often been a "pull" relationship: A commercial need exists, and it provides the impetus to expand the availability and performance of discrete display technologies while lowering their cost, size, and power requirements.

The computer industry has often driven the development of better and cheaper discrete displays. The explosion in the availability and power of personal computers and workstations in the 1970s and 1980s was a major cause of the similar explosion in the types and capabilities of discrete display devices. For every personal computer, there must be at least one monitor, a discrete display. These displays must be easy to use and set up and must run on household power. In addition, every portable computer includes a discrete display. These displays must be small, lightweight, and rugged; they must also run from batteries and consume little power. The tremendous economic incentives to develop lighter, cheaper, and better displays for the computer industry have had a vast effect on the discrete display technologies available.

Military needs, especially the needs of military aviation, have also driven the development of discrete displays. The demand for small, rugged cockpit- and helmet-mounted displays has had a major impact on discrete display technology. Military studies have also funded the basic research for several display technologies, and continue to push the state of the art in several fields.

13.1.2 Discrete Displays Vs. Continuous Displays

Before exploring further the applications and economic considerations that affect displays, one must ask why the various display technologies grouped under the term "discrete" are treated differently from other devices. As discussed in this chapter, there are significant and fundamental differences between discrete and continuous devices. Although the end uses of discrete and continuous displays are often the same, the methods used to analyze the devices are necessarily different. Discrete devices by definition employ sampling, and sampling has significant meaning when applied to information transmission and display. A large portion of this chapter is thus devoted to sampling theory and the effects of sampling on image displays.

An important part of understanding how discrete displays perform is in understanding how information sent to them is generated. Displays are rarely the devices that perform the first sampling in a system. That step is usually taken by an imaging device such as a camera tube or a charge-coupled device. A complete discussion of sampling and discrete displays must necessarily include the topic of imaging.

13.1.3 Plan of the Chapter

Section 13.2 discusses the most common discrete display technologies. This discussion includes a brief description of the underlying physical processes of each technology, but centers on the distinguishing characteristics of the technology that must be accounted for when analyzing and characterizing devices. Section 13.3 then discusses the measurements and figures of merit commonly used to characterize continuous displays, and how these can be applied to characterizing discrete displays.

Section 13.4 presents a discussion on sampling, general discrete device theory, and imaging. These topics are central to the chapter, as they identify the underlying reasons why discrete devices must be treated differently from continuous devices. Section 13.5 concludes with a summary and general guidelines on analyzing, characterizing, and using discrete display devices in systems.

13.2 DISCRETE DISPLAY TECHNOLOGIES

Most of the discrete displays discussed here are presented in much greater detail elsewhere in this volume. This section, therefore, does not seek to fully explain the intricacies of construction and operation of the various displays. Rather, the discussion is limited to the basic elements of the display technology, particularly those that affect applications of the technology and characterization techniques. Particular features of the technology that require additional characterization to fully describe a device are also called out.

13.2.1 The Cathode-Ray Tube

The cathode-ray tube (CRT) is the oldest discrete display technology available to us. It was developed early in the 20th century for instrumentation applications, such as oscilloscopes. The CRT was then adapted for use in radar displays and television in the 1930s and 1940s. Cathode-ray tubes continue to be used for these applications, as well as for other display applications for video images, instrumentation, and computer interfaces.

A simple CRT, shown in Figure 13.1, consists primarily of a large evacuated tube. The tube contains a large display screen at the viewing end, covered with an electroluminescent material. The material is often generically referred to as a phosopor because phosphorus was and is a common electroluminescent material. At the rear of the tube are one or more electron guns and two electromagnetic deflection yokes.

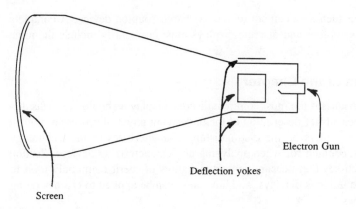

Figure 13.1 Basic CRT construction.

Electrons are produced in a continuous beam by a hot cathode in the electron gun and accelerated to a high velocity by the potential gradient in the gun. The electron beam then passes between the yokes, which deflect the beam by using electric or magnetic fields of varying intensities. The yokes act to aim the beam at a particular point on the display screen. When the electrons strike the screen, atoms in the phosphor grains are excited and produce visible light. The intensity can be adjusted by changing the number of electrons produced and their intensity. The beam can be moved to different points on the display screen by adjusting the signal applied to the deflection yokes.

The device just described is not inherently a discrete display device. The analysis of CRTs is complicated because many CRTs are used as continuous devices rather than discrete devices. The difference in classification depends on two characteristics. The first, and more obvious, characteristic is the distribution of phosphors on the display screen and the physical construction of the screen. Discrete spots of phosphors will, of course, cause the CRT to be a discrete device. The less obvious, but sometimes more important, characteristic is the nature of the electronics controlling the CRT and the nature of the information being displayed. If the signals and circuits driving the electron guns and the deflection yokes produce pixelated information on a continuous phosphor screen, the entire device must be classed as discrete.

Analog devices, such as most oscilloscopes, usually contain CRTs with a continuous coating of phosphors on the display screen. The deflection yokes scan the electron beam at a constant rate horizontally and deflect the beam vertically as driven by the oscilloscope input signal. This results in a CRT that is completely continuous in nature.

DISCRETE DISPLAY DEVICES AND ANALYSIS 499

Early monochrome television sets used the same phosphor coating as oscilloscopes, but different electronics. In these sets, the electron beam is scanned both vertically and horizontally at constant rates, sweeping out a pattern of sloping horizontal lines. This results in a display that is discrete in one direction, the vertical, but continuous in the other, the horizontal, as shown in Figure 13.2.

Most monochrome television sets produced today, as well as all color sets, are discrete in both directions. A shadow mask, shown in Figure 13.3, is used to align the electron beam or beams. The shadow mask allows the beams to illuminate only the appropriate pixels. The shadow mask arrangement is the most commonly used arrangement for televisions and computer monitors. The beam is still driven across the screen in the same pattern as with early monochrome sets, but the shadow mask performs a sampling in the horizontal direction to make the display fully discrete.

This distinction between fully continuous, half-discrete, and fully discrete is a very important one and must always be considered when analyzing a CRT system. However, CRT systems built today are predominantly fully discrete, and this chapter will devote most of its discussion to them.

There are a number of other characteristics peculiar to CRTs. Because the electron beam must be steered at large angles at the extreme edges of the display, misalignments can be common. This is especially true in color systems, with three electron beams to steer and align. The effect of this is that color response, color alignment, and intensity can vary significantly from the edges of the display to the center.

There are many phosphors that can be used in CRTs, each with its own partic-

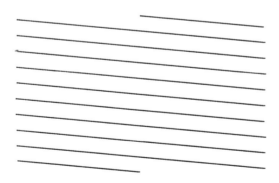

Horizontal scan lines

Figure 13.2 Half-discrete CRT display.

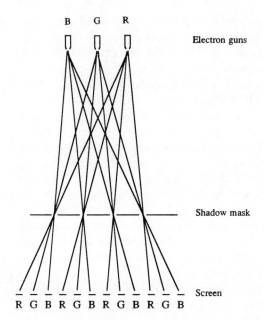

Figure 13.3 CRT shadow mask.

ular characteristics, including persistence curves, aging behavior, and color profile. In addition, the phosphor coating can vary in thickness and grain size within a pixel, between pixels in the same CRT, and between CRTs of the same model. It may take considerable effort to analyze and document the variability that will occur with a CRT display.

The shape of the tube is also important to CRT characterization. Glass technology has only recently made it practical to have a truly flat, square-cornered display. Many CRTs still use a slightly convex screen and faceplate, which can result in distortion at the corners or astigmatism.

Despite these potential drawbacks, CRTs are still very widely used. Their popularity is due primarily to the maturity of the technology and the excellent brightness, contrast, and color response provided. However, they are bulky and use considerable power at high voltages. Nevertheless, they continue to dominate the market for televisions and computer monitors.

When analyzing and characterizing CRTs, it is important to remember that they can be highly variable in response. The difference in response between two CRTs can be very noticeable, as can the difference between two regions of the same CRT. In addition, CRTs exhibit significant warm-up and aging effects due to their vacuum-tube technology.

13.2.2 The Light-Emitting Diode

The light-emitting diode (LED) was the first solid-state display device to enjoy wide commercial use. It became common in the 1960s and enjoyed considerable use in small displays such as calculators, watches, and electronic instrumentation. LEDs have not found commercial success in large image displays. Recently, a miniature display monitor for personal computers was introduced with some initial success, but the full commercial potential of the device is still unknown at this writing.

Light-emitting diodes are simple semiconductor diodes, with the exception that their construction includes luminescent materials, as shown in Figure 13.4. When a sufficient forward voltage is applied across the terminals of the LED, the device becomes active and emits light. When the voltage is removed, the device becomes inactive and stops emitting.

Light-emitting diodes have the general advantages of all semiconductor devices. They resist aging very well, and they are very rugged. They can be manufactured in large quantities easily and cheaply. However, LEDs have high power consumption for semiconductor devices. It is impractical to provide enough battery power to run an LED portable computer display, for example.

Light-emitting diodes have no significant unusual features to complicate char-

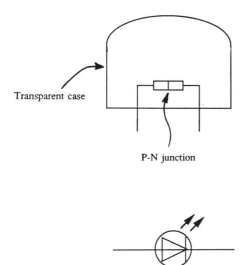

Figure 13.4 LED construction.

acterization and analysis. They are usually uniform in performance, and their characterization is fairly straightforward.

13.2.3 The Liquid-Crystal Display

Liquid-crystal displays (LCDs) are widely used in commercial applications, as well as being the subject of considerable research. Research into different LCD materials and device structures is expanding the already wide range of LCD applications. LCDs are most commonly found today in consumer electronics such as wristwatches, portable televisions, and portable computer displays.

Liquid-crystal displays use an interesting and somewhat peculiar property of certain materials, selected smectic, nematic, and cholesteric liquids. In certain liquid phases, these materials exhibit some of the electro-optic effects normally ascribed to crystals; hence the name liquid crystal. Of particular interest, the application of an electric field to these liquids causes the molecules to alter the way they rotate the polarization vector of visible light. A more detailed description of the physics of LCDs can be found in Chapter 2.

A basic LCD display consists of three layers, shown in Figure 13.5: a back polarizer, the LCD cells, and a front polarizer. Light enters the device through the back polarizer and passes through the LCD cells. Some cells will rotate the polarized light whereas others will not, depending on the electric field placed across each cell. The light then exits the device through the front polarizer. If the polarizers are arranged appropriately for the type of LCD material being used, an image will result.

The nature of LCDs introduces two problems not seen in other devices. First, the LCD modulates ambient light and does not produce any light. An auxiliary

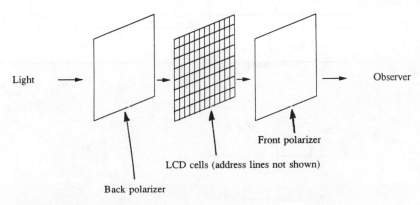

Figure 13.5 Basic LCD construction.

light source must be provided for low-light viewing, and a diffuser must be provided if ambient light is used. Second, because of the polarization necessary to use the LCD effect, the phase of the displayed image must be measured, as well as its amplitude. The LCD is the only major display technology where phase analysis and measurements are necessary. LCDs can be difficult to view at some angles and orientations because of this polarization and because of the long, rod-like molecular structure of many LCD materials.

There have been some experimental applications of LCDs in optical data processing. If the light entering the back polarizer of an LCD display has been previously modulated to hold information, the LCD array can be used as a large filter or a matrix of Boolean operator cells. Although LCD switching rates are rather low, an LCD display can provide large and relatively inexpensive processing modules for image recognition, massively parallel computing, and similar applications.

13.2.4 Other Display Technologies

There are several other types of discrete display technologies, such as ac and dc plasma displays (Slottow, 1976) and electroluminescent displays, that are not as widely used as CRTs and LCDs. These other display technologies are used primarily for special situations, especially in flat-panel displays for particular applications, many of them military. Recently these technologies have entered the portable computer market, however.

Plasma displays use a gas plasma that emits light when it is carrying an electric current. Both ac and dc versions of plasma displays are in use. Plasma displays produce a strong illumination, usually red or red-orange. Electroluminescent displays use phosphor materials, like CRTs, but excite the phosphor pixels directly with electrodes. These displays are usually monochrome, usually green or blue-green. There are no unusual characteristics to these devices, and their characterization is fairly straightforward.

13.3 CHARACTERIZATION MEASUREMENTS AND FIGURES OF MERIT

The equipment, measurements, and figures of merit used to characterize discrete displays are for the most part the same as those used for continuous displays. The experimental equipment may have to be of higher precision and will be applied somewhat differently. Many but not all of the figures of merit used on continuous displays can be used to characterize discrete displays, but the derivations must be adapted to account for sampling and other effects.

13.3.1 Measurements

Common analytical measurements in photometry, such as reflectance and polarization, are used to characterize discrete displays just as they are used to characterize continuous displays. In this regard, discrete displays are very similar to continuous displays. However, the equipment used to take the measurements may have to be more precise, as it will have to take measurements of individual pixels. It may even be necessary to make detailed measurements within a pixel to generate profiles of pixel behavior.

Many more measurements may have to be taken to characterize a discrete display than would be necessary for a similar continuous display. Enough pixels must be analyzed in separate regions of the display to derive overall performance data as well as a figure for short-range deviations. A good statistical profile should be obtained for the behavior of a random pixel from these measurements.

Of all common discrete display technologies, LCDs will probably require the most effort to make a thorough characterization by measurements. As discussed in Section 13.2.3, LCDs employ polarization as an integral part of the display process, and polarimetry is especially important for these devices. LCDs also require measurements of any diffuser used to provide uniform background illumination. If diffusion is not high enough, images of the light source will be presented to the user, a situation to be avoided.

Of course, contrast measurements should be taken on all discrete displays, just as they are on continuous displays. Contrast measurements of LCDs should be made at a number of orientations and viewing angles. Reflectance measurements are necessary to determine susceptibility to glare, both from point sources and from diffuse background light. Most systems will cover discrete display devices with a glass or plastic faceplate, and these faceplates may require antiglare coatings.

Colorimetric measurements are also important. Information that has been computer generated or computer enhanced can be understood better and more rapidly by human observers if displayed in color. However, these color choices must be psychologically and physiologically appropriate if the information is to be conveyed effectively. Colorimetry is also important to ensure that color representation of images is consistent with the original objects, that color is uniform across the display, and that device aging does not affect color balance.

Detailed time-domain photometry should also be performed, usually by applying a time-domain negative step function input to a pixel, as shown in Figure 13.6. Most applications of discrete display devices present time-dependent information. The persistence of the display device must be matched to the scanning frequency of the system and the anticipated information to be displayed. These measurements are very important in determining how well human observers will be able to use the display in its intended environment. If the persistence is too

DISCRETE DISPLAY DEVICES AND ANALYSIS

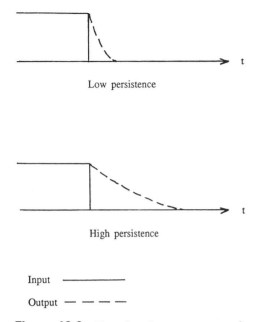

Figure 13.6 Time-domain measurements of persistency.

low, the display will appear to flicker. If the persistence is too high, rapidly moving images will smear.

13.3.2 Figures of Merit

Many of the figures of merit used to characterize continuous devices are applicable to discrete devices as well, with some caveats. These figures of merit include the optical transfer function (OTF), modulation transfer function (MTF), contrast transfer function (CTF), point-spread function, and line-spread function. All of these functions convey information about the spatial frequency response of a display device.

The OTF is a general transfer function expressing how well a sinusoidal spatial modulation is transferred through a system or device. It includes both phase and magnitude components. The MTF is the magnitude portion of the OTF. Since very few systems exhibit phase shifts of the spatial modulation, the MTF and OTF are usually synonymous, and the MTF is referenced more frequently in the literature. The CTF is similar to the MTF and OTF but uses a square-wave modulation rather than a sinusoidal modulation. Figure 13.7 shows typical MTF and CTF input and output signals for a continuous device.

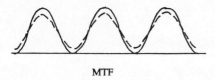

MTF

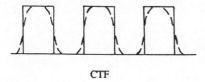

CTF

Input ────────

Output ─ ─ ─ ─

Figure 13.7 Typical MTF and CTF curves.

The OTF, MTF, and CTF can be measured experimentally by generating the appropriate input signals and measuring the device response. As discussed in detail in Section 13.4, these functions are not single-valued for discrete devices, because of sampling noise. Nevertheless, it can be practical to take measurements and compare them to expected results to identify flaws in discrete display devices.

When the pixels of a discrete device are small relative to the spatial period of the input MTF or CTF pattern, discrete device MTFs and CTFs are comparable to those for analog devices. However, as the pixel size approaches the input pattern period, the discrete MTF and CTF diverge widely from the analog MTF and CTF. The cutoff point for this behavior occurs when the pixel spacing becomes greater than approximately 20% of the input pattern period. For pixels larger than this figure, the discrete MTF can range from the predicted analog value to zero (Feltz, 1990; Feltz and Karim, 1990).

The point-spread and line-spread functions express how a device responds to impulse inputs. For discrete displays, these functions have limited utility, because they measure the response of the device to input for one pixel or one line of pixels. The theoretical output, obviously, is a uniform rectangle. The point-spread and line-spread functions are limited to describing the profiles of individual pixels or scan lines, identifying any crosstalk between pixels, and describing any defocusing (or low-pass filtering) caused by optical elements such as faceplates between the actual display and the observer.

13.4 SAMPLING AND DISCRETE DEVICE THEORY

As mentioned at the beginning of the chapter, sampling is central to a discussion of discrete display devices. A large volume of work has been devoted to the effects of sampling on electric signals, and it has been carried over into imaging systems very well. This section cannot present an exhaustive treatment of the subject, but it does present the basic characteristics of sampled systems in a general way and gives more detailed information about sampled images.

13.4.1 General Sampling Theory

Any discussion of sampling should begin with Nyquist's theorem, which describes the relationship between the frequency content of an information signal and the frequency used to sample that signal. An information signal must be sampled at a rate at least twice that of the highest frequency in the signal in order to be reproduced later from the sampled data. The minimum sampling rate for a signal is known as its *Nyquist frequency*.

Note that Nyquist's theorem specifies a frequency "at least" twice that of the highest frequency in the signal. A practical system may have to sample at a much higher rate than the Nyquist frequency in order to compensate for noise and signal degradation from other system components. This procedure is known as *oversampling*.

The need for oversampling is shown in the spectrum diagram in Figure 13.8. Sampling has the effect of duplicating the frequency spectrum of the original signal at multiples of the Nyquist frequency f_s. As the sampling rate is increased, the stop band between the duplicate spectra increases. If a signal is not sampled at a rate above the Nyquist frequency, the phenomenon of aliasing occurs.

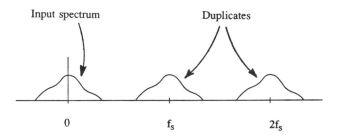

Figure 13.8 Frequency diagram of sampling.

Aliasing describes how a high-frequency component of a signal can be mapped into a low-frequency band after sampling. This is demonstrated in Figure 13.9, which shows a sinusoid being sampled at a rate below the Nyquist frequency. When an attempt is made to reconstruct the continuous waveform from the samples, the result is a low-frequency alias of the original signal. Figure 13.10 is a spectrum plot illustrating aliasing. The high frequencies of the input signal have been "folded" over the $f_s/2$ line. The frequency $f_s/2$ is often known as the *folding frequency* because of this behavior.

The numerical relationship between the original signal and the aliased result is given by Eq. (13.1). Here, f_i is the input frequency, n is the index to a particular spectrum duplicate, and f_a is the frequency of the aliased output. If a signal is grossly undersampled, several duplicate spectra can contribute to a particular alias frequency.

$$f_a = nf_s - f_i, \qquad f_s < 2f_i \tag{13.1}$$

A commonly cited example of aliasing in a discrete display is television broadcasts of sporting events. The spatial sampling rate of the television system is at times not high enough to avoid aliasing the striped shirts of referees. Because the stripes are a square-wave modulation, they have significant harmonics at high frequencies. The information presented to the user is often only the first harmonic of the stripes, which appear to an observer as repeated shadings from light to dark rather than sharp stripes. The aliased high-frequency components appear as a gradual modulation from light to dark across the shirt.

The discussion of sampling up to this point has been ideal sampling; that is, the sampling function has been a string of impulse functions. Impulse functions are used for sampling because the convolution of any function with an impulse function is the original function. It is impossible, however, to generate true impulse functions; they are only a mathematical convenience.

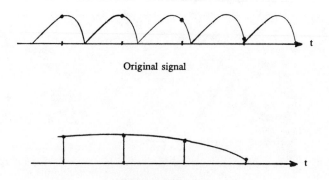

Figure 13.9 Example of aliasing.

DISCRETE DISPLAY DEVICES AND ANALYSIS

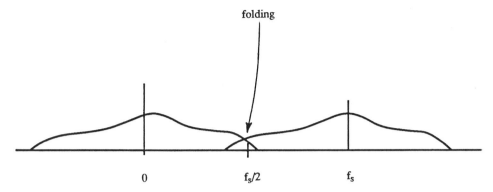

Figure 13.10 Frequency diagram of aliasing.

Real sampling uses functions that approximate impulse functions, such as narrow rectangular or triangular pulses, truncated sinc pulses, or truncated Gaussian pulses. The effect of using a nonideal sampling pulse is that the signal cannot be completely reconstructed from the samples. The convolution of a signal with a nonideal pulse, no matter how narrow, will necessarily be different from the original signal. For many systems the differences may be negligible, but not for image systems.

Most image systems use very large sampling pulses, for two reasons. First, the human visual system is not good at image reconstruction from sample points. Human observers do not react favorably to small pixels on a neutral background. Second, the dynamic range needed for accurate image reconstruction can be several orders of magnitude. Imaging devices and display devices need large pixels to overcome the resultant signal-to-noise problems. Any noise that is not signal-dependent in magnitude will overwhelm small signals, so it is desirable to have large pixels to capture as much signal energy as possible.

The error introduced into signals by sampling with nonideal pulses is thus of primary concern in systems using discrete displays. The next subsection discusses how this error can be quantified for imaging devices, such as CCDs, and display devices using the MTF.

13.4.2 Practical Sampling in Image Systems

As mentioned earlier, it is necessary to understand how discrete information enters a system in order to understand how it will be displayed. Discrete display devices are not used independently of other devices; they must be incorporated in a system to be useful. By understanding how the elements in a discrete system work together, analysis of discrete displays can be simplified.

One of the goals a designer of discrete image systems should have is to match the sampling behavior of all elements in the system. Every time a signal is reconstructed and sampled internally, the resolution of the image on the output is jeopardized. If all the elements in the system are matched appropriately, the entire system can be represented by only one sampling action, along with a number of filters and entry points for noise.

Matching the sampling behaviors means matching a number of characteristics: the pixel size, the pixel separation (or pitch), and the number of pixels, all in two spatial dimensions. It also means matching the time-domain scanning rate. If this is done correctly, the element that will be analyzed for the effects of sampling can be chosen for convenience, because all elements will have identical sampling behavior.

However, it is sometimes not feasible or desirable to match the sampling behaviors of the various elements in a system. This means that two or more sampling operations take place in the system. The first can be analyzed as occurring anywhere in the first element or group of contiguous elements with matched behavior. The output of this element or group of elements must be reconstructed into a continuous signal and then sampled by the next element or elements, which have a different sampling behavior from the first sampling. The analysis will continue in this manner until the device output is reached and the signal is reconstructed for the user.

For simplicity, however, consider the case where all the elements in a system have matched sampling behavior. For most discrete image systems, the pixels will be very large, in many cases contiguous or nearly contiguous. The error induced by sampling with these large pixels can be significant. The interested reader is referred to Feltz and Karim (1990) and Feltz (1990), which present the detailed derivation of the results discussed below.

The maximum resolution of a device is often given by the highest frequency at which the MTF is above a certain cutoff value, such as 0.7 or 0.9. The best possible resolution for a device would thus be $f_s/2$. The sampling error introduced by the large pixels commonly found in imaging devices means that the resolution of such devices is as much as an order of magnitude lower, and certainly no higher than $f_s/10$ for the above cutoff values.

Note that these figures have nothing to do with aliasing or other considerations of ideal sampling. They refer only to the effects of real sampling that cause degradation of contrast, or modulation. Even when a signal is reproduced faithfully according to a rigid reading of Nyquist's theorem, it may not be identifiable by a human observer.

13.4.3 Numerical Calculations

Actually performing detailed calculations of the response of discrete image system components can be difficult, not to mention tedious. There are several ways to perform the calculations rapidly using Fourier transform techniques on a digital computer. There are any number of commercial and public-domain software packages available to perform these calculations.

The Fourier transform is a very useful tool for performing calculations on continuous signals. An analog to this transform, the discrete Fourier transform (DFT), has been developed that can perform similar transform operations on discrete signals. The DFT can be difficult to perform computationally, but there is a way around this limitation.

If the number of points to be transformed is an integral power of 2 or some other highly composite number, the fast Fourier transform (FFT) can be used. The FFT is not a different transform from the DFT, but merely a special case of the DFT that takes advantage of some computational shortcuts. Several algorithms exist that can perform rapid FFT calculations (Brigham, 1988).

A convenient property of the various Fourier transforms is that they are separable along orthogonal axes. To take the two-dimensional FFT of an image, assemble the data in two-dimensional array. Take the FFT of each row, and replace the row in the array with its transform results. Then take the FFT of each column and replace the column in the array with its transform results. The result is a two-dimensional FFT. The order of calculation of rows and columns can be freely interchanged without affecting the end result.

Convolutions and other manipulations can be made much more conveniently in the transform domain than in the image domain. It is often best to completely transform a system or problem into the Fourier domain and work all calculations there, transforming back to the spatial domain only at the end of the calculation. The speed and ease of use of FFT algorithms has greatly enhanced the ability of designers and researchers to experiment with different system designs to achieve better overall results.

13.5 CONCLUSION

The most important information in this chapter for the professional using discrete display devices is the description of sampling in Section 13.4. Once the effects of sampling are understood and taken into account, discrete devices can be treated like other components of a system for calculating overall system performance. It is very important that sampling be accounted for accurately; it is not uncommon to find only the best-case results used. This leads to an overestimation of system capabilities, sometimes resulting in a system that cannot be used in its intended application.

References

Brigham, E. O., (1988). *The Fast Fourier Transform and Its Applications,* Prentice-Hall, Englewood Cliffs, NJ, pp. 131–166.

Feltz, J. C., (1990). Development of the modulation transfer function and contrast transfer functions for discrete systems, particularly charge-coupled devices, *Opt. Eng.* *13*(8).

Feltz, J. C., and M. A. Karim, (1990). The modulation transfer function of charge-coupled devices, *Appl. Opt.* 29(5):717–722.

Slottow, H. G., (1976). Plasma displays, *IEEE Trans. Electron Devices ED-23*: 760–772.

Other Bibliographic Sources

IEEE Transactions on Electron Devices has periodically produced special issues devoted to electronic display technologies. The August 1986 issue contains a wealth of information about technological advances that are now reaching maturity and is excellent reading. The IEEE *Spectrum* annually runs a review of technology in its January issue; there are usually several pages devoted to display technology.

14

Analytical Modeling and Digital Simulation of Scanning Charge-Coupled Device Imaging Systems

Terrence S. Lomheim and Linda S. Kalman

*The Aerospace Corporation,
Los Angeles, California*

14.1 INTRODUCTION

The imaging performance of visible scanning image sensors that employ silicon solid-state imaging focal plane devices is generally described in terms of some measure of resolving capability, which in turn is determined by the combined effects of the input scene image modulation, system modulation transfer function (MTF), and single-pixel signal-to-noise ratio (S/N). (A detailed definition of MTF is presented in Chapter 11.)

A commonly used figure of merit for film-based systems is the system's limiting resolution. This quantity is expressed as the spatial frequency corresponding to the intersection of the system MTF function and the threshold modulation (TM) or aerial image modulation (AIM) function. The MTF decreases with increasing spatial frequency, whereas the TM and AIM increase with increasing spatial frequency (Smith, 1985; Welch, 1972). The system MTF describes the ability of the sensor system to respond to sinusoidally varying spatial input as a function of frequency. The TM or AIM function displays the minimum image modulation required to produce a response in a focal-plane recording medium versus spatial frequency. The effects of S/N are hence included in the TM/AIM function.

The same basic concept can be applied to solid-state imaging systems. Here we specifically focus on sensor systems that use time-delay-and-integrate (TDI) charge-coupled device (CCD) scanning imagers (Bradley and Ibrahim, 1979; Lareau and Chandler, 1986). These devices spatially sample the input scene. The pixel in the image plane projects to a "ground footprint" in the object plane. For reference we define the ground sample distance (GSD) as the linear dimension between pixel footprints (object plane); the terminology here obviously derives from downlooking image sensor systems.

The limiting spatial resolution detectable by a spatially sampled, electro-optical (EO) imaging system is closely related to the Nyquist sampling criterion, which states that two samples are needed to unambiguously determine a spatial frequency. The inverse of the system limiting spatial frequency is called the ground-resolved distance (GRD) and is approximately equal to twice the GSD. The relationship is *approximate* because signal-to-noise ratio also has an impact on apparent resolution. For EO imaging systems an approach similar to that used in film systems can be used to determine the limiting resolution (Lyon et al., 1976); the limiting resolution is determined by the intersection of the MTF and AIM curves.

The GRD is a lumped figure of merit often used to describe EO sensor imaging performance. The Image Interpretability Rating System (IIRS) (Howes, 1987) is closely related to the GRD and was developed for military target identification purposes. Howes gives a relationship between the IIRS and the GRD; for example, a GRD of 8–16 in. corresponds to an IIRS rating of 7, a GRD of 16–30 in. to an IIRS rating of 6, etc. The GRD is an optimum figure of merit at the one spatial frequency of 1/GRD. However, real imagery consists of a complex superposition of many spatial frequencies, from a spatial frequency of zero on up.

An accepted measure of image sharpness, originally proposed by Schade (1948), is the equivalent line number N_e, also called the equivalent spatial bandwidth, representing the total area under the square of the MTF curve. Experimenters have demonstrated that subjective evaluation of image sharpness correlates well with N_e (Lloyd, 1982). However, the shape of the system MTF curve between zero and the Nyquist frequency is crucial in producing high-quality, interpretable images, particularly those containing man-made objects. This has been recognized by researchers, and improved figures of merit have been defined. Wolfe (1985a1) uses a figure of merit defined as MTF_A, which is the area between the system MTF and AIM curves. Kuttner (1983) and Rosenbruch (1983) describe image quality criteria based on a variety of different ways of "integrating over the MTF" from zero to the Nyquist frequency. These techniques represent improvements over the GRD specification but do not capture the subtleties that often arise in real TDI CCD-based EO imagery. High-fidelity imaging EO sensor simulations followed by visual analysis and photointerpretation are needed to fully assess detailed sensor designs.

In this chapter we outline the details of such an imaging sensor simulation, tailored for systems that use TDI CCD focal planes. Enough detail is included in our simulation that trade-offs between S/N ratio and resolution, and between

MODELING AND SIMULATION OF CCD IMAGING SYSTEMS 515

GSD and MTF, can be studied, and the cosmetic defects induced by uncalibrated offset and nonuniformity effects can be examined. Visual examples of these and other features are given. The simulation is built around a sensor TDI CCD MTF and signal-to-noise model. Hence, starting with a sensor design description, both simulated sensor imagery and GRD-based figures of merit can be generated for cross-reference purposes. Section 14.2 presents a detailed description of the scanning TDI CCD sensor signal and noise model, scan geometry considerations, and the assumed TDI CCD layout geometry. Section 14.3 covers the two-dimensional system MTF model. Section 14.4 outlines the simulation architecture and implementation, including key approximations and assumptions, and the details of the methods used for inserting TDI CCD nonuniformity effects and onboard data processing, such as data compression. Section 14.5 shows selected design trade-off examples using the simulation. Finally, Section 14.6 discusses the experimental hardware validation of the digital simulations.

14.2 SCANNING IMAGE SENSOR DESCRIPTION

Figure 14.1 depicts a nominal electro-optical imaging sensor system that employs a scanned TDI CCD. A portion of the ground surface is imaged by the sensor system, ultimately producing a sampled digital picture. We assume that the sun illuminates the scene to be imaged. The spectral characteristics of the scene are a function of the solar spectrum, downward atmospheric transmittance, ground spectral reflectance, and sun-to-ground geometry. The sensor system scans the image formed by the optical system across the TDI CCD focal plane. The focal plane consists of a two-dimensional matrix of $m \times n$ individual detecting elements (pixels) capable of converting the optical irradiance into a proportional electrical signal. Analog processing and multiplexing produce usable voltage signals from the TDI CCD, which are then digitized, corrected for pixel-to-pixel nonuniformity effects, compressed, and digitally multiplexed to a high-speed serial digital stream for transmission to a receiving station. Of course, the imaged scene is spectrally modified by propagation through an appropriate distance of the Earth's atmosphere and by the spectral transmittance of the optical system. Operations that are inverse to those performed by the sensor image collection process are applied to reconstruct the imaged scene.

In the remainder of this section, important details of the scanning sensor chain highlighted in Figure 14.1 will be discussed, including (1) single-pixel sensor signal and noise model, (2) sensor viewing/scan geometry considerations (whiskbroom, pushbroom scanners) and an overview of important system relationships for TDI scanners, and (3) an overview of typical TDI CCD topology and associated nonuniformity issues. These topics are covered as a background and introduction to the detailed description of the CCD sensor simulation described in Section 14.4.

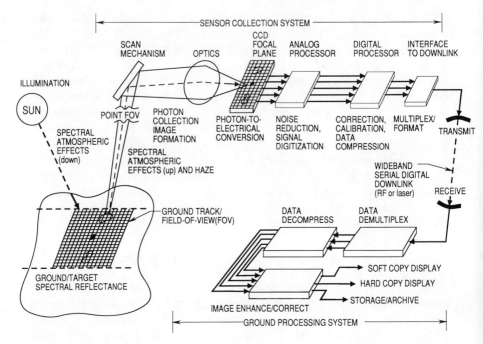

Figure 14.1 Schematic showing the signal path through sensor collection system and ground processing system. The signal-processing chain is similar for both pushbroom and whiskbroom sensor systems.

14.2.1 Sensor Signal and Noise Model

The magnitude and spectral content of the irradiance in the focused image depend on sensor design and system operating parameters. For a sufficiently small area of a solar-illuminated scene, and not considering active emission by the scene, the spectral radiance of the scene, $R_s(x, y, \lambda)$, is given by

$$R_s(x,y,\lambda) = (1/\pi)[T_A^{\text{down}}(\lambda)I_s^{\text{exo}}(\lambda)\rho(x,y,\lambda) + I_A^{\text{diffuse}}(\lambda)\,\rho(x,y,\lambda)] \qquad (14.1)$$

where $T_A^{\text{down}}(\lambda)$ is the spectral transmittance of the solar exoatmospheric spectral irradiance $I_s^{\text{exo}}(\lambda)$ down through the Earth's atmosphere to the ground surface, $\rho(x,y,\lambda)$ is the spatially varying spectral reflectance of the scene (assumed to be Lambertian), and $I_A^{\text{diffuse}}(\lambda)$ is the downwelling diffuse irradiance incident on the ground. $R_s(x,y,\lambda)$ is usually given in units of W cm^{-2} sr^{-1} µm^{-1}, and the coordinates (x,y) refer to ground spatial coordinates.

Using $R_s(x,y,\lambda)$ as the source radiance to be imaged by the sensor system, the continuous irradiance distribution imaged at the sensor focal plane, $I_{\text{fp}}(x,y,\lambda)$, is

MODELING AND SIMULATION OF CCD IMAGING SYSTEMS

$$I_{fp}(x,y,\lambda) = \Omega_{aper}T_{opt}(\lambda) \, [T_A{}^{up}(\lambda)R_s(x,y,\lambda) + R_H(\lambda)] * \text{PSF}_{opt}(x,y,\lambda) \quad (14.2)$$

where Ω_{aper} is the effective collecting solid angle of the optical aperture, $T_{opt}(\lambda)$ is the optical system spectral transmittance (including any waveband-defining filters), $T_A{}^{up}(\lambda)$ is the upward atmospheric spectral transmittance, $R_H(\lambda)$ is the contribution to the radiance arising from scattering within the atmosphere, and $\text{PSF}_{opt}(x,y,\lambda)$ is the wavelength-dependent, spatially stationary optical system point-spread function (PSF). The backscattered light, $R_H(\lambda)$, is the "haze" contribution arising from scattering from aerosols and can be considered a constant, additive term for simulations over relatively small areas. The symbol * denotes a two-dimensional convolution. This convolution quantifies, in part, the degradation in the spatial frequency information of the scene due to the imaging process.

The two-dimensional irradiance distribution is electronically captured by the TDI CCD focal plane during the scanning process. Before proceeding with the description of the electronic part of the image capture process, we review the details of image time delay and integration (TDI).

The geometric formation of an image by an $m \times 1$ line array scanner is accomplished by collecting an $m \times 1$ line image using the scanner, moving this line image one pixel width in the scan direction (perpendicular to the long dimension of the array), forming a second $m \times 1$ line image, etc., thus building up a two-dimensional image. In practice, the scan motion is continuous, whereas transducing the optical signal to the electrical domain requires integration over a discrete time period, the "integration time." Generally the scan rate is chosen such that the image is moved exactly one pixel width in one integration time (or one sample per pixel dwell). Clearly, some image smear occurs over an in-scan pixel width. Image formation using a TDI scanner is, from a geometrical viewpoint, identical to that of a line scanner described above.

In TDI the same piece of the image is successively scanned across n adjacent TDI pixels, which coherently sum the resulting signal. A column of n TDI pixels is available for each cross-scan pixel channel. Charge-coupled device technology is ideally suited for this purpose; in-scan CCD charge transport under optically transparent electrodes provides for this coherent summation directly in the analog domain. Assuming no noise correlation between pixels, the coherent signal summation scales as n, whereas the noise adds in quadrature, scaling as $\sqrt{n}$. Hence, an $m \times n$ TDI CCD scanner looks like an $m \times 1$ line scanner with its signal-to-noise ratio enhanced by the factor $n/\sqrt{n} = \sqrt{n}$. Important differences between the two exist, such as dynamic range capability and image smear effects. Smear effects associated with TDI CCDs will be discussed in Section 14.3.

The TDI CCD forms a two-dimensional sampled version of the continuous irradiance distribution described in Eq. (14.2), where the sampling distance corresponds to the pixel-to-pixel center spacing. In the in-scan direction this implies one sample per dwell. As will be seen in Sections 14.3 and 14.4, it is useful to describe the captured electronic image as a function of continuous variables (e.g., just before the two-dimensional spatial sampling process). The electronic signal response, $S_e(x,y,)$, where x and y denote continuous spatial variables, is in electrons/pixel.

$$S_e(x,y) = \frac{A_{pix} t_{int} N_{tdi}}{hc} \int_{\lambda_1}^{\lambda_2} \lambda \, \eta(\lambda) I_{fp}(x,y,\lambda) * PSF_e(x,y,\lambda) \, d\lambda \qquad (14.3)$$

A_{pix} is the area of a pixel, t_{int} is the single-pixel integration time, N_{tdi} is the number of TDI pixels (per cross-scan pixel channel), λ is the optical wavelength, $\eta(\lambda)$ is the spectral quantum efficiency of the CCD, h is Planck's constant, c is the speed of light, $PSF_e(x,y,\lambda)$ is the "electronic" point-spread function associated with the TDI CCD image capture process [i.e., conversion of the irradiance map $I_{fp}(x,y,\lambda)$ to an equivalent electronic signal map], and * denotes the two-dimensional convolution of $I_{fp}(x,y,\lambda)$ and $PSF_e(x,y,\lambda)$. Note that $S_e(x,y)$ is proportional to N_{tdi}, as discussed above. The collecting solid angle Ω_{aper} is related to key optical system parameters via

$$\Omega_{aper} = (1 - A_{obs}) \pi / 4 \, (f\#)^2 \qquad (14.4)$$

where A_{obs} is the optical area obscuration and $f\# = efl/D_{op}$ is the optical focal ratio (efl is the system effective focal length and D_{op} is the optical aperture diameter).

Combining Eqs. (14.2) and (14.3), we get

$$S_e(x,y) = \frac{A_{pix} t_{int} N_{tdi} \Omega_{aper}}{hc} \int_{\lambda_1}^{\lambda_2} \lambda \, \eta(\lambda) \, T_{opt}(\lambda) [T_A^{up}(\lambda) R_s(x,y,\lambda)$$
$$+ R_H(\lambda)] * PSF_{sys}(x,y,\lambda) d\lambda \qquad (14.5)$$

where

$$PSF_{sys}(x,y,\lambda) = PSF_{opt}(x,y,\lambda) * PSF_e(x,y,\lambda) \qquad (14.6)$$

$PSF_{sys}(x,y,\lambda)$ is the overall sensor system PSF defined at pixel coordinates (x,y) and for wavelength λ, and * denotes a two-dimensional convolution.

Given that $S_e(x,y)$ is expressed in units of electrons/pixel, it is important to provide a physical interpretation of this continuous function of x and y. First note that $S_e(x,y)$ already contains the effects of *spatial integration* over a single pixel area, A_{pix}, but not spatial sampling. The spatial variable y is obtained by transforming the temporal variable t (associated with scanning) via the scan velocity v_s, so that $y = v_s t$. If we could place a TDI CCD pixel of area A_{pix} at any (x,y) position of the irradiance distribution, $I_{fp}(x,y,\lambda)$ [Eq. (14.2)], the collected signal sample in electrons would be $S_e(x,y)$. However, the sampling positions are known, being established in the cross-scan x direction by the TDI CCD cross-scan pixel spacing, and in the in-scan, or y, direction, by the temporal sampling (e.g., one sample per dwell time, dwell time being the time it takes to scan a pixel width at the scan velocity).

This sampling can therefore be expressed mathematically as

$$S_e(x_n, y_m) = \iint \delta(x - x_n) \delta(y - y_m) S_e(x,y) \, dx \, dy \tag{14.7}$$

where $\delta(\cdot)$ is the Dirac delta function, $x_n = n\Delta x$, $y_m = m \, v_s t_{\text{dwell}}$, n is the nth cross-scan pixel, Δx is the cross-scan pixel-to-pixel center spacing, and m is the mth in-scan temporal sample. Note that the in-scan sampling distance is thus $\Delta y = v_s t_{\text{dwell}}$, where 1 sample/dwell implies $t_{\text{int}} = t_{\text{dwell}}$.

While measuring signal in electrons is a useful convention in the CCD world, signal manipulation beyond the CCD occurs in the voltage domain. The voltage level V_{sig} that is presented to the analog-to-digital converter (ADC) is given by

$$V_{\text{sig}} = G_{\text{conv}} G_{\text{ap}} S_e \tag{14.8}$$

where G_{conv} is the CCD output conversion gain typically expressed in units of μV/electron, and G_{ap} is the voltage gain (unitless) of the signal processing electronics from the CCD to the ADC. The spatial designation for S_e is now dropped because V_{sig} is a time-division-multiplexed signal; the remapping of the V_{sig} time samples to the $S_e(x,y)$ spatial samples requires specific knowledge of the CCD parallel-to-serial multiplexing approach and any subsequent multiplexing before the A/D conversion process.

The relationship between the output signal voltage V_{sig} and the number of digital counts, N_{ADC}, after its digitization is

$$N_{\text{ADC}} = V_{\text{sig}} / V_{\text{LSB}} \tag{14.9}$$

where V_{LSB} is the voltage corresponding to the least significant bit of the ADC. Typically an ADC is rated as having a maximum range and a certain resolution. For such an ADC, V_{LSB} is given by

$$V_{\text{LSB}} = V_{\text{ADC}}^{\max} / 2^n \tag{14.10}$$

where $V_{\text{ADC}}^{\max}$ is the maximum range and n is the number of bits of resolution. For example, if a 12-bit ($n = 12$) ADC with a 5 V input range ($V_{\text{ADC}}^{\max} = 5$ V) were used, then $V_{\text{LSB}} = 1.22$ mV. Hence a signal V_{sig} corresponding to 12.2 mV would produce $N_{\text{ADC}} = 10$ counts.

Equations (14.1)–(14.10) allow us to express sensor signal information in any domain that is convenient: as a spectral radiance in object space, $R_s(x,y,\lambda)$; as a focal-plane irradiance, $I_{\text{fp}}(x,y,\lambda)$; as sampled CCD signal electrons, $S_e(x_n, y_m)$; as a voltage input to an ADC, V_{sig}; or as a digital quantity, N_{ADC}.

The corresponding noise model associated with the CCD sensor has several components. In a TDI CCD sensor that uses buried-channel technology, we can restrict ourselves to noise sources that are due to (1) scene and haze photons, (2) dark current generation, (3) readout electronics, and (4) quantization effects. By noise we mean both the spatial fixed pattern and temporal characteristics. For the four sources listed above, the first and second possess spatially varying characteristics (pixel to pixel) whereas the others do not. Calibration schemes exist for removing or greatly reducing the fixed-pattern (spatially varying) part of the noise; however, the effects of the corresponding temporal noise cannot be eliminated. A discussion of these and other related gain and offset nonuniformity effects follows in Section 14.2.3.

The photon noise is due to the random arrival rates of photons (Kingston, 1979) and is represented as a Poisson process in the conversion of photons to photoelectrons. The root-mean-square (rms) value of the photon noise, $n_{ph}(x_n, y_m)$, is simply the square root of the collected scene and haze electrons,

$$n_{ph}(x_n, y_m) = [S_{e,sig}(x_n, y_m) + S_{e,haze}]^{1/2} \tag{14.11}$$

where $S_{e,sig}(x_n, y_m)$ and $S_{e,haze}$ are the scene and haze electron maps as described by Eq. (14.5). Since the scene electron map varies with image pixel position, the associated photon noise varies similarly.

Charge-coupled device dark current physically arises from thermally induced electron–hole pair generation across the silicon bandgap (Sequin and Tompsett, 1975, p. 116). The dark current manifests itself as an output dc offset and also contributes shot noise (a Poisson process) whose rms value scales as the square root of the dark-current level. At the TDI CCD output, the integrated dark-current level is given in electrons/pixel at a pixel x_n by

$$D(x_n) = D_{tdi}(x_n) + D_{mux} \tag{14.12}$$

$$D_{tdi}(x_n) = t_{int} (\Delta x_{tdi} \Delta y_{tdi}) G_{tdi}(x_n) N_{tdi}/q \tag{14.13}$$

$$D_{mux} = t_{int} (\Delta x_{mux} \Delta y_{mux}) G_{mux}/q \tag{14.14}$$

where $D_{tdi}(x_n)$ and D_{mux} are the integrated dark-current levels (electrons/pixel) associated with the TDI imaging section and on-chip CCD multiplexer, respectively. The quantity t_{int} is the single-pixel integration time (scan direction); Δx_{tdi}, Δy_{tdi} and Δx_{mux} and Δy_{mux} are the TDI and multiplexer pixel dimensions, respectively; and $G_{tdi}(x_n)$ and G_{mux} are the dark-current generation rates of the TDI and multiplexer regions, respectively (typically in units of nA/cm²). The dark-current generation rates $G_{tdi}(x_n)$ and G_{mux} are strong functions of temperature. Modest cooling of the CCD (Baker, 1980) can reduce dark current to negligible levels in many applications. The spatial variable x_n corresponds to the pixel spatial location in the cross-scan direction, indicating that the component $D_{tdi}(x_n)$ varies with pixel position in this direction only. Variation in the in-scan, or y, direction can be significantly reduced by the continuous TDI summation process. The serial multiplexer transports the $D_{tdi}(x_n)$ pattern spatially intact to the TDI CCD output. Dark current generated and collected in the serial multiplexer itself reaches an equilibrium level that is the same for each pixel and hence represents a constant level (no spatial dependence) as shown in Eqs. (14.12) and (14.14).

The shot noise from the dark current (Sequin and Tompsett, 1975, p. 116) is thus

$$n_{dc}(x_n) = [D_{tdi}(x_n) + D_{mux}]^{1/2} \tag{14.15}$$

and reflects the cross-scan pixel-to-pixel spatial dependence due to $D_{tdi}(x_n)$.

The rms readout noise, v_{ro}, accounts for the on-chip electrometer voltage noise (Kosonocky and Carnes 1972), $v_{ro,ccd}$, and any additional noise generated in the analog processor, $v_{ro,ap}$. Referring these noise sources to the TDI CCD output in electrons gives

$$n_{ro} = [1/G_{conv}] [v_{ro,ccd}^2 + (v_{ro,ap}/G_{ap})^2]^{1/2} \tag{14.16}$$

Finally, the rms quantization noise associated with the ADC process, when referred to the TDI CCD output in electrons, is (Ahmed and Natarajan, 1983)

$$n_q = V_{LSB}/G_{conv} G_{ap} \sqrt{12} \tag{14.17}$$

Note that if the voltage gain of the analog signal processor or the CCD conversion gain is made large, the relative contribution of readout and quantization noise sources, n_{ro} and n_q, can be minimized. The penalty in doing this is to potentially limit the system dynamic range by exceeding the usable input voltage range of the ADC and/or analog processor.

The total rms noise is obtained by the quadrature sum of Eqs. (14.11) and (14.15)–(14.17), or

$$n_{tot}(x_n, y_m) = [n_{ph}^2(x_n, y_m) + n_{dc}^2(x_n) + n_{ro}^2 + n_q^2]^{1/2} \tag{14.18}$$

with the indicated (x_n, y_m) pixel variations. Assuming that the fixed pattern noise due to dark current and other sources can be removed by calibration, the two-dimensional pixel-to-pixel signal-to-noise map is given by the ratio of Eqs. (14.5) and (14.18). In reality, the TDI CCD parallel/serial readout architecture makes things more complicated than this by introducing additional nonuniformity effects. The origin of these nonuniformities and their relationship to the TDI CCD topology is briefly discussed in Section 14.2.3. The method for introducing fixed pattern and temporal noise effects implied by Eqs. (14.12) and (14.18) (as well as other realistic CCD nonuniformities) into the sensor simulation is described in Section 14.4.3.

14.2.2 Sensor Scan Geometry Considerations and System Relationships

Two basic scan techniques applicable to line image sensors (including TDI CCDs) are depicted in Figures 14.2a and 14.2b: the "whiskbroom" and "pushbroom" scans. Elementary considerations for these two basic sensor approaches are discussed in order to give the reader a perspective for the extent over which the scanning system simulation described in Sections 14.3 and 14.4 are valid.

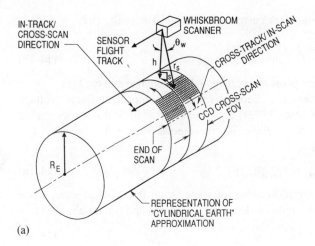

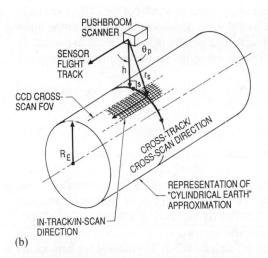

Figure 14.2 Graphic illustration of the collection mode for (a) a whiskbroom sensor and (b) a pushbroom sensor, showing projection of the collection array onto the "cylindrical Earth." S is the distance along the cylinder surface from nadir to the LOS vector.

For simplicity, we derive ground sample distance (GSD) sizes based on the "cylindrical" Earth approximation implied in Figures 14.2 and 14.3. This approximation assumes that the effects of Earth's curvature in the sensor in-track direction are negligible compared to the cross-track direction. In any realistic scanning sensor design, a more realistic and complex oblate Earth model would be required. Additionally, the interplay between the Earth's rotation and the sensor in-track motion must be modeled for a complete treatment. For aircraft-based

MODELING AND SIMULATION OF CCD IMAGING SYSTEMS

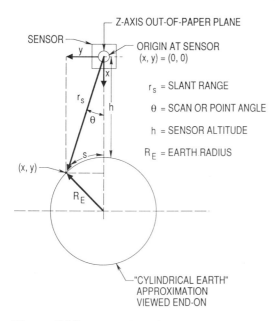

Figure 14.3 Illustration of the sensor/target geometry derived for the "cylindrical Earth" approximation.

scanning sensors the flight trajectory must be quantified, whereas for space-based systems the orbital parameters must be known in order to model the sensor in-track motion relative to the Earth's surface.

A generic whiskbroom scanner is shown in Figure 14.2a. The pertinent geometric parameters are defined by the cylindrical Earth end view in Figure 14.3. The direction labeled "in-track" refers to the direction of the sensor motion. The whiskbroom scan motion of the TDI CCD focal plane is perpendicular to the sensor "in-track" direction; hence the in-track and "cross-scan" directions are the same. It is assumed that the whiskbroom scanning sensor has at least a two-dimensional (2-D) scan capability, where one component of the scan motion is exactly antiparallel to the in-track direction, and at a rate that cancels the velocity of this motion when projected to the cylinder surface (ground). The orthogonal in-scan motion then forms a linear "strip" image along the curved part of the cylinder as shown. If this in-track velocity cancelation is not implemented, the projected strip will have an S shape rather than being linear. It should be noted that the correction for in-track motion during whiskbroom scanning is achieved by smoothly varying (e.g., scanning) the pitch angle of the sensor line of sight (LOS). (The pitch angle is measured from the sensor nadir direction to the LOS as it is pointed along the in-track direction.)

The width of the linear strip, when projected onto the cylinder, is the cross-scan field of view (FOV) and increases as a function of increasing scan angle θ_w. Both in-scan and cross-scan magnification of the strip image occur as a function of θ_w, as will be discussed below. This magnification effect (not shown in Figs. 14.2a and 14.2b) is called the "bowtie" distortion. The whiskbroom scan occurs at an angular rate of $d\theta_w/dt$ between the limits of the angle θ_w ($\pm \theta_w^{max}$), corresponding to the "end-of-scan" position in Figure 14.2a. The figure shows the scan both reversing at the end-of-scan position and advancing by one ground-projected cross-scan FOV width. This implies a bidirectional scan capability for the TDI CCD. If a unidirectional TDI CCD is used, the sensor would have to rapidly change from a pointing angle of $+\theta_w^{max}$ to $-\theta_w^{max}$ while advancing one cross-scan FOV. This implies additional complexity in executing and controlling the required two-dimensional scan.

For a spherical or oblate Earth model, another important geometric distortion must be considered. As the TDI CCD is whiskbroom-scanned, its projection onto a spherical surface rotates about the in-scan direction with increasing scan angle θ_w. Compensation of this rotation then requires a counterrotation or "yaw" correction about the LOS as a function of scan angle. In our further discussion of the whiskbroom approach we assume that the sensor scan and pointing subsystem has the 2-D capability needed to remove the S-shaped scan and yaw distortions. Only the bowtie distortion (due to magnification effects) remains.

Variants on the whiskbroom scan are possible and often desirable. The angular scan limits need not be symmetrical about the sensor nadir direction: whiskbroom scanning can have a bias to either side of nadir. Successive scan strips may be overlapping, noncontiguous, and/or of different lengths depending on the relationship between the scan rate $d\theta_w/dt$, scan limits $\pm \theta_w^{max}$, and sensor in-track velocity. Control of the LOS pitch angle (needed to eliminate S-shape scan distortion) also allows forward and back-scanning capability in the in-track direction. Clearly, the overall sophistication, programmability, and control of the scan/pointing subsystem will determine the sensor's agility and targeting capability. The simplest scan mechanism to envision is one that provides continuous in-track image coverage between the angles $\pm \theta_w^{max}$. Efficiency is maximized when the scan rate, $d\theta_w/dt$, which need not be constant, is adjusted so that the in-scan limits ($\pm \theta_w^{max}$) are covered in the time it takes the sensor to move one cross-scan FOV width. As this width varies as a function of the scan angle (magnification effects), the optimization is typically done at nadir and, consequently, scan strip overlap occurs for the off-nadir angles.

A pushbroom scanner is shown in Figure 14.2b. Figure 14.3 again defines the geometry parameters. The pushbroom is easier to implement than the whiskbroom; the sensor LOS is simply pointed at some angle θ_p, and an image is scanned out by the sensor in-track motion (flight). As such, no "active" scanning occurs. The TDI CCD focal-plane scan (in-scan for pushbroom in Fig. 14.2b) occurs in the sensor in-track direction. The width of the strip scanned out by the pushbroom is determined by the focal-plane cross-scan angular extent.

As before, variants on this simple approach are possible. For sensitivity enhancement the sensor scan rate could be decoupled from the in-track velocity by backscanning. This is done by smoothly varying (scanning) the LOS pitch angle as described above for the whiskbroom scanner. Such an approach would obviate continuous ground coverage. The pushbroom scanner also suffers from geometrically induced scan distortions. For off-nadir pointing angles, projection onto a spherical Earth causes a line in the TDI CCD image to rotate about the in-scan direction. A counterrotation or yaw correction about the LOS must be made (just as with the whiskbroom) as a function of pointing angle $\pm \theta_p$. Removal of yaw distortion, operating in flexible modes (e.g., backscanning), and LOS agility (ability to rapidly repoint the sensor LOS) require, even for a pushbroom scanner, a sophisticated 2-D pointing/scanning subsystem.

The nature of some of the geometric distortions mentioned above and simplified to the case of the cylindrical Earth approximation, can be seen by examining Figure 14.3. We first define the instantaneous field of view (IFOV) of a sensor pixel, which is assumed to be square—the angle subtended by the center-to-center spacing Δx of two adjacent pixels in the cross-scan direction, and the angle subtended by the width defined by one sample per dwell, $\Delta y = v_s t_{dwell}$, in the in-scan direction, are equal:

$$\text{IFOV} = \Delta x/\text{efl} = \Delta y/\text{efl} = \text{GSD}/r_s \tag{14.19}$$

where efl is the sensor effective focal length, GSD is the ground sample distance, and r_s is the slant range (see Fig. 14.3). This equation is valid only if the surface we are imaging is flat. Hence, it applies in the in-track, but not the cross-track, direction in Figures 14.2 and 14.3. To compute the effective cross-track GSD we must first determine the relationship between S, the distance along the cylinder surface in Figure 14.3, and the quantities h, r_s, θ, and R_E, the sensor altitude, slant range to position (x,y), sensor scan or point angle, and Earth radius, respectively. No forward or aft pointing (e.g., nonzero sensor pitch angle) is assumed. The quantity S is given by

$$S = R_E \arcsin\left[(r_s \sin \theta) / R_E\right] \tag{14.20}$$

and the slant range r_s is given by

$$r_s = (R_E + h) \cos \theta - [(R_E + h)^2 \cos^2 \theta - 2 R_E h - h^2]^{1/2} \tag{14.21}$$

The cross-track GSD (for either the pushbroom or whiskbroom scanner) at the scan or point angle θ' is

$$\text{GSD}_{ct}(\theta') = \text{IFOV} \frac{dS}{d\theta} \quad (\theta = \theta') \tag{14.22}$$

and the in-track GSD is [by Eq. (14.19)]

$$\text{GSD}_{it}(\theta') = \text{IFOV} \, r_s \quad (\theta = \theta') \tag{14.23}$$

To show an example of the impact that off-nadir pointing or scanning has on GSD, consider a sensor located at an altitude h of 500 km, capable of scanning or pointing from $-50°$ to $+50°$. We wish to compare the in-track and cross-track GSDs for $\theta' = 50°$. Using Eqs. (14.22) and (14.23), we find

$$\text{GSD}_{it}(\theta' = 50°) / \text{GSD}_{it}(\theta' = 0°) = 1.65 \tag{14.24}$$

and

$$\text{GSD}_{ct}(\theta' = 50°) / \text{GSD}_{ct}(\theta' = 0°) = 4.96 \tag{14.25}$$

Elongation of GSD for a whiskbroom scanner is valid provided the CCD cross-scan FOV represents a fairly small angle compared to the maximum scan angle, θ_w^{max}. For a pushbroom scanner this approximation is certainly valid; the angle subtended by the CCD device in the in-scan (in-track) direction is the TDI length divided by the effective focal length. With these assumptions it is clear that the GSDs for both pushbroom and whiskbroom scanners grow asymmetrically with point angle (θ_p) or scan angle (θ_w) when square pixels are used in the focal plane. The overall 2-D pixel maps generated by such scanners reflect the in-track GSD magnification effects due to increased slant range, while cross-track GSD growth [Eq. (14.22)] is due to increased slant range, obliquity of ground surface to LOS (cosine effect), and Earth's curvature in the cross-track direction. A detailed presentation of examples of the types of geometric distortions discussed above is given by Anuta et al. (1983) for the Landsat Multispectral Scanner (MSS).

It is apparent that imaging figures of merit (such as the GRD defined in Section 14.1) that are directly proportional to GSD are valid only over a narrow range of scan/point angles. Typically, system performance is quoted at nadir; the off-nadir cases are quoted in terms of the appropriate degradation factor from nadir performance. Area rate of coverage (ARC) is another important figure of merit for scanned image sensors. Since the GSD and GRD are both functions of θ_w and θ_p, ARC will also be a function of these angles. Quoting the ARC at nadir can be a little misleading because ARC increases with increasing scan/point angles and we want to maximize the ARC in most sensor applications. It may appear that off-nadir pointing or scanning (pushbroom) or large scan angles (whiskbroom) might be desirable until we recall that the increased ARC comes at the expense of degraded resolution (increased GSD and GRD).

For simplicity we define ARC *at nadir* (typically in units of km²/sec) as

$$\text{ARC} = S_G v_{scan} \tag{14.26}$$

where v_{scan} is the ground-projected scan velocity (nadir), and S_G, the cross-scan "swath" width, is

$$S_G = r_s \text{ IFOV } N_{pix,cs} \tag{14.27}$$

where r_s, the slant range, equals the sensor altitude h at nadir, and $N_{\text{pix,cs}}$ is the number of cross-scan pixel channels in the TDI CCD focal plane. The origin of v_{scan} depends on sensor type (pushbroom vs. whiskbroom) and platform (aircraft, spacecraft in circular orbit, etc.). Within each category (a whiskbroom scanner on a satellite, a pushbroom scanner on an aircraft, etc.), v_{scan} will depend on the sophistication and intent of the scan mode, from using backscanning for sensitivity enhancement (thus decoupling the sensor image scan rate from the sensor flight velocity) to the simplest case of allowing the sensor flight velocity to directly determine v_{scan}.

To meet the criterion of one sample per dwell, v_{scan} must also equal

$$v_{\text{scan}} = \text{GSD}/t_{\text{int}} \tag{14.28}$$

where t_{int} is the single-pixel integration time and is related to the scanner line rate f_{line} by

$$f_{\text{line}} = 1/t_{\text{int}} \tag{14.29}$$

The total data rate from this sensor, N_{data}, in bits per second is

$$N_{\text{data}} = N_{\text{pix,cs}} f_{\text{line}} b \tag{14.30}$$

where b represents the number of bits per cross-scan pixel.

The trade-offs and constraints implied in Eqs. (14.26)–(14.30) must be considered carefully. For example, if we consider a pushbroom scanner mounted in an aircraft flying at a fixed altitude ($r_s = h$, at nadir) and velocity v_{scan}, and wish to achieve a certain GSD during a TDI CCD operating at one sample per dwell, the parameters ARC, S_G, and N_{data} become derived parameters varying only with $N_{\text{pix,cs}}$. The integration time and hence the line rate are fixed by Eq. (14.28). Since GSD = r_s IFOV at nadir, the IFOV is also fixed.

In general, the pushbroom scanner approach is chosen for narrow-FOV, high-sensitivity, high-resolution applications, whereas the whiskbroom scanner is used in lower sensitivity, lower resolution, wider area coverage applications. At the level of the TDI CCD, the two approaches generally evoke different requirements. Focal planes for pushbroom scanners tend to be longer in total cross-scan extent and operate at lower line rates (longer integration times) with fewer stages of TDI. For whiskbroom scanners, the cross-scan extent of the focal plane is generally less and the line rates higher (shorter integration times), employing more stages of TDI to make up for the loss in sensitivity brought on by shorter integration times.

A final subject of importance in sensor scan geometry is that of image smear. The ability to compensate for the pushbroom and whiskbroom scanner distortions (which occur via projection onto an oblate Earth) is usually restricted to a single position within the TDI CCD FOV (usually the center). As we move away from this position, uncompensated image motion can occur in both the in-scan and cross-scan directions; these motions degrade the sensor resolution, and this degradation becomes worse as more stages of TDI are used in the sensor. The treatment of this degradation from the TDI CCD imaging viewpoint is covered in Section 14.3, with examples following in Sections 14.5 and 14.6. The computation of the in-scan and cross-scan uncompensated motion velocities requires use of an oblate Earth model for either the pushbroom or whiskbroom scanner.

14.2.3 TDI CCD Layout Geometry

Figure 14.4 shows the nominal geometrical layout for a scanning TDI CCD device. Examples of such an architecture are given in Bradley and Ibrahim (1979) and Lareau and Chandler (1986). Clearly, a TDI CCD is a two-dimensional array, where the device dimension in the in-scan or TDI direction is usually much shorter than in the cross-scan direction.

The charge transport in the TDI section of the CCD occurs in the in-scan direction as discussed in Section 14.2.1. Each summed image line exiting the bottom of the TDI section is then loaded in parallel into a serial CCD multiplexer, which transports this image line serially to an output electrometer (charge-to-voltage conversion circuit). Clearly, the clocking rate of the serial multiplexer must be fast enough that an image line completely exits the multiplexer before the next image line is loaded in parallel from the TDI section. This rate is proportional to the length of the serial multiplexer. For example, if a TDI CCD is operated at a parallel line rate of 20,000 lines/sec and the length of the multiplexer is 500 pixels, the required serial rate is at least 10^7 pixels/sec. If the length of the multiplexer instead were 100 pixels, the minimum serial rate would be

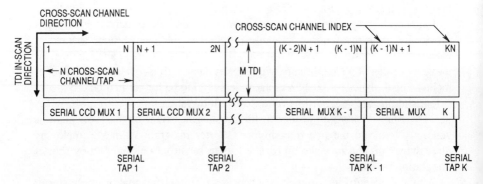

Figure 14.4 Diagram illustrating the physical layout of the front-illuminated CCD TDI linear array device.

2×10^6 pixels/sec to support the same line rate. Physical limitations on this serial rate usually dictate the use of multiple serial readout registers on a single TDI CCD chip, particularly when the total cross-scan pixel length is large. The physical limitations include considerations of on-chip power dissipation, electrometer and signal-chain noise, and analog signal-chain design constraints including ADC operating speed and power dissipation.

Each serial multiplexer therefore drives a serial output tap via an on-chip electrometer; for example, a 2048 cross-scan pixel TDI CCD with 512 pixel serial multiplexers will require four of these multiplexers, along with the electrometers and serial taps. For the purposes of this discussion, we assume that an analog signal processing chain is provided for each serial tap output, including the ADC. Sufficient differences in the electrical offset and gain characteristics associated with each serial tap and analog signal chain will produce "bandlike" imperfections in the final image. In general, we can assume that these serial-tap-induced gain and offset differences are uncorrelated. The physical distance between the on-chip CCD electrometers is usually large enough for this to be true, and there is no obvious reason for the distinct analog signal chains for each tap output to be correlated. The visual impact of such effects is covered in Section 14.4.3.

The pixel-to-pixel gain and offset nonuniformity effects obviously arise in the physical structure of the TDI CCD chip. The dominant signature of such nonuniformities comes from the TDI section. Pixel-to-pixel variations in responsivity and gain can originate from variations across the TDI region in physical pixel sizes (expected to be very small owing to excellent photolithography for VLSI production these days), variations in spectral responsivity due to layer thickness variations (Kesler and Lomheim, 1986), variations in doping density, etc. These effects generally have both a random (uncorrelated from pixel to pixel) and systematic (correlated pixel to pixel) component. Pixel-to-pixel variations in offsets arise from spatial variations in dark-current generation rate over the TDI section. These dark-current-induced offsets also have both random and systematic (correlated) components (Anon., 1987). The dark-current "patterns" that can arise are often traceable to defects and striations in the silicon boule from which the IC wafers are taken. Present CCD manufacturing technology uses epitaxial techniques to "grow" the optically active collection volume of the CCD, and these problems are greatly reduced.

It should be noted that pixel-to-pixel variations in both gains and offsets are continuous across the tap boundaries shown in Figure 14.4. The TDI section is physically contiguous across its extent in the cross-scan direction, with the vertical TDI channels being delineated by channel stops that continue uninterrupted across tap boundaries. The distinct nature of those pixels associated with a particular tap comes from the characteristics of the associated serial multiplexer, electrometer, and analog signal chain. Pixel-to-pixel variations in the in-scan direction are usually reduced by the TDI summation process. What is seen at the bottom of the TDI section is an effective gain and offset for a particular cross-scan pixel representing a corresponding TDI column. By the same token, the pixel-to-pixel signature of each serial multiplexer is also averaged out. Here the pixel-to-pixel information in the line image passes through each pixel in the serial multiplexer; hence the gain and offset of a given multiplexer are each represented by a single effective value. The nonuniformities induced in the output image will therefore have structure only in the scan direction; the pixel-to-pixel variations are usually referred to as *streaking*, while the tap-to-tap effects are referred to as *banding*.

Section 14.4.3 outlines the methodology used for including the pixel-to-pixel and tap-to-tap gain and offset nonuniformity effects in our TDI CCD sensor simulation. Examples of some of the visual effects are also displayed.

14.3 TWO-DIMENSIONAL MODULATION TRANSFER FUNCTION MODEL

The reduction of spatial frequency information in original scene spatial frequency content due to the entire electronic imaging process—that is, from a continuous input scene radiance to a continuous electronic image (one step before spatial sampling)—is expressed by the convolutions in Eq. (14.5). We can rewrite the part of Eq. (14.5) subject to wavelength integration as

$$S_e(x,y) \approx \int W(x,y,\lambda) * \text{PSF}_{sys}(x,y,\lambda) d\lambda \qquad (14.31)$$

where

$$W(x,y,\lambda) = \lambda\, \eta(\lambda) T_{opt}(\lambda) \{ [T_A^{up}(\lambda)\, T_A^{down}(\lambda) I_s^{exo}(\lambda) \rho(x,y,\lambda)\,]/\,\pi \\ + [I_A^{diffuse}\, \rho(x,y,\lambda)\, T_A^{up}(\lambda)\,]\,/\,\pi + R_H(\lambda) \} \qquad (14.32)$$

The symbols have their previous meaning and $*$ denotes the two-dimensional convolution of the continuous functions $W(x,y,\lambda)$ and $\text{PSF}_{sys}(x,y,\lambda)$. $W(x,y,\lambda)$ varies with the original input scene radiance distribution given by Eq. (14.1) at each λ and can be thought of as a two-dimensional spectral weighting function for the system PSF. Calculation of a final sampled electronic image is possible if we have access to analytical or sufficiently accurate numerical representations of $W(x,y,\lambda)$ and $\text{PSF}_{sys}(x,y,\lambda)$.

The function $\text{PSF}_{sys}(x,y,\lambda)$ is obtained by analyzing the details of the sensor system. However, the TDI CCD sensor analytical models are described in the spatial frequency domain by using the modulation transfer function (MTF). The system PSF and system optical transfer function (OTF) form a Fourier transform pair. The MTF is defined to be the modulus of the OTF (Goodman, 1968). Hence the MTF can have only positive values. In useful symmetrical incoherent imaging systems, the component OTFs are real but can take on negative values. (When assymetries are present in $PSF_{sys}(x,y,\lambda)$ the OTF's become complex.) It is a widespread practice in the CCD and imaging communities to refer to the OTF and MTF as equal even though the MTF is defined to have only positive values. We continue this practice here (in order to have consistency with the existing literature); Table 14.2 lists MTFs that have sinc functions, which clearly can take on negative values. Technically these are OTFs. The system MTF is generally more amenable to analytic description than the system PSF. Therefore, in practice, it is more straightforward to implement the convolution in Eq. (14.31) in the spatial frequency domain. A further advantage of describing the sensor degradation with MTFs is that the total sensor system MTF is a *product* of MTFs, each describing particular sensor components in the overall spatial frequency degradation.

Equation (14.31) can be written analogously in the spatial frequency domain as the product of the two-dimensional Fourier transform of $W(x,y,\lambda)$ and the sensor system MTF:

$$S_e(x,y) \approx \int FT^{-1} \{ FT [W(x,y,\lambda)] \, MTF_{sys}(f_x,f_y,\lambda) \} \, d\lambda \qquad (14.33)$$

where f_x, f_y denote spatial frequency variables in the x and y directions, and FT and FT^{-1} denote the Fourier and inverse Fourier transforms, respectively. Recall that $S_e(x,y)$ represents a continuous two-dimensional function that is then spatially sampled by the pixel grid per Eq. (14.7).

Equation (14.33) is generally not amenable to direct implementation in practical simulation work. It is challenging, if not impossible, to obtain empirical data, or to model from first principles the function $W(x,y,\lambda)$, over any significant spectral bandwidth. Approximations become inevitable; the limitations produced by these approximations and their implications are discussed in detail in Sections 14.4.1 and 14.4.2. Section 14.4 describes the architecture and implementation of a TDI CCD sensor simulation that utilizes a spatial frequency description of the sensor (MTF) and high-resolution input imagery obtained from another sensor. Nonuniformity effects such as those outlined in Section 14.2.3 and the signal and noise models described in Section 14.2.1 are also included.

We now provide a detailed description of the two-dimensional MTF model $[MTF_{sys}(f_x,f_y,\lambda)$ in Eq. (14.33)], which is appropriate for a TDI CCD sensor system in most applications. At the end of this section we discuss other MTF sources that are not included in the model and our reasons for treating them as second-order effects.

In the simulation process discussed in Section 14.4, the system MTF behavior must be understood over the visible spectral region that is applicable to silicon devices. Since TDI CCDs usually employ frontside-illuminated architectures, this response is limited to the regime of 400–1000 nm. Several of the component MTFs depend on wavelength, and the system MTF reflects this concomitant wavelength dependence. In the case of a broadband imaging application, the system MTF is polychromatic, whereas in narrowband spectral imaging (e.g., multispectral imaging), the MTF will be essentially monochromatic. The overall system MTF is a two-dimensional function of the spatial frequency variables f_x and f_y, both in units of line pairs per millimeter, where x denotes the cross-scan direction and y denotes the in-scan direction. Table 14.1 lists all the in-scan and cross-scan MTF components that we consider for our description of a scanning TDI CCD system and their specific dependencies on optical and CCD device parameters. Equations summarizing each of the component MTFs appear in Table 14.2, along with pertinent references.

The pixel spatial aperture MTF (Table 14.2A) is a product of sinc functions resulting from the assumption that the response across a pixel has a trapezoidal shape. This model assumes an area of flat response in the center of the pixel, with the response falling off linearly at either edge of the pixel to zero at the center of the adjacent pixels. At the boundary between two pixels a response of 0.5 means that a photoelectron has an equal probability of being collected by either of the two pixels. Independent pixel response functions may apply in the in-scan versus the cross-scan directions, and hence Table 14.2A contains distinct MTF equations for each.

Table 14.1 Cross-Scan and In-Scan MTF Components

f_x (cross-scan)	f_y (in-scan)
Detector spatial aperture ($\Delta x, \Delta S_x$)	Detector spatial aperture ($\Delta y, \Delta S_y$)
Optics ($\lambda, f\#, \varepsilon$)	Optics
Optical degradation (λ, σ, ϕ)	Optical degradation
Carrier diffusion [$\alpha(\lambda), L_{depl}, L_{diff}$]	Carrier diffusion
Velocity mismatch [$n_{TDI}, k_{smear}(x)$]	Velocity mismatch [$n_{TDI}, k_{smear}(y)$]

λ	wavelength	$\alpha(\lambda)$	silicon absorption coefficient
$\Delta x, \Delta y$	pixel pitch		
$\Delta S_x, \Delta S_y$	width of flat response	L_{depl}	CCD pixel depletion width
$\Delta S_x, \Delta S_y$	width of flat response	L_{diff}	CCD carrier diffusion length
$f\#$	focal ratio		
ε	optical linear obscuration	n_{ph}	number of CCD clock phases
σ	optical rms wave front error		
ϕ	optical correlation function	n_{TDI}	number of TDI stages
$k_{smear}(y) = \Delta v_y T_{int}/\Delta_y$		k_{smear}	smear coefficient
$k_{smear}(x) = \Delta v_x T_{int}/\Delta x$			
Δv_x	velocity mismatch, cross-scan		
Δv_y	velocity mismatch, in-scan		
T_{int}	pixel integration time		

Table 14.2 Component MTF Equations

A. Detector spatial aperture (trapezoidal)
   (Schumann and Lomheim, 1989)
   $$\text{MTF}_{apt}(k_x,k_y) = \text{MTF}_{apt}(k_x,\Delta x,\Delta S_x)\, \text{MTF}_{apt}(k_y,\Delta_y,\Delta S_y)$$
   $$\text{MTF}_{apt}(k,\Delta\ell,\Delta S) = \text{sinc}(k\Delta\ell)\, \text{sinc}\,[k(\Delta\ell - \Delta S\ell)]$$

B. Detector temporal aperture/velocity mismatch
   (Johnson, 1982)
   $$\text{MTF}_{int}(k_x,k_y) = \frac{\text{sinc}\,[n_{TDI}(k_x\Delta V_x + k_y\Delta V_y)T_{int}]}{\text{sinc}\,[(k_x\Delta V_x + k_y\Delta V_y)T_{int}/n_{ph}]} \times \text{sinc}\left[\frac{k_x\Delta Y + (k_x\Delta V_x + k_y\Delta V_y)T_{int}}{n_{ph}}\right]$$

C. Carrier diffusion
   (Seib, 1974)
   $$\text{MTF}_{diff}(k_x,k_y,\lambda) = \frac{1 - [\exp(-\alpha L_{dep}\ell)]/(1 + \alpha L)]}{1 - [\exp(-\alpha L_{dep}\ell)]/(1 + \alpha L_{diff})]}$$
   $$L = L_{diff}/(1 + 4\pi^2 L_{diff}^2 k^2)^{1/2} \quad k^2 = k_x^2 + k_y^2, \quad L_{diff} = \sqrt{D\tau}$$

D. Optical diffraction
   (O'Neill, 1956)
   $\text{MTF}_{opt}(k_x,k_y,\lambda)$ and $k = \sqrt{k_x^2 + k_y^2}$
   (use O'Neill formula if optics is diffraction-limited)

E. Optical degradation
   (Nicholson, 1975)
   $\text{MTF}_{od}(k_x,k_y,\lambda)$ and $k = \sqrt{k_x^2 + k_y^2}$
   (use Nicholson equations)

F. Composite system MTF:
   $$\text{MTF}_{total}(k_x,k_y) = \text{MTF}_{apt}(k_x,k_y)\, \text{MTF}_{int}(k_x,k_y)\, \text{MTF}_{diff}(k_x,k_y)\, \text{MTF}_{opt}(k_x,k_y)\, \text{MTF}_{od}(k_x,k_y)$$
   True at a single λ

---

k = spatial frequency, typically in ℓ p/mm
sinc = $(\sin \pi x)/\pi x$; x = cross-scan; y = in-scan.
$\Delta x, \Delta y$ = pixel width (x,y)
$\Delta S_x, \Delta S_y$ = pixel flat-response width (x,y)
n_{TDI} = # TDI stages
ΔV_x = velocity error, x direction
ΔV_y = velocity error, y direction
T_{Int} = integration time
n_{ph} = no. of clock phases per pixel
α = absorption coefficient of silicon at λ
$L_{dep\ell}$ = CCD depletion width
L_{diff} = diffusion length
One sample per well assumed in the in-scan direction.

In a scanned imaging system there are several sources that can contribute to MTF degradation due to uncompensated image motion. This MTF degradation is accounted for by the temporal aperture/velocity mismatch MTF term given by Table 14.2B, where Δv_x and Δv_y are the velocity errors produced by this uncompensated motion. Here, x and y correspond to in-scan and cross-scan directions, respectively.

The equation in Table 14.2B is a function of the number of TDI phases, N_{tdi}, and the number of clock phases per pixel, N_{ph}. Note that when no velocity errors are present ($\Delta v_x = \Delta v_y = 0$), this equation reduces to sinc ($k_x \Delta x T_{int} / N_{ph}$), which is often referred to as the temporal aperture MTF (Lomheim et al., 1990). This residual MTF component is due to the slight smearing caused by the fact that the scanned image moves smoothly (i.e., continuously) in the TDI direction but the CCD charge transport in the TDI direction moves in discrete steps given by the width of a phase gate. For example, in a four-phase CCD the charge packets move in discrete steps of ¼ pixel as the TDI image is built up.

Sources of uncompensated image motion derive from differences between the expected and actual scanning velocity. For a scanning system, imperfections in the scan mirror mechanism and discrete angular position control or time encoding effects introduce uncertainty into the actual scan velocity and cause deviations from the ideal velocity. Failure to precisely align the TDI columns with the scan direction due to imperfect control of the sensor platform attitude can introduce additional blurring, as subsequent TDI pixels do not view exactly the same portion of the scene as the previous line. For pushbroom sensors, the effective projected ground velocity at opposite ends of the array may differ enough to introduce noticeable smear, particularly for long arrays (several thousands of pixels long) and at large off-nadir angles where the Earth's curvature may become a significant factor. These terms can be grouped together into a single "velocity mismatch" factor that acts as an effective uncompensated angular motion over one integration period. Similar distortions can occur in whiskbroom TDI scanners as discussed in Section 14.2.2. These terms are often small over one integration period but may become increasingly important if large numbers of TDI stages are used to increase the signal-to-noise ratio. For example, an effective smear of 0.05 pixel per integration period is relatively benign for 10–20 stages of TDI, compared to other typical smear sources but may be a significant limiting factor in resolution under conditions of low illumination requiring long effective integration periods. A particular advantage of the digital simulation to be discussed in Section 14.4 is the ability to investigate the trade between S/N ratio and image smear to determine optimal operating modes and restrictions for design parameters.

Sensor platform motions can range from "low-frequency" wobble to "high-frequency" vibrations. A high-frequency vibration is defined as a motion whose period is much shorter than the exposure or integration time, while a low-frequency motion has a period much larger than one integration time. For high-frequency motions the amplitude of the vibration essentially determines the effective MTF and is treated explicitly. Trott (1960) gives the closed form for this contributor. Isolation of the sensor subsystem from the supporting platform by vibration-damping techniques usually makes this component negligible in most applications. For low-frequency motion, the effective velocity can vary during one cycle of the motion, and therefore the smear is time-dependent; that is, it varies from one integration period to the next, depending upon where in the motion cycle the integration period falls. This smear component can be accounted for by using the worst smear value, the smear occurring when the velocity is greatest. This is a pessimistic estimate, and a statistical treatment is more suitable (Wulich and Kopeika, 1987).

The CCD carrier-diffusion MTF (Table 14.2C) is assumed to be circularly symmetric and describes the effects of pixel crosstalk due to image charge generated in the field-free diffusion volume of the CCD, which then migrates by means of the random process of diffusion to the TDI pixels for collection (Seib, 1974). The wavelength dependence of the absorption coefficient of silicon gives this MTF component its spectral dependence.

Finally, the optical diffraction and optical degradation MTFs (Table 14.2D and E) are given by the O'Neill formula (O'Neill, 1956) for a circularly symmetric diffraction-limited optical system, and by the Nicholson equations (Nicholson, 1975), respectively. Clearly, this assumes ideal optics. For the case of nondiffraction-limited optics, analysis programs such as Code V (Optical Research Associates, Pasadena, CA) can be used to numerically predict the MTF, provided a detailed optical design exists. The optical diffraction MTF also exhibits wavelength dependence.

Note that many other sensor-related MTF components can be considered, such as CCD charge transfer efficiency (CTE), electronics processing, and optical defocus. In many TDI CCD scanned imaging applications, the dark-current and background flux levels combine to provide a pixel bias level of several hundred to several thousand electrons. Under these conditions CTE-induced effects can usually be ignored. For the purposes of this paper, we assume that these additional MTF components are negligible and that optical parameters such as defocus are under control. Arguello (1979) gives a detailed summary of a larger subset of component MTFs than we consider here. These MTF components can also be included in the modeling of an imaging system's capability.

It is useful to discuss the impact of atmospheric MTF effects on the sensor imaging capability. The Earth's atmosphere affects imagery obtained from overhead sensors by reducing scene contrasts and introducing a blurring effect. Reduction in contrast can arise from multiple scattering due to particulates and aerosols in the light path (commonly known as adjacency effects), as well as introduction of path radiance, which is light scattered into the sensor FOV by the atmosphere itself. The contribution due to path radiance is effectively a background radiance level upon which the scene signal is measured. Increasing path radiance decreases the S/N ratio in the same way that dark current and readout noise affect the imagery. The path radiance can be a significant portion of the signal received from the scene, particularly for large slant ranges obtained from off-nadir views and for the shorter visible wavelengths. For a realistic simulation it is important that the radiance contribution due to path radiance be accurately included in the imagery.

The "adjacency effect" arises because of scattering of radiation reflected from the surface by aerosols and particulates in the atmosphere. Horizontal scattering of light as it passes through the atmosphere causes an apparent loss of contrast. The visual effect is that dark objects adjacent to bright objects appear brighter than would otherwise be expected owing to contributions from photons coming from the bright surface that are scattered into the path from the dark surface. Modeling this process is time-consuming and requires complex radiative transfer codes. In addition, it is a nonlinear effect; the magnitude of the effect depends explicitly upon the scene content. Methods for "removing" this effect from remotely sensed imagery have been developed (Diner et al., 1989), and conceivably the effect could be modeled and incorporated into a simulation. However, the complexity of this process is beyond the scope of this work.

If the high-resolution background scene consists of data obtained from an existing sensor, then it already implicitly contains atmospheric effects related to the geometry and atmospheric conditions under which it was acquired. If those conditions are similar to those expected for the sensor being simulated, no further corrections are required. The simplest modification that can be made is the introduction of additional path radiance into the simulated image. If, for example, the high-resolution imagery is acquired using low- or mid-altitude aircraft, the path radiance is less than that expected from an orbiting platform viewed through the entire atmosphere. The difference between expected path radiance and actual implicit path radiance in the input imagery can be computed using standard atmospheric codes, and the difference added to the bakground scene on a pixel-by-pixel basis prior to signal chain computations. The correction is simple to perform and perhaps represents the most significant visual atmospheric effect.

The main contributor to reduction in MTF due to the intervening atmosphere is turbulence. Turbulence is responsible for the twinkling of stars at night and the rippling motion of images viewed through heated air rising from a hot surface such as a road. Atmospheric turbulence is manifest as a spatial variation in the index of refraction of the air layers so that light passing though the layers is refracted and images appear to move. The blurring effect caused by the apparent motion of small objects and edges can be minimized by utilizing short exposure

MODELING AND SIMULATION OF CCD IMAGING SYSTEMS

times. Analysis presented by Kopeika (1987) and Fried (1966) shows that under nominal conditions, atmospheric turbulence limits the resolution of objects viewed down through the Earth's atmosphere to objects larger than 4–6 cm. The effect of turbulence for applications considered here (GSDs on the order of 1 m) is minimal.

In the next section we introduce the methodology and architecture of a scanning TDI CCD sensor simulation that utilizes the two-dimensional MTF model described in Table 14.2.

14.4 SENSOR SIMULATION ARCHITECTURE AND IMPLEMENTATION

Figure 14.5 shows a functional flow diagram illustrating the CCD simulation process. The simulation process is broken into several steps that mimic portions of the CCD imaging chain shown in Figure 14.1. The basic processes are (1) obtaining an appropriate input scene, (2) simulating the blurring represented by the convolution of the sensor PSF with the input scene, (3) sampling the filtered scene at the appropriate ground sample distance (GSD) or sampling interval, (4)

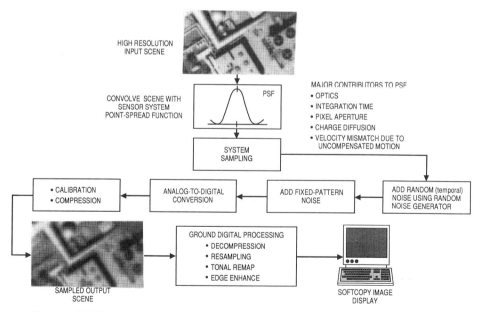

Figure 14.5 The major components of the entire image simulation procedure. The component boxes indicate independent processing modules that emulate corresponding elements in the sensor collection process and signal chain.

simulating the electronic signal chain, including effects of random noise components, fixed pattern noise, and analog-to-digital (A/D) conversion, and (5) applying digital filters, bandwidth compression algorithms, or other signal-processing procedures to the output data stream. The data stream can then be reformulated into a two-dimensional image or subjected to further processing or data extraction procedures, as appropriate.

In what follows we present the important details and approximations associated with each step in the simulation process. Specifically, in Section 14.4.1, the characteristics and limitations associated with the input scene radiance distribution are discussed; Section 14.4.2 outlines the practical implementation of the MTF, signal, and noise equations including important approximations made to the equations outlined in Sections 14.2.1 and 14.3; Section 14.4.3 discusses the implementation of the TDI CCD response and output offset nonuniformity effects based upon their physical origin in the CCD architecture (per Section 14.2.4); Section 14.4.4 discusses potential processing of the image data in the digital domain, including data calibration and data compression; and Section 14.4.5 discusses the specific computer hardware configuration used for the simulation work.

14.4.1 Input Scene Radiance Distribution

The input scene is represented as a two-dimensional array of digital values representing the in-band radiance distribution at the object plane of the sensor. One basic requirement for the input scene is that it be radiometrically realistic—for example, that mean radiances and contrasts present in the scene represent those that would actually be observed by the proposed sensor. In addition, the scene characteristics should include realistic geometric characteristics that represent the acquisition conditions of a real sensor: Image data for a nadir-looking instrument should look like nadir-looking data and not as if they were acquired by a side-looking instrument. Finally, the scene data should have an inherent resolution (IFOV) and ground sample distance (GSD) several times better than that of the sensor being simulated. This requirement facilitates the sampling/interpolation process that occurs later in the signal chain simulation, minimizes the effects of aliasing, and correctly simulates the continuum nature of the background against which a real sensor would be looking.

The scene used as input to the simulation is a discrete, digital representation of the continuous two-dimensional radiance distribution that the sensor would see during normal operation. This scene can have several forms including synthetic test data (such as an array of bar targets or other resolution targets), synthetic scene data, digitized photographic imagery, digital data from another sensor, etc. Synthetic test patterns are useful for analyzing in detail the effects of various components in the imaging chain because they allow the computation of standard quantitative measures of image quality such as the contrast transfer function (CTF), edge response function (ERF), and signal-to-noise ratio (S/N). However, if one wishes to investigate the subjective visual characteristics of imagery that would be produced by the sensor, it is necessary to run the simulation with realistic renditions of the actual scenes to be viewed by the sensor. Estimating visual quality through analysis by using test patterns may lead to a pessimistic result. Minute defects that may be very noticeable in imagery of test patterns containing high contrasts and dominated by uniform areas and sharp edges may actually not be a problem in operational imagery where the typical scene content obscures the defects. Unfortunately, although the simulation of a proposed sensor design against realistic scene images often provides the best estimate of the sensor performance, the acquisition or generation of appropriate input imagery to use in the simulation may be a difficult task.

There are several reasons for the requirement for high spatial resolution in the input imagery. The digital input radiance distribution is a discrete version of the continuous scene that the sensor would observe. Sampling the continuous scene to produce the digital scene introduces an extraneous transfer function component into the simulation process. Failure to oversample sufficiently can lead to aliasing or other visual artifacts in the simulated imagery that do not realistically represent the performance of the sensor system. In addition, the use of high-resolution background data enhances the accuracy of the interpolation process used to emulate the system sampling occurring toward the end of the simulation.

If the input scene consists of digital data derived from another sensor, then it implicitly contains both the characteristics of the scene and those of the acquisition sensor from which the data were obtained. The acquisition sensor characteristics will propagate through the digital simulation process and be included in the final product. These effects can be minimized by utilizing background imagery with higher resolution than that which the sensor being simulated would be expected to produce, as mentioned above. A rule of thumb is that an oversampling ratio (the ratio of input scene resolution to simulated system resolution) of 4 or 5 is generally adequate to correctly approximate a continuous input image. For example, simulation of a system with a proposed GSD of 10 m would require input imagery with a GSD of approximately 2 m.

There are other benefits to be derived from a large oversampling ratio. Generally, image data obtained from an existing imaging sensor are not completely free of noise. For example, various sorts of fixed pattern noise and random noise may be superimposed on the scene data. Application of the sensor MTF can be physically interpreted as an averaging or smoothing of the input image data. When adjacent pixels are averaged together to produce one simulated sensor pixel, the noise effects in the input imagery tend to be smoothed out and minimized; the more pixels averaged together, the less apparent are the characteristics of the acquisition sensor noise in the input scene data.

The second requirement on input imagery is that it radiometrically represent the true scenes against which the sensor would be looking. Mean in-band radiances and edge contrasts present in the scene should reflect those that would be obtained in actual operation. If imagery from an existing sensor is used as input to the simulation, then it is desirable that the spectral characteristics of this acquisition sensor—in particular, $T_{opt}(\lambda)$, the sensor optical transmission, and $\eta(\lambda)$, the detector spectral quantum efficiency—are similar to those of the design to be simulated. Imagery derived from a photographic source would not be an optimal candidate for simulating a CCD sensor because the spectral responses of film and silicon detectors differ significantly and the photographic imagery would not exhibit appropriate radiometric characteristics.

There are a variety of means for generating input imagery for use in simulations. The simplest procedure is to use data from an existing sensor operating in the same wavelength passband and exhibiting similar spectral response, but with higher spatial resolution, as described before. If such imagery is to be used, it is important that it be obtained under sensor viewing conditions similar to those that would be encountered with the sensor being simulated. For example, the use of imagery obtained with a nadir-looking instrument would not be a good candidate for input to a simulation of an off-nadir side-looking instrument, since the nadir-look imagery would not contain information about the vertical characteristics of terrain features. Similarly, if low-altitude aircraft imagery was used to simulate a spaceborne sensor, it would be important to modify the radiometric characteristics of the aircraft imagery to account for the increased haze caused by looking through the entire atmosphere. In those cases where the acquisition sensor (source of input scene) and simulations do not have the same spectral bandpass, a scaling factor can often be applied to the input scene data that results in relatively correct in-band radiances and contrasts.

14.4.2 MTF Signal and Noise Equations

As mentioned previously, the first step in the simulation process is to spatially blur, by convolution, the high-resolution input scene with the system PSF. Our choice is to do this convolution in the spatial frequency domain, both because the sensor system MTF can be constructed analytically (Section 14.3) and because our particular computer hardware configuration uses an array processor, which provides fast computation of two-dimensional Fourier transforms on large sections of an image (up to 512×512 pixels).

Equation (14.33) shows that this is a three-step process for each wavelength: (1) The two-dimensional Fourier transform of the input scene is computed; (2) the transformed scene is then multiplied by the two-dimensional system MTF (Table 14.2); and (3) the result is then subjected to an inverse two-dimensional

Fourier transform. It should be noted that Gaussian functions are often used to approximate sensor system PSFs (Serayfi, 1973) (hence the MTFs are also Gaussian). To evaluate the merits of a proposed sensor design it is useful to describe it in as much physical detail as possible, such as is given in the model in Table 14.2. Gaussian PSF/MTF approximations are useful for quick but less accurate estimates of system performance.

Table 14.2 makes it clear that the system MTF exhibits a wavelength dependence, and therefore the two-dimensional weighted scene radiance function [$W(x,y,\lambda)$ in Eq. (14.32)] will be subjected to different amounts of spatial blurring as a function of λ. In general, the spectral distribution of scene radiance varies from one (x,y) position to the next; for terrain backgrounds the predominant wavelengths received from water bodies differ from those received from soil. If these various terrain features occur simultaneously within the input scene, then to accurately simulate the imaging process, a weighted superposition of blurred scenes, each defined over a narrow spectral bandwidth, would be needed. This is unrealistic because *high-resolution* multispectral input scene data are hard to come by, and even if they were available, constructing a broadband simulation with narrowband multispectral scenes would prove to be computationally intensive.

We now discuss the procedure for radiometrically calibrating an input scene that is obtained from another, higher resolution sensor, which we shall refer to as the "acquisition sensor." Clearly, if the input scene is based on image data taken with the acquisition sensor, then we have direct access only to $S_e(x,y)$ as given in Eq. (14.5) [or a quantity proportional to $S_e(x,y)$]. To obtain the correct underlying scene spectral radiance, we must extract the function $R_s(x,y,\lambda)$ from the right-hand side of Eq. (14.5). This can be done accurately only if the system PSF is approximated as a product of delta functions in the x and y directions and (a) the spectral bandwidth of the acquisition sensor is sufficiently narrow or (b) the spectral functions $\eta(\lambda)$, $T_{opt}(\lambda)$, $T_A^{up}(\lambda)$, $R_s(x,y,\lambda)$, and $R_H(\lambda)$ do not vary appreciably over this bandwidth and are known.

The assumption of a delta function PSF is generally met with sufficient accuracy if we follow the "rule of thumb" mentioned in Section 14.4.1, namely, that the input scene have greater than five times the resolution desired in the final sensor simulation output. However, as mentioned above, narrowband multispectral input imagery of adequate resolution is often not available and the more available, relatively broadband imagery must be used. In such cases the spectral radiance $R_s(x,y,\lambda)$ can be determined only as an average over the spectral bandwidth.

Using the delta function PSF assumption in Eq. (14.5), the convolution of $PSF_{opt}(x',y',\lambda)$ with $R_s(x',y',\lambda)$ becomes $R_s(x,y,\lambda)$. If we also assume a narrow spectral bandwidth $\Delta\lambda$ with a center wavelength λ_c, Eq. (14.5) as applied to the input-scene acquisition sensor is

$$S_e^{aq}(x,y,\Delta\lambda) = \{A_{pix}^{aq}\ t_{int}^{aq}\ N_{tdi}^{aq}\ \Omega_{aper}^{aq}\ \eta^{aq}(\lambda_c)\ T_{opt}^{aq}(\lambda_c) / hc\}$$
$$\times [T_A^{up,aq}(\lambda_c)\ R_S^{aq}(x,y,\Delta\lambda) + R_H^{aq}(\Delta\lambda)]\ \Delta\lambda^{aq}\ \lambda_c \quad (14.34)$$

where all symbols have their previous meaning and the superscript "aq" labels the sensor and system parameters appropriate for the acquisition sensor. Other approximation methods can be used; for example, if $R_s^{aq}(x,y,\lambda)$ and $R_H(\lambda)$ are known to not vary appreciably over the spectral bandwidth $\Delta\lambda$, then we can approximate Eq. (14.34) by

$$S_e^{aq}(x,y,\Delta\lambda) = \{A_{pix}^{aq}\ t_{int}^{aq}\ N_{tdi}^{aq}\ \Omega_{aper}^{aq}\ F^{aq}(\Delta\lambda)\ R_s(x,y,\Delta\lambda)\} / hc \quad (14.35)$$

where

$$F^{aq}(\Delta\lambda) = \int_{\Delta\lambda} \lambda\ \eta^{aq}(\lambda)\ T_{opt}^{aq}(\lambda) d\lambda \quad (14.36)$$

and

$$R_s(x,y,\Delta\lambda) = T_A^{up,aq}(\lambda_c)\ R_s^{aq}(x,y,\Delta\lambda) + R_H^{aq}(\Delta\lambda) \quad (14.37)$$

$F^{aq}(\Delta\lambda)$ is the weighting function, and $R_s(x,y,\Delta\lambda)$ is the average, or "effective," scene spectral radiance function, both defined over $\Delta\lambda$. Typically, we solve Eq. (14.35) for $R_s(x,y,\Delta\lambda)$, thereby "radiometrically calibrating" our input scene with an accuracy that is limited by the indicated approximations.

Note that $S_e^{aq}(x,y,\Delta\lambda)$ is written as a continuous two-dimensional function of (x,y) as given in Eq. (14.5). In reality, the acquisition input image is sampled and would therefore be described by Eq. (14.7). However, the assumption of five times greater resolution of the input image compared to the simulated image allows us to approximate the input image as continuous for the purpose of simulation.

Having "radiometrically calibrated" the input scene via Eq. (14.35), we are now ready to use $R_s(x,y,\Delta\lambda)$ as an input to a sensor simulation. The first step is to blur the input scene by the system PSF; and as described above, our choice is to do this in the spatial frequency domain as indicated in Eq. (14.33).

$W(x,y,\lambda)$ in Eq. (14.33) can be calculated using empirical data or models that describe the spectral functions contained in the right-hand side of Eq. (14.32). Note that Eq. (14.32) contains terms that are proportional to the input scene radiance; the scene radiance levels were "calibrated" as averaged or effective quantities over the spectral bandpass of the associated acquisition sensor per Eq. (14.35). This averaging obliterates the spectral detail in $R_s(x,y,\lambda)$ in obtaining the calibrated input scene, $R_s(x,y,\Delta\lambda)$. However, it is not inconsistent to *assume* a spectral dependence for $R_s(x,y,\lambda)$ based on modeled or empirical data for each of $T_A^{down}(\lambda)$, $I_s^{exo}(\lambda)$, $I_A^{diffuse}(\lambda)$, and $\rho(x,y,\lambda)$. For example, if the scene contains a lot of uniform vegetation, the associated reflectance spectra, $\rho_{veg}(x,y,\lambda)$, could be used.

Rather than carry out the two-dimensional spatial Fourier transform at each incremental wavelength $d\lambda$ and then sum these transforms [as implied by the integration in eq. (14.33)], we make approximations that allow us to avoid this computationally intensive procedure. First we assume that the input scene spectral content does not vary as a function of scene position (this is usually valid over sufficiently small regions) and hence

$$R_s(x,y,\lambda) = R_s(x,y) \, g(\lambda) \tag{14.38}$$

where $g(\lambda)$ is the spectral function associated with all the input scene pixels. The system MTF is replaced by a weighted polychromatic MTF given by

$$\text{MTF}_{\text{sys}}^{\text{poly}}(f_x, f_y, \Delta\lambda) = \sum_i W(\lambda_i) \, \text{MTF}_{\text{sys}}(f_x, f_y, \lambda_i) \, / \, \sum_i W(\lambda_i) \tag{14.39}$$

and

$$W(\lambda_i) = \lambda_i \, \eta(\lambda_i) \, T_{\text{opt}}(\lambda_i) \, \{ \, T_A^{\text{up}}(\lambda_i) \, \rho(\lambda i) \, (1/\pi) \, [\, I_s^{\text{exo}}(\lambda_i) \, T_A^{\text{down}}(\lambda_i) \\ + \, I^{\text{diffuse}}(\lambda_i) \,] \, + \, R_H(\lambda_i) \, \} \tag{14.40}$$

where all of the symbols have their previous meaning and λ_i represents wavelength increments over $\Delta\lambda$. For our work, $\text{MTF}_{\text{sys}}^{\text{poly}}(f_x,f_y,\Delta\lambda)$ is computed using 20 equally spaced wavelengths in the bandpass $\Delta\lambda$, where $\Delta\lambda$ goes from 0.4 to 1.0 µm.

The input scene blurring referred to at the beginning of this section, and shown as the first step in Figure 14.5, is described by

$$R_s^{\text{blurred}}(x,y,\Delta\lambda) = \text{FT}^{-1} \, \{ \, \text{FT} \, [\, R_s(x,y,\Delta\lambda) \,] \, \text{MTF}_{\text{sys}}^{\text{poly}}(f_x,f_y,\Delta\lambda) \, \} \tag{14.41}$$

The operations inside the curly brace { } in Eq. (14.41) consist of the complex multiplication of the real two-dimensional array representing the function $\text{MTF}_{\text{sys}}^{\text{poly}}(f_x,f_y,\Delta\lambda)$ [as given by Eq. (14.39) and Table 14.2] with the Fourier transform of the 2-D input scene radiance image, $R_s(x,y,\Delta\lambda)$,

$$\text{FT} \, (R_s(x,y,\Delta\lambda)) = R_{s,\text{real}}(f_x,f_y) + iR_{s,\text{imag}}(f_x,f_y) \tag{14.42}$$

Figure 14.6 shows the graphical representation of the 2-D image and its Fourier transform (in the format required by our software). The pixel spacings in this input image (Δx in the cross-scan direction and Δy in the in-scan direction) are indicated, along with the equations for the spatial frequency spacings ($\Delta f_x, \Delta f_y$) and the sampling or Nyquist frequencies ($f_{\text{NYQ},x}, f_{\text{NYQ},y}$) that result after the Fourier transform. Note that the scene is chosen to have dimensions that are a power of 2 (typically 512 × 512) so we can utilize the speed of the fast Fourier transform (FFT) (Bracewell, 1986).

The spatial frequencies associated with the 2-D MTF array elements [$\text{MTF}_{\text{sys}}^{\text{poly}}(f_x,f_y,\Delta\lambda)$] correspond exactly with those of the function $\text{FT}\{R_s(x,y,\Delta\lambda)\}$. Thus the function $\text{MTF}_{\text{sys}}^{\text{poly}}(f_x,f_y,\Delta\lambda)$ is evaluated well beyond the Nyquist frequency ($f_x, f_y >> f_{\text{NYQ}}$) of the simulated sensor. Values for

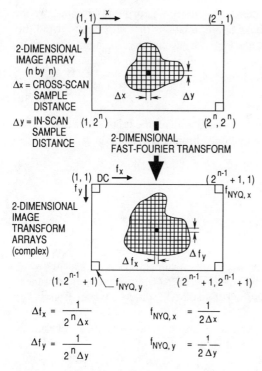

Figure 14.6 The relationship between the indices and spatial characteristics of the two-dimensional input image array and the image Fourier transform and corresponding spatial frequencies.

$\mathrm{MTF}_{\mathrm{sys}}^{\mathrm{poly}}(f_x,f_y,\Delta\lambda)$ at spatial frequencies above the optical cutoff frequency f_{co} are set to zero, because the optical system cannot physically pass those frequencies. The optical cutoff frequency is given by

$$f_{co} = 1 / (\lambda f\#) \qquad (14.43)$$

where λ and $f\#$ have their previous meanings and λ is typically chosen at the center of the system spectral bandpass, λ_c. The criterion for setting $\mathrm{MTF}_{\mathrm{sys}}^{\mathrm{poly}}(f_x,f_y,\Delta\lambda)$ to zero is based on the test

$$f > f_{co} \qquad (14.44)$$

where $f = (f_x^2 + f_y^2)^{1/2}$.

To convert the blurred scene from units of spectral radiance to a sampled image in units of electrons per pixel we use Eqs. (14.5) and (14.7). The approximations for $S_e(x,y)$ that were used in obtaining Eq. (14.34) or (14.35) are also used for the simulated sensor. Thus we have for the sampled simulated sensor output image, $S_e^{\text{sim}}(x_n,y_m)$,

$$S_e^{\text{sim}}(x_n,y_m) = \iint \delta(x-x_n)\, \delta(y-y_m) S_e^{\text{sim}}(x,y,\Delta\lambda)\, dx\, dy \tag{14.45}$$

where $S_e^{\text{sim}}(x_n,y_m)$ has units of electrons/pixel and the spatial sampling described by Eq. (14.7). $S_e^{\text{sim}}(x_n,y_m)$ represents the 2-D image output obtained after the "system-sampling" step shown in Figure 14.5. The sampling process is easily implemented as a bilinear interpolation of the blurred image, $S_e^{\text{sim}}(x,y,\Delta\lambda)$. The sample spacing in the in-scan direction is controlled by the temporal sampling, whereas the sample spacing in the cross-scan direction is controlled by the detector pitch [see discussion after Eq. (14.7)]. Note that no aggregation of pixel image values is required as part of the sampling process; integration over the TDI CCD pixel area is built into the 2-D system MTF spatial filtering process in the detector aperture function.

The decoupling of the blurring and sampling processes makes it trivial to handle such operations as temporal oversampling (e.g., multiple samples per dwell) and focal-plane architectures where the pixels are spatially arranged in nonsymmetric, overlapping patterns ("staggered arrays"). More sophisticated techniques such as bicubic interpolation can be used to perform the sampling at the expense of computer CPU time. If the input image is sufficiently oversampled to start with, the benefits to be gained from using a larger interpolation kernel over the bilinear function are minimal.

The function $S_e^{\text{sim}}(x,y,\Delta\lambda)$ appearing on the right hand side of Eq. (14.45) contains all of the approximations to our original signal equations that are required to practically implement a simulation. By analogy with Eqs. (14.34)–(14.37), we have

$$S_e^{\text{sim}}(x,y,\Delta\lambda) = \{A_{\text{pix}}^{\text{sim}}\, t_{\text{int}}^{\text{sim}}\, N_{\text{tdi}}^{\text{sim}}\, \Omega_{\text{aper}}^{\text{sim}}\, \eta^{\text{sim}}(\lambda_c)\, T_{\text{opt}}^{\text{sim}}(\lambda_c)\, R_s^{\text{blurred}}(x,y,\Delta\lambda)\, \Delta\lambda_{\text{sim}}\lambda_c\} / hc \tag{14.46}$$

or

$$S_e^{\text{sim}}(x,y,\Delta\lambda) = \{A_{\text{pix}}^{\text{sim}}\, t_{\text{int}}^{\text{sim}}\, N_{\text{tdi}}^{\text{sim}}\, \Omega_{\text{aper}}^{\text{sim}}\, F^{\text{sim}}(\Delta\lambda)\, R_s^{\text{blurred}}(x,y,\Delta\lambda)\} / hc \tag{14.47}$$

where

$$F^{\text{sim}}(\Delta\lambda) = \int \Delta\lambda_{\text{sim}}\, \lambda\, \eta^{\text{sim}}(\lambda)\, T_{\text{opt}}^{\text{sim}}(\lambda)\, d\lambda \tag{14.48}$$

The choice of using Eq. (14.46) or (14.47) depends on the choice of method of approximation as discussed previously. Equation (14.46) is appropriate in narrowband or multispectral applications; Eq. (14.47) is appropriate in broadband panchromatic sensor simulations.

The next step in the simulation is the incorporation of the sources of temporal noise. Equation (14.18) of Section 14.2.1 gives an equation for the rms noise, in electrons per pixel, that is a two-dimensional function of the final *sampled* image pixel positions, or $n_{tot}(x_n, y_m)$. The image pixel positions, (x_n, y_m), are the same as those for our working signal, $S_e^{sim}(x_n, y_m)$, as defined in Eq. (14.45). In other words, the rms noise associated with image pixel $S_e^{sim}(x_i, y_j)$ is $n_{tot}(x_i, y_j)$, where these functions are defined by Eqs. (14.45) and (14.18), respectively. Note that since analog-to-digital conversion is simulated explicitly, the quadrature noise term related to the A/D conversion process is not included at this point. The effect of that noise term is implicitly included through the integer truncation in the A/D process.

The addition of temporal noise to the simulation output can be expressed as

$$ST_e^{sim}(x_n, y_m) = S_e^{sim}(x_n, y_m) + G(n,m)n_{tot}(x_n, y_m) \tag{14.49}$$

where ST_e^{sim} is the simulation sampled output image after the addition of temporal noise, and the unitless factor $G(n,m)$ is a two-dimensional array of random uncorrelated numbers. $G(n,m)$ is obtained by randomly selecting a value from a Gaussian noise distribution whose standard deviation is unity. Specifically, a white Gaussian random number generator (standard deviation of unity) is independently exercised to obtain each value in the array, $G(n,m)$. This array is then scaled by multiplication with the noise array, $n_{tot}(x_n, y_m)$, which has the spatial effects of the signal (photon) and dark-current-dependent temporal noise source built into it.

Figure 14.7 illustrates the simulation flow successively by pictorial examples through the level of Eq. (14.49). Figure 14.7a is an overview of the high-resolution image used as input for the simulation. The image data were originally obtained by digitizing low-altitude, high-resolution aerial photographs (Snyder et al., 1982). The original digitized image was 4096 × 4096 pixels in size, with a sample distance of approximately 3 in. per pixel. Prior to use in these simulation examples, the image data were block-averaged (using a 2 × 2 window) to a resolution of 6 in. and an input size of 2048 × 2048 pixels. At this resolution, the ratio of sensor GSD to input scene GSD is 6.565 (1 m/6 in.) and satisfies the oversampling requirements for accurate simulation.

As we pointed out previously, the input imagery chosen for the simulation should be obtained with an acquisition sensor with spectral characteristics similar to those of the sensor being simulated. The choice of digitized black-and-white photographic imagery as the high-resolution scene for simulation of a silicon CCD imager does not satisfy this requirement. Furthermore, the tonal characteristics of the low-altitude imagery are not representative of those derived from a space platform, as little atmospheric haze is present in the high-resolution data. However, for the purposes of demonstration we have chosen to ignore these inconsistencies and have selected this imagery primarily for its spatial resolution characteristics and interesting features.

Figure 14.7b shows a magnified subregion of the input image at full resolution (to show the detail in the original scene), and Figure 14.7c is a plot of the in-scan

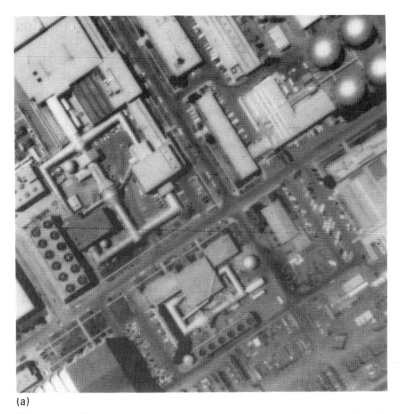

(a)

Figure 14.7 The sequence of images shows the form of the two dimensional image array as it passes through the simulation procedure illustrated in Figure 14.5. (a) Overview (every fourth line and pixel) of the high-resolution image array. (b) A section of the high-resolution input image shown at full resolution. (Image courtesy of Dr. H. Snyder, Department of Industrial Engineering and Operations Research, Virginia Tech University.) (c) Plot of the in-scan total sensor MTF and the component MTFs for the sensor parameters in Table 14.3. (d) A section of the input scene shown in (b) after blurring with the sensor 2-D MTF. (e) The blurred image sampled at the sensor GSD. (f) The final image output after the signal chain computations [image (e) with noise added and converted to digital values]. (g) A section of the simulated image (e) enlarged to the same scale as (b) for comparison.

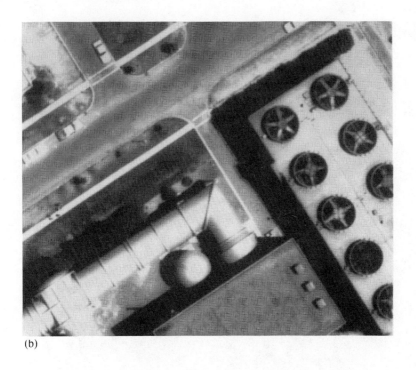

(b)

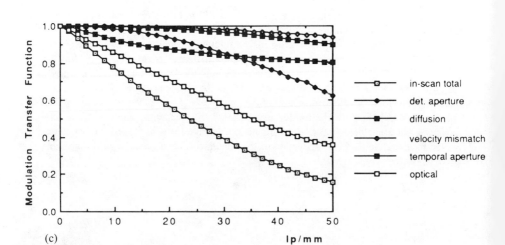

(c)

Figure 14.7 (Con't.)

MODELING AND SIMULATION OF CCD IMAGING SYSTEMS

(d)

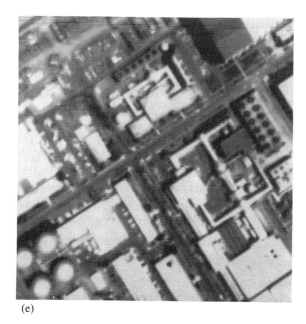

(e)

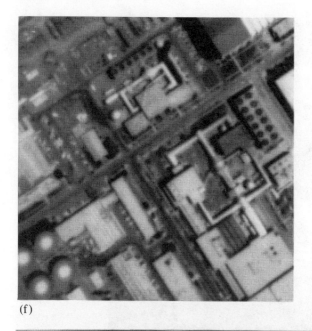
(f)

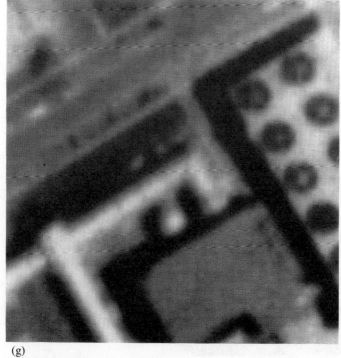
(g)

Figure 14.7 (Con't.)

system MTF (including the component MTFs based on the model listed in Table 14.2 with the specific input parameters for the example case given in Table 14.3. Figure 14.7d is a section of the input image (corresponding to Figure 14.7b) after blurring by the 2-D sensor system MTF [described in Eq. (14.41)], where the loss of resolution due to the spatial filtering clearly stands out. Figure 14.7e shows the entire simulated scene, after sampling, as described by Eq. (14.45), at a scale corresponding to Figure 14.7a. Figure 14.7f is the sampled, simulated image after addition of noise per Eq. (14.49), and Figure 14.7g is a portion of this same image magnified to the same scale as Figure 14.7b, for comparison. The effects of sampling and noise make Figure 14.7g appear grainy in comparison with Figure 14.7d. The aggregate effect of the simulation process through the level of Eq. (14.49) is clearly seen by comparing Figures 14.7b and 14.7g.

Figure 14.7g and Eq. (14.49) represent the *best possible image* that the TDI CCD sensor described in Table 14.3 could produce, that is, the theoretical limit. Systematic, spatially varying pixel gain and offset nonuniformities are not included in Figure 14.7g, nor are they reflected in Eq. (14.49). We describe our method for including these important "real-world" nonuniformity effects in the next subsection. Examples of their impact on simulated output imagery follow in Section 14.5.

14.4.3 TDI CCD Nonuniformity Effects

The output of the TDI CCD sensor simulation, which is quantified by Eq. (14.49) and displayed in Figure 14.7g, represents the best image (theoretical limit) that can be obtained for a sensor based on the parameters in Table 14.3. However, in realistic sensor systems, nonuniformities can blemish the image and easily be the dominant source of limiting signal-to-noise behavior. Hence it is important to include these effects in our simulation in as realistic a way as possible.

Sensor-based pixel response and offset nonuniformity effects can, in principle, be made negligible by hardware design (i.e., make the hardware "perfect") or by removal via sensor calibration techniques. Designing and fabricating perfectly uniform TDI CCD devices will place enormous demands on the CCD integrated circuit manufacturing technology, usually resulting in extraordinarily low device production yields. On the other hand, nonuniformity correction via flooded radiometric calibration techniques increases the complexity and operation of the sensor. By including nonuniformity effects in our simulations, we can ascertain at what levels they become important as a function of sensor, system, and operating mode parameters.

In Section 14.2.3 the physical origins of the significant pixel-to-pixel TDI CCD response and offset nonuniformities were discussed, including effects associated with TDI CCD parallel serial and multiple serial tap output structure.

Table 14.3 Example Sensor Design Parameters

Architecture	Scanning, line array, silicon CCD	
	1024 detectors × 32 TDI	
	8 taps (128 channels/tap)	
	Altitude 200 km/circular orbit	
Optics	efl	200 in. (5.077 m)
	f#	10.0
	D_{aper}	20 in.
	T_{opt}	0.7 (flat across spectral region)
	Obscuration (area)	0.15
	Waveband	0.5–0.9 μm
Detectors	Material	Silicon
	Pitch	10μm (in-scan, cross-scan)
	IFOV	5μrad
	QDE	0.35 (average across spectral region)
	Depletion width	6 μm
	Diffusion length	50 μm
	Full well capacity	125,000 electrons
Bandwidth	Integration time per TDI stage	50 μsec
	2.65 MHz/tap	
	205 Mbps (full data rate)	
A/D	5 μV/electron	
	10 bits/pixel	
	5 V maximum out	
Noise	Dark current	2 nA/cm² at 25°C
	Device noise	100 electrons, rms
Smear	0.01 μrad / sec (0.1 pixel / integration time)	
Coverage	Scan velocity	2 × 10⁴ m/sec
	Ground sample distance	1 m
	Cross-scan width	1.024 km
	Area rate of coverage (ARC)	2.048 km²/sec
	Sensor line rate	20,000 lines/sec

We now outline our method for inserting these effects into our simulation. The four principal sources of nonuniformity follow from the discussion in Section 14.2.3: (1) cross-scan pixel-to-pixel offset variations due to spatial variations in dark current, (2) pixel-to-pixel variations in photoresponse, (3) variations in the TDI CCD serial tap offset levels, and (4) variations in the serial tap gains. Figure 14.8 shows a schematic of the procedure used to include random and systematic noise effects in the simulation.

The dark-current spatial variations manifest themselves in only the cross-scan

MODELING AND SIMULATION OF CCD IMAGING SYSTEMS

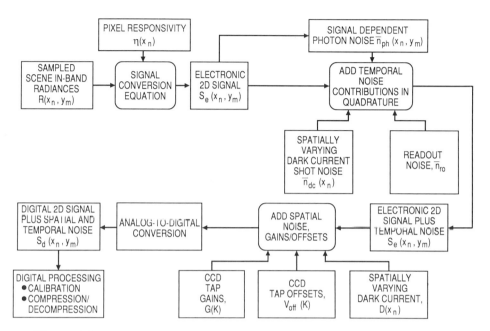

Figure 14.8 Flow diagram illustrating the component processes in the signal chain model.

direction due to the in-scan summation provided by the TDI process. The variation in dark current with cross-scan pixel position is given formally by Eq. (14.12). These variations are implemented in the simulation by

$$D(x_n) = \beta\, \alpha(x_n)\, D_{tdi} + D_{mux} \tag{14.50}$$

where D_{tdi} is an average dark-current level due to the CCD TDI region, $\alpha(x_n)$ is a unitless constant describing the cross-scan pixel-to-pixel offset spatial structure, β is a unitless amplitude scaling factor, and D_{mux} is a constant (structureless) dark-current offset. As mentioned in Section 14.2.3, dark-current generation is correlated from pixel to pixel and hence the constant $\alpha(x_n)$ must reflect this. We assume that these spatial variations are represented by a set of Gaussian random numbers, each corresponding to a cross-scan pixel position. Furthermore, we assume that these variations, when described in the spatial frequency domain, follow a spatial power spectral density (PSD) function that is proportional to $1/f$, where f is the spatial frequency, usually quoted in units of line pairs per millimeter (lp/mm). The associated structure in $\alpha(x_n)$ is typical of that observed in pixel-to-pixel dark-current nonuniformities (Anon., 1987). The technique for generating the random number set $\alpha(x_n)$ that follows a $1/f$ PSD is given in Keshner

(1982) and Barnes and Jarvis (1971). Since dark current can have only positive values, the final realization of $D(x_n)$ [Eq. (14.49)] can have only positive values. The numbers $\alpha(x_n)$ are scaled so that $\alpha^{max}(x_n) - \alpha^{min}(x_n) = 1.0$. The amplitude scaling factor β determines the peak-to-peak dark-current structure variation as a fraction of the average value, D_{tdi}. D_{mux} is usually large enough at typical CCD dark-current levels that the unphysical situation of generating negative values of $D(x_n)$ in Eq. (14.49) is avoided. By adjusting β, D_{tdi}, and D_{mux}, a quantitatively realistic cross-scan dark-current pattern can be synthesized. The size of this pattern will scale correctly with pixel integration time and number of TDI stages via Eqs. (14.12)–(14.14).

Dark-current variations, particularly at the scan rates and TDI lengths used in most pushbroom and whiskbroom scanning sensors, produce a fine striping structure that is easily obscured in image areas with any significant detail. The dark-current spatial variations are visible only when they approach the level of the photon noise in the image (or readout noise for low-light-level images). A more visible nonuniformity effect is that of pixel-to-pixel response variations that scale directly with signal level.

In our simulation, pixel-to-pixel response nonuniformity is implemented by introducing a corresponding variation in cross-scan pixel-to-pixel quantum efficiency. As with dark current, the TDI summation obliterates structure in the in-scan direction, and only cross-scan pixel-to-pixel variations remain. The implementation corresponds to replacing $\eta^{sim}(\lambda)$ by

$$\eta^{sim}(x_n, \lambda) = \eta^{sim}(\lambda) [\, 1 + \beta' \gamma(x_n)] \tag{14.51}$$

where $\eta^{sim}(\lambda)$ is the effective pixel quantum efficiency, $\gamma(x_n)$ is a unitless function that describes its spatial variations, and β' is a scaling factor.

Based on our discussion in Section 14.2.4, we are using $\gamma(x_n)$ in Eq. (14.51) to account for differences in pixel area, pixel collection efficiency, and spectral response as they manifest themselves after TDI as cross-scan pixel-to-pixel variations. In the case of pixel-to-pixel spectral response variations, Eq. (14.51) is an approximation of

$$\eta^{sim}(\lambda) [\, 1 + \beta' \gamma(x_n) \,] \approx \int \eta_n^{sim}(\lambda) \, d\lambda \tag{14.52}$$

where $\eta_n^{sim}(\lambda)$ would be the spectral response function of cross-scan pixel n after TDI. Hence, Eq. (14.51) is a phenomenological description of the aggregate of the response nonuniformity effects after TDI. As with dark-current variations, the response variations are also correlated from pixel to pixel. The response variations are introduced by choosing the $\gamma(x_n)$ to be random correlated numbers by the same method used to generate the $\alpha(x_n)$ for Eq. (14.50), and subject to $\gamma^{max}(x_n) - \gamma^{min}(x_n) = 1.0$. Two separate realizations of $\gamma(x_n)$ and $\alpha(x_n)$ are always used in each individual simulation because the two spatial effects (dark-current and response variations) do not have a common physical origin. The scaling parameter β' was used to set the size of the nonuniformity; for example, $\beta' = 0.1$ gives a peak-to-peak response nonuniformity of 10%.

Figure 14.9a shows a simulated image free of nonuniformity effects, whereas Figure 14.9b shows the same simulated image with $\beta' = 0.3$. Figure 14.9c shows a plot of $\eta^{\text{sim}}(x_n)$ versus cross-scan pixel position for 100 adjacent detectors. The correlated nature of this variation is evident in the plot. The responsivity values fluctuate about the nominal value of 0.34. The peak-to-peak spread of the individual detector responsivities is 30% of 0.34, or approximately 0.1. The value chosen for β' is somewhat higher than would be expected from modern CCD TDI devices but was chosen specifically to illustrate the responsivity variations. The response nonuniformity effect appears as correlated, high spatial frequency striping in the image and is most obvious in areas of uniform image tone. Dark-current nonuniformity is not visible in this figure; to make its effects visible would require setting its mean value and peak-to-peak variation to anomalous levels, at this scene radiance level.

The multiple serial output tap architecture of the TDI CCD serves as an additional source of gain and offset nonuniformity. As shown in Figure 14.4, each serial tap consists of a parallel-to-serial CCD multiplexer feeding a charge-to-voltage converting electrometer. Equations (14.8) and (14.9) describe the conversion of each output tap from electrons/pixel to the voltage domain, and finally to digital counts after ADC. The output of each serial tap is thus a time-division multiplexed video signal consisting of sequential voltage steps that are each proportional to the corresponding pixel charge. Further time-division multiplexing of the k serial tap video signals (see Figure 14.4) can occur in the analog or digital domains. The choice of a specific architecture depends on the system design trade-offs for the analog and digital signal processing electronics. For example, several of the k serial CCD tap output streams could be time-division multiplexed (in the analog domain) into a single higher speed video stream and then fed to an ADC for digitization. More often than not, each serial tap output is processed by a single signal processing electronics chain through A/D conversion. For this latter case we ascribe to each tap, k, an overall gain and offset that each contain the aggregate effects of the entire processing chain (from electrons/pixel at the CCD pixel level through digital counts after the ADC). Hence, tap gain variations are traceable to the on-chip (CCD) conversion gain G_{conv} and analog signal processor gain G_{ap} [Eq. (14.8)]. Similarly, offsets can originate at various points along the signal chain.

The serial tap nonuniformity effects are included in the simulation *after* implementation of the pixel-to-pixel dark-current offset and response nonuniformities. The tap offsets are best described in the voltage domain just prior to A/D conversion. The kth offset just prior to the ADC is given by

$$V_{\text{off}}(k) = \varepsilon(k) V_{\text{off}}^{\text{pp}} \qquad (14.53)$$

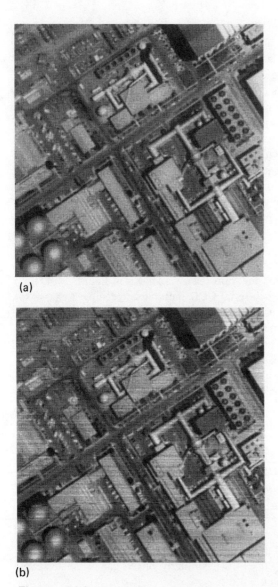

(a)

(b)

Figure 14.9 (a) The simulated output scene from Figure 14.7e, with signal-dependent noise included but no spatial nonuniformities. (b) Simulated image showing the visual striping effect arising from pixel-to-pixel variations in detector responsivity. (c) Plot of the detector responsivity as a function of cross-scan pixel position. (d) The simulated output scene from Figure 14.7e, with signal-dependent noise included but no spatial nonuniformities. (e) Simulated image showing the effect of variations in tap voltage offsets. (f) Plot of the tap voltage offsets (four taps modeled).

MODELING AND SIMULATION OF CCD IMAGING SYSTEMS

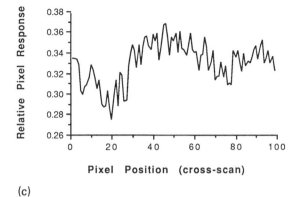

(c)

(d)

(e)

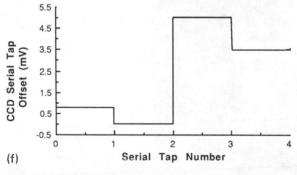

(f)

Figure 14.9 (Con't.)

where $V_{\text{off}}^{\text{pp}}$ is a scaling voltage and represents the peak-to-peak variation in the k tap offsets, and $\varepsilon(k)$ are a set of k random numbers normalized such that $\varepsilon^{\max}(k) - \varepsilon^{\min}(k) = 1.0$. The $\varepsilon(k)$ are obtained from a Gaussian number generator. No correlation between values is induced [as was done, for example, with $\gamma(x_n)$ and $\alpha(x_n)$ above] because the large physical distance between the CCD tap electrometers and the physically distinct analog signal processing chains (one per tap) do not justify correlation effects.

The two-dimensional output image (described in the voltage domain) after modification for offset effects is

$$V_{\text{sig}}^{\text{off}}(x_n, y_m) = V_{\text{sig}}(x_n, y_m) + \varepsilon(k, x_n) V_{\text{off}}^{\text{pp}} + V_{\text{bias}} \qquad (14.54)$$

where $V_{\text{sig}}(x_n, y_m)$ is obtained from Eq. (14.8). The cross-scan pixel index n and tap index k must be related by the inequality

$$N(k-1) + 1 \leq n \leq Nk \qquad (14.55)$$

where N is the total number of cross-scan pixels in a tap. For example, if $N = 250$ and $1 \leq n \leq 250$, $k = 1$, is required for the inequality in Eq. (14.55) to be satisfied. This constraint forces the correct association between cross-scan pixel location x_n and tap index k. The term V_{bias} is a voltage offset that is used to place $V_{\text{sig}}^{\text{off}}(x_n, y_m)$ within the dynamic range of the ADC. Note that the effects of signal truncation at the top end of the ADC range can easily be included in the simulation.

Variations in effective tap gains are included by substituting the factor $G_{\text{tap}}(k, x_n)$ for the product $G_{\text{conv}} G_{\text{ap}}$ in Eq. (14.8). $G_{\text{tap}}(k, x_n)$ is given by

$$G_{\text{tap}}(k, x_n) = [1 + \Gamma \Psi(k, x_n)] G_{\text{conv}} G_{\text{ap}} \qquad (14.56)$$

where Γ represents the peak-to-peak variation in the tap gains. Equation (14.56) is also subject to the constraint given by inequality (14.55). The $\Psi(k, x_n)$ are a random number set generated in a manner identical to the factors $\varepsilon(k, x_n)$ in Eq. (14.54). The realizations of $\varepsilon(k, x_n)$ and $\Psi(k, x_n)$ are, however, distinct and independent for Eqs. (14.54) and (14.56) as the gain and offset mechanisms are not related in first order. Equation (14.56) allows for variations in gain above and below the nominal tap/chain gain of $G_{\text{conv}} G_{\text{ap}}$. Physically realistic choices of Γ are on the order of a few percent and easily avoid the unphysical situation of $G_{\text{tap}}(k, x_n) < 0$.

Figure 14.9d is the nonuniformity-free reference image, and Figure 14.9e shows the effects of tap offset variations for four taps, each of which serves 256 cross-scan pixels. No other nonuniformity effects are included in Figure 14.9e. Figure 14.9f displays the relative tap offset levels as generated by Eq. (14.54). $V_{\text{off}}^{\text{pp}}$ is 5.0 mV in Figure 14.9e; this corresponds to a peak-to-peak variation of 1000 electrons, which is 10 times the rms noise floor in the image. An example image showing tap-gain variations is not included because it is difficult to see visual effects for any physically reasonable variation in tap gains. The cosmetic effect on image quality is more of an annoyance for large images. The image is degraded by banding, with each band corresponding to the number of cross-scan pixels in each tap. The abrupt discontinuous transitions at the tap boundaries can have secondary degrading effects, particularly when digital signal processing methods such as bandwidth compression are applied to the video stream before serial transmission.

The aggregate effect of sufficiently high levels of nonuniformities will clearly reduce the quality and hence utility of the image information. In principle, the nonuniformity effects can be removed by various calibration schemes. In practice, the techniques will still yield residual nonuniformities whose magnitude will depend on the sophistication and complexity of the calibration scheme. In particular, the interaction of digital data compression with tap nonuniformities (i.e., tap boundaries) may produce video anomalies that cannot be removed. In the next section, the digital signal processing functions of nonuniformity correction and data compression are discussed.

14.4.4 On-Board Digital Signal Processing

In the previous section we described our TDI CCD sensor simulation through the level of voltage inputs to the ADCs [Eqs. (14.54) and (14.56)]. After the A/D conversion process we have a two-dimensional array of digital values representing the amplitude of each pixel. We assume that fully detailed image processing and exploitation occurs in the ground processing system, as shown in Figure 14.1. The two most likely forms of digital image processing to occur on board the sensor prior to serial transmission are image uniformity correction and image bandwidth compression.

If digital data compression is used to maximize the sensor area rate of coverage, the interaction of the image data compression algorithms with the image nonuniformities may limit the effectiveness of ground-based calibration schemes. In some circumstances, it may become necessary to remove the nonuniformities prior to data compression. A number of techniques for removing spatial nonuniformities may be used, including flat-field correction, histogram normalization, and Fourier filtering.

One simple and easily implemented method for removing image nonuniformities is flat-field correction. A uniform calibrated and variable irradiance level is exposed on the TDI CCD focal plane (using some sort of optical calibration assembly in the sensor). The response of each cross-scan pixel at a given TDI level is measured at a series of flat-field irradiance levels that appropriately cover the pixel dynamic range. A measurement at a "zero" irradiance level gives the cross-scan pixel offset levels. From such data a response versus input irradiance curve can be obtained for each cross-scan pixel. This curve can be fitted to a higher order polynomial or described by a series of linear segments between calibration irradiance points, where, in either case, the resulting coefficients provide the means by which differences in pixel-to-pixel offsets and responses can be mathematically corrected (e.g., through an onboard look-up table). If the pixel responses are extremely linear over the dynamic range, the correction can be made with a few or perhaps even a single irradiance level.

The ability of the flat-fielding scheme to remove nonuniformity effects will depend on pixel linearity, the signal-to-noise ratio at each calibration point, and the total number of such points. Ultimately, these factors determine the accuracy of the computed gain and offset calibration coefficients for each pixel, and these coefficients in turn determine the "flat-field" correction capability. The result of the nonuniformity correction will fall somewhere between the "perfect" images in Figure 14.9a and the nonuniform images in Figures 14.9b and 14.9e (or some combination of these latter two).

Many image data compression techniques have been studied, including one- and two-dimensional differential pulse code modulation (DPCM) (Habibi, 1972; Ready and Spencer, 1975), vector quantization (VQ) (Nasrabadi and King, 1988), and many variations on transform coding (Chen and Smith, 1977). More straightforward methods, such as pixel aggregation and selective transmission of subportions of the TDI CCD image, can also be used to reduce the required bandwidth of the transmitted signal. Although lossless image data compression algorithms (algorithms that do not introduce visual image quality loss) have been developed, complexity often precludes their implementation in practical sensor applications. More practical candidate algorithms do produce some loss in image quality; the greater the compression the greater the loss. This loss in image quality must somehow be accounted for in the sensor design margin so the final image achieves the required quality. The digital simulation described earlier is an excellent tool for examining the impact of such losses on image quality.

Section 14.5 gives a series of TDI CCD sensor simulation examples that emphasize some important sensor trade-offs and dependencies, including anomalies induced by the interaction of tap offset nonuniformities and selected data compression algorithms.

14.4.5 Computational Considerations

A primary benefit in performing digital simulations is the ability to study the effects of changes and modifications in sensor design parameters in a timely manner. With sufficient computing capability, the simulation process can be streamlined and used to produce imagery in a quick-turnaround production mode. In addition, the simulation process can be used to generate a library of images representing system performance under varying acquisition conditions. Such a database is useful for utility analysis and signal processing algorithm development efforts that would otherwise require construction and operation of an actual testbed instrument. It is apparent from the description of the computations required by the simulation process that the utility of the simulation procedure will be greatly influenced by the hardware and software available to the user. At a minimum, the computational configuration should include a host CPU capable of performing complex operations on large arrays of floating point numbers, sufficient storage capacity for holding input and output images and two-dimensional transfer functions, and a high-quality display monitor and image processing software with which to examine the results of the simulation. Specific requirements on the hardware configuration will now be addressed and referenced to portions of the simulation process discussed earlier.

The digital simulation process can be broken down into three main functional processes: (1) computation of the two-dimensional MTF, sensor signal chain analysis, and digital signal processing; (2) Fourier filtering to simulate the spatial blurring process; and (3) display of the simulated imagery. Although intermediate processing steps fall between them, these three processes place particular requirements on the CPU speed, memory requirements, and display capabilities.

Constructing the two-dimensional MTF is a computationally intensive process. Section 14.4.2 described the procedure for computing the polychromatic MTF using a spectrally weighted sum of subinterval MTFs. Each wavelength-dependent MTF component is computed 20 times, once per wavelength interval, for each frequency (f_x, f_y), in order to generate the polychromatic value. In our facilities, the MTF computations are performed using a VAX 8200 rated at 1 MIPS. The MTF array constructed for filtering a 512×512 pixel image contains 257×257 floating point values. Generation of the 2-D MTF occupies a large fraction of the computing time required for a single simulation, and it is clearly advantageous to minimize this with a capable CPU.

Once the two-dimensional MTF is constructed, the MTF filtering process involves performing a forward two-dimensional Fourier transform of the image array, complex multiplication of the image FT with the MTF, and an inverse Fourier transform of the result. We restrict the array dimensions to be powers of 2, to take advantage of the computational shortcuts provided by the fast Fourier transform. Figure 14.10 shows a hardware configuration routinely used for performing sensor simulations. The process is enhanced through the use of an array processor that allows a 512×512 pixel array to be filtered in a few seconds. The

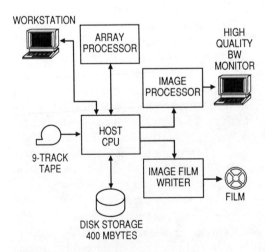

Figure 14.10 A hardware configuration well suited for performing digital simulations.

actual computational time in the array processor is minimal (on the order of 1–2 sec per Fourier transform); the majority of the time is spent performing input-output to and from the array processor. The advantage of using an array processor decreases as the capability of the host CPU increases, as ultimately the array processor functions will be limited by the speed with which data can be transferred back and forth.

The output image produced through digital simulation is smaller than the input image, owing to the oversampling requirements for the background scene. For example, an input scene 512×512 pixels in size with an inherent resolution four to five times better than the sensor system being simulated will produce images approximately 100 pixels square. Although this image size may be sufficient to obtain a general appreciation for the performance of the system, a larger image provides more contextual information and better interpretability and utility. However, generation of a simulated image with 1024×1024 pixels, for example, requires starting with a high-resolution background scene on the order of 4096×4096 pixels. As the size of the input image increases, the storage and memory requirements for the Fourier transform process also grow. At some point, depending on the configuration of the host CPU, there will simply not be sufficient memory to compute the required Fourier transform.

To accommodate the filtering of extended input images, we have developed a technique that involves filtering individual subsections of the large input image and stitching the resulting subimages back together. Figure 14.11 illustrates this process. The image is divided into subimages of appropriate size for the filtering software. The subimages overlap in both the x and y spatial directions. Each subimage is filtered separately using the 2-D MTF computed for the sensor system. (Note that the 2-D MTF need be computed only once because the same MTF is applied to the entire image.) The output image array is filled with the results of filtering the individual input subimages. The overlap regions are divided evenly and filled with values from each of the two adjoining images. The overlap is required to minimize the edge effects that inevitably accompany finite, discrete Fourier transform calculations. The amount of overlap is typically several times the effective width (in high-resolution pixel elements) of the sensor PSF. This allows for a smooth transition at the boundaries between the subimages. For example, consider an input scene 4096×4096 pixels in size, with a sampling resolution four times better than the system being simulated. An input scene of this size is typically broken into 100 subimages (a 10×10 array of overlapping 512×512 pixel subimages) and requires 100 filtering operations to fill in the output image array. The output array is then sampled at the required GSD and reduced in size to a 1024×1024 array.

In spite of attempts to minimize edge effects, "seams" in the final product are sometimes observed at the locations where output subimages are joined. This effect is most noticeable when the image content in one subimage differs signifi-

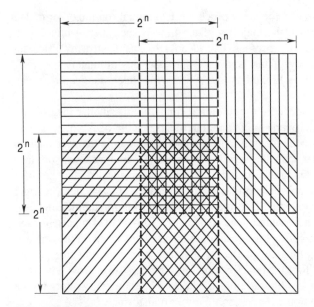

Figure 14.11 Schematic showing the subimage sectioning process used for processing large image arrays. The large image array is broken into several overlapping smaller arrays, which are individually filtered with the sensor MTF and then restitched together before sampling.

cantly from that in the adjoining subimage, particularly in areas of little contrast or detail. These seams can be further minimized by employing greater overlap of the adjoining images, or by performing a linear interpolation across the seam, using values from the two contributing images to estimate image values in the overlap region. Generally, we have found that for scenes of terrestrial backgrounds, the "seams" are either not visible or are of minor consequence in evaluating the quality or utility of the output product. The completion of a filtering operation of the size discussed in the previous paragraph typically requires several hours and is performed in batch mode. Much smaller image sizes can be used to evaluate the effects of minor changes in design parameters, but the capability to filter large images provides an output product with greatly enhanced contextual information content.

Subsequent steps in the digital simulation process involve signal conversion, addition of random and fixed pattern noise, and further digital processing such as bandwidth compression. The signal chain calculations and noise computations are not easily vectorized owing to the pixel-dependent processes, and the array processor does not provide significant computing advantage over the host CPU.

We prefer to store intermediate results after each stage of processing in order

MODELING AND SIMULATION OF CCD IMAGING SYSTEMS

to view the progress at each stage and identify the sources of various image degradations that appear in the final results. An input array consisting of 2000 × 2000 pixels (16 bits/pixel) requires 8 Mbytes of storage. Storage of the 2-D MTF and intermediate results can easily double or triple that value. Sufficient online disk storage is critical for minimizing I/O transfer time during the intermediate stages of processing and providing fast turnaround time.

Finally, in order to evaluate the system performance it is necessary to visually inspect the image. At a minimum, the hardware configuration should provide the capability to view the output imagery with a level of scrutiny at least as good as, and preferably superior to, that which would be used in practice on real imagery. Use of a high-quality monitor (e.g., 60 Hz, noninterlaced) along with image display capabilities provided by most standard image processing systems (roam, zoom, tonal remap, convolution filtering, etc.) is required for precise inspection and characterization of defects and artifacts in the imagery.

14.5 SENSOR SIMULATION: EXAMPLE CASES

In this section we present several example simulations designed to show the utility and capabilities of the digital simulation process. The baseline system parameters used for these examples are presented in Table 14.3. The following examples illustrate the utility of the simulation procedure for examining visual effects of component MTFs on image quality, effects related to focal-plane defects, and various signal-processing procedures.

One of the advantages of using a selectable TDI device is the ability to accommodate a wide dynamic range. When imaging under low illumination, additional stages of TDI can be used to increase the signal-to-noise ratio (S/N). Under more favorable illumination conditions, fewer stages of TDI are required to achieve a nominal S/N ratio. Unfortunately, increased S/N achieved this way is accompanied by an MTF degradation that increases with the number of TDI stages if velocity mismatch is present. The MTF degradation arising from velocity mismatch is given by the equation in Table 14.2B. It is clear that the magnitude of this degradation is directly dependent upon the product of the amount of velocity mismatch and the number of TDI stages.

Figure 14.12 shows the results of simulations of a system operating under low levels of illumination. For this simulation, an uncompensated motion of 0.01 rad/sec results in an image movement of 0.1 pixel during the integration time for a single TDI stage (0.1 msec). The example illustrates the trade-off between S/N and MTF smear arising from uncompensated, residual image motion, when TDI is used to control S/N. Figure 14.12b shows the simulated imagery produced using four stages of TDI. The short exposure time (product of the integration time per TDI stage and the number of stages of TDI) combined with the low illumination level results in a grainy appearance to the image. In contrast, Figure

(a)

Figure 14.12 Illustration of the trade-off between S/N ratio and image smear in TDI imaging applications. Simulation results for sensor in Table 14.3, with 10% velocity mismatch. (a) 32 stages of TDI; (b) four stages of TDI. (c) Plots of in-scan MTF curves for cases (a) and (b). The benefit of increased S/N accomplished by using more TDI is offset by the increased spatial blurring.

14.12a shows the results obtained if the total exposure time is increased by using 32 stages of TDI. The graininess representative of lower S/N ratios is reduced, but the image appears blurred, owing to the reduced MTF associated with the velocity mismatch and increased TDI. Figure 14.12c shows the overall in-scan MTF and the velocity mismatch components for the two cases. The loss in MTF due to the velocity mismatch smear is evident. Severe velocity mismatch can actually null out frequency ranges with accompanying loss of critical spatial information (Lomheim et al., 1990). The ability to simulate imagery obtained under various operating conditions, such as presented in this example, allows a system designer to make decisions concerning the utility of the sensor under a wide variety of conditions and potential operating modes.

In previous sections we have described a technique for computing a spectrally weighted polychromatic MTF to be used in simulations of broadband imaging

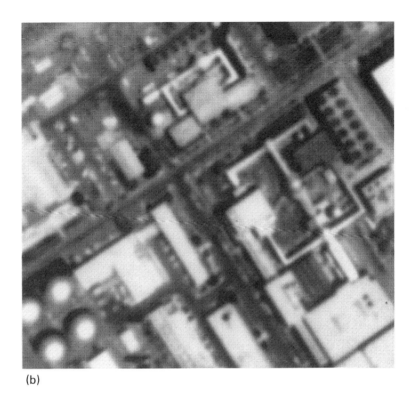

(b)

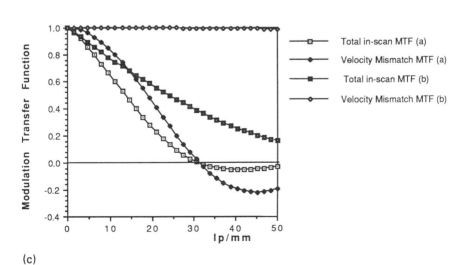

(c)

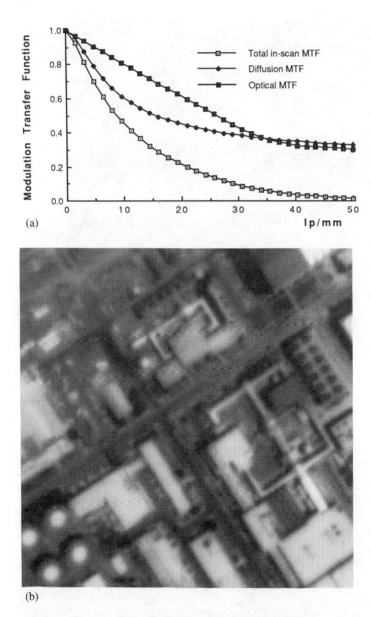

Figure 14.13 Illustration of the effects arising from spectral dependence of sensor MTF. (a) In-scan MTF and wavelength-dependent components for a system with $\lambda_c = 0.95$ μm ($\Delta\lambda = 0.1$ μm). (b) Blurred, sampled image (prior to signal-chain computations) resulting from filtering the image in Figure 17.7a with the sensor MTF shown in (a). (c) In-scan MTF and components for system centered at $\lambda_c = 0.45$ μ ($\Delta\lambda = 0.1$ μm). (d) Blurred, sampled image resulting from filtering with sensor MTF shown in (c).

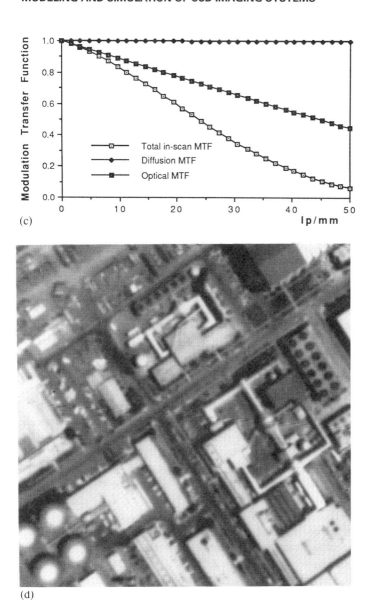

(c)

(d)

systems. The technique can be used equally well over more restricted portions of the spectrum. To demonstrate the wavelength-dependent nature of the MTF function components, we use the system design provided in Table 14.3, but with the addition of a spectral filter. The two cases we have chosen to simulate have spectral coverages of 0.4–0.5 μm ("blue") and 0.9–1.0 μm (near-IR, referred to as "red"). Figures 14.13a and 14.13c show the total in-scan MTF functions and the wavelength-dependent component MTFs for these two cases. Both the optical and carrier diffusion MTF components exhibit increased degradation for the longer wavelength (red) case, and the corresponding system MTF is severely degraded. The resulting additional image blur is clearly evident through comparison of the image examples shown in Figures 14.13b and 14.13d. Notice that although the total in-scan MTF for the "red" and "blue" cases are both less than 10% at the Nyquist frequency, the shapes of the two MTF curves are very different. The "red" MTF drops off much faster than the "blue" MTF and shows a severe loss of information in the middle spatial frequencies as well. This illustrates the problem of specifying the performance of an imaging system at only one spatial frequency (e.g., GRD).

It should be noted that the imagery in Figure 14.13 demonstrates image quality loss due to MTF processes alone; no optical signal analysis was performed for either image, and no attempt was made to simulate the actual gray tones that would be indicative of the terrain in those two wavebands. In reality, particularly for an overhead system looking through the Earth's atmosphere, the increased image quality available with the "blue" system would be offset by the reduction in S/N ratio arising from multiple scattering and attenuation processes in the atmosphere at these wavelengths.

Figure 14.14a shows a portion of the simulation example from Figure 14.9b. Nonuniformities due to variations in detector-to-detector responsivities are included in this simulation. In addition, a dark line representing a failed detector and a bright line representing a saturated detector were inserted into the image at the nonuniformity implementation step. Such defects are representative of those that might be expected in actual focal-plane arrays, and the inclusion of such defects in the simulation allows one to study the interactions of focal-plane defects and downstream processing algorithms. Figure 14.14b shows the result of compressing and decompressing (expanding) the data in Figure 14.14a using a one-dimensional DPCM algorithm. A compression ratio of 2.66 was achieved by reducing the original 10 bit pixels to 3.75 bits. The action of the compression algorithm is in the detector column direction; each column is compressed individually. Comparison of Figures 14.14a and 14.14b shows little difference. Under close examination, the image in Figure 14.14b appears slightly noisier, but the effect is essentially visually unnoticeable. Figure 14.14d shows the "difference image" created by subtracting the compressed/expanded image from the original image. Larger differences are represented as white features, while no difference is represented as black. As is typical with DPCM algorithms, the largest differences, that is, the places where the compression algorithm most often fails to reproduce the original data, are primarily along edges of high-contrast features.

The example in Figure 14.14c shows Figure 14.14a after compression and reexpansion with a two-dimensional DPCM algorithm. A compression ratio of

MODELING AND SIMULATION OF CCD IMAGING SYSTEMS 571

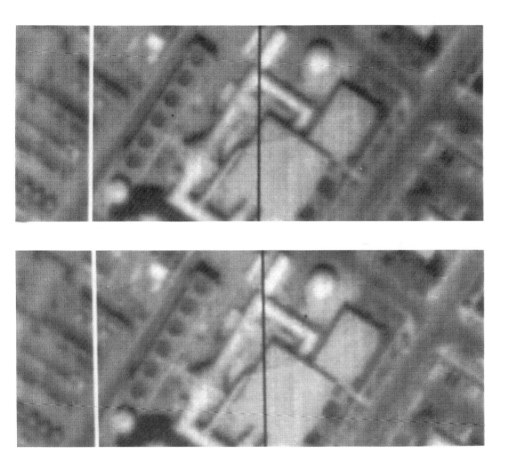

Figure 14.14 Illustrations showing the interactions of bandwidth compression algorithms with focal plane defects. (a) Enlarged portion of original simulated image from Figure 14.9b, with one failed (dark line) and one saturated (bright line) detector. (b) Figure (a) after compression with one-dimensional DPCM algorithm. (c) Figure (a) after compression with two-dimensional DPCM algorithm. Note the artifacts adjacent to the failed and saturated detectors in (c). (d) Difference image derived from images (a) and (b). (e) Difference image derived from images (a) and (c). (Larger differences show up as lighter shades; black = no difference.) (f) Histograms of the difference images.

3.64 was obtained by compressing the image to 2.75 bits per pixel (bpp). This algorithm essentially produces a single digital value for each block of 2×2 pixels in the original image, hence the two-dimensional nature of the algorithm. Figure 14.14e shows the difference image produced for Figure 14.14c. The primary visual effect of the two-dimensional algorithm is noticeable degradation of object

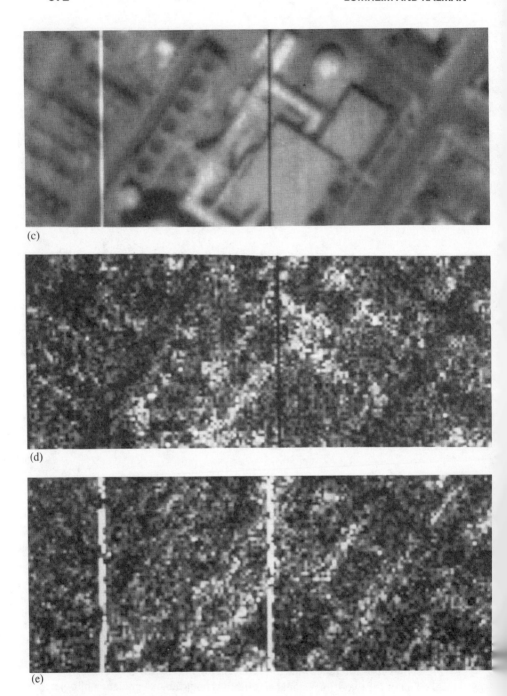

Figure 14.14 (Con't.)

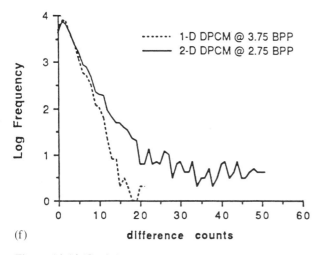

(f)

Figure 14.14 (Con't.)

edges. This is most evident alongside the saturated pixel column, where the column adjacent to the saturated column also exhibits artifacts related to the failure of the algorithm to faithfully represent abrupt edges. This artifact was not present in Figure 14.14b because the one-dimensional algorithm did not "cross" the saturated detector column. However, in the presence of a similar defect or sharp transition occurring in the in-scan direction, the one-dimensional algorithm would show similar behavior.

Figure 14.14f shows a log plot of the frequency of occurrence of difference values for the one- and two-dimensional compression examples of Figures 14.14d and 14.14e, respectively. Histograms of difference values such as that shown in Figure 14.14f illustrate the fidelity of the compression algorithms. The plot in Figure 14.14f shows that the two-dimensional DPCM algorithm produces more errors (at a given difference level) and larger differences than the one-dimensional DPCM algorithm. The higher compression ratio obtained with the two-dimensional algorithm-means that more imagery could be transmitted to the ground for a fixed data link rate, at the expense of greater loss in image quality. Utilizing onboard compression with longer line arrays can conceivably increase the area coverage rate or allow for increased dynamic range through the collection of more bits per pixel. However, the potential benefits derived from including bandwidth compression must be weighed against the potential degradations associated with the particular form of compression algorithm. This example illustrates the increased degradation that accompanies increases in compression ratio. The example also illustrates the results of potential interactions between the actual form of the compression algorithm and focal-plane nonuniformity ef-

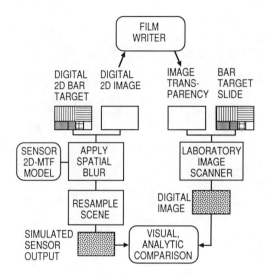

Figure 14.15 Schematic showing the conceptual method for validating the digital simulation procedure using a laboratory-based sensor.

fects. The decision of whether a compression algorithm will produce acceptable results is best determined by exercising the algorithm under the entire range of conditions and scenes expected, and visually assessing and characterizing the resulting degradations. The simulation process allows the relatively quick and inexpensive construction of a database of imagery for use in such evaluations.

14.6 EXPERIMENTAL SIMULATION VERIFICATION

Validation of the sensor simulation procedure described in Section 14.4 requires a laboratory image-scanning system along with a TDI CCD device capable of capturing and storing two-dimensional digital imagery. The scene information input to both the analytic (digital) and hardware simulation are derived from the same source, be it "bar-pattern" input scenes used for analytic and quantitative comparison of MTFs or extended scenes, where the comparisons are visual and hence more subjective. Figure 14.15 outlines the steps required in this overall comparison.

We have experimentally verified selected features in our digital simulation using a scanning TDI CCD hardware simulator developed in an electro-optics facility. A complete description of this hardware system is given in Lomheim et al. (1990). At the time of this work the digital and hardware simulations were compared by using a bar-pattern image only. This laboratory image consisted of bar-pattern sequences of increasing spatial frequency, delineated by the deposition of chrome strips on a glass slide. This chrome-on-glass transparency pro-

MODELING AND SIMULATION OF CCD IMAGING SYSTEMS

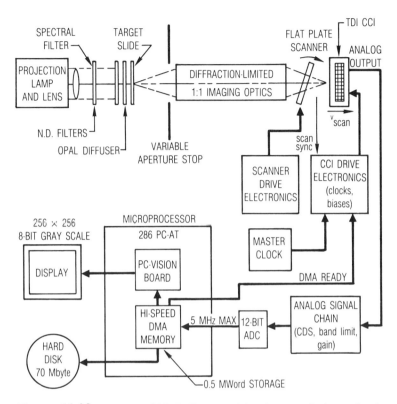

Figure 14.16 Functional block diagram of the electro-optical scanning imaging system used for characterizing TDI CCDs.

vides high optical fidelity of the ideal input pattern, compared to a film-based transparency, thereby avoiding potential nonlinearity effects. We have recently acquired some selected chrome-on-glass extended scenes with integral bar-pattern sequences along the edges, derived from a digital high-resolution input scene, and in future work we will provide the complete comparison depicted in Figure 14.15. We report here the comparative results of sensor MTF extractions using chrome-on-glass bar-pattern inputs to the hardware imager system and analogous digital bar-pattern sequences as the input to the digital simulation. Before proceeding we describe the essential features of our experimental configuration.

Figure 14.16 shows a top-level block diagram of the overall experimental system. Starting in the upper left-hand portion of this figure, it is seen that the object or target that is to be scanned across the TDI CCD is formed by back-illuminating a desired transparency. For extended target imaging characterization, a chrome-on-glass or photographic transparency of the desired image or bar pattern is back-illuminated by a projection lamp and lens assembly, which ensures uniform illumination of the slide. Intensity and optical wavelength control are obtained by inserting neutral density and spectral bandpass filters in the collimated region of the projection lamp system. The opal diffuser just behind the target slide provides uniform illumination. For point-target or impulse-response characterization experiments, the transparent target is replaced by an appropriate wavelength laser focused by a microscope objective lens through a pinhole.

The optics consists of a unity-magnification multielement refractive reimager with an adequate field of view and large front and back focal distances, and is optimized for high-resolution imaging over the visible spectral region. The reimager focal ratio (f-number) can be independently controlled by placing various diameter apertures directly behind the entrance aperture of the reimager.

For scanning an extended or point-source image across the TDI CCD we used an optical flat-plate scanner (Wolfe, 1985b). An optical ray incident on an optical flat at some angle with respect to the flat's normal exits the far side of the flat in a direction parallel to its entry, but with a lateral displacement easily calculated by Snell's law. If the optical flat is then rotated about an axis perpendicular to the flat's normal, the lateral displacement of the optical ray is linearly displaced in kind. Hence, if the flat is rotated at uniform angular velocity, the ray is displaced uniformly in time, giving a proportional linear scan motion. An entire image is scanned in this fashion, where the fixed image is located at the object plane of the reimager.

If the flat-plate angular rotation rate can be controlled precisely, the corresponding linear scan motion of the image can be controlled with the same precision. A feedback system using an encoder wheel drives a phase-locked loop circuit, and this arrangement provides precise angular velocity control (and hence image linear scan velocity control). The master clock driving the TDI CCD timing signals also drives the phase-locked loop system, which permits us to precisely match the image scan velocity with the TDI CCD electronic parallel scan velocity and thereby control the degradation effects of the temporal aperture MTF (Table 14.2B). The optics aperture control allows us to vary the optical point-spread function (PSF) and hence the optical MTF (Table 14.2D) by diffraction. The reimager was verified to have diffraction-limited behavior for focal ratios greater than $f/8$ at wavelengths of 632.2 and 832 nm. Finally, spectral bandpass filters provide color control, which affects both the optical and diffusion MTF components (Table 14.2C and D).

In addition to the experimental control of the component MTFs indicated above, a method was developed that gave us control over the phase modulation of the in-scan MTF. This was accomplished by the use of a timing circuit that temporally synchronized the image data capture down to a fraction of a pixel in the TDI scan direction. This control is important for sensor MTF extraction from bar-pattern inputs, particularly as the Nyquist frequency is approached (phase-modulation effects are important here).

MODELING AND SIMULATION OF CCD IMAGING SYSTEMS 577

The TDI CCD used in the hardware simulation was located at the focus of the optics, just behind the optical flat-plate scanner, and consisted of a 96 × 2048 pixel array (Bradley and Ibrahim, 1979) (96 TDI pixels × 2048 cross-scan pixels) with eight serial taps, each servicing a subset of 256 pixels. Image data were taken from a single serial tap, thus producing a cross-scan image width of 256 pixels. As the image was scanned across the CCD, the resulting analog serial video stream was processed by an analog signal chain that (1) provided voltage amplification as needed to preserve the noise floor of the CCD and provide an appropriate match of the CCD output voltage range to the input voltage range of the analog-to-digital converter (ADC), (2) provided signal-chain electronic bandlimiting prior to sampling to reduce the broadband signal chain noise, and (3) contained a correlated double-sampler circuit (CDS), essential for reducing the noise associated with the on-chip electrometer driving the serial tap output.

The digitized pixel stream (post-ADC) was captured at the full video rate using a dual-port solid-state memory (DPM) housed in a personal computer (PC) via direct memory access (DMA). The 2-D image (about 200 in-scan × 256 cross-scan pixels) was then transferred to an image processing and display board (located in the PC). Once transferred, the image could be manipulated using a PC-based image processing software package, including image zoom and pan, contrast enhancement, and display of line profile plots along selected slices of an image. The raw digital image data (in the DPM) could also be loaded into the PC working memory (specifically into the elements of a 2-D Fortran array) for mathematical processing required for MTF extraction.

The digital simulation technique described in this paper was used to simulate the laboratory-based sensor and produce output imagery that could be directly compared with that produced by the actual sensor. A digital image representing the bar target was constructed for use as input to the digital simulation. The sensor characteristics, as measured in the laboratory, were used with the models in Table 14.2 to create an overall sensor system MTF for filtering the bar-target image. The output images were sampled at the appropriate distances corresponding to the effective sampling distances of the CCD device used in the laboratory and the appropriate scan velocity. Cases of perfect velocity synchronization as well as velocity mismatch were simulated. The output imagery provided a digital simulation of the laboratory-derived images.

The Coltman method (Coltman, 1954) was used to extract MTF versus spatial frequency plots from the TDI CCD's response to variable-frequency bar patterns and from the simulated imagery (Griffice et al., 1990). Figure 14.17a shows in-scan composite (TDI CCD and optical) MTF versus spatial frequency plots for the composite in-scan MTF model per Table 14.2 using the pertinent hardware TDI CCD and optical parameters at a wavelength of 632.8 nm, the MTF extracted (Coltman method) from the bar-pattern responses in a digital simulation that used the same hardware parameters, and the MTF extracted (Coltman method) from bar-pattern responses in the hardware simulation. Excellent agree-

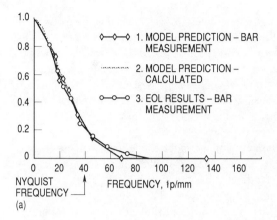

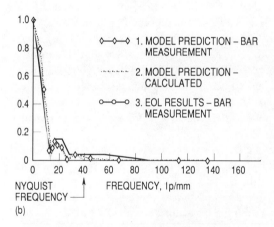

Figure 14.17 Comparison of (1) MTF derived from simulated imagery with (2) predicted system MTF computed using Table 14.2 equations and (3) MTF derived from images of bar targets. (a) No velocity mismatch; (b) 6.91% velocity mismatch.

ment between the three cases is evident. Figure 14.17b displays the same three plots as in Figure 14.17a except that here the hardware was adjusted to deliberately produce 6.91% velocity mismatch in the scan direction when using 96 stages of TDI. For this level of velocity mismatch, the velocity mismatch component of the temporal aperture MTF (Table 14.2B) dominates; the zeros in the sinc function at spatial frequencies of 15.9 and 31.8 lp/mm are verified in both the digital and hardware simulations.

Figure 14.18a is a visual display of the nominal bar-pattern response associated with the MTF plot of Figure 14.17a. Figure 14.18b shows the impact of

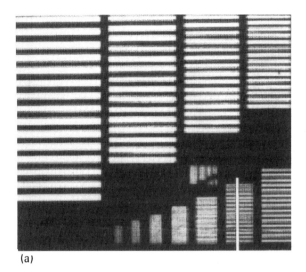

(a)

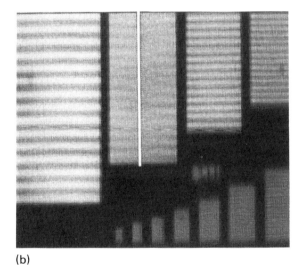

(b)

Figure 14.18 (a, b) Imagery produced using laboratory-based scanning TDI sensor (overall sensor transfer functions shown in Figure 14.17). (a) No velocity mismatch; (b) 6.91% velocity mismatch. (c,d) Imagery produced through digital simulation of laboratory-based sensor. (c) No velocity mismatch; (d) 6.91% velocity mismatch.

Figure 14.18 (Con't.)

the in-scan velocity mismatch, which was precisely chosen to obliterate the response in the second largest pattern (12.95 lp/mm), by placing the first null in the temporal aperture sinc function at the spatial frequency of this bar pattern. The precision control of the linear image scan velocity relative to the TDI CCD electronic scan rate allows us to selectively null distinct spatial frequencies. To show the concomitant effects of such an anomalous effect on extended images, the digital simulation was used with a high-resolution digital scene and with the same level of velocity mismatch as was used to create Figure 14.17b. The result is shown in Figure 14.18d; Figure 14.18c has no velocity mismatch and is included for comparison.

14.7 SUMMARY

In this work, we have presented the details of a scanned imaging sensor simulation that assumes the use of a silicon solid-state TDI CCD in its focal plane. The simulation requires a high-resolution input scene, which is then convolutionally "blurred" to the resolution of the simulated sensor using a detailed two-dimensional sensor MTF model. The simulation also includes the effects of sampling, temporal noise, realistic and systematic gain nonuniformity effects, and the effects of digital signal processing, such as digital video data compression, on final image quality. Each of these features was individually visually illustrated (e.g., output image display) using a nominal imaging sensor example.

The utility of the digital image simulation approach over evaluations based solely on commonly used figures of merit such as GRD and IIRS rating was emphasized by the example cases. The simulation was also experimentally verified using a hardware image simulator that contained a scanned TDI CCD; the simulation was adjusted to match the hardware simulators' optical and sensor parameters.

The described imaging simulation permits the detailed parametric evaluation of specific sensor designs well in advance of engineering development, including the impact of cosmetic defects induced by certain types of focal-plane nonuniformities. With an appropriate database of high-resolution imagery, this type of imaging sensor simulation could be used for the evaluation of a number of detection system types, including infrared or visible point-source detection systems and imaging multispectral systems.

Acknowledgments

This work was supported by the Aerospace Corporation company-sponsored research funds under the VISTAS and SISTER projects. We thank Charles P. Griffice and Warren F. Woodward for their contributions to the validation work described in Section 14.6, particularly the data reduction leading to the hardware

MTF plots, and Mark Vogel for supplying the specific algorithms for the one- and two-dimensional DPCM image data compression used in Section 14.5. Philip J. Peters, Edward S. Meinel, and Shahen Hovanessian are thanked for their critical review of the manuscript.

References

Ahmed, N., and T. Natarajan (1983). *Discrete Time Systems and Signals*, Reston Publishing, Reston, VA, pp. 318–321.

Anon. (1987). *CCD Data Book*, CCD Imaging Division, Fairchild-Weston Systems, Sunnyvale, CA, p. 77.

Anuta, P. E., J. L. Carr, C. D. McGillem, D. M. Smith, and T. C. Strand (1983). Data processing and reprocessing, in *Manual of Remote Sensing*, 2nd ed., Vol. I (F. C. Billingsley, Ed.), Sheridan Press, Falls Church, Virginia, Chapter 17, pp. 755–762.

Arguello, R. (1979). Image chain analysis of a high-resolution, high-speed, CCD film reader system, course notes, Session X, presented at the Tenth Annual Modeling and Simulation Conference, School of Engineering, Univ. Pittsburgh (26–27 April).

Baker, W. D. (1980). Intrinsic focal plane arrays, *Topics Appl. Phys. 38*: 39–44.

Barnes, J. A., and S. Jarvis (1971). Efficient numerical and analog modeling of flicker noise processes, NBS Tech. Note 604, Natl. Bureau of Standards, Washington, DC.

Bracewell, R. N. (1986). *The Fast Fourier Transform and Its Applications*, McGraw-Hill, New York.

Bradley, W. C., and A. A. Ibrahim (1979). 10,420 pixel focal plane with five butted 2048 by 96 element TDI CCD's, *Proc. SPIE 175*: 72–80.

Chen, W. H., and C. Smith (1977). Adaptive coding of monochrome and color images, *IEEE Trans. Commun. COM-25*:1285–1292.

Coltman, J. W. (1954). The specification of imaging properties by response to sine wave input, *Opt. Soc. Am. 44*:468–471.

Diner, D. J., J. V. Martonchik, E. D. Danielson, and C. J. Bruegge (1989). Atmospheric correction of high resolution land surface images, *Proceedings of IGARSS'89*, 12th Canadian Symposium on Remote Sensing, Vol. 2, p. 860.

Fried, D. L. (1966). Limiting resolution looking down through the atmosphere, *Opt. Soc. Am. 56*: 1380.

Goodman, J. W. (1968). *Introduction to Fourier Optics*, McGraw-Hill, New York, p. 114.

Griffice, C. P., L. S. Kalman, and W. F. Woodward (1990). Laboratory validation of E-O sensor simulation procedures, Internal Memorandum, The Aerospace Corporation, Los Angeles, CA, May.

Habibi, A. (1972). Delta modulation and DPCM coding of color signals, *Proceedings of the International Telemetry Conference 8*, Los Angeles, CA, Oct. 10–12, p. 333–343.

Howes, P. D. (1987). Es-250 reconnaissance system, *Proc. SPIE 833*: 79–84.

Johnson, J. F. (1982). Statistics of charge-coupled imager aperture MTFs, Internal Memorandum, The Aerospace Corporation, Los Angeles, CA, July.

Keshner, M. S. (1982). $1/f$ noise, *Proc. IEEE 70*:212–218.

Kesler, M. P., and T. S. Lomheim (1986). Spectral response nonuniformity analysis of charge-coupled imagers, *Appl. Opt.* 25:3653–3663.

Kingston, R. H. (1979). *Detection of Optical and Infrared Radiation* (Springer Ser. Opt. Sci., Vol. 10) Springer-Verlag, New York, pp. 10–16.

Kopeika, N. S. (1987). Imaging through the atmosphere for airborne reconnaissance, *Opt. Eng.*, 26:1148.

Kosonocky, W. F., and J. E. Carnes (1972). Noise sources in charge-coupled devices, *RCA Rev.* 33:91–92.

Kuttner, P. (1983). Review on methods for simplification of optical performance criteria, *Proc. SPIE 467*: 56–61.

Lareau, A., and C. Chandler (1986). Advanced CCD reconnaissance detector, *Proc. SPIE 694*: 124–129.

Lloyd, J. M. (1982). *Thermal Imaging Systems*, Plenum, New York.

Lomheim, T. S., L. W. Schumann, R. M. Shima, J. S. Thompson, and W. F. Woodward (1990). Electro-optical hardware considerations in measuring the imaging capability of scanned time-delay-and-integrate charge-coupled imagers, *Opt. Eng.* 29:911–927.

Lyon, R., R. E. Othmer, and I. W. Doyle (1976). Computerized performance predictions for a solid state reconnaissance camera, *Proc. SPIE 79*: 216–227.

Nasrabadi, N. M., and R. A. King (1988). Image coding using vector quantization: a review, *IEEE Trans. Commun. COM-36*: 957–971.

Nicholson, D. S. (1975). Estimating optical system performance from a statistical description of its probable manufacturing errors, *Proc. SPIE 54*: 163–168.

O'Neill, E. L. (1956). Transfer function for an annular aperture, *Opt. Soc. Am.* 46:285–288.

Ready, P. J., and D. J. Spencer (1975). Block adaptive DPCM transmission of images, *NTC'75 Conf. Rec. 2*: 22-10–22-17.

Rosenbruch, K. J. (1983). Optical quality criteria as a means of a simplified description of image behavior, *Proc. SPIE 467*: 66–72.

Schade, O. H., Sr. (1948). Electro-optical characteristics of television systems, *RCA Rev.* 9: 5–37.

Schumann, L. W., and T. S. Lomheim (1989). Modulation transfer function and quantum efficiency correlation at long wavelengths (greater than 800 nm) in linear charge-coupled imagers, *Appl. Opt.* 28:1701–1709.

Seib, D. H. (1974). Carrier diffusion degradation of modulation transfer function in charge-coupled imagers, *IEEE Trans. Electron Devices ED-21*:210–217.

Sequin, C. H., and M. F. Tompsett (1975). *Charge Transfer Devices*, Academic, New York.

Serayfi, K. (1973). *Electro-Optical Systems Analysis*, Electro-Optical Research Company, Los Angeles, pp. 179–181.

Smith, W. J. (1985). Optical design, in *The Infrared Handbook*, rev. ed., (W. L. Wolfe and G. Zissis, Eds.), Office of Naval Research, Washington, DC, p. 8–29.

Snyder, H. L., M. E. Maddox, D. I. Shediry, J. A. Turpin, J. J. Burke, and R. N. Strickland (1982). Digital image quality and interpretability: data base and hardcopy studies, *Opt. Eng.* 21:14.

Trott, T. (1960). The effects of motion in resolution, *Photogrammetric Eng.* 26:819–827.

Welch, R. (1972). The prediction of resolving power of air and space photographic systems, *Image Technol. 14*:25–32.
Wolfe, W. L. (1985a). Imaging systems, in *The Infrared Handbook*, rev. ed. (W. L. Wolfe and G. Zissis, Eds.), Office of Naval Research, Washington, DC, pp. 19-19–19-20.
Wolfe, W. L. (1985b). Optical-mechanical scanning techniques and devices, in *The Infrared Handbook*, rev. ed. (W. L. Wolfe and G. Zissis, Eds.), Office of Naval Research, Washington, DC, pp. 10-18–10-19.
Wulich, D., and N. S. Kopieka (1987). Image resolution limits resulting from mechanical vibrations, *Opt. Eng. 26*:529–533.

15
Display and Enhancement of Infrared Images

Jerry Silverman and Virgil E. Vickers

*Rome Laboratory,
Hanscom Air Force Base, Massachusetts*

15.1 INTRODUCTION

15.1.1 Nature of Infrared Imagery

High quality infrared (IR) imagery from current staring focal plane arrays has now reached or exceeded TV resolution. For PtSi-based Schottky barrier IR cameras (Shepherd, 1988), minimum resolvable temperatures below 0.02°C have been achieved, and arrays as large as 512 × 512 are commercially available. Hence, the processing, display, and enhancement of high-resolution wide-dynamic-range staring IR imagery, whether for soft-copy display or in real-time hardware embodiments, is becoming an important topic in image processing, but one still in the early stages of development (Silverman et al., 1990).

At the outset, we review the differences between IR images and the more familiar baseline of visible imagery (Silverman et al., 1992). In visible images, objects reflect the light of a source or sources to a sensor. In ideal (passive) IR images, the objects emit IR radiation as determined by their absolute temperature; in effect, we view a temperature profile of the scene. However, the monotonic relationship between object brightness and scene temperature can be perturbed by the presence of an IR source such as the sun, which gives a more visual look to daytime IR imagery. Figure 15.1 contrasts day and night IR images of the same scene, the latter probably closer to an ideal thermal profile. Lesser perturbations in the brightness versus temperature relationship arise from the deviations of real objects from ideal blackbodies as well as from the interaction be-

Figure 15.1 Day and night versions of the same scene.

DISPLAY AND ENHANCEMENT OF IR IMAGES 587

tween the spectral response of the camera and the spectral content of the radiating scene.

Another important distinction lies in the inherently low contrast of IR images compared to visible light images. The typical IR image is dominated by the background radiation at the average scene temperature, as most of the objects will be at ambient temperature and will radiate at roughly the same intensity, leading to a typical variation of a factor of 2 from lowest to highest signal. Consequently, plots of the number of pixels at various signal levels, the raw image histograms, typically have one high and narrow main peak due to this background radiation (several main peaks if more than one background is present such as sky and ground). Temporal noise as well as small contrast variations within the background will spread out the main peak(s) somewhat but generally leave more concentrated histograms than are found for visible images. Frequently, the "targets" or objects of interest are on a small number of pixels and are warmer and hence separated in gray level from the background levels; this leads to a low-level trailing edge in the histogram. The image histograms play a major role in the algorithms of Section 15.2, and examples and further discussion are found there. A second consequence of the low contrast of IR images is the importance of noise sources, such as spatial noise (Mooney et al., 1989), which are generally insignificant in the visible.

A subtle point that should be clarified is that this limited contrast does not preclude imagery of such wide dynamic range as often to exceed the 8 bit sensitivity of high-quality monitors, and the latter sensitivity is effectively further reduced by the limitations of the inherent gray-scale sensitivity of the eye. Images used in this chapter were taken with PtSi Schottky infrared cameras operated in the 3–5 μm band, with noise characteristics that have been carefully analyzed and measured (Mooney et al., 1989; Murguia et al., 1990). Infrared image dynamic ranges depend on weather, time of day, detector array technology, camera design, and image content. For PtSi cameras, raw signal levels at the upper end of representative ranges will span about 1000–2000 ADU (analog to digital units) after digitizing single frames to 12 bits. Since a typical noise level is 5 ADU [see Murguia et al. (1990) for details], a usable dynamic range of up to 200–400 levels is often encountered. We are faced, therefore, with a classical problem in image display: the disparity between the image dynamic range and the smaller dynamic range of the monitor/eye display system.

Ideally, in assessing the relative efficacy of several alternative techniques for display or enhancement, one should work as closely as possible with the type of imagery, display hardware, ambient light conditions, and so on specific to the application(s) in question. However, in an imperfect world, the ideal is not always practical. Hence, in evaluating and comparing algorithms for general-purpose applications, we have adhered to the following philosophy. We have employed an extensive set of locally taken imagery: indoor, outdoor, day and night scenes as diverse as possible. Whatever general conclusions are advanced in the following sections about the relative merits of one algorithm versus another, it is usually not difficult to find an image or image type that belies any particular conclusion. Therefore, we believe it most useful to emphasize how the various algorithms typically interact with IR imagery. We hope the reader will thereby gain the insight to choose algorithms based on his or her application and expected image set.

15.1.2 Display Scales, Algorithms, and Applications

With the philosophy just stated in mind, let us turn to the question of the gray scale used for the final 8 bit display, a matter less mundane than one might imagine, as we have found a strong interplay between the display algorithms and the gray scale. The "default" gray scale on our Sun workstation monitors is a linear "colormap" between the display value i and the luminance command value for the red, green, and blue components of monitor intensity:

$$\text{red}[i] = \text{green}[i] = \text{blue}[i] = i \qquad (15.1)$$

where i, red, green, and blue all range from 0 to 255. This default scale on our monitors (and we suspect similarly on other monitors) is unbalanced and suboptimum in that it is too sensitive at the bright end and not sensitive enough at the dark (zero) end. Although one could argue that this is desirable for many IR applications where the information of interest tends to be at the hot (bright) end, we would counter that such bias if desired is better introduced into the algorithms rather than into the display scales.

Thus we have assumed that features of interest in the imagery—averaged, so to speak, over many images and applications—are equally likely to occur in any range in the final display. We sought such balance, as well as an optimum number of discernible shades of gray, by going to the form

$$\text{red}[i] = \text{green}[i] = \text{blue}[i] = \gamma(i) \qquad (15.2)$$

where the gamma function for the monitor (Briggs, 1987) represents the desired mapping from display value to screen luminance command value.

Although formal algorithmic procedures for determining $\gamma(i)$ are available (Briggs, 1981), we have found the following simple procedure adequate. Using software-generated standard bar patterns with a range of spatial frequencies, we set display contrasts at $\Delta i = 2$ or 3. The bar pattern backgrounds were set from dark to light in gradual increments. Working in a darkened room (which means, strictly speaking, that the scale should always be used in a darkened room—needless to say, it wasn't), one of us mapped the above display contrasts into luminance command contrasts that were comfortably perceptible for the larger bar patterns and just perceptible for the smallest pattern. The gamma function thus derived is shown in Figure 15.2 (the optimum mapping varies slightly from

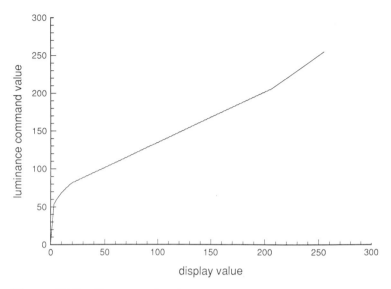

Figure 15.2 The gamma function.

monitor to monitor). Figure 15.3 shows the difference this gamma function makes, compared with the default gray scale, using both a real daytime image and a simulated image of uniform blocks going from 0 to 255 in unit steps. Note the difference in balance and sensitivity between the scales.

For the gamma-corrected scale, when i is between 35 and 175, a change of 2 is just perceptible, while in the regions above and below these limits, a change of 3 is required. We estimate that roughly 110 shades of gray are discernible out of the 256 nominal levels. Although the use of pseudocolor is beyond the scope of this chapter, we note in passing that a color scale with about 200 discernible levels has been designed for use with IR images (Silverman et al., 1990).

15.2 GLOBAL MONOTONIC DISPLAY ALGORITHMS
15.2.1 Introduction

To display our images, we need to map from the raw recorded signal (digitized to 12 bits for the images used to illustrate this chapter) to 8 bit values. Algorithms for this purpose can be divided into a global monotonic group treated in this section, and all others treated in Section 15.3; Table 15.1 lists the major algorithms considered in this chapter. More specifically, we now consider global mappings (no influence of local context) in which the radiometric trend from low

Figure 15.3 Examples showing effect of gamma function correction. (a, c) Displays with default gray scale; (b, d) with gamma function correction.

to high in the recorded image is retained in the displayed image (monotonic). We found the division of algorithm type between Sections 15.2 and 15.3 meaningful for IR images; it is typically ignored for visible images, for which raw signal levels depend strongly on natural and artificial light sources in the vicinity (Schreiber, 1978; see his Figure 1) and for which the intimate familiarity of the human brain with such imagery allows for flexible interpretation, so that we are not disturbed by deviation from monotonicity.

These global monotonic algorithms are subdivided into direct-scaling and related algorithms and histogram-based algorithms. The fundamental distinction here is between the linear or piecewise linear mapping of the scaling algorithms versus the nonlinear mapping of the histogram-based algorithms. In the former, one is reserving dynamic range in the display for empty regions, if present, within the span of the raw signal histogram, whereas in the latter, unoccupied levels are "squeezed" out of the display.

DISPLAY AND ENHANCEMENT OF IR IMAGES

Table 15.1 Major Algorithms Considered in This Chapter

Algorithm	Acronym	Section introduced
Direct scaling	DS	15.2.2
Histogram equalization	HE	15.2.3
Histogram projection	HP	
Undersampled projection	UP	
Threshold projection	TP	
Plateau equalization	PE	
Local range modification	LRM	15.3.2
Overlapping projection	OP	
Sliding projection	SP	
Raw modulo	RM	15.3.3
Modulo projection	MP	
Weak sinc sharpening	WS	15.3.4
Strong Gaussian sharpening	SG	
Medium Gaussian sharpening	MG	
Weak Gaussian sharpening	WG	

We next introduce a basic image set chosen to illustrate the operation of the various display algorithms. A common practice, particularly in the literature on restoration and for visual imagery, is to employ some familiar standard images previously used by other workers. Aside from the practical matter of the absence of such standard IR images, we reiterate that specific images can be quite misleading or at least unrepresentative of general trends. For example, we firmly contend that histogram equalization, a standard technique discussed in Section 15.2.3, is an ineffective algorithm for IR images. Yet it is not hard to find images for which the technique is quite satisfactory.

How, then, does one handle this difficulty and still retain coherence and some degree of brevity? We ask the reader to accept on faith that the small set of three images that will be used as a common thread throughout the rest of the chapter to illustrate trends and conclusions are fairly representative of the range of possibilities for the (literally) hundreds of IR images used to test the many algorithms. Where needed to make a special point, or hopefully to prevent boredom, other images will be interjected at certain junctures.

The three images are a sunny day image of Canada geese on a grassy background, a fairly complex night image of an airport scene, and a staged indoor scene of wide dynamic range with hot and cold cups (low-contrast details on each cup), between which is a set of bar patterns. The digitized raw signal histograms of the image set are presented in Figure 15.4. The geese image exemplifies a very common type of IR image with a histogram much like the prototype described in Section 15.1.1: a very concentrated main peak from the grassy back-

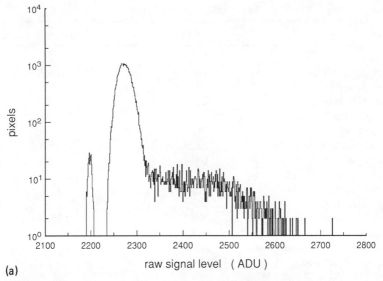

(a)

Figure 15.4 Raw signal histograms for the three standard images, (a) geese; (b) airport; (c) cups.

ground and a trailing edge on the high side of the peak arising from warmer objects that occupy fewer pixels. The airport scene is a representative night image for mild clear weather (a noisier cold night landscape scene will be used as well). The complex histogram of the two-cup image has portions related to the cold cup (ice water), bar patterns, background, and warm cup (hot water). This staged image is rich in specific local details and is interesting for another reason as well. The first two images are typical of most of our surveyed images in that they are "grabbed" single frames that were one-point-corrected by camera electronics (Murguia et al., 1990). The noise of such images includes the temporal noise of the single frame and the residual spatial noise associated with an imperfect correction. The two-cup image is produced from three direct (uncorrected) images; each image is an average over 256 successive measured frames and hence has negligible temporal noise. The three direct images are high- and low-temperature uniform scenes and the direct cup scene itself, and the final image is the two-point-corrected result (Murguia et al., 1990), which has lower residual spatial noise than one-point corrected images (Mooney et al., 1989). Hence, this image has low-contrast details and very little noise. (It will be compared later with a single-frame, one-point-corrected similar image.) The portion of its histogram above 2600 is an artifact of the interaction between the two-point correction and the bad pixels on the top few rows and contains no information, but it can affect the operation of display algorithms.

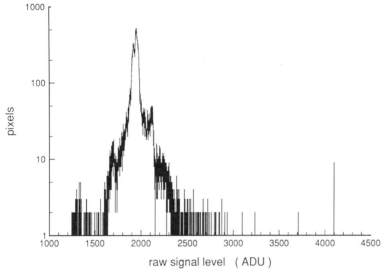

(b)

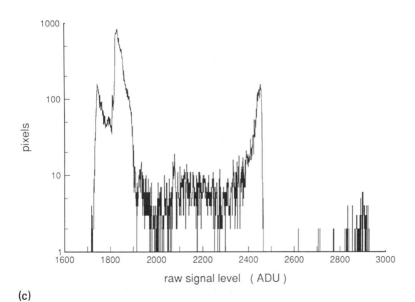

(c)

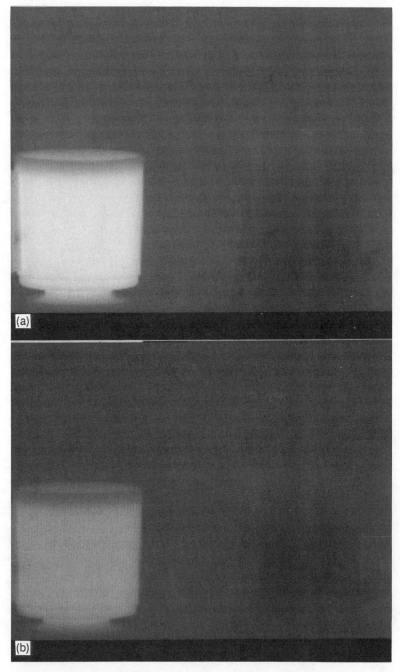

Figure 15.5 Direct-scales (DS) display with two choices for white level: (a) 2500; (b) 2900. Cf. Figure 15.4c.

15.2.2 Direct Scaling and Related Linear Mappings

Direct scaling (DS), if carried out interactively in software, is similar to manual offset/gain adjustment (contrast and brightness) of "live" camera imagery as guided by the eye. In the interactive mode, the observer views the histogram and chooses a black-and-white level whose span is then linearly mapped into the full display dynamic range. By and large, the obvious level choices usually provide a display very similar to that given by the histogram projection algorithm discussed in the next subsection, despite the linear/nonlinear distinction between the two mappings. In some cases, the optimum choices are not so obvious, as in the direct-scaled displays of the two-cup image in Figure 15.5, which take white as 2500 or 2900, respectively. The second choice wastes display dynamic range with no increase in scene content.

However, a more fundamental problem with DS, particularly for hardware implementation as an automatic offset/gain control to replace manual adjustment, is the difficulty of finding a robust automatic counterpart of the interactive process, because of scene and application dependencies. One simple candidate technique is to determine the black and white levels with symmetrical criteria by "integrating" to some fraction of the total area of the histogram plot from the bottom (black) or top (white) end. Figures 15.6 and 15.7 show the results of such an algorithm for the geese and airport images with choices of 0.05%, 0.1%, and 1.0%, respectively, of the area (values are percents of the pixels below the black level and above the white level). Also included is the computationally simple but naive choice in which the lowest and highest occupied raw signal levels are taken as black and white, respectively; the presence of a few unreliable pixels typically makes such a high/low scaling a poor choice. The optimum of the integration procedure is often at the 0.1% level; going beyond the optimum, as in the fourth picture of the sequences of Figures 15.6 and 15.7, tends to increase overall contrast but leads to a too high black level or too low white level, giving poor grayscale resolution at the low or high end (the latter here).

To increase the utility and robustness of a DS algorithm, particularly for hardware implementation, one would need to incorporate a more sophisticated analysis of the histogram and allow for unsymmetrical criteria for black and white choices. In our opinion, the results would still fall short in robustness for many image types compared with the results of the algorithms described in the next section.

A further extension of the DS concept would be a piecewise linear transformation, for example, of the interactive type referred to as "function processing" (Woods and Gonzalez, 1981) but restricted to monotonic mappings for the purposes of this section. Carefully probing the raw signal data of the two-cup image and after some trial-and-error attempts, we came up with the three-piece linear mapping shown in Figure 15.8, affording the display in Figure 15.9. This is superior to any of the other global monotonic algorithms on this image, although

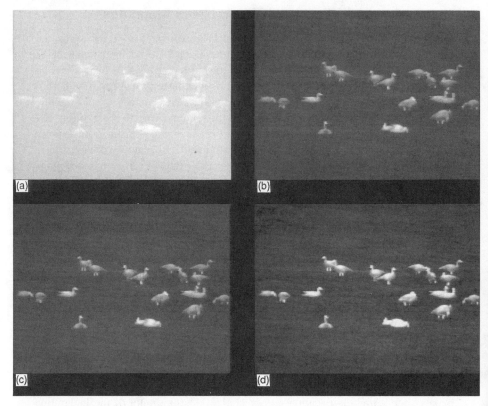

Figure 15.6 Examples of DS displays with various choices for black/white levels: (a) high/low; (b) 0.05%; (c) 0.1%; (d) 1.0%. See text.

the hybrid algorithms described in the next subsection come close. Since this technique is customized to each image and requires substantial operator intervention, we submit that it is more of a "process" than an "algorithm" and not practical for general hardware or software use.

15.2.3 Histogram-Based Nonlinear Mappings

By histogram-based algorithms, we refer generally to nonlinear mappings governed by transformations from the raw signal histogram to some desired final display histogram. The prototype for such methods is the well-known technique of *histogram equalization* (HE) described in many texts on image processing (e.g., Pratt, 1978). As the name indicates, the desired final form is a uniform histogram distribution. Although several variations of HE have been proposed

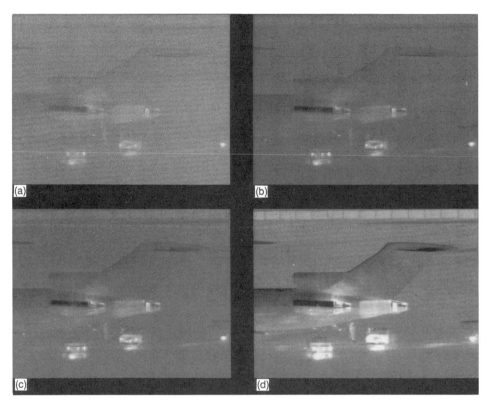

Figure 15.7 Examples of DS displays with various choices for black/white levels: (a) high/low; (b) 0.05%; (c) 0.1%; (d) 1.0%.

(Hummel, 1977), including hyperbolic and exponential distributions as desired goals, we believe that such refinements are essentially related to shifts in the gamma function of the gray-scale display (Section 15.1.2), which can be treated at a "pre-algorithmic" stage. Hence, our treatment in this section will focus on HE and a newer polar opposite to it called *histogram projection* (HP), as well as hybrids of the two.

Histogram equalization often gives excellent results on visible imagery and is claimed to be optimum from the standpoint of information theory (Tom and Wolfe, 1982). Generally it has been used to redisplay 8 bit data on an 8 bit scale. For the present purpose, mapping 12 bit IR data to 8 bits, the algorithm has major problems.

To implement HE for the discrete case, one converts the histogram of the

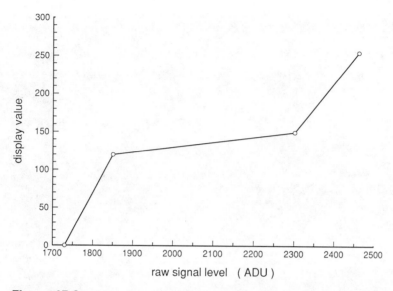

Figure 15.8 Custom piecewise-linear mapping function for cups image.

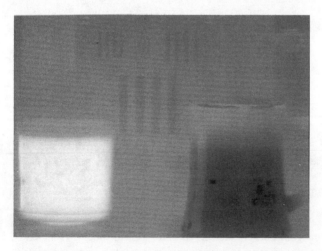

Figure 15.9 Display using piecewise-linear mapping shown in Figure 15.8.

DISPLAY AND ENHANCEMENT OF IR IMAGES

starting (raw) data to a cumulative distribution function, $F[i]$, which rises in discrete jumps from 0 to 1 and specifies what *fraction of pixels* are at or below the raw signal level i. Figure 15.10 shows $F[i]$ for the geese image. Note the large jump in $F[i]$ within the histogram peak (refer back to Fig. 15.4a). For display on an 8 bit scale, pixels at raw level i are mapped into

$$\text{Display value} = 255 \, F[i] \tag{15.3}$$

The resulting display histogram and displayed image are shown in Figures 15.11 and 15.12, respectively. The transformation of Eq. (15.3) produces an approximately uniform display histogram by fusing sparsely occupied adjacent raw signal levels while reserving a greater display dynamic range for signal levels with high pixel counts (places where $F[i]$ has large jumps). Empty levels can be created on the display scale (Fig. 15.11) between densely occupied adjacent raw signal levels. Referring back to the starting histogram (Fig. 15.4a), we see that the small leading peak at 2200 and the long trailing stream beyond 2300 retain their identity in the display histogram over a very narrow display range at the dark and light ends.

We reiterate that the geese scene is representative of a common type of IR image in which most of the pixels are on a background (ground here) with relatively little detail or variety, while a minority of pixels are on smaller objects of interest whose signal levels are separated from this background. As a measure of

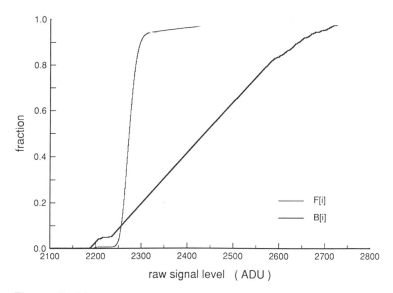

Figure 15.10 Cumulative distribution functions for geese image.

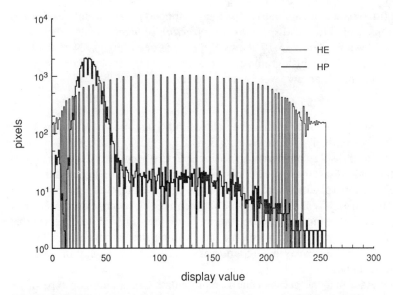

Figure 15.11 Display histograms for histogram equalization (HE) and histogram projection (HP) for geese image.

the degree of histogram concentration (Fig. 15.4a), we note that 25% of the total of 451 occupied raw signal levels (the presence of at least one pixel at a given signal level defines occupancy) account for 95% of the pixels. For images of this common type, HE assigns most of the display levels to the small variations in the background and its associated noise. Raw signal levels of the small objects, if outside this background peak, are compressed into a few display levels. (Hence the geese are displayed as white blobs starkly separated from the background but without the internal thermal detail contained in the raw data.)

This incompatibility between IR images and the HE technique was pointed out by Dion and Cantella (1984) as follows: "Since it operates on the basis of probability of occurrence, a small object within a given field-of-view can ordinarily be de-emphasized if its detail lies at an infrequently occurring level." Their solution was to precede HE with a high-pass filter. However, one can then no longer guarantee a global, monotonic mapping, and furthermore the tendency of HE to amplify the background noise is increased. Another solution first reported in 1988 (Silverman and Mooney, 1988) is a *histogram projection* (HP) algorithm whose guiding principle is diametrically opposite to HE: Display dynamic range is assigned equally to each signal level present, regardless of how many pixels occupy that level.

DISPLAY AND ENHANCEMENT OF IR IMAGES

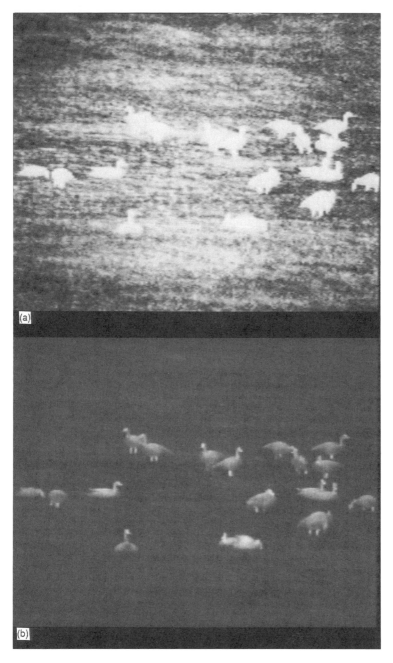

Figure 15.12 Comparison of HE and HP displays. (a) HE; (b) HP.

To perform HP, one need only compute an occupancy (binary) histogram and then order the occupied raw signal levels from 1 to N from lowest to highest, with N the total number of levels occupied. The cumulative distribution corresponding to $F[i]$, now called $B[i]$, represents the *fraction of occupied levels* at or below level i. $B[i]$ rises from 0 to 1 in discrete uniform steps of $1/N$ and is also illustrated in Figure 15.10.

An expression similar to Eq. (15.3) can be used to describe the 8 bit display of HP:

$$\text{Display value} = \lfloor 256\,(B[i] - 1/N) \rfloor \tag{15.4}$$

where $\lfloor\ \rfloor$ represents truncation to the next lower integer. The uniform step rises in $B[i]$ in effect assign equal display space to each occupied level. A more accurate description of HP can be based on thinking of the N occupied levels as linearly mapped or "projected" into the 8 bit display scale. If N is greater than 256 levels, neighboring occupied levels in the raw signal are fused to the same display value on the 8 bit scale by the compression factor $N/256$. In a typical software implementation using integer arithmetic, one could write for each pixel

$$\text{Display value} = \lfloor 256\,(n - 1)/N \rfloor \tag{15.5}$$

where n is the order number (from 1 to N) of the pixel's occupancy level.

The display of the geese image according to HP is also seen in Figure 15.12, and the corresponding display histogram in Figure 15.11. The close correspondence between this display histogram and the original (Fig. 15.4a) is typical of the algorithm. Indeed, thinking in terms of a transform driven by a desired final histogram, one seeks with HP to project the original histogram (excising empty levels) into the available display space; corresponding features such as peaks become higher, of course, if N is greater than 256. The natural, although somewhat dark-level, view of the background and the excellent resolution in gray scale of the smaller, warmer objects are characteristic of this algorithm.

The HP display is typically indistinguishable from the best DS result, despite the nonlinear/linear difference between the two mappings, as in the comparison of Figure 15.13 for the airport image (the keen-eyed viewer might spot some differences such as in the airplane windows). With the increasing sensitivity of IR imagers, the influence of unoccupied levels within the linear span of the image signal content may well become more important and the payoff from the nonlinear feature more apparent. In any case, the HP algorithm is easier to implement in real time and is more robust than the DS algorithm.

Some further comparisons of the displays generated by HE and HP are given in Figures 15.14–15.17. For the airport image, the displays are more complementary in their strong and weak points with the usual superior gray-scale resolution of small-scale objects in the HP result (the person and the vehicle grilles) but with a broader delineation and hence clearer spatial sense between fore-

DISPLAY AND ENHANCEMENT OF IR IMAGES

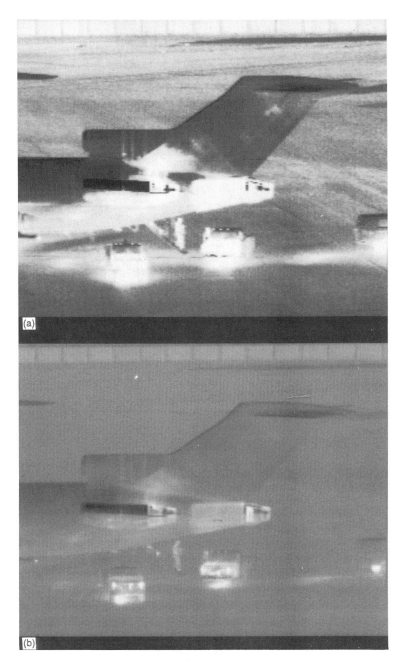

Figure 15.13 Optimum direct scaling compared with histogram projection. (a) DS at 0.1%; (b) HP.

Figure 15.14 Comparison of HE and HP displays. (a) HE; (b) HP.

DISPLAY AND ENHANCEMENT OF IR IMAGES

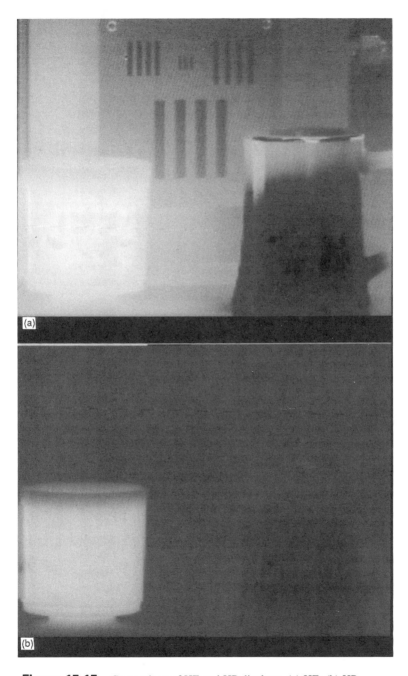

Figure 15.15 Comparison of HE and HP displays. (a) HE; (b) HP.

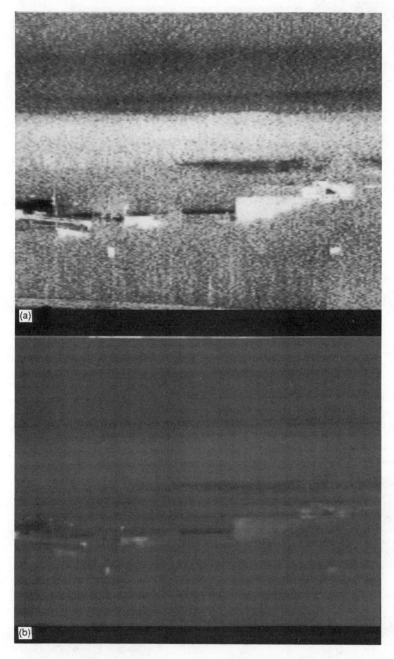

Figure 15.16 Comparison of HE and HP displays. (a) HE; (b) HP.

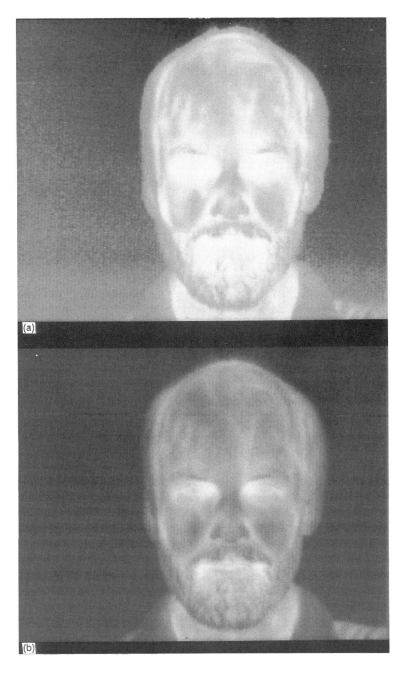

Figure 15.17 Comparison of HE and HP displays. (a) HE; (b) HP.

ground and background in the HE display. The face tends to favor the HP result, but if we had less neutral backdrop and more face in the field of view, the two displays would be more comparable. The cold winter-night landscape (Fig. 15.16) is a striking example of the interaction of temporal noise at the background levels with the two algorithms; the tendency of HE to amplify such noise is horrendous here, to the extent that image recognition is almost destroyed. Finally, the two-cup image displays represent one of a small number of cases where HE gives the better overall display. In these instances, some hybrid form of the two algorithms is usually superior to either; this brings us to the next topic in this subsection.

The allocation of display dynamic range according to histogram height, the prominent effect of HE, can be beneficial when used in a weaker mode than occurs in HE. Images that especially benefit from such weighting, such as the two-cup image, have rather complex multipeaked histograms and little information of interest in sparsely occupied raw histogram levels. For example, as described above, the occupied levels above 2600 in the histogram of the cup image (Fig. 15.4c) are artifacts. These "eat up" dynamic range in the HP display.

Our first attempt to fuse the two algorithms was literally a "hybrid" process in which a weighted combination of the cumulative distribution functions of each is used:

$$\text{Display value} = W \lfloor 256 \, (B[i] - 1/N) \rfloor + (1 - W) \, 255 \, F(i) \qquad (15.6)$$

Values of W from 0.9 to 0.7 often give the best result, that is, "mixing-in" between 10 and 30% of the HE weighting effect. Figure 15.18 shows the displays resulting from a 25%/75% mix of HE/HP for the airport and two-cup images. Since such a hybrid procedure reintroduces and even accentuates the computational complexity of the HE algorithm, we sought alternatives that hybridize the results without such increased complexity. Three are discussed here; two are basically variations on HP called *undersampled projection* (UP) and *threshold projection* (TP), and one is really a variation on HE, *plateau equalization* (PE).

Before examining these three algorithms, we emphasize that each of them depends on a single parameter that introduces the assignment of dynamic range on the basis of histogram height in a gradual and controlled manner. As implemented in software on actual imagery, they generate similar sets of displays going from the HP result to close to the HE result. There are, however, subtle differences between the three techniques that a simulated test pattern image will clarify below. Further, in a hardware real-time embodiment, their noise characteristics should differ (see Section 15.4 for further discussion).

Undersampled projection (UP) is the simplest way to gradually introduce the weighting characteristic of HE—one that in fact further reduces the computational overhead. By calculating a binary histogram based only on every second, fourth, eighth, etc. pixel, one gradually increases the display allocation of more

DISPLAY AND ENHANCEMENT OF IR IMAGES

Figure 15.18 Hybrid displays using Eq. (15.6) with $W = 0.75$. (a) Airport; (b) cups.

heavily occupied levels. In effect, this is occurring on a probabilistic basis, by not detecting sparsely occupied levels; this leaves more display dynamic range for the occupied levels. The displays for ¼ and ⅛ undersampling for the airport scene and for ¹⁄₁₆ and ¹⁄₃₂ for the two-cup image are given in Figure 15.19. The display histograms for HP and the more undersampled UP are compared in Figure 15.20. Note the effect on display range allocation of increasing the weighting given to pixels that are at frequently occurring levels.

An alternative to undersampling is to require a threshold number of occupancies—2, 4, 8, etc.—before deeming a level "occupied" (TP). Threshold projection shifts the dynamic range allocation similarly to UP, by preferential detection of more densely occupied levels, but operates in a deterministic mode rather than a probabilistic one. On actual imagery, very similar results are obtained (Fig. 15.21; display histograms, not shown, are much like those in Fig. 15.20).

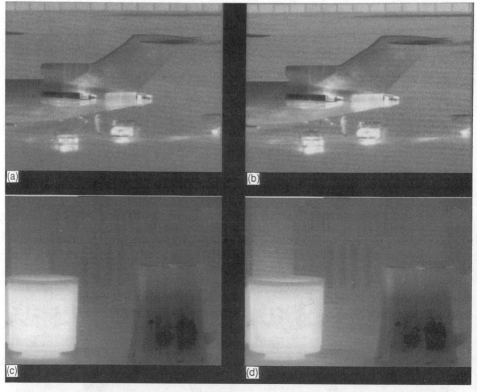

Figure 15.19 Examples of UP displays. (a) Airport at ¼; (b) airport at ⅛; (c) cups at ¹⁄₁₆; (d) cups at ¹⁄₃₂.

DISPLAY AND ENHANCEMENT OF IR IMAGES

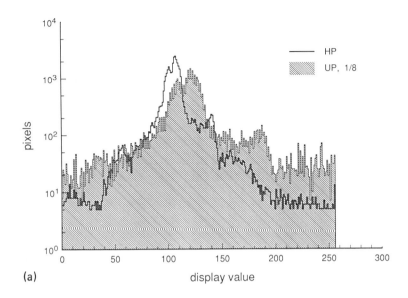

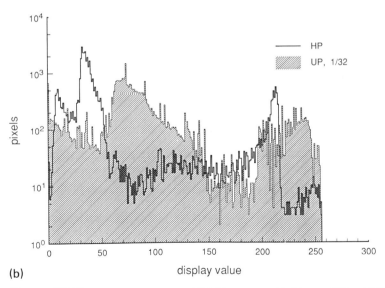

Figure 15.20 Examples of HP and UP histograms. (a) Airport, HP and UP at $1/8$; (b) cups, HP and UP at $1/32$.

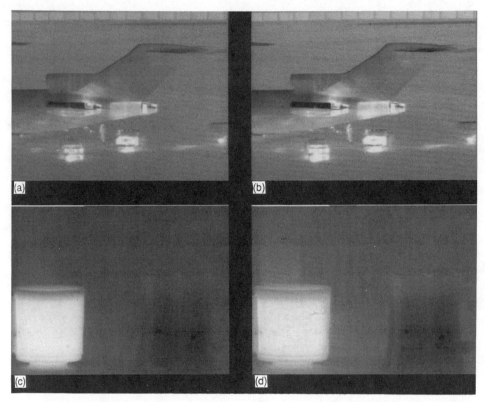

Figure 15.21 Examples of TP displays with threshold at (a) 2; (b) 4; (c) 4; (d) 8 pixels.

In principle, a better way to blend the benefits of projection and equalization is to perform HE with a cutoff saturation level or plateau imposed on the histogram distribution (PE). Typically, a plateau of 10–20 counts is optimum for the 160 × 244 arrays used to generate the imagery here; the desirable plateau would increase with total pixel count. In PE, one generates the $F[i]$ function used in Eq. (15.3) as before from the full histogram but in effect suspends counting the occupancies for a given level when and if the plateau is reached. All levels at or above this plateau are given equal weight as in projection, and levels below are weighted (in compressed form) as in equalization. Typical results are shown in Figure 15.22, with Figure 15.23 comparing the display histograms for HP and PE. Note the close similarity with Figures 15.19 and 15.20.

The subtle differences in principle (though usually not in practice on current IR imagery) between UP, TP, and PE are best revealed through a simulated test pattern such as one whose raw histogram is given in Figure 15.24. A two-level

DISPLAY AND ENHANCEMENT OF IR IMAGES

Figure 15.22 Examples of PE displays with plateau at (a) 10 pixels; (b) 20 pixels.

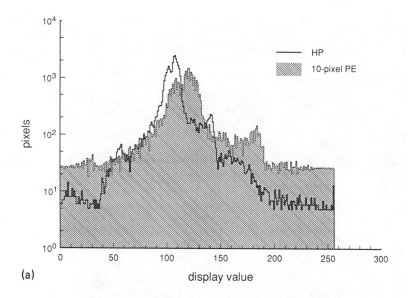

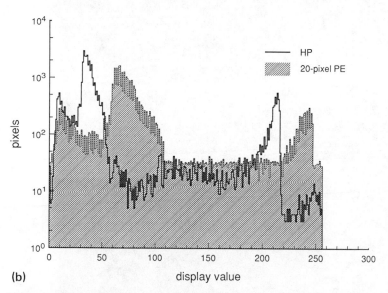

Figure 15.23 Examples of HP and PE histograms. (a) Airport, HP and PE with 10 pixel plateau; (b) cups, HP and PE with 20 pixel plateau.

DISPLAY AND ENHANCEMENT OF IR IMAGES

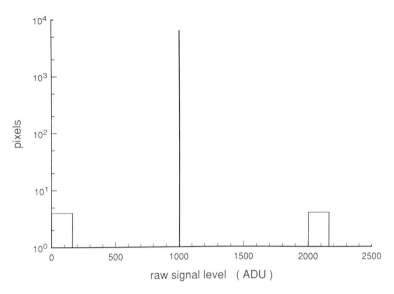

Figure 15.24 Histogram for test pattern.

(1000 and 1004) low-contrast checkerboard pattern in the center of the image leads to the histogram peak of 3200 at about level 1000. The intensity-graded wavy line signals at the top and bottom of the image lead to the square peaks of height 4. These arise from 4 pixels each at levels 1–162 (top) and at levels 2001–2162 (bottom). In all, there are 327 occupied levels. As could be foreseen, HE brings out the checkerboard pattern but loses the gray-scale resolution of the wavy lines (Fig. 15.25), whereas HP optimizes the latter but loses the contrast in the checkerboard. The computationally intensive hybrid procedure (75% projection) and PE (40 pixel plateau) afford optimum displays (Figs. 15.26a, 15.26b), bringing out the checkerboard pattern while largely retaining the gray-scale gradations of the wavy lines. Undersampled projection and threshold projection (Figs. 15.26c, 15.26d) shift dynamic range by not detecting levels that are actually present. In the former, segments of the wavy lines are not graded but lumped into fixed display values, whereas in the latter, when the minimum threshold condition exceeds 4 the checkerboard pattern is vividly brought out but each wavy line has been reduced to a fixed display value.

We conclude this section with the following observation. Our survey of global monotonic algorithms was initially motivated by a pressing need for a real-time automatic contrast control to replace the manual offset/gain controls on IR cameras, which require frequent adjustment. The HP algorithm has successfully provided the requisite automated and optimized display (see Section 15.4).

Figure 15.25 Comparison of HE and HP displays of test pattern. (a) HE; (b) HP.

DISPLAY AND ENHANCEMENT OF IR IMAGES

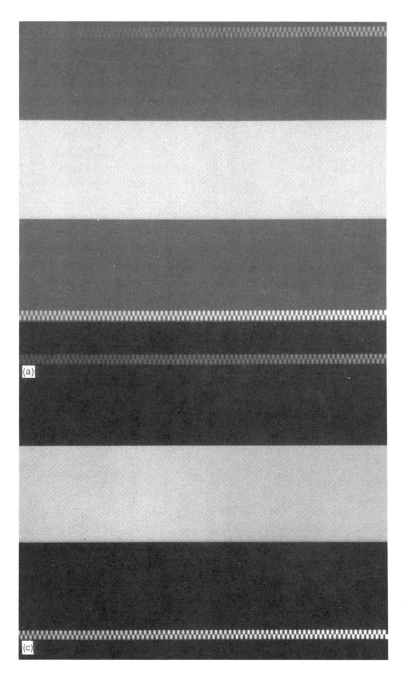

Figure 15.26 Examples of hybrid algorithm displays of test pattern. (a) Eq. (15.6), $W = 0.75$; (b) PE, 40 pixel plateau; (c) UP, $\frac{1}{16}$; (d) TP, 8 pixel threshold.

15.3 LOCAL CONTRAST ENHANCEMENT

15.3.1 Introduction

A familiar experience to any operator of an infrared camera is that in bringing out local details by adjusting one portion of the image for optimum contrast, one will obliterate other parts of the image. As a software example, Figure 15.27 highlights the rudder of the airplane image and the warm cup portion of the cup image by the assignment of the raw signal span of these parts of the images to the whole display dynamic range. The purpose of the algorithms considered in this section is to perform this enhancement function automatically and simultaneously for all parts of the image with a minimum generation of artifacts.

To accomplish such a locally enhanced display, one must depart from global, monotonic mappings. Low-contrast details such as those on the cups are frequently not discernible in such mappings. In the HP algorithm, for example, the total number of occupied levels N is a measure of the signal dynamic range. N is 451, 1282, and 877 levels for the geese, airport, and cup images, respectively. When one is mapping on the order of 1000 levels of information into 256 nominal levels of display (about 100 discernible levels of gray), about every four adjacent occupied levels will be fused to the same display value, and raw signals eight or more levels apart can end up as indistinguishable adjacent display values, though a representative noise level is only 3–5 (Murguia et al., 1990) for the 12 bit raw signal. Clearly, real information can be lost in such global, monotonic mappings. [In on-line imagery, the spatial and temporal averaging performed by the eye (Mooney, 1991) would lower the quoted noise levels slightly.]

In the next three subsections we describe three distinct categories of algorithms for locally enhanced display.

15.3.2 Local Implementation of Global Algorithms

Many locally adaptive enhancement methods described in the literature take advantage of the obvious fact that the dynamic range of a subimage is typically less than that of the total image. For example, in the HP algorithm, the degree of contrast hinges on the number of occupied levels. To the degree that the local occupancy differs from the total, one can increase the display contrast by a local application. In Figure 15.28, the results in applying the HP procedure to four, eight, and 16 disjoint subimages, respectively, are shown for the cups image. As the number of subimage divisions increases, the degree of contrast expansion does also, but so does the number of luminance "seams" at subimage borders; this tends to distract the eye and has the potential of destroying information. Although this is the simplest way to apply a global algorithm locally, better results are achieved (at greater computational cost) by using sliding or overlapping windows or interacting subimages. Both overlapping- and sliding-window im-

Figure 15.27 Examples of mapping part of raw signal range to full 8 bit display range: (a) 1800–1900; (b) 2300–2460.

DISPLAY AND ENHANCEMENT OF IR IMAGES

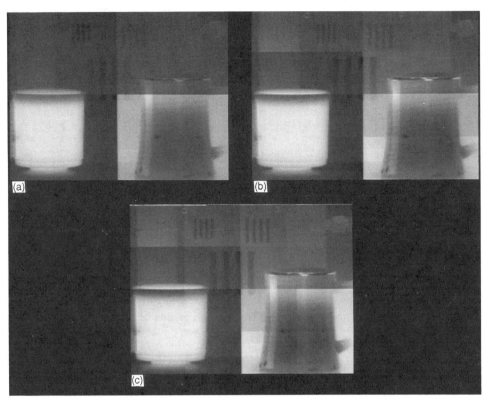

Figure 15.28 Results of applying HP to disjoint subimages. (a) Four subimages; (b) 8 subimages; (c) 16 subimages.

plementations of HP will be described below. Tom and Wolfe (1982) describe local implementation of HE with a sliding window.

A clever example of local implementation of a global algorithm of the linear type described in Section 15.2.2 is local range modification (LRM) of Fahnestock and Schowengerdt (1983) (see also Schowengerdt, 1983, pp. 66–67). The image is partitioned into disjoint subimages. The maximum and minimum signal levels of each region are determined. One then assigns to the corner of each subimage the maximum (minimum) of all the maxima (minima) of the subimages that share this corner. Despite the disjointness of the subimages, allowing interaction of neighboring regions through the assigned corner values is similar in effect to using overlapping windows. For each pixel, a local maximum and minimum are computed from a bilinear interpolation of the corner values and then used to define a linear local contrast stretch. We have implemented versions of this algorithm for IR images, with some results given in Figures 15.29 and 15.30.

For the airport image, LRM yields a display similar to local overlap projection described shortly (see Fig. 15.33). Results are shown for block sizes 20 × 61 pixels (32 total subimages) and 20 × 31 pixels (64 total subimages). The false changes of luminance are artifacts that are also found in local overlap projection. A more serious weakness of the LRM procedure is its sensitivity to the influence of outlying or fallacious pixels such as are found in the first few rows of the cups image or to very strong edges in the image such as the cup boundaries. Large regions can be "whited" or "blacked" out because of an inappropriate minimum or maximum in the local stretch equation. In Figure 15.30, two versions of the LRM algorithm are applied with 20 × 31 block size. The versions differ in regard to the detection and attempted rectification of the effects of anomalous pixel values. (Details are not important here. We simply emphasize the great difficulty in making the LRM algorithm robust to such severe artifacts over a range of image types.) If one applies these same two LRM variations to a single-frame, one-point-corrected version of the cups image (Fig. 15.31), different but equally severe artifacts are present. [This alternative cup image, referred to hereafter as the "noisy" cup image, is more representative of the majority of our images in its degree of temporal and spatial noise (see end of Section 15.2.1) and is used in the remainder of this section in addition to the first cups image to illustrate the noise-amplifying effects of local enhancements.]

The LRM algorithm inspired a somewhat analogous but more successful local implementation of a global algorithm, namely a local form of HP based on overlapping windows (*overlap projection*, OP). Overlap projection is implemented after first (conceptually) adding four phantom rows to the image to give 160 columns × 248 rows of pixels. (These pixels have widths twice their height.) The image is then divided into 80 disjoint subimages (16 × 31 pixels). To eliminate or reduce the luminance seams at subimage boundaries, one applies the HP transformation within a sliding window (size 32 × 62 pixels) that encompasses four subimages (Fig. 15.32). The HP transformation is performed for each of the 63 distinct positions taken by the window as it translates by half its linear dimension in each direction—thus the overlap. For the four corner subimages or regions, a unique transformation is defined for each pixel. For the 28 edge regions, two transformations are defined, and for the 48 interior regions four transformations are defined for each pixel. For the pixels in the edge and interior regions, the final display value is generated by linear and bilinear interpolations, respectively, of the window-defined transformations.

Results of OP are shown in Figures 15.33–15.35. Generally, the algorithm is quite successful in bringing out low-contrast features such as structural details on the airplane rudder, designs and lettering on the cups, and veins on the hand. False luminance-change artifacts are present, especially noticeable for the cups

DISPLAY AND ENHANCEMENT OF IR IMAGES

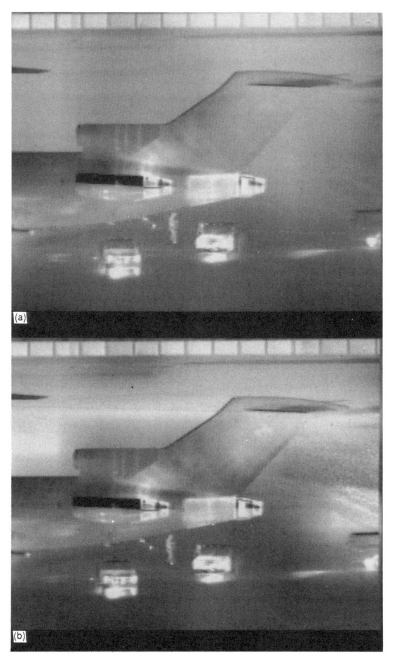

Figure 15.29 Examples of LRM displays. (a) 20 × 61 block size; (b) 20 × 31.

Figure 15.30 Examples of displays from two versions of LRM, 20 × 31 block size. See text.

DISPLAY AND ENHANCEMENT OF IR IMAGES

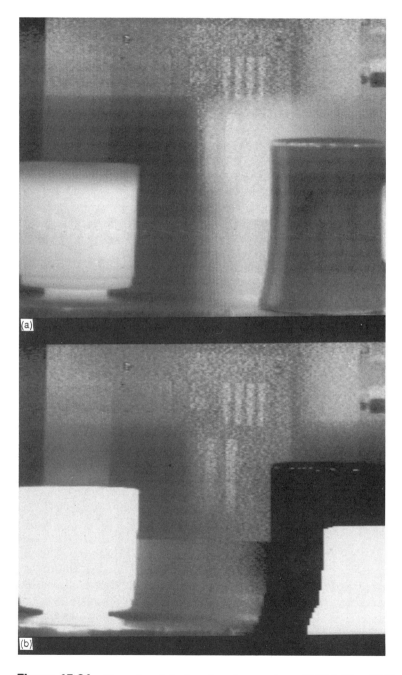

Figure 15.31 Examples of displays from two versions of LRM, 20 × 31 block size, for noisy cup image. See text.

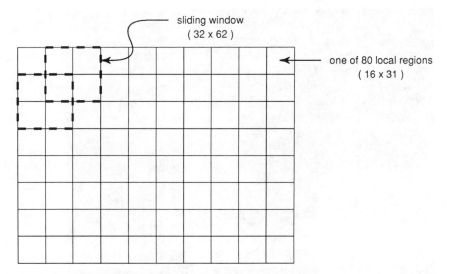

Figure 15.32 Schematic of the OP technique.

image. However, rarely do the luminance effects destroy information as radically as in the LRM algorithm. For the hand and Boston skyline images, HP and OP are directly compared. For the latter image, the many levels occupied by the clouds (this image has a total of 2025 out of 4096 possible levels occupied!) compress the global display of the buildings into the low end of the display scale. Hence the local implementation brings out much lost detail. The hand image is a graphic example of the difference between a global monotonic display and a locally adaptive one.

An alternative means of applying local implementation of HP is by a sliding-window approach (SP). Using window sizes of 11×11, 21×21, 31×31, and 41×41 pixels, we compute the HP transformed display for the pixel centered in this window as well as that of the pixel directly one row below. One then slides the widow to center the next pair of pixels and repeats the computation. (If the pixels were square, one would do four pixels at a time.) The computationally intensive SP algorithm is a very strong contrast enhancer (Fig. 15.36) that brings out much noise as well. It is most useful as a supplement to OP in the smaller window size versions. As its degree of enhancement lessens to approach that of the OP procedure (31×31 and 41×41 sizes), SP has more serious artifacts that are more likely to destroy information.

Two major drawbacks of local implementations of HP are, first, the computational speed and memory requirements of a real-time implementation, and, second, the haphazard interaction between the image and the degree of enhancement

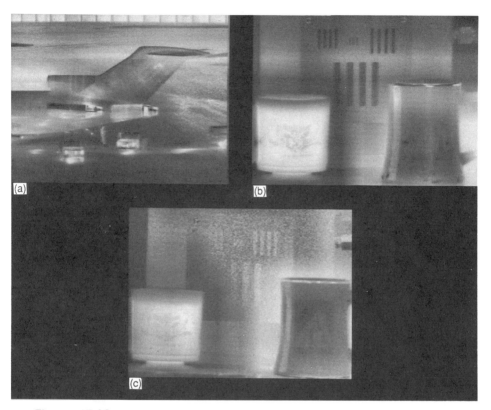

Figure 15.33 Examples of applying OP to three standard images: (a) airport; (b) cups; (c) noisy cups.

afforded by the procedure. The latter reflects the circumstantial dependence of the ratio of local versus global numbers of occupied levels. These drawbacks are circumvented by the algorithms described next.

15.3.3 Modulo Processing

The mappings described in this subsection (Silverman et al., 1990) grew out of the well-known technique of sawtooth scaling, "often used to produce a wide dynamic range image on a small dynamic range display" (Pratt, 1978, pp. 309–312). For 8 bit displays, one reduces the raw signal level modulo 256; that is, one keeps the remainder (0–255) after division by 256. The problem with such a sawtooth mapping (Fig. 15.37; the mapping shown for a raw signal range of 10 bits can, of course, be continued to arbitrary size) is the discontinuities at multiples of 256 in the raw signal, which lead to sudden black/white or white/black

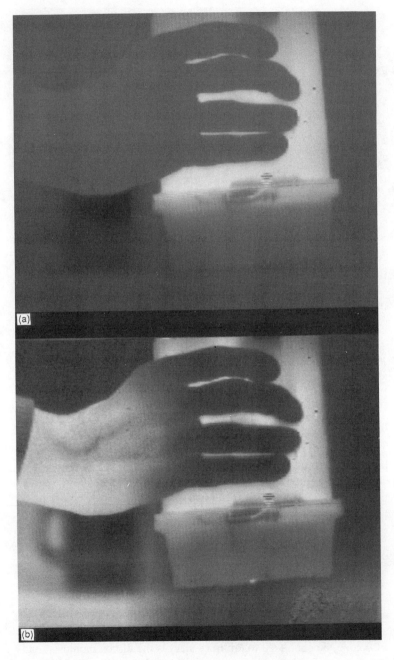

Figure 15.34 Comparison of HP and OP. (a) HP; (b) OP.

DISPLAY AND ENHANCEMENT OF IR IMAGES

Figure 15.35 Comparison of HP and OP. (a) HP; (b) OP.

Figure 15.36 Examples of SP displays of cup images using various window sizes: (a) 11 × 11; (b) 21 × 21; (c) 31 × 31; (d) 11 × 11 (noisy cups).

transitions in the display. The simple modification (labeled as "modulo" in the figure, although strictly speaking the sawtooth mapping is the mathematical definition of modulo) to a continuous triangular mapping with the same periodicity increases the utility of the procedure. The asymmetry of the sawtooth mapping is replaced by the symmetry of the modulo mapping (mirror planes at multiples of 128 in the raw signal). In terms of the modulo-reduced value m, which ranges from 0 to 255, the revised mapping is given by

$$\text{Display value} = \begin{cases} 2m, & m < 128 \\ 2(256 - m) - 1, & m \geq 128 \end{cases} \quad (15.7)$$

Equation (15.7) generates the displays of our (now four) standard images shown in Figure 15.38. This algorithm is referred to as *raw modulo* (RM), as the modulo property takes us close to the raw signal data themselves. As an image-

DISPLAY AND ENHANCEMENT OF IR IMAGES

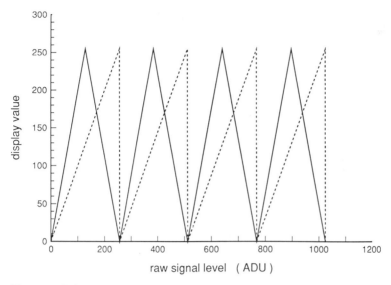

Figure 15.37 Sawtooth (dashed line) and modulo (solid line) mappings.

independent global many-to-one mapping (in contrast to the other algorithms in the previous and next subsections), it can be simply implemented in table lookup format, while affording a strong and relatively artifact-free form of local contrast enhancement.

The main drawback of the RM technique occurs whenever a local low-contrast feature straddles a mirror plane in the mapping (cf. the lettering in the lower left-hand corner of the original cup image in Fig. 15.38) where the deviation from monotonicity garbles the display, or whenever regions with high spatial frequency information map into one or more periods in the display, leading to a too rapidly changing, confusing display (cf. the vehicle grilles in the airport scene). Two simple adjuncts to RM can address these problems and increase the utility of the procedure. The first fix is to allow division of the raw signal data by factors of 2, 4, etc. before performing the modulo reduction; this is tantamount to increasing the period of the mappings in Figure 15.37 by the same factors. The result of division by 2 and 4 for the airport scene (Fig. 15.39) provides more "readable" displays. A sequence of RM displays "toned down" by increasing factors of 2, as in the face image (Fig. 15.40), often provides a useful survey of the information content of an image. The mirror-plane artifact stems from the arbitrary "phase" in the raw signal (additive constant) with respect to the modulo reduction. The simple fix is to shift the raw signal data by values such as ± 32, 64, 128 before the modulo step (Fig. 15.41). Note the improved reading of the lettering in the lower left-hand portion of the warm cup in the modulo -32 ver-

Figure 15.38 Examples of RM displays of the four standard images.

sion and the clearer view of the plant logo and "coffee connection" letters in the modulo +128 version.

The RM algorithm involves no histogram processing, is simple and effective, but ignores the dynamic range requirements of the particular image. A *modulo projection* (MP) technique is a more elaborate algorithm designed to adapt to the image dynamic range as measured by the total number of occupied signal levels N. The following version was "tuned" to the present IR images; in principle it can be adjusted for other kinds of imagery. If N is less than 512, one applies the RM algorithm except that for additional contrast enhancement the raw signal data are doubled (if $N < 100$) or multiplied by 1.5 (if $100 \leq N < 512$) before the modulo 256 reduction. For $N \geq 512$, the occupied levels are ordered from 1 to N (as in the HP technique) and m in Eq. (15.7) is taken as the modulo-reduced occupancy number rather than the modulo-reduced signal level. For still higher dynamic range images ($N \geq 800$), neighboring occupied levels are coalesced to

DISPLAY AND ENHANCEMENT OF IR IMAGES

Figure 15.39 Examples of RM with raw signal divided by (a) 2; (b) 4.

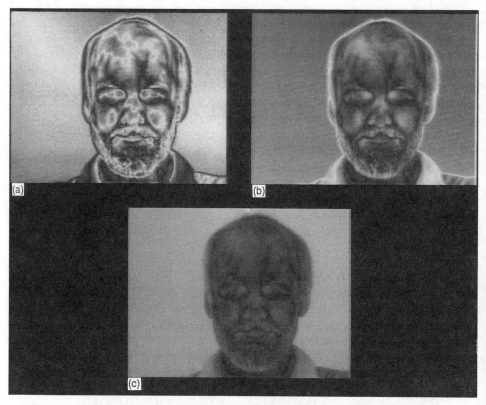

Figure 15.40 Examples of RM with raw signal divided by (a) 1; (b) 2; (c) 4.

some extent—but by a factor of 2 less than occurs automatically in using the HP algorithm—before the modulo 256 reduction.

Figures 15.42 and 15.43 show enhancements produced by the MP algorithm. As the images in these figures have $N > 512$, MP acts largely as a toned-down version of RM. Comparing the airport scenes in Figures 15.38 and 15.43, the MP version is rather similar but superior to the RM divided by 4 (Fig. 15.39) display, with sharper structural details on the plane rudder. With very low noise imagery such as the original cups image, the strongly contrast-enhancing effect of RM works well. For the noisy cups scene, MP gives a clearer view of the warm cup, while RM reveals more of the bar patterns and cold cup details, along with more temporal noise. For the hand image (Fig. 15.42; compare Fig. 15.34), both modulo algorithms bring out the veins and sleeve cuff details (note the reverse contrast change in the veins in Fig. 15.42), with better detail in the fingers than the OP algorithm gives. The OP display does retain a better sense of the

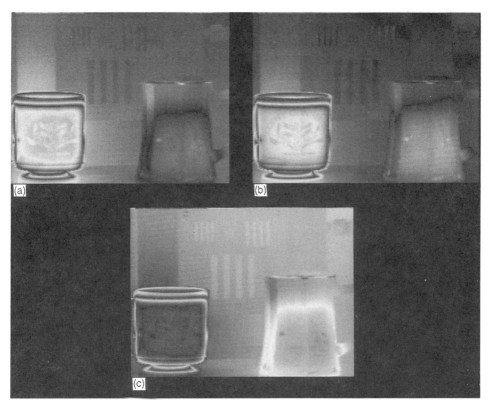

Figure 15.41 Examples of RM with raw signal shifted by (a) 0; (b) −32; (c) +128.

global thermal sense of the raw signal—this is typically true. More comparisons among the algorithms for local contrast enhancement will be given in Section 15.3.5.

15.3.4 High-Frequency Enhancement

Up to now, we have concentrated exclusively on the spatial domain (SD) point of view. The Fourier domain is now widely used in the analysis and filtering of multidimensional signals (Dudgeon and Mersereau, 1984) because of the wide availability of the FFT routine for computing the discrete Fourier transform (DFT). For images (2-D signals), the spatial frequency Fourier domain (SFD) viewpoint is widely employed for the "restoration" problem in image processing, but in general only lip service is paid to the SFD in the "enhancement" problem in image processing. [An exception is the excellent book by Wahl (Wahl, 1987)

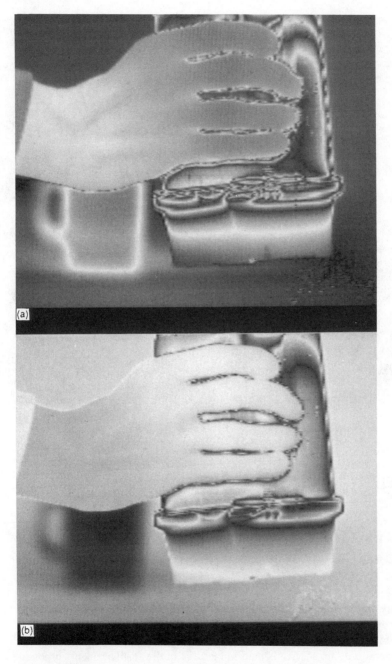

Figure 15.42 Comparison of RM and MP. (a) RM; (b) MP.

DISPLAY AND ENHANCEMENT OF IR IMAGES

Figure 15.43 Examples of MP displays. (a) Airport; (b) cups; (c) noisy cups.

in which the DFTs of images are used throughout to underscore trends and basic principles.]

The algorithms treated in this subsection, high-frequency enhancement with linear filters, can be implemented in either the SD or SFD domains. To our knowledge, there are only two standard algorithms designed strictly for SFD implementation. One is homomorphic filtering (Wahl, 1987, pp. 84–86), in which the signal is modeled as the product of a high-frequency reflectance component and a low-frequency luminance component. One converts the product to a sum through the log function, enhances the (now) additive high-frequency component in the SFD, and exponentiates the inverse-transformed result to recover the processed image. We have tried this algorithm on our IR images without success—not surprisingly, as the underlying model is not suitable for IR images. The second SFD-specific procedure is "alpha rooting" (Dudgeon and Mersereau, 1984, pp. 124–126), in which one takes the alpha root ($\alpha < 1$) of the magnitude of the DFT but retains the phase. Again, we found poor results in applying this technique to IR images.

As any algorithm for increasing local contrast more or less involves enhancing higher spatial frequencies at the expense of lower ones, it seems natural to examine this problem in the SFD. Figure 15.44 shows the DFTs of the airport image after display with the HP, OP, RM, and MP algorithms, respectively. [We are using 128 × 256 point FFTs on our images by truncating the first and last 16 columns and mirror-expanding the last 12 rows of our 160 × 244 images. More specifically, Figure 15.44 shows log-like displays of the squared magnitude of the DFT with the center of symmetry indicating the (0, 0) spatial frequency. See Wahl's 1987 text for further details.] Although the three local algorithms have the general effect of enhancing high spatial frequencies, they do so in a nonlinear

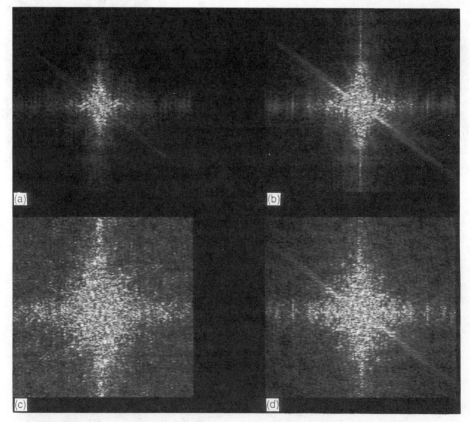

Figure 15.44 DFTs of the airport scene after applying (a) HP; (b) OP; (c) RM; (d) MP.

DISPLAY AND ENHANCEMENT OF IR IMAGES

and spatially adaptive fashion that would have no counterpart in the SFD. For example, the degree of high-frequency enhancement with the OP algorithm varies locally with the ratio of local to global number of occupied levels—a signal-dependent and operator-uncontrolled process. One would like to accomplish such high-frequency enhancement in a controlled and directed manner.

The goal of many high-frequency enhancements is to improve subjective image quality from the standpoint of psychophysics by accenting edges, called *edge crispening* (Pratt, 1978, pp. 322–326) or (the term we prefer) *image sharpening*. A typical convolution mask for this purpose is

$$\frac{1}{7} \begin{bmatrix} -1 & -2 & -1 \\ -2 & 19 & -2 \\ -1 & -2 & -1 \end{bmatrix} \qquad (15.8)$$

Sharpening the raw signal data of the airport image with this mask and then mapping into 8 bits with HP gives the display of Figure 15.45a. The thermal span (monotonicity) of the image is largely intact, but only slight contrast enhancement results. One can implement a graded set of similar operations by means of the equation

$$P'_c = P_c + a(P_c - \bar{P}_{n \times n}) \qquad (15.9)$$

where P_c and P'_c are the initial and final raw signal values, respectively, of each pixel centered in an $n \times n$ square neighborhood (n odd). a is a small positive integer that controls the degree to which the difference between the central pixel and the $n \times n$ neighborhood average $\bar{P}_{n \times n}$ is amplified. The choice of $a = 2$, $n = 3$ (Fig. 15.45b) gives virtually the same result as the mask in Eq. (15.8), while a choice such as $a = 4$, $n = 9$ (Fig. 15.45c, hereafter referred to as the WS algorithm for weak sinc sharpened), although more blurry, is beginning to exhibit the degree of local contrast enhancement sought.

We arrived at further improved masks for local contrast enhancement by "tuning" a start from Eq. (15.9) in the SFD and implementing the result in the SD. Applying the WS algorithm to an "impulse" image and computing the DFT of the sharpened result (Fig. 15.46a), one obtains the transfer function of this filter. As expected, it is a two-dimensional sinc function aligned along the axial directions with the requisite number of side lobes from the 9×9 neighborhood. We next converted this function into an equivalently strong, circularly symmetric Gaussian (exponential) filter by using the three-parameter form suggested by Wahl (1987, p. 85),

$$G(r) = \begin{cases} 1, & r = 0 \\ \lambda_1 - (\lambda_2 - \lambda_2)\exp(-r^2/\lambda_3^2), & \text{otherwise} \end{cases} \qquad (15.10)$$

where r is the distance in frequency space from the (0, 0) frequency. Equation (15.10) was strengthened and adjusted in the SFD by using it to filter the DFT of

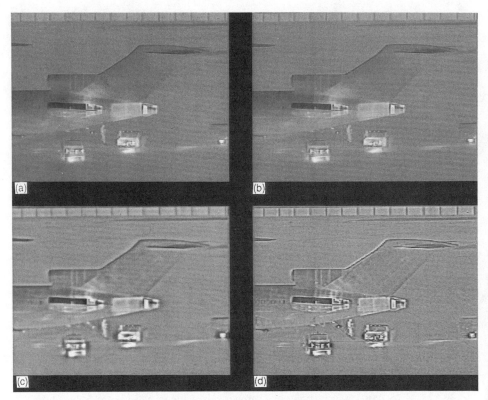

Figure 15.45 Examples of HP displays after high-frequency enhancement with (a) Eq. (15.8) mask; (b) Eq. (15.9), $a = 2, n = 3$; (c) Eq. (15.9), $a = 4, n = 9$ (WS); (d) MG mask.

representative images and inverse-transforming the result to inspect the filtered image. The final Gaussian G was then transformed back into an (SD) convolution mask. We then approximated this mask using integers, shortened the mask extension, and toned down the effect slightly by trial and error. What finally emerged are the following three Gaussian convolution masks, referred to as strong, medium, and weak (SG, MG, WG), respectively,

$$SG: \begin{bmatrix} 0 & -1 & -2 & -1 & 0 \\ -1 & -2 & -3 & -2 & -1 \\ -2 & -3 & 37 & -3 & -2 \\ -1 & -2 & -3 & -2 & -1 \\ 0 & -1 & -2 & -1 & 0 \end{bmatrix} \quad (15.11)$$

DISPLAY AND ENHANCEMENT OF IR IMAGES

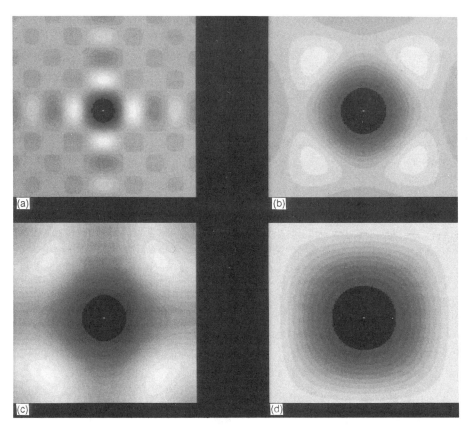

Figure 15.46 Filter transfer functions for (a) WS; (b) SG; (c) MG; (d) WG.

$$MG: \begin{bmatrix} 0 & 0 & -1 & 0 & 0 \\ 0 & -1 & -2 & -1 & 0 \\ -1 & -2 & 17 & -2 & -1 \\ 0 & -1 & -2 & -1 & 0 \\ 0 & 0 & -1 & 0 & 0 \end{bmatrix} \quad (15.12)$$

$$WG: \begin{bmatrix} -1 & -2 & -1 \\ -2 & 13 & -2 \\ -1 & -2 & -1 \end{bmatrix} \quad (15.13)$$

whose filter functions are compared to the starting WS filter in Figures 15.46b–15.46d. The MG mask with power of 2 coefficients and the small 3 × 3 WG mask were designed for ease of implementation.

Figure 15.45d completes the sequence of sharpening comparisons of the airport scene with the use of the MG mask. Excellent local contrast enhancement is achieved with a more "natural look" than with the OP, RM, or MP algorithms (see Figs. 15.33, 15.38, and 15.43). The DFTs of the airport displays in Figures 15.45b–15.45d are shown in Figure 15.47. Comparison with Figure 15.44 indicates that the sharpening filters produce a more structured operation in frequency space, with greater deemphasis of the low-range and midrange spatial frequencies.

The WS filter [$a = 4$, $n = 9$ in Eq. (15.9)] is compared in operation to the three Gaussian masks on both cup images in Figures 15.48 and 15.49. The weak or medium masks are often optimum, as in this case, for noisier images, while the SG mask, or even stronger versions, match up well with very low noise im-

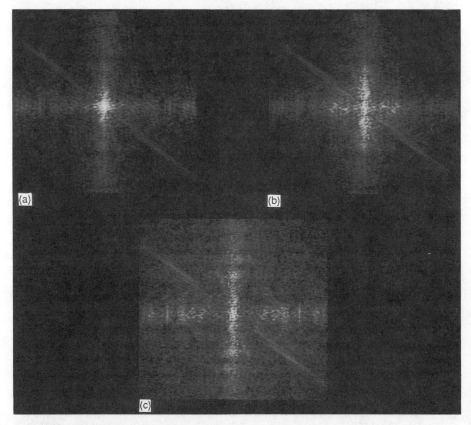

Figure 15.47 DFTs for airport displays after high-frequency enhancement with (a) Eq. (15.9), $a = 2$, $n = 3$; (b) WS; (c) MG.

DISPLAY AND ENHANCEMENT OF IR IMAGES

Figure 15.48 Examples of HP displays after high-frequency enhancement with (a) WS; (b) WG; (c) MG; (d) SG.

ages. Especially for such low noise images, the Gaussian masks can reveal high-frequency information missed by the previous algorithms, as for example in the license plate numbers on the automobile in Figure 15.50 (a 256 frame average and hence a low temporal noise image).

These masks—or for that matter any specific high-frequency enhancement algorithm—have two drawbacks. First, they amplify both residual spatial noise (as in the horizontal lines on the warm cup in Fig. 15.49) and temporal noise (as in the bar patterns in the same image or the background in the face image of Fig. 15.51). Second, if the low-contrast information is not high spatial frequency, such as the veins in the hand image (Fig. 15.51), modulo and local techniques (Figs. 15.42 and 15.34b) do a better job of enhancement.

A basic difference between the algorithms in this subsection and all previous algorithms should be underscored. Here we are not mapping from raw signal to

Figure 15.49 Examples of HP displays of noisy cups after high-frequency enhancement with (a) WS; (b) WG; (c) MG; (d) SG.

display values but are rather preprocessing the raw signals. Hence we can join any of the previous algorithms in tandem with one of these high-frequency enhancements. The examples shown so far have all been sharpening/HP. A very effective combination (results shown in Fig. 15.52) is to sharpen, for example, with the WS algorithm and then map into an 8 bit display with the OP algorithm. This affords a stronger, more locally balanced, contrast enhancement than just OP, but with much reduced luminance artifacts (compare to Figs. 15.33 and 15.50). Apparently, the preprocessing with sharpening equilibrates to some degree the set of local numbers of occupied levels, giving smoother transitions between regions upon using the OP procedure (see Section 15.3.2).

15.3.5 Conclusions

The comments scattered throughout this section on the pros and cons of the various techniques for local contrast enhancement are gathered together and

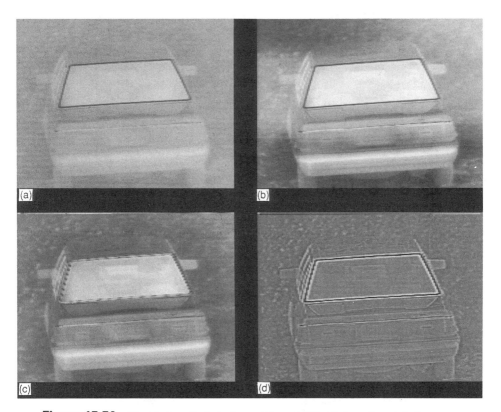

Figure 15.50 Comparison for bringing out license plate digits and car interior among (a) HP; (b) OP; (c) RM; (d) SG/HP.

categorized in Table 15.2. We summarize some broad conclusions from this table.

The three types of local enhancement algorithms—local implementation of global algorithms, modulo processing, and high-frequency sharpening—are all useful and sometimes complementary "software tools," which afford a comparable and effective degree of contrast enhancement, although one can find images or image types that match particularly well to each category. Local techniques like OP and SP are less predictable in their effects, with their circumstantial dependence on the number of locally occupied levels. The high-frequency enhancement techniques rank high with respect to absence of artifacts, other than edge overshoots, and in "natural look." One might well ask, What is a natural look for IR imagery? We suspect that large regions that retain the monotonic thermal sense of the global mappings of Section 15.2 look more natural. Hence, modulo processing does not rank high in this respect and tends to look confusing, particularly to inexperienced observers.

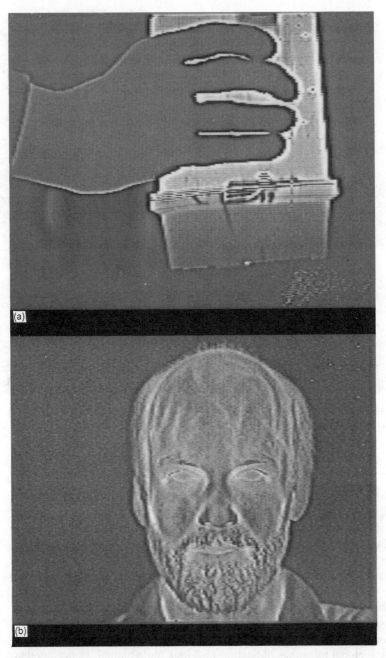

Figure 15.51 Examples of drawbacks of high-frequency enhancement. (a) SG/HP; (b) MG/HP.

DISPLAY AND ENHANCEMENT OF IR IMAGES

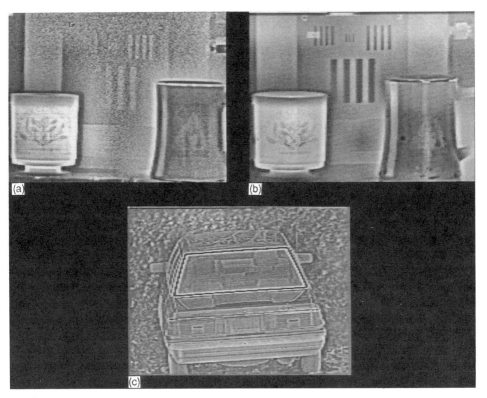

Figure 15.52 Examples of high-frequency enhancement followed by OP. (a) WS/OP; (b) WS/OP; (c) SG/OP.

If one turns to issues central to real-time operation in hardware on IR cameras, then RM processing, which entails simple global, signal-independent transformations, offers a computationally cheap form of contrast enhancement. Adding the two "bells and whistles" described in Section 15.3.3 (division of the raw signal by 2, 4, etc., and additive shifts in the raw data scale) would increase the utility of such a camera adjunct.

15.4 HARDWARE IMPLEMENTATIONS AND FUTURE WORK

Many IR imaging applications such as low-altitude night navigation, target tracking, and autonomous landing systems require a real-time automatic contrast control that provides optimized display imagery. While manual offset/gain adjustment on laboratory-designed cameras generally affords an excellent view of the

Table 15.2 Comparison of Major Algorithms for Local Contrast Enhancement

Algorithm	Strength of enhancement	Natural look/ thermal span	Main artifacts	Ease of real-time implementation	Comments
OP	Volatile but usually strong	Fair	False luminance changes; volatile	Very difficult	"Volatile" due to image-dependent variation of local occupancy
SP	Very strong	Poor	Worse than OP, depending on window size	Extremely difficult	Windows 11×11, 21×21, 31×31, 41×41
RM	Strong	Poor	Display sense reversals; gray scale arbitrary	Very simple	Raw signal compression and level shifts as useful adjuncts
MP	Moderate	Poor	Display sense reversals; gray scale arbitrary	Simple	Adapts to dynamic range of image
WS	Weak	Good	Edge overshoots	Moderate	
SG	Very strong				
MG	Strong	Good	Edge overshoots	Moderate	Only enhances high-freq. detail; amplifies temporal and residual spatial noise
WG	Moderate				

IR scene content (in the hands of a skilled operator), frequent readjustment is required as the camera is panned or as the IR content of the scene changes. Real-time hardware implementations of the HP algorithm for this purpose have now been incorporated into several cameras designed in-house and afford a very useful alternative to the offset/gain controls.* In similar active areas of development, several U.S. companies have already implemented HP or are now implementing it, typically in the UP or TP variation, for the same real-time function. So far, a fixed parameter in implementing UP or TP has been used. However, an implementation of UP with a programmable parameter (ranging, say, from HP itself to $\frac{1}{32}$ undersampling) would allow for some adjustments to the application by providing online flexibility in the dynamic range mapping.

Aside from issues of computational complexity, other issues arise when real-time implementations are considered, such as interactions with temporal noise and mean display level flicker. A problem noticed in some of the HP implementations is the suspected presence of some frame-to-frame flicker noise, particularly in blander scenes (small total number of occupied levels N). Referring back to Eq. (15.5), we recall that the final display value of each pixel depends on its order number from 1 to N in the hierarchy of occupied levels. Even in a stable scene, noise variations from frame to frame, especially in their effect on sparsely occupied levels, can cause shifts in pixel order numbers. The most noticeable effects would arise from changes at the low or dark end (say, as "detected" occupied levels disappear or are created) because such changes tend to affect the order number as a constant shift of a large majority of the pixels. The resulting rapid changes in the mean level of the display can become perceptible. Noise simulations have been performed, and several fixes are being considered (Mooney and Ewing, private communication, 1990).

We conclude this treatment of the display of IR images with a caveat. We have surveyed algorithms in two broad categories: "standard" algorithms used in other contexts (HE, LRM, homomorphic filtering, etc.) which we have tested on IR images, and algorithms newly devised for application to IR imagery (HP, OP, RM, etc.). The imagery driving our work and underlying this survey was based exclusively on PtSi staring technology and taken with cameras designed in our laboratory. More and more, IR imagery of this caliber is coming into the hands of foreign and domestic industrial companies that make PtSi cameras (Shepherd, 1988), as well as imagery from other technologies such as InSb and HgCdTe. As staring IR imagery from other cameras, technologies, and wavelength regions (in particular, the 8–12 μm atmospheric window) becomes available, it will be desirable to revisit our surveyed algorithms. We therefore expect modifications and expansions of the perhaps somewhat parochial point of view of our chapter as

---

*The first such implementation was done in conjunction with the Hughes Aircraft Company; details are available upon request from the authors.

more standard algorithms are tried on IR imagery or as new algorithms come along. Conversely, we anticipate possible use of some new algorithms such as HP, RM, or MP on other types of imagery with similar dynamic range requirements, such as medical imagery.

Acknowledgements

We are grateful to Dr. J. M. Mooney, who contributed to much of the original work reported in this chapter, for his insightful reading of the manuscript. We are also grateful to Dr. F. D. Shepherd, who has encouraged and supported our work since its inception. Finally, although the list is too long to mention individual names, we wish to thank the many scientists past and present who worked in our laboratory on the conception, fabrication, design, and development of PtSi infrared cameras. Without their efforts, we would not have had this high-quality imagery in our hands.

References

Briggs, S. J. (1981). Photometric technique for deriving a "best gamma" for displays, *Opt. Eng. 4*: 651.
Briggs, S. J. (1987). Soft copy display of electro-optical imagery, *Proc. SPIE 762*: 153–170.
Dion, D. F., and M. J. Cantella (1984). Real-time dynamic range compression of electronic images, *RCA Eng. 29*: 42.
Dudgeon, D. E., and R. M. Mersereau (1984). *Multidimensional Digital Signal Processing*, Prentice-Hall, Englewood Cliffs, NJ.
Fahnestock, J. D., and R. A. Schowengerdt (1983). Spatially variant contrast enhancement using local range modification, *Opt. Eng. 22*: 378.
Hummel, R. (1977). Image enhancement by histogram transformation, *Comp. Graph. Imag. Processing 6*: 184.
Mooney, J. M. (1991). Effect of spatial noise on the minimum resolvable temperature of a staring sensor, *Appl. Opt., 30*: 3324.
Mooney, J. M., F. D. Shepherd, W. S. Ewing, J. E. Murguia, and J. Silverman (1989). Responsivity nonuniformity limited performance of infrared staring cameras, *Opt. Eng. 28*: 1151.
Murguia, J. E., J. M. Mooney, and W. S. Ewing (1990). Evaluation of a PtSi infrared camera, *Opt. Eng. 29*: 786.
Pratt, W. K. (1978). *Digital Image Processing*, Wiley, New York, pp. 307–344.
Schowengerdt, R. A. (1983). *Techniques for Image Processing and Classification in Remote Sensing*, Academic, New York.
Schreiber, W. F. (1978). Image processing for quality improvement, *Proc. IEEE 66*: 1640.
Shepherd, F. D. (1988). Silicide infrared staring sensors, *Proc. SPIE 930*: 2–10.
Silverman, J., and J. M. Mooney (1988). Processing of IR images from PtSi Schottky barrier detector arrays, *Proc. SPIE 974*: 300–309.

Silverman, J., J. M. Mooney, and V. E. Vickers (1990). Display of wide dynamic range infrared images from PtSi Schottky barrier cameras, *Opt. Eng. 29*: 97.

Silverman, J., J. M. Mooney, and F. D. Shepherd (1992). Infrared video cameras, *Sci. Am.*, *266*, no. 3: 78–83.

Tom, V. T., and G. J. Wolfe (1982). Adaptive histogram equalization and its applications, *Proc. SPIE 359*: 204–209.

Wahl, F. M. (1987). *Digital Image Signal Processing*, Artech House, Norwood, MA.

Woods, R. E., and R. C. Gonzalez (1981). Real-time digital image enhancement, *Proc. IEEE 69*: 643.

16

The Human Factor Considerations of Image Intensification and Thermal Imaging Systems

Clarence E. Rash

*U.S. Army Aeromedical Research Laboratory,
Fort Rucker, Alabama*

Robert W. Verona

*Universal Energy Systems,
Fort Rucker, Alabama*

16.1 INTRODUCTION

16.1.1 Background

Humans have always sought to enhance their ability to see during periods of low illumination and in adverse weather. Historically, natural nighttime illumination from the moon and stars was augmented with artificial illumination to enable people to perform tasks or to provide surveillance capability. This was a simple and relatively inexpensive means to achieve an enhanced nighttime capability. Technologists then developed alternative means of permitting people to enhance their natural night vision in a passive manner when artificial illumination either was not cost-effective (e.g., lighting inland waterways) or compromised the task (e.g., conducting covert night surveillance). Even though law enforcement agencies have made extensive use of these newer passive systems, the military has been the primary impetus for the continuing research and development of passive night vision devices (NVDs). While there are many roles and resulting requirements for NVDs, military aviation requirements are the most demanding. Consequently, aviation requires the most sophisticated of these passive night vision devices.

Independent of the nighttime imaging system technology used, some of the "natural fidelity" of the external scene is lost in the imaging process. The specific characteristics of each system determine the nature of the presented image and consequently can significantly affect user performance. Compared to unaided night vision, modern night imaging systems enhance some visual information while degrading other information. Generally, however, the user operates with fewer visual cues than are available in daylight—a handicap that may not be obvious to the inexperienced user.

16.1.2 Night Imaging Systems

A simplified block diagram of a night imaging system is shown in Figure 16.1. The two basic sections are the sensor and the display. Currently, the two technologies used for passive night vision sensors are image intensification and thermal imaging (forward-looking infrared, FLIR). Image intensifiers (I^2) amplify (intensify) reflected or emitted light so the eye can more readily see a poorly illuminated scene. They depend on the presence of some minimum amount of light in order to produce a usable image. This is analogous to using a microphone, amplifier, and speaker to allow the ear to more easily hear a faint sound. In both cases some of the "natural fidelity" may be lost in the amplification process. The intensified image resembles a black-and-white television image, except that it is in shades of green (due to the selected display phosphor) instead of shades of gray. The physics of the I^2 tube is described in detail in Chapter 1.

The second type of night imaging system uses a thermal imager. This type of sensor does not depend on levels of ambient light, but on temperature differences, based on infrared (IR) radiation generated by the scene. The FLIR sensor can be designed to "see" radiation in either the 3–5 or 8–12 μm spectral range. All objects radiate measurable amounts of IR energy in this spectral range, and the atmosphere is more transmissive in these regions. Infrared imaging is discussed in greater detail in Chapter 15.

16.1.3 Helmet-Mounted Devices

Some night imaging systems are mounted fully, or in part, on the user's head. This is particularly true in the aviation environment, where the system is attached to, or integrated into, the aviator's helmet. When the helmet is used as a platform for the presentation of flight imagery and symbology, certain characteristics become more important with regard to performance and safety. The visual input

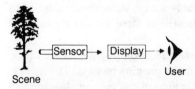

Figure 16.1 Simplified block diagram for a night vision system.

provided by these devices does not approach that experienced using the unaided eye during periods of daylight illumination. Compared to unaided daylight flight, many visual parameters—for example, acuity, field of view, color, and depth perception—are understandably compromised when night vision devices are used. In addition, the mounting of these devices on the helmet increases the hazards associated with excessive weight and pronounced center-of-gravity changes.

16.2 IMAGING SYSTEM THEORY
16.2.1 Image Intensifiers

Image intensifiers belong to a class of devices known as *image tubes*. These devices combine a photoemissive surface (sensor) with a luminescent screen (display) to form a self-contained device. An image intensifier is an image tube that amplifies ambient light. By definition, it does not induce significant changes in the spectral range covered (Levi, 1968). When such a change does occur, as in the conversion of an infrared scene into a visible image, the device is known as an image converter. Unfortunately, as early first-generation devices have been improved upon up to the current third generation, this distinction has been lost. All three generations are called image intensifiers even though they also function by definition as image converters.

In first-generation image intensifiers, the selected scene is focused onto a photosensitive surface called the ***photocathode***. First-generation tubes use an S-20 multialkali photocathode. Electrons in proportion to the amount of incident light are emitted and accelerated via an electric field toward a ***phosphor screen***. Electroluminescence of the phosphor is responsible for the final conversion step where photons (in the form of visible light) are again produced. The high light gain is achieved by cascading optically coupled stages from phosphor screen to photocathode. The observer views the amplified image of the scene on the final phosphor screen.

Second-generation tubes are significantly smaller and lighter than first-generation tubes. They use a ***microchannel plate*** (MCP) in conjunction with the S-20 photocathode to produce the required light amplification. The MCP is a thin wafer of tiny glass tubes that channel the electrons to the phosphor screen. As in first-generation tubes, scene energy is focused onto the photocathode. Now, however, emitted electrons impinge on the MCP. As the electrons pass through the glass tubes, they strike an emissive material coating the channel walls. The channels are tilted slightly so the electrons strike the walls many times, increasing tube gain. Thousands of electrons exit the MCP for each electron produced at the photocathode. In addition to increased gain, the presence of the MCP minimizes contrast reduction imposed by bright light sources in the image intensifier's field of view. The MCP imposes bright source localization. Individual channels can saturate without causing the entire device to saturate. However, local area contrast degradation can still occur. A bright light source produces high electron densities at the MCP and phosphor screen. These high densities sometimes produce a "halo" around the image of the light source.

The photocathode, MCP, and phosphor screen of second-generation devices are in "proximity focus." The source of the electrons is very close to the phosphor screen, so the electrons do not diverge and cause the image to blur. Often, the phosphor screen is deposited on a fiber-optic inverter. This inverter rotates the final image 180°, providing an erect image as required for most applications.

Third-generation image intensification devices are similar in design to second-generation ones, but with 5 to 7 times the sensitivity and triple the lifespan. The major difference is the use of a gallium arsenide (GaAs) photocathode. Compared to the S-20 photocathode, the GaAs cathode has enhanced sensitivity beyond 55° nanometers (nm) and extends the tube's range into the near-infrared. The MCP in the third-generation tubes also has a metal oxide film that serves as a barrier to prevent cathode destruction from positive ion bombardment.

16.2.2 Thermal Imagers

Passive night vision systems based on thermal imaging technology operate by detecting infrared emission of objects in the scene. No universal definition exists for "infrared energy." For imaging, it is generally accepted as thermally emitted radiation in the 1–20 μm (1 μm = 10^{-6} m) region of the electromagnetic spectrum (Fig. 16.2). Currently, most thermal imaging is performed in the 3–5 or 8–12 μm region. These regions are somewhat dictated by the IR transmittance windows of the atmosphere (see Section 16.8.1).

Thermal imaging theory is based on the fact that every object emits radiation. This radiated energy is a direct result of the vibration of the molecules making up the object. An object's temperature is a measure of its vibrational energy. Hence, the higher the temperature of a body, the greater its radiated energy. In turn, the temperature of a body is determined by several factors, including (1)

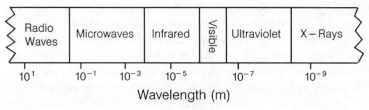

Figure 16.2 The electromagnetic spectrum.

HUMAN FACTOR CONSIDERATIONS

the object's recent thermal history, (2) the reflectance and absorptance characteristics of the object, and (3) the ambient temperature of the object's surroundings.

An object's recent thermal history includes its exposure to external thermal sources, for example, direct sunlight and other surrounding objects, and the presence of internal thermal sources, such as engines. Figure 16.3 depicts a simplified scenario of a boulder sitting in the open under direct sunlight. The boulder is absorbing energy radiated by the sun, the amount being dependent on the reflectance and absorptance characteristics of the boulder's exterior surface. The boulder also is absorbing energy radiated by other objects in the scene, which in this simplified scenario consists only of the surrounding atmosphere and the ground. Energy may also be acquired from the boulder's *physical contact* with the atmosphere if the temperature of the atmosphere is higher than the temperature of the boulder. At the same time, the boulder is emitting energy in an amount related to its temperature. If the boulder is at a higher temperature than the atmosphere and ground, it also will be losing energy due to contact with the surrounding atmosphere and ground. If the net effect of all of this energy flux is an increase, then the total molecular vibrational energy (and therefore the temperature of the boulder) will increase. Conversely, if the energy flow results in a net decrease in total energy, then the temperature of the boulder will decrease. For example, at night, when the boulder's primary external energy source (the sun) is no longer available, the boulder's net energy transfer will be negative, and its temperature will decrease. However, at any given time, the boulder and the atmosphere can be represented by temperature values T_b and T_a, respectively. These values generally are different and will change as a function of time.

Figure 16.3 Simplified scenario of boulder (T_b) and surrounding atmosphere (T_a).

The simple scenario discussed above can be expanded by recognizing the complex nature of real objects. If a more realistic object, such as a vehicle, is investigated, then several other factors must be considered. A vehicle has geometric features and is manufactured from several different materials. These materials have different reflectance and absorptance characteristics. This will result in different parts of the vehicle being at different temperatures. The vehicle's geometric features, such as sides, front and back, and top, can result in nonuniform solar heating. The vehicle also has a major internal source of thermal energy, its engine. Our simplified picture of an object at a single uniform temperature must be replaced by one in which the object consists of a multitude of temperature values, resulting in many different levels of energy emission.

Thermal imaging sensors form their image of the external world by collecting energy from individual segments of the scene. This may be accomplished by using a single detector (or row of detectors) that scans over the scene, building its image one part at a time. An alternative technique is to use a matrix of detectors, with each one collecting energy from a different part of the scene (Fig. 16.4). The size of each detector's collection angle defines the smallest area of the scene that can be imaged, that is, the resolution of the sensor. The output of each detector is related to the amount of energy emitted from a small part of the scene. The overall result is a two-dimensional energy emission profile of the scene. To be able to discriminate between two segments of the scene, or between two objects, the two objects must be at two different emission levels, and the sensor must be able to discriminate between the two levels.

Current military FLIR systems use a "common module" developed by the U.S. Army Center for Night Vision and Electro-Optics Laboratory to ensure design compatibility and to reduce production costs: This design employs a parallel scan of 60, 120, or 180 detectors arranged in a single vertical row. An optomechanical system is used to scan the outside scene across the detector array. As

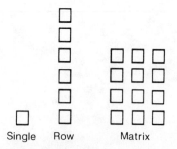

Figure 16.4 Single detector and row or matrix of detectors.

implied by the name, the common-module FLIR design divides the sensor package into separately functional assemblies. This parallel scan modular imaging approach provides the advantages of higher sensitivity, simpler scan mechanism, and higher reliability, compared to alternative systems. The goal of the device determines the complexity. A hand-held observation device would require the fewest number of detectors; a tank sight needs a moderate number; a pilotage system has the highest detector requirement. In addition, interlacing detector scanning doubles the effective number of detectors.

One disadvantage to most thermal imagers is the requirement that they be "cooled" prior to effective use. This process may take from 5 to 30 min. The optimal cooldown temperature is a function of the detector material and the selected spectral sensitivity range. Cooling is required to improve the thermal imager's signal-to-noise ratio. Cooling can be accomplished with liquid nitrogen or very high pressure air (>6000 psi) in what is termed an open-cycle cooler because the nitrogen or air is expelled after the cooling process. Closed-cycle cryogenic coolers that continuously recirculate the refrigerant material are used on most military thermal imagers, particularly those fixed on vehicles with mercury cadmium telluride (HgCdTe) detectors operating in the 8–12 μm region. Thermoelectric cooling is used on other types of detector materials such as indium antimonide, which operates in the 3–5 μm region. Indium antimonide detectors are used frequently in hand-held or small weapon sighting imaging devices. Although newer ceramic detector materials that do not require cooling have been developed in recent years, they have not reached a stage of fielding competency.

16.3 MILITARY APPLICATIONS

16.3.1 Military Requirements

The nature of warfare and its corresponding peacetime training demands match well the operational capabilities of passive night vision systems. The ability to conduct nighttime operations and to provide undetected surveillance are major goals in most military scenarios. Although each branch of the armed forces often has unique mission requirements, most current night vision devices share common designs and are based on the principles of either image intensification or thermal imaging.

Nighttime tasks requiring night vision systems can be grouped generally into three categories: pilotage, navigation, and targeting. The U.S. Navy, Air Force, and Marines have fielded primarily I^2 technology devices for pilotage and thermal imaging devices for navigation and targeting. The U.S. Navy's and the U.S. Marines' fixed-wing communities have a third-generation I^2 pilotage system known as CATS-EYES. The rotary-wing communities of these branches and the U.S. Air Force's fixed-wing community use another third-generation I^2 system known as the Aviator's Night Vision Imaging System (ANVIS). The U.S. Army uses I^2 devices and thermal systems for pilotage, navigation, and targeting.

16.3.2 Night Vision Goggles and Aviator's Night Vision Imaging System

Military aviation has fielded two night vision systems based on the image intensifier tube. The earliest version is known as the AN/PVS-5 series night vision goggles (NVG) and is based on second-generation image intensifier tubes. The current version, which utilizes third-generation tubes, is known as the AN/AVS-6 Aviator's Night Vision Imaging System (ANVIS) (Fig. 16.5).

The original version of the AN/PVS-5 NVG was a full-face configuration designed for use by ground troops. The faceplate provided support for the I^2 tubes and contained their emitted light. However, this design had several deficiencies with respect to aviation, the major one being total occlusion of the aviator's peripheral vision. This lack of peripheral vision was a major safety concern to the

Figure 16.5 Aviator's night vision imaging system (ANVIS).

HUMAN FACTOR CONSIDERATIONS

aviation community. In 1982, a modified faceplate version of the NVG, shown in Figure 16.6, was developed as an interim measure until the ANVIS could be fielded (McLean, 1982).

The NVG and ANVIS systems are imaging devices that amplify low-level ambient light reflected from objects. An amplified image is presented on a phosphor screen (Verona, 1985). Both systems use two image intensifier tubes to form a binocular device that is attached to the aviator's helmet.

The second-generation NVGs have a typical system gain of 2000–3000 and a peak display luminance between 0.3 and 0.9 foot lamberts. The tubes in the third-generation ANVIS operate similarly to the second-generation ones, but they have greater sensitivity and resolution, operate over a slightly different spectral range, and have a peak display luminance of between 0.7 and 2.2 fL. For cockpit lighting compatibility, the interior surfaces of the objective lenses of ANVIS are coated with a dielectric film (a minus-blue filter) that rejects wavelengths less than 625 nm. From the user's point of view, the major difference between NVGs and ANVIS is that ANVIS, with third-generation tubes, is usable during periods of lower light levels where starlight is the only source of illumination. (Note: There is a three-decade difference between the illumination levels of full moon and starlight.)

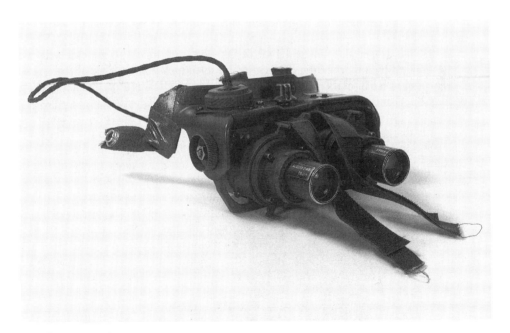

Figure 16.6 AN/PVS-5 night vision goggles (cut a way version).

16.3.3 Pilot's Night Vision System

In the Army's newest production aircraft, the AH-64 (Apache) attack helicopter, thermal imaging is used for both pilotage and targeting. The targeting sensor system is known as the Target Acquisition and Designation System (TADS), and the pilotage sensor system is known as the Pilot Night Vision System (PNVS). The PNVS provides imagery to a helmet-mounted display (HMD) called the Integrated Helmet and Display Sighting System (IHADSS) (Fig. 16.7). Both the TADS and the PNVS use thermal imaging sensors that are mounted on the nose of the aircraft and operate in the 8–12 μm spectral range (Fig. 16.8).

16.3.4 Integrated Helmet and Display Sighting System (IHADSS)

The IHADSS was developed specifically for the U.S. Army AH-64 attack helicopter. The system is designed around a helmet referred to as the Integrated Helmet Unit (IHU). Along with sight sensor electronics, the following components are included: visor assemblies (clear and tinted), monocular optical relay unit (known as the Helmet Display Unit, HDU), miniature cathode-ray tube (CRT), and communication and video cables. The function of helmet-mounted display components of the IHADSS is to provide night vision information to the aviator for the purposes of nap-of-the-earth (NOE) pilotage, target acquisition and identification, and weapons aiming and to provide daytime flight symbology (Walker et al., 1980).

In the basic operation of the IHADSS, an electronic image of the external scene, formed by a thermal imaging sensor mounted on the nose of the aircraft, is converted into a light image (symbology overlaying FLIR imagery) on the face of the CRT. This image is relayed optically through the HDU and reflected off a beamsplitter, also known as a combiner, into the pilot's eye (Fig. 16.9). Therefore, it is through the HDU that the pilot receives primary visual data to fly the aircraft. Infrared detectors mounted in the IHU allow the aircraft's imaging sensor to be slaved to the pilot's head movements. Aircraft parameters symbology, along with the sensor video, is presented to the pilot by means of the HDU. In addition, target acquisition and weapons information also can be displayed. The display system is designed so the image of the 30° vertical by 40° horizontal field of view (FOV) of the sensor subtends a 30° × 40° field at the pilot's eye (equivalent to viewing a 21 in. television screen from a distance of 28 in.). This provides an imaging system of unity magnification. This field of view is controlled by the pilot's line of sight and has a field of regard of ±90° in azimuth and +40° to −70° in elevation.

The IHU is custom fitted with pads to provide a stable platform for the HDU. The display has a 10 mm exit pupil in order to provide for some eye position tolerance and to accommodate eye rotation.

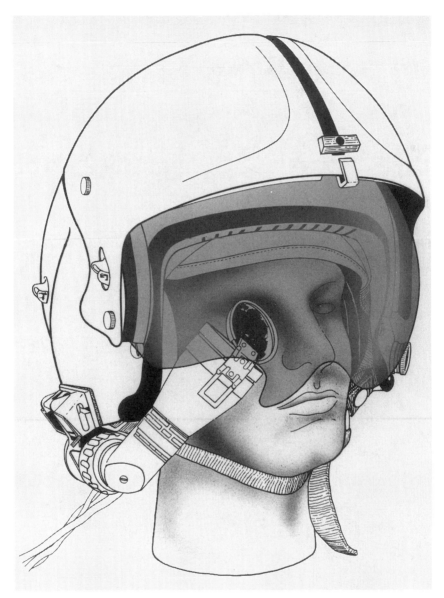

Figure 16.7 The Integrated Helmet and Display Sighting System (IHADSS).

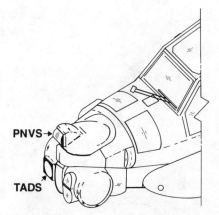

Figure 16.8 The positions of the PNVS and TADS sensors on the AH-64.

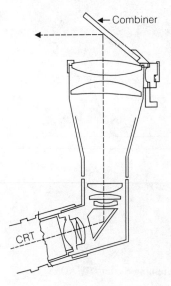

Figure 16.9 The helmet display unit (HDU), consisting of CRT and relay optics.

HUMAN FACTOR CONSIDERATIONS

The IHADSS represents a tremendous transition in helmet sophistication. The IHU in the IHADSS plays a crucial role in linking the pilot and aircraft. Aviator performance and safety are highly dependent on the transfer of the sensor information to the eye through the HDU. With the advent of the IHADSS helmet, Army aviation has moved from an era of the "slap-on, cinch-up" helmet to one where the helmet is a precision-tuned piece of equipment that requires special considerations and care. The purpose of this helmet extends beyond that of protection, to include providing a platform for presentation of high-resolution infrared imagery, sighting direction and movement commands for a gimbaled cannon and the PNVS sensor, and flight imagery and weapons delivery information (symbology).

16.4 BIODYNAMIC CONSIDERATIONS

16.4.1 Head-Supported Weight

The placing of night vision devices on the helmet has significantly increased the amount of weight that must be supported by the head. The head-supported weights of the cutaway version of the NVG and ANVIS when worn with the standard U.S. Army SPH-4 aviator's helmet are 6.8 lb (3.1 kg) and 5.9 lb (2.7 kg), respectively. These weights include typical values of counterweights used by aviators to offset center-of-gravity (CG) shifts. The head-supported weight of the IHADSS is 4.0 lb (1.8 kg). This weight includes the helmet (Integrated Helmet Unit, IHU), Helmet Display Unit (HDU), and miniature CRT.

The effects of placing additional weight on the aviator's head generally can be grouped into two areas: mission effectiveness and crash dynamics. Increased head-supported weight can affect mission effectiveness either directly (via physical effects) or indirectly (via fatigue). The physical effect of increased inertia alone causes rapid lateral head movements to be slowed and delayed (Gauthier et al., 1986). These inertial effects are seen at levels of head-supported weight (4.4 lb) similar to those added by the IHADSS. Limited research has been conducted to document the fatigue factor associated with increased head-supported weight. One study, conducted in 1968 by the U.S. Army Human Engineering Laboratory, found that a total head-supported weight in excess of 5.3 lb (2.4 kg) degraded the performance of complex sighting tasks (U.S. Army Human Engineering Laboratory, 1968). This degradation manifested itself in slower head motions, most likely the result of muscle strain. Fatigue in the head/neck muscles can slow reaction times associated with movements of these muscle groups. In situations where the primary pilotage imagery input is controlled by head movement, this slowing of reaction time creates a dangerous condition and also may contribute to decreased maneuvering accuracy, increasing the risk of an accident. However, the quantitative relationship between weight and performance degradation is not fully documented.

The effect of increased head-supported weight in crash dynamics is a direct result of the additional mass. For the 50th percentile male head, the head and neck weight is 11.7 lb. In the case of AN/PVS-5 NVG mounted on the SPH-4 helmet with counterweights, an additional 6.8 lb results in a 58% increase in head-supported weight and accompanying G force in a crash. Adding IHADSS (4.0 lb) results in a 34% increase in head/neck weight. This increased G loading will contribute further to head and neck muscle fatigue during maneuvers of low to moderate accelerations (<3G). However, of most concern is the additional amount of G force that will act during crashes. In the absence of external forces, and neglecting forces due to the neck muscles, the total internal force exerted on the skull base during acceleration is approximately equal to the product of the total mass of the helmeted head and the head acceleration. Thus, for a given head acceleration, the larger the mass (head-supported weight), the larger the force and the greater the risk of injury.

To help reduce the hazard of increased head/neck mass in an accident sequence, most current night vision devices (and their accompanying counterweights) are designed to "break away" when they encounter high accelerative forces. For example, the ANVIS should break away from the aviator's helmet at a head acceleration of 10–15 G. This design feature reduces the total head/neck/HMD mass during the critical period of high accelerations.

16.4.2 Center of Gravity

Both NVG and ANVIS are attached to the front of the helmet. The IHADSS helmet display unit is attached to the right side of the IHADSS helmet. All of these systems cause a shift in the location of the center of gravity of the head/neck/HMD system and result in an asymmetrical loading of the head/neck system. The consequences of this CG shift are similar to those of increased head-supported weight, affecting both crash kinematics and mission effectiveness. The offset center of gravity creates a moment arm by which the weight of the head/neck/HMD system produces a torque on the head and neck musculature. The effort required to balance this torque contributes to fatigue. In addition, the presence of the torque increases injury risk during a crash sequence.

Shifting the head/neck CG forward 10 cm has been shown to reduce neck isometric endurance following 5 or 35 min of dynamic (lateral) neck exercise (Phillips and Petrofsky, 1984). However, using a smaller CG shift (2.5–5.0 cm) and lighter head-supported weight (3 lb), a physiologically optimal CG position was found to be either forward or lateral (Phillips and Petrofsky, 1983). These conclusions are in conflict with a more recent study in which aviators preferred rearward and vertical CG shifts to forward or lateral shifts (Butler et al., 1987). The current ANVIS and NVG have forward CG shifts; IHADSS has a forward and right lateral shift. Although there is disagreement about which type of CG offset is best tolerated, any amount of offset most likely will increase fatigue, which will further degrade performance during extended missions.

With respect to the dynamics of a crash, if no external forces are present and the effects of the neck muscles are again neglected, the torque at the skull base is approximately equal to the product of the mass of the head/neck/HMD system, the linear acceleration, and the distance from the head/neck center of gravity to the head/neck/HMD center of gravity. Thus, for a given head combination of head-supported weight and acceleration, the greater the CG offset, the greater the torque and the risk of injury.

16.4.3 Analysis

As can be deduced from the above discussions, the effects of head-supported weight and a shift in the CG are not independent. An increase in either parameter increases the torque or bending stresses in the neck due to maneuver flight loads and/or crash loads. Although complaints about weight and CG asymmetry do not predominate in surveys of AH-64 pilots (Hale and Piccione, 1989; Rash and Martin, 1987a; Crowley, 1991), helmet designers should strive to maintain the head/night vision device CG as close as possible to that of the head/neck CG to reduce fatigue and risk of injury (Glaister, 1988).

Even though upper bound values for weight and CG have not been established analytically, empirically aviators have provided subjective data that should be used as a guideline. These aviators consider AN/PVS-5 NVG without counterweights an unacceptable configuration for rotary-wing aircraft. The ANVIS and IHADSS meet with greater approval with helicopter pilots, but objections have been raised by pilots who fly high-performance aircraft. There also seems to be a consensus that some adverse weight effects can be minimized by bringing the HMD CG in line with that of the head/neck system, the implication being that the CG parameter is more critical to user acceptance than the weight parameter. This observation is substantiated by aviators who typically add counterbalancing weights to their helmets to counteract the adverse CG of AN/PVS-5 NVG (Jones, 1983).

16.5 SENSOR AND DISPLAY CONSIDERATIONS

As depicted in Figure 16.1, imaging systems can be represented as comprising two basic sections: the sensor and the display. When convolved with the characteristics of the human operator, the operating characteristics of these two sections define the upper performance of the man–machine system.

16.5.1 Sensor Parameters

All sensors share some common operational parameters. Other parameters are sensor-dependent. However, because the two classes of imaging systems being discussed in this chapter are based on divergent technologies, even common parameters require different approaches to their explanation.

Image Intensifiers

Measurable sensor parameters that affect the performance of the image intensifiers in I^2 night vision systems include spectral response, signal-to-noise (S/N) ratio, equivalent background input (EBI), sensitivity, and gain. Also, there are defects in the image intensifier tubes that can degrade their performance. Some of the tube characteristics are fixed at the time of manufacture, and others change during the lifetime of the tube. When the tube is manufactured, it has a certain spectral response based on the chemical composition of its photocathode. Second- and third-generation tubes have the spectral response shown in Figure 16.10. Second-generation tubes are responsive to wavelengths between 380 and 850 nm, which includes all of the visible spectrum. Third-generation intensifiers are responsive to wavelengths between 550 and 950 nm, which includes only a

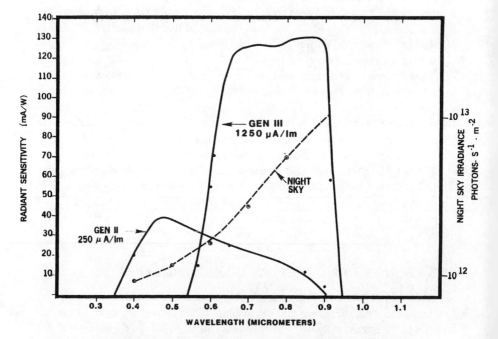

COMPARISON OF SENSITIVITIES BETWEEN GEN II AND GEN III PHOTOCATHODES

Figure 16.10 Comparison of sensitivities between second- and third-generation image intensifier tubes.

HUMAN FACTOR CONSIDERATIONS

portion of the visible spectrum but extends further into the near-infrared. Even though objects appear to have the same shape when viewed through second- and third-generation tubes, they may have entirely different intensities. For example, dirt roads against green grass may appear very dark through second-generation tubes but appear bright through third-generation tubes. This occasional contrast reversal when comparing second- and third-generation images is a function of the difference in the spectral sensitivity between the I^2 generations and the scene reflectivity. The contrast, and therefore the visibility, of objects must be related to a device of a specific generation. A computer model to predict scene contrasts for I^2 devices has been developed by the U.S. Army (Decker, 1989).

The contrast and resulting image quality also are a function of the signal-to-noise ratio of the intensifier tube. Signal is the information transmitted through the intensifier. Noise is unwanted disruption in the signal. Some effects of visual noise are discussed later in Section 16.6.2. The S/N ratio decreases as the image intensifier tube ages. Second-generation tubes have lower S/N ratios than third-generation tubes. The S/N ratio of second-generation tubes continue to decrease throughout their useful life. (The laboratory life time is cited as 2000–4000 hr but may be operationally closer to 1000 hr.) The S/N ratio of third-generation tubes tends to remain constant over their 7500 hr life and then falls off rapidly. End of life is defined as the point when the S/N ratio falls below a prescribed value.

Another important parameter affecting image quality and a major limitation of I^2 NVDs is equivalent background input (EBI). EBI is a measure of the output luminance of an image intensifier with no input, the analog of dark current in photodiodes. The EBI parameter is important because if it is too high the contrast of the intensified scene image may be decreased to a level unacceptable to the user. Contributors to this background screen luminance include thermal dark current, secondary electron emission from ion impact on the photocathode, and output light backscatter (Levi, 1980).

The sensitivity of an image intensifier, whether second- or third-generation, is established when it is manufactured. The sensitivity of second-generation intensifiers gradually decreases over time. Third-generation intensifiers maintain a relatively constant sensitivity over time. The eventual loss of sensitivity results in a weakened signal, thus a lower S/N ratio, which signifies end of life.

The gain of an image intensifier is defined as the ratio of signal (light) in to signal (light) out. Tube gain is controlled primarily by the tube's power supply. Increasing the voltage between elements in the intensifier tube increases its gain. The close proximity of the elements in the tube limit the voltage that can be applied before breakdown or arcing will occur. When the tube is new, increasing the gain increases both the signal level and the noise level. At some point near its end of life when the sensitivity of the tube is low, increasing the gain increases the noise more than the signal, thus decreasing the S/N ratio.

Other sensor characteristics affecting image quality are inherent tube defects. Some minor defects are normal and may be present when the tubes and systems are accepted from the manufacturer; others develop during the life of the device. Some black spots, white spots, fixed pattern noise, and distortion are permissible in new image intensifiers. The tube contractual specifications define the limits of acceptable defects.

A black spot is an inactive region in any of the various elements of the tube—MCP, photocathode, phosphor screen, etc. The size, location, and number of black spots are measured and compared with the specification to determine if they are acceptable. Spots in the central area of the image large enough to block objects from view, such as other aircraft, are unacceptable.

Bright spots usually identify emission points where there is a high energy density caused by impurities or surface irregularities. This type of defect can degrade contrast and lead to premature tube or power supply failure. Power supply current sensing circuits can be fooled by the localized high currents and misadjust the tube's operating parameters. The emissions usually burn off, resulting in black spots discussed above. Some emissions are "spoiled," that is, converted into black spots, by a process of applying laser radiation during final manufacturing testing. Bright spots are not acceptable, but some black spots are acceptable.

Fixed pattern noise may appear as a faint geometrical pattern of the internal fiber optics. If the presence of the pattern interferes with the aviator's ability to use the I^2 NVD, the tube is rejected. Finally, excessive distortion of the internal fiber-optic inverter is measured. Unacceptable distortion levels can cause flat surfaces to appear to have depressions or bulges. Tubes exhibiting this level of distortion are obviously unacceptable for flight.

Thermal Imagers

Sensor design parameters as well as user-adjustable sensor controls affect the performance of thermal imaging systems. The design parameters include sensitivity, signal-to-noise ratio, component time constants, spectral response, and resolution. User-adjustable sensor controls include gain and bias level. However, many of these parameters are interrelated. In complicated imaging systems such as the common-module PNVS, it is difficult to describe individual control operation and effects. What is more important is an overview of how the sensor affects the quality of the "picture" of the outside world presented to the aviator.

For a thermal sensor to be able to image a scene of the outside world, it must be able to respond to the energy being emitted by objects in the scene. The spectral response of the sensor, or that part of the energy spectrum over which it can collect energy, can be defined as the ratio of the sensor's output signal to the amount of collected energy, as a function of the wavelength. This response is determined primarily by the choice of detector material. One of the most popular

HUMAN FACTOR CONSIDERATIONS 671

detector materials, and the one used in the PNVS, is mercury cadmium telluride (HgCdTe). The response for a specific HgCdTe detector varies with the chemical formulation, the mechanism of energy conversion (photoconductive or photovoltaic), and the system's operating temperature. A typical spectral response for the AH-64 PNVS system detector is shown in Figure 16.11.

A sensor's ability to "see" (or resolve) detail is often presented as the dominant parameter in determining the quality of the "picture" obtained. The smallest segment of the scene that can be imaged is a measure of the spatial resolution of the sensor. This can be defined as the solid angle over which the detector can collect energy. For a single detector, this is usually expressed as the subtended angle representing the detector's instantaneous field of view (IFOV). However, in the more complex common-module PNVS, the spatial resolution obtained is determined by the interrelationship of the single-detector IFOV, the number of detectors and their geometry, the scan method, and the scheme for digital sampling of the detector's analog output. Also, in scan-type thermal imaging systems, the vertical resolution and the horizontal resolution are different.

Finer features of a scene can be detected if they are of sufficient size and there is adequate contrast between the feature and its background. The threshold contrast depends on size. Hence, resolution, by itself, does not guarantee preservation of detail. Contrast transfer is another important parameter. Figure 16.12a shows a scene containing a horizontal row of trees. Assume that these trees are of the same width and are separated by a distance that is equal to their width. Also assume that each tree is identical in its emitted energy and is viewed by the detector as an object of temperature T_1. Let the background of the scene be at

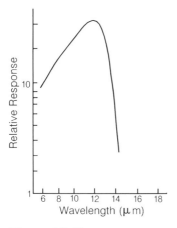

Figure 16.11 Spectral response of typical HgCdTe detector.

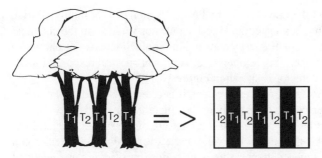

Figure 16.12 (a) Visual and (b) thermal representations of a scene consisting of a row of trees.

temperature T_2. This leads to a simplified representation of the scene, as presented in Figure 16.12b.

To understand how resolution affects scene imaging, allow the detector to be placed at different ranges from the transformed scene (Fig. 16.13). As the range decreases, the amount of the scene from which the detector collects energy, that is, the scene area within the detector's IFOV, becomes smaller. At the greatest range, the detector is collecting energy from a large part of the scene. At the closest range, the detector is collecting energy from only a portion of one of the bars representing a single tree. At the farthest range, the detector is collecting energy from multiple bars (trees). Consider the detector's output at these two extreme ranges as the detector scans across the scene. As the detector scans at the closest range, its output will be at its maximum value when its IFOV is filled fully with a target bar (Fig. 16.14a), a lesser value when the IFOV is filled partially by a target bar and a background bar (Fig. 16.14b), and its minimum value when the IFOV is filled fully with a background bar (Fig. 16.14c). The representative output of the detector is shown in Figure 16.14d.

Two important concepts are demonstrated in Figure 16.14. First, the output signal undergoes a modulation (a change in amplitude) that generally follows the increasing and decreasing temperatures (emitted energy) of the scene pattern. The frequency of the bars in the scene and the maximum and minimum values (which determine contrast) are retained in the output signal. Second, the sharp transition in the scene between a target and a background bar is deemphasized in the detector's output. This deviation from a completely faithful representation of the scene occurs as the detector's IFOV is collecting energy simultaneously from both a target and a background bar. During this period, the detector's output value falls somewhere between the values obtained for target and background bar alone.

At the farthest range, the IFOV is collecting energy from a part of the scene

HUMAN FACTOR CONSIDERATIONS

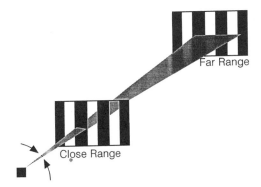

Figure 16.13 The intersection of the detector's instantaneous field of view with the scene at different ranges.

containing several target and background bars during the entire scan (Figs. 16.15a–16.15c). As the detector scans, the output signal (Fig. 16.15d) varies little in amplitude and does not undergo a modulation at the frequency in the scene. Neither the scene frequency nor the contrast is accurately reproduced; the individual trees may not be distinguishable on the display.

In the explanation above, as the scene is placed at increasing ranges, the number of bars within the IFOV (target spatial frequency) increases. As this spatial frequency increases, the modulation of the scene, as reproduced in the image, decreases. If the modulation of the image, compared to the actual scene modulation, is graphed as a function of increasing spatial frequency, a curve similar to the one in Figure 16.16 will be obtained. This curve is known as the *modulation transfer function* (MTF) and is a figure of merit for comparing detectors.

Complex real-world scenes are a composite of varying spatial frequencies and contrasts, unlike the single-frequency, high-contrast scene discussed in the above example. Therefore, the image formed by the detector will not reproduce faithfully all scene information. The contrast of the higher spatial frequencies may be particularly degraded, causing loss in scene detail. Unfortunately, the detector is not the only system component that has an MTF. Further degradation of the scene may result from the MTF of the optics and the display. These component MTFs are cascaded to provide an end-to-end system MTF. A component MTF value (for a given spatial frequency) is always less than 1, and a system MTF value is always less than any of the component values.

Another figure of merit used to compare the performance of thermal sensors is *minimum resolvable temperature* (MRT). The MRT is a measure of sensitivity. Often, it is defined incorrectly as the minimum temperature difference the sensor can resolve. Actually, it is not a measure of temperature sensitivity, but of energy sensitivity (relating to the material *and* its temperature). The concept of MRT,

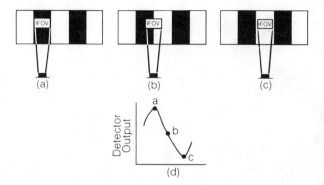

Figure 16.14 Detector scanning of a close-range scene and its representative output.

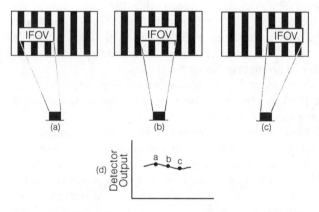

Figure 16.15 Detector scanning of a far-range scene and its representative output.

while an important laboratory parameter, has no practical significance to the user except for system comparison. In general, the lower the MRT, the better the sensor can discriminate between objects in a scene.

Two adjustable sensor controls are available to the user. These are *gain* and *bias level*. Gain relates to the range of thermal energy levels being detected, and bias relates to the mean energy level. These controls are intended to allow the user to optimize the sensor's performance. In operation, these adjustments affect the IR detector output signal as it is passed to the next sensor stages. Proper settings of these controls, which are highly dependent on environmental conditions, optimize the dynamic range (ratio of maximum to minimum signal levels)

HUMAN FACTOR CONSIDERATIONS

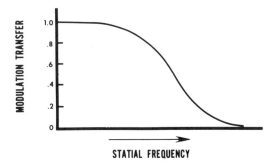

Figure 16.16 Typical modulation transfer function (MTF) curve.

of the transferred detector signal. Improper detector settings, for a given scene and environment, will result in loss of scene detail and a degraded image. The optimal settings for these controls change with environmental conditions and scenes. Since they are interactive, as are the brightness and contrast controls on the display, an iterative, time-consuming process is required to optimally set them, only to have them require additional adjustment as the environment or scene background changes. Many attempts to automate these controls have failed. The topic of the dynamic range problem is discussed in the display parameter section (Section 16.5.2).

A third control over the PNVS sensor output, but one that does not actually affect sensor operation, is the *FLIR polarity* switch. This switch converts the polarity of the sensor's output from "white hot" to "black hot." This refers to the presentation of the imagery on the display. In the white hot mode, objects emitting the greatest amount of energy appear "whiter" (actually "greener" for the IHADSS display phosphor) than objects of less emission. Conversely, in the black hot mode, objects emitting the greatest energy appear "blacker." A noticeable difference between these two modes is the appearance of the sky background. In the white hot mode, the sky will appear darker than the horizon. Aviators appear to indiscriminately switch between modes, selecting the "best" image, with no definable criteria for selecting one mode over another. The ability to switch polarity is particularly useful when objects are located under tree cover. In the white hot mode, the objects of interest are bright objects being viewed against the bright trees and ground. For *certain gain and level settings*, switching to "black hot" may cause the targets to stand out as dark objects while the background remains relatively bright. Users may frequently switch polarity in order to optimize picture quality.

An additional effect on sensor performance is that of the cooling system. If the cooling system in a thermal sensor has not had time to adequately cool the detector or has exceeded its operating lifetime (1000–2000 hr) and is unable to provide the designed operational temperature, the resulting sensor output will be "noisy" and result in degraded operation.

16.5.2 Display Parameters

The purpose of the display is to present a final visual image of the scene to the user. Most currently fielded night imaging systems accomplish this by means of phosphor displays. (Future systems may make greater use of developing flat-panel technologies.)

Image intensification devices (AN/PVS-5 and ANVIS) are direct look-through systems where the detector-generated electrons are imaged (via the fiber-optic inverter) directly onto the phosphor display screen. In these devices, display parameters and their effects on image quality are limited to the operating characteristics of the phosphor. The phosphor selection for the display is critical and must be optimized for its intended use. Each phosphor exhibits specific physical characteristics. In general, each phosphor emits a unique spectrum of light when activated by electrons. In the image intensifier, the phosphor is activated by a two-dimensional array of electrons. Each pixel in the image is an element of the continuous electron array. Pixel brightness is determined by the rate of electron flow.

Rise time and persistence are critical parameters that affect how long it takes for the phosphor to radiate light at 90% of maximum luminance after being exposed to the electron beam and how long it takes for the light to fall to 10% of its maximum luminance when the electron beam is removed, respectively. The phosphor's luminous efficiency specifies the ratio of luminous energy output for a specified energy input.

The PNVS/IHADSS is a remote-view system using temporally synchronized scanning techniques to maintain scene spatial integrity. The PNVS sensor scans the outside scene, converting it into an electric video signal suitable for input to a video display.

By definition, the video display converts the electrical representation of the scene generated by the sensor into a two-dimensional image that can be viewed by the eye. The video display is typically a cathode-ray tube, and the image is similar to that produced on a black-and-white television. A modulated beam of electrons is scanned very rapidly over a phosphor screen. The beam produces a tiny modulated dot of light that generates the two-dimensional, illuminated visual image.

The quality of the imagery is determined by the scene's characteristics, the sensor's operating parameters, the intervening electronics, and the display's operating parameters. The role of the sensor's parameters was discussed in Section 16.5. The characteristics of the scene depend on its spatial frequency content and environmental conditions (see Section 16.8.1). Display parameters that affect the

HUMAN FACTOR CONSIDERATIONS 677

quality of CRT images include line rate, screen phosphor, spot size and shape (electronic focus), maximum luminance (brightness), dynamic range, gray scale, resolution, and display MTF. For helmet-mounted displays, user-adjustable controls often include optical focus, brightness, and contrast. Many of these parameters are interrelated. Additional adjustment controls for electronic focusing, positioning, and sizing of the CRT image are present, but typically they are not designed for routine adjustment.

In the United States, a commercial television (TV) picture is generated from 525 horizontal scan lines. Each TV picture or frame is presented every $1/30$ sec. To minimize visual flicker in the display, every other line (half the picture or field) is presented every $1/60$ sec. The number of discrete horizontal scan lines determines the maximum vertical resolution of the display. In a 525-line scan system, only about 490 lines are active, that is, present visual information to the viewer. A vertical line from the top to the bottom of the display would consist of 490 vertical dots, one on each scan line. Regardless of size, every display has 490 scan lines. Consequently, there is no more information on a 5 ft screen than there is on a 5 in. screen. Other common line rates used for special-purpose television systems are 875 and 1024.

The PNVS and TADS use the Department of Defense "common-module" thermal imaging system and operate at an 875 line horizontal scan rate to improve the apparent vertical resolution with about 817 active lines (information lines). The vertical rates are the same, $1/30$ sec per vertical frame and $1/60$ sec per vertical field. Some parameters are traded off for the 875 line system compared to the 525 line system. The electron beam moves faster in an 875 line system compared to a 525 line system, but the faster beam must have higher energy to produce the same luminance.

The number of gray shades is the number of visually distinct luminous steps from black to white a display can reproduce. One gray-scale step is defined as $\sqrt{2}$ or (1.414) times as bright as its predecessor. This means that the theoretical number of gray-scale steps a display can reproduce can be calculated given the luminance values of the darkest and brightest areas of the image. Better displays can reproduce a larger gray scale; 10 or more steps are desirable for good image reproduction. Displays that are not very dark in the least bright areas typically cannot reproduce an acceptable gray scale. Light scatter inside the CRT can significantly reduce the number of gray-scale steps available at high luminance levels. Fiber-optic faceplates sometimes are used to reduce light scatter. The larger the gray scale, the smoother the transitions from light to dark areas and the better the overall picture contrast.

The maximum operating luminance of a display is critical if the display is going to be used in high ambient light environments. However, the maximum operating luminance parameter by itself can be misleading. Gray scale and luminance are interrelated parameters and must be specified at the same operating condition. For example, if a display is going to be used only at night, then the number of gray-scale steps should be specified at a luminance in the range of 4–40 fL at the eye. If the display is to be used in a daylight environment, the number of gray-scale steps should be specified in the range of hundreds of footlamberts at the eye. It should be noted that developing flat-panel technology displays require more gray steps owing to the discrete nature of their pixels.

By definition, the imaging system consists of the display and the sensor. The display works together with the sensor to present the image of the outside scene. The dynamic range parameter exemplifies this relationship. The dynamic range of the video display is limited compared to that of the thermal sensor. The sensor is capable of sending about 30 gray-scale steps (distinguishable levels of brightness) to the display. However, the display is capable of presenting only about 10 gray-scale steps to the eye. To illustrate this concept, allow each of the levels of the sensor's output signal to represent 1°C difference in the temperature of a simple object. A 30-step gray scale would allow the sensor to produce an output signal representing a range of 1–29°C, each degree representing a distinguishably higher signal level. This 30-level signal would be sent to the display, where only 10 levels can be displayed. By adjusting the sensor's gain and level controls and the display's brightness and contrast controls, any 10° range could be displayed. If the details in the object, perhaps represented by the energy levels associated with the temperature range of 5–14°C were of interest, then the display could be set up to show the 5°–14° (energy-level) range. However, any similar objects with temperatures (energy levels) above 14°C would not be discernible, being presented at the maximum luminance level associated with the upper (14°C) level. A tank with engine, drive wheels, and exhaust at temperatures between 20°C and 30°C would look like a white blob; only details that were in the 5–14°C range would be distinguishable as shades of gray on the display. Details represented by energy levels of the 5°C level and below would be black.

Automatic controls within the thermal sensor are designed to minimize the potentially dangerous effects of the display's limited dynamic range and the resulting "blooming" or whiting-out of the display. A problem occurs when the sensor is looking at the relatively warm ground below the horizon and the cold sky above the horizon, a large temperature difference. The pilot needs to see the horizon, but even more important, the pilot needs to see details of the ground and objects around him that are much warmer than the cold sky. The automatic controls work relatively well for large temperature differences that are on different scan lines, but not so well when the large temperature differences are on the same scan line. This situation can occur when the aircraft banks and the sensor sees the cold sky and the warm ground on the same scan line.

Spot size is the size of the electron beam footprint on the phosphor screen, measured at its 50% output luminance points. The spot size determines the maximum resolution that can be expected from the CRT. This concept is similar to

HUMAN FACTOR CONSIDERATIONS

drawing with a small, thin drafting pencil or a big, thick carpenter's pencil—finer detail can be drawn with the thinner lead. The size of the electron beam increases as its energy increases, and the larger the beam footprint, the poorer the limiting resolution of the display. A general relationship can be defined for line rate, luminance, resolution, and spot size. As the line rate increases from 525 to 875, the electron beam must move approximately 67% faster to draw the greater number of lines. The beam has less dwell time to activate the phosphor, thus producing a lower luminance image at 875 than at 525. A higher energy electron beam will increase the luminance but will result in a larger spot size and decreased resolution.

The modulation transfer function (MTF) is used as a measure of a display's efficiency in presenting information at various spatial frequencies just as it was a measure of the sensor's efficiency. Modulation contrast is measured at several spatial frequencies starting at about 5 cycles per display width (five alternating black and white bars across the display) to a spatial frequency with a modulation contrast of less than 2%. The modulation contrast reading at 5 cycles per display width will provide an indication of the number of gray scales the display can present. A display must have approximately 93% modulation contrast at 5 cycles per display width to reproduce 10 gray scales (Task and Verona, 1976). The modulation contrast reading of less than 2% shows the maximum *horizontal* resolution of the display; that is, the display will not be able to reproduce information above that spatial frequency. The display's maximum *vertical* resolution is limited by the number of vertical scan lines.

An understanding of the display parameters discussed above is essential to the understanding of the AH-64 IHADSS and panel-mounted head-down displays. The AH-64 has a sophisticated video system with two thermal imagers (PNVS for pilotage and TADS for targeting), a day television sensor, two symbology generators, a video tape recorder, and four video displays (two IHADSS displays and two panel-mounted displays). In the IHADSS, the 1 in. CRT and relay optics, referred to as the Helmet Display Unit (HDU), as shown in Figure 16.9, are mounted on the right side of the aviator's helmet (Fig. 16.7). The HDU, the helmet, and additional electronics collectively are referred to as the Integrated Helmet and Display Sighting System (IHADSS). The IHADSS display is designed to provide a virtual image of a one-to-one presentation of the 30° vertical by 40° horizontal field of view provided by the sensor. The line-of-sight direction for the PNVS sensor is controlled by the head position of the aviator, which is continuously monitored by infrared detectors mounted in the helmet. Processing electronics of the IHADSS convert this information into drive signals for the PNVS. The result is a visually coupled system in which the PNVS is slaved to the aviator's head motion. In addition to the PNVS or TADS imagery, symbology representative of various aircraft operating parameters, such as altitude, heading, and torque, can be presented on the HDU (Fig. 16.17).

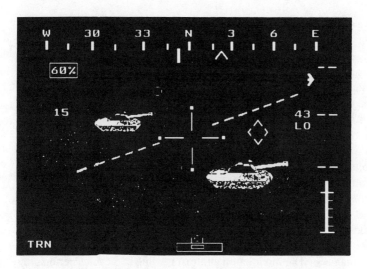

Figure 16.17 Artist's depiction of FLIR imagery and flight symbology.

The thermal image from the PNVS or TADS can be presented to the pilot on a miniature (1 in. diameter) CRT in the HDU, shown in Figure 16.18, or on a 5 in. panel-mounted display. The image generated on the helmet-mounted 1 in. CRT is viewed through magnifying relay optics and a see-through beamsplitter (combiner). The magnifying optics increases the 1 in. CRT image to an apparent size equivalent to that of a 21 in. display viewed at a distance of 28 in. This results in a 40° horizontal by 30° vertical image that corresponds to the FOV of the sensor and provides a total system magnification of unity. The 5 in. direct-view panel-mounted display appears as a 7° horizontal by 5° vertical image. The same information is present on both displays, but the panel display *appears* to have a better image because it is eye-limited (smaller detail than the eye can see) and the HDU is display-limited (the eye could see more if the display could present more). In addition, the contrast provided by the panel display will be better because it is a direct-view image, not a HDU see-through virtual image.

The optical beamsplitter (combiner) shown in Figure 16.9 is a delicate and critical component of the HDU. The combiner is made of a 50% neutral density filter coated with a dielectric thin-film stack. The dielectric coating reflects 80% of the light from the P43 phosphor to the eye while attenuating 90% of light with the same wavelengths as the P43 phosphor that passes through the neutral density substrate. Smudges, fingerprints, scratches, and other distracting features on the combiner may draw the eye's attention and focus to the combiner rather than to the image projected from the combiner. The combiner must be kept free of distracting marks. The see-through feature of the combiner is intended to provide a

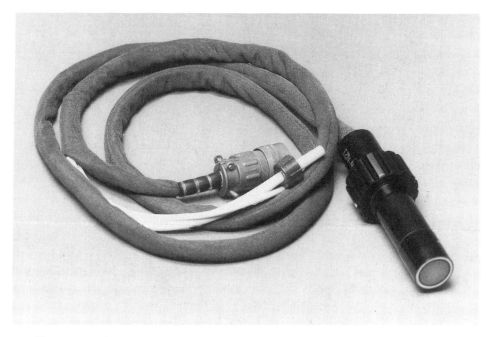

Figure 16.18 Miniature (1-in. diameter) cathode-ray tube used in the IHADSS.

measure of registration between the display image and the outside world. One disadvantage is that bright light sources, when viewed through the combiner, degrade the imagery contrast.

In the IHADSS, an electronically generated gray scale can be displayed to aid in setting up of the user's brightness and contrast controls (Fig. 16.19). This setup is valid for the sensor only if the sensor video output matches the same range as the display's gray-scale video signal. If the sensor video level is lower than the maximum gray-scale level, the resulting sensor video looks washed out and generally lacks contrast. If the sensor video level is higher than the maximum gray-scale level, the resulting sensor video will have too much contrast and will lack detail in the shadows.

The scene information acquired by the sensor is presented as brightness levels on the display. The minimum and maximum brightness levels that can be presented determine the available contrast and shades of gray. The IHADSS is capable of presenting to the eye highlight brightness levels of 4–150 fL. At night, FLIR imagery brightness is typically 8–10 fL. As the CRT ages, the phosphor becomes less efficient and its brightness drops. If a higher brightness setting is used as compensation, it results in increased electron beam size and lower horizontal resolution.

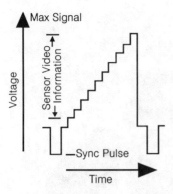

Figure 16.19 Gray-scale video signal used in the setup of the user's brightness and contrast controls.

Among the user adjustments on the HDU is optical focus. This adjustment allows the user to set the semitransparent display image at optical infinity so no change in accommodation is necessary when the user switches attention from distant real objects to the display's virtual image. The user looks at a distant object and adjusts the optical focus so the sensor image is focused at the same point as the distant object. One apparent disadvantage of this display focus approach is the indication that the display eye tends to focus on the HDU beamsplitter (combiner) rather than at optical infinity. Recent studies have suggested a relationship between this misaccommodation and underestimations of size and distance (Iavecchia et al., 1988). In addition, a 1988 survey of 52 AH-64 aviators identified problems relating to size and distance perception. Sixty-five percent of the survey respondents indicated that objects viewed on the HDU were perceived as being smaller and farther away than they actually were. During certain phases of flight, such as landing approaches, these misperceptions may seriously affect the aviator's ability to maintain a proper approach angle or avoid obstacles (Hale and Piccione, 1989).

The above problem relates to the eye's accommodation, or focusing, point. There is another problem that is associated with the mechanical focusing of the HDU. This focusing is achieved by the rotation of a knurled ring located at the rear of the HDU barrel. The focus can be adjusted over a range of $+3$ to -6 diopters. In 1989, a study was conducted to measure the HDU focus adjustment settings of 20 AH-64 aviators. Measurements were taken just prior to takeoff. Ninety percent of the aviators were found to have focus settings of -0.5 diopter or greater (more negative). The range of focus settings was 0 to -5.25 diopters with a mean of -2.25 diopters. The required positive accommodation by the aviator's eye to offset these negative focus settings is likely a source of headaches and visual discomfort during and after extended periods of flight (Behar and

HUMAN FACTOR CONSIDERATIONS

Rash, 1990). Aviators can increase their accommodation workload inadvertently by misadjusting the optical focus. Then they force their visual system to accommodate to a display image that is abnormally close; this is in addition to the normal crewstation and distant real objects accommodation changes.

Additional user adjustments of CRT image orientation, position, and size also can significantly affect performance if done incorrectly. The image orientation is controlled by the rotation of the CRT with respect to the optical axis of the HDU. If the image rotation is improperly adjusted, the pilot may experience a conflict between the symbology and his otolith-derived sense of gravity. A head tilt may develop to compensate for this mismatch, creating a situation analogous to the leans, a common vestibular illusion (Gillingham and Wolfe, 1985; Crowley, 1991). Misadjustments of position and size are addressed in discussions of field of view and visual fields (Section 16.6.3).

When the two applications of a phosphor display, as discussed above, are compared, a number of differences between the directly viewed I^2 display and the IHADSS raster-scanned display can be noted. The I^2 display is a continuous display. In other words, all picture elements (pixels) are presented successively by a simultaneous flow of electrons to each pixel as modulated by the input scene. There is no image breakup or flicker because the image is continuously refreshed. The horizontal and vertical resolution are the same because the I^2 tube is rotationally invariant, that is, there is no defined top, bottom, left, or right. The raster-scanned display may produce images from a thermal imager, as with the PNVS, but may also be used to produce electronically generated symbology or images from a remotely located image intensifier coupled to a video camera. In this case, there must be a definitive spatial relationship with the scene through a one-to-one mapping with the sensor. In the scanned display, where a single electron beam creates the image on the phosphor screen, the horizontal and vertical resolutions usually are different. In scanned systems there is a continuous image along the horizontal axis and a sampled image along the vertical axis. Because of the vertical sampling, there are usually fewer pixels in a raster-scanned display than in a continuous display.

The scanned display does offer the advantage of providing an electronic interface where additional image signals can be added (video mixing) and presented simultaneously on the display. Currently, optical overlay techniques must be used to add additional images to a direct-view image intensifier. These overlays may be combined in object space or image space. However, when combined, each image must be of equivalent energy and combined at the same optical distance. An exception is when one of the overlays is symbology, which should be 20–100% brighter.

Finally, phosphor persistence characteristics, discussed earlier, are critical to the temporal performance of both continuous and scanned displays. Smearing of the image will occur if the rate at which a pixel is required to change its luminous output is greater than the phosphor decay time (persistence).

16.5.3 Image Quality

The process of transferring the information content of the outside scene to the user culminates with the final transfer to the human eye. An important intermediate step is the formation of the image on the display. The ability of the user to interpret and make decisions based on this image—that is, user performance—depends on the quality of this image. This "image quality" results from the individual and interactive characteristics of the sensor and display, as discussed above (Sections 16.5.1 and 16.5.2) and the ambient conditions (Section 16.8.1).

There are two general categories of image quality measures: subjective and objective. The subjective determination of image quality relies on an observer's assessment of a reproduced image. If a sensor and display are being evaluated, the sensor views a standard pattern and the observer views the final display image to make the subjective quality judgment. The Electronic Industries Association (EIA) resolution chart, which was specifically developed for use with television systems, and the 1951 U.S. Air Force tribar target are examples of standard patterns. The wedge-shaped patterns of the EIA chart (Fig. 16.20a) are used to subjectively determine the limiting resolution of the television system. The point at which the normally individually distinguishable converging lines are no longer distinct is the limiting resolution. A numbered scale is printed along the wedge pattern so a numeric value can be related to the subjective limiting resolution.

Similarly, the Air Force tribar target (Fig. 16.20b) is divided into seven groups of six bars. Each set of bars represents a specific spatial frequency. Three of the six bars are oriented horizontally, and three are oriented vertically. The limiting resolution of the imaging system is subjectively determined by selecting the smallest set of bars that can be resolved as three distinct bars.

These subjective assessment methods can be used effectively to evaluate the performance of an imaging system under a specific set of operating conditions. However, as the human observer serves as the measuring instrument, the assessments tend to vary from observer to observer. The objective approach to image quality assessment is more consistent in measuring the parameters of an imaging system, but the problem is to relate these measured parameters to the human's performance.

One of the most widely used objective measures of image quality is the modulation transfer function (MTF). As discussed previously (Section 16.5.2), the MTF of an electro-optical (E-O) system is a measure of the E-O system's capability to transfer contrast from the input scene to the output image as a function of spatial frequency. As any object can be analyzed as a sum of spatial frequencies, the fidelity of the image reproduction is dependent on the capability of the E-O system to reproduce individual spatial frequencies of a given contrast. Typically, as the spatial frequency increases, the reproduction capabilities of the E-O system decrease. In visual terms, the bright areas are not as bright and the dark areas are not as dark as in the original object.

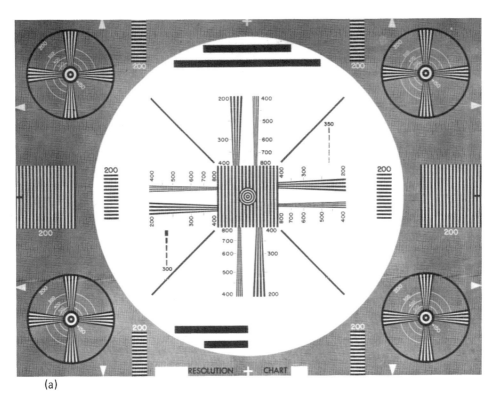

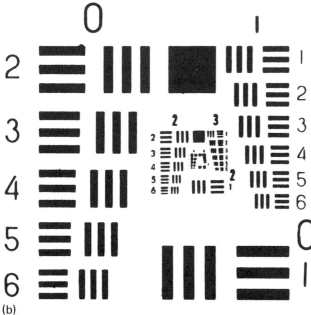

Figure 16.20 Standard patterns of (a) Electronic Industries Association (EIA); (b) U.S. Air Force.

The overall MTF of an E-O system is the product of the system component MTFs. A poor choice of a single component can have a devastating effect on the overall system performance. Similarly, using an expensive component with performance far beyond that of other components may not significantly improve the system's performance. Thus far, we have discussed only the static cases where the imagery remained stationary. Another dimension can be added to the MTF curve that depicts the degrading effects of dynamic imagery. The temporal characteristics of the E-O components effect the modulation transfer in this dimension. This is discussed in more detail in Section 16.7.1.

As stated earlier, the objective E-O system measurements are more consistent than the subjective measurements but are also more difficult to relate directly to the human observer. Ideally, knowing the measured objective performance of an E-O system should enable one to predict the system's performance with a human observer. In practice, this last step is somewhat illusive. The utility of the MTF and other objective measures have been more beneficial in comparing the performance of like systems and a quality control technique than as a means of accurately predicting the capability of a human observer using an E-O system to perform a specified task.

In addition to the MTF, other figures of merit (FOMs) have been developed to compare the performance of various E-O sensors, displays, and systems (Task, 1979). However, the various manufacturers typically do not provide like hardware specifications that permit performance comparisons across manufacturers. A discussion of these FOMs is outside the scope of this survey chapter, but it is important to note that they do exist, each with its inherent benefits and limitations. The FOMs can be used to compare hardware performance, but they are not generally useful until adopted by a standardization organization and legitimatized as a measurement and specification standard similar to the EIA resolution chart.

16.6 VISUAL PERFORMANCE ISSUES

16.6.1 Visual Acuity

Visual acuity is a measure of the ability to resolve fine detail. Snellen visual acuity commonly is used and expressed as a comparison of the distance at which a given set of letters is correctly read to the distance at which the letters would be read by someone with clinically normal eyesight. A value of 20/80 indicates that an individual reads at 20 ft letters that normally can be read at 80 ft. Normal visual acuity is 20/20. As measured through imaging systems, visual acuity is a subjective measure of the operator's visual performance using these systems. The acquisition of targets is a primary performance task. For this task, a reduced acuity value implies that the observer would achieve acquisition at closer distances. However, providing an acuity value for thermal or I^2 imaging systems is difficult. For thermal systems, the parameter of target angular subtense is confounded by the emission characteristics of the target. For I^2 systems, resolution is cited in lieu of visual acuity. Resolution is an objective measure of the capability of an I^2 NVD to distinguish a separation between two objects. Night vision device procurement, test, and end-of-life specifications are defined in terms of resolution rather than visual acuity. However, for comparison with other systems, Snellen visual acuity with the AH-64 PNVS/IHADSS is given as 20/60 (Greene, 1988). Snellen visual acuity with the AN/PVS-5 NVG is 20/50 under optimal conditions (high contrast and scene luminance). Snellen visual acuity with the AN/AVS-6 ANVIS is 20/40 under optimal conditions. But these optimal conditions will seldom be encountered by an aviator in the real world.

Optimal I^2 resolution (and hence visual acuity) is obtained under high light level conditions with high-contrast targets. The resolving power of the I^2 NVD decreases with light level because the noise in the intensified image increases. The high light level resolution limit was stated earlier as equating to a visual acuity range of 20/50 to 20/40. The low light level resolution is not limited but continues to decrease with decreasing light levels. Remember, image intensifiers are light amplification devices that require a minimum level of illumination to function effectively. Although there may be measurable resolution at very low light levels, the AN/PVS-5 NVGs lose their operational effectiveness at about starlight, as visual acuity approaches 20/100, and ANVIS lose their operational effectiveness at overcast starlight.

In order for the user to take full advantage of the I^2 system's full resolution capability, the eyepiece lenses in front of each eye must be adjusted to suit the wearer's eyes. Each eyepiece lens is independently adjustable over a range from $+2$ to -6 diopters. This adjustment may eliminate the need for some aviators to wear their spectacles when using I^2 NVDs. Aviators still may require spectacles to correct astigmatism or to see items in the crewstation under the I^2 NVDs. The IHADSS helmet display unit does not have a conventional eyepiece. However, an optical focus of the final imagery ($+2$ to -6 diopters) is achieved by an in/out translation of the CRT within the HDU barrel.

16.6.2 Presentation Mode

Modes of visual information presentation used by night imaging systems are monocular, biocular, and binocular. The human visual system is binocular. A binocular system receives two visual inputs from two sensors that are slightly displaced in space. This configuration is used in the AN/PVS-5 series NVGs and AN/AVS-6 ANVIS, where two separate image intensifier tubes, one tube for each eye, are the sources of the visual input. A variation in this design is a biocular system in which one visual input is presented to both eyes using mirrors or prisms. This configuration is used in the AN/PVS-7 series NVGs where one image intensifier is the source of visual input and the same image is presented to both eyes. The AN/PVS-7 may be used by military ground troops and aircraft crewmembers but is not authorized for use by aviators.

Image intensifiers present "noisy" images to the eyes. This noise appears as scintillations, commonly called *sparkles* or *snow*. The noise increases as the ambient light level decreases. The noise is more predominant with second-generation image intensifiers than with the more sensitive third-generation image intensifiers. These scintillations block detailed information in the image. Since the intensity and location of the noise varies with time, it is unlikely that a scintillation will appear at the same place at the same time in both tubes. The observer therefore benefits from having two independent (uncorrelated) images in a binocular system compared to the duplication of the (correlated) image in a biocular system. Therefore, there is a theoretical improvement in perceived image quality due to visual noise reduction in the binocular system compared to a biocular system (Davon, 1976).

Binocular systems also present a brighter image to the viewer than biocular systems because the luminance output of the image intensifier in a typical biocular system is divided, with half presented to each eye. An additional reliability benefit, particularly in aviation, of binocular systems is redundant image intensifiers. Increased cost, weight, and complexity are obvious detractors of binocular systems compared to biocular systems.

Binocular systems might be expected to provide the additional visual capabilities of depth perception and stereopsis such as with the unaided eyes. Depth perception is the ability of a viewer to judge the relative location of objects in space. Stereopsis is the ability of the observer to judge depth based purely on the differences in the two retinal images caused by the separation of the two eyes. However, the ANVIS and NVG binocular feature provides less stereo resolution than unaided binocular vision at the same ambient illumination.

Depth perception with NVGs is approximately equal to that of the monocular unaided eye (Wiley et al., 1976). Size consistency, overlay, interposition, motion parallax, shadows, and convergence of lines are monocular cues used for depth perception. Resolution also plays an important role in depth perception. The viewer can extract better monocular cues from a scene using an I^2 NVD with better resolution, thus enhancing the viewer's ability to judge relative distances. The ANVIS provides better depth perception cues than NVGs because ANVIS provides better resolution. Accommodation, convergence, and divergence are weak binocular cues used for unaided eye depth perception that are unavailable to night vision device users.

As stated above, stereoscopic vision results from images collected from two slightly different perspectives, such as with the two displaced eyes. Stereoscopic vision is used primarily in eye–hand coordination tasks and is most effective at ranges of less than 50 ft. Experiments with binocular and biocular NVGs (AN/PVS-5 series vs. AN/PVS-7 series) have demonstrated that the image disparity obtained by the separation of image input in the binocular system does not provide effective stereoscopic vision. There was no statistically significant performance difference between the binocular and biocular systems (Wiley, 1989). Neither does ANVIS offer effective stereoscopic vision.

The aviator's ability to judge depth using I^2 NVDs is critical, particularly during confined area maneuvers such as in parking areas and landing zones, and during hovering, bob-ups, and nap-of-the-earth flight. At best, the depth perception cues available are equivalent to performing these maneuvers with one eye during the day.

In the human visual system, the eyes work together and change their focus together from about 6 in. to infinity. This variable-focusing ability can occur consciously or unconsciously and is quite rapid. The objective focus on a night vision device is quite different. The objective lenses on each channel are focused manually independently from about 10 in. to infinity (distances greater than 50 ft are equivalent to infinity for I^2 NVDs). The aviator must view an object more than 20 ft away when focusing the objective lenses for distant viewing. Misadjusting the objective lenses will greatly reduce the ability of the aviator to view distant objects clearly. The objective lens of one or both channels must be adjusted manually to view objects closer than 20 ft through the I^2 NVD. The aviator does not need to refocus the I^2 NVD for viewing inside the crewstation because of the look-under capability of the I^2 NVDs.

The IHADSS used with the PNVS thermal imaging system on the U.S. Army's AH-64 helicopter employs a monocular presentation. Imagery is presented only to the right eye. During the development of the IHADSS, there were two major concerns with the proposed monocular display format: eye dominance and binocular rivalry. Eye, or "sighting," dominance refers to a tendency to use one eye in preference to the other during monocular viewing (Beaton, 1985). Critical cost and weight considerations favored a monocular display format for the IHADSS, which logically would be located on the right side of the helmet because most of the population is right-eye-dominant. However, there were serious questions as to whether a left-eye-dominant pilot could learn to attend to a right-eye display. In fact, there is evidence linking sighting dominance with handedness and various facets of cognitive ability, including tracking ability and rifle marksmanship performance (Crowley, 1989). One study addressing head aiming and tracking accuracy with helmet-mounted display systems indicated eye dominance to be a statistically significant factor (Verona et al., 1979). However, this small amount of research is far from compelling, and the IHADSS is produced to fit the right-eyed majority. Since the AH-64 has been fielded, there have been no reports in the literature addressing the influence of eye dominance on IHADSS targeting accuracy or AH-64 pilot proficiency.

Probably a more troublesome phenomenon is binocular rivalry, which occurs when the eyes receive dissimilar input. This ocular conflict apparently is resolved by the brain by suppressing one of the images (Bishop, 1981). The IHADSS presents the eyes with a multitude of dissimilar stimuli: color, resolution, field of view, motion, and brightness. Aviators report difficulty making the necessary attention switches between the eyes, particularly as a mission progresses (Bennett and Hart, 1987). For example, the relatively bright green phosphor in front of the right eye can make it difficult to attend to a darker visual scene in front of the left eye. Conversely, if there are bright city lights in view, it may be difficult to shift attention away to the right eye (Hale and Piccione, 1989). AH-64 pilots report occasional difficulty in adjusting to one dark-adapted eye and one light-adapted eye (Crowley, 1991). It may be hard to read instruments or maps inside the cockpit with the unaided eye because the PNVS eye "sees" through the instrument panel or floor of the aircraft, continuously presenting the pilot with a conflicting outside view. In addition, attending to the unaided eye may be difficult if the symbology presented to the right eye is changing or jittering (Crowley, 1991). Some pilots resort to flying for very short intervals with one eye closed, an extremely fatiguing endeavor (Hale and Piccione, 1989; Bennett and Hart, 1987). The published user surveys generally agree that the problems of binocular rivalry tend to ease with practice—although under conditions of a long, fatiguing mission, particularly if there are system problems (display focus or flicker, poor FLIR imagery, etc.), rivalry is a recurrent pilot stressor (Hale and Piccione, 1989; Bennett and Hart, 1987). It is likely that sighting dominance interacts with binocular rivalry, affecting a pilot's ability to attend to one or the other eye.

An apparent disadvantage of a monocular display such as the IHADSS is the complete loss of stereopsis (visual appreciation of three dimensions during binocular vision). Stereopsis is thought to be particularly important in tactical helicopter flying, as the terrain is invariably within the 200 m limit of effective stereo vision in this mode of flight (Tredici, 1985). However, monocular depth cues (e.g., retinal size, motion parallax, interposition, and linear perspective) generally are acknowledged to be more important for routine flying. A 1989 study of visual acuity and stereopsis with night vision goggles (a second-generation binocular I^2 system) found stereopsis with this system to be greatly reduced (Wiley, 1989). Aviators using monocular pilotage systems can improve their nonstereo depth perception with training, although the degraded acuity inherent in these systems will adversely affect the perception of even monocular depth cues to some extent. This is reflected frequently in aviator surveys (Hale and Piccione, 1989; Crowley, 1991). However, with practice, most AH-64 aviators are able to fly competently throughout the night nap-of-the-earth (NOE) environment.

Currently, it is in vogue to suggest that the next generation of HMDs should deliver imagery to both eyes, instead of one as in the IHADSS. This can be accomplished in two ways: binocularly (each eye is presented with a distinct image, and two sensors or two sensor perspectives are used) or biocularly (both eyes are presented with identical images from the same sensor). Theoretically,

HUMAN FACTOR CONSIDERATIONS

advantages and disadvantages can be cited for both of these display modes. However, it should be noted that AH-64 aviators report using the eye left free by the IHADSS's monocular mode for a variety of functions, including reading instruments and maps within the cockpit, cross-checking range and size information derived from the PNVS, and maintaining color vision and dark adaptation in one eye (Hale and Piccione, 1989; Crowley, 1991; Hart, 1988). It would seem advisable to at least allow the aviator the option of selecting a monocular format should it be deemed necessary.

16.6.3 Field of View and Visual Fields

The human eye has an instantaneous field of view (FOV) that is roughly oval and typically measures 120° vertically by 150° horizontally. Considering both eyes together, the overall FOV covers approximately 120° vertically by 200° horizontally (Zuckerman, 1954). The size of the FOV provided by an imaging system is determined by trade-offs among various sensor and display parameters including size, weight, placement, and resolution.

In both NVGs and ANVIS, the FOV of a single image intensifier tube is a circular 40°. The tubes have a 100% overlap; hence the combined field is also a circular 40°. A pictorial representation of the 40° FOV for ANVIS and NVG is provided in Figure 16.21.

The circular 40° FOV seems small when compared to the overall FOV of the eyes, but the reduction is not so significant considering the multiple visual obstructions (e.g., armor, support structures, glareshield) normally present in military aircraft cockpits. Still, the aviator must use continuous head movements in a scanning pattern to help compensate for the limited I^2 NVD FOV.

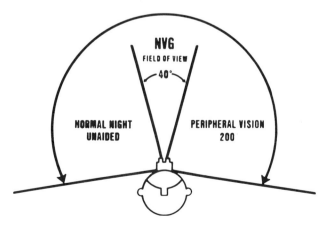

Figure 16.21 Instantaneous field of view for the human visual system and current I^2 helmet-mounted displays (HMDs).

The 40° FOV of the I² NVD is a theoretical value based on the user being able to place his or her eye into the exit pupil of the I² NVD optics. Variations in head anthropometry and the wearing of a protective mask or corrective lenses may prevent proper eye placement and result in a reduced FOV. Improper adjustment of the helmet attachment also can preclude the aviator from achieving a full 40° FOV. These small FOV losses are not always obvious to the aviator.

Current I² designs extend the visual field in addition to that provided through the I² NVD optics. Instruments and other objects within the crewstation can be viewed by virtue of the I² NVD's limited look-under, look-around capabilities. However, the eyes are adapted to the light emitted by the image intensifiers.

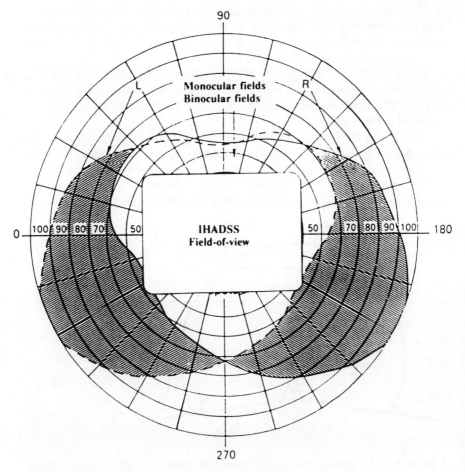

Figure 16.22 Pictorial representation of IHADSS 30° × 40° field of view.

HUMAN FACTOR CONSIDERATIONS

Supplemental lighting at a level equivalent to the output of the I^2 NVD is necessary for the aviator to see detailed scenes such as maps and dial legends.

The IHADSS FOV is rectangular, measuring 30° vertically by 40° horizontally (Fig. 16.22). Although the HDU does physically obstruct unaided lateral visibility to the right, the IHADSS provides an unimpeded external view throughout the range of PNVS movement ($\pm 90°$ azimuth and $+20°$ to $-45°$ elevation). However, the AH-64 pilot is trained to use continuous scanning head movements to compensate for the limited FOV. A potentially disorienting effect occurs when the pilot's head movements exceed the PNVS range of movement—the image suddenly stops, but head movement continues. This could be misinterpreted by the pilot as a sudden aircraft pitch or yaw in the direction opposite the head movement. If there are lights visible to the unaided eye (or through the combiner), diplopia (double vision) usually results.

The IHADSS is designed to present the sensor's FOV in such a manner that the image on the combiner occupies the same area in front of the eye, resulting in a one-to-one representation of the outside world (i.e., no magnification or reduction). However, to achieve this design goal, the pilot must place his eye within the exit pupil of the HDU optics. (The exit pupil of an optical system is a small volume in space where the user must place his eye in order to obtain the full available field of view.) The major determinant of whether this can be attained is the physical distance between the eye and the combiner. Variations in head and facial anthropometry greatly influence the ability of the aviator to comfortably achieve a full FOV. Some aviators report discomfort due to pressure against the zygomatic arch (cheekbone) (Rash and Martin, 1987a), and many report difficulty seeing all the provided symbology (Hale and Piccione, 1989). The interposition of chemical protective masks and/or spectacles (either for laser protection or for correction of refractive error) increases the eye–combiner distance, further reducing the likelihood that the pilot will see the full FLIR image or symbology display (Rash and Martin, 1987b; McLean and Rash, 1984). Improper adjustment of the HDU/helmet attachment bracket also can prevent the aviator from achieving the design FOV.

The effects of reduced FOV on aviator performance are not understood fully. The task of determining the minimum FOV required to fly is not a simple one. First, the minimal FOV required is highly task-dependent. Consider the different sensory cues used for high-speed flight across a desert floor (narrow FOV) versus a confined-area hovering turn (wide FOV). Second, the FOV required to maintain orientation depends on workload. A small attitude indicator bar (or cue), occupying only a few degrees of the visual field, does not provide much information to the peripheral retina, which normally mediates visual information regarding orientation in the environment (Gillingham and Wolfe, 1985). Acquiring this orientation information from the central (foveal) vision requires more concentration and renders the pilot susceptible to disorientation should his attention be diverted to other cockpit tasks for even a brief period. Third, with helmet-mounted displays such as the IHADSS, any reduction in FOV also may deprive the pilot of critical flight symbology.

To compensate for the FOV problems cited above, some AH-64 pilots resort to using the CRT horizontal and vertical size controls to reduce the overall size of the image (Hale and Piccione, 1989). This allows the aviator to view all of the imagery and symbology, but the sensor's 30° × 40° FOV now occupies a smaller area on the combiner than it needs to provide a one-to-one scene representation. As this reduced image can cause problems with distance and size perception, it is not recommended.

16.7 SYSTEM FACTORS

16.7.1 Temporal Characteristics

Discussions in the previous sections have addressed parameters that are related primarily to the *spatial* characteristics of night imaging systems. However, the temporal characteristics of the system also can affect performance, especially in a dynamic environment (Rash and Verona, 1987). Thermal imaging systems have time constants associated with the detector, the scanning mechanism, and the display. The important time constant with I^2 devices is related to the display phosphor. For thermal systems, the dynamic environment may introduce additional temporal factors such as sensor gimbal jitter, head motion in visually coupled systems, and relative target–sensor motion.

In thermal systems, where the sensor is composed of multiple detectors, an individual detector's time constant determines the detector's speed of response to temperature (energy) changes in a scene segment. In a static environment, in which the detector continuously images the same scene segment, the detector time constant's contribution to the temporal characteristics of the sensor is minimal. Rapid temperature changes are not routine events in the real world. However, in a dynamic environment or in a scanning imaging system, the detector is continually imaging different scene segments.

In visually coupled display systems, the interface between the pilot's head movements and the corresponding sensor movements is an additional potential source of problems. Any latency between the movement of the head and the movement of the sensor must be reduced to an imperceptible level. The PNVS gimbal with its 120°/sec maximum velocity is responsive and approaches the desired level of imperceptible latency (see Section 16.7.3). However, the communication of the helmet sight command signals to the PNVS gimbal generates a perceptible but acceptable latency. The TADS with its 60°/sec velocity is appreciably slower and, although acceptable as a pilotage backup and targeting/navigation system, is unacceptable as a primary pilotage system.

Rapid head movements in visually coupled systems generate a rapidly moving scene on the display. The head movement rates greatly exceed the nominal relative movement rates that are observed between an aircraft and a moving ground object. This results in more demanding time constants associated with both sensors and displays.

Phosphor persistence is often the most important display parameter that affects the temporal response of a phosphor-based display. Excessive persistence reduces modulation contrast and causes the reduction of gray scale in a dynamic environment in which there is relative motion between the target and the sensor (Rash and Becher, 1983). Persistence effects may cause the loss of one or more gray-scale steps. This may be of minor concern at low spatial frequencies where there are many gray-scale steps. However, where there is only enough modulation contrast to present only one or two gray-scale steps under the static condition, the loss of one gray-scale step at high spatial frequencies would be significant.

This effect is well demonstrated in the history of the phosphor selection for the IHADSS. A P1 phosphor (24 msec persistence) initially was selected to satisfy the high-luminance daytime symbology requirement. After initial flight tests, the CRT phosphor was changed to the shorter persistence P43 (1.2 msec) from the more efficient P1 because of the image smearing reported. The test pilots reported that tree branches seemed to disappear as pilots moved their heads in search of obstacles. It was determined that the long persistence of the P1 phosphor was responsible for the phenomenon. ANVIS use a P20 phosphor with a 350 μsec persistence, but are converting to P43 or P22 because of the environmental dangers of P20 production.

The current PNVS electro-optical multiplexer, a Department of Defense common module that is used to convert the mechanically scanned thermal detector outputs to a video signal, also introduces a significant delay time to the video image. Next-generation devices will most likely be replaced with solid-state multiplexers with improved delay times.

16.7.2 Sensor Location

Before night imaging systems were available, an aviator's primary visual sensors were his eyes. His experience in the perception and interpretation of visual input is referenced to the eyes' position in the head. This point of perception is left intact with the use of head-mounted I^2 devices. However, when flying the AH-64, the primary visual input for night and foul weather flight is the PNVS sensor. This sensor is located in a nose turret approximately 20 ft forward and 3 ft below the pilot's design eye position. This exocentric positioning of the sensor can introduce problems of apparent motion, parallax, and incorrect distance estimation (Brickner, 1989). However, this mode of sensor location does provide the advantage of unobstructed visual fields. The aviator's field of view is no longer affected by the physical obstructions of the aircraft frame. The PNVS sensor provides the aviator with the capability to look through the floor of the aircraft, a definite advantage when landing in an uncleared area, where the sensor can be used over the full field of regard. However, this field of regard is affected by the attitude of the aircraft.

The design for the next-generation military aviation imaging system calls for the integration of FLIR and I^2 sensors. Because of the weight and size characteristics of FLIR technology, the FLIR's position will remain exocentric. However, the I^2 sensor has two location options. It can be collocated with the FLIR sensor on the nose of the aircraft, or it can be helmet-mounted. If both sensors are exocentrically located, only the basic concerns of this mode of location, as listed above, require consideration. However, if the I^2 sensor is helmet-mounted, there may be problems associated with the mixed location modes and the resultant switching of visual reference points.

16.7.3 Head–System Interface

By virtue of their design, helmet-mounted displays are mounted totally, or in part, on the aviator's helmet. Current I^2 devices are totally (sensor and display) head-mounted. For these devices, the major interfacing problem is the requirement for a proper and stable fit. Slippage resulting in the inability to maintain the eyes in the system's exit pupils must be avoided. This slippage usually is induced by the center-of-gravity offset present in these designs.

For the IHADSS, only the system's display section is helmet-mounted. The sensor section is integrated with the helmet in that the direction of the sensor line of sight is controlled by head movement. To ensure optimal system performance, proper interfacing of the helmet and its attached display with the aviator's head is critical. The criteria for proper interfacing include the static placement and stability of the eye into the exit pupil of the HDU optics and the dynamic transformation of aviator head movement to sensor movement.

The AH-64 aviator receives primary sensory data through the HDU to fly the aircraft. To receive the total imagery available on the HDU, the aviator must adjust the helmet and HDU to match the position of the exit pupil of the HDU optics to his right eye. In addition, the helmet must remain stable, maintaining this exit pupil position in the presence of head movements and aircraft vibration. The helmet has become a critical piece of equipment, requiring special consideration and care. One of these special considerations is the fitting process.

The basic fitting process involves numerous steps, including, but not limited to, adjustments to the suspension system, proper location and alignment of the HDU, and final trimming of the helmet visor to accommodate the HDU when in the operating position. The objectives of the fitting procedure are to (1) obtain a comfortable, stable fit of the IHU (helmet) that will enable the aviator to achieve the maximum field of view provided by the HDU and (2) achieve boresight, which permits accurate engagement of weapons systems (Honeywell, 1985).

Rash et al. (1987) evaluated the U.S. Army fitting program for the IHADSS. Several important lessons were learned during this evaluation. For the first time, the impact that head anthropometry has on helmet fit was recognized. Not only are there problems associated with one or more extreme head dimensions, but there are additional problems related to head irregularities such as one ear lower than the other, tapering forehead, and bulges. All of these variations increase the detailed attention required to provide the aviator with a comfortable and stable helmet fit.

Aviator facial anatomy also is crucial to optimal HDU interface. If the aviator's eye is not located in the exit pupil along the optical axis but is some distance behind it, a "knothole effect" is experienced. The field of view provided is decreased in a manner similar to that experienced when a person looking through a knothole begins to move away from the knothole. A protruding cheekbone or deeply sunken eyes can prevent the HDU from being positioned close enough to obtain the full field of view. Even a small displacement can substantially reduce the available field of view. If, due to anthropometric and facial anatomic irregularities, the aviator is unable to achieve full field of view, he may attempt to position the HDU to select what he considers to be the critical portion of the imagery and/or symbology for the task at hand. A good indication of poor or difficult fit is the extension of the combiner. A good fit is indicated by a combiner extension distance of a ¼ in. or less. As previously discussed (Section 16.6.3), aviators also may resort to adjusting the size of the CRT image in order to view all of the provided symbology.

Helmet-mounted imaging systems such as the PNVS/IHADSS use the aviator's head as a control device. Head position is employed to produce drive signals that slave the sensor's gimballed platform to aviator head movements. As described in Section 16.3.4, infrared detectors mounted on the helmet continuously monitor the head position of the aviator. Processing electronics of the IHADSS convert this information into drive signals for the PNVS gimbal. This control system is a closed-loop servosystem that uses the natural visual and motor skills of the aviator to remotely control the sensor and/or weapon.

As previously mentioned, one of the most important operating parameters of visually coupled systems is the sensor's maximum slew rate. The inability of the sensor to slew at velocities equal to those exhibited in the aviator's head movements would result in (1) significant errors between where the aviator thinks he is looking and where the sensor is actually looking and (2) time lags between the head and sensor lines of sight. Medical studies of head movements have shown that normal adults can rotate their heads $\pm 90°$ in azimuth (with neck participation) and $-10°$ to $+25°$ in elevation without neck participation. These same studies showed that peak head velocity is a function of movement displacement; that is, the greater the displacement, the greater the peak velocity, with an upper limit of 352°/sec (Allen and Webb, 1983; Zangemeister and Stark, 1981). However, these studies were laboratory-based and do not reflect the velocities and accelerations indicative of a helmeted head in military flight scenarios. In support of the AH-64 PNVS development, Verona et al. (1986) investigated single pilot head movements in a U.S. Army JUH-1M utility helicopter. In this study, head position data were collected during a simulated mission in which four JUH-1M pilot subjects fitted with a prototype IHADSS were tasked with searching for a threat aircraft while flying a contour flight course (50–150 ft above ground level). The acquired head position data were used to construct frequency histograms of azimuth and elevation head velocities. Although velocities as high as 160 and 200°/sec in elevation and azimuth, respectively, were measured, approximately 97% of the velocities were found to fall between a range of 0–120°/sec. This conclusion supported the PNVS design specification maximum slew rate of 120°/sec. It also lends validity to aviator complaints that the TADS sensor (with a maximum slew rate of 60°/sec) is too slow to be used for pilotage.

16.8 EXTERNAL FACTORS

16.8.1 Environmental Conditions

Environmental conditions can affect greatly the performance of night imaging systems. Since I^2 and thermal imaging systems use divergent technologies and have different spectral responses, the effects of the environment on each are different.

For I^2 NVDs, environmental factors affecting performance include level of illumination, weather, cloud cover, and obscurants. Some environmental conditions can enhance and others can degrade I^2 NVD performance. As stated earlier, I^2 NVDs are amplification devices. A minimum amount of energy is necessary for acceptable performance. Major sources of this energy are the moon, stars, and artificial lighting.

Energy from natural sources such as the moon and stars and from many artificial sources such as nearby towns or tactical flares reflects from objects. The composition, surface characteristics, and environmental conditions determine the amount and spectral distribution of the reflected energy that reaches the I^2 NVD. The I^2 NVD's objective lens characteristics determine the amount and spectral distribution of the energy that is transmitted to the photocathode (light-sensitive portion of the image intensifier tube) of the I^2 NVD. The chemical composition, age, and voltage of the photocathode determine the characteristics of the reflected energy that finally is presented to the viewer.

The moon and stars emit a significant amount of energy that cannot be seen by the human eye but can be seen by the I^2 NVDs. The stars are a uniform omnidirectional source of energy. The moon, like the sun, is a point source of

HUMAN FACTOR CONSIDERATIONS 699

energy that casts shadows as it moves across the sky from moonrise to moonset. The percent of moon illumination, rise and set times, and the maximum moon angle above the horizon are data available from the weather service for a specific geographical area.

In the aviation environment, the direction of flight with respect to the location of the moon, sky glow, and ground lighting is critical to I^2 NVD performance. Flying toward light sources will reduce the effectiveness of the I^2 NVD, similar to flying toward the sun during daylight. Flying away from the light sources may enhance the I^2 image. Shadows cast by terrain features and manmade objects from the external lighting are important considerations for aviators. Most wires are not detectable through I^2 NVDs, but often their support poles and towers provide cues to the existence and path of the wires. The configuration of the external lighting with respect to the aircraft's direction of flight can enhance or mask the shadow cues.

Atmospheric conditions also can have a significant effect on I^2 performance. The energy reflected from objects is scattered by water vapor and particulate matter suspended in the air. Optimal performance is obtained on a dry, clear night. The presence of precipitation (e.g., rain and snow) and obscurants (e.g., fog, dust, and smoke) degrade the performance to varying degrees. In general, atmospheric conditions that degrade unaided visual performance will also degrade the performance of I^2 NVDs.

In the high-speed environment of aviation, users with minimal I^2 experience often overextend themselves by failing to recognize the gradual degradation in their night vision device's performance due to deteriorating atmospheric conditions. One feature of I^2 NVDs is that they maintain a constant average display luminance over a wide range of input energy. Therefore, it may not be obvious to the aviator that conditions are deteriorating. The experienced I^2 NVD aviator has learned that deteriorating environmental conditions are indicated by an increase in image noise. High visual noise levels indicate that the device is operating at its performance limits.

There are also potential hazards associated with flying in clouds. Thin, wispy clouds visible to the naked eye may not be seen when using an I^2 NVD. The presence of cloud cover can increase or decrease the amount and distribution of predicted energy available to I^2 NVDs. Cloud cover may attenuate the energy from natural sources but enhance the amount of energy available from artificial sources, such as sky glow from towns, shopping centers, and highway lighting.

Modern artificial lights—for example, high-efficiency sodium and mercury vapor lamps—by design, produce a large percentage of their energy in the visual portion of the spectrum. Second-generation image intensifiers are sensitive to all of the visual spectrum and can see all of the artificial light, but the third-generation image intensifiers see only a part of the visible spectrum and hence only a part of the artificial light. Therefore, energy generated by artificial sources may be more beneficial to second-generation I^2 NVDs than to third-generation devices.

Military use of I^2 devices usually is strictly controlled for nonoperational (e.g., training) flights. Current Army policy restricts I^2 NVD flights to periods of natural illumination that meet or exceed the lunar conditions of 23% fractional illumination and 30° altitude above the horizon. These conditions can be waived when the aircraft is equipped with an artificial infrared search or landing light.

When selecting the correct lamp wattage and dispersion pattern to be used with an infrared filter in the search or landing light with the I^2 NVD, it is critical to consider the environment in which the aircraft will be flown. Some lights generate large-area footprints of low intensity, and others generate small footprints of light with high intensity. In either case, there is a tendency to become channelized by the infrared light, similar to driving at night with headlights. The aviator becomes less aware of the periphery and significantly limits the peripheral visual scans.

Thermal imaging systems, such as the PNVS, are capable of providing acceptable pilotage imagery in a wide range of environments (deserts, swamps, mountain areas, etc.) and weather conditions (fog, snow, etc.). However, their effectiveness can be limited by the operating environment and prevailing

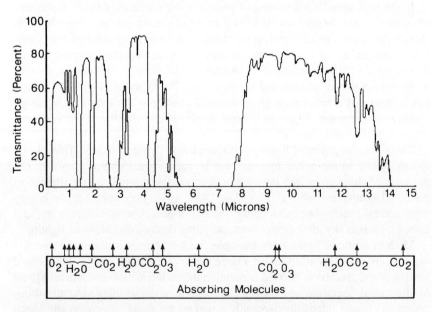

Figure 16.23 Spectral transmittance of the Earth's atmosphere.

weather. The atmosphere absorbs, emits, and scatters IR radiation and often is the major driver in system performance. The choice of detectors, with respect to their spectral response, is governed to a degree by the transmittance of the atmosphere. A plot of the transmittance of the earth's atmosphere is depicted in Figure 16.23. This figure demonstrates the effect of three major IR radiation absorbers: water vapor (at 2.7, 3.2, 6.3, and 11.9 μm), ozone (at 4.8, 9.6, and 14.2 μm), and carbon dioxide (at 2.7, 4.3, 12.6, and 15.0 μm). It can be seen in this plot that certain open "windows" of transmittance exist. Part of one window includes the visible spectrum (0.4–0.7 μm) used by the human visual system. The effectiveness of the 3–5 and 8–12 μm response ranges of the HgCdTe detector becomes obvious when the remaining windows are studied.

In addition to absorption of IR radiation by the atmosphere, there is scattering by the various atmospheric molecules. The scattering of the IR radiation further attenuates the IR signal and contributes to the IR background noise. The overall attenuation, a summation of scattering and absorption, is expressed by the extinction coefficient. These coefficients can be used to compare how the atmosphere will transmit IR radiation for various atmospheric conditions. In Table 16.1, transmissivities for some of these conditions (calculated by the Beer–Lambert law) are presented for ranges of 1 and 10 km. Most of the conditions in Table 16.1 are related to moisture. However, additional elements such as dust and smoke (obscurants) affect the composition and density of the atmosphere and therefore the IR radiation transmission (Pratt and Reid, 1985).

As discussed above, the atmospheric transmission as affected by the environmental conditions attenuates the IR signal. In addition to this effect, these conditions reduce the solar heating of objects (targets), thereby reducing their thermal signatures.

Table 16.1 Atmospheric Transmissivity

Condition	Extinction coefficient	Transmissivity (%)	
		At 1 km	At 10 km
Very clear and dry	0.05	95	61
Haze	0.11	90	33
Light snow	0.51	60	<1
Moderate rain	0.69	50	<1
Heavy rain	1.39	25	<1
Light fog	1.90	15	<1
Heavy fog	9.20	<1	<1
Heavy snow	9.20	<1	<1

Note: Extinction coefficients are expressed in km^{-1} for the 8–12 μm spectral region.

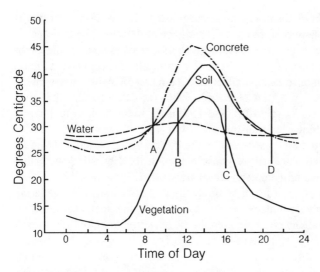

Figure 16.24 Representative 24-hr thermal history curves for soil, water, vegetation, and concrete.

The role of the sun and existing environmental conditions are responsible for one phenomenon, referred to as *thermal crossover*, that is unique to thermal imaging. Figure 16.24 shows representative 24 hr thermal history curves for four materials: soil, water, vegetation, and concrete (Pratt and Reid, 1985). Just after midnight, all are emitting more energy than they are absorbing (due to the absence of the sun). Each material's temperature is different (the vegetation has the lowest temperature, and the water has the highest), and all temperatures are still slowly decreasing. As sunrise (assumed to be 0600) approaches and occurs, the temperature of each material begins to rise, at a different rate for each substance. By approximately 0900, the temperatures of three of the substances (soil, water, and concrete) reach nearly the same value (point A) and the thermal sensor may be unable to discriminate among them. Point A is a crossover point for these three materials. As the day proceeds, the materials continue to increase in temperature. At point B, the water and the vegetation reach a crossover point. At these crossover points, the relative order of the temperature values reverses. For example, prior to point B, the temperature of the water is higher than that of the vegetation. Following point B, the temperature of the vegetation is higher than that of the water. On the display of the imaging system, where the different temperatures (actually energy levels) are represented by different levels of brightness, the materials undergoing the crossover *reverse* contrast. Where the water may have been "brighter than" the vegetation, it is now "darker than" the vegetation.

HUMAN FACTOR CONSIDERATIONS

Over a 24 hr period, crossover for the representative materials and thermal histories shown occurs twice (points A and D, B and C). However, there is no crossover point for the vegetation with either the soil or the concrete. In the real world, the presence and frequency of crossover points for any two substances are dependent on geographic location, season, weather, and many other factors. In the 1988 AH-64 aviator survey, 98% of the respondents reported instances where the FLIR image was degraded to the extent that mission completion was compromised (Hale and Piccione, 1989). Most often, this was a result of IR crossover.

16.8.2 Internal and External Light Sources

The use of either I^2 or thermal night imaging systems requires the presence of external signal energy. In general, the higher the level of the signal energy, the better the system performance. However, *intense* sources of energy within the spectral response region of the sensor most often degrade performance. In addition, energy sources in proximity to the user may degrade total system performance (as with totally head-mounted I^2 NVDs) or display performance (as with helmet-mounted displays). Equally important is the related issue of how certain external sources are imaged by the systems.

Internal lighting conditions are of most importance when I^2 devices are being used. Once the user is assured that the I^2 NVD is operating properly and the environmental conditions will permit I^2 NVD operations, the next major concern is the immediate lighting environment in which the I^2 NVD will be used. Nowhere is this more important than in aviation use. Proper internal crewstation lighting is critical to the overall effectiveness of the I^2 NVD. Because I^2 NVDs based on image intensification operate on the principle of light amplification, bright lights emitting energy in that portion of the electromagnetic spectrum in which these devices are sensitive produce severe veiling glare that can obscure the overall image and degrade the performance of the I^2 NVD (Keane, 1984). To prevent this, and to protect the image intensifier assembly from permanent phosphor burns, both NVGs and ANVIS are equipped with a bright source protection (BSP) circuit, which operates as an automatic gain control. The BSP circuit decreases the sensitivity of the image intensifier tubes when they are exposed to bright lights emitting energy in the ANVIS/NVG-sensitive portion of the electromagnetic spectrum. The net result is a reduced response to energy originating from the external scene, effectively reducing the aviator's capability to view outside the cockpit.

Numerous methods—the use of low-reflectance black paint, light louvers, filters, low-intensity lamps, etc.—have been attempted to provide "compatible" cockpits (Holly, 1980). However, only limited success has been achieved using any of these methods. Since second-generation intensifiers "see" the same energy as the human eye, it is difficult to produce light levels adequate to view instruments with the naked eye without severely degrading the night vision device, particularly at low exterior illumination levels. A design solution to this problem is incorporated in the ANVIS. The interior surfaces of the objective lenses in the ANVIS have been coated with a dielectric film (a "minus-blue filter") that would reject wavelengths shorter than 625 nm. This design was planned to make the ANVIS compatible with cockpits having blue-green crewstation lighting. However, programs to convert current aircraft to blue-green lighting have not been fully implemented. Therefore, some aircraft are currently being flown without compatible lighting as specified in MIL-L-85762 (Department of Defense, 1986). Incorporating this type of filter in second-generation devices would degrade their performance to an unacceptable level. The third-generation devices are robust enough to accept the filter without significant performance degradation.

Flares, rocket motors, strobe lights, lightning, and other bright light sources may temporarily affect the performance of the I^2 NVD. The interruption may last only a few seconds, but if unexpected, the interruption of the image presentation may be quite disconcerting. A bright flash will cause the image intensifier to saturate, evidenced by a temporary, but complete, loss of detail in the intensified image. Because saturation limits the light passed to the eye, recovery from a flash through an image intensifier is much more rapid than recovery from a flash to the unaided dark-adapted eye.

Because the output image of the I^2 NVD is presented to the viewer on a green phosphor screen, all color information from the surrounding environment is converted to shades of green analogous to the imagery in black-and-white television. This is particularly important when considering color-coded information from external light sources, such as aircraft navigation and tower lights. The lights on an approaching and departing aircraft look the same. Color-coded taxiway and runway lights also appear the same through an I^2 NVD.

Intensity and distance cues are easily confused at night, particularly with I^2 NVDs. A light source can appear the same to an aviator whether it is small and close or large and far away. Lights from other aircraft can blend easily into ground lights. Midair collisions may have resulted because aircrews were not able to differentiate lighted aircraft in their flight path from background lights.

Anticollision strobe lights improve the conspicuousness of aircraft, but not without trade-offs. Often, the high-intensity xenon strobe lights interfere with the I^2 NVD and distract the aviators because their light reflects off the rotor blades and nearby objects in terrain flight. The front 180° of the anticollision light is covered with tape to prevent the unwanted reflections. During formation flight, only the trailing aircraft can use an anticollision light because other aircraft's strobe lights would be in the view of aviators in the rear of the formation. The upper position lights have their lower hemisphere taped, and the lower position lights have their upper hemisphere taped to reduce unwanted external light from entering the crewstation. Some newer aircraft use small near-infrared lights, visible only with I^2 NVDs, to covertly mark their position without interfering with the I^2 NVD.

Light from internal and external sources is scattered by dirt, grease, moisture, and abrasions on the windscreen and reduces the contrast of the intensified scene. Clean, dry, and unabraded windscreens are essential to optimum I^2 NVD performance. The night driving analogy is again appropriate. Oncoming headlights through a dirty, wet, pitted windshield cause the same visual degradation.

Unlike systems that use the principle of image intensification, thermal imaging systems are less sensitive to degrading effects of internal crewstation and environmental energy sources (Verona and Rash, 1989; Rash and Verona, 1989). However, as discussed above, the performance of thermal systems can be degraded by some sources, most of them natural. Performance degradation also can be induced by manmade sources, including specially designed countermeasures that target thermal imagers. Whereas the effects of internal and external energy sources are associated primarily with the sensor, the display is where these effects are perceived by the aviator.

Physically large areas of very hot or very cold temperatures that are much hotter or colder than the area of interest and are in the FOV of the sensor tend to confuse the automatic level and gain controls. In trying to cope with the large temperature range, these controls may mask the area of interest. This may happen naturally with the cold sky region above the horizon. It is more difficult to introduce very cold manmade regions, but it is relatively easy to artificially create large, very hot regions by lighting large-area fires. A similar problem may be encountered when viewing a landing zone at a close distance with a thermal sensor when several aircraft have landed previously and are heating the landing zone with their exhausts. Much of the detail in the scene may be masked by the hot exhaust.

In the air, formation flying can be troublesome to a thermal imager when the hot exhaust of a leading aircraft is viewed against a cold sky background. This effect may be even more troublesome when there are several large, hot exhausts against a cold sky background. The forward aircraft detail may be seriously degraded—enough to mask unexpected or rapid changes in attitude or the initiation of a maneuver by a lead aircraft.

A special group of external sources that are a concern to thermal imaging systems is lasers. Any laser that operates within the spectral response region of the thermal sensor (8–12 μm) potentially can blind or destroy an unprotected sensor. Carbon dioxide (CO_2) lasers operating at 10.6 μm fall within this region. The neodymium-YAG laser used in the AH-64 laser rangefinder operates at 1.06 μm and cannot be seen by the PNVS sensor. Lasers can be aimed at the sensor or can be used as a covert pointer or designator when aimed at a target. The night vision device can "see" laser radiation (in the IR) the naked eye is unable to see. Neodymium-YAG lasers are particularly dangerous to the human observer because the laser energy at this wavelength can pass easily through the eye's cornea display permit the bright light from the flare to degrade its contrast to the eye. These effects are more apparent when the flare is in the field of view of the sensor. Looking away from the flares will reduce the flare's effects.

The display can be degraded by internal and external visible light sources, particularly high-intensity sources in the crewmember's area of interest. Even though the thermal sensor may not "see" the light source, if the light is in the crewmember's line of sight, it will degrade the contrast of the display imagery.

16.9 PHYSICAL AFTEREFFECTS

With any new technology or device, there is a concern with short-term and long-term health hazards (aftereffects) associated with use. For helmet-mounted devices, these concerns can be grouped into those related to visual effects and those related to the muscular-skeletal system.

Two studies, by Glick and Moser (1974) and Moffitt et al. (1986), looked at complaints of afterimages by AN/PVS-5 night vision goggle users. The 1974 study investigated reports of transient (1–2 min) "brown-and-white" color vision following the use of ANVIS and AN/PVS-5 NVG. Pre- and post-wear color perception tests showed no permanent change in color vision. The short-term effect on color vision was concluded to be a normal physiological phenomenon, afterimages resulting from visual adaptation to a relatively intense monochromatic (green) light source, the phosphor display. The 1986 study similarly concluded that the green phosphor in the AN/PVS-5 NVG is capable of generating chromatic afterimages that can temporarily affect discrimination of white and green light sources.

A comprehensive visual assessment of AH-64 aviators was conducted in 1989 (Behar et al., 1990) in response to concerns over long-term effects of using the IHADSS helmet-mounted display. The study performed an epidemiological survey of reported visual problems among AH-64 aviators, conducted a clinical and laboratory evaluation of the refractive and visual status of a sample group of these aviators, and investigated the occurrence of improper focusing of the helmet display unit.

The survey found that over 80% of the aviators registered at least one visual complaint (visual discomfort, headache, double vision, blurred vision, disorientation, or afterimages) associated with use of the IHADSS. Many comments provided within the survey indicated that symptoms occurred during or following long flights and/or flying with poor quality or improperly focused displays. The battery of vision tests (visual acuity, contrast sensitivity, color vision, depth perception, binocular rivalry, and clinical tests of refraction, accommodative function, and oculomotor status) failed to relate to any visual complaint index. Any differences in performance or status between the left (unaided) or right (display) eye were small in all cases. It was concluded that the likely source of the reported visual discomfort and headache was the positive accommodation induced by the aviators' failure to achieve proper focusing of the display.

Less information is available on muscular-skeletal aftereffects. Some studies (Anderson, 1988; Schall, 1989) have pointed to the high potential of nonejection cervical spine injuries due to $+G_z$ forces in high-performance aircraft with standard aviation helmets. It is a logical assumption that the increased head-supported weight of HMD systems would exacerbate this potential. Anecdotal data from aviator debriefings often cite incidences of stiff or sore necks following flights with AN/PVS-5 NVDs that are not properly counterweighted.

16.10 SUMMARY

Night imaging systems provide the ability to operate effectively at night and in adverse weather. Although these devices do not change night into day, they do significantly enhance night operations compared to the unaided eye. The quality of the image presented to the user is impacted by the operating characteristics of the system's optics, detector, and display. With night vision devices, compromises must be accepted in many visual parameters such as acuity, field of view, and depth perception, and the effects of these compromises on performance must be understood. In addition, among the most significant factors influencing image quality, and hence user performance, are the energy levels associated with the scene targets and background and the environmental effects on these levels.

Acknowledgments

We wish to thank James H. Brindle, John S. Martin, William E. McLean, Roger W. Wiley, and Robert J. Whitcraft for their guidance and assistance in the preparation of this work.

The views, opinions, and/or findings contained in this paper are those of the authors and should not be construed as an official Department of the Army position unless so designated by other official documentation.

References

Allen, J. H., and R. C. Webb (1983). Helmet mounted display feasibility model, Naval Training Equipment Center, NAVTRAEQUIPCEN IH-338, Orlando, FL.

Anderson, H. (1988). Neck injury sustained during exposure to high *G*-forces in the F-16B, *Aviat., Space, Environ. Med. 59*: 356–358.

Beaton, A. (1985). *Left Side, Right Side: A Review of Laterality Research*, Yale Univ. Press, New Haven.

Behar, I. and C. E. Rash (1990). Diopter focus adjustment of Apache IHADSS, *U.S. Army Aviat. Dig.*, January: 14–15.

Behar, I., R. W. Wiley, R. R. Levine, C. E. Rash, D. J. Walsh, and R. L. S. Cornum (1990). Visual survey of Apache aviators (VISAA), U.S. Army Aeromed. Res. Lab., USAARL Rep. 90-15, Fort Rucker, AL.

Bennett, C. T., and S. G. Hart (1987). PNVS-related problems: pilots reflections on visually coupled systems, 1976-1987, NASA-Ames Research Center, Moffett Field, CA.

Bishop, P. O. (1981). Binocular vision, in *Adler's Physiology of the Eye*, 7th ed. (R. A. Moses, Ed.), C. V. Mosby, St. Louis, pp. 575-648.

Brickner, M. S. (1989). Helicopter fights with night vision goggles—human factors aspects, NASA Tech. Memo. 101039, Moffett Field, CA.

Butler, B. P., R. E. Maday, and D. M. Blanchard (1987). Effects of helmet weight and center-of-gravity parameters in a vibration environment, U.S. Army Aeromed. Res. Lab., USAARL LR-87-12-4-5, Fort Rucker, AL.

Crowley, J. S. (1989). Cerebral laterality and handedness in aviators: performance and selection implications, USAFSAM-TP-88-11, School of Aerospace Medicine, USAF Systems Command, Brooks AFB, TX.

Crowley, J. S. (1991). Night vision devices and in-flight illusions: aviator experience, USAARL Rep. 91-15, U.S. Army Aeromed. Res. Lab., Fort Rucker, AL.

Davon, H. (1976). *The Physiology of the Eye*, Academic, New York, p. 254.

Decker, W. M. (1989). Predicting the performance of night vision devices using a simple contrast model, *Helmet-Mounted Displays*, Proc. SPIE 1116, pp. 162-169.

Department of Defense (1986). Military specification: Lighting, aircraft, interior, night vision imaging system compatible, MIL-L-85762, Dept. of Defense, Washington, DC.

Gauthier, G. M., B. J. Martin, and L. W. Stark (1986). Adapted head- and eye-movement responses to added head inertia, *Aviat., Space, Environ. Med. 57*: 336-342.

Gillingham, K. K., and J. W. Wolfe (1985). Spatial orientation in flight, in *Fundamentals of Aerospace Medicine* (R. L. Dehart, Ed.), Lea & Febiger, Philadelphia, pp. 299-381.

Glaister, D. H. (1988). Head injury and protection, in *Aviation Medicine*, 2nd ed. (J. Ernsting and P. King, Eds.), London, pp. 174-184.

Glick, D. D., and C. E. Moser (1974). Afterimages associated with using the AN/PVS-5 night vision goggle, U.S. Army Aeromed. Res. Lab., USAARL LR 75-1-7-1, Fort Rucker, AL.

Greene, D. A. (1988). Night vision pilotage system field-of-view (FOV)/resolution tradeoff study flight experiment report, U.S. Army Night Vision Lab., NV 1-26, Fort Belvoir, VA.

Hale, S., and D. Piccione (1989). Pilot assessment of AH-64 helmet mounted display system, Essex Corporation, Alexandria, VA.

Hart, S. G. (1988). Helicopter human factors, in *Human Factors in Aviation* (E. L. Weiner and D. C. Nagel, Eds.), Academic, San Diego, pp. 591-638.

Holly, F. F. (1980). A night vision goggle compatible lighting system for Army aircraft, U.S. Army Aeromed. Res. Lab., USAARL LR 80-4-2-2, Fort Rucker, AL.

Honeywell, Inc. (1985). Integrated Helmet and Display Sighting System (IHADSS) helmet fitting procedures, Rep. 46220-1, St. Louis Park, MN.

Iavecchia, J. H., H. P. Iavecchia, and S. N. Roscoe (1988). Eye accommodation to head-up virtual images, *Hum. Factors 30* (6): 689–702.
Jones, V. P. (1983). Night vision goggle counterbalance system, *U.S. Army Aviat. Dig.*, May: 12–14.
Keane, J. (1984). Forward looking infrared and night vision goggles, a pilot's perspective related to JVX, Naval Air Development Center, Warminster, PA.
Levi, L. (1968). Luminescence and applications, in *Applied Optics*, Vol. 1 (S. S. Ballard, Ed.), Wiley, New York, p. 266.
Levi, L. (1980). Photoelectric and thermal detectors, in *Applied Optics*, Vol. 2 (S. S. Ballard, Ed.), Wiley, New York, p. 550.
McLean, W. E. (1982). Modified faceplate for the AN/PVS-5 night vision goggles, U.S. Army Aeromed. Res. Lab., USAARL Rep. 83-1, Fort Rucker, AL.
McLean, W. E., and C. E. Rash (1984). The effect of modified spectacles on field-of-view of the helmet display unit of the Integrated Helmet and Display Sighting System, U.S. Army Aeromed. Res. Lab., USAARL Rep. 84-12, Fort Rucker, AL.
Moffitt, K., J. Cicinelli, and S. P. Rogers (1986). Chromatic aftereffects caused by night vision goggles, Tech. Rep. 625-9, U.S. Army Avionics Res. and Development Activity, Fort Monmouth, NJ.
Phillips, C. A., and J. S. Petrofsky (1983). Neck muscle loading and fatigue: systematic variation of headgear weight and center-of-gravity, *Aviat., Space, and Environ. Med. 54* (10): 901–905.
Phillips, C. A., and J. S. Petrofsky (1984). Cardiovascular responses to isometric neck muscle contractions: results after dynamic exercises with various headgear loading configurations, *Aviat., Space, Environ. Med. 55*: 740–745.
Pratt, J. B., and R. Reid (1985), *AH-64 AQC PNVS Handbook*, Directorate of Training and Doctrine, Fort Rucker, AL.
Rash, C. E., and J. Becher (1983). Preliminary model of dynamic information transfer in cathode-ray-tube displays, *Conf. Proc. IEEE Southeastcon*, pp. 166–168.
Rash, C. E., and J. S. Martin (1987a). A limited user evaluation of the Integrated Helmet and Display Sighting System, U.S. Army Aeromed. Res. Lab., USAARL Rep. 87-10, Fort Rucker, AL.
Rash, C. E., and J. S. Martin (1987b). Effects of the M-43 chemical protective mask on the field-of-view of the helmet display unit of the Integrated Helmet Display and Sighting System, U.S. Army Aeromed. Res. Lab., USAARL LR-87-10-2-5, Fort Rucker, AL.
Rash, C. E., and R. W. Verona (1987). Temporal aspects of electro-optical imaging systems, *Imaging Sensors and Displays*, Proc. SPIE 765, pp. 22–25.
Rash, C. E., and R. W. Verona (1989). Cockpit lighting compatibility with image intensification imaging systems: issues and answers, in *Helmet-Mounted Displays* (J. T. Carollo, Ed.), Proc. SPIE 1116, pp. 170–174.
Rash, C. E., J. S. Martin, D. W. Gower, Jr., J. R. Licina, and J. V. Barson (1987). Evaluation of the U.S. Army fitting program for the Integrated Helmet and Display Sighting System, U.S. Army Aeromed. Res. Lab., USAARL Rep. 87-8, Fort Rucker, AL.
Schall, D. G. (1989). Non-ejection cervical spine injuries due to $+G_z$ in high performance aircraft, *Aviat., Space, Environ. Med. 60*: 445–456.
Task, H. L. (1979). An evaluation and comparison of several measures of image quality

for television displays, Aerospace Med. Res. Lab., AMRL-TR-79-7, Wright-Patterson Air Force Base, OH.
Task, H. L., and R. W. Verona (1976). A new measure of television display quality relatable to observer performance, Aerospace Med. Res. Lab., AMRL-TR-76-73, Wright-Patterson Air Force Base, OH.
Tredici, T. J. (1985). Ophthalmology in aerospace medicine, in *Fundamentals of Aerospace Medicine* (R. L. Dehart, Ed.), Lea & Febiger, Philadelphia, pp. 465–510.
U.S. Army Human Engineering Laboratory (1968). Aircraft crewman helmet weight study: progress report No. 1, U.S. Army Human Eng. Lab., Aberdeen Proving Ground, MD.
Verona, R. W. (1985), Image intensifiers: past and present, AGARD-CP-379, Advisory Group for Aerospace Res. and Development, Neuilly-sur-Seine, France.
Verona, R. W., and C. E. Rash (1989). Human factors and safety considerations of night vision systems flight, in *Display System Optics II* (H. M. Assenheim, Ed.), Proc. SPIE 1117, pp. 2–12.
Verona, R. W., J. C. Johnson, and H. Jones (1979). Head aiming/tracking accuracy in a helicopter environment, U.S. Army Aeromed. Res. Lab., USAARL Rep. 79-9, Fort Rucker, AL.
Verona, R. W., C. E. Rash, W. E. Holt, and J. K. Crosley (1986). Head movements during contour flight, U.S. Army Aeromed. Res. Lab., USAARL Rep. 87-1, Fort Rucker, AL.
Walker, D. J., R. W. Verona, and J. H. Brindle (1980). Helmet-mounted display system for attack helicopters, *Tech. and App.*, Vol. 2, No. 3, pp. 129–130.
Wiley, R. W. (1989). Visual acuity and stereopsis with night vision goggles, U.S. Army Aeromed. Res. Lab., USAARL Rep. 89–9, Fort Rucker, AL.
Wiley, R. W., D. D. Glick, C. T. Bucha, and C. K. Park (1976). Depth perception with the AN/PVS-5 night vision goggle, U.S. Army Aeromed. Res. Lab., USAARL Rep. 76–25, Fort Rucker, AL.
Zangemeister, W. H., and L. Stark (1981). Active head rotations and eye-head coordination, *Ann. NY Acad. Sci.* *374*: 541–559.
Zuckerman, J. (1954). *Perimetry*, Lippincott, Philadelphia, p. 237.

17
Color Control in Digital Displays

Celeste McCollough Howard

*University of Dayton Research Institute,
Aircrew Training Research Division, Armstrong Laboratory,
Williams Air Force Base, Arizona*

17.1 INTRODUCTION

Information concerning color control in digital displays has appeared in publications addressed to audiences in widely varying specialties, including color science, optics, visual physiology, psychophysics, color perception, display engineering, computer graphics, electronic image processing, and machine vision. This interdisciplinary interest is not limited to display color; Watson (1990) has summarized digital video applications of concepts developed to describe the perception of space, form, and motion as well as color. This chapter surveys the literature pertaining to display color, in order to identify those concepts and techniques that have achieved general recognition across the several user communities. The chapter focuses on control of color in displays and in full-color reproductions of displays; Chapter 5 in this volume describes techniques for reproducing color displays in black and white.

Color science provides a general foundation for the developments reviewed here. The most complete handbook on color science is Wyszecki and Stiles (1982). Briefer treatments can be found in Hunt (1987), Sproson (1983), Foley et al. (1990, Chapter 13), Hall (1988, Chapter 3), Meyer and Greenberg (1987), Gershon (1985), and Bumbaca and Smith (1987). Color science has achieved its greatest success in the area of *color specification*, the objective description of color based upon psychophysical data from color-matching and color discrimination experiments. Recent articles on color control in digital displays show an increasing trend toward use of the descriptive systems recommended by the International Commission on Illumination (CIE).

However, a great deal of work remains to be done in the area of *color appearance*. Color reproduction technology has not yet advanced to the point of using *perceptual* rather than *colorimetric* standards in judging the quality of images, although the desirability of perceptual standards has been recognized (Stone et al., 1986; Laihanen, 1989). Wyszecki (1986) has carefully defined the terms that play a role in the study of color appearance; these terms include hue, lightness, brightness, chroma, saturation, and colorfulness. His discussions of color adaptation, contrast, object-color constancy, and uniform color scales provide a good basis for understanding current research in this active area. No method for predicting all color appearance variables has yet developed far enough to achieve recommendation by the CIE.

Device-independent color rendering has become generally recognized as a necessity for computer-based color printing of pictures or graphic designs. It is not easy to achieve correspondence between the "soft proof" on a monitor and the eventual "hard copy" from a printer. The CIE systems, all of which are based on *XYZ* tristimulus values, provide a common color description language that is independent of the particular display device. Use of a device-independent color space can help in the effort to achieve WYSIWYG (What You See Is What You Get) (Holub et al., 1988a; MacDonald, 1990). With increasing use of multiple-device displays to provide a wider field of view, device independence has also become a requirement in the real-time simulation of real or virtual environments for pilot or driver training. Choice of a device-independent color space has frequently led to improvement in image processing for purposes of visual enhancement or efficient electronic transmission. As a consequence, relatively few recent articles describe color processing in the traditional red-green-blue (*RGB*) cubical color space of cathode-ray tube (CRT) primaries.

17.2 IDEAL METHODOLOGY OF COLOR CONTROL

Figure 17.1 is based upon a diagram by Hall (1988, p. 117), which compares the "ideal methodology for image storage and display" with "current practice." Hall's representation of current practice is duplicated in Figure 17.1, but his diagram of ideal methodology has been modified to include additional stages. Instead of beginning at the image-computation step, the ideal methodology as shown here starts with the image color data, which may come from three different kinds of sources: (1) reflectance and illumination data describing the physical objects and lighting of the scene to be simulated; (2) *RGB* values from a CRT

Ideal Methodology

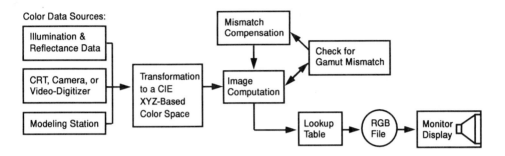

Current Practice

Figure 17.1 Ideal methodology for image computation and display, contrasted with current practice. [Adapted from Hall (1988), p. 117. "Current practice" section reproduced by permission of Springer-Verlag, New York.]

color table, color camera, or video-digitizer; or (3) *XYZ* tristimulus values selected at a CIE-based color-editing station. Image data from these sources then pass through a second stage in which they are transformed into one of the CIE perceptual color spaces before they undergo image computation in stage 3. For applications that require the output to be displayed in hard copy or on one or more different display devices, there is a processing loop within which the image calculations are checked and corrected for possible gamut mismatches between the design device and the output device(s). Digital *RGB* (or CMYK) values tailored to the intended display (or printer) are generated in stage 4 by way of a look-up table (LUT). This LUT replaces the traditional (and misleading) application of "gamma correction." The resulting digital values may then be stored before being sent to the output device.

The ideal methodology contrasts sharply with what Hall recognizes as current practice, in which all computations are carried out in the *device-dependent RGB* space and stored in an *RGB* file before being "gamma-corrected" on the way to the output device. The disadvantages of current practice can be made clear by examining successive stages in the ideal methodology.

17.3 SOURCES OF COLOR IMAGE DATA

17.3.1 Illumination and Reflectance Data

Physical data, such as object surface reflectances and spectral energy distributions (SEDs) of illuminants in a scene, can be directly transformed into CIE color spaces using standard colorimetric equations. Therefore, these data may constitute the best and most natural source of color image data, as several writers have suggested (Hall, 1988, pp. 58–62; Gershon and Jepson, 1989; Brainard and Wandell, 1990). Use of such data is clearly in line with the current trend toward building graphics images as objects in a three-dimensional environment rather than as regions in a two-dimensional picture.

Much of the raw material for an object-centered physical data approach is already available. Information about the spectral energy in natural daylight and other sources of radiant energy can be found in Wyszecki and Stiles (1982). Meyer and Greenberg (1986) cite references giving spectral reflectances of building materials and interior paints, as well as SEDs of common indoor lighting sources. They have used SEDs of the sky at sunset, as a function of angular position above the horizon, to generate the colors for a city skyline silhouetted against a sunset sky. Maloney (1986) shows how data on the spectral reflectance of different types of terrain and vegetation [such as the data provided in Krinov (1953)] may be used to obtain three or four basis functions that are sufficient to describe all naturally occurring reflectance or illuminance functions.

Color selection based on physical data can improve both efficiency and realism in simulator displays designed for mission rehearsal. Imagery simulating real-world geographical environments can be developed directly from Defense Mapping Agency terrain elevation data, which includes specification of the materials making up terrain and cultural features in particular locations. If reflectances of these materials are incorporated into the database-generation process, color selection occurs automatically as terrain elevation data are brought into the database. Expressing illumination and reflectance data as basis functions further increases computational efficiency and speed.

17.3.2 *RGB* Data from CRT, Camera, or Video-Digitizer

RGB color data, from whatever source, must first be transformed into a device-independent color space before image computation begins. Such transformations always rest upon actual physical measurements (calibration) of the input device. When the input device is a CRT, it is necessary to measure both luminance and chromaticity of each primary (R, G, and B) independently at eight to 16 voltage levels. Other display devices may require different and perhaps more extensive measurements; for light-valve projectors (LVPs), it is desirable to measure a number of *combinations* of R, G, and B at different voltage levels. For evidence that calibration measurements are really necessary, even for very well-behaved CRTs, see Virgin et al. (1986), Hartmann and Madden (1987), Brainard (1989), Post and Calhoun (1989), DuFlon (1990), Engeldrum and Ingraham (1990), and Lucassen and Walraven (1990). CRTs with self-calibrating features have begun to appear (McMillan, 1990).

Lee (1988), relying on work by Maloney (1986) and Maloney and Wandell (1986), reports a method for obtaining CIE color data from a video-digitizer, given that the digitizer itself has been calibrated and the illuminant SED is known. Wandell (1986) has described the derivation of CIE color data from color photographs (such as photographic reproductions of paintings) or from video-camera *RGB* output when the sensitivities of the three camera sensors are known. The Stanford Color Analysis Package (CAP) (Brainard and Wandell, 1990) uses linear models to extract surface reflectance functions from camera data.

17.3.3 Interactive Color Editors in Perceptual Color Space

Within the past few years, a number of interactive color editors have been developed that incorporate device-independent color information. Meyer and Greenberg (1986, 1987) describe a program called Chameleon, which was developed primarily for instructing computer graphics students about color science. Chameleon can display monitor-gamut colors in relation to any of several CIE color spaces, but it permits only limited color editing. Tektronix has developed a color space (TekHVC) based on the CIELUV space (see Section 17.4.3) for use as an interface between CRT design displays and color printers (Taylor et al., 1989). This space has become an integral part of the Tektronix Color Management System (TekCMS), which includes an interactive color interface (ICI) and software for incorporating calibration files specific to individual input and output devices. A Macintosh version of this color editor has been available for some time as TekColor; TekCMS has recently been incorporated as an enhancement to X Windows.

Several other device-independent editing systems are being offered, including a prepress system for gravure printing (Schreiber, 1986) that employs CIE-traceable "absolute" data throughout image processing. The Alvey Color Appearance Model (ACAM), under development in England (MacDonald, 1990; Luo et al., 1991a, 1991b), goes beyond CIE-recommended systems to include features of R. W. G. Hunt's model of color vision for predicting color appearance (Hunt, 1987, 1991). The Logical Visual Display at MIT's Media Laboratory (Jacobson and Bender, 1989) incorporates information from the Munsell color order system (see Section 17.4.5) into a two-dimensional array of choices for interactive color editing.

Other efforts of this kind are under way in most companies with an interest in color hard copy. For the most part, these developments are not yet well represented in the published literature, and no attempt will be made to compare or evaluate them. It is important, however, to recognize that color-editing programs operating only in *RGB* device-dependent color space are becoming appropriately obsolete.

17.4 COLOR SPACES

The variety of color spaces mentioned in the literature may well be bewildering. This section will deal first with those color spaces that are transformations of the *XYZ* space recommended by the CIE in 1931; equations defining these transformations are listed in Table 17.1. All of these spaces are device-independent; they differ primarily in the extent to which they successfully model psychophysical and perceptual data. The section will conclude with brief descriptions of non-

Table 17.1 Colorimetric Equations[a]

$x = X/(X + Y + Z)$	(1)
$y = Y/(X + Y + Z)$	(2)
$z = Z/(X + Y + Z)$	(3)
$u = u' = 4X/(X + 15Y + 3Z)$	(4)
$v = 6Y/(X + 15Y + 3Z)$	(5)
$v' = 9Y/(X + 15Y + 3Z)$	(6)
$L^* = 116(Y/Y_n)^{1/3} - 16 \qquad Y/Y_n > 0.008856$	(7)
$L^* = 903.3(Y/Y_n) \qquad Y/Y_n \leq 0.008856$	
$u^* = 13L^*(u' - u'_n)$	(8)
$v^* = 13L^*(v' - v'_n)$	(9)
$C^*_{uv} = (u^{*2} + v^{*2})^{1/2}$	(10)
$h_{uv} = \arctan(v^*/u^*)$	(11)[b]
$s_{uv} = 13[(u' - u'_n)^2 + (v' - v'_n)^2]^{1/2} = C^*/L^*$	(12)
$\Delta E^*_{uv} = [(\Delta L^*)^2 + (\Delta u^*)^2 + (\Delta v^*)^2]^{1/2}$	(13)
$a^* = 500[(X/X_n)^{1/3} - (Y/Y_n)^{1/3}]$	(14)[c]
$b^* = 200[(Y/Y_n)^{1/3} - (Z/Z_n)^{1/3}]$	(15)[c]
$C^*_{ab} = (a^{*2} + b^{*2})^{1/2}$	(16)
$h_{ab} = \arctan(b^*/a^*)$	(17)[b]
$\Delta E^*_{ab} = [(\Delta L^*)^2 + (\Delta a^*)^2 + (\Delta b^*)^2]^{1/2}$	(18)

[a] Subscript *n* indicates that the value is for the reference white, usually D65.
[b] *h* lies between 0° and 90° if numerator and denominator are both positive; between 90° and 180° if numerator is positive and denominator is negative; between 180° and 270° if both numerator and denominator are negative; between 270° and 360° if numerator is negative and denominator is positive.
[c] If any of the ratios X/X_n, Y/Y_n, or Z/Z_n is equal to or less than 0.008856, its term in Eqs. (14) and (15) is replaced by $7.787R + 16/116$, where *R* represents the ratio. The cube root operation is applied *only* to ratios whose value exceeds 0.008856.

CIE color spaces, most of which are derivatives of the cubical *RGB* space. Although some of these spaces model perceptual dimensions more effectively than others, all of them are device-dependent, and they should be recognized as having very limited usefulness today.

17.4.1 1931 *Y,x,y* Space

In the original planning of color television, the color output of any CRT device was engineered to be traceable to the *XYZ* tristimulus values derived from color-matching data. These data were obtained experimentally from human observers looking at small test fields (2° of visual angle). The values of *X*, *Y*, and *Z* can be calculated for any real color by using radiometric data and the standard observer's three color-matching functions ($\bar{x}$, $\bar{y}$, and $\bar{z}$). Since the *Y* tristimulus value corresponds to the photometric luminance of the color (in candelas per square meter, or nits), it is customary to regard *Y* as the luminance dimension of a three-dimensional color space.

Chromaticity information is described by *x*, *y*, and *z* [Table 17.1, Eqs. (1)-(3)], and the chromaticity of all real colors can be plotted as *x* and *y* coordinates ($0 < x < 0.735$; $0 < y < 0.834$) in a two-dimensional space, the CIE 1931 2° Chromaticity Diagram. In this diagram, mixtures of any two real colors lie on a straight line joining the chromaticity coordinates of those colors. Moreover, the relative position of a mixture on that line reflects the proportions of the two colors in the mixture. According to some authors (e.g., Hunt, 1987, p. 66), color diagrams in which mixtures of two colors *do not* lie on a straight line should not be called "chromaticity diagrams" (see Section 17.4.3).

In 1960 the CIE supplemented the 2° chromaticity diagram by adopting new color-matching data obtained from observers who viewed 10° test fields. Display users working with field sizes of 4° or greater should employ the 10° color-matching functions in determining the tristimulus values of their color image data. Wyszecki and Stiles (1982) provide tables of color-matching functions at 1 and 10 nm intervals (2°, pp. 725–737; 10°, pp. 738–749); Hunt (1987) presents both 2° and 10° functions at 5 nm intervals in a single table (pp. 182–183).

17.4.2 1976 *Y, u', v'* Color Space

The *x,y* chromaticity diagram is "perceptually nonuniform." Two colors separated by a distance *d* in one region (for example, blue) will appear perceptually much more different from each other than two colors separated by the same distance in other regions (particularly in the green). A linear transformation of *x* and *y* to *u* and *v* [Table 17.1, Eqs. (4) and (5)] improved the uniformity somewhat, and after 1964 some workers used *Y*, *u*, *v*, instead of *Y*, *x*, *y* as a color space. The tristimulus values *XYZ* were replaced by tristimulus values *UVW*, with $W = Y$ representing luminance.

At the 1975 CIE meeting, the value of v was increased by a factor of 1.5 [Table 17.1, Eq. (6)] to improve the uniformity, and Yuv became $Yu'v'$. The $u'v'$ chromaticity space is represented in the CIE 1976 UCS (Uniform Chromaticity Space) diagram. Chromaticity coordinates in u' and v' can be computed from either the 2° or 10° functions; for large-screen displays, the 10° functions should be used. Since $u'v'$ space is a linear transformation of xy space, the 1976 UCS diagram also meets the linear-mixture criterion and qualifies as a true chromaticity diagram.

17.4.3 CIELUV and CIELAB Color Spaces

In 1976, the CIE also recommended two color spaces, CIELUV and CIELAB, for use in the representation of three-dimensional *color appearance*. Robertson (1990) has explained the reasons behind the recommendation of two spaces rather than one. Briefly stated, each of the two spaces proposed for consideration had too many proponents to be rejected, yet too few to be chosen overwhelmingly. Folklore abounds concerning the proper conditions for using CIELAB rather than CIELUV (or vice versa). According to Robertson, industries concerned with small color differences between object colors tend to use CIELAB. However, Pointer (1981) presents data showing that neither color space is significantly better than the other for representing color-difference data in a uniform manner. MacDonald (1990) accepts a common view that CIELUV should be used for "self-luminous objects or additive sources," while CIELAB is appropriate for "objects viewed by reflective light." Carter and Carter (1983) and Post (1984) offer guidance concerning the appropriate reference white for CIELUV description of self-luminous displays.

According to Wyszecki (1986), no certain evidence yet exists to settle the CIELUV/CIELAB issue. Brill and Derefeldt (1991) and Tajima (1983) recommend the use of CIELUV for electro-optical displays, and Sproson (1983) favors CIELUV for "film and TV applications." Silverstein et al. (1986) use CIELUV in implementing an algorithm for selecting a set of optimally discriminable display colors. CIELUV space qualifies by the linear-mixture criterion as a chromaticity diagram, while CIELAB space does not; mixtures of two colors in CIELAB space tend to lie along curved lines (see Figure 17.6). Wyszecki (1986) and Robertson (1990) see this difference as a possible advantage for CIELUV in digital display technology.

Gentile et al. (1990a, 1990b) use CIELUV for solving gamut mismatch and quantization problems; Dalton (1990) advocates CIELUV space for quantization and halftoning, emphasizing that the availability (in both CIELUV and CIELAB) of a separate hue dimension has advantages for processing algorithms that preserve hue. TekHVC space (Taylor et al., 1989) is essentially a rescaled version of CIELUV space in which the 0° position has been shifted to coincide with a line between reference white and "unique red."

On the other hand, Spooner (1990) favors CIELAB for use in prepress proofing systems, claiming that "equally perceived color differences plot to approximately equal distances in $L^*a^*b^*$ but not in $L^*u^*v^*$ or $Yu'v'$." The extent of this difference can be seen in Figures 1 and 2 of Billmeyer (1988, p. 142), which show CIELUV and CIELAB diagrams of Munsell and OSA colors spaced at experimentally determined equal-appearing intervals. Ikegami (1989) and Stone (1990) also use CIELAB as an intermediary between RGB and the printer subtractive primaries cyan, magenta, yellow, and black (CMYK).

This report will stress the *similarities* of CIELUV and CIELAB rather than their differences. Both systems represent an advance from $u'v'$ space toward a better representation of *color appearance* data, and both systems permit the derivation of *color appearance descriptors* from CIE tristimulus variables. This advance is most clearly seen by examining the conversion of luminance Y, a psychophysical variable, to lightness L^*, a psychological (perceptual) variable [Table 17.1, Eq. (7)]. Use of a cube-root lightness metric is perhaps the least controversial feature of the CIELUV and CIELAB transformations. Schuchard (1990) has shown that a cube root of Y function is a good approximation to the data on brightness discrimination over the range of luminances usually represented on CRTs (about -2 to 2 log cd/m²). It is less satisfactory below and above this range.

Since L^* is a better approximation to the perceptual dimension of lightness than Y, CIELUV and CIELAB make it possible to describe and visualize the *gamut of possible colors* for any display device at different levels of *apparent lightness*. Figure 17.2 shows the gamut boundary colors (in CIELUV) for a typical CRT (right side) and a single LVP (SLVP) display (left side). L^* is represented on the vertical axis at the center of both figures; achromatic colors (the gray scale) are located on this axis from $L^* = 0$ (black) to $L^* = 100$ (maximum white). Because the SLVP has a relatively luminous "dark field" even in the absence of any color signal, its gamut contains no colors with L^* below 30. Consequently, the contrast available within an SLVP display is much less than the contrast within a CRT display.

Cross sections of the CIELUV space at constant L^* are *linear transformations of $u'v'$ space*, in which the $u'v'$ distance of a color from the reference white (usually D65; see Brill and Derefeldt, 1991) is weighted by the lightness L^* of that color [Table 17.1, Eqs. (8) and (9)]. This weighting reflects the fact that *apparent colorfulness increases as lightness increases*; colors with the same u',v' coordinates will appear more "colorful" at higher lightness levels.

Two measures of colorfulness are available in CIELUV, *chroma* and *saturation* (see Wyszecki, 1986, Chp. 9, p. 4). Chroma (the degree to which a chromatic color differs from an achromatic color *of the same lightness*) is represented as the distance from the achromatic L^* axis, weighted by L^* [Table 17.1, Eq. (10)]. Saturation (the degree to which a chromatic color differs from an achromatic color *regardless of their lightness*) is the distance s without L^* weighting

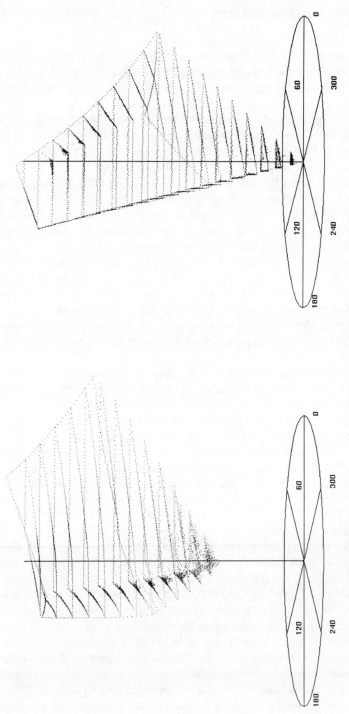

Figure 17.2 Gamuts of a typical CRT monitor (right) and a single light-valve projector (left), shown in CIELUV color space with D65 as reference white. A gamut is represented by a maximum of 7200 points, each being that point of greatest chroma C^*_{uv} in one of 7200 categories (360 in hue angle h_{uv} × 20 on lightness L^*). The light valve gamut contains no points in categories with $L^* < 30$.

COLOR CONTROL IN DIGITAL DISPLAYS

[Eq. (12)]. Equation (13) is offered as a measure of color difference; one ΔE^*_{uv} unit represents approximately one just noticeable difference (JND).

The gamuts in Figure 17.2 can therefore also be described with reference to *polar coordinates* C^*_{uv} and h_{uv} [Table 17.1, Eqs. (10) and (11)]. The perceptual dimension of hue h_{uv} is represented as a hue angle with red near the origin (0°), green in the region near 150°, and blue near 225°. The three-dimensional gamuts shown in side view in Figure 17.2 appear in top view as inserts in Figure 17.7, with the CRT again on the right. Viewed thus in polar coordinates, the CRT and SLVP gamuts have red primaries similar in hue but differing in chroma. The green primaries of these devices differ in hue, but both are definitely yellow-green. The blue primaries are at hue angles near 255°, which corresponds to a violet-blue.

Figures 17.3 and 17.4 illustrate the representation of the same CRT gamut in

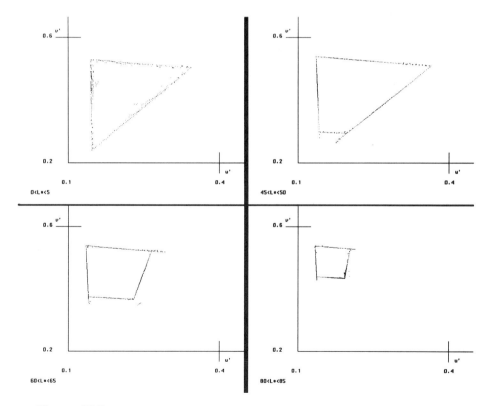

Figure 17.3 The CRT gamut from Figure 17.2, shown in CIE 1976 Uniform Chromaticity Space. Four lightness levels are shown: $0 < L^* < 5$ (upper left), $45 < L^* < 50$ (upper right), $60 < L^* < 65$ (lower left), and $80 < L^* < 85$ (lower right).

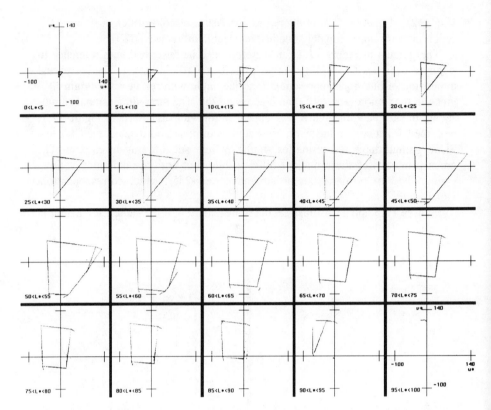

Figure 17.4 The CRT gamut from Figure 17.2, with the 20 lightness levels shown separately in order of increasing L^* from top left to bottom right.

$u'v'$ and u^*v^* space, respectively. Figure 17.3 shows the $u'v'$ gamut at four lightness levels. For L^* between 0 and 5, the points of maximum chromaticity form a triangle; this triangle is the familiar "monitor gamut," which is often drawn without mention of the lightness dimension. As the remaining panels in Figure 17.3 show, the full extent of this triangular gamut can actually be achieved only at the lowest lightness level.

In order to show more than a few lightness levels, Gentile et al. (1990a) present device gamuts in an array similar to Figure 17.4, where boundary colors for each L^* category are shown separately. The 20 levels of L^* for the CRT in Figure 17.2 appear here in a 5 × 4 array.

Figure 17.5 shows the gamut boundary colors, calculated in CIELAB, for the same SLVP and CRT whose CIELUV gamuts appear in Figure 17.2. To illustrate the difference between CIELUV and CIELAB, Figure 17.6 displays L^* level

COLOR CONTROL IN DIGITAL DISPLAYS

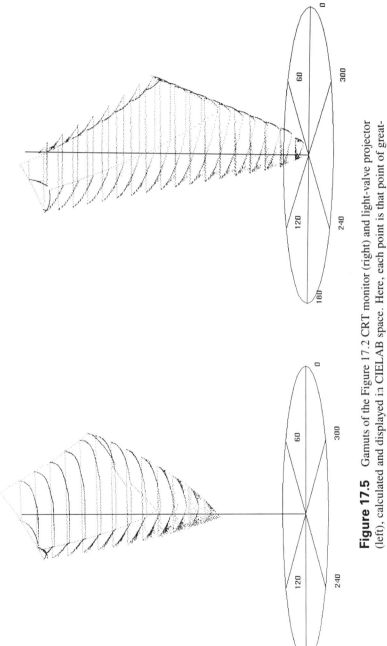

Figure 17.5 Gamuts of the Figure 17.2 CRT monitor (right) and light-valve projector (left), calculated and displayed in CIELAB space. Here, each point is that point of greatest chroma C_{ab}^* in one of 7200 categories (360 in hue angle h_{ab} × 20 in lightness L^*).

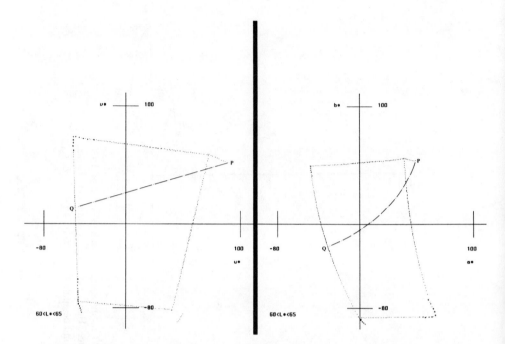

Figure 17.6 One lightness level (60 < L^* < 65) from the CRT gamut shown in Figures 17.2 and 17.5, shown in u^*v^* (left) and a^*b^* (right) coordinates. The 11 points from P to Q represent the positions in CIELUV or CIELAB space, respectively, of mixtures of the colors at P and Q. Such mixtures lie along a straight line in CIELUV, but they generally lie along a curved line in CIELAB.

60–65 for the CRT in both u^*v^* and a^*b^* diagrams, taken from the data for Figures 17.2 and 17.5. Two equiluminant colors P (with XYZ = 39.9, 30, 3.25) and Q (with XYZ = 19.6, 30, 27.4) are shown in each of these gamuts and connected by a line passing through nine mixtures of these two colors. Note that the mixtures lie along a straight line in CIELUV but along a curved line in CIELAB. Since a^* and b^* [Table 17.1, Eqs. (14) and (15)] represent *nonlinear* transformations from XYZ, CIELAB does not meet the linear-mixture criterion and cannot serve as a chromaticity diagram in the usual sense. CIELAB has measures for chroma and hue [Eqs (16) and (17)] that are analogous to those measures in CIELUV space, but it lacks the saturation measure s. Its color difference measure ΔE^*_{ab} [Eq. (18)] parallels ΔE^*_{uv} for CIELUV.

17.4.4 Color Appearance Models

Although CIELUV and CIELAB move from color description toward the representation of color appearance, neither system attempts to deal with the full range of factors affecting color appearance. Two of these factors are particularly relevant for digital displays. The effect of mean display luminance on *apparent brightness* of saturated colors (the classical Purkinje effect) has created problems for photometry in the mesopic region (between 0.001 and 10 nits) that have not yet been resolved (CIE, 1989). Moreover, in a self-luminous display with mean luminance around 10 nits, it is possible to have reds, magentas, and blues that appear much brighter, relative to other colors, than similarly colored objects in daylight. These colors may then appear to "glow" (fluorence or *Farbenglut*; see Wyszecki, 1986, chp. 9, p. 7). Taylor and Murch (1986) report data typical for CRTs; their observers chose a video blue of 2.7 nits (and a red of 4.7 nits) as equal in apparent brightness to a white of 10 nits.

Level of illumination *in the display surround* has measurable effects on the colorfulness of regions in an image (Hunt, 1987, 1991; Rhodes, 1989). A display seen in a dim or dark surround will appear less bright and less colorful than the same display seen in a normally illuminated surround, even though the display is shielded from ambient light. A white border is often included as part of a video display when direct comparisons are to be made between the display and a photograph viewed in room illumination (Gorzynski and Berns, 1990).

Two color appearance models have been proposed (Hunt, 1987, 1991; Nayatani et al., 1990) that include parameters to adjust for the effects of image and surround luminance, simultaneous lightness and color contrast, and chromatic adaptation. Both models are being evaluated experimentally (Billmeyer, 1988; Nayatani et al., 1990; Luo et al., 1991b), but neither has yet been officially recommended by the CIE. The ACAM color-editing system (MacDonald, 1990; Luo et al., 1991a, 1991b) is the first interactive color editor to incorporate surround effects from such a model.

17.4.5 Munsell, OSA, and NCS Color Order Systems

Color systems made up of reflective color samples have a long history and abiding usefulness. Hunt (1987) provides excellent color illustrations of the three most widely used systems (Munsell, OSA, and NCS). The Munsell samples have been chosen to represent approximately equal intervals in hue, value (lightness), and chroma. These samples are arranged in a three-dimensional space similar to CIELUV or CIELAB, in which value [closely represented by L^*; see Hunt (1985)] is drawn on the vertical axis with hue and chroma represented in polar coordinates within horizontal planes. Munsell samples can be described in CIE Y, x, y coordinates (Wyszecki and Stiles, 1982, pp. 840–852) and transformed to other CIE spaces. Pointer (1981) and Billmeyer (1988) show Munsell constant-value diagrams in CIELUV and CIELAB spaces, with Billmeyer finding that the diagrams appear more nearly circular in the SVF color space proposed by Seim and Valberg (1986) than in either CIELAB or CIELUV.

The Logical Visual Display at MIT's Media Lab (Jacobson and Bender, 1989) translates Munsell color space into CIE coordinates by using the conversion tables such as those found in Wyszecki and Stiles (1982). These coordinates are then converted into device *RGB* by using calibration data. Thus the system incorporates Munsell information on the extent of chroma available at each hue and value level, so that the number of available colors in the editing space can be proportional to the number of distinct steps in a perceptual color space. This procedure yields a two-dimensional color space with a "tulip" shape. Hansen's (1990) discussion of the Bender–Jacobson approach includes color illustrations.

The OSA color tiles have been developed more recently as another attempt to fill color space with samples separated by equally perceptible intervals. The OSA system avoids the problems inherent in a polar coordinate system by adopting a regular rhombohedral lattice geometry. Both the Munsell and OSA systems have been discussed in detail by Wyszecki (1986).

The Swedish Natural Color System (NCS) is based on three opponent-color dimensions (black/white, red/green, and blue/yellow). Hedin and Derefeldt (1990) describe a color graphics editing system (Palette) based on the NCS system. Its user interface has some similarity to the TekICI; with the addition of device characterization data, Palette could be used like TekCMS to perform device-independent color rendering between electronic display units and hard copy.

These three color order systems, together with the less well known German system, DIN (see Wyszecki and Stiles, 1982, pp. 509–511), are all designed for applications that involve reflective samples. All can be converted to CIE *XYZ* tristimulus values. They are of interest to video display users principally as reference systems in which the perceptual distances between neighboring color samples have been experimentally determined to be approximately equal.

17.4.6 *YIQ* and *YUV*

YIQ, like its European counterpart *YUV* (not to be confused with *Yuv* or *Yu'v'*, dealt with in Section 17.4.2), is a transformation of *RGB* space into three mutually orthogonal dimensions. Scheibner (1970) clarifies the German language equivalent terms and concepts. Buchsbaum (1987) discusses the development of *YIQ* as a method of adding a two-dimensional color signal (I, Q) to the existing monochromatic intensity signal (Y) without exceeding the available bandwidth. Although this solution to technical problems was designed to take advantage of known properties of human color vision, the I and Q color signals do not have perceptual meaning. Moreover, since *YIQ* is a transformation of *RGB* space, it is device-dependent.

The equations normally given for transformations between *RGB* and *YIQ* (Sproson, 1983; Pratt, 1991) are *only approximate* when the standard NTSC primary chromaticities and gamma are used instead of values determined from the individual device. However, because the *YIQ* transformation separates luminance and chromaticity, *YIQ* space has been found suitable for image-processing applications where luminance alone is to be manipulated without changing chromaticity (Mitra et al., 1989). The outcome of such processing remains device-dependent unless the image data are later transformed into CIE space.

17.4.7 "Hexcone" and "Double Hexcone" Color Spaces

Color editing in *RGB* space has impressed some authors as psychologically unnatural or counterintuitive, and several have proposed transforming *RGB* data into a three-dimensional space corresponding to the perceptual dimensions of hue, lightness, and saturation. Smith (1978) helped to popularize a transformation from *RGB* to a hexcone model (HSV) with dimensions of hue, saturation, and value (lightness), and Rogers (1985) proposed a double hexcone (HSL). Both hexcone models represent the *RGB* primaries and their complementaries (CMY) as vertices of a hexagon. Another HSL space replaces the hexagon with a circle to form a double cone (Joblove and Greenberg, 1978). In 1978 Tektronix introduced a version of this model, called HLS (Murch, 1987, p. 23), of which at least nine variations eventually developed. Joblove and Greenberg also suggested an HSL cylinder, which expands the base and top of the double cone into hue circles. Hall (1988, p. 46) presents diagrams of HSV, HSL, and HLS spaces.

These models have been offered to facilitate color selection in graphics design applications, and choice among them has been based on local preference. Schwarz et al. (1987) report an experimental comparison of the speed and accuracy of subjects using different color-editing models for a color-matching task. Their experiment included *RGB*, *YIQ*, HSV, CIELAB, and an "opponent color" model. "Subjects using the *RGB* color model matched quickly but inaccurately compared with those using the other models. . . . Users of the HSV color model were the slowest . . . but were relatively accurate" (Schwarz et al., 1987, p. 123). Kotera and Kanamori (1989, p. 257) propose using a transformation from *RGB* to HLS in order to achieve "perceptual color control" while adjusting "color tone," then transforming the image data from HLS to hardcopy CMY for a sublimation transfer color printer. Ware and Cowan (1990) have proposed an *RGYB* space for color editing; in this two-dimensional space, the opponent colors are displayed at opposite corners of a square.

Finally, an equilateral triangle derived by passing a plane through the *R*, *G*, and *B* corners of the *RGB* cube occasionally appears in digital display literature. Smith (1978) calls this the "triangle" model and recognizes its potential for separating hue, saturation, and lightness dimensions. It should not be confused with the "Maxwell triangle" (Wyszecki and Stiles, 1982, p. 121), for which the *R*, *G*, and *B* axes are actual primaries from color-matching experiments. Huntsman (1989) uses the triangle transformation to derive his "planar-vector based color space for graphic arts." Strickland et al. (1987) and Bockstein (1986) use it to separate brightness, hue, and saturation components for digital color image enhancement (see Section 17.5.3).

All the color spaces discussed in this section are entirely device-dependent, and they give no information about the relative gamuts of different display devices.

17.5 IMAGE COLOR COMPUTATIONS

17.5.1 Computations Intended to Manipulate Lightness

Many of the computations carried out in the Image Calculation and Processing stage of Figure 17.1 are intended to affect primarily the lightness dimension, leaving hue and saturation unchanged. Computations to achieve shading, antialiasing, and texture are of this sort. If image computations are carried out in *RGB* space, it often happens that hue and saturation do not remain constant as lightness is increased or decreased (Dalton, 1990). Therefore, it is advantageous to perform these computations in a color space that separates luminance and chromaticity dimensions. Although *YIQ* space has occasionally been used for this reason (Mitra et al., 1989), an *XYZ*-based color space such as CIELUV or CIELAB provides the additional advantage of device independence. Dalton (1990) considers HSL and HSV spaces better than *YIQ* because they possess a separate hue dimension, but he recommends CIELUV for its device independence and greater perceptual uniformity.

Antialiasing algorithms rely on luminance averaging to obtain intermediate gray values. When these algorithms are applied in *RGB* space, the resulting gray values reflect the nonlinearity of the luminance–voltage function of the display device. Since linearity of gray-scale values improves edge smoothing and apparent resolution (Covington, 1990; Jacobsen, 1990), *RGB* users have been advised to use gamma correction. However, the ideal methodology obtains linearity in the image-processing stage by using Y or L^* as the luminance variable and postponing the *RGB* transformation to a much later stage. Gentile et al. (1990b) found CIELUV L^* superior to *RGB* for halftoning by error diffusion algorithms. Jacobsen's image quality ratings (Jacobsen, 1990) indicate that the cube root variable L^* may perform just as well as linear Y when the entire luminance ramp has at least 16 gray levels.

17.5.2 Compensation for Gamut Mismatch

Any three-primary display device has a limited gamut of realizable colors, and printer gamuts do not match electronic display gamuts. Pointer (1980) provides CIELUV and CIELAB comparisons of printer and CRT gamuts; Stone's CIELAB comparisons also reveal the differences in L^* (Stone, 1990). Adjustments must be made to compensate for gamut mismatch. Such compensation is also required when imagery designed on a CRT is to be displayed by an LVP; the degree of gamut mismatch between these two devices is documented here in Figures 17.2, 17.5, and 17.7.

Gentile et al. (1990a) compared several approaches to gamut mismatch compensation, obtaining observer ratings of match quality between each output image and the original picture. Except for a clipping algorithm applied in *RGB*, all their algorithms operate on CIELUV variables. The results indicate that CIELUV-based algorithms provide a better basis than *RGB* for gamut mismatch compensation. Image-dependent compression techniques were preferred over image-independent methods, which do not consider the color gamut of the particular image being processed. Clipping algorithms were preferred to compression methods. "The most preferred technique was that of clipping with constant lightness and hue, implemented to maintain these components exactly while clipping chroma to the border of the realizable output gamut" (Gentile et al., 1990a, p. 181). Clipping of both hue and chroma also received good ratings, as long as lightness was preserved.

Lindbloom, on the other hand, prefers to "treat hue as being more important than saturation, which in turn is more important than lightness" (Lindbloom, 1989, p. 122). His method, which he has applied to color monitors, film recorders, electronic prepress systems, and color hardcopy devices, preserves the high luminance resolution that is present at low luminances in *RGB* space by "companding" the *RGB* variables before transforming them into $L^*u^*v^*$ for mismatch compensation.

Gamut mismatch computations appear as an internal loop in the ideal methodology of Figure 17.1. These computations are transparent to the user when the image is edited on an electronic display using a color-editing program that has access to calibration data for both editing and output devices. By making use of a display based on a perceptual color space with lightness, hue, and saturation dimensions, the user is enabled to "adjust the psychophysical variables" (MacDonald, 1990) and let software take care of the transformation to physical variables.

Dalton (1990, p. 183) provides diagrams from the TekColor interface for the Macintosh. Figure 17.7 shows a similar interface from the Color Modeling Workstation under development for flight simulation research at Armstrong Laboratory, Williams Air Force Base. This color modeling program contains some TekCMS components; additional components have been designed to accommodate SLVPs as output devices. Each "hue leaf" shows the gamut boundaries for both display devices and the location of colors being edited. Regions outside the area common to both gamuts are regions of gamut mismatch.

Not all device-independent color editors display multiple gamuts. The display described by Schreiber (1986) simply alerts the user to gamut mismatch by

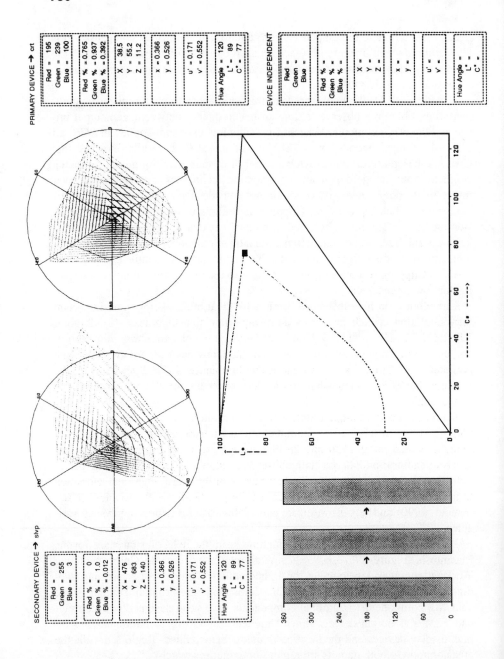

"blinking" any color that the program recognizes as being outside the bounds of the output device gamut.

17.5.3 Other Image Computations

The image-computation stage may also include image processing to improve perceptibility of information in the image, to compress the information for electronic transmission, or to "segment" the image into figure-ground units for machine vision. Most authors recognize, explicitly or implicitly, the dominant role of luminance contrast in the perception of high spatial frequency edge information. Trends toward use of a uniform chromaticity space have appeared in all these applications, but the use of *RGB* space continues to dominate image processing.

Mitra et al. (1989) find *YIQ* superior to *RGB* for digital processing of color images; histogram equalization applied to *Y* alone is more efficient and less likely to produce false color than when it is applied to *R*, *G*, and *B*. A desire to deal separately with saturation led Strickland et al. (1987) to base their computations on the triangle model (see Section 17.4.7). Because the center of this triangle lies on the luminance diagonal of the *RGB* cube, the distance of any color from this diagonal provides a measure of saturation that can be used in image enhancement. By adding the spatial frequency information carried by saturation to the information carried by luminance, they were able to enhance some edges and create others, putting more detail into dark areas of images representing outdoor scenes. Mayo (1989) applied pseudocolor to bring out the low spatial frequency information in monochrome medical ultrasound images, where high-frequency luminance variations ("speckle") tend to obscure the medically significant low-frequency variation.

"Color compression" sometimes means only reducing the total number of colors to a set whose size is limited by the frame buffer. Wan et al. (1990) report a variance-based method of color selection applied in *RGB* space; they comment that the same method should give better results in a UCS. The Palette program (Hedin and Derefeldt, 1990) is explicitly designed for selecting a limited number of colors in NCS color space. McFall et al. (1989) use *YUV* (the European ver-

Figure 17.7 Sketch of the user interface display from the Color Modeling Workstation under development at Armstrong Laboratory. The lower part of the display shows the gamut boundaries of a single hue leaf at 120° (in the yellow/green) for a CRT (solid line) and an SLVP (dash-dot line). The upper part of the display shows the cursor location color descriptors for the CRT (right side) and SLVP (left side), alongside CIELUV gamuts of each device in polar L^*C^*h coordinates (bounding circle drawn at $C^* = 150$). These gamuts appear in side view in Figure 17.2; here they are seen from directly above the top of the L^* axis.

sion of *YIQ*) to make color selection in a space that separates luminance and chrominance.

Sigel et al. (1990) applied subsampling for data compression to four images, using *YUV*, CIELUV, CIELAB, and HLS color spaces. *YUV* gave the best results, HLS the worst. Tsai (1989) describes color-picture compression architecture based on *YIQ*. Mitchell et al. (1989) use a transform similar to *YIQ*, with one luminance and two chrominance channels.

Authors writing on the segmenting of color images for machine vision take for granted that the segmenting should be done according to a model that simulates human color vision. Murano et al. (1989) extract hue and saturation data after *R*, *G*, and *B* images from a video camera have been used to identify a moving object. The color information is obtained through a look-up table and passed on with 5 bits allocated to hue and 3 to saturation. Celenk (1990) performs image segmentation in CIELAB space, using circles and their radii as approximations to Munsell space. Bumbaca and Smith (1987) describe a real-time computer color vision system using their own version of a UCS with one achromatic luminance channel and two opponent-color chromatic channels, but they offer no data for comparing it with other uniform spaces.

17.6 CONVERTING UCS LUMINANCE–CHROMATICITY VALUES TO DIGITAL COMMANDS

In the ideal methodology, the image-computation stage yields values of those variables that have been employed in the computations, preferably UCS values of $Yu'v'$, $L*u*v$, or $L*a*b$. For additive output devices, these values must be passed to a program that converts them to *RGB* digital commands in the domain of the intended display. For subtractive devices, such as printers, the values are converted to the CMYK domain of the printer that is to be used.

A great deal has been written about the conversion from continuous variables in uniform chromaticity space to integer variables in *RGB* space. Consensus is developing for the use of look-up tables rather than gamma correction functions in this conversion; Catmull (1979) explains early methods for developing such tables. This section will deal first with procedures for generating the look-up table.

17.6.1 Look-up Tables for CRT Output Devices

Matrix equations for conversion between *XYZ* tristimulus values and *RGB* luminance appear in many articles and textbooks, often accompanied by a chromaticity matrix based on NTSC standards and a gamma value near 2. Such conversions are bound to result in color errors unless calibration data from the particular CRT are used. Even with device calibration, the matrix method can work only when phosphor chromaticity does not change with voltage and when the power supply is adequate to guarantee gun independence (Cowan and Rowell, 1986). Meyer (1990) has shown why it is also necessary to adjust the monitor for proper gun balance, so that white chromaticity does not vary with luminance. However, not one of a set of eight projection CRTs in the Armstrong Laboratory's DART (Display for Advanced Simulation and Training) could be adjusted to meet Meyer's criterion for gun balance. All eight projectors have gray scales that become greenish below 0.25 V because the luminance output falls off faster for the red and blue components than for the green.

Naturally, effort should be made to obtain monitors that reach these standards if possible. The inevitable shortcomings can then be dealt with by calibration and computation (Brainard, 1989; Neri, 1990). Post and Calhoun (1989) compared several methods for converting from *XYZ* to digital *RGB* and found that their best method had two important features. First, it dealt with the possible dependence of phosphor chromaticity upon voltage by changing the chromaticity matrix according to voltage level (the "variable chromaticity" feature). Second, it created a luminance–voltage LUT for each phosphor by piecewise linear interpolation between the eight or 16 voltage levels actually measured radiometrically on the CRT. This method (PLVC, piecewise linear interpolation with variable chromaticity) gave smaller luminance and chromaticity errors than piecewise linear interpolation with constant chromaticity, linear regression, or the use of either a logarithmic or power function to represent the luminance–voltage relation.

We have used their method of *XYZ* to *RGB* conversion, coupled with an automatic iterative "measure and adjust" procedure (Howard, 1990) suggested by Post (personal communication), to derive different *RGB* command codes that will produce matching colors in each of the DART's eight display windows. We have used the same iterative program on a daily basis to maintain a set of 72 color stimuli, presented on a single CRT, within much narrower tolerance limits for vision research over a period of almost 3 months.

For CRTs viewed in ambient room lighting, the look-up table should be constructed using the luminance and chromaticity of the ambient light as a fourth factor; Holub et al. (1988b) and Howard (1988) describe this procedure.

17.6.2 Look-Up Tables for Devices Other than CRTs

Chromaticity-based matrix methods that work for the CRT do not work for LVPs. Howard (1988) found that inclusion of the dark-field luminance and chromaticity as a separate factor (analogous to ambient light on a CRT) improves LVP color prediction by the PLVC matrix method, but large luminance and chromaticity errors still remain. For device-independent color rendering on an SLVP, we have turned to a different calibration procedure for characterizing the device. Sisson and Pierce (1992) measured two LVPs (an SLVP and a "multiple" LVP) at 512 combinations of eight command levels for each primary; the data include 20 replications of these sets of 512 measurements. A weighted least squares polynomial fit to one set of 512 points predicts LVP output with reasonable accuracy. Table 17.2 presents a sample of Y, u', v' values for which predictions were generated by three methods: polynomial, PLVC, and PLVC with dark field as a separate component. Luminance and chromaticity errors are acceptable in all luminance ranges only with the polynomial method. That method was used in computing the SLVP gamuts shown in Figures 17.2, 17.5, and 17.7, and it will be employed in the Color Modeling Workstation for device-independent color transformations between the modeling station and LVP displays.

Greenberg (1991) notes that flat-panel displays are the most rapidly growing type of electronic displays and that "LCD color measurement is not at all like CRT color measurement." Issues in the design of color matrix displays (CMDs) have been discussed by Jacobsen (1990) and by Silverstein et al. (1990). Users of these new display technologies should anticipate that they may be no more amenable to traditional CRT computation methods than LVPs have been.

17.6.3 Color Resolution: How Much Is Enough?

Nonlinearity of the luminance–voltage relation produces variation in the amount of luminance resolution available over the operating range of each CRT phosphor. Cowan (1983) analyzes the resolution ranges that are likely to produce discreteness artifacts. Murch and Weiman (1990) calculate the effective number of bits at any voltage level as the base 2 logarithm of L_{max}/L_i, where L_{max} is the maximum luminance of the display and L_i is the incremental luminance of one digital step at that voltage level. For their Tektronix monitor with a 12 bit video controller, the effective number of bits varied from 13.8 at 0.36 V to 10.4 at 0.98 V. Some authors take explicit steps to retain the relatively high luminance resolution available at low luminance levels, Lindbloom (1989) by "companding" *RGB* input to the processing stage, Gentile et al. (1990b) by implementing the *RGB* to CIELUV transformation with an 18 bit look-up table.

It is important to realize that the current practice of using *RGB* digital values during image computation and then passing them through a "gamma correction" on the way to the display *sacrifices* the luminance resolution bonus available in the low-luminance region of CRTs. This bonus is also sacrificed when image computation is done on digital values that have *already been gamma-corrected* (a procedure that Covington (1990) recommends for improving antialiasing). Some advanced graphics workstations (such as the Silicon Graphics 4D IRIS) implement a default gamma correction that must be disabled for careful color editing. Table 17.3 (upper half) shows the resolution in bits for four voltage regions when the CRT from Figure 17.2 is uncorrected and when it is corrected for gammas of 1.2, 1.4, and 2.0. As the lower half of Table 17.3 shows, the story is very different for the SLVP, which has luminance resolution that increases slightly with increasing voltage.

COLOR CONTROL IN DIGITAL DISPLAYS 735

Table 17.2 Accuracy of Luminance and Chromaticity Prediction for a Single Light-Valve Projector

Digital code	Measured color output			Error in prediction (polynomial)		Error in prediction (standard matrix)		Error in prediction (matrix + dark field)	
	u'	v'	Y(nits)	Y (%)	$u'v'$ dist.	Y (%)	$u'v'$ dist.	Y (%)	$u'v'$ dist.
0,9,0	0.186	0.508	80	−3.4	0.0102	136	0.0144	0.4	0.0005
9,0,9	0.215	0.422	87	−0.1	0.0051	124	0.0337	0.6	0.0019
0,9,9	0.169	0.433	94	−3.7	0.0043	115	0.0289	0.1	0.0004
9,9,0	0.232	0.514	99	−4.2	0.0033	110	0.0254	0.4	0.0009
27,0,0	0.311	0.509	106	2.5	0.0031	102	0.0618	−0.2	0.0002
9,9,9	0.206	0.448	113	−4.8	0.0025	95	0.0176	0.2	0.0014
27,9,91	0.252	0.467	145	−2.5	0.0041	75	0.0237	0.8	0.0029
0,54,54	0.151	0.401	253	0.3	0.0015	45	0.0243	2.1	0.0003
0,9,255	0.128	0.351	290	−0.5	0.0011	38	0.0355	1.3	0.0011
191,0,191	0.233	0.379	374	1.2	0.0033	74	0.0354	44.7	0.0312
191,9,9	0.370	0.514	382	−0.5	0.0022	31	0.0453	2.9	0.0092
255,9,9	0.380	0.519	433	−1.1	0.0021	28	0.0465	2.9	0.0121
136,54,90	0.246	0.437	451	0.01	0.0010	42	0.0054	18.4	0.0065
191,54,191	0.221	0.412	487	1.3	0.0029	60	0.0249	37.6	0.0253
54,136,90	0.186	0.450	544	−0.5	0.0005	30	0.0085	9.9	0.0075
0,191,191	0.148	0.435	663	−0.02	0.0020	23	0.0089	6.3	0.0008
27,191,27	0.196	0.542	689	−0.6	0.0010	22	0.0070	3.8	0.0028
191,191,0	0.271	0.548	780	0.4	0.0013	18	0.0125	4.6	0.0009

Table 17.3 Color Resolution for CRT and Single LVP[a]

Digital command region	Uncorrected for gamma	Corrected for gamma 1.2	Corrected for gamma 1.4	Corrected for gamma 2.0
Typical CRT monitor				
32–40	14.9	12.1	10.7	9.1
40–48	13	11.2	10.1	8.7
128–136	8	7.9	7.6	7.3
240–248	6.6	6.6	6.7	6.8
Single light-valve projector				
40–48	7.6	7.8	7.8	7.8
128–136	7.8	7.9	7.9	7.9
240–248	9	9	9	9

[a] All values in bits.

Variable luminance resolution can have important consequences for image quality. Gentile et al. (1990b), Hall (1988), and Mulligan (1986) point out that low luminance resolution may result in noticeable quantization artifacts in image areas with low-contrast patterns. Images processed with error diffusion algorithms (Foley et al., 1990, p. 572) received higher acceptability ratings from observers than images processed with ordered dither (Gentile et al., 1990b). DuFlon (1990) describes how a CRT display with a gamma of 1.8 will make shadow areas in an image scanned from a photograph have a "muddy filled-in appearance," while shadow areas in an image captured from a video frame will appear too bright. It is important to know the luminance–voltage function in order to take advantage of its resolution consequences as well as to compensate for its nonlinearity.

17.7 SUMMARY

This survey of the literature on digital display color control leads to the following conclusions:

1. There is a strong movement away from the use of device-dependent *RGB* and *YIQ* color spaces for all color image applications *except* image data compression for signal transmission.
2. For color image computations, perceptual color spaces are favored over *RGB* for their use of perceptually meaningful dimensions, such as luminance and chromaticity or lightness, hue, and saturation.
3. Among perceptual color spaces, the trend is toward using device-independent variables derived from CIE *XYZ* tristimulus values rather than

device-dependent variations of HSV, HSL, HLS, or an *RGB* triangle. The CIE spaces are favored also for their approximate *perceptual uniformity*.
4. Calibration of individual display and hardcopy devices is an essential component of successful color rendering.
5. Color-editing programs are becoming available that enable the user to detect and correct gamut mismatch problems during color selection and editing. Such programs may also use device calibration data to make the final device-dependent conversion to digital command codes for storage or transmission to the output display controller.
6. Use of illumination and reflectance data is recommended by many authors as the preferred basis of color selection for computer-generated imagery. Camera and digitizer data can also be made device independent through calibration of these input devices.
7. Conversion of image data from continuous variables in a device-independent domain to a device-specific integer domain should be done as late as possible in the processing sequence. Such conversion should be planned as a look-up table rather than as a functional transformation.
8. Color graphics designers should keep in mind these perceptual factors affecting displays: (a) the sensitivity of human vision to small luminance steps in low-luminance regions, (b) the greater apparent lightness of saturated colors at low mean luminance levels, and (c) the decreased saturation of all colors in images that are viewed in dim surrounds.

References

Billmeyer, F. W. J. (1988). Quantifying color appearance visually and instrumentally, *Color Res. Appl. 13*: 140–145.

Bockstein, I. M. (1986). Color equalization method and its application to color image processing, *J. Opt. Soc. Am. 3A*: 735–737.

Brainard, D. C., and B. A. Wandell (1990). Calibrated processing of image color, *Color Res. Appl. 15*: 266–271.

Brainard, D. H. (1989). Calibration of a computer controlled color monitor, *Color Res. Appl. 14*: 23–34.

Brill, M. H., and G. Derefeldt (1991). Comparison of reference-white standards for video display units, *Color Res. Appl. 16*: 26–30.

Buchsbaum, G. (1987). Color signal coding: color vision and color television, *Color Res. Appl. 12*: 266–269.

Bumbaca, F., and K. C. Smith (1987). Design and implementation of a colour vision model for computer vision applications, *Comput. Vision, Graphics, Image Processing 39*: 226–245.

Carter, R. C., and E. C. Carter (1983). CIELUV color-difference equations for self-luminous displays, *Color Res. Appl. 8*: 252–253.

Catmull, E. (1979). A tutorial on compensation tables, *Comput. Graphics 13*: 1–7.

Celenk, M. (1990). A color clustering technique for image segmentation, *Comput. Vision, Graphics, Image Processing 52*: 145–170.

CIE (1989). *Mesopic Photometry: History, Special Problems and Practical Solutions*, CIE Publ. 81, Bureau Central de la CIE, Paris.

Covington, M. A. (1990). Smooth views, *Byte 5*: 279–283.

Cowan, W. B. (1983). Discreteness artifacts in raster display systems, in *Colour Vision: Physiology and Psychophysics* (J. D. Mollon and L. T. Sharpe, Eds.), Academic, New York, pp. 145–153.

Cowan, W. B., and N. Rowell (1986). On the gun independence and phosphor constancy of colour video monitors, *Color Res. Appl. 11*: S34-S38.

Dalton, J. C. (1990). Visually optimized color image enhancement, *Proc. SPIE 1250*: 177–189.

DuFlon, R. (1990). Color display systems achieve color consistency, *Comput. Technol. Rev. 10*: 113–115.

Engeldrum, P. G., and J. L. Ingraham (1990). Analysis of white point and phosphor set differences of CRT displays, *Color Res. Appl. 15*: 151–155.

Foley, J. D., A. van Dam, S. K. Feiner, and J. F. Hughes (1990). *Computer Graphics: Principles and Practice*, Addison-Wesley, Reading, MA.

Gentile, R. S., E. Walowit, and J. P. Allebach (1990a). A comparison of techniques for color gamut mismatch compensation, *J. Imaging Technol. 16*: 176–181.

Gentile, R. S., E. Walowit, and J. P. Allebach (1990b). Quantization and multilevel halftoning of color images for near original image quality, *Proc. SPIE 1249*: 249–260.

Gershon, R. (1985). Aspects of perception and computation in color vision, *Comput. Vision, Graphics, Image Processing 32*: 244–277.

Gershon, R., and A. D. Jepson (1989). The computation of color constant descriptors in chromatic images, *Color Res. Appl. 14*: 325–334.

Gorzynski, M., and R. S. Berns (1990). The effects of ambient illumination and image color balance on the perception of neutral in hybrid image display systems, *Proc. SPIE 1250*: 111–118.

Greenberg, D. P. (1991). More accurate simulations at faster rates, *IEEE Comput. Graphics Appl. 11*: 23–29.

Hall, R. (1988). *Illumination and Color in Computer Generated Imagery*, Springer-Verlag, New York.

Hansen, R. (1990). Breaking the color barrier, *Comput. Graphics World 13*(7): 39–48.

Hartmann, W. T., and T. E. Madden (1987). Prediction of display colorimetry from digital video signals, *J. Imaging Technol. 13*: 103–108.

Hedin, C. E., and G. Derefeldt (1990). Palette—a color selection aid for VDU images, *Proc. SPIE 1250*: 165–176.

Holub, R., W. Kearsley, and C. Pearson (1988a). Color systems calibration for graphic arts: I. Input devices, *J. Imaging Technol. 14*: 47–52.

Holub, R., W. Kearsley, and C. Pearson (1988b). Color systems calibration for graphic arts: II. Output devices, *J. Imaging Technol. 14*: 53–60.

Howard, C. M. (1988). Display characteristics of example light-valve projectors, Air Force Human Resources Lab. TR-88-44, Air Force Systems Command, Brooks Air Force Base, TX.

Howard, C. M. (1990). An automated method of device-independent color rendering, *Proc. 1990 Image V Conf.*, pp. 270–273.

Hunt, R. W. G. (1985). Perceptual factors affecting colour order systems, *Color Res. Appl. 10*: 12–19.

Hunt, R. W. G. (1987). *Measuring Colour*, Ellis Horwood, Chichester, England.

Hunt, R. W. G. (1991). Revised colour-appearance model for related and unrelated colours, *Color Res. Appl. 16*: 146–165.

Huntsman, J. R. (1989). A planar-vector based color space for graphic arts color analysis and reproduction, *Color Res. Appl. 14*: 240–259.

Ikegami, H. (1989). New direct color mapping method for reducing the storage capacity of look-up table memory, *Proc. SPIE 1075*: 26–31.

Jacobsen, A. R. (1990). Determination of the optimum gray-scale luminance ramp function for anti-aliasing, *Proc. SPIE 1249*: 202–213.

Jacobson, N., and W. Bender (1989). Strategies for selecting a fixed palette of colors, *Proc. SPIE 1077*: 333–341.

Joblove, G. H., and D. P. Greenberg (1978). Color spaces for computer graphics, *ACM Comput. Graphics 12*: 20–25.

Kotera, H., and K. Kanamori (1989). The new color image processing techniques for hardcopy, *Proc. SPIE 1075*: 252–259.

Krinov, E. L. (1953). Spectral reflectance properties of natural formations, Tech. Translation TT439 (E. Belkov, Tr.), National Research Council of Canada, Ottawa.

Laihanen, P. (1989). Optimization of digital color reproduction on the basis of visual assessment of reproduced images, *Proc. SID 30*(3): 183–190.

Lee, R. L. J. (1988). Colorimetric calibration of a video digitizing system: algorithm and applications, *Color Res. Appl. 13*: 180–186.

Lindbloom, B. J. (1989). Accurate color reproduction for computer graphics applications, *Comput. Graphics 23*: 117–126.

Lucassen, M. P., and J. Walraven (1990). Evaluation of a simple method for color monitor recalibration, *Color Res. Appl. 15*: 321–326.

Luo, M. R., A. A. Clarke, P. A. Rhodes, A. Schappo, S. A. R. Scrivner, and C. J. Tait (1991a). Quantifying color appearance. Part I. LUTCHI color appearance data, *Color Res. Appl. 16*: 166–180.

Luo, M. R., A. A. Clarke, P. A. Rhodes, A. Schappo, S. A. R. Scrivner, and C. J. Tait (1991b). Quantifying color appearance. Part II. Testing colour models performance using LUTCHI colour appearance data, *Color Res. Appl. 16*: 181–197.

MacDonald, L. (1990). Color perception in imaging and graphics, *Adv. Imaging 5*(3): 56–58.

McFall, J. D., J. L. Mitchell, and W. B. Pennebaker (1989). Displaying photographic images on computer monitors with limited colour resolution, *Proc. SPIE 1075*: 179–184.

McMillan, T. (1990). Desktop color publishing, *Comput. Graphics World 13*(1): 37–42.

Maloney, L. T. (1986). Evaluation of linear models of surface reflectance with small numbers of parameters, *J. Opt. Soc. Am. 3A*: 1673–1683.

Maloney, L. T., and B. A. Wandell (1986). Color constancy: a method for recovering surface spectral reflectance, *J. Opt. Soc. Am. 3A*: 29–33.

Mayo, W. T. (1989). Using color to represent low spatial frequencies in speckle degraded images, *Proc. SPIE 1077*: 137–145.

Meyer, G. W. (1990). The importance of gun balancing in monitor calibration, *Proc. SPIE 1250*: 69–79.

Meyer, G. W., and D. P. Greenberg (1986). Color education and color synthesis in computer graphics, *Color Res. Appl. 11*: S39-S44.

Meyer, G. W., and D. P. Greenberg (1987). Perceptual color spaces for computer graphics, in *Color and the Computer* (H. J. Durrett, Ed.), Academic, New York, pp. 83–100.

Mitchell, J. L., W. B. Pennebaker, and C. A. Gonzales (1989). The standardization of color photographic image data compression, *Proc. SPIE 1075*: 101–106.

Mitra, S. K., I. Zarrinnaal, and Y. Wang (1989). Digital processing of color images, *Proc. SPIE 1077*: 132–136.

Mulligan, J. B. (1986). Minimizing quantization errors in digitally-controlled CRT displays, *Color Res. Appl. 11*: S47-S51.

Murano, T., S. Sasaki, and T. Ozaki (1989). Time-varying color image processing system based on a reconfigurable pipeline architecture, *Proc. SPIE 1075*: 18–25.

Murch, G. (1987). Color displays and color science, in *Color and the Computer* (H. J. Durrett, Ed.), Academic, New York, pp. 1–25.

Murch, G., and N. Weiman (1990). Assessing visual grey scale sensitivity on a CRT, *Proc. SPIE 1249*: 214–223.

Nayatani, Y., T. Mori, K. Hashimoto, K. Takahama, and H. Sobagaki (1990). Comparison of color-appearance models, *Color Res. Appl. 15*: 272–284.

Neri, D. F. (1990). Color CRT characterization presented solely in terms of the CIE system, *Percept. Motor Skills 71*: 51–64.

Pointer, M. R. (1980). The gamut of real surface colors, *Color Res. Appl. 5*: 145–155.

Pointer, M. R. (1981). A comparison of the CIE 1976 colour spaces, *Color Res. Appl., 6*: 108–118.

Post, D. L. (1984). CIELUV/CIELAB and self-luminous displays: another perspective, *Color Res. Appl. 9*: 244–245.

Post, D. L., and C. S. Calhoun (1989). An evaluation of methods for producing desired colors on CRT monitors, *Color Res. Appl. 14*: 172–186.

Pratt, W. K. (1991). *Digital Image Processing*, 2nd ed. Wiley, New York.

Rhodes, W. L. (1989). Digital imaging: problems and standards, *Proc. SID 30*(3): 191–195.

Robertson, A. R. (1990). Historical development of CIE recommended color difference equations, *Color Res. Appl. 15*: 167–170.

Rogers, D. F. (1985). *Procedural Elements for Computer Graphics*, McGraw-Hill, New York.

Scheibner, H. (1970). Chrominanz und Farbart (Chromatizität) beim Farbfernsehen, *Opt. Acta 17*: 143–150.

Schreiber, W. F. (1986). A color prepress system using appearance variables, *J. Imaging Technol. 12*: 200–211.

Schuchard, R. A. (1990). Review of colorimetric methods for developing and evaluating uniform CRT display scales, *Opt. Eng. 29*: 378–384.

Schwarz, M. W., W. B. Cowan, and J. C. Beatty (1987). An experimental comparison of RGB, YIQ, LAB, HSV, and opponent color models, *ACM Trans. Graphics 6*: 123–158.
Seim, T., and A. Valberg (1986). Towards a uniform color space: a better formula to describe the Munsell and OSA color scales, *Color Res. Appl. 11*: 11–24.
Sigel, C., R. Abruzze, and J. Munson (1990). Visual artifacts in chromatically subsampled images, *J. Opt. Soc. Am. 7A*: 1969–1975.
Silverstein, L. D., J. P. Lepkowski, R. C. Carter, and E. C. Carter (1986). Modeling of display color parameters and algorithmic color selection, *Proc. SPIE 624*: 26–35.
Silverstein, L. D., J. H. Krantz, F. E. Gomer, Y. Y. Yeh, and R. W. Monty (1990). Effects of spatial sampling and luminance quantization on the image quality of color matrix displays, *J. Opt. Soc. Amer. 7A*: 1955–1968.
Sisson, N., and B. Pierce (1992). Predicting light-valve luminance and chrominance, Technical Report in preparation at Armstrong Laboratory.
Smith, A. R. (1978). Color gamut transform pairs, *Comput. Graphics 12*: 12–19.
Spooner, D. L. (1990). Derivation of the color gamut plots of prepress proofing systems, *Proc. SPIE 1250*: 80–89.
Sproson, W. N. (1983). *Colour Science in Television and Display Systems*, Adam Hilger, Bristol, U.K.
Stone, M. C. (1990). Color printing for computer graphics, in *Computer Graphics Techniques: Theory and Practice* (D. F. Rogers and R. A. Earnshaw, Eds.), Springer-Verlag, New York, pp. 79–127.
Stone, M. C., W. B. Cowan, and J. C. Beatty (1986). A description of the color-reproduction methods used for this issue of *Color Research and Application, Color Res. Appl. 11*: S83-S88.
Strickland, R. N., C.-S Kim, and W. F. McDonnell (1987). Digital color image enhancement based on the saturation component, *Opt. Eng. 26*: 609–616.
Tajima, J. (1983). Uniform color scale applications to computer graphics, *Comput. Vision, Graphics, Image Processing 21*: 305–325.
Taylor, J. M., and G. M. Murch (1986). The effective use of color in visual displays: text and graphics applications, *Color Res. Appl. 11*: S3-S10.
Taylor, J. M., G. M. Murch, and P. A. McManus (1989). TekHVC™: a uniform perceptual color system for display users, *Proc. SID 30*: 15–21.
Tsai, Y. T. (1989). Real-time architecture for error-tolerant color picture compression, *Proc. SPIE 1075*: 140–147.
Virgin, L., G. Murch, B. TenKate, and P. McManus (1986). Colorimetric calibration and specification of CRT systems, *SID 86 Dig. 17*: 334–337.
Wan, S. J., P. Prusinkiewicz, and S. K. M. Wong (1990). Variance-based color image quantization for frame buffer display, *Color Res. Appl. 15*: 52–58.
Wandell, B. A. (1986). Color rendering of color camera data, *Color Res. Appl. 11*: S30-S33.
Ware, C., and W. Cowan (1990). The RGYB color geometry, *ACM Trans. Graphics 9*: 226–232.
Watson, A. B. (1990). Perceptual-components architecture for digital video, *J. Opt. Soc. Am. 7A*: 1943–1954.

Wyszecki, G. (1986). Color appearance, in *Handbook of Perception and Human Performance* (K. R. Boff, L. Kaufman, and J. P. Thomas, Eds.), Wiley, New York, Chapter 9.

Wyszecki, G., and W. S. Stiles (1982). *Color Science: Concepts and Methods, Quantitative Data and Formulae*, Wiley, New York.

18

The Human Factors of Helmet-Mounted Displays and Sights

Maxwell J. Wells

Logicon Technical Services, Inc.
Dayton, Ohio

Michael Haas

Harry G. Armstrong Aerospace Medical Research Laboratory,
Wright-Patterson Air Force Base, Ohio

18.1 INTRODUCTION

Since their inception in the mid-1960s, helmet-mounted displays (HMDs) have been the cause of both excitement and frustration—excitement because of the possibilities offered by a display that could be coupled directly to the head, and frustration because of the difficulties associated with constructing and using such a device. The potential applications of HMDs seem endless and include improved person–machine interface, enhanced visual performance at night and in low visibility, increased naturalness in the presentation and display of information, sensor fusion, increased situation awareness, and access to virtual worlds. However, insufficient design specification, complicated display technologies, and unrealistic and premature expectations have sometimes been the reality. Fortunately, the potential uses for the displays, and the enthusiasm of their advocates, have ensured sufficient resources for the development and maturation of the technology. With this maturation has come increased utilization and the realization that there are many human factors issues associated with the use of HMDs that must be addressed if these devices are to meet their expectations. Perhaps more than any other electro-optical device, HMDs have challenged the human factors community to extend and utilize their body of knowledge. The purpose of this chapter is to document the status of research and understanding of the human factors issues of HMDs.

The study of human factors is the study of human characteristics that are applicable to the design of systems and devices of all kinds. Therefore, very few topics fall outside the scope of human factors. This chapter is organized into four main categories: visibility, comfort, fidelity, and intuitiveness. This is an arbitrary taxonomic structure, imposed on a complex set of issues in an attempt to simplify their presentation. There is, by necessity, some repetition of factors between categories. However, any taxonomy will exhibit this feature, which reflects the fact that the issues are not mutually exclusive. Some topics are not included in this chapter either because they fall outside the range of our expertise or because the human factors issues are not specific to HMDs. Two such topics are safety and maintainability.

The term *helmet-mounted display* describes a device in which a display is attached to the head by means of a helmet. However, as the applications for these devices transition into the civilian arena, the term *head-mounted display* may become more appropriate. In either case, the acronym HMD applies. The terms are used interchangeably in this chapter. A display is a device that presents information to an observer. Usually, helmet-mounted display refers to a visual display; however, it could be appropriate to refer to a helmet-mounted auditory or tactile display. This chapter addresses only the human factors issues of visual displays. A head- or helmet-mounted sight refers to a system comprising a reticle attached to the head and a transducer for measuring head orientation. The term *head-coupled system* is used to describe a system consisting of a head position transducer and a device that is pointed by the output from that transducer. An example of a head-coupled system is the Apache integrated helmet and display sighting system (IHADSS), in which a forward-looking infrared sensor (FLIR) is pointed by using head position. In the IHADSS, the output from the FLIR is viewed on the HMD. However, it is not necessary to have an HMD for the system to be head-coupled. An example of such a system is a head-steered infrared missile seeker, in which the only feedback about seeker orientation is provided open-loop, by the orientation of the head-mounted reticle.

Finally, reference is made to helmet-mounted simulators, in particular, the visually coupled airborne systems simulator (VCASS). This consists of a non-see-through, binocular, helmet-mounted display, a helmet position sensor, and an image generator (Fig. 18.1). The device was constructed to test the feasibility of some of the proposed uses for HMDs (Furness and Kocian, 1986). VCASS presents the wearer with a field of view 120° wide by 60° high, of computer-generated imagery. Head movement information is used by the image generator to produce an image appropriate to where the wearer's head is pointing. In addition, VCASS consists of a mock-up of an aircraft cockpit, with controls that also provide information to the image generator. Sitting inside the cockpit, wearing the helmet, an observer can be presented with simulated moving scenes as if he or she were actively flying an aircraft.

HUMAN FACTORS OF HMDS

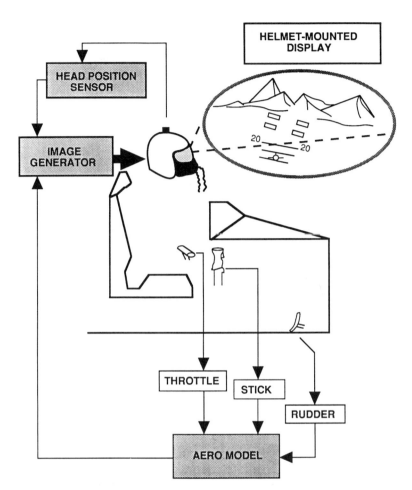

Figure 18.1 Diagram of the Visually Coupled Airborne Systems Simulator (VCASS), an example of a helmet-mounted simulator.

18.2 VISIBILITY

18.2.1 Light management

Light management refers to the methods by which the images on the display are made visible by controlling their luminance and contrast. Many of the questions regarding the visibility of HMDs can be answered using results from studies conducted with video display terminals (VDTs). For example, the ANSI standard for VDTs (ANSI, 1988) specifies that "Either the character or its background, whichever is of higher luminance, shall be able to achieve a luminance of 35 cd/m² (10 foot-lamberts) or more." The ANSI requirements for contrast are that "Character luminance modulation shall be equal to or greater than 0.5. A luminance modulation of at least 0.75 is preferred." Luminance modulation (M) is defined as $M = (L_{max} - L_{min})/(L_{max} + L_{min})$, where L_{max} is the higher luminance of the characters or the background and L_{min} is the lower luminance of the two.

The range of luminance in the environments in which HMDs may be used varies from less than 0.001 FL (0.003 cd/m²) on a moonless night to 20,000 FL (68,500 cd/m²) during the day with clouds in the sky (Buchroeder, 1989). Because some of the required light levels on the HMD exceed those that it is currently possible to produce, it is necessary to determine the type of information to be displayed (imagery, alphanumerics, or both) and the level of ambient illumination in which the display will be used (e.g., night or day, inside or outside).

The luminance requirements for alphanumerics are less stringent than those for imagery (Task, 1991). This is because the visual system needs to perceive, and the display needs to produce, only two shades of gray (on and off). High ambient illumination may wash out several shades of gray present in imagery, with subsequent loss of interpretability. Solutions to washout include reducing background scene luminance, either totally, with a non-see-through combiner, or partially, with a combiner with low or variable transmittance. A completely opaque combiner effectively occludes a part of the field of view (see Section 18.2.2), and a variable transmittance combiner that is too dark may also produce a performance decrement. Individual differences between observers also dictate the provision of readily accessible controls for luminance and contrast.

18.2.2 Field of view

When discussing field of view (FOV), the term *instantaneous FOV* is used to describe the FOV of a sensor (the eye or an electro-optical sensor) at any instant, that is, with no sensor movement. The instantaneous, achromatic FOV of both eyes combined is approximately $\pm 60°$ vertically and $\pm 100°$ horizontally (Boff and Lincoln, 1988). It is not possible, with current technology, to replicate the FOV of the eyes with an HMD. Even if it were, there would be a shortfall when the eyes moved in the head. Therefore, an important human factors issue is the minimum FOV required to achieve acceptable performance.

The importance of FOV size depends on the task for which the device will be used. Figure 18.2 outlines some applications depending on whether the HMD is see-through or opaque and whether its primary function is the display of imagery, symbology, or alphanumerics.

Two factors should be considered in discussing field of view. These are the

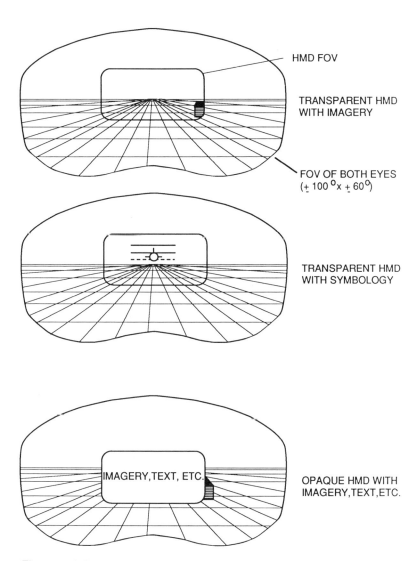

Figure 18.2 An illustration of transparent and opaque HMDs superimposed on the instantaneous FOV of both eyes.

field of view afforded by the HMD (the intradisplay FOV) and the field of view outside of the HMD (the paradisplay FOV). This latter aspect has sometimes been ignored in display design and evaluation. In most applications, the limits of the paradisplay FOV are those of the eyes' instantaneous FOV. However, within that paradisplay FOV there will be occluded areas. These may be produced by the intradisplay FOV or by HMD supports, oxygen, or gas masks, or other protective gear. In the case of an HMD with an opaque combiner, the occluded area may be large. An application where the intra- and paradisplay FOVs are the same size is VCASS, where the apparatus is arranged to occlude everything except the information on the HMD. An example of the use of this arrangement is described in subsequent sections.

Where a person is looking (gaze position or line of sight) is a combination of head position and eye-in-head position. Normal visual inspection utilizes a synergistic coordination between the head and the eyes (Gresty, 1974; Barnes, 1979). In many instances, the mobility of the head and eyes can compensate for intrusions into the paradisplay FOV or limitations in the intradisplay FOV. However, an HMD with a small FOV forces the wearer to reduce the amount of eye-in-head motion during head movements in order to see the information on the display. A small FOV also requires the wearer to make more head movements in order to view the world. Similarly, a display that obscures a large or important portion of the paradisplay FOV forces more head movement in order to see around the blind spot.

See-Through HMD for Imagery

Potential applications include those where images are overlaid on, and incorporated into, the real world. Examples are presentation of a radar image at the angular position of an object beyond visual range (egocentric radar), previewing possible modifications to an existing building, and visualizing data on tectonic stress. In an experiment that simulated one such application, Wells and Venturino (1990) had subjects search for and then shoot three, six, or nine targets. The targets were placed in random locations above the subject, along with computer-generated terrain. Both were viewed on a helmet-mounted simulator with a FOV 120° wide by 60° high. However, the targets were visible only when they were located in the central portion of the display, in an area that varied, in different presentations, from 120° wide to 20° wide. The FOV in which the terrain was visible simulated the paradisplay FOV, and the FOV in which the targets were visible simulated the intradisplay FOV. The observer had to locate the targets and then monitor them for changes in shape. A target that changed shape constituted a threat and had to be shot within 5 sec, or it would shoot the observer. Reducing the size of the intradisplay FOV (the area on which the targets were visible) reduced performance (Fig. 18.3). Subjects were threatened for longer times, and shot fewer targets, with smaller FOVs than with larger FOVs. For a simple task

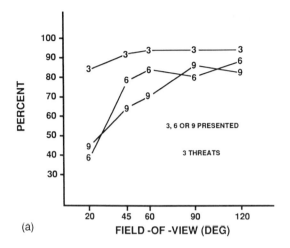

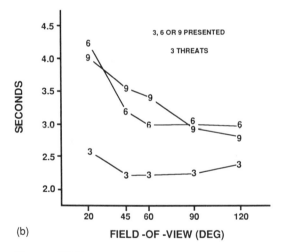

Figure 18.3 Mean performance as a function of FOV size and number of targets, from 17 subjects using a simulated, transparent HMD. (a) Threats hit; (b) time threatened.

(three targets), a FOV of 20° was adequate, whereas for the most complex task (nine targets), a FOV of 60° yielded better performance.

The Wells and Venturino experiment simulated a generic air-to-air application in which objects beyond visual range are made visible by suitable processing of radar images. The paradisplay FOV size was limited by the experimental apparatus to 120° wide by 60° high, and the intradisplay FOV size varied between 20° and 120° in width. Other applications may be more or less sensitive to the size of both intra- and paradisplay FOVs. In some applications it may be desirable to occlude paradisplay FOV, similar to applying blinders, to concentrate attention on the portion of the FOV over which the HMD is superimposing information.

See-Through HMD with Symbology and Alphanumerics

Applications of see-through HMDs with symbology and alphanumerics include portrayal of flight-critical information and cuing about target location. The advantage of an HMD over a fixed-position HUD, for this type of information portrayal, is that information can be made available irrespective of head position. The "all-aspect HUD" has been demonstrated to be capable of allowing users to spend more time with their heads pointing outside the cockpit during certain critical phases of a mission (Osgood et al., 1991).

The minimum HMD FOV size for information-presentation applications is similar to the FOV sizes for the fixed-position HUDs that they replace (15–30°). The type of information being presented (symbology and alphanumerics) generally requires the wearer to use foveal vision. Increasing the area over which the information is displayed (e.g., by increasing the FOV size) therefore increases the amount of eye movement required to view the information. Based on research with panel-mounted displays, information relevant to one task should be presented in an area subtending about 9° at the eye (Baker, 1962; Enoch, 1959).

In aircraft control applications, restrictions to paradisplay FOV should be minimized. Important events occur outside the HMD FOV. If the design of the display necessitates obscuring a part of the natural FOV, portions above the line of sight are considered to be more important to flying performance (Dryden, 1991). Other applications may impose other restrictions.

Opaque HMD with Imagery

One of the applications driving HMD technology is the use of head-coupled systems for night flying. Such a system might consist of an HMD, a head position sensor, and a head-steered electro-optical sensor (FLIR or low-light-level television). Current (or soon-to-be-current) systems do not have opaque combiners, but discussion of these systems is included in this section because the imagery they present is displayed against a (mostly) black and/or empty background.

In the night vision application, all of the information from outside the cockpit is derived from the restricted FOV of the HMD. Despite this limited FOV, it appears that some tasks can be accomplished adequately. For example, in a simulated night attack, a 10°-dive bomb delivery was performed with FOVs of 20°, 30°, 40°, 60°, and 80° (Osgood and Wells, 1991). Subjects flew toward a target whose position was indicated on their HUDs. They then executed a pop-up maneuver, acquired the target visually on their HMD, and aimed and released a bomb. In all conditions, there was an offset between the indicated target location (indicated on the HUD) and the actual location. The offset varied from 1000 ft (304.8 m) to 6000 ft (1828.7m) and forced the subjects to search for the target during the pop-up maneuver. It was possible to perform the task with all FOV sizes, but with larger FOVs targets were acquired sooner. Target acquisition precipitated a chain of events that culminated in bomb release. Bombs were released at higher altitudes, with less possibility of shrapnel damage to the aircraft, when the targets were sighted earlier. There was no significant difference in the time to find targets with FOVs larger than 40° in the large-offset conditions (see Fig. 18.4). Further analysis of the data suggests that the subjects who acquired the targets early made changes to their flight profile during ascent that allowed them to aim at the target earlier on the descent (Wells and Osgood, 1991). However, extrapolation of these results to other flying tasks must be accomplished with caution.

Restrictions to the paradisplay FOV in night vision applications may affect performance. Reports by pilots who have flown with head-steered FLIR with their paradisplay FOV occluded (daylight test flights with the head covered) report unease with the visual arrangement (Giles, 1991). There are at least two explanations for this. The first is that there is a loss of visual information from outside the cockpit. The second is that there is a loss of non-instrument information from inside the cockpit. It is unclear how much information, if any, is available from outside the cockpit at night, and how beneficial that information might be. Indeed, there are reports of outside-the-cockpit information being detrimental to performance at night (e.g., Kraft, 1978). Information from inside the cockpit may help inform the pilot of the position of his head with respect to the aircraft. Without this information, there may be a tendency for disorientation (see Section 18.5.2). For whatever reason, it would appear that occlusion of the FOV outside the HMD should be avoided in this type of application.

Opaque HMD with Symbology and Alphanumerics

The advent of small, inexpensive head-mounted displays [e.g., the Private Eye, (Peli, 1990)] makes the arrangement of an opaque HMD for displaying symbology and text a practical possibility. Applications include personal computer terminals, fax machines, and electronic reference manuals. For these applications, a FOV similar to those used in the applications that the device replaces would seem appropriate.

The Private Eye has a FOV 21° wide by 14° high. With an opaque HMD, increasing the display FOV increases the intrusion into the paradisplay FOV.

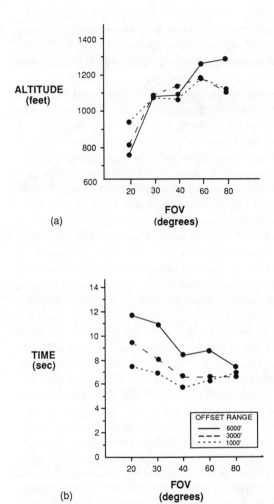

Figure 18.4 Mean times after the start of the pop maneuver targets were sighted (upper) and altitudes at which bombs were released (lower) by five pilots during a simulated air-to-ground attack. The attacks were conducted as if at night, using a head-steered FLIR.

Depending on the application, this may aid or hinder performance. For example, the Private Eye provides a monocular image and may produce binocular rivalry (Peli, 1990; also see Section 18.2.4). In this case, occlusion of the other eye (decreasing the paradisplay FOV) may be beneficial.

18.2.3 Vibration

Helmet-mounted displays are particularly susceptible to the effects of whole-body vibration. For example, Furness (1981) reported that, at some frequencies, the reading error produced with a panel-mounted display was produced with HMD using one-tenth the amplitude of vibration. Wells and Griffin (1987a) reported that the number of numerals read correctly from an HMD in a helicopter decreased from 2.4 sec^{-1} while stationary on the ground to 1.0 sec^{-1} during in-flight vibration. The reason for the vibration-induced decrement in performance is relative motion between the line of sight of the eye and the optical axis of the HMD. Rotational oscillation of the head causes vibration of the HMD, but the eyes, under the influence of the vestibular ocular reflex (VOR), remain space-stable (Benson and Barnes, 1978; Wells, 1983). The VOR, which normally serves to keep images stable on the retina during body movement and vibration, acts to degrade performance with HMDs.

Vibration is also a factor in the use of helmet-mounted sights and head-coupled systems, where head movement is used to direct weapons, sensors, and other systems. Under normal circumstances a person can aim his/her head at a stationary target with pitch and yaw axis errors as small as 0.1° root mean square (Wells and Griffin, 1987b). Tracking moving targets with the head is easily learned (Wells and Griffin, 1987c) and, depending on the difficulty of tracking the target motion, can be accomplished successfully. Whole-body vibration disrupts both head aiming and head tracking. With random vibration, aiming at a stationary target is disrupted by the vibration-induced head motion (vibration breakthrough). However, the decrement in head tracking during vibration is greater than the sum of the decrement caused by tracking and the decrement caused by vibration breakthrough (see Fig. 18.5). It is likely that the additional decrement results from attempts to reduce the error between the head-mounted reticle and the target, which, due to lags in the response of the head, result in greater error.

Solutions to the problems caused by vibration should, in principle, begin at the source, with engineering solutions to reduce the amplitude of vibration of the vehicle or avoidance of high-vibration flight regimes. If these solutions are not feasible or possible, reductions in vibration transmission to the operator may be accomplished by seat design or placement. Another approach has been active image stabilization for helmet-mounted displays (Wells and Griffin, 1984; Velger and Merhav, 1986), in which the display image was deflected to match the line

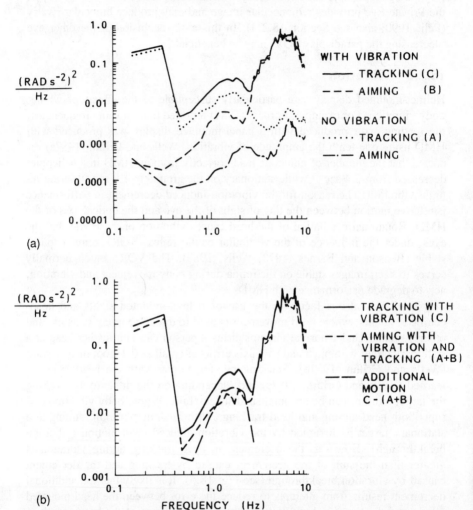

Figure 18.5 Power spectra of pitch axis head motion while aiming at or tracking a target, with and without random whole-body vibration. Mean data for six subjects.

of sight of the eye. In this approach, accelerometers are attached to the display and the image is space-stabilized. To be successful, this technique requires that the stabilization system distinguish between vibration-induced head motion and voluntary head motion. One method, used successfully in flight trials (Wells and Griffin, 1987a), used adaptive filters that altered their high-pass characteristics in response to the signals from the accelerometers. A diagram of the adaptive filter is shown in Figure 18.6 (Lewis et al., 1987), and performance with the system is shown in Figure 18.7.

Increasing the size of the alphanumerics can also reduce the effect of vibration (Lewis and Griffin, 1979), but at the cost of decreased information density. Some background luminance conditions have also been shown to affect reading performance during vibration (Furness, 1981). Generally, increasing the background luminance improves performance. The worst conditions for reading during vibration are bright letters against a black background (Wells and Griffin, 1987a).

The frequencies of vibration that cause most problems are those in the 2–10 Hz range, where there is sufficient transmission of vibration to the head to cause rotational head displacement (Lewis and Griffin, 1980). Vehicles whose vibration spectra have sufficient energy in this range of frequencies include off-road vehicles and both fixed-wing and rotary-wing aircraft (Boff and Lincoln, 1988). With fixed-wing aircraft, vibration is typically turbulence-induced and random, with periods of low vibration amplitude interspersed with periods of high vibration amplitude. Rotary-wing aircraft are also subjected to random turbulence, and in addition there can be significant components at the rotor frequency, around 5 Hz (Jackson and Grimster, 1972) . With both types of aircraft, certain phases of a flight—for example, transition to hover in a helicopter or taking off and landing in a fighter—may generate more vibration, possibly with detrimental effects on the use of the HMD.

18.2.4 Binocular rivalry

When the eyes are presented with different scenes, the image from one eye may be totally or partially suppressed. This is called retinal rivalry. It may occur, for example, with a monocular HMD of the type used in the U.S. Army's AH-64 helicopter, where the output from a head-steered FLIR is viewed by the right eye and the cockpit instruments are viewed by the left eye. Users of the systems have reported difficulty making the necessary attention switches between the eyes, particularly as a mission progresses (Bennett and Hart, 1987).

Binocular rivalry is affected by a large number of factors including ocular dominance, voluntary control, and the luminance, color, and contrast of the images (Levelt, 1968; Laycock, 1976). Suppression may be total, where the image from the suppressed eye is not perceived, or partial, so that only parts of the

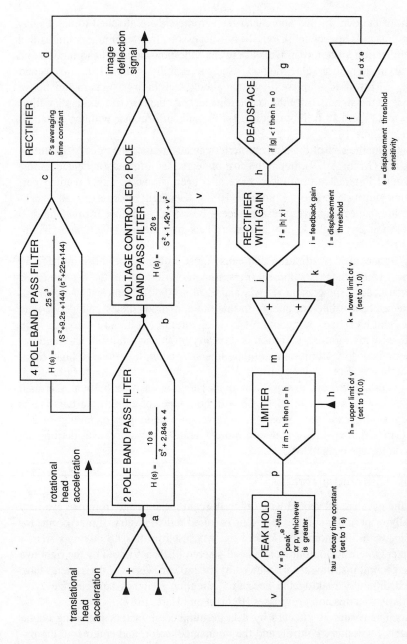

Figure 18.6 Schematic diagram of an adaptive HMD image deflection system. The bandpass filters are equivalent to combined integrators and high-pass filters.

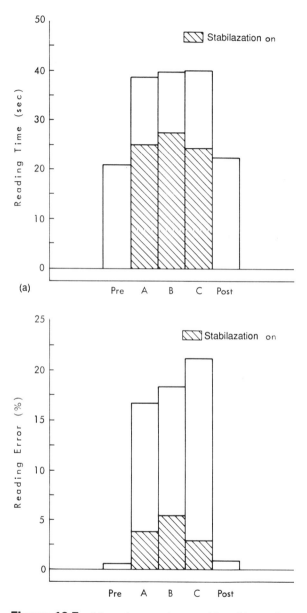

Figure 18.7 Mean times and errors (six subjects, four replications per subject) for reading an array of 50 numerals on a helmet-mounted display with and without stabilization of the image. Pre, preflight; A, nil wind hover; B, left at 15 knots; C, 50 knots at <30 m; Post, postflight.

image from each eye contribute to the total percept (Arditi, 1986). The suppressed image may alternate with the perceived image over a period of 1–4 sec. It has been argued that, since normal binocular viewing can present the eyes with conflicting visual information, and since we normally perceive a single image, suppression may be a mechanism for reducing image disparities (Arditi, 1986). This argument conflicts with the conventional wisdom that the disparate views from each eye are fused into one image.

18.2.5 Interference

At issue is the effect of images displayed on an HMD on the visibility of objects in the real world. Interference might arise, for example, with an all-aspect HUD, in which task-critical information may be in front of the eyes irrespective of head position. An HMD image may interfere with the perception of images from the outside world by completely obscuring those images or by reducing their contrast. These problems can be minimized or overcome by the provision of suitable controls to turn off the display or reduce its brightness, and by training.

Other factors to consider are the effects of using spectrally selective combiners, or displaying color, on the perception of scenes from the outside world. For example, it is not uncommon to use an HMD combiner that transmits all wavelengths except those emitted by its CRT, which it reflects to the eye. This has the advantage of minimally decreasing the transmission of light from the outside world while maximizing the visibility of the information on the CRT. However, if there are panel-mounted CRTs that use the same phosphors as those on the HMD, the information on those displays will not pass through the combiner.

Typically, the combiner may filter out a narrow portion of the visible electromagnetic spectrum, and so the effect on the perception of the color of the outside world will be limited. This may not be the case if the combiner is tuned to reflect to the eye more than one wavelength—for example, for displaying more than one color or for protection against lasers—or if a colored visor is used. The effect of a colored (yellow) visor on color perception was reported by Post (1992), who tested color-naming performance on a CRT. Although the full range of chromaticities was on the display, the subjects, looking through a yellow visor, saw them all as green, yellow, or orange; that is, they saw no blue. The reduction was caused by the yellow visor removing the blue from the phosphor emission, leaving only the weak green emitted by the blue phosphor. The visor thus converted the RGB three-color CRT into a two-color CRT. This may have implications for the perception of color-coded information on a panel-mounted display.

There is another potential problem with color HMDs. Color from a displayed image may combine with color from the outside world to produce a false percept of one or both images.

18.2.6 Monocular vs. Binocular vs. Biocular

In a monocular HMD, one image source (e.g., a CRT) and one set of optics present an image to one eye. A biocular HMD has one image source, the image from which is relayed to both eyes via two sets of optics. A binocular HMD utilizes two image sources and two sets of optics and presents slightly different images to the two eyes.

Monocular displays have the advantage of low weight and small volume. They are also simpler to integrate within or on the helmet, and they require alignment with only one eye. Monocular displays also restrict less of the paradisplay FOV, as one eye has almost unrestricted view of the environment. This may be advantageous, for example, when the display blanks out owing to failure of the CRT or the sensor, or for some other reason, during use of an airborne head-coupled system at night. Vision from the unrestricted eye could be used to transition to instrument flight using the panel-mounted displays. The unaided eye can also maintain dark adaptation and color vision and thus may be used to routinely perform a variety of tasks in the cockpit (Rash et al., 1990). Disadvantages of a monocular HMD include potential problems with binocular rivalry, lack of stereopsis, and reduced intradisplay FOV size. Also, two eyes are better at discerning small contrast differences, that is, at seeing low-contrast objects.

Biocular displays and binocular displays present images to both eyes, but biocular displays are lighter. The assumed advantage of viewing an image with both eyes is sometimes expressed as $\sqrt{2}$, based on the Green and Swets (1966) model of binocular summation. According to that model, the binocular response is obtained by sampling from two independent noisy channels. Because the signal is correlated in the two eyes, the resultant sampling distribution will have the summed signal strength of the monocular channels. However, the uncorrelated noise sometimes adds and sometimes subtracts in the two channels, increasing only as the square root of the sum of the squares. Thus, if the signal and noise are equal in the two eyes, the predicted binocular improvement is $\sqrt{2}$, or about 41% (Arditi, 1986). When little noise is present (high signal/noise ratio) the $\sqrt{2}$ rule is inapplicable. It should be noted that the actually observed advantages of using two eyes are nearly always small, exhibiting large intersubject variability.

The disadvantages of biocular displays, relative to monocular displays, include increased weight and size of the added optics and the need to align the images for both eyes. Adjustment of the optics must include provision for interpupillary distance of observers, and alignment of the optical axes of the two channels for divergence, convergence, and vertical disparity.

Binocular displays exhibit the advantage of presenting images to both eyes, as do biocular displays, and in addition they can provide binocular disparities.

provide binocular disparities. These disparities, or differences in the images from the two channels, allow the visual system to use stereopsis to transform the monocular images into an impression of solid three-dimensional space. The disadvantages of binocular displays are similar to the disadvantages of biocular displays (increased weight and size, adjustments for two eyes). In addition, in order to provide retinal disparities, there is a need for two sensors to provide images from two different locations.

The perception of depth is considered important in some aspects of flight and is aided by stereopsis. However, there are visual cues other than stereopsis by which observers can achieve depth perception, including motion parallax, object interposition, linear perspective, and retinal image size, all of which can be achieved monocularly (or with a biocular display). Furthermore, stereoscopic cues are most effective for objects up to 30 ft (9 m) from the observer (Stevens, 1982) and lose much of their effectiveness beyond about 200 ft (61 m) (Tredici, 1985). Thus, although stereopsis is useful for flight close to the ground (e.g., helicopters), it is less useful for jet aircraft, where the locus of attention is far forward of the aircraft and flight is seldom at altitudes as low as 200 ft (61 m).

18.2.7 Spatial Resolution

Resolution is an objective measure typically used in conjunction with other parameters, such as luminance or contrast ratio, to fully describe the quality of a display. Subjective measures of image quality are correlated with these objective measures. While the conventional definition of resolution is associated with the spatial characteristics of a display, temporal resolution characteristics must also be considered when dealing with helmet-mounted displays. Temporal resolution is discussed in Section 18.4.3.

Information on an HMD may be computer-generated or derived from other sources such as a video camera or infrared sensor. Whatever the source, the input to the HMD system has a limited information content. This input is passed through display electronics, which change it to a format that the image source on the HMD can accept. The image source transforms the electronic format to a light format, which is passed through the HMD optics. The HMD optics further transform the light. Thus, the final spatial resolution of the HMD begins with the information content of the input and is modified at the HMD image source, display electronics, and optics.

The effects of HMD spatial resolution can be considered for two types of visual information: alphanumerics (and symbology) and imagery. There are specifications for the resolution required for alphanumerics for applications such as text processing, data entry, and data enquiry. It can be assumed that these specifications also apply to symbology. The ANSI standard (ANSI, 1988) for

HUMAN FACTORS OF HMDS 761

VDTs uses the modulation transfer function area (MTFA). The MTFA is an area on a graph of luminance modulation (on the ordinate) against spatial frequency (on the abscissa). It is the area above the eye's threshold of detectability (the contrast threshold function, CTF) that is also below the modulation transfer function (MTF) of the display. The resultant area is in units of modulation (a scalar) × cycles/degree. The standard recommends a value of at least 5.

For imagery, the applications determine the HMD resolution requirements. For example, if the task is to locate a target, the spatial resolution of a scene need not be very high to perform the task. However, if the wearer must identify that object, the required spatial resolution is greater. In order to address some of these questions, a flight simulation was performed at McDonnell Douglas in 1990 in which a FLIR scene was used in the performance of a night close-air-support mission. This study was performed to establish the operational effects of HMD field-of-view size and resolution. The FLIR was simulated with several fields of view, ranging from 20° to 60° circular, both as a HUD image (fixed to the boresight of the aircraft) and as an HMD image (line of sight slaved to the pilot's head). Two levels of resolution were used, a 512 line format at all FOV sizes and a 1024 line format at all FOV sizes. With both line formats, as the field of view increased, the angular resolution (lines per degree) decreased. The hypothesis underlying the study was that performance of the task would be limited by field of view in the 20°–30° range (i.e., that the resolution loss going from 20° to 30° would not affect performance) but that resolution loss would limit performance with the larger FOVs. Performance measures included bomb miss distance and exposure to surface-to-air threats. The results showed decreasing operator performance as the FOVs were increased. Best performance was obtained with the 20° FOV. The explanation for this somewhat unexpected result is that the mission, a gravity weapon delivery, was resolution-limited. The target had to be detected and identified from far away, when the aircraft was high and had not yet been committed to a descent toward the target. At those distances, angular resolution with the larger FOVs was inadequate. The pilot could not see the targets well enough with the larger FOVs.

18.3 COMFORT

18.3.1 Fit

There is an increasing requirement to integrate a number of electro-optical components on, or into, the pilot's helmet. These include HMDs and night vision goggles. This trend introduces new problems regarding helmet fit. Traditional methods of fitting helmets to individuals involved assigning one of three or four helmet shells, based on head size, as measured by some combination of head breadth, width, and circumference. The shells were designed to accommodate the majority of head sizes. Differences between the size and shape of the head and the size and shape of the shell were accounted for by the use of foam pads or adjustable straps. Although the helmets usually fitted snugly, two individuals with the same head size, but different head shape, fitted into the helmets differently. The addition of helmet-mounted optics has removed some of this latitude associated with helmet fit. This is illustrated by Figure 18.8. The illustration on the left shows the outlines of the heads of two subjects, aligned so that the pupils are superimposed (the pupil locations are shown by the X's labeled "P"). The two horizontal lines pass through the pupils and are parallel with the Frankfort plane. The same outlines are shown on the right with the crania aligned. The heads are the same size but are shaped differently. Thus, with the crania aligned, the locations of the faces with respect to the tops of the skulls are different. The same-size helmet shell would align differently with the two heads. An HMD adjusted to fit one subject would not fit the other subject. It may be possible to make the helmet and/or the HMD attachment more adjustable, but at the expense of added helmet weight.

Helmet fit has implications for the amount of helmet-on-head movement that will occur during vibration or high-g maneuvering. During aircraft vibration, most head and helmet movement is in the pitch axis. With conventional helmet fasteners, it is impossible to remove all helmet movement in that axis because of

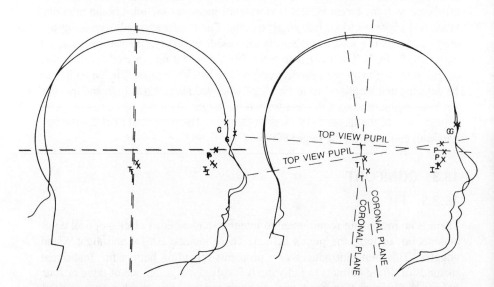

Figure 18.8 Profiles of the heads of two subjects arranged so that the pupils (P) overlay (left) or so the crania are aligned (right). [From Robinette and Whitestone (in press).]

the movement of the skin on the scalp. Ideally, a properly fitting helmet will reduce the amount of hysteresis, so that once displaced, the helmet-mounted optics will return to, or close to, their former position in front of the eye. During high-g maneuvers, an ideal helmet would be rigid in the z axis and would not have a tendency to rotate on the head. Neither of these requirements is currently achieved (Kennedy and Kroemer, 1973; Jarrett, 1978).

The added complications in helmet design produced by intersubject differences in head size and shape, eye position, and line of sight are compounded by intrasubject differences between the eyes (see Section 18.2.6).

18.3.2 Display Color

The color of the symbology or background on an HMD scene may affect the comfort and/or performance of the user. For example, research at the Royal Aircraft Establishment investigated the effects of the colors of the primary flight display. The colors tested were red, brown, blue, cyan, amber, white, magenta, and green. Magenta was found to be the most uncomfortable color (Caldow, 1984). The reasons for this visual discomfort are unclear, but this effect should be considered in the design of HMDs, particularly when they may subtend a large portion of the FOV for long periods.

There are recommendations for color use on video display terminals that may also apply to HMDs (ANSI, 1988). The use of pure blue on a dark background is not recommended for text, for thin lines, or for high-resolution information because the normal eye is blue-blind in the central fovea. Simultaneous presentation of pure blue and pure red may result in chromostereopsis, the apparent difference in depth between two objects or areas of different color lying in the same plane. A similar effect can be obtained with red and green and with blue and green. The effect is thought to result from different dispersions of colored light due to chromatic aberration of the eye. The ANSI standard also contains recommendations for choosing colors for color coding and for color contrast.

Also of relevance to display color is the issue of color blindness. About 8% of males and 1% of females have color deficiencies. The ANSI standard recommends that pure red should be avoided, arguing that individuals with the most common form of color deficiency (red–green), experience difficulty distinguishing colors that differ only in the amount of red or green. The standard recommends that colors used in VDTs should differ in the amount of blue as well as in the amount of red or green.

18.3.3 Thermal

One of the possible consequences of adding HMDs to a helmet is that the heat produced by the CRTs may affect the thermal comfort of the wearer. This effect may be enhanced by integrated helmets that totally enclose the head and face.

Although this may appear to be of only nuisance value, designers of integrated helmets may find better acceptance of their products if consideration is given to this basic requirement.

18.3.4 Movement restrictions

Mass, Volume, and Center of Gravity

Helmet and HMD mass is of primary importance to the acceptability and utility of head-coupled systems. In the laboratory, excessive helmet mass can be compensated for by suspending the helmet from the ceiling. For example, the VCASS helmet weighs approximately 11.5 lb (5.2 kg) but is counterbalanced by a constant-force spring to have an apparent weight on the head of approximately 4 lb (1.8 kg). Although this does not reduce the mass or inertia of the helmet, it does increase the duration for which subjects can use the device. In airborne applications, helmet suspension is a less feasible solution, and a limit on helmet mass is necessary. In high-performance aircraft, the helmet will be required to undergo rapid movement, with varying and high-amplitude g forces. The total weight on the head that a user can tolerate, even for short-duration missions, is quite limited. Furthermore, in the event of ejection, helmets are meant to provide protection, but an excessively massive helmet may become a liability.

Another possible effect of adding mass to the head may be changes in the acceleration or velocity of head movement, with subsequent effects on target acquisition times. In order to quantify this, Wells and Griffin (1987b) had six subjects acquire and track a target that was offset by 60° to the side of, or above, the subjects. Adding up to 0.7 kg (1.5 lb) as much as 250 mm (9.8 in.) in front of the helmet tended to increase the time to acquire targets, but the effect was small. In this experiment, the total head-supported mass was approximately 5.5 lb (2.5 kg), but there was no simulation of the varying g forces that are found in flight. There is a need for in-flight data with varying g forces to quantify the effects of head-supported mass on head movement, and to quantify the effects on performance of the kinds of head-aiming tasks that will be performed with HMDs. Currently, the only estimates of head movement, during realistic flight tasks were obtained during simulated air-to-air engagements. These data show that most of the time the head moved very little but that there were periods when the head moved up to 600°/sec and over large amplitudes (see Fig. 18.9 and Table 18.1).

It has been argued that the effects of increased head-supported weight on the aviator can be separated into the effects on crash kinematics and the effects on mission effectiveness (Rash et al., 1990). However, to this should be added the effects of weight on comfort. It would appear that pilots are prepared to endure increased discomfort, up to a point, and even increased risk of injury during ejection, if the head-coupled system sufficiently increases mission effectiveness. The problem is that this point is poorly defined and varies considerably among pilots. In part, this may be because there is a lack of systematic evidence on which to base any decision. Also, the issue of discomfort may be one with which pilots are ill at ease. It may be considered a weakness to admit that a helmet feels uncomfortable or that discomfort will affect performance. However, two points should be borne in mind. The first is that helmet discomfort may affect the perceived utility of an HMD; that is, an otherwise surmountable shortcoming with the head-coupled system may be deemed unacceptable if the helmet hurts. This could delay acceptance of the systems in the operational community. The second point is that the distraction caused by helmet discomfort may add to the numerous other stressors in the cockpit, and a system that was designed to aid performance may have the opposite effect.

Research on the effects of helmet weight has investigated the effects on muscle fatigue (Phillips and Petrofsky, 1983) and head movement (U.S. Army Human Engineering Laboratories, 1968; Gauthier et al., 1986; Wells and Griffin, 1987b). The currently acceptable maximum weight of the helmet plus accessories appears to be about 4.5 lb (2 kg) (Glaister, 1988; Fulghum, 1990).

Cables, Attachments, and Personal Equipment
Restrictions to movement caused by cables and other attachments may affect both the utility of the HMD and the comfort of the wearer. The effects of these attachments are often unnoticed during normal flight operation because any restriction to head movement is compensated for by eye movement. However, unrestricted head movement takes on an added importance with head-coupled systems. For example, during a target acquisition task using a helmet-mounted sight, Wells and Griffin (1987b) found an increase in target acquisition time at some target positions because the stole of the life jacket impeded head motion. Also, it should be borne in mind that cables fastened to the helmet add to the weight supported by the head. A large fiber-optics cable can add appreciable weight, as can heavily insulated high-voltage cables.

18.3.5 Display Location

Where the HMD should be located in the visual field is dependent on the type and size of display (see Section 18.2.2), the task to be performed, and the comfort of the viewer. Data exist on the preferred viewing angles for panel-mounted displays. These range from 10° below the horizontal (VanCott and Kinkade, 1972) to 40° ± 20° below the horizontal (Kroemer and Hill, 1986). Recommendations for lateral displacement range from 15° to the left or right (Department of Defense, 1981) to between 30° and 40° to the left or right (Farrell and Booth, 1984). Very few data exist to guide designers on the placement of HMD images. One study conducted by Katsuyama et al. (1989) had subjects perform a dual

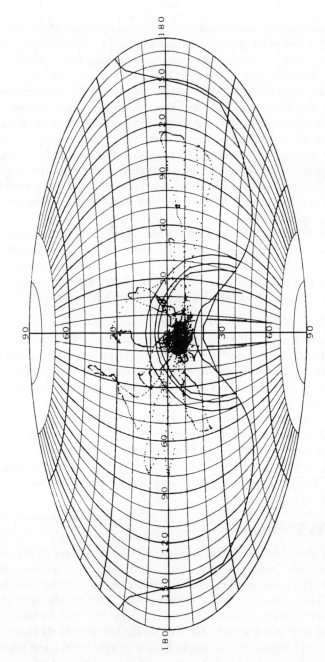

Figure 18.9 Head displacement during an air-to-air engagement. The engagement lasted 193 sec. Each point represents a sample of head orientation. Data were sampled at 20 Hz. [From Wells and Haas (1990).]

Table 18.1

	DISPLACEMENT (deg)			VELOCITY (deg/sec)			ACCELERATION (deg/sec^2)		
	el	az	radial	el	az	radial	el	az	radial
MAXIMUM	89	180	148	344	601	554	2452	4753	7772
MINIMUM	-77	-179	0	-251	-547	-563	-2332	-4746	-4481

Maximum and minimum head movements from 28 air-to-air engagements conducted in a simulator [From Wells and Haas (1990)].

task, with the primary task presented on a panel-mounted display in front of the subject and the secondary task presented on a miniature simulated helmet-mounted display placed at various locations. Their results suggest that the head-mounted information should be presented 15° below the primary viewing area. However, in the laboratory experiment the subjects' heads were held stationary with a bite bar. Normal head movements required in operational situations may require somewhat different HMD locations.

18.4 FIDELITY

18.4.1 Apparent Size and Distance

Pilots viewing imagery on an HMD often misperceive the size and/or distance of objects within the scene. For example, Hale and Piccione (1990) found in a survey that 65% of AH-64A pilots reported that objects appeared smaller and further away than they actually were. Furthermore, 43 of the 53 pilots (82%) indicated problems maintaining a proper approach angle to landing points because of inaccurate judgments of distance. The problem has been termed *accommodation micropsia* (apparent shrinking of objects) and accommodation macropsia (apparent enlarging of objects). The topic has been discussed extensively in the literature and has been outlined elegantly by Roscoe (1977, 1987), and most recently by Hale (1990). The following is a summary of the issues.

The apparent size of a visual object is not simply a function of the angle that that object subtends at the eye, that is, image size on the retina. Thus, although a 6 ft (1.8 m) man at 10 ft (3 m) subtends twice the visual angle as a man the same size at 22 ft (6.7 m), we do not perceive the closer man as being twice as tall as the further man. Generally, we perceive both men to be the same size but at different distances. This perception of object size as invariant with distance, despite variation of the retinal image size, is termed *size constancy*. There are a number of factors that contribute to size constancy including, the richness of the visual scenery, the size of the visual surroundings, the location of the image on the retina, and the state of accommodation of the eye. If some, or all, of these factors are missing or distorted, our ability to accurately interpret object size from the retinal image can be reduced. When this happens, our perception of the size, or the distance, of the object becomes distorted. Manipulation of the factors involved in size constancy is often used in psychophysics experiments that investigate visual performance. Everyday occurrences also provide opportunities for erroneous perception. For example, an object floating in the sky can be perceived as the Goodyear blimp in the distance or a car dealer's promotional balloon one block away. One of the more famous examples is that the full moon close to the horizon looks appreciably larger than the full moon high in the sky, although both subtend about the same visual angle.

The perceived size of objects has been shown to be strongly correlated to the accommodation of the eye. When the eye is over accommodated (focused at a distance closer than the viewed object), objects at a distance greater than the point of accommodation appear smaller and/or further away. Roscoe (1977) provides the following demonstration of this phenomenon. Close one eye, focus your open eye on your thumb held at arm's length, observe a more distant object such as a window or a picture on a wall, and, while continuing to focus on your thumb, draw it toward you and observe the change in size of the window or picture. [Another example, for those with access to an HMD, is as follows: Observe the apparent size of a word on the HMD while you walk toward, or away from, a wall about 15 ft (4.6 m) away. The word will appear to shrink and expand, presumably because of your accommodation reflex being driven by your distance from the wall.]

The available evidence suggests that subjects over accommodate when viewing collimated displays. Hence, objects viewed on an HMD (giving a collimated image) appear smaller and may be interpreted as being further away than they actually are.

18.4.2 Accommodation/Convergence

With normal vision, the eyes accommodate (focus) to image objects sharply on the retina, and diverge or converge to align the eyes' lines of sight with the object. Both responses of the eye are coupled. In the absence of any changes in the stimulus for accommodation, changes in convergence produce changes in accommodation. Similarly, changes in accommodation are accompanied by changes in convergence. If one eye is covered, the covered eye makes an inward (or convergent) motion when accommodation is increased, and an outward (divergent) motion when accommodation is relaxed (Boff and Lincoln, 1988). However, when images are viewed on a binocular or biocular HMD, these two ocular responses need to be uncoupled. The HMD presents images at optical infinity, and any accommodative changes brought about by convergence are inappropriate, as they reduce retinal image sharpness. The effects of these changes on performance warrant further investigation.

18.4.3 Temporal Resolution

The speed with which the HMD system can change/update moving imagery is one aspect of its temporal resolution. As with spatial resolution, the temporal resolution of the HMD system is limited by the image source. For a raster-scanned electro-optical sensor, this may be the frame rate. For a computer image generator, this may be the scene update rate. (Data exist for frame rate requirements, drawn from the literature related to television and VDT use [see, e.g., Kiver and Kaufman (1983)].) In contrast to spatial resolution (see Section 18.2.7), the passage of the information through the display electronics and optics has little effect on this aspect of temporal resolution.

The frame rate of the HMD source is another aspect of the system's temporal resolution. For CRT image sources, the image update rate may be that of a standard television system. In the United States, standard television has a 2:1 interlace, with a field rate of 30 Hz and a frame rate of 60 Hz. This effective frame rate of 60 Hz, coupled with the temporal integration of the eye, is sufficient to produce a flicker-free television picture (for most observers, under most conditions). However, Crookes (1957) describes a phenomenon that may have implications for the perception of HMD images generated by raster-scanned systems. The reason for the phenomenon, an apparent change in shape of the television picture, is shown for a left-to-right eye movement in Figure 18.10. Normally, if the eye passes over a brightly illuminated picture in a dark room, the perceived picture may be blurred. This is because each part of the retina over which the image passes is stimulated successively by different parts of the picture. However, with a raster-scanned image source, the point of light produced by the scanning electron beam that generates the picture is sufficiently fast that there is not the same blurring on the retina, even during rapid eye movements. During head movement with an HMD, it is likely that there will be a substantial amount of eye movement. Also, it might be expected that the perception of any eye-movement-induced distortion of an image would be most likely when the integration time of the eye is long (long persistence). Long integration time is associated with low retinal illuminance (de Lange, 1958), in other words, conditions during which the HMD is likely to be used. Precisely how any distortion in the retinal image will be perceived is difficult to predict. In addition to the parallelogram distortion illustrated in Figure 18.10, there may be image compression and elongation during downward and upward eye movements. It is also possible that the two fields of an interlaced frame will be torn apart during eye movements.

An additional temporal resolution factor with HMDs is the effect of the head-coupled control loop. Images that move in response to head motion will be affected by the characteristics of the head tracking system. Consideration must be

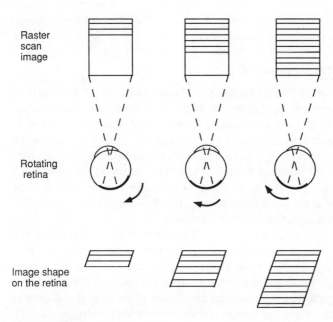

Figure 18.10 The shape of a raster-scanned picture on the retina during rapid eye movement. [Adapted from Crookes (1957).]

given to both the update rate and lag of the tracking device. The effects of update rate and lag may vary, depending on the application and on whether or not the user receives competing sensor feedback. For example, in one application explored with VCASS, a virtually imaged cockpit was meant to remain stable with respect to the pilot's head movements (Osgood and Wells, 1991). Despite an update rate as low as 8 Hz and a lag of approximately 120 msec, the simulation was adequate. Partly, the adequacy of the simulation was achieved because the HMD was opaque and the computer-generated images were not viewed in conjunction with the real world. When the VCASS optics were made transparent, so that the subject could see the computer-generated images superimposed over the real world, the lag and update rate of the images became unacceptable. One reason for the acceptability of the VCASS simulation may have been that there appears to be a certain amount of "dead zone" associated with the perception of head position (see Section 18.5.2). It is also possible that, in the absence of real-world images, the adaptability of the head and eye movement systems (e.g., Melville Jones et al., 1984) compensated for the changes in control necessary to perform with the slow update rate and long delay in the computer-generated images.

Other applications may have the wearer of the HMD using a virtual visual representation of his or her hand to interact with controls. In the absence of suitable virtual tactile feedback, it has been proposed that the person interact with real controls. In this case, any lag in the visual feedback (virtual images of the hand) would compete with the tactile feedback (touching the real control). There is very little guidance in the literature as to the boundaries of what is acceptable.

18.4.4 Illusions

Pilots wearing HMDs are subject to the same illusions as those not wearing HMDs (e.g., Harding and Mills, 1983). In addition, there are problems related to the use of electro-optical sensors and problems with the use of the head. The former category includes loss of orientation over large bodies of water (insufficient contrast between the sea and the sky with IR sensors), poor orientation in poor contrast conditions, and confusion of ground lights with stars. Problems in the latter category include head drift (see Section 18.5.2) and the possibility of disorientation if the cockpit is not correctly mapped for use with an electromagnetic sensor or if the head tracking system "tumbles."

18.4.5 Acceptable Tolerances

Perceptual Mismatch

Perceptual mismatch refers to differences between the movement or position of objects viewed on the display and the movement or position of those objects as perceived by other means. For example, consider a pilot using the IHADSS system while hovering in a helicopter. If the sensor lags the head, at the end of a rapid head movement the HMD may still show a moving scene. However, the pilot's vestibular and proprioceptive systems would indicate a stationary head. In this case the perceptual mismatch would be between the movement of the scene as viewed on the HMD and the scene movement that would be expected with a stationary head. Another example would be the same arrangement viewed during the day with a see-through HMD. If both a real and a displayed object were visible to the eye and there was a misalignment between the sensor and the head, the HMD image and the directly viewed object would not superimpose (Fig. 18.11).

It is probably impossible to realize a system that does not produce some perceptual mismatch. The human factors challenge is to design systems that reduce the amount and the impact of the mismatch. Perceptual mismatch can result in motion sickness, loss of simulation fidelity, and degraded performance at some tasks. For example, So (1991) found that with a head tracking task, delays of greater than 60 msec between the head-controlled reticle and the head had a significant degrading effect on tracking performance. In Section 18.4.3 it is reported that with perceptual mismatch a simulation on VCASS showed unacceptable lags. The solution adopted was to reduce the perception of real-world objects by making the HMD opaque. This reduced the perceptual mismatch and resulted in a usable simulation.

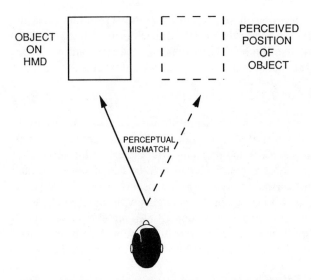

Figure 18.11 Perceptual mismatch occurs when the movement or position of an object viewed on the display differs from the movement or position of the object as perceived by other means.

Tolerances for perceptual mismatch are difficult to quantify. Just noticeable differences in movement or position are small. However, there may be a difference between what is noticeable and what is tolerable. Also, the nature of the task may dictate what is tolerable. This is an area in which more research is needed.

Optical Distortion

Some types of distortion within the HMD optical system may be tolerable (e.g., Fantone, 1990). Distortion outside the system, for example, angular deviation (prismatic shifting) caused by the canopy, may be intolerable. Optical deviation is produced by curved canopies, which displace light differently at different canopy points. Curved canopies are required for good aerodynamics. The induced deviation is minimized, as far as possible, in canopy manufacture and by calibration after manufacture. Prior to the introduction of helmet-mounted sights, the deviation problem was relevant to the area behind the HUD. Canopy deviation with HMDs is a problem only when the operator is working open-loop, for example, when a sighting reticle is being used to aim a missile and the operator has no feedback about where the missile is pointing. Clubine (1982) produced a model for permitting computer correction for canopy deviation for all points on the canopy, for helmet-mounted sights when they are being used in the open-loop configuration. Maximum deviation in an F-16 cockpit is about 7.1 mrad in elevation and 3.9 mrad in azimuth.

Alignment

With binocular or biocular displays, the observer is presented with an image for each eye. If there is misalignment between the optical axes of the system, the observer may have difficulty fusing the two images, as the images won't fall upon corresponding parts of the two retinas. For moderate misalignment, fusion is still possible, but it requires effort and may cause eyestrain and headache and even nausea. As the misalignment worsens, fusion becomes more difficult until, with sufficient misalignment, fusion breaks down and the observer sees two images (diplopia). The effort to maintain fusion has a cumulative effect. Equipment that appears to be adequate for short-term use may prove to be unacceptable over longer durations. Self (1986) provides a comprehensive review of the literature relating to binocular alignment and formulates a set of optical limit specifications for HMDs. His conclusions include the following: Perfection is unnecessary and too expensive; there are significant individual differences in sensitivity and tolerance to misalignment; the small samples of observers do not allow formulation of percentile tolerances; tolerances for comfortable use are tighter than tolerances based on the ability to fuse images; tolerances that used real-world (complex) scenes are tighter than tolerances obtained using simple images; multidigit values, cited in the literature, implying high accuracy, are sometimes the result of multiplying one- or two- digit values by a conversion factor and carrying the answer out to too many digits.

18.4.6 Binocular Overlap

In binocular systems, there is a trade-off between total intradisplay FOV and binocular overlap. For example, VCASS uses two helmet-mounted displays, each with an 80° FOV, to produce a total horizontal FOV of 120° with 40° of overlap (Fig. 18.12). An increase in horizontal FOV could be achieved by reducing the amount of overlap. As discussed in Section 18.2.6, there may be applications in which the advantages of binocular overlap could be sacrificed for a larger FOV. In order to investigate whether changing the binocular overlap affects performance, Tsou et al. (1991) conducted an experiment with professional car drivers on a test track. The drivers were fitted with vision-occluding goggles that restricted their FOV and also altered the amount of overlap. This resulted in what the authors call a "distortion-free, eye-limited viewing condition." Measurements of driving time, course error, and head movement were made as the drivers negotiated an obstacle course. The experiment was a pilot study conducted with two subjects. The results revealed differences between FOV sizes but no differences between the different amounts of overlaps.

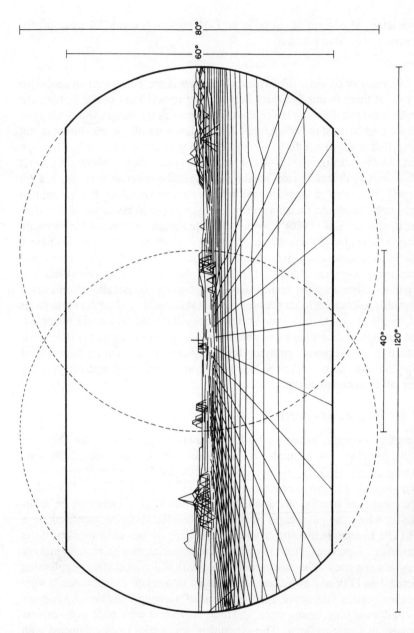

Figure 18.12 The arrangement of the two helmet-mounted displays in VCASS that gives a total instantaneous field of view 120° wide by 60° high.

18.5 INTUITIVENESS

18.5.1 Information Formatting

Information Content

In the display of information, there are demands for ease of interpretability, high information density, multiple functions, and compelling features. Often these demands are partially, or totally, mutually exclusive. For this reason, the design of displays occupies the border between art and science. There are numerous design principles to which the display designer can refer. Some of these are described in texts by Tufte (1983, 1990) and Laurel (1990). These principles include variations in object and text size and color, use of blank space, data chunking, use of icons, layering and separation, variations in object and text brightness, and the use of metaphor. These principles apply equally well to the display of information on HMDs, but some peculiarities associated with HMDs require special consideration. Principal among these is that the HMD is inescapable. Information presented on the display will be in some portion of the FOV at all times. This means, for example, that flashing a display element to attract attention may be effective, but it may also be annoying. For the same reason, an overly compelling display may cause the wearer to lose contact with other, non-display information. With see-through displays, consideration must be given to the obscuration of objects in the real world by high information density on the HMD. The fact that the HMD moves with the head also has implications for the presentation of attitude information, as discussed later.

Recognition of Reality

Information presented on the HMD is compelling. In a simulator, this feature may add to the reality of the simulation. However, in an aircraft, confusion between reality and displayed information can be fatal. Evidence of this type of confusion was observed during trials conducted in the mid-1970s in the United Kingdom (Karavis and Clarkson, 1991). In these trials, pilots were presented with a head-mounted reticle and a HUD. On several occasions pilots attempted to control their flight path during simulated low-level missions by raising their line of sight instead of applying control input.

Even with the less compelling HUD, and in a real aircraft, pilots can become confused. An example is the following account of a person flying an F-16 equipped with LANTIRN (Low Altitude Navigation and Targetting Infra Red system for Night) with automatic terrain following.

> My sortie was a typical low level . . . day training mission. I was number 2 on a low level mission northwest of Luke AFB, Arizona. After three legs of the low level, I experienced a . . . failure. . . . I removed my vision restricting device and decided to finish the profile as a day VFR low level to South

TAC range. I descended into a valley, checked my gas, my timing, and pressed on towards the upcoming mountain ridge. Shortly thereafter, I got the scare of my life. I was not flying this airplane. I was waiting for it to fly itself over the ridge! I was lucky I recognized it early and was able to safely clear the rocks. But what had happened? (Slocum, 1991).

There is the potential for HMDs to contribute to such a compelling mental representation of a situation that the wearer may respond inappropriately to the primary task of flying the aircraft. For example, a scene from a missile-mounted video camera, viewed on an HMD, may confuse the wearer sufficiently to cause disorientation. As systems become more sophisticated and HMDs are used more often, other subtle sources of confusion are likely to occur. The incidents described above exemplify the need for an integrated systems approach to the design of cockpit displays.

Head vs. Aircraft vs. World Stabilization

Mounting a display on the head allows attitude information to be presented in front of the eye, irrespective of head pointing angle. This offers potential benefits for off-boresight aircraft control, whereby a pilot, looking outside the cockpit, can control the aircraft by referring to the attitude information on the HMD. However, head movement introduces another variable into the display of attitude information, as illustrated by Figure 18.13. The column on the left of the figure shows four views of an outside-in display, that is, a display showing aircraft attitude as if the pilot were located outside of, and behind, the aircraft and looking in. Irrespective of where the pilot points his/her head, this type of display will show the same information. The second column from the left shows four views of an inside-out display, that is, a display that superimposes the artificial horizon on the real horizon. The inside-out display uses the metaphor of a "hole" through which the pilot can see the real horizon. Therefore, changes in orientation of the aircraft result in changes in the position of the horizon in the display.

To be consistent with the metaphor, the "hole" should move with the head. This is shown in the third column from the left, with the wearer's head pointing to the right. With this arrangement, the display gives the same information when the aircraft is straight and nose high and the pilot's head is pointing to the right (column 3, row 3), as when the aircraft is level and rolled right and the pilot's head is pointing forward (column 2, row 2). This is a potential source of confusion. To minimize this confusion, a hybrid display of outside-in and inside-out information was formulated. The column on the right of Figure 18.13 shows the hybrid display with the wearer's head pointing to the right. The elements of the display are illustrated in Figure 18.14, and an experiment that tested the utility of this display is reported by Osgood et al. (1991).

How to display off-boresight information requires careful consideration and may well be the source of considerable debate, similar to that concerning inside-

HUMAN FACTORS OF HMDS

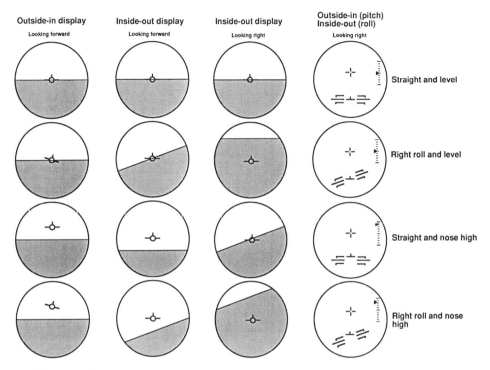

Figure 18.13 Outside-in, inside-out and hybrid displays. A complete explanation is provided in the text. [From Geiselman (1991).]

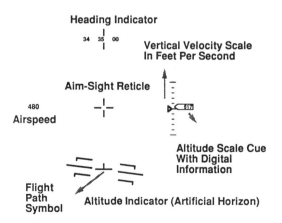

Figure 18.14 A hybrid inside-out/outside-in display used for presenting attitude information off-boresight.

out vs. outside-in attitude displays for panel-mounted applications (Roscoe, 1968).

18.5.2 Aiming with the Head

Helmet-mounted displays and helmet-mounted sights are frequently described as being intuitive, and it is assumed that this intuitiveness will improve performance. Studies have shown that the addition of a head tracking device to tactical cockpits will increase overall weapons system effectiveness by more efficiently coupling the pilot to the aircraft avionics (Arbak, 1989). Also, Barnes (1989) presents a vivid picture of many uses of HMDs in air-to-air and air-to-ground operations in an F-16. These include keeping track of other aircraft during tactical intercepts by using pointing symbology driven by the avionics, presentation of own-ship energy state information during off-boresight visual searches, and aiding target and waypoint detection as well as designation of navigation system updates. However, it is worth remembering that aiming with the head is not as natural as it may at first appear. Normally, a combination of head and eye movements is used to acquire visual targets. The change from the unconscious use of the head and eyes to the use of the head for aiming is achieved easily, but there are some performance implications.

There is evidence that the addition of an HMD into the FOV affects eye movement. Stern et al. (1988) investigated head and eye movement using a 12° FOV see-through HMD in a flight simulator modeling a strategic bomber. They found unusual head and eye movement patterns not seen in other laboratory settings. For example, head movements lagged eye movements by 100–200 msec in the flight simulator, as opposed to 50 msec in laboratory settings. Also, compensatory eye movements, which normally compensate for head movement during the acquisition of visual targets, were frequently accomplished before the head motion began.

Without feedback of head position, the head can drift by up to 3° in 2 min (see Fig. 18.15). This head drift has implications for the use of an HMD for flying at night and may explain some of the reports of misaligned landings that occur with pilots who are unfamiliar with the system (Dryden, 1991). Figure 18.16 illustrates one interpretation of what may be happening. The drawing on the left shows the situation where the pilot's head is aligned with the aircraft and the aircraft is aligned with the runway for a normal landing. The middle drawing shows what happens if the pilot's head drifts to the right and he is unaware of this displacement. The sensor, which is aligned with his head, provides sufficient visual information to allow the pilot to correctly deduce his direction of travel, using the optic flow field (Warren, 1990). However, if he assumes that his head is aligned with the aircraft, then the visual information is appropriate to the aircraft being yawed to the right. The drawing on the right shows the result of this misperception. The pilot maneuvers the aircraft so that his head (and the sensor)

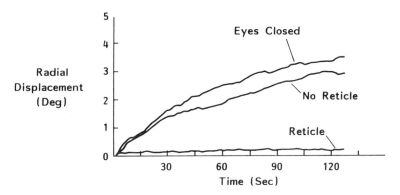

Figure 18.15 Head displacement from straight ahead (0°) over time, under three visual conditions. [From Wells and Griffin (1987b).] Subjects were instructed to aim their head at a stationary target. Mean data from 12 subjects.

point straight down the runway. In doing so he must make the vehicle yaw to the left. The aircraft now approaches the runway yawed to the left, which may result in increasing control problems as the aircraft gets closer to the ground.

18.5.3 Cross Spectral Interpretation

The images presented on HMDs are often from head-steered thermal sensors. The sensors, and their associated electronics, transduce energy in the infrared part of the electromagnetic spectrum and turn it into energy in the visible portion of the spectrum. The result is that differences in thermal energy are seen on a monochrome display as differences in gray shade. The polarity of the system determines whether white or black (or green or black for an HMD with a green phosphor) represents hot or cold. These images are different from images obtained from a visual sensor, lacking, among other things, shadow and texture (Brickner 1989, Foyle et al., 1990). The ability to switch the polarity of the image is important for allowing the users to obtain the most intelligible image. With displays where colors are used to indicate temperature or intensity, differences from visual sensor images may be very large and may pose other problems of interpretation.

Acknowledgments

We are grateful to our colleagues who, through conversation, editorial revision, guidance, and example, contributed significantly to the preparation of this chapter. These include Jeff Craig, Bob Eggleston, Dean Kocian, Nick Longinow, Wayne Martin, Robert Osgood, Dave Post, Brad Purvis, Herschel Self, Lee Task, and Rik Warren. We also benefited immeasurably from our interviews with Joe Bill Dryden and Keith Giles, test pilots for General Dynamics. Jeff Schmidt,

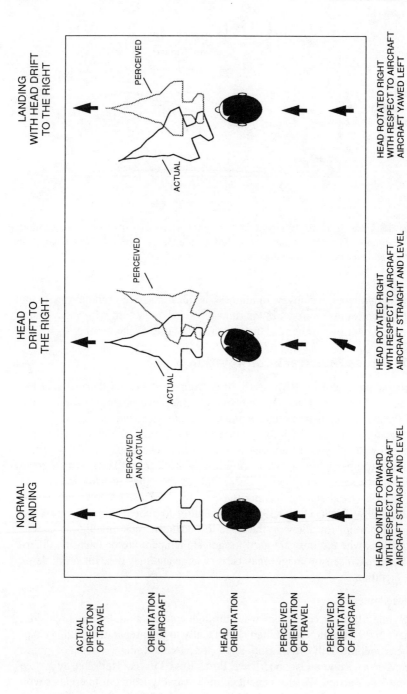

Figure 18.16 One possible consequence of head drift is a pilot is using a head-steered sensor and has no feedback of head orientation. See text for details.

Nancy Dolan, and Eric Geiselman helped in the preparation of some of the illustrations. Don Polzella and Mike Gravelle of the Crew System Ergonomic Information Analysis Center (CSERIAC) helped to obtain some of the references.

References

ANSI (1988). American National Standard for Human Factors Engineering of Visual Display Terminal Work Stations, ANSI/HFS Standard 100-1988, Human Factors Society, Santa Monica, CA.

Arbak, C. J. (1989). Utility of a helmet-mounted display and sight, *Proc. Helmet Mounted Displays*, Society of Photo-Optical Instrumentation Engineers, Bellingham, WA, pp. 138–141.

Arditi, A. (1986). Binocular vision, in *Handbook of Perception and Human Performance*, Vol. 1 (K. R. Boff, L. Kaufman, and J. P. Thomas, Eds.), 23.1–23.41. Wiley, New York.

Baker, C. (1962). *Man and Radar Displays*, Pergamon, New York.

Barnes, G. R. (1979). Vestibulo-ocular function during co-ordinated head and eye movements to acquire visual targets. *Journal of Physiology, 289*, 127–147.

Barnes, W. J. (1989). Tactical applications of the helmet-mounted display in fighter aircraft, *Proc. Helmet Mounted Displays*, Society of Photo-Optical Instrumentation Engineers, Bellingham, WA, pp. 149–160.

Bennett, C. T., and S. G. Hart (1987). PNVS-related problems: pilots reflections on visually coupled systems, 1976–1987, Working paper, NASA-Ames Research Center, Moffett Field, CA. Cited in Rash, C. E, R. W. Verona, and J. S. Crowley (1990). U.S. Army Aeromedical Research Laboratory Report No. 90–10. Fort Rucker, AL.

Benson, A. J., and G. R. Barnes (1978). Vision during angular oscillation—the dynamic interaction of visual and vestibular mechanisms, *Aviation Space Environ. Med. 49*: 340–345.

Boff, K. R., and J. E. Lincoln (1988). *Engineering Data Compendium*, Harry G. Armstrong Aerospace Medical Research Laboratory, Wright Patterson Air Force Base, OH.

Brickner, M. S. (1989). Helicopter flights with night vision goggles: human factors aspects (Technical Memorandum No. 101039). NASA.

Buchroeder, R. A. (1989). Helmet-mounted displays, Tutorial Short Course Notes T26, SPIE Symposia on Aerospace Sensing.

Caldow, R. (1984). Airborne electronic colour displays—a review of U.K. activities since 1981, *Sixth Advanced Airborne Display Symposium*, Naval Air Test Center, Patuxent River, MD, pp. 45–65.

Clubine, W. R. (1982). Modeling the helmet-mounted sight system, unpublished M.Sc. Thesis, Air Force Institute of Technology.

Crookes, T. G. (1957). Television images, *Nature 179*: 1024–1025.

de Lange, J. (1958). Research into the dynamic nature of the human fovea–cortex systems with intermittent and modulated light. 1. Attenuation characteristics with white and colored light, *J. Opt. Soc. Am. 48*: 777–784.

Department of Defense (1981). Human engineering design criteria for military systems, equipment and facilities, MIL-STD-1472C, Department of Defense, Washington, DC.

Dryden, J. W. (1991). Personal communication. (General Dynamics test pilot with over 200 hr experience in an F-16 equipped with an HMD and head-steered FLIR.)

Enoch, J. M. (1959). Effect of the size of a complex display upon visual search, *J. Opt. Soc. Am. 49*: 280–286.

Fantone, S. (1990). Visual imaging system design—the human factor, *Photonics Spectra*, October: 123–126.

Farrell, R. J., and J. M. Booth (1984). *Design Handbook for Imagery Interpretation Equipment*, D180-19063, Boeing Aerospace Company, Seattle.

Foyle, D. C., Brickner, M. S., Staveland, L. E., and Sanford, B. D. (1990). Human object recognition as a function of display parameters using television and infrared imagery. In *SID International Symposium XXI*, pp. 269–272.

Fulghum, D. A. (1990). Navy orders contractors to stop work on ejectable night vision helmets, *Aviation Week Space Technol.*, 67–68 December.

Furness, T. A. (1981). The effects of whole-body vibration on the perception of the helmet-mounted display, Unpublished Doctoral Thesis, Univ. Southampton.

Furness, T. A., and D. F. Kocian (1986). Putting humans into virtual space, *Soc. Comput. Simulation Ser. 16*: 214–230.

Gauthier, G. M., B. J. Martin, and L. W. Stark (1986). Adapted head- and eye-movement responses to added head inertia, *Aviation, Space Environ. Med. 57*: 336–342.

Geiselman, E. E. (1991). Personal communication.

Giles, K. (1991). Personal communication. (General Dynamics test pilot with over 200 hr experience in an F-16 equipped with an HMD and head-steered FLIR.)

Glaister, D. H. (1988). Head injury protection, in *Aviation Medicine*, 2nd ed. (J. Ernsting and P. King, Eds.), 174–184 London.

Green, D. M., and J. A. Swets (1966). *Signal Detection Theory and Psychophysics*, Wiley, New York.

Gresty, M. A. (1974). Coordination of head and eye movements to fixate continuous and intermittent targets, *Vision Res. 14*: 395–403.

Hale, S. (1990). Visual accommodation and virtual images—a review of the issues, U.S. Army Engineering Lab. Tech. Note 3-90, Aberdeen Proving Ground, MD.

Hale, S., and D. Piccione (1990). Pilot performance assessment of the AH-64 helmet display unit, U.S. Army Engineering Lab. Tech. Note 1-90, Aberdeen Proving Ground, MD.

Harding, R. M., and F. J. Mills (1983). Function of the special senses in flight, *Br. Med. J. 286*: 1728–1731.

Jackson, C. E., and W. J. Grimster (1972). Human aspects of noise and vibration in helicopters, *J. Sound Vibration 20*(3): 343–351.

Jarrett, D. N. (1978). Helmet slip during simulated low-level high speed flight, Royal Aircraft Establishment Tech. Rep. 78078.

Karavis, A., and G. J. N. Clarkson (1991). The design and evaluation of a fast-jet helmet mounted display, *Proc. AGARD Symposium on Helmet Mounted Display and Night Vision Goggles* 2.1–2.8.

Katsuyama, R. M., E. P. Rolek, S. Johnson, and D. L. Monk (1989). Effects of miniature CRT location upon primary and secondary task performances, AAMRL-TR-89-018.

Kennedy, K. W., and K. H. E. Kroemer (1973). Excursion of head, helmet and helmet attached reticle under $+G_z$ forces, AAMRL-TR-72-127.
Kiver, M., and M. Kaufman (1983). *Television Electronics—Theory and Practice*, Van Nostrand Reinhold, New York.
Kraft, C. (1978). Psychophysical contributions to air safety, in *Psychology from Research to Practice* (H. Pick, H. W. Leibowitz, and J. F. Singer, Eds.), Plenum, New York.
Kroemer, K. H. E., and S. G. Hill (1986). Preferred line of sight angle, *Ergonomics* 29(9): 1129–1134.
Laurel, B. (Ed.) (1990). *The Art of Human–Computer Interface Design*, Addison-Wesley, Reading, MA.
Laycock, J. (1976). A review of the literature appertaining to binocular rivalry and helmet mounted displays, RAE Tech. Rep. 76101.
Levelt, W. J. M. (1968). *On Binocular Rivalry*, Mouton, The Hague.
Lewis, C. H., and M. J. Griffin (1979). The effect of character size on the legibility of a number display during vertical whole-body vibration, *J. Sound Vibration 67*: 562–565.
Lewis, C. H., and M. J. Griffin (1980). Predicting the effects of vibration frequency and axis and seating condition on the reading of numeric displays, *Ergonomics 23*: 485–507.
Lewis, C. H., A. F. Higgins, M. J. Wells, and M. J. Griffin (1987). Stabilization of head-coupled systems for rotational head motion, Final Report for USAF contract T33615-82-C-0513.
Melville Jones, G., A. Berthoz, and B. Segal (1984). Adapted modification of the vestibulo-ocular reflex by mental effort in darkness, *Brain Res. 56*: 149–153.
Osgood, R. K., and M. J. Wells (1991). The effect of field-of-view size on performance of a simulated air-to-ground night attack, *Proc. AGARD Symposium on Helmet Mounted Displays and Night Vision Goggles, 517*: 10.1–10.7.
Osgood, R. K., E. E. Geiselman, and C. S. Calhoun (1991). Attitude maintenance using an off-boresight helmet-mounted virtual display, *Proc. AGARD Symposium on Helmet Mounted Displays and Night Vision Goggles, 517:* 14.1–14.7.
Peli, E. (1990). Visual issues in the use of a head mounted monocular display, *Opt. Eng. 29*: 883–892.
Phillips, C. A., and J. S. Petrofsky (1983). Neck muscle loading and fatigue: systematic variation on headgear weight and center-of-gravity, *Aviation, Space Environ. Med. 54*(10): 901–905.
Post, D. L. (1992). Applied color-vision research, in *Color in Electronic Displays* (H. Widdel and D. L. Post, Eds.), 137–173 Plenum, New York.
Rash, C. E., R. W. Verona, and R. S. Crowley (1990). Human factors and safety considerations of night vision systems flight using thermal imaging systems, USAARL Rep. 90-10.
Robinette, K. M. and J. J. Whitestone (in press). Two methods for characterizing the human head for the decision of helmets. AL-TR-92–xxxx.
Roscoe, S. N. (1968). Airborne displays for flight and navigation, *Human Factors 10*: 321–332.
Roscoe, S. N. (1977). How big the moon, how fat the eye?, Tech. Rep. AFOSR-77-2.

Roscoe, S. N. (1987). The trouble with HUDs and HMDs, *Human Factors Soc. Bull.* *30*(7): 1-2.

Self, H. C. (1986). Optical tolerances for alignment and image differences for binocular helmet-mounted displays, AAMRL-TR-86-019.

Slocum, D. (1991). The F-16 and LANTIRN, *Flying Safety* April: 5-7.

So, R. H. (1991). Effects of time delays on head tracking performance and the benefits of lag compensation by image deflection, *Proc. AIAA Flight Simulation Technologies Conference and Exhibit* 124-130.

Stern, J. A., R. Goldstein, and D. N. Durnham (1988). An evaluation of electro-oculographic, head movement, and steady state evoked response measures of workload in flight simulation, AAMRL-TR-88-036.

Stevens, K. (1982). Computational analysis. Implications for visual simulation of terrain, in *Vision Research for Flight Simulation* (W. Richards and D. Dismokes, Eds.), National Academy Press, Washington, DC.

Task, H. L. (1991). Optical and visual considerations in the specification and design of helmet mounted displays, *Proc. SID Int. Symp.*, pp. 297-300, Playa del Rey, CA.

Tredici, T. J. (1985). Ophthalmology in aerospace medicine, in *Fundamentals of Aerospace Medicine* (R. L. Dehart, Ed.), 465-510 Lea and Febiger, Philadelphia.

Tsou, B. H., B. M. Rogers-Adams, and C. D. Goodyear (1991). The evaluation of partial binocular overlap on car maneuverability—a pilot study, *Proc. Fifth Annual Space Operations, Applications, and Research Symposium*.

Tufte, E. R. (1983). *The Visual Display of Quantitative Information*, Graphics Press, Cheshire, CT.

Tufte, E. R. (1990). *Envisioning Information*, Graphics Press, Cheshire, CT.

U.S. Army Human Engineering Laboratories (1968). Aircraft crewman helmet weight study, Progress Report No. 1, Aberdeen Proving Ground, MD.

VanCott, H. P., and R. G. Kincade (1972). *Human Engineering Guide to Equipment Design*, rev. ed., U.S. Govt Printing Office, Washington, DC.

Velger, M., and S. Merhav (1986). Reduction of biodynamic interference effects in helmet-mounted sights and displays, *Proc. 22nd Annual Conference on Manual Control*, Dayton, OH.

Warren, R. (1990). Preliminary questions for the study of egomotion, in *Perception and Control of Self-Motion* (R. Warren and A. H. Wertheim, Eds.), 3-32 Erlbaum, Hillsdale, NJ.

Wells, M. J. (1983). Vibration-induced eye motion and reading performance with the helmet-mounted display, Unpublished Doctoral Thesis, Univ. Southampton.

Wells, M. J., and M. J. Griffin (1984). Benefits of helmet-mounted display image stabilization under whole-body vibration, *Aviation Space Environ. Med.* *55*: 13-18.

Wells, M. J., and M. J. Griffin (1987a). Flight trial of a helmet-mounted display image stabilization system, *Aviation Space Environ. Med.* *58*: 319-322.

Wells, M. J., and M. J. Griffin (1987b). Performance with helmet-mounted sights, Inst. Sound and Vibration Res. Tech. Rep. 152.

Wells, M. J., and M. J. Griffin (1987c). A review and investigation of aiming and tracking performance with head-mounted sights, *IEEE Trans. Syst. Man, Cybern, SMC-17*: 210-221.

Wells, M. J., and M. Haas (1990). Head movements during simulated air-to-air engagements, *Proc. Helmet Mounted Displays II*, Society of Photo-Optical Instrument Engineers, 246–257 Bellingham, WA.

Wells, M. J., and R. K. Osgood (1991). The effects of head and sensor movement on flight profiles during simulated dive bombing, *Proc. 35th Annual Scientific Meeting of the Human Factors Society*, 22–26 Santa Monica, CA.

Wells, M. J., and M. Venturino (1990). Performance and head movements using a helmet-mounted display with different sized fields-of-view, *Opt. Eng.* 29: 870–877.

19
Perceptual Effects of Spatiotemporal Sampling

Julie Mapes Lindholm

*University of Dayton Research Institute,
Aircrew Training Research Division, Armstrong Laboratory,
Williams Air Force Base, Arizona*

19.1 INTRODUCTION

With current electro-optical display technology, complex "real-time" images are created by a raster-scan process. An input signal modulates an electron beam while it traces a fixed pattern of horizontal lines (the raster) from left to right, from top to bottom. The electron beam, in turn, either excites phosphor in a cathode-ray tube (CRT) or writes a diffraction grating on the oil film of a light valve projector.

The scanning process may be either interlaced or noninterlaced (progressive). When it is interlaced, a frame (all the horizontal lines) is divided into two fields, one consisting of odd-numbered lines and the other of even-numbered lines. The odd-numbered lines are scanned during one refresh period, the even-numbered lines during the next. In contrast, when it is noninterlaced, all the lines in a frame are scanned during each refresh period.

The form and motion percepts supported by the resulting images do not always correspond to the forms and motions of the objects that are represented. In certain cases, a lack of correspondence mirrors the perceptual illusions or distortions that would occur if the original scene were viewed. In most cases, however, nonveridical perception is attributable to the imaging technology.

The input to a raster display device can be modeled as a sampled version of a continuous function and the output as the result of a reconstruction process. Sampling theory can therefore provide insight into the percepts that result from observation of display images.

19.2 SAMPLING THEORY

A sampled version of a continuous function $f(x,y,t)$ is obtained if it is multiplied by a three-dimensional array of impulse functions with spacing Δx, Δy, Δt. The Fourier transform of this sampled function is a periodic replication of the transform of the original function (Fig. 19.1). These replicas are separated in the f_x, f_y, f_t frequency space by the sampling rates $1/\Delta x$, $1/\Delta y$, $1/\Delta t$, respectively (Gonzalez and Wintz, 1987; Jain, 1989; Rosenfeld and Kak, 1976; Stanley et al., 1984).

If the continuous function is bandlimited, its Fourier transform is zero outside a bounded region (the *region of support*) in the frequency space. Sampling the function at greater than twice these maximum frequencies, that is, at greater than the function's *Nyquist rates*, ensures that none of the replicas of the sampled function will intrude into the region of support for the original function (Fig. 19.1b). The original function can then be recovered from the sampled function by a low-pass filter, that completely attenuates the replicas introduced by sampling and passes the baseband (i.e., original) spectrum without attenuation (Jain, 1989).

If a function is sampled at less than its Nyquist rate for a particular dimension, frequency components representing the original function will extend *above* and corresponding components of the appropriately spaced replicas will extend *below* half the sampling rate (Fig. 19.2). This phenomenon is known as *aliasing*; half the sampling rate is referred to as the *folding frequency* (Stanley et al., 1984).

A function that is sampled at less than its Nyquist rates cannot be recovered by subjecting the sampled function to a low-pass reconstruction filter with cutoff frequencies equal to half the sampling rates. The high frequencies in the original function will be lost, and aliased frequencies will distort the function that is passed. To prevent aliasing, the original function must be filtered before it is sampled to ensure that it does not contain frequencies that are higher than the folding frequencies. Appropriate filtering during reconstruction will then result in a lower resolution but otherwise distortion-free copy of the original function.

The importance of sampling at the Nyquist rates can, however, be overstated. The sampling-induced replicas may overlap with (and thus modify) the baseband spectrum (Fig. 19.2b). Alternatively, they may merely introduce new, nonzero frequencies below the folding frequencies (Fig. 19.2a). In the latter case, a carefully selected nonrectangular filter could both remove the aliased frequencies and preserve the high frequencies of the original function.

19.3 DISPLAY INPUT

The source of the input signal for a raster display can be either a television camera or a computer image generator (IG). The continuous image that is sampled by the former is typically the spatiotemporal distribution of light in a "real-

PERCEPTUAL EFFECTS OF SPATIOTEMPORAL SAMPLING

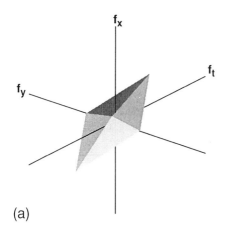

(a)

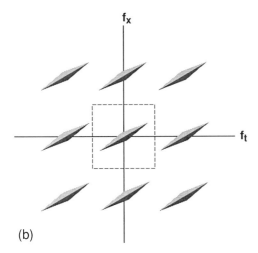

(b)

Figure 19.1 Spectral consequences of sampling a continuous function. (a) The region of support of the Fourier transform of a hypothetical, continuous function. (b) The projection on the $f_t f_x$ plane of the support of the Fourier transform of a sampled version of the function. The replicas occur at multiples of the temporal and spatial sampling frequencies. The dashed line demarks the spatiotemporal frequency space defined by half the sampling rates.

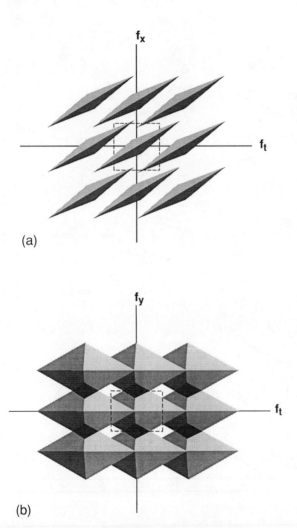

Figure 19.2 Spectral consequences of sampling at less than the Nyquist rates. In both graphs the replicas occur at multiples of the sampling rates and the dashed line demarks the frequency space defined by half the sampling rates. (a) The projection on the $f_t f_x$ plane of the support of the Fourier transform of a three-dimensional function that has been sampled at less than its Nyquist rates. Although the sampling-induced replicas extend into the frequency space defined by half the sampling rates, there is no spectral overlap. (b) The projection on the $f_t f_y$ plane of the region of support depicted in (a). Note that the replicas on the f_y dimension overlap the baseband spectrum.

world" scene. The continuous image that is sampled by an IG is derived from a database that contains a symbolic model of some region of the world. In neither case is the spatial frequency content of the image bandlimited. Moreover, since the temporal frequency of a moving spatial sinusoid equals the product of its spatial frequency and its velocity, the temporal frequency content of a time-varying image is also unlimited (Watson et al., 1986). If aliasing is to be avoided, the original image must be filtered prior to sampling.

The sensors in a video camera integrate the input light over regions of space and time, thereby serving as low-pass (albeit nonoptimal) presampling filters in both the spatial and temporal domains. Although this filtering does not completely eliminate aliasing, it does substantially reduce its amplitude.

Image generation systems sample their unfiltered, data-based scenes at infinitesimal points in space and time. To reduce spatial aliasing, the more advanced IGs have implemented algorithms that approximate the spatial filtering of a video camera (Foley et al., 1990). Rather than sampling at the centers of the displayable picture elements (pixels), each pixel is subdivided, and the scene is sampled at the centers of the subpixels. The digital value for a displayable pixel is a weighted combination of the values of a number of subpixels. Current IG systems have not, however, implemented any procedures to reduce temporal aliasing. A pixel's content is determined by the scene at a single point in time.

The spatiotemporal sampling grids of video cameras and IGs also differ. The sensors in the former are scanned sequentially. The spatially ordered video signal therefore contains information from successive, partially overlapping temporal intervals. In contrast, an IG samples the spatial array simultaneously; each pixel in a two-dimensional spatial image represents the same point in time.

19.4 DISPLAY OUTPUT

Regardless of the source of the input signal for a raster display, the postsampling filter is usually, although not necessarily, limited to the spatiotemporal characteristics of the display device itself. In a CRT, spatial and temporal filtering are due primarily to the intensity distribution of the electron spot and the temporal response of the phosphor, respectively. The former, which is approximately Gaussian, serves as a reasonable postsampling filter if its size is optimally adjusted. In contrast, very little temporal filtering is provided by current CRT phosphors, which typically decay to 10% of their maximum intensity in less than a millisecond. Although the intensity distribution of the effective spot of a light valve is not Gaussian, it does substantially attenuate high spatial frequencies. Light valves also attenuate high temporal frequencies to some extent. The diffraction grating on the oil film decays relatively slowly, with the luminance output sometimes as high as 40% of its peak value at the end of a 16.7 msec field (G. Reining, personal communication, April 1991).

19.5 HUMAN VISUAL SYSTEM

A visual stimulus (e.g., a display image) is filtered by the optics of the eye before it reaches the retina. The resulting retinal image is then sampled spatially by an array of photoreceptors. Signals from the photoreceptors are processed by successive levels of the visual system.

The spatial frequency limit imposed by the point-spread function of the eye is estimated to be approximately 60 cycles/deg at the fovea (Westheimer, 1986). This cutoff frequency is maintained out to eccentricities of about 10° (Jennings and Charman, 1981). In contrast, cone density decreases dramatically from the central fovea to an eccentricity of 10° (Curcio et al., 1987). The spatial sampling rate of the cones is high enough to prevent spatial aliasing only at the central fovea (Wilson et al., 1990), where the spacing between rows of the hexagonal cone mosaic is approximately 30 arc min (Curcio et al., 1987; Williams, 1988).

Psychophysical and physiological data suggest that the signals from the photoreceptors are processed by a number of neural mechanisms or "channels" that act in parallel. Each channel is driven by photoreceptors in a limited region of the retina (its "receptive field") and is broadly tuned with respect to both spatial and temporal frequency. [For recent reviews of the literature on spatial and temporal sensitivity, see Olzak and Thomas (1986) and Watson (1986), respectively.] Although these channels show considerable variation in the spatial frequency to which they are most sensitive, there is evidence that their temporal tuning is either low-pass, up to about 10 Hz, or bandpass, centered at about 10 Hz (Anderson and Burr, 1985; Kulikowski and Tolhurst, 1973). The temporal low-pass mechanisms are predominant for higher spatial frequency channels, whereas the temporal bandpass mechanisms are predominant for low spatial frequency channels (Anderson and Burr, 1985; Tolhurst, 1973). Moreover, the bandpass mechanisms appear to be direction-selective and hence velocity-tuned (Burr et al., 1986; Levinson and Sekuler, 1975; Watson, 1986).

The overall spatiotemporal contrast sensitivity function of the visual system presumably reflects the combined operation of all such channels and describes the system's transfer function (at the visibility threshold). At high temporal and spatial frequencies, the effects of the two dimensions are nearly separable; that is, the spatiotemporal contrast sensitivity function equals the product of the spatial contrast sensitivity function and the temporal contrast sensitivity function (Watson, 1986). However, spatial and temporal contrast sensitivity are not separable over the full range of visible frequencies; sensitivity falls off when the temporal and spatial frequency are both low (Kelly, 1985; Koenderink and van Doorn, 1979; Robson, 1966).

The exact shape of the spatiotemporal contrast sensitivity function varies with the mean luminance, orientation, width, and retinal location of the grating stimulus. The highest resolvable spatial frequency falls between 30 and 60 cycles/

deg (Wilson et al., 1990), the highest resolvable temporal frequency between about 50 and 70 Hz (Farrell et al., 1987; Watson, 1986).

Finally, it should be noted that studies of the spatiotemporal filtering characteristics of the visual system are usually conducted with the observer fixating a stationary mark. Although fixation does not eliminate all eye movements (Riggs et al., 1954), the results with fixation and with the image artificially stabilized on the retina are similar (Tulaney-Keesey, 1982). In any case, the intent of such research is to minimize the effects of eye movements and to describe the response of the visual system when the retinal image and the stimulus are in close correspondence. However, natural time-varying images usually involve object motion, and under unrestricted viewing conditions a moving object tends to elicit smooth pursuit eye movements. Such eye movements have two important effects. First, pursuit affects the trajectory and thus the temporal frequency content of the retinal image. Second, the extraretinal signals associated with pursuit affect the percept that results from a given retinal image.

19.6 IDEAL DISPLAY IMAGE

Because the filtering characteristics of the human visual system are independent of the spatial and temporal sampling rates of the display system, sampling at the (image-defined) Nyquist rates is neither necessary nor sufficient to prevent perception of sampling artifacts. Whereas aliased frequencies will not be problematic if the human visual system is insensitive to them, nonaliased frequencies in the replicas will be problematic if they are passed by the human visual system as well as by the display. Moreover, when a continuous image is filtered prior to sampling, problems can arise even if the presampling filter prevents aliasing and the postsampling filter removes the spectral replicas. A nondistorted low-pass version of the original image may not provide a perceptually acceptable level of resolution.

Ideally, the continuous image would be filtered to exclude only those spatiotemporal frequencies that exceed the limits of the human visual system. If this low-pass signal were then sampled in excess of its Nyquist rates, the sampled image could be displayed without a postsampling filter. The displayed image would contain all of the perceptible information in the original image, and the human visual system would not pass the spectral replicas.

This ideal cannot be consistently met in the spatial domain because the spatial frequency content of the retinal image depends upon the size of the display and the distance from which it is observed. For most viewing conditions, however, the spatial sampling rates of current noninterlaced, high-resolution systems are adequate. Moreover, although the spatial pre- and postsampling filters are not ideal, they can be assumed to be so for the purposes of this chapter. Perceptual distortions attributable to spatial aliasing are usually minimal.

Whereas display size and viewing distance have no direct effect upon the temporal frequency content of a retinal image, the oculomotor behavior of the observer does. As will be discussed in the following sections, the unpredictability of this behavior precludes specification of the retinal image that will result from a given display image and thus of the optimal temporal presampling filter and sample rate.

19.7 IMAGE MOTION

If a two-dimensional spatial form is translating over a uniform field at a constant velocity v, the region of support of the Fourier transform of the space–time image is an oblique plane that goes through the origin of the f_x, f_y, f_t frequency space. The spatial frequencies within this plane are determined by the spatial representation of the object and are independent of the speed and direction of motion. The temporal frequency of each spectral component equals the negative of the dot product of its two-dimensional spatial frequency vector and the two-dimensional velocity vector (Watson and Ahumada, 1985).

Figure 19.3 illustrates the time–space contrast distribution and the spatiotemporal frequency spectrum of a vertical bar moving from left to right at a constant velocity. (Because the spatial form is assumed to be constant in the vertical direction and the motion is horizontal, the y dimension can be omitted from these representations.) The region of support of the spectrum has a slope of $-1/v$ and extends beyond the spatiotemporal frequency window that is depicted. Its sinc-like cross section reflects the width of the bar.

The contrast distribution and spectrum for a time-sampled version of a moving bar are illustrated in Figure 19.4. In the time–space plot, successive samples are separated on t by the sampling period Δt and on x by $\delta x = v \cdot \Delta t$; in the frequency plot, the spectral replicas are spaced at multiples of $1/\Delta t$ on f_t and cross the f_x axis at multiples of $1/\delta x$ (Fahle and Poggio, 1981; Hsu, 1985; Watson et al., 1986).

If the bar in Figure 19.3a is taken to subtend a visual angle of 12.5 arc min and to be translating at $10°$ sec^{-1}, then Figure 19.4a represents a sampling rate of 60 Hz, and the displayed spectra (Figs. 19.3b and 19.4b) each cover ± 120 Hz horizontally and ± 12 cycles/deg vertically. Clearly, for this velocity, the spectral replicas introduced by a 60 Hz sampling rate would contain spatiotemporal frequencies that are well within the passband of the human visual system.

19.8 WINDOW OF VISIBILITY

In this regard, Watson et al. (1986) proposed that sampled and continuous motion will be indistinguishable only if the spectral replicas fall outside the "window of visibility," which they defined as a rectangular frequency space bounded by the

PERCEPTUAL EFFECTS OF SPATIOTEMPORAL SAMPLING

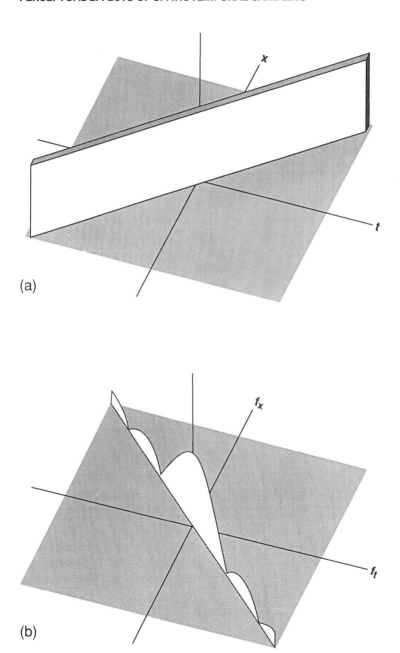

Figure 19.3 The time–space contrast distribution and spatiotemporal frequency spectrum of a vertical bar moving from left to right at a constant velocity.

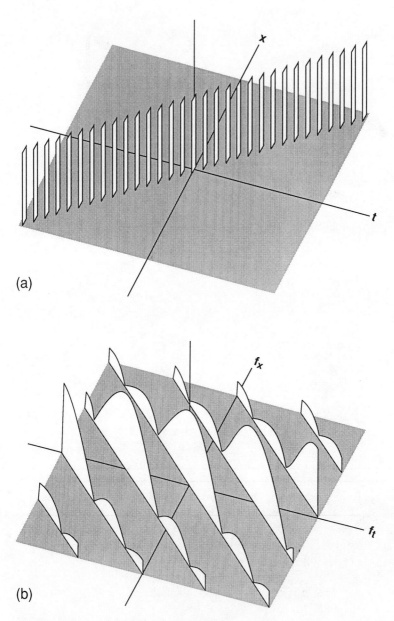

Figure 19.4 The contrast distribution and spectrum of a time-sampled version of a vertical bar moving horizontally at a constant velocity.

spatial and temporal frequency limits (f_{sl} and f_{tl}, respectively) of the human visual system. For an object moving at a constant velocity v, the sampling rate at which the first spectral replicas just touch the window of visibility (i.e., the critical sampling rate, f_{tc}) is shown to be a linear function of that velocity:

$$f_{tc} = f_{tl} + v f_{sm} \tag{19.1}$$

where f_{sm} is the maximum effective spatial frequency in the continuous image. This spatial frequency is given by f_{sl} or, if lower, the highest spatial frequency (in the direction of motion) of the continuous image.

Despite its simplification of the spatiotemporal frequency sensitivity of the human visual system, this model accounted quite well for the results of several psychophysical experiments (Watson et al., 1986). For example, two observers attempted to distinguish between (essentially) continuous and sampled versions of a narrow vertical line moving horizontally. As predicted, the critical sampling rate increased linearly with velocity for both observers. Although the slopes of the two functions differed substantially (6 and 13 cycles/deg) and, as estimates of f_{sl}, were quite low, the two intercepts were similar (about 30 Hz) and provided reasonable estimates of f_{tl}. A sampling rate of 60 Hz was inadequate for all but the lowest velocities (less than 5°/sec for one observer and 2°/sec for the other).

19.9 BREAKDOWN OF APPARENT MOTION

Watson and his colleagues (1986) did not describe their observers' percepts when the sampling rate was less than that needed for the perception of continuous motion. Our research with a 60 Hz display indicates, however, that as velocity is increased the percept changes from that of an object in continuous motion to that of a row of stationary objects, with the separation between objects equal to δx. For velocities between these two extremes, some sense of movement is retained, but it is not so much a property of the object as of a light sweeping across a row of objects, several of which appear to be illuminated at any point in time. [Fahle and Poggio (1981) offer a similar description.]

To examine whether percepts of this type can be accounted for by the filtering characteristics of the visual system, we applied a variety of low-pass temporal filters to the Fourier transforms of sampled images. (These filters, which differed in their high-frequency falloff, were all causal—that is, their impulse responses equalled zero when t was less than zero.) The filtered spectra were then each subjected to an inverse Fourier transform. In the resulting time–space contrast distributions, the intensity of each sample decayed slowly over time. (The duration of this impulse response varied with the high-frequency falloff of the filter.) Thus, the spatial pattern at any one point in time consisted of several bars separated by δx. These bars differed in intensity, decreasing from the most recent to the earliest presentation. When δx was less than the width of the bar, the overlapping portions of the bar were of greater intensity than the nonoverlapping portions. The number of bars above an arbitrary threshold increased with the duration of the impulse response.

Thus, when the replicas in a sampled version of a moving image contain spatiotemporal frequencies to which the human visual system is sensitive, there is general agreement between the percept of an observer and the image that would be passed by a low-pass temporal filtering mechanism. Research has shown, however, that when a target appears to occupy multiple locations simultaneously, the *number* of locations visible at one time, and thus the effective duration of persistence in the human visual system, increases with target velocity (Farrell, 1984; Hogben and DiLollo, 1985). Moreover, objects that are or appear to be in continuous motion often exhibit little if any motion blur—and thus little if any retinotopically mapped persistence (Burr, 1980, 1981; Hogben and DiLollo, 1985). These findings have been interpreted as evidence that the temporal low-pass channels, which respond to stationary stimuli, are inhibited by activation of the velocity-tuned, temporal bandpass channels (Burr, 1980; Burr et al., 1986). Alternative interpretations involve some form of inhibition based on spatial proximity (DiLollo and Hogben, 1985; Farrell et al., 1990).

19.10 TEMPORAL ANTIALIASING

For a system with a fixed sampling rate f_{ts}, the only way to keep the spectral replicas from extending into the window of visibility would be to filter the continuous image prior to sampling. An appropriate filter would ensure that

$$vf_{sm} \leq f_{ts} - f_{tl} \tag{19.2}$$

Given that vf_{sm} is a *temporal* frequency, the most straightforward filter would operate directly in the temporal domain.

If it is assumed that the window of visibility has a temporal frequency limit of 30 Hz, a system with a 60 Hz sampling rate should have a presampling filter that eliminates temporal frequencies greater than 30 Hz (i.e., the difference between f_{ts} and f_{tl}). Conveniently, in this case the postsampling filter provided by the human visual system equals half the sampling rate. Because of this relationship, the prescribed presampling filter eliminates only those frequencies that would be aliased and preserves all of the spatiotemporal frequencies to which the visual system is sensitive.

However, the temporal frequency sensitivity of the visual system is not, in general, limited to 30 Hz. If a higher limit, say 50 Hz, were estimated or assumed, then the spectral replicas introduced by a 60 Hz sampling rate would be visible unless the presampling filter eliminated all temporal frequencies greater than 10 Hz. But such a filter would eliminate many spatiotemporal frequencies to which the visual system is sensitive. The resulting display image would be unlikely to provide an acceptable level of resolution. To keep the replicas outside the window of visibility while simultaneously retaining (in the spectral line representing continuous motion) all of the spatiotemporal frequency information to which the observer is sensitive, the sampling rate would have to be at least 100 Hz and the presampling filter would have to pass, unattenuated, all temporal frequencies less than 50 Hz and fully attenuate all temporal frequencies greater than 50 Hz.

This argument is, of course, merely hypothetical. It is an empirical question whether an observer would be able to distinguish between the unfiltered version of a continuous image and the display version produced by a system with an adequate sampling rate (i.e., $>2f_{tL}$) and a proper presampling filter (i.e., one that attenuated temporal frequencies greater than f_{tL}).

Preliminary work in our laboratory, with a 60 Hz system, indicates that application of a temporal low-pass filter prior to sampling can restore apparent motion to high-velocity images. With this sampling rate and the algorithms we have implemented, however, the spatial form appears to be elongated.

19.11 VISUAL PURSUIT

The appropriateness of a temporal presampling filter depends upon a high correspondence between the display and retinal images and thus upon a constant fixation. This is unlikely under free viewing conditions. Smooth pursuit is elicited by sampled versions of an object moving at a constant velocity even when the spectral replicas contain spatiotemporal frequencies to which the visual system is highly sensitive. Indeed, while the perceptual mechanisms that interpret the retinal image may be able to discriminate sampled and continuous motion whenever the sampling-induced replicas are visible, our research indicates that the oculomotor system is insensitive to this information. Stated more positively the oculomotor system can extract the constant-velocity line from quite complex spectra.

If the velocity of the eyes were to exactly match the velocity of the target in the original image, then the target form in the corresponding display image would be repeatedly imaged on the same retinal location. The region of support of the Fourier transform of this retinal image would consist of vertical lines at multiples of the sampling rate (Hsu, 1985). Each spatial frequency in the baseband would be associated with a temporal frequency of zero. The replicas introduced by sampling would be visible (and then as flicker) only if the sampling rate were less than the temporal frequency limit of the visual system.

The pursuit system is not, however, highly accurate (Hallett, 1986; Wetzel, 1988), and the retinal image that results from tracking a constant-velocity target will usually be characterized by an inconstant residual velocity. To examine the frequency content of continuous retinal images of this type, several hypothetical pursuit-dependent trajectories were subjected to Fourier transforms. In the resulting spectra, the low and intermediate spatial frequencies were predominantly, although by no means exclusively, associated with a temporal frequency of zero. (The energy concentrated at $f_t = 0$ decreased as the variation in velocity increased.) The energy for the higher spatial frequencies was spread more or less evenly over the full range of temporal frequencies.

Given the high temporal frequencies in these hypothetical retinal images, a sampling rate of 60 Hz would result in aliasing. This aliasing could not be prevented by a presampling filter; the temporal frequency content of the retinal image would be determined by the accuracy of pursuit, not by the velocity of the target. Thus, the retinal image that would have resulted from pursuit of the continuous image is probably unrecoverable. Information concerning such a retinal image is presumably important if the perceptual system incorporates precise information about the movement of the eyes.

Although there is no evidence that aliasing in a pursuit-dependent retinal image affects the resulting percept, image perception during visual tracking (and presumably during fixation) is dependent upon extraretinal signals associated with movements of the eyes. If an observer tracks a 60 Hz display of an object translating at a relatively high velocity, an unblurred representation of the object appears to move continuously. It is as if visual pursuit were perfect, with the perceived spatial form corresponding to the stationary retinal image and the perceived motion of that form corresponding to the motion of the eyes. In fact, our data indicate that a temporal presampling filter, in this case, serves only to degrade the spatial percept.

At sufficiently high velocities, however, the perception of continuous motion does break down even under tracking instructions. The image will then appear to consist of a row of objects separated by δx. This breakdown in apparent motion is clearly related to a decline in pursuit accuracy. Moreover, as for the comparable situation during fixation, application of a temporal presampling filter results in a display image that supports the perception of motion although the spatial percept extends over multiple locations on the sampled trajectory.

19.12 UPDATE RATE

Smooth pursuit eye movements of an approximately constant velocity are also sometimes elicited by sampled (and probably continuous) versions of images with inconstant velocity profiles. This property of the oculomotor system is of particular importance for display images created by systems in which the update rate of the IG is less than the refresh rate of the display device. In such images, a representation of a moving target is presented r times at every sampled position, where r is the ratio of the refresh rate to the update rate. Thus, if the continuous image is moving at a constant velocity v, the displacement of the display image

PERCEPTUAL EFFECTS OF SPATIOTEMPORAL SAMPLING

will be in accord with a velocity of zero for $r - 1$ and a velocity of rv for 1 out of every r periods.

Our research indicates that the perceptual consequences of such images vary with a number of interdependent factors, including the duration of the motion sequence and the velocity of the target (Lindholm, 1988, 1992; Lindholm and Askins, 1989). In general, however, perception of a single object in "jerky" motion occurs only when the motion sequence is very short or when the observer attempts to maintain a steady fixation. When the observer is encouraged to track the target and is given time to do so, the velocity of the eyes approximates the velocity of the continuous image. As illustrated in Figure 19.5, the resulting spatial percept corresponds to the spatial form that would be repetitively "painted" on the retina if the pursuit velocity were precisely v [see also Hempstead (1966); Szabo (1983)]. Such multiobject percepts appear to move at a constant velocity, although they may also be characterized by flicker or some form of internal movement.

Figure 19.6 illustrates the Fourier transform of the display image that would result if the continuous image of Figure 19.3a were sampled at 30 Hz and displayed at 60 Hz ($r = 2$). Note that the region of support of this spectrum consists of lines at multiples of the update rate (i.e., 30 Hz). Although these lines all have

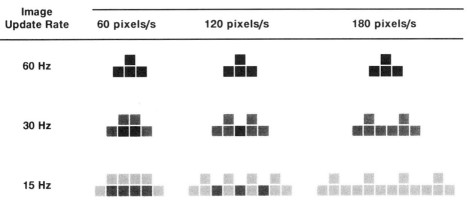

Figure 19.5 Illustration of the effects of image update rate on the perceived form and contrast of a moving object. The spatial percept depicted for the 60 Hz update rate corresponds to the form of the object in the continuous image. When the update (sampling) rate of the IG is less than the 60 Hz rate of the display, a multiobject form is perceived during ocular pursuit. Note that the spacing of the components in the multiobject percepts equals the distance that the eyes would have traveled during a refresh period if the velocity of the eyes exactly matched the velocity of the target in the continuous image.

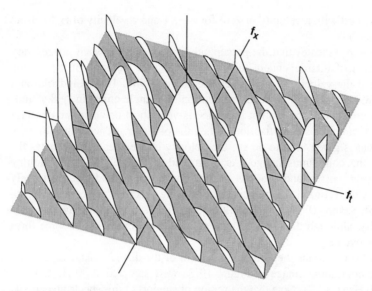

Figure 19.6 Spectrum of the display image that would result if the continuous image of a moving bar (see Fig. 19.3a) were sampled at 30 Hz and displayed at 60 Hz.

a slope of $-1/v$, they do not all have the same spatial spectrum; the spectrum of the constant-velocity line is replicated at even-numbered multiples of the update rate (i.e., at multiples of the display rate), whereas a different spectrum is replicated at odd-numbered multiples of the update rate. Neither spatial spectrum matches that in Figures 19.3b and 19.4b. In particular, then, the spatial form defined by the constant-velocity line in Figure 19.6 is not a single vertical bar. Rather, as demonstrated by an inverse Fourier transform of the spectrum of this line, it is the dual-object form that would be perceived during visual tracking.

19.13 INTERLACING

In the discussion so far, noninterlaced sampling and display grids have been assumed. With current technology, however, the scanning process is usually interlaced. Figure 19.7 illustrates the spectral consequences of interlacing when the input device (either a video camera or an IG) samples once every field (Amanatides and Mitchell, 1990; Jain, 1989). If $1/\Delta t$ is taken to represent the field rate and $1/\Delta y$ the line rate, then the spectrum of an interlaced display differs from that of a noninterlaced display by the addition of spectral replicas at $(f_t, f_y) = (m/2\Delta t, n/2\Delta y)$, where m and n are both odd integers. Any perceptual aberrations that are limited to interlaced displays result from these extra replicas (ERs).

PERCEPTUAL EFFECTS OF SPATIOTEMPORAL SAMPLING

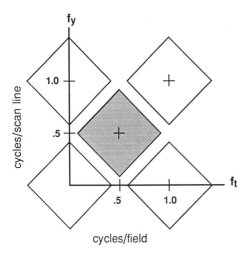

Figure 19.7 Projection on the f_t, f_y plane of the support of the Fourier transform of an interlaced image. The shaded replica would not be present if the system were noninterlaced.

The diagonal sampling grid of an interlaced system is more efficient than a rectangular sampling grid if the pre- and postsampling filters pass comparable, diamond-shaped spectra (Jain, 1989). On the other hand, if the spatiotemporal filter of the human visual system can be approximated by a rectangular "window of visibility," as suggested by Watson et al. (1986) and assumed in the preceding discussion, an interlaced sampling grid is inappropriate.

Koenderink and van Doorn (1979) depicted the spatiotemporal contrast sensitivity of the human visual system as a function of log frequency. In their Figure 3, the isosensivity contours near the limits of sensitivity do indeed delimit an approximately rectangular region [see also Kelly (1985)]. When these contours are replotted in a linear space, however, the region of sensitivity is roughly diamond-shaped. Thus, an interlaced sampling grid appears to be well suited to the spatiotemporal filter of the human visual system.

Nonetheless, interlacing poses special problems. The appropriate presampling filter is clearly more complicated for a diagonal sampling grid than for a rectangular one, as are the spectral consequences of a filter of the wrong shape or size. The diagonally positioned ERs also increase the importance of a good match between the sampling rates of the imaging system and the window of visibility.

Even if the presampling filter prevented overlap of the replicas, portions of the ERs could be visible if $1/2\Delta t$ and $1/2\Delta y$ (in cycles per degree) were not greater than the temporal and spatial frequency limits, respectively, of the (diamond-shaped) window of visibility. Moreover, these ER components could be of high amplitude. The intensity distribution of the display's electron spot would be expected to provide little attenuation of components with spatial frequencies less than ±0.5 cycle/scan line.

As illustrated in Figure 19.7, spectral components in the continuous image that are near half the sampling rate for one dimension will be replicated in the first ERs (i.e., those for which m and n equal ±1) as components near half the sampling rate for the other dimension. In particular, with a 60 Hz interlaced display, energy near (0 Hz, ±0.5 cycle/scan line) will be replicated as energy near (±30 Hz, 0 cycle/scan line).

A human observer perceives flicker when presented with a spatiotemporal frequency of (30 Hz, 0 cycle/deg) unless the amplitude of luminance modulation is low (Farrell et al., 1987). In other words, 30 Hz is less than the maximum temporal frequency of the window of visibility, and if the first ERs in an interlaced display extend near (±30 Hz, 0 cycle/scan line), they are likely to be visible. Indeed, when the spectrum of a continuous, static image contains sufficient energy in spatial frequencies near half the vertical sampling rate, the interlaced display image appears to flicker (Amanatides and Mitchell, 1990).

Such flicker can be eliminated if the presampling filter substantially attenuates vertical frequencies near 0.5 cycle/scan line. To do so, a box filter, for example, must span two pixels vertically—that is, the values of the subpixels in an area equal to twice the line height must be averaged to obtain a pixel's value (Amanatides and Mitchell, 1990). Although such a filter has the deleterious effect of reducing vertical resolution, it does eliminate some of the more serious artifacts attributable to interlacing.

An adequate spatial filter can also eliminate some of the perceptual artifacts associated with the interlaced display of an image with vertical motion. Although these problems could as well be handled by an appropriate temporal filter, real-time IG systems do not filter at all temporally, and the temporal filtering provided by a video camera is independent of spatial frequency. [But see Glenn and Glenn (1987) for a description of a video system in which the temporal presampling filter and sampling rate vary with spatial frequency.]

Consider the case where the box filter of an IG spans only one pixel and the continuous image contains a moving object (or critical component of an object) whose height is less than or equal to the width of a scan line. Assume that at $t = 0$ such an object is centered on a scan line in the field that is *not* being sampled. If the object moves vertically an odd number of scan lines per field time, its position will consistently correspond to that of an even scan line whenever the odd scan lines are sampled and to that of an odd scan line whenever the even scan lines are sampled. Such a moving object will never be part of the display image. Similarly, if the object moves an even number of scan lines per field, it will consistently fall on either an odd- or an even-numbered scan line and therefore will be displayed only every other field.

In the frequency domain, the absence of an object in the display image corresponds to complete overlap of the baseband and ERs: If the object moves up one scan line per field, the ERs centered at (-30 Hz, 0.5 cycle/scan line) and (30 Hz, -0.5 cycle/scan line) will completely overlap the baseband; if the object moves three scan lines per field, the ERs centered at (-90 Hz, 0.5 cycle/scan line) and (90 Hz, -0.5 cycle/scan line) will provide the overlap. The "blinking" of an object that moves two scan lines per field occurs because the replicas centered at (-30 Hz, 0.5 cycle/scan line) and (30 Hz, -0.5 cycle/scan line) cross the f_t axis at $+30$ Hz and -30 Hz, respectively. (The components in the replicas that cross the f_t axis correspond to the 0.5 cycle/scan line component of the baseband.)

In addition to these artifacts, which can be eliminated by limiting the vertical frequency content of the image prior to sampling, replicas of the (0 Hz, 0 cycle/scan line) component can cause perceptual distortions during ocular pursuit if (0 Hz, 0.5 cycle/scan line) is within the window of visibility. As discussed in Section 19.11, tracking causes the baseband and the sampling-induced replicas of the retinal image spectrum to tilt in accord with the velocity of the eyes. (Perfect tracking undoes the spectral tilt due to object motion.) If the replicas are not centered on the f_t axis, pursuit also causes the locations of the replicas to shift. For example, if the eyes were moving up at a velocity of 1 scan line per field (60 scan lines/sec), the ER centered at (30 Hz, 0.5 cycle/scan line) would be shifted to (0 Hz, 0.5 cycle/scan line). Thus, the (0 Hz, 0 cycle/scan line) component of the original image would be replicated, in the retinal image, at (0 Hz, ± 0.5 cycle/scan line)—that is, as a stationary 0.5 cycle/scan line grating.

A 0.5 cycle/scan line grating corresponds to the pattern that would be repetitively imaged on the retina if an observer were accurately tracking the interlaced display of a (not too small) bright object moving vertically an odd number of scan lines per field. Therefore, if this spatial frequency is perceptible, such an object may appear as a moving horizontal grating of 0.5 cycle/scan line (Szabo, 1983).

This effect is not limited to moving objects. The interlaced display of a uniformly bright field contains energy not only at (0 Hz, 0 cycle/scan line) but also, in the first ERs, at (± 30 Hz, ± 0.5 cycle/scan line). Ongoing research in our laboratory indicates that if smooth pursuit eye movements can be elicited by any object (e.g., a small black dot) moving an odd number of scan lines per field, the surrounding stationary bright field will look like a moving grating. A similar, albeit more complicated, effect is seen with some light valve projectors, which reverse the phase of one or two of the diffraction gratings on each successive field (G. Reining, personal communication, April 1991).

19.14 CONCLUSION

The form and motion percepts that result from observation of raster display images often differ substantially from the percepts that would result from observation of the original scene. Some of these perceptual artifacts can be understood by consideration of (1) sampling theory and the spatiotemporal frequency content of display images; (2) the filtering characteristics of the human visual system when the eyes are stationary; (3) the response of the oculomotor system to moving stimuli; and (4) the effects of pursuit eye movements on the retinal image and on the interpretation of that retinal image by the perceptual system. This understanding should, in turn, shape the further development of imaging technology.

Acknowledgment

I thank Tim Askins for software support, Margie McConnon for figure preparation, Marge Keslin for manuscript editing, and especially Norwood Sisson for many helpful discussions of sampling theory and signal processing. Our work was funded by Air Force Contract F33615-90-C-0005 to the University of Dayton Research Institute. A version of this chapter is also being published as an Armstrong Laboratory Technical Report.

References

Amanatides, J., and D. P. Mitchell (1990). Antialiasing of interlaced video animation, *Comput. Graphics 24*: 77–85.
Anderson, S. J., and D. C. Burr (1985). Spatial and temporal selectivity of the human motion detection system, *Visual Res. 25*: 1147–1154.
Burr, D. C. (1980). Motion smear, *Nature 284*: 164–165.
Burr, D. C. (1981). Temporal summation of moving images by the human visual system, *Proc. Roy. Soc. Lond. B211*: 321–339.
Burr, D. C., J. Ross, and M. C. Morrone (1986). Seeing objects in motion, *Proc. Roy. Soc. Lond. B227*: 249–265.
Curcio, C. A., K. R. Sloan, O. Packer, A. E. Hendrickson, and R. E. Kalina (1987). Distribution of cones in human and monkey retina: individual variability and radial asymmetry, *Science 236*: 579–582.
DiLollo, V., and J. H. Hogben (1985). Suppression of visible persistence, *J. Exp. Psychol. Hum. Percept. and Perform. 11*: 304–316.
Fahle, M., and T. Poggio (1981). Visual hyperacuity: spatiotemporal interpolation in human vision, *Proc. Roy. Soc. Lond. B213*: 451–477.
Farrell, J. E. (1984). Visible persistence of moving objects, *J. Exp. Psychol. Hum. Percept. Perform. 10*: 502–511.
Farrell, J. E., B. L. Benson, and C. R. Haynie (1987). Predicting flicker thresholds for video display terminals, *Proc. SID 28*: 449–453.
Farrell, J. E., M. Pavel, and G. Sperling (1990). The visible persistence of stimuli in stroboscopic motion, *Vision Res. 30*: 921–936.

Foley, J. D., A. van Dam, S. K. Feiner, and J. F. Hughes (1990). *Computer Graphics: Principles and Practice*, Addison-Wesley, Reading, MA.

Glenn, W. E. and K. G. Glenn (1987). HDTV compatible transmission system, *J. Soc. Motion Pict. Telev. Eng., 96*: 242–246.

Gonzalez, R. C., and P. Wintz (1987). *Digital Image Processing*, Addison-Wesley, Reading, MA.

Hallett, P. E. (1986). Eye movements, in *Handbook of Perception and Human Performance*, Vol. 1, Sensory Processes and Perception (K. R. Boff, L. Kaufman, and J. P. Thomas, Eds.) Wiley, New York, Chapter 10.

Hempstead, C. F. (1966). Motion perception using oscilloscope display, *IEEE Spectrum, 3*: 128–135.

Hogben, J. H., and V. DiLollo (1985). Suppression of visible persistence in apparent motion, *Percept. Psychophys. 35*: 450–460.

Hsu, S. C. (1985). Motion-induced degradations of temporally sampled images, Unpublished Master's Thesis, Mass. Inst. Technology, Cambridge, MA.

Jain, A. K. (1989). *Fundamentals of Digital Image Processing*, Prentice-Hall, Englewood Cliffs, NJ.

Jennings, J. A. M., and W. N. Charman (1981). Off-axis quality in the human eye, *Vision Res. 21*: 445–456.

Kelly, D. H. (1985). Visual processing of moving stimuli, *J. Opt. Soc. Am. 2*: 216–225.

Koenderink, J. J., and A. J. van Doorn (1979). Spatiotemporal contrast detection threshold surface is bimodal, *Opt. Lett. 4*: 32–34.

Kulikowski, J. J., and D. J. Tolhurst (1973). Psychophysical evidence for sustained and transient detectors in human vision, *J. Physiol. 249*: 519–548.

Levinson, E., and R. Sekuler (1975). The independence of channels in human vision selective for direction of movement, *J. Physiol. 250*: 347–366.

Lindholm, J. M. (1988). Temporal factors affecting perception of raster graphics images, *Bull. Psychonom. Soc. 26*: 495.

Lindholm, J. M. (1992). Temporal and spatial factors affecting the perception of computer-generated imagery (AL-TR-1991-0140), Aircrew Training Res. Div., Armstrong Laboratory, Williams AFB, AZ.

Lindholm, J. M., and T. M. Askins (1989). Temporal-to-spatial conversion in apparent motion, *Bull. Psychonom. Soc. 27*: 494.

Olzak, L. A., and J. P. Thomas (1986). Seeing spatial patterns, in *Handbook of Perception and Human Performance, Vol 1, Sensory Processes and Perception* (K. R. Boff, L. Kaufman, and J. P. Thomas, Eds.), Wiley, New York, Chapter 7.

Riggs, L. A., J. C. Armington, and F. Ratliff (1954). Motions of the retinal image during fixation, *J. Opt. Soc. Am. 44*: 315–321.

Robson, J. G. (1966). Spatial and temporal contrast sensitivity functions of the visual system, *J. Opt. Soc. Am. 56*: 1141–1142.

Rosenfeld, A., and A. C. Kak (1976). *Digital Picture Processing*, Academic, New York.

Stanley, W. D., G. R. Dougherty, and R. Dougherty (1984). *Digital Signal Processing*, Reston, Reston, VA.

Szabo, N. S. (1983). Digital image anomalies: static and dynamic, in *Computer Image Generation* (B. J. Schachter, Ed.), Wiley, New York, pp. 125–135.

Tolhurst, D. J. (1973). Separate channels for the analysis of the shape and movement of a moving visual stimulus, *J. Physiol. 231*: 385–402.

Tulunay-Keesey, U. (1982). Contrast thresholds with stabilized and unstabilized targets, *J. Opt. Soc. Am. 72*: 1284–1286.

Watson, A. B. (1986). Temporal sensitivity, in *Handbook of Perception and Human Performance*, Vol. 1, Sensory Processes and Perception (K. R. Boff, L. Kaufman, and J. P. Thomas, Eds.), Wiley, New York, Chapter 6.

Watson, A. B., and A. J. Ahumada, Jr. (1985). Model of human visual-motion sensing, *J. Opt. Soc. Am. A2*: 322–342.

Watson, A. B., A. J. Ahumada, Jr., and J. E. Farrell (1986). Window of visibility: a psychophysical theory of fidelity in time-sampled visual motion displays, *J. Opt. Soc. Am. A3*: 300–307.

Westheimer, G. (1986). The eye as an optical instrument, in *Handbook of Perception and Human Performance*, Vol. 1, Sensory Processes and Perception (K. R. Boff, L. Kaufman, and J. P. Thomas, Eds.), Wiley, New York, Chapter 4.

Wetzel, P. A. (1988). Error reduction strategies in the oculomotor control system, Doctoral Dissertation, Univ. Illinois-Chicago.

Williams, D. R. (1988). Topography of the foveal cone mosaic in the living human eye, *Vision Res. 28*: 433–454.

Wilson, H. R., D. Levi, L. Maffei, J. Rovamo, and R. DeValois (1990). The perception of form: retina to striate cortex, in *Visual Perception: The Neurophysiological Foundations*, (L. Spillmann and J. S. Werner, Eds.) Academic, New York, Chapter 10.

20
Electro-Optic Displays—The System Perspective

Donald L. Moon

*The University of Dayton,
Dayton, Ohio*

20.1 INTRODUCTION

Electro-optic (E-O) displays are among the most used and most useful sensors in almost all systems where a man-machine interface is required. The number of applications of these displays, in one technological form or another, is growing in what seems to be an unbounded manner. For instance, we find E-O displays in all computer systems, automotive systems, tank systems, aircraft systems (fixed-wing and rotary-wing), naval systems, etc., in both military and civilian applications. The number of different implementation technologies that are being used and developed is large. That is, we have liquid crystals, plasma, electroluminescence, vacuum fluorescence, light-emitting diodes, etc. Also, we have flat-panel, CRT, and flat-panel CRT embodiments of these technologies. In addition, these displays have a large number of functions. We have panel displays, head-up displays, horizontal situation displays, night vision displays, projection displays, stereoscopic displays, holographic displays, three-dimensional displays, and many others.

When one considers all of these applications, technologies, and functions and the fact that many applications will accommodate many technologies and multiple functions, the problem of "display integration" becomes considerable. How do all the displays in a given system operate together in light of human factors and image quality? Can "integration technologies" such as computers, busing structures, and software be organized in such a way as to enhance the image quality and information quality of the overall system? Even though these questions may (seem to) be rhetorical in some sense, their answers are being actively pursued in the research and engineering communities. A significant amount of work toward answering these questions has been expended in recent years, and as yet there are not clear answers. This chapter is dedicated to that work and therefore attempts to present some of the ideas, definitions, and system concepts that are allowing display systems engineering and architectures to evolve to answer these questions.

20.2 DEFINITION OF IMAGE QUALITY

The visual quality of an image on a display surface, as the image is perceived by the human eye, is generally referred to as the *image quality*. Therefore, the parameters that characterize a display are those metrics that are identifiable as impacting on the way the human eye will perceive an image in terms of aesthetics, dynamics, and information content (Moon, 1988).

Ideal image quality would be defined for a display as follows: The display would (1) modulate the ambient light when it is abundant but emit bright light in the dark; (2) be capable of producing saturated colors at will; (3) be visible from all angles; (4) have high resolution; (5) respond in microseconds but retain the image indefinitely if so desired; (6) have a contrast ratio of at least 50:1 and 64 levels of gray; and (7) consume negligible power at low voltage. Unfortunately, no single technology exists that can meet all of these ideal characteristics. As a consequence, image quality is generally very closely aligned with the application environment for which a specific display is intended, and accordingly the image quality is enhanced by optimization of the parameters most important to that application.

In order for a display to exhibit good image quality, certain requirements with respect to legibility, luminance, and contrast must be adhered to by the display regardless of the technology used. A general discussion of these image quality requirements and their measurable features is presented in the following paragraphs.

20.2.1 Legibility

The first requirement of a visual display device is that the displayed information must be "seen"—easily, accurately, and without ambiguity—under the conditions of use. This quality can be summarized as *legibility*. A great deal of human factors research has been conducted on this subject, with error rate and speed of recognition as the principal measurement parameters. The situation is complicated by the fact that prior training, environmental distractions, and cumulative fatigue can directly influence the results, independent of the quality of the display (Conrac, 1980).

It is clear, however, that three major factors—contrast, luminance, and size of the graphic elements—all affect legibility. In order to provide the maximum amount of information within a display of a given size, it is an advantage to keep the size of the graphic elements (e.g., line width, character height) to a minimum established primarily by the viewing distance. But the legible minimum is directly affected by contrast and luminance. With a given graphic-element size, legibility can be significantly enhanced by increasing either or both of these variables.

The human eye (or more precisely, the eye–brain system) is a differential input device. It is designed to gain information by interpreting the differences detected in color values and luminance levels. Hence, the use of contrasting colors and luminance contrast, which can be generally defined as the difference between the color or luminance that represents information and the corresponding values for the "background" or non-information areas of the display surface, is important.

20.2.2 Luminance

Increased luminance contributes to display legibility by increasing the acuity of the eye—the ability to separate fine detail such as individual dots or lines on the display surface. An "average observer" under average lighting conditions has an angular acuity of approximately 1 min of arc. For a typical viewing distance of 18 in., this means that the lower separable limit for dot diameters or line widths is on the order of 0.05 in. Above-average luminance levels can reduce this dimension by nearly half, but lower luminance levels have the opposite effect. At very low levels, the minimum separable display-element size may increase to several minutes of arc.

Such precision may seem irrelevant for most display applications. Increased visual acuity directly enhances, however, the "crispness" of the boundary lines that define, for example, an alphanumeric character. This in turn increases the legibility of each character or graphic element, but only if adequate contrast between the image and the background is also maintained. If the luminance levels of both are increased simultaneously (as occurs when ambient light is reflected off the entire display surface), the legibility will actually decrease.

This paradox is partially the result of a second characteristic of the human eye. Increased levels of luminance may increase the acuity of the eye, but they reduce the all-important ability to discern luminance differences. The minimum discernible luminance difference is, over a broad range, a fixed percentage or fraction of the average luminance level. The higher the average luminance, therefore, the less sensitive the eye is to luminous differences.

20.2.3 Contrast

The direct relationship between luminance levels and the ability to perceive luminance differences explains why the contrast of a display is normally expressed as a ratio. The absolute luminance values of the image and background surfaces of a display are, within reasonable limits, not nearly as important as their "relative" luminance values. In fact, the contrast ratio of a display device is probably the most important single characteristic contributing to the legibility of a display and its "looking-good" appearance.

Unfortunately, the contrast ratio is also the most difficult display parameter to predict with precision, and this difficulty is compounded when an attempt is made to compare different display technologies or even different devices within the same technology. In theory, the contrast ratio of a display device can be simply calculated as the maximum luminance divided by the minimum luminance. (It should be noted that this is not a universally accepted definition.)

The effective maximum and minimum luminance values may not, however, be easily obtained. Many manufacturers of display devices publish only the "intrinsic" contrast ratios of their displays, without taking into account the effect of reflected ambient light. The environmental conditions in which the displays may be used are too variable to anticipate. This leaves it up to the display-system designer or user to predict (or suffer with) the "extrinsic" contrast ratio, which can be significantly lower. The extrinsic contrast ratio can be calculated, of course, for any known ambient conditions, but this statement presumes that the display-system designer also knows the reflective and diffusing qualities of the display surface. Such information is rarely provided by the published literature for a display device and may require specific tests by either the manufacturer or the potential user.

To complicate matters further, a number of other contrast calculation conventions have been established, some of which result in values identified as "contrast" rather than "contrast ratio." But the two terms are generally used interchangeably in the same written material. It is advisable, therefore, to determine the ways in which contrast ratios have been calculated before using them for comparison. If the display consists of discrete elements, the contrast specifications may also differentiate between display-element "on" and "off" conditions, with separate contrast ratios for each situation.

Another way to specify contrast is to state the "shades of gray" that the device can display. As noted earlier, the minimum discernible luminance difference is a function of the luminance level—approximately 3% for most observers. The shades-of-gray convention uses an arbitrary multiplier of 1.41 (the square root of 2) for each gray-scale step. For example, 1 shade-of-gray translates to a minimum contrast ratio of 1; 2 translates to 1.41; 3 translates to 2.0; 4 translates to 2.8; etc.

All of the contrast considerations discussed up to this point have related to the display itself, including the effect of reflected ambient light on the display contrast. There is, however, a secondary ambient-light effect that must be taken into account. Light reflected off the surroundings or directed into the viewer's eyes can alter the luminance-adaptation level of the eyes, and this, too, can directly affect the legibility of the display.

The *surround contrast* is defined as the ratio between the luminance level of the surroundings and the background luminance of the display. The ideal surround contrast ratio is 1:1. If the surrounding luminance is markedly lower than that of the display background, the effect on legibility is relatively minor, although eyestrain may result. But as the surround contrast ratio increases toward 10:1, which can easily happen in an outdoor environment, the display contrast ratio must be at least doubled to maintain an adequate level of viewing comfort and legibility.

20.3 MEASUREMENT OF IMAGE QUALITY

Monochromatic images can be represented as a two-dimensional light emission (or reflection) function, $f(x,y)$. The value of the function is proportional to the intensity at each (x,y) location. If the image is approximately continuous, as in a photograph, then x and y can be considered to be continuous variables. On the other hand, if one is referring to most flat-panel displays or digitially addressed CRTs, then x and y are discrete variables. Digital images are discrete in intensity as well as in spatial coordinates. The (x,y) elements of a digital image are called image elements, picture elements, pixels, or pels. To be compatible with digital computer programs and storage conventions, digital image sizes are usually a power of 2 such as 512 × 512, with 8, 16, 32, or 64 intensity levels (Tennas, 1985).

Regardless of whether the displayed image is continuous (analog) or discrete (digital), the intensity dimension must ultimately be measured or calculated in luminance units in order to relate the appearance of the image to the visual system of the observer. Appropriate units are candelas per square meter (cd/m^2) in the SI system or foot-lamberts (fL) in the English system. All useful measures of image quality can be calculated from the $f(x,y)$ data measured from the image.

In measurement terms, "image quality" has been used in two general contexts: (1) that dealing with physical measures of the image itself with little or no regard for the ability of the observer to obtain information from the image and (2) that dealing with perceived or measured quality from the human observer, sometimes with little regard for the physical characteristics of the image. Both concepts of image quality are critical for an understanding of the importance of clearly measuring image quality in any design context.

20.3.1 Physical Measures of Image Quality

Physical measures of image quality attempt to define or describe pertinent image statistics relative to a baseline or ideal image. For example, some physical image metrics relate intensity distributions of an image to assumed ideal image intensity distributions or relate an original image to a degraded version of the image such that the differences in the statistical intensity distributions are a measure of the degradation or the reduced image quality. This type of metric is usually described as a "pixel error" metric.

Another set of physical metrics of image quality are based on the modulation transfer function (MTF), a measure of the displayed contrast in an image as a function of the size of objects in the image.

An important distinction with regard to the physical measures of image quality is that they do not attempt to relate the measured quantities directly and empirically to the visual performance of the user of the image. That is, the emphasis is placed solely upon the physical measurement, and the validity of the measurement, calculation, or construct is assumed from some statistical characterization of the user or the visual system rather than from a direct experimental test of the perceived quality of the image with a user population or sample.

20.3.2 Behaviorally Validated Measures of Image Quality

Behaviorally validated measures of image quality, on the other hand, strongly emphasize the visual performance or perception of the user and relate this performance empirically to physical characteristics of the image. Most of the behavioral measures of image quality are simpler in form than are the purely physical measures, perhaps because of the less sophisticated level of mathematical interest of the researchers performing psychophysical experiments or perhaps more because of the awareness of the limited precision of perceptual measurement and the desire to maintain a comparable level of sophistication in the measurements of both the physical image and the perceptual response to the image.

A careful evaluation of the literature of image quality convinces the evaluator that it is both meaningful and useful to conduct and apply research that relates physical measures of image quality to user performance. A theme of this chapter is that useful measures of image quality must contain both repeatable and design-relevant physical descriptions of the image as well as suitable behavioral measures by which to assess the validity of the measure.

20.4 METRICS OF IMAGE QUALITY

In this section, the purpose is to delineate the metrics that affect the quality of images that are displayed on an electro-optical display. This delineation will be concerned, primarily, with the most important metrics that are generally associated with CRT displays and flat-panel displays.

The use of CRTs as a display medium has been thoroughly investigated, and the variables of these displays that influence image quality are well known. When considering flat-panel or matrix displays, a likely first step would be to identify analogous variables between CRTs and matrix displays. One might then hypothesize, on the basis of the extent of the CRT literature, that the variables that influence performance with CRTs would similarly influence matrix displays. Following up on this, however, may not be a productive strategy because matrix displays differ fundamentally from CRTs in a number of ways. For example, the distribution of luminance on a CRT is spatially continuous in one dimension and discontinuous in the other dimension, while matrix displays, by their very nature, have a discrete distribution of luminance in both spatial dimensions. Consequently, it is probably appropriate that proper consideration of matrix displays requires that these displays be considered as a unique set. In any case, since any display operates in both the temporal and spatial domains, as a starting point we will assume that each of the two types of displays requires consideration of a distinct set of metrics for image quality. Later, we will see that this assumption may not be extremely important and, in most cases, is not necessary.

20.4.1 CRT Display Image Quality

The variables that affect the quality of a CRT image may be categorized into three groups: geometric, electronic, and photometric (Task, 1979). Within these three groups we find the elements listed in Table 20.1

The most important of these metrics with respect to image quality are luminance (brightness), contrast, frame rate, field rate, resolution, spot size and shape, and modulation transfer function. These parameters are briefly discussed below.

Brightness

Recommending an optimum brightness for an image is difficult. Typical peak brightness values range from 30 to 100 fL, with a minimum of 10 fL recom-

Table 20.1 Variables That Affect CRT Image Quality

Geometric	Electronic	Photometric
Viewing distance	Bandwidth	Luminance
Display size	Dynamic range	Gray shades
Aspect ratio	Signal/noise ratio	Contrast radio
Number of scan lines	Frame rate	Halation
Interlace ratio	Field rate	Ambient
Illuminance		MTF
Scan line spacing		Color
Linearity		Resolution, spot size and shape, luminance, uniformity

mended under even the best viewing conditions (Goldmark, 1949), for example, low ambient lighting, no glare sources, and high contrast.

Contrast

The use of the term *contrast* in CRT work refers to the ratio of the maximum to the minimum luminance on the screen. Contrast ratios of 50:1 to 100:1 are typically adequate for effortless viewing (Goldmark, 1949). The greater the contrast ratio of a CRT, the greater the number of quantized levels possible. These levels of luminance steps are referred to as *gray-scale* steps, generally accepted as equal logarithmic steps that increase by a factor equal to $\sqrt{2}$ between steps. The number of gray scale levels possible introduces a concept known as the *dynamic range* of the system. For example, a system having 14 gray levels would possess a dynamic range of $(\sqrt{2})^{14} = 128$. The contrast capability of the system can be measured by the number of gray-scale steps that can be reproduced.

Frame Rate and Field Rate

The *frame rate* of a CRT system is the frequency at which the video picture is fully updated, whereas the *field rate* is the frequency at which a full picture is written on the screen. The continuity of the picture is dependent on the field rate, which must be high enough to represent motion in an "apparently" continuous manner while avoiding the presence of flicker.

A field rate of 60 $\sec^{-1}$ is considered to be sufficient to fuse discrete movements of an object, providing the object does not move across the frame at too excessive a rate. It can be pointed out here, however, that a selected field rate does imply a maximum object/sensor relative velocity above which motion may be marked by visible jerks in the progress of the object across the screen.

Resolution

Spatial resolution is perhaps one of the most important parameters in determining the picture quality of an imaging system. As the size of the picture elements decreases, the sharpness of the picture increases to some limiting value determined by pixel size. The ability of an imaging system to represent the fine detail in a scene is referred to as the *resolution* of the system. In television, the resolution is measured by the maximum number of adjacent parallel lines that can be produced in the image. The term *TV lines of resolution* generally refers to the maximum number of horizontal lines that can be accommodated within the vertical height of the display. This is, of course, dependent on the number of scan lines being used in the imaging sensor.

Next we consider the horizontal resolution. We start with the simple statement that a measure of horizontal resolution is the maximum number of parallel vertical bars that can be accommodated within the width of the display. The horizontal resolution is, therefore, the maximum number of vertical lines (or bars) that can

E-O DISPLAYS—THE SYSTEM PERSPECTIVE

be produced on the display. Since each black and white (dark and bright) line or bar pair can be regarded as a full cycle of the signal voltage, the term *spatial frequency* is often used to represent the number of bright and dark pairs of bars present on a display. The spatial frequency can be expressed in terms of cycles per display width or cycles per unit distance across the display (e.g., cycles/mm or cycles/in.).

Spot Size and Shape

Related somewhat to resolution is the size and shape of the spot formed on the phosphor screen by the electron beam. In analyzing the anatomy of the CRT spot, both the current distribution in the electron beam and the spreading of the light on the phosphor screen are points of consideration (Goldmark, 1949). The current is highest in the center of the spot and decreases in a Gaussian manner. Current is plotted as a function of distance in standard deviations out from the center of the spot. Since in a Gaussian distribution the current never decreases to zero, it is customary to choose some point on the curve to serve as an arbitrary measure of the spot's diameter. One commonly chosen value is the 50% point on the curve.

The lack of a finite diameter for the light distribution of a spot on the phosphor screen raises a question concerning the obvious overlapping of the Gaussian scan lines. The luminance value at a single point, even without consideration of scattered or ambient light, is the mathematical summation of the contributions of every scan line above and below it (Levi, 1968).

The term *spot size* in the above discussion has been used to convey two distinct, but related, descriptions of the CRT spot. One refers to the geometric shape of the spot, the other to the distribution of the light within the spot. The latter is more properly referred to as the *point-spread function*. So, more precisely, the CRT spot can be described as circular in shape with a Gaussian point-spread function.

Modulation Transfer Function

The last measure of CRT image quality to be presented is the one that is currently most used, the modulation transfer function (MTF). For displays where the information to be presented is either rapidly changing or of high density, it is advantageous and necessary to know how accurately the CRT responds to the modulation of the electron beam. As applied to CRTs, the MTF measures the sine wave frequency response of the CRT system from electronic input to visual output (Jenness et al., 1976). In more precise terms, the MTF of the display system indicates the system's capability to transfer contrast from an input screen to the output image as a function of spatial frequency. The MTF can be obtained experimentally by first modulating the electron beam with a specific frequency sinusoidal signal. The resulting sinusoidal spatial luminance pattern on the phosphor screen is then scanned by a photometer, and the amplitude of the luminance pattern is recorded. The modulating signal is progressively increased in frequency by repeating the measurements. The obtained amplitude values are normalized to the low-frequency values. A plot of the normalized values as a function of spatial frequency is the MTF. A detailed methodology for obtaining the luminance profile on the face of a CRT can be found in Virsu and Lehlio (1975).

For an MTF to validly describe a system, the response of the system must be uniform through the field of view (homogeneous) and in all directions (isotropic), and the response must be independent of input levels, that is, possess the properties of a linear system. CRT systems approximate all of these properties except one: the CRT system is anisotropic. The imagery presented on the CRT is the result of continuous sampling in the horizontal direction but discrete sampling in the vertical direction. This departure from a linear system strictly means that two MTFs (one vertical and one horizontal) are required. However, the horizontal MTF is the more commonly utilized figure of merit.

Several "MTF-derived" metrics have been proposed from the research community. The most important of these should be included in any serious discussion of image quality and are so included here by reference only: equivalent passband (Linfoot, 1960), Strehl intensity ratio (Linfoot, 1960), MTF area (Snyder, 1980; Boroughs et al., 1967; Gutman et al., 1979), gray shade frequency product (Task and Verona, 1976), and integrated contrast sensitivity (Van Meeteran, 1973).

20.4.2 Flat-Panel Display Image Quality

The variables that affect the quality of a matrix display (MD) image are generally the same as for CRT displays except for some unique spatial parameters. These are presented in the following paragraphs along with a short discussion of temporal and luminance variables.

Spatial Variables

There are several spatial variables such as emitter size, spacing, percent active area, shape, packing, and gradients. Each of these is described below.

Emitter Size. This is the physical extent of the light-emitting surface. For square emitters, size is given by the edge dimensions; for circular emitters, the diameter is used. For emitters without a square luminance profile—emitters with an edge luminance gradient—the 50% luminance point defines the size. Size may also be expressed in terms of visual angle (VA) if the viewing distance d is known. Visual angle is given by VA $= 2 \arctan(h/2d)$, where h is the height of the viewed image at d.

Emitter Spacing. This is the distance between centers of adjacent emitters. Spacing may also be described in terms of visual angle.

Percent Active Area. This is the amount of the total display surface that emits

E-O DISPLAYS—THE SYSTEM PERSPECTIVE

light. For square emitters, percent active area is given by (emitter size/emitter spacing)^2 × 100. For other shapes, the total active area must be computed on the basis of the emitter size and number of emitters and this result divided by the total available display surface area.

Emitter Shape. The emitters in an MD potentially can be any geometric shape. Shape is an important variable because it will influence how the emitters are packed on the display. In addition, shape affects the maximum possible percent active area for a given display.

Emitter Packing Format. This is a classification of the geometry of the emitters in the matrix. In orthogonal packing the centers of emitters in adjacent rows and columns are in register. To an extent, emitter shape dictates packing format. For example, circular emitters would pack "best," in the sense that the percent active area is maximized in a rhombic matrix, with the centers of adjacent emitters forming the vertices of an equilateral triangle.

Edge Luminance Gradient. This is the luminance profile of the emitter. Typically emitters have a sharp luminance edge (or a square luminance profile). Alternatively, the luminance distribution may taper off gradually. The latter instance is similar to the Gaussian luminance profile of a CRT spot.

Temporal Variables

A matrix display also has variables in the temporal domain. Present and projected matrix displays do not emit light continuously but rather emit pulses at some frequency. The duration of these pulses specifies the "on" time, and the frequency of the pulses gives the *refresh rate* of the display. The ratio of the on time to the period, where the period is the reciprocal of the refresh rate, defines the *duty cycle*.

Luminance Variables

Finally, luminance (L) parameters are important to matrix displays. The *emitter luminance* specifies the output of a given emitter on the display. The *inactive area luminance* (obviously germane only to displays with less than 100% active area) may also vary. Together, these variables define the maximum possible contrast, $(L_{max} - L_{min})/L_{min}$, which will vary between $-\infty$ and $+\infty$. Alternatively, the two luminance values may be used to express the maximum possible modulation using the relationship $(L_{max} - L_{min})/(L_{max} + L_{min})$, which ranges from 0 to 1.

20.5 DISPLAY SYSTEMS

20.5.1 Military Aircraft Cockpit Displays

Perhaps the problem that most drives the requirements in military aircraft is the full situation awareness problem, that is, the desire of the pilot to have 4 Π steradian visibility in the region of flight. The beyond-visual-range situational awareness problem solution requires the "fusion" or "integration" of radar, electronic warfare, the Joint Tactical Information Distribution System (JTIDS) navigation, and electronic map information on a large display. This would allow the pilot to look at a single source to "get the Big Picture" (Adam, 1990).

From the early 1950s to the late 1980s, the evolution of fighter cockpit displays was typically as follows:

Period	Aircraft	Display size and number
1950s	F-4	3in. diameter (1)
1960s	F-15	4 in. × 4 in. square (1)
1970s	F-18	5 in. × 5 in. square (2)
		5.5 in. × 5.5 in. square (1)
1980s	F-15E	5 in. × 5 in. square (1)
		6 in. × 6 in. square (2)

Generally speaking, until the latter part of this period, the displays tended to be stand-alone, that is, no sharing of capability, no fusion of data—no integration. The search for a better solution to the situation awareness problem and the increased capability of both sensors and computers is leading to larger display surfaces. Predictions are that flat-panel matrix technologies will deliver surface areas that are 10–15 times what is available using CRT technology today.

During the latter part of the 1980s, newer aircraft cockpits began to enjoy the availability of larger surface displays as well as the results of better understanding of the requirements for solution of the situation awareness problem. Today's typical cockpit is compared to what are predicted to be the characteristics of "Cockpit 2000" in Figure 20.1.

Cockpit 2000 is predicted to have about twice the display area of current fighters and to differ from today's cockpit in two important aspects: (1) A helmet sight and display will provide all normal head-up display (HUD) functions on the helmet visor with the added benefit of off-axis target designation, and (2) the 10 in. × 10 in. global situation display will be not only larger but also more productive than any three of the small multifunction displays currently in use.

Beyond the year 2000, an increase in display technology research and development will eventually provide flat matrix display panels with large surface areas, high brightness, high resolution, and long life. These large "Big Picture" displays will provide 10 times the display area of today's CRTs, allowing plan and perspective views, split screen, and movable inserts. A helmet sight and display, voice command, and touch-sensitive surfaces will provide pilot interface with the weapons system. In short, the "Big Picture" provides the pilot with full control over the configuration and content of almost 400 in.^{2} of display surface

E-O DISPLAYS—THE SYSTEM PERSPECTIVE

Figure 20.1 Cockpit 2000 solves the two most pressing cockpit problems: tactical and global situational awareness.

to match the mission whether it be air-to-air, air-to-surface, navigation, terrain following/terrain avoidance, or system status. Manned simulations have shown a 100% increase in the situational awareness of pilots using the "Big Picture" over those using a conventional cockpit with its two or three small multifunction displays (CRTs).

20.5.2 Commercial Aircraft Cockpit Displays

Commercial airlines have accepted the new, simplified "glass cockpits," with displays generated on the faces of multiple-color cathode-ray tubes, as a means of standardizing crew levels at one pilot and one first officer (*Aviation Week and Space Technology*, April 11, 1988). Although the new "glass cockpit" technology represents a remarkable advance in pilot–vehicle interface techniques, the displays utilized are largely electronic renditions of the electromechanical (round dial) instruments that they replaced (Hatfield, 1990).

The current state of the art, as well as a potential future direction, in advanced cockpit technology is indicated in Figure 20.2. Representing the state of current cockpit technology is the MD11 transport aircraft cockpit. Here, a clean, uncluttered pilot interface is provided by an "all-glass" cockpit configuration, developed jointly by Douglas and Honeywell (Morello, 1989). Display information is computer-generated on full-color CRT displays.

The displays are form-factor D units, each having a 6.5 in. × 6.5 in. display surface. They are arranged six across in a side-by-side configuration. Flight control information is presented on the two outside primary flight displays, and engine-monitoring/alert information and system status information is presented on the two center displays. Navigational information, in the form of charts

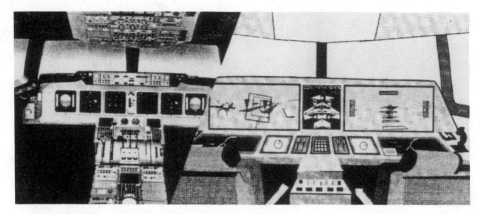

Figure 20.2 State-of-the-art transport cockpit (MD 11) and an envisioned future cockpit.

(maps) or horizontal situation indicators, is presented on the remaining two displays. Pilots interface with the aircraft control system and the navigational system primarily via the glareshield control panel and the navigation control-display units (multifunction keyboard/display units shown in the center console). The MD11 flight management system provides lateral and vertical guidance with laser inertial reference system capabilities. The aircraft makes extensive use of reliable digital avionics and automation to aid the pilots in flight management, aircraft control, and onboard systems monitoring. The centralized fault display system provides built-in test equipment capabilities.

Glass cockpits have emerged in other commercial aircraft such as the Boeing B757, B767, and B747–400; the Gulfstream IV; and the Airbus Industries A310 and A320. In these cockpits, as in the case of the MD11, the conventional electromechanical flight instruments have been replaced with color CRTs driven by modern processors. Although these modern transport all-glass cockpits represent a significant advance in capabilities, crews have complained of too much complexity and too high a workload during operations under 10,000 ft altitude (Wiener, 1989). This complaint is associated primarily with the crew's programming of the flight management system (requiring extensive "head-down" time segments) during descent phases of the mission. This "head-down" task, with its associated high workload, is especially undesirable during the descent phase of the mission, because it is during this phase that air traffic becomes most congested and "out-the-window" scanning for other traffic is a routine operational procedure.

Clearly, there are areas in which current transport cockpit technology can be improved and future transports will have unique cockpit requirements. As we move toward future transports and their associated cockpits, there are many technologies that must be developed and exploited. A vision for the future transport cockpit is shown by Figure 20.2, an advanced cockpit technology concept emanating from aeronautical human factors research. Depicted here is an advanced, "all-glass" flight deck that is unusually clean and uncluttered and that makes use of large-screen, integrated, pictorial display technology and human-centered design techniques. This concept and the technology that needs to be developed to make it feasible will be explored in a later section. First, technologies that are already under development will be discussed.

The rapid advance of display media, graphics and pictorial display, and computer technologies, along with human factors methodologies is making possible important advances in next-generation cockpits. These technologies have the potential to enable the construction of cockpits with a smaller crew workload during critical mission phases and improved crew situational awareness, safety, and operational efficiency. The following sections discuss some of these technologies.

A major thrust that is under way in the research and development community is the replacement of color CRT display technology with flat-panel display technology. The main design goal of these flat-panel displays is to minimize depth, weight, and power consumption while improving reliability and sunlight viewability.

The color CRT provides the advantages of low cost (because of its maturity) and high-resolution display. However, it has the disadvantages of large depth and ungraceful degradation. Further, it is susceptible to washout under high levels of ambient light.

The most promising full-color flat-panel technology seems to be the active-matrix liquid-crystal display. One such device, made by General Electric, has a 6.25 in. × 6.25 in. usable screen area and is capable of high-resolution (512 × 512 color pixels and 1024 × 1024 monochrome pixels) graphics and/or video with 16 gray-scale levels (Credelle, 1987 and Haim, 1988).

Boeing has plans to replace color CRT display technology with color flat-panel display technology when they initiate production of the B777 long-haul transport. The chosen flat-panel technology for this application is color active-matrix LCD technology.

NASA Langley Research Center's research interests presently include two thrusts in the area of large-screen, pictorial displays for future transport cockpits. These efforts are predicted on (1) the observation that color electronic display media for cockpits have been steadily increasing in size to provide additional display space with its associated flexibility, and (2) the assumption that that trend will continue. The research is aimed at determining the benefits and limitations of large-screen media and the new pictorial formats that they may enable, as well as providing guidelines for the application of these new formats. The two basic thrusts are multiwindow reconfigurable displays and panoramic displays. Although, initially, these display concepts are being studied separately, it is felt that they will ultimately be merged into a concept that involves both windowing and wide-screen panoramic display.

Several types of display technologies, coupled with advances in real-time graphics generators, could enable the type of advanced-concept cockpit depicted in Figure 20.2, wherein total integration of the crew's information requirements is achieved through panoramic, wide-field-of-view, integrated pictorial displays. These technologies include ultralarge-screen flat panels, edge-abutted large-screen flat panels, or projection display technology (using compact folded projection optics), such as LCD light valves or laser projection technology. Figure 20.3 depicts some of the operational modes of large-screen display systems being developed at LARC.

20.5.3 Automobile Display Systems

During the last decade, electro-optic displays have found increasing use in the passenger compartment of new automobiles and, as automobile design procedures take more of a systems approach, display applications in automobiles are expected to proliferate at an even heavier rate (Rodda, 1987; Ross, 1988; Rivard, 1987; Jurger, 1991). Historically, display technologies in automobiles have typically evolved as indicated in Table 20.2.

The need to convey information to an automobile's driver through instrumentation was apparent even in the early years of automotive development. In Chev-

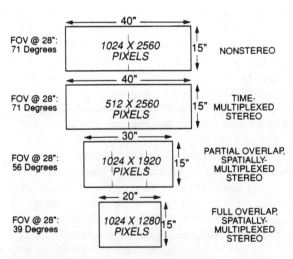

Figure 20.3 Operational modes of large-screen panoramic display system being developed at Langley Research Center.

Table 20.2 The Evolution of Automobile Display Technology

	1975–1985	1985–1990	Beyond 1990
Mechanical	D-Arsonval movement	Electronically controlled, air core	Electronically controlled, air core
Fixed format	Vacuum-fluorescent, light-emitting diode	Higher resolution vacuum-fluorescent, liquid crystal	High-resolution vacuum fluorescent, liquid crystal
Variable format	None	Cathode-ray tube, liquid-crystal dot matrix	Cathode-ray tube, liquid-crystal dot matrix, electroluminescent dot matrix, programmable vacuum-fluorescent dot matrix, addressable displays, projected head-up displays, holographic displays

rolet's first year of production, 1912, its Classic Six model featured no less than two gauges and two odometers. All of these were, of course, mechanical.

During the following decades the number of readouts increased, and they became segregated into what are now called primary and secondary displays. Primary displays are placed directly in front of the driver and present essential information: vehicle speed, engine parameters, and critical warnings. Secondary displays convey information of lesser importance, such as that relating to heating and air conditioning, the radio, and redundant or less important vehicle-parameter readouts. Secondary displays have typically been located in the dash near the car's midline between the driver and front-seat passenger.

As time went on, the increasing number of mechanical readouts were supplemented by electrical ones, such as resistive fuel-level gauges, but the first electronic display did not appear until the 1978 Cadillac Seville. This vehicle featured three digital displays for the fuel gauge, speedometer, and trip computer, using gas-plasma technology (which was changed to vacuum-fluorescent technology soon after the vehicle was introduced).

Liquid crystals were first utilized for a primary display in the 1984 Chevrolet

Corvette. Three separate twisted-nematic LCDs were used for the speedometer, driver information center, and tachometer. Corvette designers based their selection on the LC's design flexibility and color capability as well as its consistency with the high-tech theme of the newly redesigned Corvette.

More recent milestones in the short history of electronic automotive displays are the red-green-blue CRT in the 1985 Toyota Soarer (optionally available in Japan) and the first production use of a monochromatic CRT in the 1986 Buick Riviera. The Toyota's CRT was placed directly in front of the driver. The Riviera's was, and is, a combined secondary display incorporating touch-screen controls. Like the Corvette, the Riviera was a newly redesigned vehicle, and the unusual instrumentation helped create a unique image for it.

But despite these well-known innovations and the scattering of American and Japanese models that have electronic displays as either standard or optional equipment, purely mechanical instruments—those with pointer-and-dial readouts driven by flexible mechanical shafts—still make up 50% of the primary instrument clusters integrated into cars sold in the North American market. Electromechanical instruments—pointer-and-dial readouts driven by electronic signals generated by transducers—constitute a further 32%, leaving only 14% of the market to be divided among all the purely electronic technologies. Of these, a significant majority are vacuum fluorescent. Liquid crystals are next, followed by a small number of CRT and LED displays. Most of the electronic displays have been used as primary instrumentation—the readout has been placed directly in front of the driver.

There is significant optimism that the use of electro-optical displays will continue to increase. Several additional factors tend to convert the optimism about electronic displays to a conviction among most people in the industry that their ascendancy is inevitable. In the past, the information displayed in automobiles has been segmented into speed, fuel level, engine parameters, and time. Increasingly, in the future, information will consist of integrated and derived parameters presented in graphic form, as in navigation and fault diagnosis (Fig. 20.4).

The problem of limited dashboard "real estate" for displays and controls is already a problem in today's vehicles. With more information to be displayed, reconfigurable displays will become all but a necessity except in the simplest vehicles.

Pointer-and-dial instruments work well for primary information. With their graphics capability and reconfigurability, electronic displays are particularly desirable for secondary displays. The trend, even in fairly basic automobiles, toward a growing ratio of secondary to primary display function also favors the growth of electronics.

Innovative locations for displays are under serious consideration at all major manufacturers. Head-up displays, displays in the steering-wheel hub, in the rearview mirror, and in the headliner, as well as the now common placement in the

E-O DISPLAYS—THE SYSTEM PERSPECTIVE

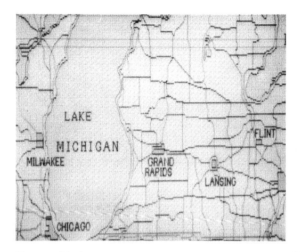

Figure 20.4 The transition from discrete to integrated information display in automobiles, as in this prototype navigation display, is expected to accelerate the move to electronic displays.

dashboard and console, will frequently place a premium on thin packages, light weight, and low power.

Integration, reconfigurability, the increase of the number of secondary display functions, and innovative instrument placement favor electronic displays generally and LCDs in particular. This is not to say that LCDs are likely to be the favored technology in all applications; many different display technologies will continue to grow and survive. But the capabilities of LCDs when compared to the automotive environment, engineers' goals, and stylists' goals indicate a good "fit."

Collision-avoidance systems combining radar, laser, visual, infrared, and ultrasonic sensing are being developed to monitor the space surrounding a vehicle. Such systems will monitor not only ahead of the vehicle to detect a rapidly closing obstacle and a possible collision but also the driver's blind spots at the car's rear corners for safer lane changes in heavy traffic. In its simplest form, the system would analyze what it perceives and give an appropriate warning to the driver. More advanced systems would actively control the throttle and brakes in response to a sensed possible collision.

The passenger compartment, and especially the driver's controls, will also receive much attention. Displays will present information in a way that greatly reduces the driver's work.

Such essential information as vehicle speed will be provided continuously, possibly by a head-up display (Fig. 20.5). Other information will show on a

Figure 20.5 One type of head-up display uses holographic techniques to project key driver information onto a car's windshield. In a simple version (a), only the speedometer readout is projected in front of the driver; a more advanced version (b) projects all the conventional instrument readouts; an even more advanced kind (c) projects an infrared image of the scene (generated below the instrument panel) directly in front of the driver, with the image focused at infinity to aid the driver at night or in fog. Finally (d), an image of a navigational map could be projected along the car midline between driver and passenger. (Illustrations courtesy of Ralph V. Wilhelm, Jr. of Delco Electronics Corp., Kokomo, IN.)

multifunction panel that employs one of a number of display technologies, such as liquid crystal, vacuum fluorescence, or light-emitting diode. Performance data will show whenever the system senses the unusual, and a warning device will bring those data to the driver's attention. The driver will also be able to choose the most convenient and agreeable format for displaying any array of information.

The system will display maintenance information automatically, notifying the driver what servicing is needed and when it should be done. A speech-recognition system might make it easier to control the entertainment system, call

E-O DISPLAYS—THE SYSTEM PERSPECTIVE

up driver information, format the display, or dial numbers on a cellular telephone.

Because of what appears to be an unsolvable traffic congestion problem using standard design methodologies, a new total system design methodology is developing. This new technique is heavily weighted toward the use of communications and computer systems and is referred to as an Intelligent Vehicle Highway System (IVHS). A key element in any such new system will rely on the availability of display technologies to support driver situation awareness around the automobile. One such system that is being developed is the Ali-Scout system (Fig. 20.6).

The Ali-Scout system (now renamed Euro-Scout) is a three-in-one system that handles advanced traffic management, route navigation, and driver information. Developed by Siemens Automotive, an operating unit of Siemens AG, Munich, it uses infrared transmitters and receivers to transfer navigation information between roadside beacons and onboard displays in appropriately equipped vehicles.

The system's infrastructure equipment includes infrared beacon transmitter-receivers located on traffic lights and linked to a central control facility where computers originate the system guidance information.

The in-vehicle unit includes an input keyboard, a small arrow indicator guidance display, and a voice messaging system. Unseen by the driver are an onboard computer for the dead-reckoning navigation system and an infrared transmitter-receiver.

The transmitter-receiver picks up traffic data as it passes a beacon and also sends data back to the central computer, where they are factored into the degree of route congestion and travel time for all suitably equipped vehicles on all beacon-to-beacon links. The computer also factors in the historical congestion profiles, including pedestrian traffic for that intersection. The information can then be used to adjust traffic light timing to smooth overall flow throughout the system.

Ali-Scout does not use map displays but does store an up-to-date map of the city in its central computer. It continually computes a driver's minimum trip time route and transmits that information to his car.

Another IVHS system that's being developed is Trafficmaster, from General Logistics PLC, Luton International Airport, Bedfordshire, England, which is a real-time traffic information system that provides motorists with up-to-the-minute information on the speed, direction, and length of any traffic backup. It enables drivers to assess the situation and take avoiding action, if desired, but does not suggest what evasive action to take.

The system uses infrared sensors mounted on highway bridges at approximately 3 km intervals that log the speed of traffic passing below them. Each unit

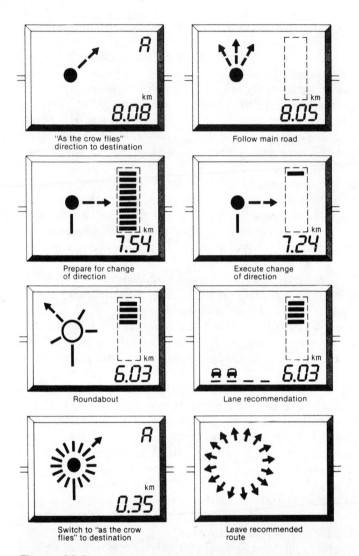

Figure 20.6 In-vehicle display gives the driver a number of informational displays. Explicit directions—get in the right lane, turn right, and so on—will be displayed for getting to the destination the driver has input via the keyboard. When the vehicle is outside the digital map area, an arrow is displayed showing "as the crow flies" direction to the destination, but the driver has to determine which roads to take.

E-O DISPLAYS—THE SYSTEM PERSPECTIVE

illuminates two detection zones with encoded streams of infrared light. If the speed drops below a preset threshold of 40 km/hr, the sensors relay that information over a radio link to the control center. From there the information is transmitted by means of a radio paging network to a receiver mounted on the dashboard of the vehicle.

The receiver displays the information in map form, giving the driver an audible and a visual signal as soon as updated information is received. The display then zooms in to a close-up map of the area where the problem is occurring; a flashing block shows the location, speed, and direction of the holdup while the number of blocks denotes the length of the backup. Information is updated every 3 min.

These IVHS systems for future automobiles will require more displays, and an integrated system of displays will be needed to increase their availability.

20.5.4 Integrated Display Systems

It seems clear that future systems will be highly integrated whether the system application is air-, land-, or sea-based. In addition, the display subsystems will be integrated along with other specific function subsystems. The architectural feature that all of these integrated systems share is a data bus communication scheme that is computer controlled via relatively sophisticated software. Generalized examples of such systems (Rodda, 1987; Spitzer, 1987) are shown in figure 20.7 for automobiles and in Figures 20.8 and 20.9 for aircraft. Given that these types of integrated display systems will evolve, it is likely that a much more critical and comprehensive evaluation of image quality will be needed.

20.6 A MORE UNIVERSAL DEFINITION OF IMAGE QUALITY

Given the current diversity of display device technologies and systems, image quality metrics, image quality measurement techniques, and the variability of human vision systems, it can be concluded that we need a better approach for studying image quality. The "new" approach must minimally depend on technology and must maximally depend on the human vision system and its environment. Several possible integrated tasks for such an approach are presented in the following paragraphs.

20.6.1 Empirical Task

All of the metrics of image quality have one thing in common: They are based upon some theoretical approach to the notion of image quality and quantification of the visual system, and lead directly to a model of image quality based upon that theoretical approach. A totally different approach is to offer no theory or

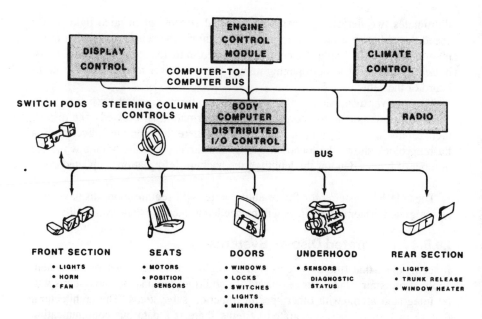

Figure 20.7 Data bus communications.

concept and simply determine empirically which concepts best predict observer performance, letting the resulting pool of predictions define quantitatively what is meant by "image quality." Such an approach was taken by Snyder and Maddox (1978) for digitally addressed displays using a stepwise linear multiple regression technique.

Using three different tasks, they performed experiments that varied the structure of the display in terms of pixel size, shape, contrast, spacing, and the like. They collected observer performance on two different search tasks and a reading task and correlated these performance measures with a variety of physical measurements of geometric and photometric characteristics of the image. In this statistical approach all known variables are permitted to enter into a linear prediction equation, and the computed result is a "model" that defines the best predictive combination of any or all of the variables. Their empirical model, which has subsequently been cross-validated, predicts 50% of the variance for the search task and 52% of the variance in the reading task. Of perhaps more interest are the combinations of variables that entered into the prediction equations. These predictor variables are almost entirely modulation and MTFA type measures and generally support the results previously obtained for these types of image quality measures.

E-O DISPLAYS—THE SYSTEM PERSPECTIVE

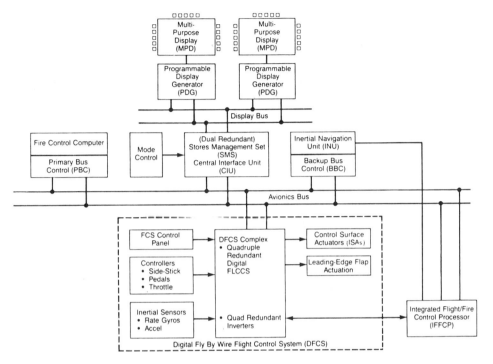

Figure 20.8 F-16 Digital Avionics System architecture. [From Ramage et al. (1983). © American Institute of Aeronautics and Astronautics; reprinted with permission.]

Their model equations represent the best empirically derived measures of image quality for digital displays, for the purpose of display design specification. They do not deal directly with the recommended dynamic range of a given image or any other image-specific parameters as do some of the measures described in the literature. Thus, these equations are useful to the designer in optimizing displays, particularly for the presentation of alphanumeric information. They are of no help in describing or suggesting processing algorithms to make a literal image more interpretable.

20.6.2 Psychological Task

Any image can be described as a two-dimensional modulation of luminance and color across space. A psychological approach to image quality considers the detectability and discriminability of these images in terms of the spatial, temporal, and color-sensitive mechanisms of perception (Rigowitz, 1985).

Display technologies bombard phosphors with electrons or excite gases to

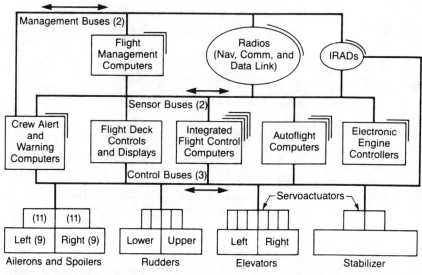

Figure 20.9 Proposed architecture for transport aircraft. [From Tagge et al. (1985).]

emit photons. Although these technologies vary greatly, the same language can be used to describe the images they create. In all cases, the picture created is a two-dimensional variation of luminance and color across space. The fact that the images created by diverse methods can all be described with the same vocabulary marks the first step in developing a technology-independent approach to characterizing and improving image quality. The second step involves the realization that the way an image "looks" depends on the visual system looking at it. The luminance of an object, for example, can be measured quite easily. The apparent "darkness" or "lightness" of that object, however, depends on the workings of human perception. The Hermann grid, for example, is an array of black squares. Differences in perceived luminance at certain areas of the grid cannot be measured by a photometer; they are not present in the physical stimulus. We see the nonexistent gray circles in the grid because of the way our visual system processes adjacent areas of dark and light. The science that studies the mechanisms that cause us to see the physical world as we see it is *visual psychophysics*.

It has been known since Fourier's time that a complex periodic signal can be represented as the sum of its sinusoidal components. This theorem, originally developed to describe the decomposition of heat waves, has been broadly applied to electric waves, auditory waves, and visual images. Any complex visual image can be represented as the sum of sinusoidal modulations of luminance (i.e., sine wave gratings) that vary in their spatial frequency, contrast, phase, and orientation. This provides a convenient way of representing images. This characterization of visual images is particularly interesting to the visual scientist because both electrophysiological and psychophysical evidence support the hypothesis that special mechanisms exist in vision that are particularly sensitive, "selectively" sensitive, to narrow ranges of spatial frequency information. It is as if we had specialized spatial frequency filters.

This spatial frequency framework can also be used to characterize interactions in our response to spatial patterns that vary in their temporal or chromatic composition. It can be shown that our temporal response depends on spatial frequency. Spatial frequency channels sensitive to low spatial frequencies—that is, to the low-resolution parts of an image—have a transient response; channels sensitive to high spatial frequencies—that is, to the high-resolution parts of an image—have a sustained response. We have higher spatial resolution for luminance modulations than for modulations of pure color. Color-sensitive mechanisms are tuned to large areas.

In this framework, then, we can express the images produced on CRT and matrix displays in terms of sinusoidal variations in luminance and wavelength, and consider these modulations the input to specialized visual filters.

Independent of the technology producing them, images can all be characterized in terms of their two-dimensional spatial frequency spectra. Represented this way, the complex, varied images produced by various processes can all be considered within the same framework. Most important, this particular framework allows us to address "image quality" in terms of the response properties of human visual mechanisms.

20.6.3 Direct Task

In Task and Verona's work (Verona et al., 1979) a direct method for measuring CRT image quality has evolved. Their technique is based on the modulation transfer function (MTF) theory and human psychophysical data. The measurements consist of *directly* recording the modulation contrast available on the CRT display as a function of spatial frequency. An electronic sine wave generator produces a sine wave intensity pattern on the face of the CRT display. The display luminance distribution is scanned using a telephotometer or microphotometer, depending on the size of the display. The modulation contrast of the display is obtained from the photometer scan for several spatial frequencies. The resulting graph showing modulation versus frequency is defined as the sine wave response (SWR) curve of the display.

Since human vision is not linearly related to modulation, it is desirable to transform the modulation axis to another parameter that is linearly related to vision. This can theoretically be accomplished by transforming the modulation contrast to the square root of two incremented gray shades. The resulting gray shade response (GSR) indicates how many gray shades are visible as a function of spatial frequency.

Therefore, they have defined a new single display quality metric using the GSR curve of a display. The measure is derived from the modulation transfer function area (MTFA) concept and is defined as the area between the visual threshold curve and the GSR. This area is referred to as the gray-shade frequency product or GFP. A brief study was performed to determine the correlation of GFP with performance in a target recognition task. The results for three display conditions indicate that the GFP is at least as good a measure of display quality as the MTFA.

Unfortunately, it is not clear how useful the direct method described above is for discrete-type displays. Therefore, research needs to be pursued that would investigate the possibility of extending the direct method to technologies other than CRT displays. It seems reasonable that this extension might depend on square-wave functions as a basis (i.e., instead of, or in addition to, the sine wave functions used for the CRT direct method). Walsh functions and their derivatives (or subsets) have all the "interesting" mathematical characteristics exhibited by sine waves.

20.6.4 Summary

It appears that there will be a need for a more universal definition of image quality than is currently available in keeping with the expected diversity of display technology and integration in the future. The discussion here has presented the basis for such a definition by indicating what seem to be the necessary ingredients for a universally usable definition. The research community, of course, must continue its evolutionary work.

References

Adam, G. (1990). Tactical cockpits—the coming revolution, Eleventh Annual IEEE/AESS Symposium, Nov. 28, pp. 15–23.

Boroughs, Fallis, Warnock, Britt (1967). Quantitative determination of image quality, Boeing Company Rep. D2-114058-1.

Conrac (1980). *Raster Graphics Handbook*, Conrac Corporation, Covina, CA.

Credelle, T. L. (1987). Avionic color liquid crystal displays—recent trends, SAE Paper No. 871790, Aerospace Technology Conf. and Exposition, Long Beach, CA, Oct. 5–8.

Goldmark, P. C. (1949). Brightness and contrast in television, *Electr. Eng.* 68: 170–175.

Gutman, Snyder, Farley, Evans (1979). An experimental determination of the effect of image quality on eye movements and search for static and dynamic targets, AFAMRL-TR-79-51.

Haim, E., F. Luo, G. Biermann, and M. Cushing (1988). One million pixel full-color

Liquid crystal display for avionics applications, 8th Digital Avionics Systems Conf., San Jose, CA, Oct. 17–20, 1988.

Hatfield, J. (1990). Advanced cockpit technology for future transport aircraft, IEEE/AESS Symposium, Nov. 28, 1990, pp. 77–87.

Jenness, J. R., W. A. Eliot, and J. A. Ake (1976). Intensity ripple in a raster generated by a Gaussian scanning spot, *J. Soc. Motion Picture Telev. Eng.* 76: 544–550.

Jurger, R. K. (1991). Smart car and highways go global, *IEEE Spectrum*, May: 26–36.

Levi, L. (1968). *Applied Optics—A Guide to Modern Optical System Design*, Wiley, New York, Vol. I.

Linfoot, E. H. (1960). *Fourier Methods in Optical Image Evaluation*, Focal Press, London.

Moon, D. L. (1988). Image quality for discrete-element displays: variables, metrics and measurements, *Image Processing, Analysis, Measurements, and Quality*, SPIE Vol. 901, pp. 161–170.

Morello, S. A. (Compiler) (1989). *Aviation Safety/Automation Program Conference*, NASA Conf. Publ. 3090, sponsored by NASA HQ, Virginia Beach, VA, Oct. 11–12, 1989.

Ramage, J. K., et al. (1983). Development of the AFTI/F-16 Triplex Digital Flight Control System, AIAA paper 83–2278-CP.

Rigowitz, B. E. (1985). A psychological approach to image quality, *Proc. SPIE 549*: 9–13.

Rivard, J. G. (1987). The automobile in 1997, *IEEE-Spectrum, October*: 67–71.

Rodda, W. J. (1987). Automotive electronics, *IEEE Potentials, February*: 18–21.

Ross, D. A. (1988). Automotive LCD's, *SID-Inf. Display, February*: 12–15.

Snyder, H. L. (1980). Human visual performance and flat panel display image quality, VPI and State Univ. Tech. Rep. HFL-80–1.

Snyder, H. L., and M. E. Maddox (1978). Information transfer from computer-generated, dot-matrix displays, VPI and State Univ. TR-HGL-78–3.

Spitzer, G. R. (1987). *Digital Avionics Systems*, Prentice-Hall, Englewood Cliffs, NJ.

Tagge, G. E., L. A. Irish, and A. R. Bailey (1985). System study for an integrated digital/electric aircraft (IDEA), NASA Contractor Rep. 3840 with Suppl. 1.

Task, H. L. (1979). An evaluation and comparison of several measures of image quality for television display, Wright Patterson AFBase, OH, Aerospace Res. Lab., AMRL-TR-79–7.

Task, H. L., and R. W. Verona (1976). A new measure of television display quality relatable observer performance, AFAMRL-TR-76–73.

Tennas, L. E. (Ed.) (1987). *Flat-Panel Displays and CRTs*, Van Nostrand Reinhold, New York.

Van Meeteran, A. (1973). *Visual Aspects of Image Intensification*, Institute for Perception TNO, Soesterberg, The Netherlands.

Verona, R. W., H. L. Task, V. C. Arnold, and J. H. Brindle (1979). A direct measure of CRT image quality, U.S. Army Aeromed. Res. Lab. USAARL Rep. 79–14.

Virsu, V., and P. K. Lehlio (1975). A microphotometer for measuring luminance distributions on a CRT, *Behavior Res. Methods Instrum.* 7:29–33.

Wiener, E. L. (1989). Human factors of advanced technology ("glass cockpit") transport aircraft, NASA Contractor Rep. 177528, Contract NCC2-377, Ames Research Center, June 1989.

Index

Accommodation, 419, 768
Active-matrix, 69
Active-matrix LCD, 69, 95, 162
  display performance, 91
Addressing, 54
  electron beam, 56
  matrix, 54
  multi-dimensional, 133
  thermal, 59
Algorithm, Display, 589
Amorphous silicon image sensors, 96
Antialiasing, 798
Automobile displays, 824
Automotive HUD, 407

Beer-Lambert law, 699
Binocular
  disparity, 219, 688
  overlap, 341, 773
  parallax, 291, 346

[Binocular]
  performance, 424
  rivalry, 755
  vision, 417, 421
Biocular, 759
  display optics, 417, 426
  viewers, 430
    catadioptric, 440
    flight simulator, 438
    Fresnel, 442
    panaromic, 438
    relaxed-view, 433
    stereoscopic, 443
Birefringence, 36
Blending, 280

Cathode ray tubes, 10, 732
Center of gravity, 664
Charge-coupled device, 511
Cholesteric, 20
  devices, 45

Chromaticity, 732
CIE, 712
Circular dichroism, 45
Collimation, 340
Color, 142, 187
  discrimination, 328
  emitters, 142
  field-sequential, 144
  filters, 143
Color appearance, 712, 724
Color control, 711
Color image decoding, 196, 206
Color image encoding, 191, 204
Color resolution, 734
Color space, 716, 727
  double hexcone, 727
  hexcone, 727
Color vision, 324
Colorimetric equations, 716
Combiners, 372
Contrast enhancement, 617
Contrast, 811
Contrast ratio, 349
Contrast transfer function, 503
Contrast threshold function, 469
Convergence, 276, 768
Cosine grating, 453
Cross-spectral interpretation, 779
Crosstalk, 304
CRT, 10, 732
  dynamic imaging, 462
  flat, 156
  MTF, 459
  phosphors, 374
  projection displays, 217
  stereoscopic, 293
  refresh rate, 305

Dichroic dyes, 53
Dielectric response, 31
Diplopia, 420

Display algorithms, 589
Display contrast ratio, 349
Display parameters, 674
Display scales, 586
Displays
  automobile, 824
  biocular, 417
  CCD, 511
  discrete, 493
  electrochromic, 178
  electroluminescent, 149
  electrophoretic, 178
  flat-panel, 121
  gas-discharge, 165
  gas-electron-phosphor, 171
  head-up, 337
  holographic head-up, 337
  nondiscrete, 447
  peripheral vision, 311
  photoluminescent, 170
  pin diode, 83
  projection, 211
  stereoscopic, 291
  system, 809
  vacuum fluorescent, 154
Dynamic imaging system, 457
Dynamic modulation transfer function, 475
Dynamic scattering, 44
Dynamic threshold, 51

Edge crispening, 637
Edge response, 455
Electrochromic displays, 178
Electroluminescent displays, 149
Electromechanically focused image tube, 2
Electron beam addressing, 56
Electrophoretic displays, 178
Encoding of color, 191, 193

INDEX

Ferroelectric liquid crystal, 24, 49
  distorted helix, 52
  soft-mode, 52
Field-of-view, 341, 689, 746
Flat CRTs, 156
Flat-panel display, 818
Flicker, 328
FLIR, 656
Flooding frequency, 788
Fourier transform, 189
Fraction of pixels, 597
Frequency domain filtering, 190

Gamut mismatch, 728
Gas-discharge displays, 165
Gas-electron-phosphor displays, 171
Ghosting, 304
Gray scale, 135

Halftoning, 139
Head-coupled system, 744
Head-up display, 337
Helmet-mounted devices, 652, 743
High-definition TV, 180
High frequency enhancement, 633
Histogram, 588
  equalization, 594
  projection, 595
HMD, 743
Holographic combiners, 372
Holographic diffraction theory, 362
Holographic displays, 337
Holographic optical elements, 370
Holography, 362
  angular characteristics, 364
  spectral characteristics, 364
Horizontal parallax, 302
HUD, 337, 407
  combiner, 360

[HUD]
  optics, 352
  pilot display unit, 390
  processor, 383
Human factors, 651, 743
  binocular overlap, 341, 773
  binocular parallax, 291, 346
  binocular rivalry, 755
  vibration, 735
Human visual system, 792
Hybrid field effect, 41

Image
  blending, 280
  degradation, 481
  motion, 794
  intensifier, 2, 651
  quality, 682, 810
  restoration, 483
  scaling, 306
  sensor
    direct addressed, 102
    matrix-addressed, 103
    photoconductor-based, 97
    photodiode-based, 99
    scanning, 513
  sharpening, 637
  tubes, 653
Impulse response, 448
Infrared imagery, 583

Laser projection displays, 255
LCD, 19, 69, 159
  active matrix, 71
  thin-film transistor, 73
LED, 145
Legibility, 810
Light-emitting diodes, 145
Light management, 745

Light valve projection displays, 230
Linear system, 448
Liquid crystal, 19
  nematic, 20
    supertwisted, 42
    lower-twisted, 42
    ferroelectric, 24
    polymer-dispersed, 46, 165
Liquid crystal display, 19
Liquid crystal light valve projection displays, 254
Liquid crystal modulator, 296
Liquid crystal properties
  applied field effects, 32
  dielectric response, 31
  disclinations, 27
  elasticity, 25
  electrochemistry, 34
  stability, 35
  surface alignment, 28
  viscosity, 26
Luminance, 732, 811

Magnification, 16
Malcolm horizon, 313
Matrix addressing, 54
Metal-insulator-metal diode, 70, 79
MIM diode LCD, 80
Microchannel plate, 6, 653
Minimum resolvable temperature, 583, 671
Modulation transfer function, 9, 57, 215, 671
  mathematical definition, 452
  two-dimensional, 528
Modulation transfer function area, 469
Modulo projection, 630

Monocular, 759
MTF, 452, 817
  degradation, 464

Nematic, 20
Night vision devices, 651
Night vision goggles, 658
Nyquist rates, 788

Oil-film light valve projection displays, 230
Optical distortion, 772
Optical mode interference, 43
Optical transfer function, 503
Optically-addressed LCLV, 246

Perceptual color space, 715
Peripheral vision displays, 311
Phosphor, 12, 14, 653
Phosphor displays, 456
Phosphor persistence, 692
Photoaddressing, 56
Photocathode, 653
Photoconductor-based sensor, 97
Photodiode-based sensor, 99
Photoluminescent DC gas discharge displays, 170
p-i-n diode displays, 83
Pixel error, 471
Plasma switching, 164
Plateau equalization, 606
Polymer-dispersed liquid crystals, 46, 165
Poole-Frenkel equation, 79
Projection display, 211
  active-matrix light valve, 240
  laser, 255
  liquid-crystal light valve, 238

INDEX

[Projection display]
    oil-film light valve, 230
Pulse-width modulation, 137
    screen, 265

Refractive HUD, 395
Refresh rate, 305
Relaxed view bioculars, 433
Resolution, 4, 16, 760
    CRT, 16
    gray-scale, 139
    intensifier, 4

Sampling, 787, 788
Scanning, 128
    image sensor, 513
    line-at-a-time, 130, 173
    point-by-point, 129
    scanning and flicker, 128
    three-dimensional, 131
Sensor noise model, 514
Sensor parameters, 665
Sensor signal model, 514
Shift invariance, 450
Smectic, 22
Sparkles, 685
Spatial disorientation, 317, 320
Spatial light modulator, 59
Spatiotemporal sampling, 787
Static MTF, 453
Stereogram, 292
Stereoscopic bioculars, 443
Stereoscopic displays, 291
    CRT, 293
    off-axis, 297
    on-axis, 299
    time-multiplexed, 294
    time-parallel, 294
Streaking, 528

Superposition, 450
Surface-stabilized device, 49

Temporal antialiasing, 798
Thermal addressing, 59
Thermal crossover, 699
Thermal imaging systems, 651, 654
Thermally-addressed LCLV, 251
Thin-film transistor, 73
Three-dimensional display, 291
Threshold projection, 606
Time-multiplexed stereoscopic display, 294
Time-parallel stereoscopic display, 294
Tricolor grating, 191, 203
Twisted nematic, 21, 39, 69

Undersampled projection, 606

Vacuum fluorescent displays, 154
Vestibular system, 320
Vibration, 753
Visibility, 745
Visual
    aberration, 428
    acuity, 684
    fields, 689
    perception, 323
    pursuit, 799
    surfaces, 428
    system, 317

White-light image processing, 187
Window of visibility, 794